.TO ASI Series

ınced Science Institutes Series

ies presenting the results of activities sponsored by the NATO Science Committee, ı aims at the dissemination of advanced scientific and technological knowledge, ı view to strengthening links between scientific communities.

.eries is published by an international board of publishers in conjunction with the) Scientific Affairs Division

'e Sciences ıysics	Plenum Publishing Corporation London and New York
athematical .d Physical Sciences	D. Reidel Publishing Company Dordrecht, Boston, Lancaster and Tokyo
ehavioural and Social Sciences ıgineering and aterials Sciences	Martinus Nijhoff Publishers The Hague, Boston and Lancaster
omputer and Systems Sciences :ological Sciences	Springer-Verlag Berlin, Heidelberg, New York and Tokyo

ı C: Mathematical and Physical Sciences Vol. 183

Clifford Algebras and Their Applications in Mathematical Physics

Clifford Algebras and Their Applications in Mathematical Physics

edited by

J. S. R. Chisholm

and

A. K. Common

Mathematical Institute,
University of Kent, Canterbury, Kent, U.K.

D. Reidel Publishing Company

Dordrecht / Boston / Lancaster / Tokyo

Published in cooperation with NATO Scientific Affairs Division

Proceedings of the NATO and SERC Workshop on
Clifford Algebras and Their Applications in Mathematical Physics
Canterbury, U.K.
15-27 September, 1985

Library of Congress Cataloging in Publication Data

NATO and SERC Workshop on Clifford Algebras and Their Applications in Mathematical Physics (1st: 1985: Canterbury, Kent)
Clifford algebras and their applications in mathematical physics.

(NATO ASI series. Series C, Mathematical and physical sciences; vol. 183)
"Proceedings of the NATO and SERC Workshop on Clifford Algebras and Their Applications in Mathematical Physics, Canterbury, U.K., 15–27 September 1985"—T.p. verso.
"Published in cooperation with NATO Scientific Affairs Division."
Includes index.
1. Clifford algebras—Congresses. 2. Mathematical physics—Congresses. I. Chisholm, J. S. R. (John Stephen Roy) II. Common, A. K. (Alan K.) III. North Atlantic Treaty Organization. Scientific Affairs Division. IV. Title. V. Series: NATO ASI series. Series C, Mathematical and physical sciences; no. 183.
QC20.7.C55N37 1985 512′.57 86–15406
ISBN 90–277–2308–7

Published by D. Reidel Publishing Company
P.O. Box 17, 3300 AA Dordrecht, Holland

Sold and distributed in the U.S.A. and Canada
by Kluwer Academic Publishers,
101 Philip Drive, Assinippi Park, Norwell, MA 02061, U.S.A.

In all other countries, sold and distributed
by Kluwer Academic Publishers Group,
P.O. Box 322, 3300 AH Dordrecht, Holland

D. Reidel Publishing Company is a member of the Kluwer Academic Publishers Group

---

Printed in The Netherlands

TABLE OF CONTENTS

Foreword ix

W. K. Clifford 1845-1879 xiii

GENERAL SURVEYS

D. HESTENES* / A Unified Language for Mathematics and Physics 1

P. LOUNESTO* / Clifford Algebras and Spinors 25

CLASSIFICATION OF CLIFFORD ALGEBRAS

J. ŁAWRYNOWICZ and J. REMBIELIŃSKI / Pseudo-Euclidean Hurwitz Pairs and Generalized Fueter Equations 39

A. DIMAKIS / A New Representation for Spinors in Real Clifford Algebras 49

R. ABLAMOWICZ and P. LOUNESTO / Primitive Idempotents and Indecomposable Left Ideals in Degenerate Clifford Algebras 61

SPIN GROUPS

A. MICALI* / Groupes de Clifford et Groupes des Spineurs 67

P. ANGLES / Algebres de Clifford $C^+_{r,s}$ des Espaces Quadratiques Pseudo-Euclidiens standards $E_{r,s}$ et structures correspondantes sur les espaces de Spineurs Associes. Plongements Naturels des Quadratiques Projectives Reelles $\tilde{Q}(E_{r,s})$ attachees aux Espaces $E_{r,s}$ 79

J. A. BROOKE / Spin Groups associated with Degenerate Orthogonal Spaces 93

A. MICALI and C. MALLOL / Algebres de Clifford Separables II 103

G. KIENTEGA / Sur une Question de Micali-Villamayor 109

CLIFFORD ANALYSIS

R. DELANGHE* and F. SOMMEN / Spingroups and Spherical Monogenics 115

J. RYAN / Left Regular Polynomials in Even Dimensions, and Tensor Products of Clifford Algebras 133

F. SOMMEN / Spingroups and Spherical Means 149

F. BRACKX and W. PINCKET / The Biregular Functions of Clifford Analysis: Some Special Topics 159

L. V. AHLFORS / Clifford Numbers and Möbius Transformations in R^n 167

MATHEMATICAL PHYSICS

R. W. TUCKER* / A Clifford Calculus for Physical Field Theories 177

V. SOUČEK* / Generalized C-R Equations on Manifolds 201

J. BUREŠ / Integral Formulae in Complex Clifford Analysis 219

G. E. SOBCZYK* / Killing Vectors and Embedding of Exact Solutions in General Relativity 227

Z. OZIEWICZ / From Grassmann to Clifford 245

I. M. BENN / Lorentzian Applications of Pure Spinors 257

E. KÄHLER / The Poincaré Group 265

P. R. HOLLAND / Minimal Ideals and Clifford Algebras in the Phase Space Representation of Spin-½ Fields 273

K. BUGAJSKA / Some Consequences of the Clifford Algebra Approach to Physics 285

PHYSICAL MODELS

M. DRESDEN* / Algebraic Ideas in Fundamental Physics from Dirac-algebra to Superstrings 293

J. G. TAYLOR / On Two Supersymmetric Approaches to Quantum Gravity: Clifford Algebra Degeneracy v Extended Objects 313

D. HESTENES* / Clifford Algebra and the Interpretation of Quantum Mechanics 321

M. F. ROSS / Representation-free Calculations in Relativistic Quantum Mechanics 347

J. P. CRAWFORD / Dirac Equation for Bispinor Densities 353

R. S. FARWELL and J. S. R. CHISHOLM / Unified Spin Gauge Theory Models 363

J. McEWAN / U(2,2) Spin-Gauge Theory Simplification by Use of the Dirac Algebra 371

F. D. SMITH, Jr. / Spin(8) Gauge Field Theory 377

A. O. BARUT / Clifford Algebras, Projective Representations and Classification of Fundamental Particles 385

G. DIXON / Fermionic Clifford Algebras and Supersymmetry 393

H. JOOS / On Geometry and Physics of Staggered Lattice Fermions 399

J. KELLER / A System of Vectors and Spinors in Complex Spacetime and Their Application to Elementary Particle Physics 425

D. W. EBNER / Spinors as Components of the Metrical Tensor in 8-dimensional Relativity 435

W. M. PEZZAGLIA, Jr. / Multivector Solution to Harmonic Systems 445

F. G. MORRIS and K. R. GREIDER / The Importance of Meaningful Conservation Equations in Relativistic Quantum Mechanics for the Sources of Classical Fields 455

ELECTROMAGNETISM

E. F. BOLINDER* / Electromagnetic Theory and Network Theory using Clifford Algebra 465

B. JANCEWICZ / Remarks on Clifford Algebra in Classical Electromagnetism 485

K. IMAEDA / Quaternionic Formulation of Classical Electromagnetic Fields and Theory of Functions of a Biquaternion Variable 495

G. A. DESCHAMPS and R. W. ZIOLKOWSKI / Comparison of Clifford and Grassmann Algebras in Applications to Electromagnetics 501

GENERALISATIONS OF CLIFFORD ALGEBRA

A. CRUMEYROLLE* / Symplectic Clifford Algebras 517

P.-E. HAGMARK and P. LOUNESTO / Walsh Functions, Clifford Algebras and Cayley-Dickson Process 531

T. T. TRUONG / Z(N)-Spin Systems and Generalised Clifford Algebras 541

A. K. KWAŚNIEWSKI / Generalized Clifford Algebras and Spin Lattice Systems 549

A. RAMAKRISHNAN / Clifford Algebra, its Generalisations and Physical Applications 555

J. HELMSTETTER / Application of Clifford Algebras to *-products 559

M. IMAEDA / On Regular Functions of a Power-associative Hypercomplex Variable 565

J. TIMBEAU / On a Geometric Torogonal Quantization Scheme 573

INDEX 583

*Invited Speaker

FOREWORD

William Kingdon Clifford published the paper defining his "geometric algebras" in 1878, the year before his death. Clifford algebra is a generalisation to n-dimensional space of quaternions, which Hamilton used to represent scalars and vectors in real three-space: it is also a development of Grassmann's algebra, incorporating in the fundamental relations inner products defined in terms of the metric of the space. It is a strange fact that the Gibbs-Heaviside vector techniques came to dominate in scientific and technical literature, while quaternions and Clifford algebras, the true associative algebras of inner-product spaces, were regarded for nearly a century simply as interesting mathematical curiosities. During this period, Pauli, Dirac and Majorana used the algebras which bear their names to describe properties of elementary particles, their spin in particular. It seems likely that none of these eminent mathematical physicists realised that they were using Clifford algebras. A few research workers such as Fueter realised the power of this algebraic scheme, but the subject only began to be appreciated more widely after the publication of Chevalley's book, 'The Algebraic Theory of Spinors' in 1954, and of Marcel Riesz' Maryland Lectures in 1959. Some of the contributors to this volume, Georges Deschamps, Erik Folke Bolinder, Albert Crumeyrolle and David Hestenes were working in this field around that time, and in their turn have persuaded others of the importance of the subject.

Since 1960, a number of groups and individuals have begun independent work on Clifford algebras. Like many mathematical physicists, we ourselves have used the Dirac and Pauli algebras for a long time; but only during the past decade have we worked more generally with Clifford algebras, together with Dr. Ruth Farwell and Dr. John McEwan. We realised that many mathematicians and mathematical physicists working in the field were unaware of each others' existence, and that many theoretical physicists did not know of the immediate relevance of Clifford algebras to modern physical theories. So in 1983 we applied to N.A.T.O. and to the S.E.R.C. for financial support for a Workshop on Clifford algebras. We gathered together the names of most of those working in the field and invited senior figures to form an International Advisory Committee. Nearly all of these accepted enthusiastically, and with this Committee's active support and advice we were able to draw up a widely representative programme of invited speakers, and to obtain the funds we needed to hold the Workshop in Canterbury. We were also very pleased that all of those invited spoke at the Workshop and have contributed to this volume.

Those attending the Workshop represented a broad spectrum of interests in various branches of mathematics, mathematical and theoretical physics, and electrical engineering. This wide range of interests is reflected in the diversity of topics covered in these proceedings, and in the variety of notations and attitudes adopted by contributors. The main problem for participants at the Workshop was to assimilate a great deal of new material, often presented using unfamiliar notations and phraseology. This difficulty was overcome by careful explanations of fundamentals by a number of speakers, and through the organisation of informal seminars by Workshop participants, stimulated by Dr. Ian Porteous. Consequently, the contents of these proceedings not only give an account of recent research work, but also review many of the fundamental areas of development and application of Clifford algebras. We are particularly grateful to the invited speakers for the care with which they have formulated and presented their contributions.

The talks at the Workshop were grouped according to subject-matter; in organising the proceedings we have again grouped together invited and contributed papers whose contents are related. Naturally, some papers could have been placed in a different section: we have tried to choose the most appropriate place for each one. We have included a small number of papers from researchers who, for reasons beyond their control, were unable to attend the Workshop. We are grateful to Professor Edwin Power of University College, London for contributing an account of Clifford's life and work. We were pleased to welcome Professor Power and Professor John Adams of Trinity College, Cambridge, as guests at the Workshop dinner, representing the two institutions where Clifford worked. We are grateful to the authors of contributions for helping us to keep to an exacting timetable for publication, and for their efforts in producing good camera-ready copy. We have checked each paper in an effort to eliminate errors. The success of both the Workshop and the proceedings has depended upon the enthusiasm and cooperation of the participants, many of whom have refereed several papers; we are pleased that both of these qualities have been unbounded. One notable contribution to the Workshop to which we must draw attention was a demonstration by Professor Pertti Lounesto of a computer programme using Clifford algebras to manipulate vectors and tensors.

The Workshop would not have taken place without the generous support of N.A.T.O. and the S.E.R.C. Mathematics Committee; this volume will provide ample justification for this support. The official backing and the advice of the International Advisory Committee has been most helpful, and we are particularly grateful to Professor Brackx and Dr. Tucker, members of the local committee. The social success of the Workshop depended upon the excellent organisation of several events by Dr. John McEwan, and upon the continual interest and help of our ladies Pauline McEwan, Ellen Common and Monty Chisholm. Several of our academic colleagues helped us with transport to social events in our homes. A major factor in the success of the Workshop was the cooperation and concern of all of the staff of Darwin College; in particular, the quality of the meals was remarkably good, drawing praise from the most fastidious of European palates. We thank the Master of Darwin and the Vice-Chancellor of the University for their active support and interest in the Workshop. Finally, we express our sincere thanks to Mrs. Sandra Bateman, secretary to the Workshop, for her careful work and cooperation before and during the Workshop, and for her patience in making final corrections to all the papers published in this proceedings.

We are pleased that the participants at the Workshop see it as the first in a regular series of meetings. Professor Micali has undertaken to arrange the second meeting at Montpellier in 1987 or 1988; we anticipate a very happy and stimulating reunion with all the friends and colleagues who made the first Clifford Algebra Workshop such a pleasant and memorable meeting.

Roy Chisholm   Alan Common

March, 1986

Yours most truly
W. K. Clifford

WILLIAM KINGDON CLIFFORD: 1845-1879

W. K. Clifford was born in Exeter the county town of Devon on 4th May, 1845. He was the son of a bookseller and as a boy lived near the cathedral and went to a local school until he was 15 years old. At 15 he went to Kings College, London where his exceptional powers in mathematics were first noted. At 18 he entered Trinity College Cambridge and published his first two papers during his freshman year at Cambridge. He graduated winning a Smith prize and being second Wrangler in 1867. In the following year he was elected a Fellow of Trinity. Three years later, at the age of 26, Clifford was appointed Professor of Applied Mathematics and Mechanics at University College, London. He remained in this chair until his untimely death from tuberculosis in Madeira, where he had gone in the hope of recovery, on 3rd March, 1879. A tragedy that such a genius should die at 33 years of age.

Council members of University College, London do not think it against the spirit of confidentiality to open the archives of UCL to publish a reference they received in 1870. It was written by Clerk Maxwell (a man not noted for exuberant praise) and says:-

"The pecularity of Mr. Clifford's researches, which in my opinion points him out as the right man for a chair in mathematical science, is that they tend not to the elaboration of abtruse theorems by ingenious calculations, but to the elucidation of scientific ideas by the concentration upon them of clear and steady thought. The pupils of such a teacher not only obtain clearer views of the subjects taught, but are encouraged to cultivate in themselves that power of thought which is so liable to be neglected amidst the appliances of education".

As we know Clifford was appointed despite his youth. He was probably pleased to leave Cambridge. He had become strongly anti-ecclesiastical during his time at University. He has become deeply involved in the spirited arguments within the Spencer-Darwin evolution controversies that dominated intellectual life at that time. This could well have been due to a reaction to the establishment atomosphere during his childhood at Exeter. His father's main sales were of devotional and ecclesiastical books in the cathedral close and Clifford senior was a magistrate. William grew up as a devoted and earnest Anglican. It was during his late teens that he became a non-believer and one with a special dislike of establishments of all kinds.

Other than his Anglican upbringing there are few records about his childhood. One story is that at the age of 6 on a visit to London for the Great Exhibition in 1851, he calculated how many sword edges could be fitted around the circumference of the wheel of the GWR coach on which he was travelling. He enjoyed kite flying - a pursuit continued into adult life. One of the many problems he submitted to the "Educational Times" (a publication with a section somewhat like the problem section of the American Mathematical Monthly) concerns the intrinsic equation for a kite string under the action of wind. He submitted questions and solutions to this problem section all his life (most are included in his collected works), in fact during his first year at Cambridge he submitted 16 such problems to the "Educational Times" in addition to those two mathematical papers. Although he became a very sick man, previously to his illnesses he was very active. He enjoyed gymnastics and had a very high power to weight ratio. As a boy he was prouder of his gymnastic feats than his academic ones. He claimed to have invented a 'movement' which he called the 'corkscrew'. This was performed with a long vertical pole which he ran towards and then grabbed high up and finally swung around and around as he descended - the aim being to decrease the pitch as much as possible by increasing the number of turns in the descent. Of course there are right handed spin(or)s and left handed spin(or)s! Reading his famous paper entitled "Further note on biquaternions" we can see Clifford projecting into right and left corkscrews and developing all the calculus of the Dirac algebra of spinors.

While a Fellow at Trinity, Clifford attended the annual meeting of the British Association for the Advancement of Science held at his home city in 1869. In those days this was the leading scientific meeting in the United Kingdom. J.J. Sylvester was the chairman and it is the tradition that the chairman's addresses were a survey of the whole of their subjects: with a special emphasis on new discoveries since the previous meetings. Sylvester gave a learned and witty defence of mathematics against attacks by T.H. Huxley and others. Huxley accused mathematics of being dull and unimaginative being a mere procession of syllogisms from self evident truths. Sylvester's rebuttal included a vivid counterexample in the use of the highest order in geometry. He told the story of Gauss, who used to say that he had laid aside several questions which he had treated analytically, and hoped to apply to them the imaginative methods of geometry in a future state of existence when his conceptions of space should have been amplified and extended. For as we can conceive of beings like infinitely attenuated bookworms in an infinitely thin sheet of paper which possess only the notion of space in two dimensions, so we may imagine beings capable of realizing space of four or more dimensions. Sylvester reported that:-

"Mr. W.K. Clifford has indulged in some remarkable speculations as to the possibility of our being able to infer that our space of three dimensions is in the act of undergoing in space of four dimensions (space as inconceivable to us as our space to the supposititious bookworm) a distortion analogous to the rumpling of the page to which that creature's powers of direct perception have been postulated to be limited".

Clifford's famous abstract to the Cambridge Philosophical Society followed in a few months. It is clear that he was well aware of the empirical basis of geometry as it applies to the continuum in which we live. With the dynamic possibility that the curvatures represent motion of matter we clearly have a precursor and a prophetic vision of Einstein's general theory. Clifford is the founding father of geometrodynamics. This is the marriage of geometry with dynamics so that the geometry of space becomes a dynamical object subject to equations of motion. It is worthwhile repeating, for a new generation of theoretical scientists, this remarkable abstract: so they may see how ideas, now commonplace, were immensely imaginative and powerful steps in the past.

"I wish to indicate a manner in which these speculations (-that there are different kinds of space and that we can find out by experience to which of these kinds of space in which we live belongs-) may be applied to the investigation of physical phenomena. I hold in fact:

(1) That small portions of space are in fact of a nature analogous to little hills on the surface which is on the average flat; namely, that the ordinary laws of geometry are not valid to them.

(2) That this property of being curved or distorted is continually being passed on from one portion of space to another after the manner of a wave.

(3) That this variation of the curvature of space is what really happens in that phenomenon which we call motion of matter.

(4) That in the physical world nothing else takes place but this variation, subject (possibly) to the laws of continuity."

In passing it is of interest to note the other preoccupations at the Exeter meeting of the British Association. There is nothing new under ...

(i) The pressures of taxation on the populace;
(ii) Comparison of brain size of European and African;
(iii) Maxwell appeals that we go metric;
(iv) Stokes comments on the rapid progress of science and the difficulty of keeping up;
(v) Stokes is sorrowful over the relations between government and science: especially because the treasury refuses money for a statue to Faraday;
(vi) Teaching of mathematics in schools is very bad: committee set up including Cayley, Sylvester and Clifford.

Sylvester was a great friend of Clifford, he was also professor (of Physics!) at University College before going to the United States. They shared a joke about certain other scientists to whom they gave the name "Faranights". Mr. Faranight differs from Mr. Faraday as night does from day: he does science for commercial gain and personal aggrandisement.

Clifford was a most successful professor. He was interested in all educational problems: one of his essays, entitled "Virchow and the Teaching of Science", concerns that ever-present dilemma, the gulf between 'being told' and 'finding out'. Of course, Clifford came down on the side of the angels in the form of Pestalozzi principles. He was an excellent teacher and was very much liked by his students: his syllabuses in applied mathematics could well stand today in the classical branches of our subject. One innovation was to teach the women and men together. Univerity College was first to have any women students but before Clifford they were taught in separate classes. In fact their classes changed on the half hour with the men changing on the hour: to prevent intermingling! Clifford was conscious of educational claims outside the narrow university mores: he gave courses, for example, to the Ladies of Kensington and at Shoreditch College. An analogy would be the Ladies of Belle Air and Watts College. In a letter to his fiancee he wrote:-

"... there is room for some earnest person to go and preach around in a simple way the main straightforward rules that society has unconsciously worked out and that are floating about in the air; to do as well as possible what one can do best; to work for the improvement of the social organization; to seek earnestly after truth and only to accept provisionally opinions one has not inquired into; to regard men as comrades in work and their freedom as a sacred thing; in fact, to recognize the enormous and fearful difference between truth and falsehood, right and wrong, and how truth and right are to be got at by free inquiry and the love of our comrades for their own sakes and nobody else's. The world is a great workshop where we all have to do our best to make something good and beautiful with the help of others. Such a preaching to the people of the ideas taught by the great Rabbis was (as near as we can make out) the sort of work that Christ did; but he differed from the Rabbis and resembled all other Jewish prophets in not being able to stand priests."

Much of Clifford's contemporary prominence was not due to mathematics but to his attitude to religion. Animated by an intense love of truth and devotion to public duty, he waged war on such ecclesiastical systems as seemed to him to favour obscurantism and to put claims of sect above those of human society. The alarm of the establishment was greater in those days as theology was still unreconciled with the Darwin theory; Clifford was regarded as a dangerous champion of the antispiritual tendencies then imputed to modern modern science. His ideas were powerful and his presentation equally so: these appeared in many well-known periodicals of various philosophical, political and scientific societies and also in several discourses and orations. In his polemical writings and speeches Clifford ventured into psychology, education, morals and ethics. He founded the Republican society. For those mathematicians who do not know about this aspect of his life they have a treat in store if they read his collected works, outside mathematics, "Lectures and Essays" published in two volumes by Macmillan: these writings exceed his mathematical collected works.

Clifford married Ethel Lucy in 1875 and they had two daughters. He contracted pulmonary tuberculosis and made several visits to sunnier climates than that of London in the 1870's. These did initially seem to halt the disease but only temporarily. He died with his girls too young to know their father. He left them a great legacy in the example of his life and also with a delightful children's story "The Giant's Shoes". He wrote the inscription for his tomb:-

I was not, and was conceived;
I loved, and did a little work;
I am not, and grieve not.

It is appropriate to end with a quotation from Clifford's essay "The ethics of belief":-

"No one man's belief is a private matter which concerns him alone. Our lives are guided by that general conception of the course of things which has been created by society for social purposes. Our words, our phrases, our forms and processes and modes of thought, are common property, fashioned and perfected from age to age; an heirloom which every succeeding generation inherits as a precious deposit and a sacred trust to be handed on to the next one, not unchanged but enlarged and purified. Into this, for good or ill, is woven every belief of every man who has speech of his fellows. An awful privilege, and an awful responsibility that we should help to create the world in which posterity will live".

Clifford at the blackboard

WORKSHOP ON "CLIFFORD ALGEBRAS" UNIVERSITY OF KENT AT CANTERBURY · SEPTEMBER

KEY TO PHOTOGRAPH OF WORKSHOP PARTICIPANTS

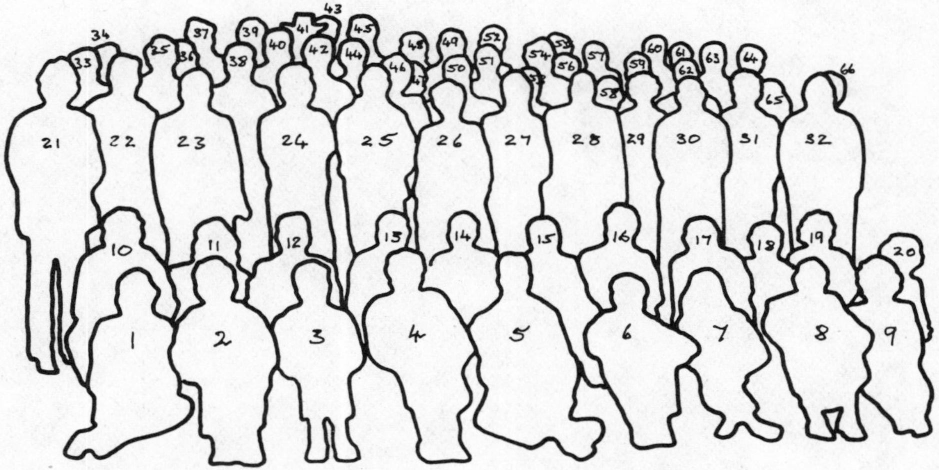

1. Mrs. Ellen Common
2. Prof. Nuri Kuruoglu
3. Prof. Bernard Jancewicz
4. Prof. Erik Folke Bolinder
5. Prof. Bill Pezzaglia
6. Prof. Marvin Ross
7. Mrs. Carol Ross
8. Prof. Jack Powles
9. Mrs. M. Chisholm
10. Dr. Aristophanes Dimakis
11. Mrs. Aristophanes Dimakis
12. Dr. Alan Common
13. Prof. Gr. Tsagas
14. Prof. Kuni Imaeda
15. Prof. Ken Greider
16. Prof. Roy Chisholm
17. Dr. Geoffrey Dixon
18. Prof. Vera Figueredo
19. Mrs. Bunty Deschamps
20. Mrs. Charlotte Kaehler
21. Dr. Jim Crawford
22. Prof. Josep Parra Serra
23. Prof. Julian Lawrynowicz
24. Dr. John Ryan
25. Prof. Richard Delanghe
26. Mrs. Regina Arcuri
27. Mr. Ricardo Marques Pereira
28. Prof. Alladi Ramakrishnan
29. Mme. Pierre Angles
30. Prof. Georges Deschamps
31. Dr. Ruth Farwell
32. Prof. Artibano Micali
33. Dr. Frank Sommen
34. Prof. Max Dresden
35. Dr. Willy Pincket
36. Mrs. Berthe Dresden
37. Dr. Richard Ziolkowski
38. Dr. Ian Porteous
39. Prof. Vladimir Soucek
40. Prof. Freddie Brackx
41. Mr. Tony Smith
42. Mr. Michael Friday
43. Prof. Garrett Sobczyk
44. Mr. David Curley
45. Prof. Pertti Lounesto
46. Prof. Jarolim Bures
47. Dr. Krystyna Bugaska
48. Prof. George McVittie
49. Prof. David Hestenes
50. Prof. Albert Crumeyrolle
51. Mrs. Nancy Hestenes
52. Prof. John Taylor
53. Prof. Lars Ahlfors
54. Prof. Jaime Keller
55. Dr. Robin Tucker
56. Prof. Krystof Kwasniewski
57. Dr. Dieter Ebner
58. Mrs. Erna Ahlfors
59. Dr. Pierre Angles
60. Prof. Bernard Goldschmidt
61. Dr. John McEwan
62. Prof. Hans Joos
63. Prof. James Brooke
64. Dr. Tuong Truong
65. Dr. Mari Imaeda
66. Prof. Erich Kaehler

A UNIFIED LANGUAGE FOR MATHEMATICS AND PHYSICS

David Hestenes
Physics Department
Arizona State University
Tempe, Arizona 85287
U.S.A.

ABSTRACT. To cope with the explosion of information in mathematics and physics, we need a unified mathematical language to integrate ideas and results from diverse fields. Clifford Algebra provides the key to a unified Geometric Calculus for expressing, developing, integrating and applying the large body of geometrical ideas running through mathematics and physics.

INTRODUCTION

This first international workshop on Clifford Algebras testifies to an increasing awareness of the importance of Clifford Algebras in mathematics and physics. Nevertheless, Clifford Algebra is still regarded as a narrow mathematical specialty, and few mathematicians or physicists could tell you what it is. Of those who can, the mathematicians are likely to characterize Clifford Algebra as merely the algebra of a quadratic form, while the physicists are likely to regard it as a specialized matrix algebra. However, the very composition of this conference suggests that there is more to Clifford Algebra than that. Assembled here we have mathematicians, physicists and even engineers with a range of backgrounds much wider than one finds at a typical mathematics conference. What common insights bring us together and what common goal do we share? That is what I would like to talk about today.

The fact that Clifford Algebra keeps popping up in different places throughout mathematics and physics shows that it has a universal significance transcending narrow specialties. I submit that Clifford Algebra is as universal and basic as the real number system. Indeed, it is no more and no less than an extension of the real number system to incorporate the geometric concept of direction. Thus, a Clifford Algebra is a system of directed numbers. I have argued elsewhere [1] that Clifford Algebra is an inevitable outcome of the historical evolution of the number concept. Its inevitability is shown by the fact that it has been rediscovered by different people at many different times and places. You might prefer to say "reinvented"

J. S. R. Chisholm and A. K. Common (eds.), Clifford Algebras and Their Applications in Mathematical Physics, 1–23.

instead of "rediscovered," but with your leave, I will not attempt to distinguish between inventions and discoveries here.

Everyone here knows that the Pauli and Dirac algebras are Clifford Algebras invented by physicists to solve physical problems without knowing about Clifford's results. The same can be said about the invention of the Clifford Algebra of fermion creation and annihilation operators. I have personally witnessed several rediscoveries of Clifford Algebra which are not apparent in the literature. Two were enountered in papers I refereed. A more spectacular and fruitful rediscovery was made by the noted expert on combinatorial mathematics, Gian-Carlo Rota. He contacted me to find out how he could secure a copy of my book Spacetime Algebra; when I asked him why he was interested, he handed me a couple of his articles. These articles developed a mathematical system which Rota claimed to be ideal for the theory of determinants. I was astounded to discover that, except for trivial differences in notation, it was identical to the formulation of determinant theory which I had worked out with Clifford Algebra. In developing these articles, Rota and his coworker Stein had essentially rediscovered Clifford Algebra, though Rota himself already knew Clifford Algebra very well. Of course, Rota must have suspected the connection and not taken the time to clear it up. The reinvention was not superfluous, however, because it produced a number of elegant results which I was able to incorporate in a broader treatment of Clifford Algebra [2].

I am sure that members of this audience can tell about other rediscoveries of Clifford Algebra. I think the discoveries of the complex numbers and quaternions should be included in this list, for I will argue later that they should be regarded as Clifford Algebras. Moreover, we shall see that even Clifford may not have been the original discoverer of Clifford Algebra.

The phenomenon of multiple discovery shows us that the development of Clifford Algebra is not the work of isolated individuals. It is a community affair. Excessive emphasis on priority of scientific discovery has contributed to the mistaken belief that scientific progress is primarily the work of a handful of great men. That belief does not take adequate account of the context of their discoveries, as indicated by the fact that the same scientific discovery is frequently made by two or more people working independently. Sociologist Robert Merton [4] has documented a large number of such multiple discoveries and argued convincingly that, contrary to conventional wisdom, multiple discoveries are far more common than discoveries by one person alone. Moreover, the more important the discovery, the more likely it is to be multiple. There is no better example than the multiple discoveries of Clifford Algebra.

The number is extremely large when all the significant variants of Clifford Algebra are included. This goes to show that genuine creative powers are not limited to a few men of genius. They are so widespread that discovery is virtually inevitable when the context is right. But the context is a product of the scientific community. So scientific discovery should not be regarded as an independent achievement by an individual. In the broad sense of discovery, Clifford Algebra should be seen as a common achievement of the scientific community.

I believe the phenomenon of multiple discovery is far more pervasive than historical accounts suggest. I hold with the Swiss psychologist Jean Piaget that the process of learning mathematics is itself a process of rediscovery. A similar view was expressed by the logician Ludwig Witgenstein, asserting in the preface to his famous Tractatus Logico-Philosophicus that he would probably not be understood except by those who had already had similar thoughts themselves. From that perspective, everyone at this conference is a codiscoverer of Clifford Algebra. That view may seem extreme, but let me point out that it is merely the opposite extreme from the view that mathematical discoveries are made by only a few men of genius. As usual, we find a deeper truth by interpolating between the extremes. Though we have many common discoveries, some of us had the benefit of broader hints. Mathematics progresses by making rediscovery easier for those who are learning it.

There may be no branch of mathematics in which more multiple discoveries can be found than in the applications of Clifford Algebra. That is because Clifford Algebra arises in so many "hot spots" of mathematics and physics. We will have many opportunities to note such discoveries, many of them by participants in this workshop. In fact, I don't believe I have ever attended a workshop with so many multiple discoverers.

The subject of spinors, about which we will be hearing a lot, may be the hot spot which has generated the largest number of multiple discoveries. I have not met anyone who was not dissatisfied with his first readings on the subject. The more creative individuals, such as the participants in this conference, are likely to try to straighten out matters for themselves. So off they go generating rediscoveries, most likely to be buried in the vast, redundant and muddled literature on spinors. One of the jobs of this conference should be to start straightening that literature out. Let us begin by noting that Professor Kähler [5] was the first among us to study the geometrical significance of representing spinors as elements of minimal ideals in the Dirac algebra.

Now I should explain my reason for bringing up the phenomenon of multiple discoveries. Though multiple discoveries may be celebrated as evidence for widespread creative powers of mankind, I want to emphasize that they are also symptoms of an increasingly serious problem, namely, the breakdown of communication among the different branches of science and mathematics. Needless duplication of effort, including multiple discoveries, is an obvious consequence of this breakdown. Moreover, progress will be seriously impeded when important results in one field are neither rediscovered nor transmitted to other fields where they are applicable.

During the last thirty years, the mathematics curriculum in universities of the United States has grown progressively less relevant to the training of physicists, so physics departments have taught more and more of the mathematics physicists need. This is only one of many signs of the growing gulf between physics and mathematics. But mathematics itself is becoming progressively more fragmented. Mathematicians in neighboring disciplines can hardly talk to one another. To mention only one example which I know very well, my father is an expert in the calculus of variations, but he has never communicated significantly with differential geometers, even in his own department, though their fields are intimately related. He is impatient with their jargon, for whenever he took the trouble to translate it, he found, they were saying something he already knew. So why bother? On the other hand, modern geometers tend to regard the calculus of variations as out of date, though they seldom reach its level of rigor and generality; even less do they realize that the field is still growing vigorously.

The language barriers between different branches of mathematics makes it difficult to develop a broad grasp of the subject. One frequently hears the complaint that mathematics is growing so rapidly that one cannot hope to keep up, let alone catch up. But how much of it is real growth, and how much is rediscovery or mere duplication of known results with a different nomenclature? New branches are being created, and names, definitions and notations are proliferating so profusely that the integrity of the subject is threatened. One gets the impression sometimes that definitions are being created more rapidly than theorems. Suffering from a Babylon of tongues, mathematics has become massively redundant.

As an example of redundancy even in traditional mathematics consider the following list of symbolic systems:

synthetic geometry	matrix algebra
coordinate geometry	Grassmann algebra
complex analysis	Clifford Algebra
vector analysis	differential forms
tensor analysis	spinor calculus

Each of these systems provides a representation of geometric concepts. Taken together, therefore, they constitute a highly redundant system of multiple representations for geometric concepts. Let us consider the drawbacks of such multiple representation:

(a) Limited access. Most mathematicians and physicists are proficient in only a few of these ten symbolic systems. Therefore, they have limited access to results developed in a system with which they are unfamiliar, even though the results could be reformulated in some system which they already know. Moreover, there is a real danger that the entire mathematical community will lose access to valuable mathematical knowledge which has not been translated into any of the current mathematical languages. This appears to be the case, for example, for many beautiful results in projective geometry.

(b) Cost of translation. The time and effort required to translate from one symbolic system to another is unproductive, yet translation is common and even necessary when one system is too limited for the problem at hand. For example, many problems formulated in terms of vector calculus or differential forms can be simplified by re-expressing them in terms of complex variables if they have 2-dimensional symmetry.

(c) Deficient integration. The collection of ten symbolic systems is not an integrated mathematical structure. Each system has special advantages. But some problems call for the special features of two or more different systems, so they are especially unwieldy to handle. For example, vector analysis and matrix algebra are frequently applied awkwardly together in physics.

(d) Hidden structure. Relations among geometric concepts represented in different symbolic systems are difficult to recognize and exploit.

(e) Reduced information content (high redundancy), is a consequence of using several different symbolic systems to represent a coherent body of knowledge such as classical mechanics.

It goes without saying that elimination of these drawbacks would make mathematics easier to learn and apply. The question is, can it be done? And how can it be done without impairing the richness of mathematical structure? There is good reason to believe that it can be done, because all ten of the symbolic systems in our list have a common conceptual root in geometry. Historically, each system was developed to handle some special kinds of geometrical problem. To some extent, the creation of each system was a historical accident. But, from a broad historical perspective, each creation can be seen as an offshoot of a continuous evolution of geometric and algebraic concepts. It is time we integrated that accumulated experience into a unified mathematical system.

This presents us with a fundamental problem in the design of mathematics: How can we design a coherent mathematical language for expressing and developing the full range of geometric concepts? In other words, how can we design a comprehensive and efficient geometric calculus? That is the problem I would like to discuss with you today. Before we get down to specifics, however, I would like set the problem in a general context.

The information explosion presents us with a staggering problem of integrating a wealth of information in science and mathematics so it can be efficiently stored, retrieved, learned and applied. To cope with this problem, we need to develop a Metamathematical Systems Theory concerned with the design of mathematical systems. The theory should explicate the principles and techniques of good mathematical design, including the choice of notations, definitions, axioms, methods of proof and computation, as well as the organization of theorems and results. I submit that one of the secrets of mathematical genius is being privy to powerful mathematical design principles which are not explicitly formulated in the literature. By formulating the principles explicitly, they can be more readily taught, so people don't have to keep rediscovering them. Moreover, the principles can then be criticized and improved. We cannot afford the laissez-faire proliferation of mathematical definitions and notations. This is not to say that freedom in mathematical design should be in any way restricted. But we need rational criteria to distinguish good design from poor design. Custom is not enough.

In computer science the theory of software systems design is undergoing vigorous development. The design of mathematical systems is no more than software systems design at the highest level. So the theories of mathematical design and software design should merge continuously into one another. They should also merge with a new branch of applied mathematics called "Mathematical Systems Theory."

Returning to the specific problem of designing a Geometric Calculus, we note that this should take us a long way towards a general theory of mathematical design, for the corpus of geometric concepts underlies a major portion of mathematics, far exceeding the ten symbolic systems mentioned above. Actually, the design and development of Geometric Calculus is far along, far enough, I think, so it can be claimed without exaggeration to apply to a wider range of mathematics and physics than any other single mathematical system. I have been working on the development of Geometric Calculus for about 25 years, and a significant portion of the results are published in references [1], [2] and [3]. I am keenly aware that most of the results are translations and adaptations of ideas developed by others. I have endeavored to incorporate into the system every good idea I could find. But many ideas would not fit without some modification, so the process involved many decisions on the details of mathematical design. Although we cannot hope to cover many of the details, I would like to review with you today some of the most important ideas that have gone into the design of Geometric Calculus, and outline a program for further development of the language. As I see it, many participants at this workshop have already contributed to the development of Geometric Calculus, whether they have thought of their work that way or not. Anyway, the future development will require contributions from people with diverse backgrounds, such as we have here. I hope many of you will join me to make this a community program.

1. DIRECTED NUMBER SYSTEMS

The first step in developing a geometric calculus is to encode the basic geometric concepts in symbolic form. Precisely what geometric concepts are basic is a matter of some dispute and partly a matter of choice. We follow a preferred choice which has emerged from generations of mathematical investigation. We take the concepts of magnitude and direction as basic, and introduce the concept of vector as the basic kind of directed number.

What does it mean to be a directed number? The answer requires an algebraic definition and a geometric interpretation. Directed numbers are defined implicitly by specifying rules for adding and multiplying vectors. Specifically, we assume that the vectors generate an associative algebra in which the square of every vector is a scalar. Multiplication is thus defined by the rules, for any vectors a, b, c,

$$a(bc) = (ab)c \tag{1.1}$$

$$a(b + c) = ab + ac$$
$$(b + c)a = ba + ca \tag{1.2}$$

$$a^2 = \text{scalar} \tag{1.3}$$

Assuming that the scalars are real numbers, the sign of the scalar a^2 is called the signature of the vector a. So the signature of a vector may be positive, negative or zero.

These simple rules defining vectors are the basic rules of Clifford Algebra. They determine the mathematical properties of vectors completely and generate a surprisingly rich mathematical structure. But the feature making Clifford Algebra stand out among algebraic systems is the geometric interpretation which can be assigned to multiplication. Although the interpretation is not part of the formal algebraic system, it determines how the system is developed and applied, in particular, how it is applied to physics. The interpretation turns Clifford Algebra into the grammar for a language describing something beyond itself.

The geometric interpretation is most apparent when the product ab is decomposed into symmetric and antisymmetric parts by writing

$$ab = a\cdot b + a\wedge b, \tag{1.4}$$

where

$$a\cdot b = \frac{1}{2}(ab + ba) \tag{1.5}$$

$$a\wedge b = \frac{1}{2}(ab - ba) \tag{1.6}$$

We may regard (1.5) and (1.6) as definitions of subsidiary products, a symmetric inner product $a\cdot b$ and an antisymmetric outer product $a\wedge b$. The product $a\cdot b$ is just the usual scalar-valued inner product on a vector space, and its geometric interpretation is well known to everyone here. The quantity $a\wedge b$ is neither vector nor scalar; it is called a bivector (or 2-vector) and can be interpreted as a directed area, just as a vector can be interpreted as a directed length. The outer product $a\wedge b$ therefore describes a relation between vectors and bivectors, two different kinds of directed numbers. Similarly the inner product $a\cdot b$ describes a relation between vectors and scalars. According to (1.4), these are combined in a single geometric product ab, which is itself a kind of composite directed number, a quantity which completely describes the relative directions and magnitudes of the vectors a and b. More details on the interpretation of the geometric product and its history are given in [1].

The inner and outer products were originated by H. Grassmann as representations of geometric relations. I learned from Professor Bolinder only a few years ago that, late in his life, Grassmann [6] added the inner and outer products to form a new product exactly as expressed by (1.4). Thus, Grassmann discovered the key idea of Clifford Algebra independently of Clifford and evidently somewhat before him. This has been overlooked by historians of mathematics, who have dismissed Grassmann's later work as without interest.

Generalizing (1.6), for the antisymmetrized product of k vectors $a_1, a_2, \ldots, a_k$, we write

$$A_k = a_1 \wedge a_2 \wedge \ldots \wedge a_k \tag{1.7}$$

This is exactly Grassmann's k-fold outer product. Unless it vanishes it produces a directed number A_k called a k-vector, which can be interpreted as a directed k-dimensional volume. I will refer to the integer k as the grade of the k-vector. I have adopted the term "grade" here mainly because it is monosyllabic, and there are some problems with alternative terms such as "dimension". The relative direction of a vector a and a k-vector A_k is completely characterized by the geometric product

$$aA_k = a \cdot A_k + a \wedge A_k \tag{1.8}$$

where, generalizing (1.5) and (1.6),

$$a \cdot A_k = \frac{1}{2}(aA_k + (-1)^{k+1} A_k a) \tag{1.7}$$

is a (k-1)-vector and

$$a \wedge A_k = \frac{1}{2}(aA_k + (-1)^{k} A_k a) \tag{1.8}$$

is a (k+1)-vector. Thus, the inner product is a grade lowering operation while the outer product is a grade raising operation. The inner product differs from a scalar product in that it is not necessarily scalar-valued. Rather, the inner product (1.7) is equivalent to the operation of contraction in tensor algebra, since k-vectors correspond to antisymmetric tensors of rank k.

This brings up an important issue in mathematical design. Clifford algebras are sometimes defined as certain ideals in tensor algebras. There is nothing logically wrong with this, but I submit that it is better mathematical design to reverse the process and introduce tensor as multilinear functions defined on Clifford algebras. Two reasons for this. First, the geometric product should be regarded as an essential part of the definition of vectors, since it is needed for the interpretation of vectors as directed numbers. Second, Clifford algebras are more versatile than tensors in applications, as I believe is born out in practice.

I should point out that if we regard the geometric product as essential to the concept of vector, as required by the most basic design principle of geometric calculus, then we must distinguish vector spaces on which the product is defined from linear spaces on which it is not. We must eschew the standard use of the term "vector" for elements of arbitrary linear spaces. If a precedent is needed for this, recall that when Hamilton introduced the term "vector" he definitely regarded multiplication as part of the vector concept.

The notation for Clifford Algebras is not yet standardized in the literature, so for the purposes of this conference, let me recommend one employed by Porteous [7] and Lounesto [8]. Let $\mathcal{R}^{p,q}$ denote a vector space of dimension $n = p + q$ over the reals consisting of a p-dimensional subspace of vectors with positive signature orthogonal to a q-dimensional subspace of vectors with negative signature. In this context, "orthogonality" of vectors means a vanishing inner product. Let $\mathcal{R}_{p,q}$ denote the Clifford Algebra generated by the vectors of $\mathcal{R}^{p,q}$, and let $\mathcal{R}^k_{p,q}$ be the $\binom{n}{k}$-dimensional subspace of k-vectors in $\mathcal{R}_{p,q}$. Elements of the 1-dimensional subspace $\mathcal{R}^n_{p,q}$ are called pseudoscalars of $\mathcal{R}^{p,q}$ or $\mathcal{R}_{p,q}$. An arbitrary element of $\mathcal{R}_{p,q}$ is called a multivector or Clifford number.

For the important case of a vector space with Euclidean (positive) signature, it is convenient to introduce the abbreviations $\mathcal{R}^n = \mathcal{R}^{n,o}$, $\mathcal{R}_n = \mathcal{R}_{n,o}$ and $\mathcal{R}^k_n = \mathcal{R}^k_{n,o}$. An n-dimensional vector space $\mathcal{C}^n$ over the complex numbers generates a Clifford algebra $\mathcal{C}_n$ with k-vector subspaces $\mathcal{C}^k_n$.

I am of half a mind to outlaw the Complex Clifford Algebras altogether, because the imaginary scalars do not have a natural geometric interpretation, and their algebraic features exist already in the real Clifford Algebras. However, there is already a considerable literature on complex Clifford Algebras, and they do have some formal advantages. For example, a student of mine, Patrick Reany, has recently shown [9] that the solutions of any cubic equation over the complex numbers are contained in the solutions X of the simple cubic equations $X^3 = C$, where X and C are elements of the complex algebra $\mathcal{C}_1$. This opens up new possibilities for the theory of algebraic equations. Note that $\mathcal{C}_1$ is the largest Clifford Algebra for which all the elements commute. But note also that $\mathcal{C}_1$ can be regarded as a subalgebra of the real Euclidean algebra $\mathcal{R}_3$, with the pseudoscalar $\mathcal{R}^3_3$ playing the role of imaginary scalars.

Clifford Algebras are worth studying as abstract algebraic systems. But they become vastly richer when given geometrical and/or physical interpretations. When a geometric interpretation is attached to a

Clifford Algebra, I prefer to call it a Geometric Algebra, which is the name originally suggested by Clifford himself. In this case, $R_{p,q}$ is to be regarded as a generalized number system, a system of directed numbers for making geometrical representations and computations. To emphasize the computational features of the number system, I like to call $R^{p,q}$ arithmetic (p,q)-space.

2. GEOMETRIC FUNCTION THEORY

Having identified Geometric Algebra as a suitable grammar representing basic geometrical relations, the next step in developing a Geometric Calculus is to use this grammar to construct a theory of functions for representing complex geometrical relations and structures. I like to call this construction Geometric Function Theory. Of course we want to integrate into the theory the rich body of results from real and complex analysis. This presents us with a definite design problem. I hope to convince you that this problem has a beautiful solution which points the direction to a powerful general theory. The key to the solution is an analysis of the geometric structure implicit in complex variable theory.

It is customary for mathematicians and physicists to regard complex numbers as scalars, mainly I suppose because it is an algebraic completion of the real number system. From the broader perspective of geometric algebra, however, this custom does not appear to be a good decision in mathematical design. Nor is it consistent with the geometrical interpretation of complex numbers which motivated the historical development of complex analysis and is crucial to many of it applications. From the beginning with multiple discovery of complex numbers by Wessel and Gauss, the unit imaginary i was interpreted as the generator of rotations in a 2-dimensional space. Moreover, development of the subject was stimulated by physical interpretations of complex functions as electric fields, magnetic fields or velocity fields of fluid flow in a physical plane.

Despite the great richness and power of complex function theory, it suffers from a defect in design which makes it difficult to see how the theory should be genralized beyond two dimensions. Fortunately, the defect is easily repaired. The defect is an ambiguity in geometrical representation. Throughout the subject, complex numbers play two distinctly different geometrical roles which are not differentiated by the mathematical formalism. On the one hand, a complex number may represent a point in a plane; on the other hand it may represent a rotation-dilation. Moreover, the representation of points has the defect of singling out a preferred direction (the real axis) with no geometrical or physical significance in applications.

The defect can be removed by regarding the "plane" of complex numbers as the even subalgebra R_2^+ of the geometric algebra R_2 for the "real plane" R^2. Then we can write $R_2 = R^2 + R_2^+$, exhibiting the entire algebra R_2 as the sum of the real and complex planes. The crucial points of this construction is that the unit imaginary i is now to be interpreted as a bivector, the unit pseudoscalar for the real plane R^2, but it is also the generator of rotations in the real plane. For it is easily proved any vector in R^2 by 90°. Parenthetically, let me note that we could have achieved the same result by using the geometric algebra $R_{0,2}$ instead of $R_2 = R_{2,0}$. But then the vectors would have negative square, which might be a drawback or an advantage, depending on what you want to do with it. The advantage is that $R_{0,2}$ can be identified with the quaternions and the even subalgebra of R_3 .

A linear mapping of the real plane $\{x\} = R^2$ onto the complex plane $\{z\} = R_2^+$ is defined by the equation $z = xe_1$, where e_1 is a fixed unit vector specifying the direction which is mapped into the real axis. By the way, the same kind of linear mapping of vectors into the even subalgebra is of great importance in all Clifford Algebra, because it defines a linear relation between Clifford Algebras of different dimension. I call it a <u>projective</u> <u>mapping</u>, because it provides an algebriac formulation for the basic idea of introducing homogeneous coordinates in projective geometry. We shall encounter it again and again in this conference.

We can use this projective mapping between the real and complex plane to map functions of a complex variable to functions on the real plane where there is no preferred real axis. As an important example, we note that for an analytic function F(z), Cauchy's Theorem

$$\oint dzF(z) = 0 \qquad \text{maps to} \qquad \oint dxf(x) = 0, \tag{2.1}$$

where $f(x) = e_1F(xe_1)$ is a vector field on the real plane. Note that $dx = dze_1$ is a vector-valued measure for the curve over which the integral is taken. I believe that this is the key to the great power of complex analysis, and we can transfer this power to all of real analysis if we use directed measures in defining multiple integrals. Let me explain how.

As you know, Cauchy's Theorem can be derived from Green's Theorem which, using geometric algebra can be written in the form

$$\int_A dA\partial f = \oint_{\partial A} dxf. \tag{2.2}$$

Here dA is an element of directed area, that is, a bivector-valued measure, so it can be decomposed into $dA = i|dA|$ where i is the unit bivector. Also, $\partial f = \partial_x f(x)$ is the derivative of f with respect to the vector x. The operator $\partial = \partial_x$ is algebraically a vector, because x is a vector. Therefore, we can use (1.6) to write $\partial f = \partial \cdot f + \partial \wedge f$, which is a decomposition of the vector derivative into divergence and curl. Note that $dA \wedge \partial = 0$, so $dA\partial = dA \cdot \partial$ has the same grade as dx, which, of course, is essential if (2.2) is to make sense.

It is fair to ask how the derivative with respect to a vector should be defined. This is a design problem of the utmost importance, because $\partial = \partial_x$ is the fundamental differential operator of Geometric Calculus. The solution is incredibly simple and can be found by applying one of the most basic principles of mathematical design: Design your definitions to put the most important theorems in the simplest possible form. Accordingly, we define the differential operator ∂ so that (2.2) is true. We simply define ∂f at x by taking the appropriate limit of (2.2) as the set A shrinks to the point x. This took me some time to realize, as I originally defined ∂ differently and then derived (2.2) with $dA \cdot \partial$ on the left without realizing how to get rid of the dot (Ref [3]). Note that this definition of the derivative is completely coordinate-free.

Equation (2.2) is the most important theorem in integral calculus. It applies not only to the real plane but to any two dimensional surface A in $R^{p,q}$ with boundary ∂A, with possible exceptions only on null surfaces. Indeed, it applies without change in form when A is a k-dimensional surface, if we interpret dA as a k-vector-valued measure on A and dx as a (k-1)-vector-valued measure on ∂A. It thus includes the theorems of Gauss and Stokes, and it generalizes the so-called generalized Stoke's theorem from the theory of diffrential forms. To relate (2.2) to differential forms when A is a 2-dimensional surface, we simply take its scalar part. From the right side of (2.2) we get a 1-form $\omega = dx \cdot f$ and from the left side we get the exterior derivative $d\omega = dA \cdot (\partial \wedge f)$. So we get the generalized Stokes' theorem

$$\int_A d\omega = \int_{\partial A} \omega \tag{2.3}$$

The trouble with this is it is only half of (2.2). The other half is the bivector equation

$$\int_A dA\partial \cdot f = \int_{\partial A} dx \wedge f. \tag{2.4}$$

It is true that this can also be expressed in terms of differential forms by introducing "dual forms", but the damage has already been done. By splitting (2.2) into parts we have diminished the power of the directed measure characteristic of complex analysis. With differential forms one cannot even write Cauchy's theorem (2.1) as a single formula. Differential forms bury the directed measure which is explicit in (2.2).

Equation (2.2) is so important that it should ot be named after Stokes or any other person. It is the result of contributions by many people. In the 1-dimensional case when A is a curve with endpoints a and b, (2.2) reduces to

$$\int_a^b dx\partial f = \int_a^b dx\cdot\partial f = \int_a^b df = f(b) - f(a). \qquad (2.5)$$

This is widely known as <u>the fundamental theorem of integral calculus.</u> That is a suitable title for its generalization (2.2). There is more about this theorem in Ref. [2].

Returning to our reformulation of complex analysis on the real plane, we see that Cauchy's theorem (2.1) follows from the fundamental theorem (2.2) if and only if

$$\partial f = 0. \qquad (2.6)$$

This implies that the divergence $\partial\cdot f$ and curl $\partial\wedge f$ vanish separately. It is, of course, just a coordinate-free form of the Cauchy-Riemann equations for an analytic function. Indeed, (2.6) is well-defined for any k-dimensional surface A in $R^{p,q}$ on which ∂ is defined, and (2.2) then gives a corresponding generalization of Cauchy's Theorem. Therefore, (2.6) is a suitable generalization of the definition of an analytic function to k-dimensional surfaces. Moreover, the main theorems of analytic function theory generalize along with it. If the variable x is a point in R^n, you might like to write $\partial_x = \nabla$, since ∇^2 is the n-dimensional Laplacian. Moreover, $\nabla f = 0$ implies $\nabla^2 f = 0$, from which one can conclude that the Cartesian components of an analytic function are harmonic functions.

If f is analytic except at a pole x' with residue q in A, then $\partial f = 2\pi q\delta(x - x')$, and substitution into (2.2) gives the "residue theorem"

$$\oint dxf = 2\pi q \int_A i|dA(x)|\ \delta(x - x') = 2\pi iq. \qquad (2.7)$$

I mention this to point out that the mysterious i in the residue theorem comes from the directed area element. This shows us immediately the generalization of the residue theorem to higher dimensions will replace the i by a unit k-vector.

Now I want to show that our reformulation of complex variable theory in terms of geometric algebra gives us someting more than the conventional theory even in two dimensions. One more reason for regarding $\partial = \partial_x$ as <u>the</u> fundamental differential operator is the fact that it has an inverse (or <u>antiderivative</u>) ∂^{-1}. Thus, if f(x) is an unknown vector (or multivector) field whose derivative $j = \partial f$ is known, then f can be obtained from $f = \partial^{-1} j$ even when $j = 0$. Of course ∂^{-1} is an integral operator and boundary values of f must be specified to get a unique inverse. An explicit expression for ∂^{-1} can be derived from a modified form of the fundamental theorem (2.2). I give only the result which is derived in Ref. [2]:

$$f(x') = \frac{1}{2\pi} \int_A |dA| \frac{1}{x - x'} \partial f(x) + \frac{1}{2\pi i} \oint \frac{1}{x' - x} dx f(x). \quad (2.8)$$

For an analytic function $\partial f = 0$ in A, this reduces to Cauchy's integral formula, where $e_1(x' - x)^{-1} = (z' - z)^{-1}$ will be recognized as the "Cauchy kernel". Thus, (2.8) is a generalization of Cauchy's integral formula to functions which are not analytic. Conventional complex variable theory does not find this generalization, because it is self-limited to the use of line integrals without area integrals. But the generalization (2.8) is valuable even in analytic function theory, because the area integral picks up the contributions from poles and cuts.

The generalization of (2.8) to higher dimensions is straightforward. One simply replaces the Cauchy kernel with an appropriate Green's function and performs directed integration over the specified surfaces. I will leave this topic to Professor Delanghe, who is one of the multiple discoverers of generalized Cauchy integral formula.

With the basic principles and theorems for vector differentiation and geometric integration in hand, the main business of Geometric Function Theory is ready to begin, namely, to develop a rich theory of special functions generalizing and integrating the magnificent results of classical function theory. This is an immense undertaking which will require an army of workers. Many participants at the conference have already contributed to the subject. One of the most elegant and valuable contributions is the generalization of the Moebius transformation by Professor Ahlfors, which we will hear about tomorrow. This fits into the program of Abel and Jacobi to develop a complete theory of algebraic functions and their integrals. However, geometric algebra gives us a much broader concept of algebraic function.

Let me suggest that "Geometric Function Theory" is a better name for this field than "Hypercomplex Function Theory" or other alternatives, because it lays claim to territory in the center of mathematics. A more esoteric name suggests that the field belongs on the periphery. I hope I have convinced you that in order to integrate complex analysis smoothly into a general geometric function theory, one should regard complex numbers rather than scalars. In this general theory the artificial distinction between real and complex analysis disappears completely, and hypercomplex analysis can be seen as multidimensional real analysis with an appropriate geometrical tool.

3. SPACETIME CALCULUS

One of the main design principles for Geometric Calculus is that the language should be efficient in all practical applications to physics. For this reason, I have endeavored to incorporate into the language definitions, notations and conventions which are as close to standard usage in physics as is consistent with good mathematical design. That accounts for some of my deviation from perfectly good notations that mathematicians prefer. The calculus is quite well developed for all branches of physics. As an example with important implications, let me outline its application to classical relativistic electrodynamics. I aim to show three things. First, the calculus provides a compact coordinate-free formulation for the basic equations of the subject, namely, the electromagnetic field equations and energy momentum tensor as well as the equation of motion for charged particles. Second, the calculus is an efficient computational tool for deriving consequences of those equations. Third, the classical formulation articulates smoothly with the formulation of quantum relativistic quantum mechanics.

To begin, we formulate a model of physical spacetime in terms of geometric algebra. When the vectors of $\mathcal{R}^{1,3}$ are employed to designate points of flat spacetime, I call the geometric algebra $\mathcal{R}_{1,3}$ which it generates the spacetime algebra (STA), because I interpret it as a complete encoding of the basic geometrical properties of spacetime. It provides us with an arithmetic of spacetime directions and magnitudes. I leave it to Garret Sobczyk to discuss the generalization of this approach to curved spacetime.

For the derivative with respect to a spacetime point x, I like to use the notation $\Box = \partial_x$ to emphasize its special importance, and because its square is the usual d'Alembertian, for which $\Box^2$ is a common notation. Of course $\Box$ is a vector differential operator which can be defined in the same coordinate-free fashion we have already discussed.

An electromagnetic field is represented by a bivector-valued function $F = F(x)$ on spacetime. The field produced by a source with spacetime current density $J = J(x)$ is determined by Maxwell's field equation

$$\Box F = J. \tag{3.1}$$

The field is found in essentially the same way we got Cauchy's integral formula, so let us write

$$F = \Box^{-1} J. \tag{3.2}$$

The Green's functions and physical boundary conditions needed to determine $\Box^{-1}$ have been thoroughly studied by physicists, mostly in a somewhat different language.

Comparison of our formulation (3.1) of Maxwell's equation with conventional formulations in terms of differential forms or tensors is easy, because the latter formalisms can be incorporated in Geometric Calculus simply by introducing a few notations. To do that, we define an electromagnetic 2-form ω by

$$\omega = \omega(dx\wedge dy) = (dx\wedge dy)\cdot F. \tag{3.3}$$

This is the projection of the electromagnetic field F onto an arbitrary directed area element $dx\wedge dy$. To get the tensor components of F, we choose an orthonormal basis $\{\gamma^\mu\}$ for $R^{1,3}$. I use the notation γ^μ which is commonly employed for the Dirac matrices to emphasize the fact that they are different mathematical representations of exactly the same physical information. I will elaborate on that point in my next lecture. Now, for the tensor components we get

$$F^{\mu\nu} = (\gamma^\nu\wedge\gamma^\mu)\cdot F = \omega(\gamma^\nu\wedge\gamma^\mu) \tag{3.4}$$

To relate the different formulations of the field equations, we use the decomposition $\Box F = \Box\cdot F + \Box\wedge F$ of the derivative into divergence and curl. The tensor form is obtained trivially by computing components with respect to a basis, using $\Box = \gamma^\mu\partial_\mu$, for example. So we can concentrate on the formulation with differential forms. The curl is related to the exterior derivative d of a differential form by

$$d\omega = dV\cdot(\Box\wedge F), \tag{3.5}$$

where $dV = dx\wedge dy\wedge dz$ is an arbitrary directed volume element. To relate the divergence to the exterior derivative we first relate it to the curl. This can be done by introducing the duality operation which, in the spacetime algebra, is simply multiplication by the unit pseudoscalar $i = \gamma_0\gamma_1\gamma_2\gamma_3$. We exploit the associative and distributive laws as follows:

$$\Box(Fi) = (\Box F)i$$

$$\Box\cdot(Fi) + \Box\wedge(Fi) = (\Box\cdot F)i + (\Box\wedge F)i.$$

Separately equating vector and trivector parts, we obtain

$$\Box\cdot(Fi) = (\Box\wedge F)i, \tag{3.6a}$$

$$\Box\wedge(Fi) = (\Box\cdot F)i. \tag{3.6b}$$

We need only the latter equation. Defining a dual form for ω by $^*\omega = (dx\wedge dy)\cdot(Fi)$, we find the exterior derivative

$$d^*\omega = dV\cdot(\Box\wedge(Fi)) = dV\cdot((\Box\cdot F)i) \qquad (3.7)$$

From the trivector Ji we get a current 3-form $^*\alpha = dV\cdot(Ji)$, which is the dual of a one form $\alpha = J\cdot dx$. Now we can set down the three different formulations of the field equations for comparison:

$$\Box\cdot F = J \qquad \partial_\mu F^{\mu\nu} = J \qquad d^*\omega = {}^*\alpha$$

$$\Box\wedge F = 0 \qquad \partial_{[\alpha}F_{\mu\nu]} = 0 \qquad d\omega = 0$$

I leave it for you to decide which form is the best. Of course, only the STA form on the left enables us to unite the separate equations for divergence and curl in a single equation. This unification is nontrivial, for only the unified equation (3.1) can be inverted directly to determine the field as expressed by (3.2).

It should be obvious also that (3.1) is invariant under coordinate transformations, rather than covariant like the tensor form. The field bivector F is the same for all observers; there is no question about how it transforms under a change of reference system. However, it is easily related to a description of electric and magnetic fields in a given inertial system in the following way:

An inertial system is determined by a single unit timelike vector γ_0, which can be regarded as tangent to the worldline of an observer at rest in the system. This vector determines a split of spacetime into space and time, which is most simply expressed by the equation

$$x\gamma_0 = t + \underset{\sim}{x}, \qquad (3.8)$$

where $t = x\cdot\gamma_0$ and $\underset{\sim}{x} = x\wedge\gamma_0$. This is just another example of a projective transformation like the one encountered in our discussion of complex analysis. It is a linear mapping of each spacetime point x into a scalar t designating a time and a vector $\underset{\sim}{x}$ designating a position. The position vectors for all spacetime points compose a 3-dimensional Euclidean vector space R^3. I denote the vectors in R^3 with boldface type to distinguish them from vectors in $R^{1,3}$. The equation $\underset{\sim}{x} = x\wedge\gamma_0$ tells us that a vector in R^3 is actually a bivector in $R^2_{1,3}$. In fact, R^3 consists of the set of all bivectors in $R^2_{1,3}$ which have the vector γ_0 as a factor. Algebraically, this can be characterized as the set of all bivectors in $R^2_{1,3}$ which anticommute with γ_0. This determines a unique mapping of the electromagetic bivector F into the geometric algebra $\mathcal{R}_3$ of the given inertial system. The space-time split of F by γ_0 is obtained by decomposing F into a part

$$\underset{\sim}{E} = \frac{1}{2}(F - \gamma_0 F \gamma_0) \tag{3.9a}$$

which anticommutes with γ_0 and a part

$$i\underset{\sim}{B} = \frac{1}{2}(F + \gamma_0 F \gamma_0) \tag{3.9b}$$

which commutes with γ_0, so we have

$$F = \underset{\sim}{E} + i\underset{\sim}{B}, \tag{3.10}$$

where, as before, $i = \gamma_0\gamma_1\gamma_2\gamma_3$ is the unit pseudoscalar. Although $i\underset{\sim}{B}$ commutes with γ_0, $\underset{\sim}{B}$ must anticommute since i does. Therefore, we are right to denote $\underset{\sim}{B}$ as a vector in $\mathcal{R}^3$. Of course, $\underset{\sim}{E}$ and $\underset{\sim}{B}$ in (3.10) are just the electric and magnetic fields in the γ_0-system, and the split of F into electric and magnetic fields will be different for different inertial systems. The geometric algebra generated by $\mathcal{R}^3$ is identical with the even subalgebra of $\mathcal{R}_{1,3}$, so we write

$$\mathcal{R}_3 = \mathcal{R}^+_{1,3}. \tag{3.11}$$

Moreover, (3.10) determines a split of the bivectors in $\mathcal{R}_{1,3}$ into vectors and bivectors of $\mathcal{R}_3$, as expressed by writing

$$\mathcal{R}^2_{1,3} = \mathcal{R}^1_3 + \mathcal{R}^2_3. \tag{3.12}$$

This spit is not unique, however, as it depends on the choice of the vector γ_0.

A complete discussion of space-time splits is given in Refs. [4] and [10]. My object here was only to explain the underlying idea. Physicists often write $E_k = F_{ko}$ to identify the electric field components of $F_{\mu\nu}$ without recognizing that this entails a mapping of spacetime bivectors into space vectors. We have merely made the mapping explicit and extended it systematically to space-time splits of all physical quantities. Finally, I should point out that the purpose of a spacetime split is merely to relate invariant physical quantities to the variables employed by a particular observer. It is by no means necessary for solving and analyzing the basic equations. As a rule, it only complicates the equations needlessly. Therefore, the best time for a split is usually after the equations have been solved.

Now we turn to the second major component of electrodynamics, the energymomentum tensor. Geometric algebra enables us to write the energymomentum tensor T(n) for the electromagnetic field F in the compact form

$$T(n) = \frac{1}{2} Fn\tilde{F}, \tag{3.13}$$

where the tilde indicates "reversion" and $\tilde{F} = -F$. For the benefit of mathematicians in the audience, let me say that the T(n) is a vector valued linear function on the tangent space at each space time point x describing the flow of energymomentum through a hypersurface with normal n. The coordinate-free form (3.13) of the energymomentum tensor is related to the conventional tensor form by

$$T^{\mu\nu} = \gamma^\mu \cdot T(\gamma^\nu) = \frac{1}{2}\langle\gamma^\mu F \gamma^\nu \tilde{F}\rangle = -\langle\gamma^\mu \cdot F\gamma^\nu F\rangle - \frac{1}{2}\langle\gamma^\mu\gamma^\nu F^2\rangle$$

$$= -(\gamma^\mu \cdot F)\cdot(\gamma^\nu \cdot F) - \frac{1}{2}\,\gamma^\mu\cdot\gamma^\nu\langle F^2\rangle = F^{\mu\alpha}F_\alpha{}^\nu + \frac{1}{2}g_{\mu\nu}F_{\alpha\beta}F^{\alpha\beta}, \tag{3.14}$$

where the angular brackets denote "scalar part."

The algebraic form (3.13) for T(n) is not only more compact, it is easier to apply and interpret than the tensor form on the right side of (3.14). To demonstrate that, let us solve the eigenvector problem for T(n) when $F^2 \neq 0$. In that case, it is easy to show (Ref. [3]) that F can be put in the canonical form

$$F = fe^{i\phi} \tag{3.15}$$

where f is a timelike bivector with

$$f^2 = [(F\cdot F)^2 + |F\wedge F|^2]\,, \tag{3.16}$$

and ϕ is a scalar determined by

$$e^{i\phi} = \left[\frac{F^2}{f^2}\right]^{\frac{1}{2}} = \frac{(F\cdot F + F\wedge F)^{\frac{1}{2}}}{[(F\cdot F)^2 + |F\wedge F|^2]^{\frac{1}{4}}} \tag{3.17}$$

Since the pseudoscalar i commutes with all bivectors but anticommutes with all vectors, substitution of (3.15) into (3.13) reduces T(n) to the simpler form

$$T(n) = -\tfrac{1}{2}fnf. \tag{3.18}$$

This is simpler because f is simpler, in general, than F. It enables us to find the eigenvalues by inspection. The bivector f determines a timelike plane. Any vector n in the plane satisfies the equation $n\wedge f = 0$, which implies that $nf = -fn$. On the other hand, if n is orthogonal to the plane, then $n\cdot f = 0$ and $nf = fn$. For these two cases, (3.18) gives us

$$T(n) = \pm \frac{1}{2} f^2 n. \tag{3.19}$$

Thus T(n) has a pair of doubly degenerate eigenvalues $\pm f^2/2$, where f^2 can be expressed in terms of F by (3.16). To characterize the eigenvectors, all we need is an explicit for f in terms of F. From (3.15) and (3.17) we get

$$f = Fe^{-i\phi} = \frac{F(F\cdot F - F\wedge F)^{\frac{1}{2}}}{[(F\cdot F)^2 + |F\wedge F|^2]^{\frac{1}{2}}} \tag{3.20}$$

This invariant of the field can be expressed in terms of electric and magnetic fields by substituting (3.10).

This method for solving the eigenvector problem should be compared with the conventional matrix method using the tensor form (3.14). The matrix solution worked out by Synge [11] requires many pages of text. Obviously, our method is much simpler and cleaner. There is a great lesson in mathematical design to be learned from this example. In contrast to matrix algebra, geometric algebra enables us to represent linear transformations and carry out computations with them without introducing a basis. To my knowledge it is the only mathematical system with this capability. Surely as a principle of good mathematical design we should endeavor to extend this capability as far as possible. Accordingly, I suggest that the mathematical community should aim to develop a complete theory of "geometric representations" for linear functions, that is, representations in terms of geometric algebra. The geometric representation theory for orthogonal functions is already well developed, and members of this audience are familiar with its advantages over matrix representations. The example we have just considered shows that there are comparable advantages in the representation of symmetric functions. Ref. [2] carries the theory quite a bit further. But geometric representation theory is still not as well developed as matrix representation theory. Indeed, physicists are so conditioned to matrix algebra that most of them work with matrix representations of Clifford Algebra rather than directly with the Clifford Algebra itself. I submit that is would be better mathematical design to turn things around, to regard geometric algebra as more fundamental than matrix algebra and to derive matrix representations from geometric representations. In other words, matrix algebra should be subservient to geometric algebra, arising whenever it is convenient to introduce a basis and work with lists and arrays of numbers or linear functions.

Now let us turn to the third major component of electrodynamics the equation of motion for a charged particle. Let $x = x(\tau)$ be the world line of a particle parametrized by proper time τ. The invariant velocity of the particle is a timelike unit vector $v = \dot{x} = dx/d\tau$. For a test particle of rest mass m and charge q in an electromagnetic field F, the equation of motion can be written in the coordinate-free form

$$m\dot{v} = qF\cdot v. \tag{3.21}$$

Set $m = q = 1$ for convenience.

Equation (3.21) is not very different from the usual covariant tensor formulation, which can be obtained by taking its components with respect to the basis $\{\gamma^\mu\}$. But I want to show you how to solve it in a way which can't even be formulated in tensor analysis. First we note that it implies $d(v^2)/d\tau = 2\dot{v}\cdot v = 2(F\cdot v)\cdot v = 2F\cdot(v\wedge v) = 0$. Therefore, the condition $v^2 = 1$ is a constant of motion, and we can represent the change in v as the unfolding of a Lorentz transformation with spin representation $R = R(\tau)$, that is, we can write

$$v = R\gamma_0\tilde{R}, \tag{3.22}$$

where R is an element of $R^+_{1,3}$ for which $R\tilde{R} = 1$. Then we can replace (3.21) by the simpler spinor equation of motion

$$\dot{R} = \tfrac{1}{2}FR. \tag{3.23}$$

For a uniform field F, this has the elementary solution

$$R = e^{F\tau/2}\,R_0, \tag{3.24}$$

where R_0 is a constant determined by the initial conditions. This gives us an explicit expression for the proper time dependence of v when substituted into (3.22). However, the subsequent integration of $\dot{x} = v$ to get an explicit equation for the particle history is a little tricky for arbitrary F. The solution is obtained in Ref. [2], which also works out the solutions when F is a plane wave or a Coulomb field.

The spinor R can be regarded as an integrating factor which we introduced with equations (3.22) and (3.23) merely to simplify the original equation of motion (3.21). However, there is reason to believe that something much deeper is involved. We can define a comoving frame $\{e_\mu = e_\mu(\tau)\}$ attached to the particle history by the equations

$$e_\mu = R\gamma_\mu \tilde{R} \tag{3.25}$$

for $\mu = 0,1,2,3$. For $e_0 = v$ this agrees with (3.22). Now here is the amazing part, if we regard the particle as an electron and identify e_3 with the direction of the electron spin, then the spin precession predicted by (3.24) agrees exactly with the prediction of the Dirac theory, except for the tiny radiative corrections. Moreover, the solution of (3.23) for a plane wave is identical to the corresponding solution of the Dirac equation (known as the Volkov solution in the literature). The relation of the Coulomb solution of (3.23) to the Coulomb solution of the Dirac equation has not been investigated. But we have enough to assert that the simple "classical equation" (3.23) has an intimate relation to the Dirac theory. I will say more about this in my next lecture.

4. ALGEBRAIC STRUCTURES

Now let me return to the general theme of unifying mathematics. We have seen how real and complex analysis can be united in a general geometric function theory. The theory of linear geometric functions which we discussed later should be regarded as a branch of the general theory concerned with the simplest class of geometric functions, however, for reasons given in Ref. [2], it should be integrated with the theory of multilinear functions from the beginning. Reference [2] also shows how the theory of manifolds and differential geometry can be incorporated in geometric function theory.

There is one other branch of mathematics which should be integrated into the theory -- Abstract Algebra. Like linear algebra it can be incorporated in the theory of geometric representations. Thus, an abstract product $A \circ B$ will be represented as a bilinear function $a \circ b = f(a, b)$ in the geometric algebra. There will be many such representations, but some will be optimal in an appropriate sense.

Why should one bother with geometric representations? Why not be satisfied with the pure abstract formulation? There are many reasons. Representations of the different algebras in a common language should make it easier to establish relations among them and possibly uncover hidden structure. It should facilitate applications of the algebraic structures to fields, such as physics, which are already formulated in terms of geometric algebra. It will make the great computational power of geometric algebra available for integrating and analyzing algebraic structures. Best of all, it should be mathematically rich and interesting. To convince you of this, let me show you a geometric representation of the octonian product. We represent the octonians as vectors a, b, . . . in $\mathcal{R}^8$. Then, with the proper choice a multivector P in $\mathcal{R}_8$, the octonian product is given by the geometric function

$$a \circ b = \langle abP \rangle_1 , \qquad (4.1)$$

where the subscript means vector part. I will not deprive you of a chance to find P for yourself before it is revealed by Kwasniewski in his lecture. Let me only give you the hint that it is built out of idempotents of the algebra.

Of course the theory of geometric representations should be extended to embrace Lie groups and Lie algebras. A start has been made in Ref. [2]. I conjectured there that every Lie algebra is isomorphic to a bivector algebra, that is, an algebra of bivectors under the commutator product. Lawyer-physicist Tony Smith has proved that this conjecture is true by pointing to results already in the literature. However, the task remains to construct explicit bivector representations for all the Lie algebras. The task should not be difficult, as a good start has already been made.

I must stop here, because my time as well as my vision is limited. The other speakers will show the great richness of Clifford Algebras approached from any of the three major perspectives, the algebraic, the geometric, or the physical.

REFERENCES

[1] D. Hestenes, *New Foundations for Classical Mechanics*, Reidel Publ. Co., Dordrecht/Boston (1985).

[2] D. Hestenes and G. Sobczyk, *Clifford Algebra to Geometric Calculus*, Reidel Publ. Co., Dordrecht/Boston (1984).

[3] D. Hestenes, *Spacetime Algebra*, Gordon and Breach, N.Y. (1966).

[4] Robert K. Merton, *The Sociology of Science*, U. Chicago Press, Chicago (1973).

[5] E. Kähler, *Rendeconti di Mathematica* **21** (3/4) 425 (1962).

[6] H. Grassmann, *Math Am.* **XII**, 375-386 (1877).

[7] I. Porteous, *Topological Geometry*, Van Nostrand, London (1969).

[8] P. Lounesto, *Found. Phys.* **11**, 721 (1981).

[9] P. Reany, 'Solution to the cubic equation using Clifford Algebra' (submitted to Am. J. Phys.).

[10] D. Hestenes, *J. Math. Phys.* **15**, 1768 (1974).

[11] J.L. Synge, *Relativity, the Special Theory*, North Holland, Amsterdam (1956).

[12] D. Hestenes, *J. Math. Phys.* **15**, 1778 (1974).

CLIFFORD ALGEBRAS AND SPINORS

Pertti Lounesto
Institute of Mathematics
Helsinki University of Technology
SF-02150 Espoo
Finland

ABSTRACT. A historical review of spinors is given together with a construction of spinor spaces as minimal left ideals of Clifford algebras. Spinor spaces of euclidean spaces over reals have a natural linear structure over reals, complex numbers or quaternions. Clifford algebras have involutions which induce bilinear forms or scalar products on spinor spaces. The automorphism groups of these scalar products of spinors are determined and also classified.

1. HISTORY

Spinors appeared for the first time when E. Cartan classified simple Lie algebras. Lie algebras of orthogonal type had irreducible representations which could not be obtained by outer products of the euclidean spaces. Tensor methods had to be supplemented by spinor representations. For Lie algebras of type B_ℓ, there is a single spinor space, while for Lie algebras of type D_ℓ, there are two spinor spaces. The Dynkin diagrams of these spinor spaces are given in Figure 1.

B_ℓ: $SO(2\ell+1)$

D_ℓ: $SO(2\ell)$

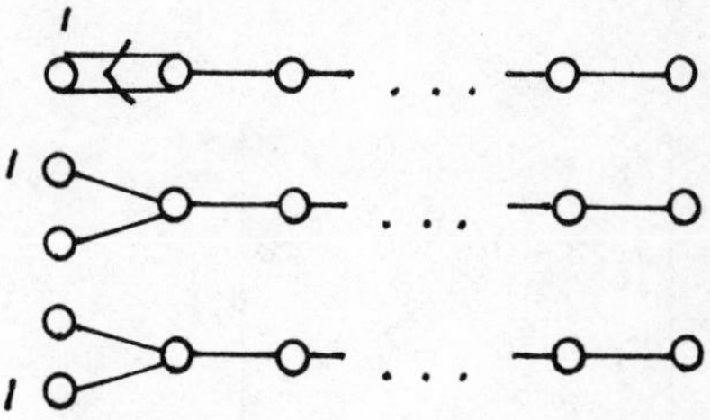

Figure 1. Dynkin diagrams for spinor spaces of Lie algebras B_ℓ and D_ℓ.

Later physicists employed spinor representations of $A_1 \simeq B_1$ or $SU(2)/\{\pm 1\} \simeq SO(3)$ in conjunction with electron spin. Pauli matrices

J. S. R. Chisholm and A. K. Common (eds.), Clifford Algebras and Their Applications in Mathematical Physics, 25–37.

$$\sigma_1 = \begin{pmatrix} 0 & 1 \\ 1 & 0 \end{pmatrix}, \quad \sigma_2 = \begin{pmatrix} 0 & -i \\ i & 0 \end{pmatrix}, \quad \sigma_3 = \begin{pmatrix} 1 & 0 \\ 0 & -1 \end{pmatrix}$$

became famous: they were used to rotate a vector $x = x_1\sigma_1 + x_2\sigma_2 + x_3\sigma_3$ around an axis $a = a_1\sigma_1 + a_2\sigma_2 + a_3\sigma_3$ by the formula

$$x \to s^{-1}xs, \qquad s = e^{ia/2}.$$

The angle of rotation was the length $|a|$ of a. A spinor

$$\psi = \begin{pmatrix} \psi_1 & 0 \\ \psi_2 & 0 \end{pmatrix}; \qquad \psi_1, \psi_2 \in C$$

transformed according to the rule

$$\psi \to s\psi.$$

While a spinor ψ made one full turn, a vector x made two full turns. For instance, if you hold a plate on your fingertips, and you rotate the plate one full turn, your arm is in a different position; you can make a second full turn to restore your arm.

Relativity and spin could be described by one equation, the Dirac equation, involving Dirac matrices (representing vectors)

$$\gamma_k = \begin{pmatrix} 0 & \sigma_k \\ \sigma_k & 0 \end{pmatrix}; \qquad k = 1,2,3; \qquad \gamma_4 = \begin{pmatrix} i & 0 \\ 0 & -i \end{pmatrix}$$

and Dirac spinors

$$\varphi = \begin{pmatrix} \varphi_1 & 0 & 0 & 0 \\ \varphi_2 & 0 & 0 & 0 \\ \varphi_3 & 0 & 0 & 0 \\ \varphi_4 & 0 & 0 & 0 \end{pmatrix}; \qquad \varphi_1, \varphi_2, \varphi_3, \varphi_4 \in C.$$

A vector $x = x_1\gamma_1 + x_2\gamma_2 + x_3\gamma_3 + x_4\gamma_4$ had a quadratic form $x^2 = x_1^2 + x_2^2 + x_3^2 - x_4^2$, but for a spinor φ the mapping

$$\varphi \to |\varphi_1|^2 + |\varphi_2|^2 - |\varphi_3|^2 - |\varphi_4|^2$$

was more interesting. This aspect will be scrutinized in Sections 3, 4 and 5.

In 1964 spinors were applied to celestial mechanics, when P. Kustaanheimo [10] introduced his spinor regularization. It was already earlier known that if the orbit of a planet is in the complex plane and if the center of gravity is in the origin, then taking a square root regularizes the motion of the planet, see Figure 2.

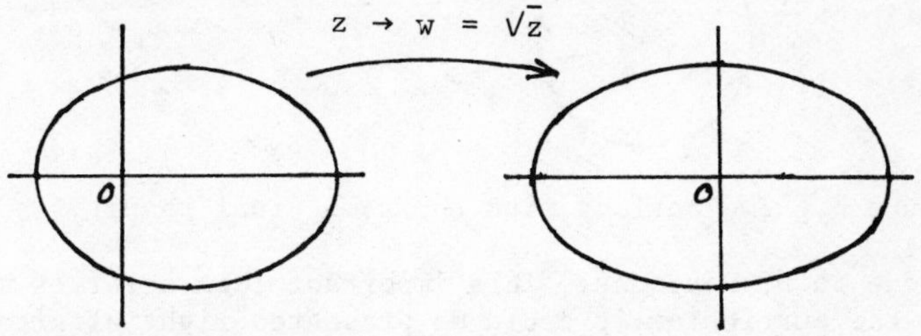

Figure 2. Regularization of Kepler motion.

For a long time it was a problem how to generalize this regularization of Kepler motion from the complex plane to the three-dimensional space. A solution was given by spinor regularization, a method which sends an even element

$$u = u_0 + u_1\sigma_2\sigma_3 + u_2\sigma_3\sigma_1 + u_3\sigma_1\sigma_2$$

of the Clifford algebra R_3 to a vector

$$x = \bar{u}\, c\, u \qquad (c \text{ is a fixed vector})$$

in R^3. Here it is essential that a non-zero u is in the Clifford group Γ_3.

2. VISUALIZATION OF SPINORS

There have been some attempts to visualize spinors as real geometrical objects [4], [8]. The following illustrates the fact that a vector makes two whole turns while a spinor makes only one turn. First, the famous equation

$$e^{i\pi} = -1$$

is connected to the fact that the two elements (sometimes called spinors)

$$\pm s = \pm e^{ia/2}$$

represent the same rotation - only the actions are different, see Figure 3.

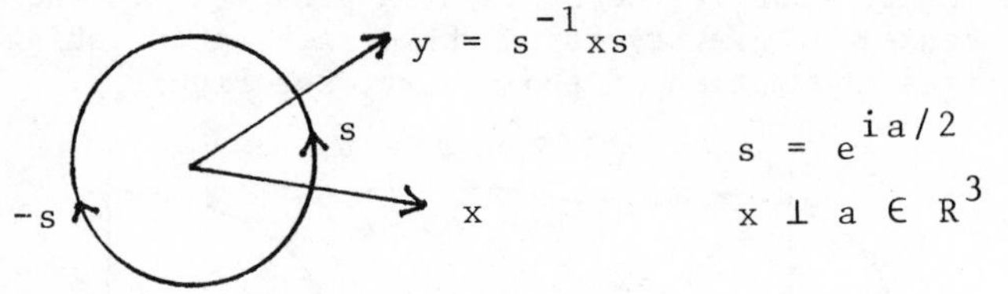

Figure 3. Two actions with the same final result.

This idea is due to D. Hestenes. This important idea deserves more attention; in the curriculum it could be presented right after assigning different numbers (mod 2π) to same oriented angle.

Secondly, the fact that a vector makes two whole turns, while a spinor makes only one turn, can be illustrated by the soap-plate trick: hold the plate on your fingertips and make one whole turn twice.

3. LINEAR STRUCTURE ON SPINOR SPACE

Spinors are elements of minimal left ideals of Clifford algebras. The Clifford algebra $R_{p,q}$ on an n-dimensional vector space $R^{p,q}$ with quadratic form

$$x^2 = x_1^2 + \ldots + x_p^2 - x_{p+1}^2 - \ldots - x_{p+q}^2 \qquad (n = p+q)$$

is the associative algebra over R containing a copy of R and a copy of $R^{p,q}$ so that

(i) the square of a vector is equal to its quadratic form
(ii) the Clifford algebra $R_{p,q}$ is generated as an algebra by the vector space $R^{p,q}$ (but not by any proper subspace of $R^{p,q}$).

Table 1 gives matrix algebras as isomorphic images of Clifford algebras $R_{p,q}$ up to $n = p+q < 8$.

Table 1. Clifford algebras $R_{p,q}$ for $p+q < 8$.

p+q \ p-q	-7	-6	-5	-4	-3	-2	-1	0	1	2	3	4	5	6	7
0								R							
1							C		${}^2R$						
2						H		$R(2)$		$R(2)$					
3					${}^2H$		$C(2)$		${}^2R(2)$		$C(2)$				
4				$H(2)$		$H(2)$		$R(4)$		$R(4)$		$H(2)$			
5			$C(4)$		${}^2H(2)$		$C(4)$		${}^2R(4)$		$C(4)$		${}^2H(2)$		
6		$R(8)$		$H(4)$		$H(4)$		$R(8)$		$R(8)$		$H(4)$		$H(4)$	
7	${}^2R(8)$		$C(8)$		${}^2H(4)$		$C(8)$		${}^2R(8)$		$C(8)$		${}^2H(4)$		$C(8)$

In order to fix a minimal left ideal S of $R_{p,q}$ choose a primitive idempotent e of $R_{p,q}$. Then $S = R_{p,q}e$. Recall that an element e of a ring A is called <u>idempotent</u> if $e^2 = e$. An idempotent is <u>primitive</u> if it is not a sum of two non-zero annihilating idempotents. Two idempotents e and f are mutually <u>annihilating</u> if $ef = 0 = fe$. A subring S of a ring A is a <u>left ideal</u> if $ax \in S$ for all $a \in A$ and $x \in S$. A left ideal is <u>minimal</u> if it does not contain properly any non-zero left ideals. A primitive idempotent e of $R_{p,q}$ can be written in the form

$$e = \frac{1}{2}(1+e_{T_1})\frac{1}{2}(1+e_{T_2})\ldots\frac{1}{2}(1+e_{T_k}),$$

where $e_{T_1}, e_{T_2}, \ldots, e_{T_k}$ is a set of commuting elements with square 1 in the canonical basis for $R_{p,q}$, generating a group of order 2^k, $k = q-r_{q-p}$. Recall the Radon - Hurwitz number r_i for $i \in Z$ given by recursion formula $r_{i+8} = r_i+4$ and Table 2.

Table 2. Radon - Hurwitz number r_i.

i	0	1	2	3	4	5	6	7
r_i	0	1	2	2	3	3	3	3

<u>Example</u>: In the case $R^{2,1}$ we have $k = 1-r_{-1} = 1 - (r_7-4) = 2$. Therefore the idempotent $e = \frac{1}{2}(1+e_1)\frac{1}{2}(1+e_{23})$ in $R_{2,1}$ is primitive. □

The division ring $K = eR_{p,q}e$ is isomorphic to R or C or H according as $(p-q) \bmod 8$ is $0,1,2$ or $3,7$ or $4,5,6$. The mapping

$$S \times K \to S, \qquad (\psi, \lambda) \to \psi\lambda$$

defines a right K-linear structure on the spinor space $S = R_{p,q}e$.
Example The Clifford algebra R_3 is isomorphic to the algebra of complex 2×2-matrices $C(2)$. Its minimal left ideal S is

$$S \ni \psi = \begin{pmatrix} \psi_1 & 0 \\ \psi_2 & 0 \end{pmatrix}, \qquad \psi_1, \psi_2 \in C$$

and the division ring $K \simeq C$ is

$$K \ni \lambda = \begin{pmatrix} \alpha & 0 \\ 0 & 0 \end{pmatrix}, \qquad \alpha \in C.$$

□

To find a matrix representation for a Clifford algebra $R_{p,q}$ choose a primitive idempotent e in $R_{p,q}$ and multiply it from the left by some elements e_T in the canonical basis for $R_{p,q}$ so that you get a basis for S.
Example From Table 1, $R_{3,1} \simeq R(4)$. Now, $k = 1-r_{-2} = 1 - (r_6-4) = 2$. Therefore, the idempotent

$$e = \frac{1}{2}(1+e_1)\frac{1}{2}(1+e_{24})$$

is primitive. Form a basis for S

$$\begin{aligned} f_1 &= e &&= \tfrac{1}{4}(1+e_1+e_{24}+e_{124}) \\ f_2 &= e_2e &&= \tfrac{1}{4}(e_2-e_{12}+e_4-e_{14}) \\ f_3 &= e_3e &&= \tfrac{1}{4}(e_3-e_{13}-e_{234}+e_{1234}) \\ f_4 &= e_{23}e &&= \tfrac{1}{4}(e_{23}+e_{123}-e_{34}+e_{314}). \end{aligned}$$

In this basis the matrics of e_1, e_2, e_3, e_4 are

$$e_1 = \begin{pmatrix} 1 & 0 & & \\ 0 & -1 & & \\ & & -1 & 0 \\ & & 0 & 1 \end{pmatrix}, \quad e_2 = \begin{pmatrix} 0 & 1 & & \\ 1 & 0 & & \\ & & 0 & 1 \\ & & 1 & 0 \end{pmatrix}, \quad e_3 = \begin{pmatrix} & & 1 & 0 \\ & & 0 & -1 \\ 1 & 0 & & \\ 0 & -1 & & \end{pmatrix}, \quad e_4 = \begin{pmatrix} 0 & 1 & & \\ -1 & 0 & & \\ & & 0 & 1 \\ & & -1 & 0 \end{pmatrix}.$$

□

4. SCALAR PRODUCTS OF SPINORS

The Clifford algebra $R_{p,q}$ has three important involutions, similar to complex conjugation. The first, called main involution, is the isomorphism $u \to u'$ obtained by replacing each e_i by $-e_i$, thereby replacing each u in k-vector space $R^k_{p,q}$ by $u' = (-1)^k u$. Clearly, $(ab)' = a'b'$. The second involution, called reversion, is an anti-isomorphism $u \to u^*$ obtained by reversing the order of factors in each $e_{\alpha_1} e_{\alpha_2} \ldots e_{\alpha_k}$ to get $e_{\alpha_k} \ldots e_{\alpha_2} e_{\alpha_1}$. Therefore, each u in $R^k_{p,q}$ is replaced by $u^* = (-1)^{[k/2]} u$. It is obvious that $(ab)^* = b^* a^*$. The third involution, called conjugation, is a combination of the two others, and we shall write $\bar{u} = u'^* = u^{*\prime}$.

It often happens that for ψ and φ in S the expressions $\psi^*\varphi$ or $\bar{\psi}\varphi$ are in K. This defines a scalar product of spinors.

Example The idempotent $e = \frac{1}{2}(1+e_3)$ is primitive in R_3. Spinor space S has a basis

$$f_1 = e = \frac{1}{2}(1+e_3)$$
$$f_2 = e_1 e = \frac{1}{2}(e_1 - e_{31})$$

in which matrix representations of e_1, e_2, e_3 are

$$e_1 = \begin{pmatrix} 0 & 1 \\ 1 & 0 \end{pmatrix}, \quad e_2 = \begin{pmatrix} 0 & -i \\ i & 0 \end{pmatrix}, \quad e_3 = \begin{pmatrix} 1 & 0 \\ 0 & -1 \end{pmatrix}.$$

The division ring $K = eR_3e$ is isomorphic to the complex field C so that

$$1 \simeq e = \begin{pmatrix} 1 & 0 \\ 0 & 0 \end{pmatrix}, \quad i \simeq e_{123} = \begin{pmatrix} i & 0 \\ 0 & 0 \end{pmatrix}.$$

Now, the reversion means hermitian conjugation (transposition and complex conjugation)

$$\psi^* = \begin{pmatrix} \psi_1 & 0 \\ \psi_2 & 0 \end{pmatrix}^* = \begin{pmatrix} \bar{\psi}_1 & \bar{\psi}_2 \\ 0 & 0 \end{pmatrix},$$

and so

$$\psi^*\varphi = \begin{pmatrix} \bar{\psi}_1 & \bar{\psi}_2 \\ 0 & 0 \end{pmatrix} \begin{pmatrix} \varphi_1 & 0 \\ \varphi_2 & 0 \end{pmatrix} = \begin{pmatrix} \bar{\psi}_1\varphi_1 + \bar{\psi}_2\varphi_2 & 0 \\ 0 & 0 \end{pmatrix}$$

belongs to the division ring K. □

Often the expressions $\psi^*\varphi$ or $\bar{\psi}\varphi$ do not belong to the division ring K. Then there is needed an element u in $R_{p,q}$ bringing $u\psi^*\varphi$ or $u\bar{\psi}\varphi$ into K.

Example Consider again $R_3 \simeq C(2)$. Then for

$$u = \begin{pmatrix} a & b \\ c & d \end{pmatrix}, \qquad \bar{u} = \begin{pmatrix} d & -b \\ -c & a \end{pmatrix}$$

and for

$$\psi = \begin{pmatrix} \psi_1 & 0 \\ \psi_2 & 0 \end{pmatrix}, \qquad \bar{\psi} = \begin{pmatrix} 0 & 0 \\ -\psi_2 & \psi_1 \end{pmatrix}.$$

So

$$\bar{\psi}\varphi = \begin{pmatrix} 0 & 0 \\ -\psi_2 & \psi_1 \end{pmatrix} \begin{pmatrix} \varphi_1 & 0 \\ \varphi_2 & 0 \end{pmatrix} = \begin{pmatrix} 0 & 0 \\ \psi_1\varphi_2-\psi_2\varphi_1 & 0 \end{pmatrix}$$

does not belong to K, but

$$e_{31}\bar{\psi}\varphi = \begin{pmatrix} \psi_1\varphi_2-\psi_2\varphi_1 & 0 \\ 0 & 0 \end{pmatrix}$$

is in K and $(\psi,\varphi) \to e_{31}\bar{\psi}\varphi$ defines a scalar product of spinors (known as symplectic product). □

In case the Clifford algebra is a sum of two simple ideals, that is, $R_{p,q}$ for $(p-q) \bmod 4 = 1$, it is reasonable to replace the spinor space S by S+S' and the division ring K by K+K'.

5. AUTOMORPHISMS OF SCALAR PRODUCTS

Example The Clifford algebra $R_{2,1}$ is isomorphic to ${}^2R(2)$. The idempotent $e = \frac{1}{2}(1+e_1)\frac{1}{2}(1+e_{23})$ is primitive in $R_{2,1}$. The basis elements

$$f_1 = \frac{1}{4}(1+e_1+e_{23}+e_{123})$$
$$f_2 = \frac{1}{4}(e_2-e_{12}+e_3-e_{13})$$

for $S = R_{2,1}e$ are such that

$$f_1^*f_1 = 0 \qquad f_1^*f_2 = 0$$
$$f_2^*f_1 = 0 \qquad f_2^*f_2 = 0$$

and

$$\bar{f}_1 f_1 = 0 \qquad \bar{f}_1 f_2 = f_2$$
$$\bar{f}_2 f_1 = -f_2 \qquad \bar{f}_2 f_2 = 0.$$

So the automorphism groups of scalar products $\psi^*\varphi$ and $\bar{\psi}\varphi$ of spinors ψ and φ in S are $GL(2,R)$ and $Sp(2,R)$ respectively. However, if you consider $S+\underline{S}'$ instead of S, then the automorphism group of the scalar product $\bar{\psi}\varphi$ of spinors ψ and φ in $S+S'$ is ${}^2Sp(2,R) = Sp(2,R)\times Sp(2,R)$. □

The automorphism groups of scalar products of spinors are given in Table 3.

Table 3. Automorphism groups of spinor spaces.

$\psi^*\varphi$

p+q \ p-q	-7	-6	-5	-4	-3	-2	-1	0	1	2	3	4	5	6	7
0								O(1)							
1							O(1,C)		^{2}O(1)						
2						SO*(2)		O(1,1)		O(2)					
3					GL(1,H)		U(1,1)		GL(2,R)		U(2)				
4				Sp(2,2)		Sp(2,2)		Sp(4,R)		Sp(4,R)		Sp(4)			
5			Sp(4,C)		^{2}Sp(2,2)		Sp(4,C)		^{2}Sp(4,R)		Sp(4,C)		^{2}Sp(4)		
6		Sp(8,R)		Sp(4,4)		Sp(4,4)		Sp(8,R)		Sp(8,R)		Sp(4,4)		Sp(8)	
7	GL(8,R)		U(4,4)		GL(4,H)		U(4,4)		GL(8,R)		U(4,4)		GL(4,H)		U(8)

$\bar{\psi}\varphi$.

p+q \ p−q	−7	−6	−5	−4	−3	−2	−1	0	1	2	3	4	5	6	7
0								O(1)							
1							U(1)		GL(1,R)						
2						Sp(2)		Sp(2,R)		Sp(2,R)					
3					^{2}Sp(2)		Sp(2,C)		^{2}Sp(2,R)		Sp(2,C)				
4				Sp(4)		Sp(2,2)		Sp(4,R)		Sp(4,R)		Sp(2,2)			
5			U(4)		GL(2,H)		U(2,2)		GL(4,R)		U(2,2)		GL(2,H)		
6		O(8)		SO*(8)		SO*(8)		O(4,4)		O(4,4)		SO*(8)		SO*(8)	
7	^{2}O(8)		O(8,C)		^{2}SO*(8)		O(8,C)		^{2}O(4,4)		O(8,C)		^{2}SO*(8)		O(8,C)

In fact, Table 3 lists the groups

$$G = \{s \in R_{p,q} \mid s^*s = 1\}$$

and

$$G = \{s \in R_{p,q} \mid \bar{s}s = 1\}$$

for $p+q < 8$. Table 3 continues with a periodicity of 8. The scalar product of spinors is

symmetric or antisymmetric
non-degenerate
positive definite $\begin{cases} \text{in } R_n & \text{for } \psi^*\varphi \\ \text{in } R_{0,n} & \text{for } \bar{\psi}\varphi \end{cases}$

and if the scalar product is not positive definite then it is neutral for all pairs (p,q) except $(0,1)$, $(0,2)$, $(0,3)$ for $\psi^*\varphi$ and $(1,0)$ for $\bar{\psi}\varphi$.

Example The Dirac algebra is the algebra of complex 4×4-matrices $C(4)$. It is isomorphic to the Clifford algebra $R_{4,1}$ and contains the Clifford algebra $R_{3,1}$ on the Minkowski space $R^{3,1}$. Let e_0, e_1, e_2, e_3, e_4 be a basis of $R^{4,1}$ so that

$$e_0^2 = 1, \quad e_1^2 = e_2^2 = e_3^2 = 1, \quad e_4^2 = -1$$

and consider $R^{3,1}$ as the space spanned by e_1, e_2, e_3, e_4. The extra dimension e_0 can be used to describe the conformal transformations of the Minkowski space $R^{3,1}$. Denote

$$i = -e_0e_1e_2e_3e_4 .$$

The idempotent

$$e = \frac{1}{2}(1-e_{034})\frac{1}{2}(1+e_{0123}) = \frac{1}{2}(1+ie_{12})\frac{1}{2}(1-ie_4)$$

is primitive in $R_{4,1}$. The minimal left ideal $S = R_{4,1}e$ of $R_{4,1}$ consists of Dirac spinors. While singling the primitive idempotent e and minimal left ideal S we have fixed

e_4 an observer (time-like vector)

e_{12} a quantization plane of spin.

The division ring $K = eR_{4,1}e$ is isomorphic to the complex field C, with ie as the imaginary unit. The right K-linear space S has a basis

$$\begin{aligned}
f_1 &= f &&= \frac{1}{4}(1+ie_{12}-ie_4+e_{124})\\
f_2 &= \frac{1}{2}(e_1+ie_2)e_{1234}f &&= \frac{1}{4}(ie_{23}-e_{31}+e_{234}+ie_{314})\\
f_3 &= e_{1234}f &&= \frac{1}{4}(e_3+ie_{123}-ie_{34}+e_{1234})\\
f_4 &= \frac{1}{2}(e_1+ie_2)f &&= \frac{1}{4}(e_1+ie_2-ie_{14}+e_{24})
\end{aligned}$$

where $i = e_{01234}^{-1}$ as before. In this basis

$$e_0 = ie_{1234} = \begin{pmatrix} 0 & i \\ -i & 0 \end{pmatrix}$$

$$e_1 = \begin{pmatrix} 0 & \sigma_1 \\ \sigma_1 & 0 \end{pmatrix}, \quad e_2 = \begin{pmatrix} 0 & \sigma_2 \\ \sigma_2 & 0 \end{pmatrix}, \quad e_3 = \begin{pmatrix} 0 & \sigma_3 \\ \sigma_3 & 0 \end{pmatrix}$$

$$e_4 = \begin{pmatrix} i & 0 \\ 0 & -i \end{pmatrix}.$$

The scalar product of spinors is given by

$$\bar{\psi}\varphi = (\bar{\psi}_1\varphi_1 + \bar{\psi}_2\varphi_2 - \bar{\psi}_3\varphi_3 - \bar{\psi}_4\varphi_4)e$$

and the automorphism group of the scalar product is U(2,2). □

6. REVIEW ON LITERATURE

Scalar products of spinors have been discussed by I.R. Porteous [12] and P. Lounesto [11]. Good papers to start studying Clifford algebras and spinors are D. Hestenes [9], M. Riesz [13] and A. Crumeyrolle [3]. C. Chevalley goes further in the algebra of spinors.

Physicists have used Clifford algebras and spinors in relativity [6], twistor theory [1], supersymmetry [7], rigid body rotation [14] and to determine spinor structures on manifolds [5].

REFERENCES

[1] R. Ablamowicz, Z. Oziewicz, J. Rzewuski: 'Clifford algebra approach to twistors'. J. Math. Phys. 23 (1982), 231-242.

[2] C. Chevalley: The Algebraic Theory of Spinors. Columbia University Press, New York, 1954.

[3] A. Crumeyrolle: Algèbres de Clifford et Spineurs. Université Paul Sabatier, Toulouse, 1974.

[4] D.W. Ebner: 'A purely geometrical introduction of spinors in special relativity by means of conformal mappings on the celestial sphere'. Ann. Physik (7) 30 (1973), 206-210.

[5] R. Geroch: 'Spinor structure of space-times in general relativity I'. J. Math. Phys. 9 (1968), 1739-1744.

[6] K.R. Greider: 'A unifying Clifford algebra formalism for relativistic fields.' Found. Phys. 14 (1984), 467-506.

[7] Z. Hasiewicz, A.K. Kwasniewski, P. Morawiec: 'Supersymmetry and Clifford algebras.' J. Math. Phys. 25 (1984), 2031-2036.

[8] H. Hellsten: 'On the visual geometry of spinors and twistors.' In P.G. Bergmann, V. de Sabbata (ed.): Cosmology and Gravitation. Spin, Torsion, Rotation, and Supergravity, pp. 457-465. Plenum Press, New York, 1980.

[9] D. Hestenes: 'Vectors, spinors, and complex numbers in classical and quantum physics.' Amer. J. Phys. 39 (1971), 1013-1027.

[10] P. Kustaanheimo, E. Stiefel: 'Perturbation theory of Kepler motion based on spinor regularization.' J. Reine Angew. Math. 218 (1965), 204-219.

[11] P. Lounesto: 'Sur les idéaux à gauche des algèbres de Clifford et les produits scalaires des spineurs.' Ann. Inst. H. Poincaré. Sect. A 33 (1980), 53-61.

[12] I.R. Porteous: Topological Geometry. Van Nostrand - Reinhold, London, 1969. Cambridge University Press, Cambridge, 1981.

[13] M. Riesz: Clifford Numbers and Spinors. The Institute for Fluid Dynamics and Applied Mathematics (nowadays Institute for Physical Science and Technology), University of Maryland, 1958. (This excellent account on Clifford algebras can be obtained from Professor John Horvath, Department of Mathematics, University of Maryland, College Park, MD 20742.)

[14] M.D. Vivarelli: 'Development of spinor descriptions of rotational mechanics from Euler's rigid body displacement theorem.' Celestial Mech. 32 (1984), 193-207.

PSEUDO-EUCLIDEAN HURWITZ PAIRS AND GENERALIZED FUETER EQUATIONS*

Julian Ławrynowicz
Institute of Mathematics
Polish Academy of Sciences
PL-90-136 Łódź, Poland

and Jakub Rembieliński
Institute of Physics
University of Łódź
PL-90-236 Łódź, Poland

ABSTRACT. In [15] the authors have reformulated, in the language of Clifford algebras, the *Hurwitz problem* [10] of finding all the pairs of positive integers (n, p) and all the systems

$$c^k_{j\alpha} \in \mathbb{R}, \quad j, k = 1, \ldots, n; \quad \alpha = 1, \ldots, p, \quad p \le n,$$

such that the set of bilinear forms

$$\eta_j = x_\alpha c^k_{j\alpha} y_k$$

satisfies the condition

$$\Sigma_j \eta_j^2 = \Sigma_\alpha x_\alpha^2 \Sigma_k y_k^2.$$

They have introduced the notion of the *Hurwitz pair* of two real unitary vector spaces V and S. They have studied the dependence of these pairs on induced symplectic decomposition and arrived in a natural way at a *complex geometry induced by a given symplectic decomposition* with a distinguished fixed direction in S, that introduced a kind of anisotropy. Further investigations in this direction have been done in [11-13, 16].

The present research is connected with an analogue of the quoted construction if the vector spaces V and S in question are equipped with *non-degenerate pseudo-euclidean real scalar products* with standard properties. This enables the authors to introduce and study some *generalized Fueter equations*, extending or improving certain results on holomorphy due to F. Brackx, R. Delanghe, V. Souček, A. Vaz Ferreira, and others.

---

* This paper is in its final form and no version of it will be submitted for publication elsewhere.

J. S. R. Chisholm and A. K. Common (eds.), Clifford Algebras and Their Applications in Mathematical Physics, 39–48.

INTRODUCTION

The origin of the investigations reported here has been described in the Abstract. The basic equation generating *the complex geometry induced by a given symplectic decomposition* has been chosen in [15] as $Jf = f \cdot n$, where J denoted the linear endomorphism corresponding to a complex structure of the complexified space V (in this paper we use the left-hand multiplication convention because we will be studying the generalized Fueter equations). Further investigations in this direction include [11-13, 16]:

a) the infinite-dimensional case with an application to the *spontaneous symmetry breaking* [16],

b) a remarkable relationship between the existence of solutions of certain *soliton* (Kadomtsev-Petviašvili) *equations* and the complex geometry in question [11],

c) a study of a *field equation* of the Dirac type with further development of the method of isospectral deformations [12],

d) an application to *quasiconformal mappings* with obtaining, as a particular case, parametrization theorems [4, 14, 16].

An analogue of the quoted construction, introduced and studied here, concerns the situation in which the vector spaces V and S in question are equipped with non-degenerate *pseudo-euclidean real scalar products*. In particular, the introduction of some *generalized Fueter equations* enables us to extend or improve certain results on holomorphy due to Brackx, Delanghe, and Sommen [6], Souček [19], Bartik, Vaz Ferreira, Markl, and Souček [5] as well as others. Similar attempts, but leading to different geometrical or physical interpretations, have already been made in [1-3] and [7, 8]. In addition to several mathematical problems arising, the investigations have an immediate interpretation in various physical theories, in particular in the *Kałuża-Klein theories* (cf. e.g. [17]).

1. CHOICE OF AXIOMS

The first step to be taken is an optimal choice of axioms. Let (ε_α) be a pseudo-orthonormal basis in S. We suppose that

$$\eta \equiv [\eta_{\alpha\beta}] \equiv [(\varepsilon_\alpha, \varepsilon_\beta)_S] = \mathrm{diag}(1, \ldots, 1, -1, \ldots, -1). \quad (1)$$

Furthermore, we assume *symmetry* for the scalar product $(a, b)_S$ in S, while the scalar product $(f, g)_V$ in V is assumed to be *symmetric or antisymmetric*. Precisely, for $f, g, h \in V$; $a, b, c \in S$ and $\alpha, \beta \in \mathbb{R}$ we assume

$$(a, b)_S \in \mathbb{R}, \qquad (f, g)_V \in \mathbb{R},$$
$$(b, a)_S = (a, b)_S, \qquad (g, f)_V = \eta (f, g)_V, \quad \eta = 1 \text{ or } -1,$$
$$(\alpha a, b)_S = \alpha (a, b)_S, \qquad (\alpha f, g)_V = \alpha (f, g)_V,$$
$$(a, b+c)_S = (a,b)_S + (a,c)_S; \quad (f, g+h)_V = (f,g)_V + (f,h)_V.$$

The most important axiom is an analogue of the *Hurwitz condition* [15] which is of the form

$$(a, a)_S (f, g)_V = (a \cdot f, a \cdot g)_V \quad \text{for } f, g \in V \text{ and } a \in S \tag{2}$$

or, equivalently,

$$C_\alpha \bar{C}_\beta + C_\beta \bar{C}_\alpha = 2\eta_{\alpha\beta} I_n, \quad \alpha, \beta = 1, \ldots, \dim S, \tag{3}$$

where

$$C_\alpha = [c^k_{j\alpha}], \; \bar{C}_\alpha = \kappa C^T_\alpha \kappa^{-1}, \; c^k_{j\alpha} = (\varepsilon_\alpha e^k, e_j)_V, \tag{4}$$

$$\kappa = [\kappa_{jk}] = [(e_j, e_k)_V], \; e^k = (\kappa^{-1})^{kj} e_j, \tag{5}$$

(e_j) being an arbitrary basis in V, and I_n the identity $n \times n$-matrix. To be precise, we suppose that there exists a linear mapping $\cdot : S \times V \to V$ (*multiplication* of elements of S by elements of V) with the properties:

(i) $(a + b) \cdot f = a \cdot f + b \cdot f$ and $a \cdot (f + g) = a \cdot f + a \cdot g$ for $f, g \in V$ and $a, b \in S$;

(ii) the *generalized Hurwitz condition* (2);

(iii) there exists the unit element ε_o in S with respect to the multiplication $\cdot$: $\varepsilon_o \cdot f = f$ for $f \in V$.

If $\cdot$ does not leave invariant proper subspaces of V, the corresponding pair (V, S) is said to be *irreducible*. In such a case we call (V, S) a *pseudo-euclidean Hurwitz pair*.

2. CLASSIFICATION OF PSEUDOEUCLIDEAN HURWITZ PAIRS

The next step is to prove two lemmas:

<u>LEMMA 1</u>. *The problem of classifying the pseudo-euclidean Hurwitz pairs* (V, S) *is equivalent to the classification problem for real Clifford algebras* $C^{(r,s)}$ *with generators* γ_α *imaginary and antisymmetric or symmetric according to* $\alpha \leq r$ *or* $\alpha > r$, *given by the formulae*

$$i \gamma_\alpha C_t = C_\alpha, \quad \alpha = 1, \ldots, r + s + 1; \; \alpha \neq t; \tag{6}$$
$$C_t \bar{C}_t = \eta_{tt} I_n, \quad t \text{ fixed},$$

the matrices C_α *being determined by* (1) *and* (3)-(5). *The relationship is given by the formulae* (6) *and*

$$\gamma_\alpha \gamma_\beta + \gamma_\beta \gamma_\alpha = 2 I_n \hat{\eta}_{\alpha\beta}, \quad \alpha,\beta = 1,\dots,r+s; \; \alpha,\beta \neq t; \tag{7}$$

$$\bar{\gamma}_\alpha = -\gamma_\alpha, \quad \mathrm{re}\,\gamma_\alpha = 0, \quad \alpha = 1,\dots,r+s, \quad \alpha \neq t, \tag{8}$$

where

$$\hat{\eta}_{\alpha\beta} = (1/\eta_{tt})\,\mathrm{diag}\,(\underbrace{1,\dots,1}_{r\ \text{times}}, \underbrace{-1,\dots,-1}_{s\ \text{times}}).$$

LEMMA 2. *The pseudo-euclidean Hurwitz pairs* (V, S) *are of bidimension* (n, p), $n = \dim V$, $p = \dim S = r + s + 1$ *with*

$$n = \begin{cases} 2^{[\frac{1}{2}p - \frac{1}{2}]} & \textit{for} \quad r - s \quad 6,7,0 \pmod 8, \\ 2^{[\frac{1}{2}p + \frac{1}{2}]} & \textit{for} \quad r - s \quad 1,2,3,4,5 \pmod 8, \end{cases} \tag{9}$$

where [ ] *stands for the function integer part.*

Proofs. Our reasoning is completely analogous to that given in [15] (Lemmas 1 and 2) in the euclidean case. It is based on the following table, where $\mathbb{M}(N, \mathbb{F})$ denotes the algebra of $N \times N$-dimensional matrices over $\mathbb{F} = \mathbb{R}, \mathbb{C}$ or $\mathbb{H}$ (the real, complex or quaternion complex fields), and $\ell = [\frac{1}{2}r + \frac{1}{2}s]$:

r − s mod 8	$C^{(r,s)}$ isomorphic to: r + s = m	Dimension of the representation space	The Majorana representation: real (+), imaginary (−)
0	$\mathbb{M}(2^{\frac{1}{2}m}, \mathbb{R})$	2^ℓ	+; -
1	$\mathbb{M}(2^{[\frac{1}{2}m]}, \mathbb{R}) \oplus \mathbb{M}(2^{[\frac{1}{2}m]}, \mathbb{R})$	$2^{\ell+1}$	+; -
2	$\mathbb{M}(2^{\frac{1}{2}m}, \mathbb{R})$	2^ℓ	+; the imaginary Majorana representation can be constructed after doubling the dimension of the representation space
3	$\mathbb{M}(2^{[\frac{1}{2}m]}, \mathbb{C})$	2^ℓ	the real and imaginary Majorana representations can be constructed after doubling the

r - s mod 8	$C^{(r,s)}$ r + s = m isomorphic to:	Dimension of the representation space	The Majorana representation: real (+), imaginary (-)
			dimension of the representation space
4	$\mathbb{M}(2^{\frac{1}{2}m-1}, \mathbb{H})$	2^{ℓ}	as above
5	$\mathbb{M}(2^{[\frac{1}{2}m-1]}, \mathbb{H}) \oplus \oplus \mathbb{M}(2^{[\frac{1}{2}m-1]}, \mathbb{H})$	$2^{\ell+1}$	-; the real Majorana representation can be constructed after doubling the dimension of the representation space
6	$\mathbb{M}(2^{\frac{1}{2}m-1}, \mathbb{H})$	2^{ℓ}	as above
7	$\mathbb{M}(2^{[\frac{1}{2}m]}, \mathbb{C})$	2^{ℓ}	as above

3. THE CASES $r - s \equiv 4 \pmod 8$ AND $r - s \equiv 7 \pmod 8$

In each of the cases determined by (9) we have to prove the existence of the matrix $\kappa = [(e_j, e_k)_V]$ corresponding to (6), i.e. such that

$$\kappa \gamma_{\alpha}^{T} \kappa^{-1} = -\gamma_{\alpha}, \quad \alpha = 1,\dots,p; \quad \alpha \neq t, \tag{10}$$

and determine the matrix C_t in (6). Let

$$\gamma_{\alpha} = \check{\gamma}_{\alpha}, \quad \alpha = 1,\dots,r; \quad \hat{\gamma}_{\beta} = \gamma_{r+\beta}, \quad \beta = 1,\dots,s.$$

By Lemma 1, (10) is equivalent to

$$\kappa \check{\gamma}_{\alpha} = \check{\gamma}_{\alpha} \kappa, \quad \alpha = 1,\dots,r; \quad \kappa \hat{\gamma}_{\beta} = -\hat{\gamma}_{\beta} \kappa, \quad \beta = 1,\dots,s.$$

Consider the real matrices

$$A = (-i)^{r} \check{\gamma}_1 \check{\gamma}_2 \dots \check{\gamma}_r, \quad B = (-i)^{s} \hat{\gamma}_1 \hat{\gamma}_2 \dots \hat{\gamma}_s.$$

Clearly,

$$A^T = (-1)^{\frac{1}{2}r(r+1)} A, \quad B^T = (-1)^{\frac{1}{2}s(s-1)} B$$

and

$$A \check{\gamma}_{\alpha} = (-1)^{r-1} \check{\gamma}_{\alpha} A, \quad B \check{\gamma}_{\alpha} = (-1)^{s} \check{\gamma}_{\alpha} B,$$

$$A \hat{\gamma}_{\beta} = (-1)^{r} \hat{\gamma}_{\beta} A; \quad B \hat{\gamma}_{\beta} = (-1)^{s-1} \hat{\gamma}_{\beta} B.$$

Now, according to (4), it is natural to consider, subsequently, eight cases $r - s \equiv 0,\dots,7 \pmod 8$. Since in this paper we are limited to a few pages, we have to restrict ourselves to two cases which seem most interesting to us, namely $r - s \equiv 4 \pmod 8$ and $r - s \equiv 7 \pmod 8$.

The first case is representative of the cases $r-s \equiv 1, 2, 3, 4, 5 \pmod 8$. It yields the case of *euclidean Hurwitz pairs:* $(n, r, s) = (8, 4, 0)$, considered in [11-16], and the case $(8,0,4)$, which is quite interesting for the physical reasons because they provide an interpretation in the *Kałuża-Klein theories*. Following our table, we have to deal with two irreducible copies of the Clifford algebra $C^{(r,s)}$. Analysing the possibility of constructing the matrix κ in $C^{(r,s)}$ and with help of embedding $C^{(r,s)}$ in $C^{(r,s+1)}$ and $C^{(r,s+2)}$ we arrive at

PROPOSITION 1. *In the case* $r - s \equiv 4 \pmod 8$ *the only possible pseudo-euclidean Hurwitz pairs are those satisfying one of the following four sets of conditions:*

$$\begin{cases} r = 4(k+k_o)+2, & \kappa = iA\check{\gamma}_{r+1} \Rightarrow \kappa^T = \kappa \quad \textit{or} \\ s = 4(k-k_o)-2, & \kappa = B \Rightarrow \kappa^T = -\kappa; \end{cases}$$

$$\begin{cases} r = 4(k+k_o)+3, & \kappa = A \Rightarrow \kappa^T = \kappa \quad \textit{or} \\ s = 4(k-k_o)-1, & \kappa = A\check{\gamma}_{r+1}\check{\gamma}_{r+2} \Rightarrow \kappa^T = -\kappa; \end{cases}$$

$$\begin{cases} r = 4(k+k_o)+4, & \kappa = B \Rightarrow \kappa^T = \kappa \quad \textit{or} \\ s = 4(k-k_o), & \kappa = iA\check{\gamma}_{r+1} \Rightarrow \kappa^T = -\kappa; \end{cases}$$

$$\begin{cases} r = 4(k+k_o)+5, & \kappa = A\check{\gamma}_{r+1}\check{\gamma}_{r+2} \Rightarrow \kappa^T = \kappa \quad \textit{or} \\ s = 4(k-k_o)+1, & \kappa = A \Rightarrow \kappa^T = -\kappa, \end{cases}$$

where => *is an abbreviation for "implies",* k *and* k_o *are integers, and* $k \geq 0$.

The second case is representative of the cases $r-s \equiv 6, 7, 0 \pmod 8$. It includes the case $(n, r, s) = (8, 7, 0)$ determining the *octonion algebra* (cf. [9]). Following our table, we choose $C^{(r,s)}$ isomorphic to the matrix algebra ober $\mathbb{C}$ and thus obtain

PROPOSITION 2. *In the case* $r - s \equiv 7 \pmod 8$ *the only possible pseudo-euclidean Hurwitz pairs are those satisfying one of the following two sets of conditions:*

$$\begin{aligned} r &= 4(k+k_o)+5, & \kappa &= A \Rightarrow \kappa^T = -\kappa \quad \textit{or} \\ s &= 4(k-k_o)-2, & \kappa &= B \Rightarrow \kappa^T = -\kappa; \end{aligned}$$

$$\begin{aligned} r &= 4(k+k_o)+7, & \kappa &= A \Rightarrow \kappa^T = \kappa \quad \textit{or} \\ s &= 4(k-k_o), & \kappa &= B \Rightarrow \kappa^T = \kappa, \end{aligned}$$

where again k *and* k_o *are integers, and* $k \geq 0$.

By the assumed property (iii) of (V, S), I_n can be decomposed as $\Sigma_1^p a^\alpha C_\alpha$ for some $a^\alpha \in \mathbb{R}$. On the other hand, by (3), C_α are determined up to a pseudo-orthogonal transformation satisfying $\mathcal{O}^T \eta \mathcal{O} = \eta$. Therefore, after suitable rotations, we can choose $a_\alpha = 0$ for $\alpha \neq t$, $a^t = 1$ and for instance $t = r + 1$, so that, taking into account the relations (7) in Lemma 1, we arrive at

COROLLARY 1. *Without any loss of generality, in Lemma* 1 *and afterwards, we may set* $C_t = I_n$ *and* $t = r + 1$, *so that* $\eta_{tt} = 1$ *and* $\hat{\eta}_{\alpha\beta} = \eta_{\alpha\beta}$ *for* $\alpha, \beta \neq t$.

The above results should be compared with those on universal Clifford algebras for non-degenerate orthogonal spaces $\mathbb{R}^{p,q}$ ([20] and [19], pp. 270-272). In analogy to [15, 16] we can now develop a *pseudo-euclidean complex geometry induced by a given symplectic decomposition of* V with a distinguished fixed direction in S. The results will be published in a separate paper.

4. GENERALIZED FUETER EQUATIONS

Given a pseudo-euclidean Hurwitz pair (V, S), let us consider a continuously differentiable V-valued mapping f with a domain in S. It can be represented as $f(x^\alpha \varepsilon_\alpha) = f^k(x) e_k$. Thus it seems natural to us to define the *generalized Fueter operator* $\bar{D}$ as

$$\bar{D} = \varepsilon_\alpha \partial^\alpha, \qquad \partial^\alpha = \partial/\partial x_\alpha, \qquad x_\alpha = \eta_{\alpha\beta} x^\beta, \tag{11}$$

$\eta_{\alpha\beta}$ being defined in (1). A mapping f introduced above is said to be *regular* in its domain if it satisfies

$$\bar{D} f = 0 \qquad \text{(the } \textit{generalized Fueter equation}\text{)} \tag{12}$$

Since there exists only the left-side multiplication of the elements of S by those of V, we can define only the left-regular mappings, called *regular* in short.

Now, with the help of the Stokes theorem we can prove an analogue of the *Cauchy integral theorem*. To this end let us introduce the hypersurface element $dv = dv^\alpha \varepsilon_\alpha$ by $dv = * dx$, where $dx = dx^\alpha \varepsilon_\alpha$ and the star $*$ denotes the Hodge $*$ operator. With a standard definition of the k--dimensional cycle homologous to zero with respect to a domain in S, we have

THEOREM 1. *If* f *is a regular mapping in a domain* $\Delta \subset S$ *and* Σ *is a* $(p-1)$*-dimensional cycle homologous to zero with respect to* Δ, *then*

$$\int_{\Sigma} dv\, f = 0 \tag{13}$$

P r o o f. Formula (13) is a consequence of the Stokes formula for differentiable chains:

$$\int_{\Sigma} dv\, f = \int_{\Sigma} dv^{\alpha}(\varepsilon_{\alpha} f) = \int_{\Omega} d\omega\, \bar{D} f = 0,$$

where $\partial\Omega = \Sigma$ and $d\omega$ denotes the volume element in S.

Furthermore, we get

THEOREM 2. *Any generalized Fueter operator* $\bar{D}$ *has its conjugate* $D = \bar{\varepsilon}_{\alpha}\partial^{\alpha}$ *where* ε_{α}, $\alpha = 1,\ldots,p$, *are defined by* $\bar{\varepsilon}_{\alpha} e_j = \bar{c}^{k}_{j\alpha} e_k$ *with the property: if* f *is regular, then* Df *is regular, so* $\bar{D}Df = \bar{D}\,Df = 0$.

The importance of the Cauchy theorem, which is quite trivial in this context, relies upon the generalized Fueter equation (12) that can be transformed to a suggestive form

$$[\Sigma_{\alpha \neq r+1}(-i\gamma_{\alpha}\partial^{\alpha}) + I_n\partial^{r+1}]\Psi = 0, \tag{14}$$

resembling the *Dirac equation* and this enables us to give physical applications. In particular, there is a connection with the *soliton equations* [12]. To be precise, we obtain

THEOREM 3. *A continuously differentiable* V*-valued mapping* $f = f^k e_k$ *with a domain in* S *is a solution of* (12) *if and only if the related mapping* $\Psi = \kappa(f^1,\ldots,f^n)^T$ *is a solution of* (14).

P r o o f. By Lemma (1), the equation (12) is equivalent to

$$\bar{C}_{\alpha}\partial^{\alpha}\Psi = 0 \quad \text{with} \quad \bar{C}_{\alpha} = i\,C_{r+1}\bar{\gamma}_{\alpha} \quad \text{for} \quad \alpha \neq r+1,$$
$$\text{and} \quad \bar{\gamma}_{\alpha} = -\gamma_{\alpha},$$

the latter being clearly equivalent to (14).

COROLLARY 2. *The equivalence of the generalized Fueter equation* (11) *and the Dirac-like equation* (14) *is independent of the existence of the unit element in* S.

COROLLARY 3. *Any Dirac-like equation* (14) *implies the continuity equation*

$$\partial^{r+1}\rho + \partial^{\alpha} j_{\alpha} = 0,$$

where

$$\rho = (\bar{\Psi}, \Psi)_V, \qquad j_{\alpha} = -i\,\bar{\Psi}\gamma_{\alpha}\Psi, \qquad \bar{\Psi} = \Psi^T\kappa^{-1}.$$

Following [5], we can now define and investigate the *index of a point with respect to a cycle*. The index may be used for obtaining the *homological form of the Cauchy integral formula*, including applications and a reformulation of the results in the physical language. The precise formulation will be given in a separate paper.

References

[1] ADEM, J.: 'Construction of some normed maps', *Bol. Soc. Mat. Mexicana* (2) 20 (1975), 59-75.

[2] ———: 'On maximal sets of anticommuting matrices', *ibid.* (2) 23 (1978), 61-67.

[3] ———: 'On the Hurwitz problem over an arbitrary field I-II', *ibid.* (2) 25 (1980), 29-51 and (2) 26 (1981), 29-41.

[4] AHLFORS, L.V.: *Lectures on quasiconformal mappings*, Van Nostrand, Princeton 1966.

[5] BARTÍK, V., A. VAZ FERREIRA, M. MARKL, and V. SOUČEK: 'Index and Cauchy integral formula in complex-quaternionic analysis', *Simon Stevin Quart. J. Pure Appl. Math.*, to appear.

[6] BRACKX, F., R. DELANGHE, and F. SOMMEN: *Clifford analysis*, Research Notes in Math. 76, Pitman, 1979.

[7] CRUMEYROLLE, A.: 'Construction d'algebres de Lie graduées orthosymplectiques et conformosymplectiques minkowskiennes', in: *Seminar on deformations, Łódź-Warsaw 1982/84, Proceedings* (Lecture Notes in Math.), Springer, Berlin-Heidelberg-New York-Tokyo, to appear.

[8] ———: 'Constante de Planck et géométric symplectique', *ibid.*, to appear.

[9] HASIEWICZ, Z. and A.K. KWAŚNIEWSKI: 'Triality principle and G_2 group in spinor language', *J. Math. Phys.* 26 (1985), 6-11.

[10] HURWITZ, A.: 'Über die Komposition der quadratischen Formen', *Math. Ann.* 88 (1923), 1-25; reprinted in: A. HURWITZ, *Mathematische Werke II*, Birkhäuser Verlag, Basel 1933, pp. 641-666.

[11] KALINA, J., J. ŁAWRYNOWICZ, and O. SUZUKI: 'A differential-geometric quantum field theory on a manifold. II. The second quantization and deformations of geometric fields and Clifford groups', *Publ. R.I.M.S. Kyoto Univ.*, to appear.

[12] ———, ———, and ———: 'A field equation defined by a Hurwitz pair', Proc. of the 12th Winter School on Abstract Analysis, Srnî (Bohemian Forest), 1985, *Suppl. Rend. Circ. Mat. Palermo* 2 (5) (1985), to appear.

[13] KALINA, J., J. ŁAWRYNOWICZ, and O. SUZUKI: 'Partial differential equations connected with some Clifford structures and the related quasiconformal mappings', Proc. Conf. "Metodi di Analisi Reale nelle Equazioni alle Derivate Parziali", Cagliari 1985, to appear.

[14] ŁAWRYNOWICZ, J. in cooperation with J. KRZYŻ: *Quasiconformal mappings in the plane. Parametrical methods* (Lecture Notes in Math. 978), Springer, Berlin-Heidelberg-New York-Tokyo 1983, VI + 177 pp.

[15] —— and J. REMBIELIŃSKI: 'Hurwitz pairs equipped with complex structures', in: *Seminar on deformations, Łódź-Warsaw 1982/84, Proceedings* (Lecture Notes in Math. 1165), Springer, Berlin-Heidelberg-New York-Tokyo (1985) 184-195.

[16] —— and ——: 'Supercomplex vector spaces and spontaneous symmetry breaking', in: *Seminari di Geometria 1984*, CNR ed Università di Bologna, Bologna, 24 pp., to appear.

[17] LEE, H.C. (ed.): *An introduction to Kaluza-Klein theories*, World Scientific, Singapore 1984.

[18] PORTEOUS, I.R.: *Topological geometry*, 2nd ed., Cambridge Univ. Press, Cambridge 1981.

[19] SOUČEK, V.: 'Biholomorphicity in quaternionic analysis. Complex quaternionic analysis, connection to math. physics. Cauchy integral formula', in: *Seminari di Geometria 1984*, CNR ed Università di Bologna, Bologna 1984, pp. 147-171.

[20] WALL, C.T.C.: 'Graded algebras, antiinvolutions, simple groups and symmetric spaces', *Bull. Am. Math. Soc.* 74 (1968), 198-202.

A NEW REPRESENTATION FOR SPINORS IN REAL CLIFFORD ALGEBRAS

A. Dimakis
Institute of Theoretical Physics
University of Göttingen
Bunsenstr. 9
3400 Göttingen
Federal Republic of Germany

ABSTRACT. A new representation of a real Clifford algebra by itself is presented, which does not use onesided minimal ideals. This generalizes the "operator form" for Dirac spinors introduced by D. Hestenes. The spinor spaces obtained thus are related to a Clifford subalgebra of the parent algebra. Classification of real Clifford algebras and interior products of spinors together with their isometry groups are discussed.

1. INTRODUCTION

Clifford algebras provide a link between the two fundamental representations of orthogonal groups: tensors and spinors. On the one hand Clifford algebras decompose into parts which are tensor representations of the orthogonal groups related to them. These are antisymmetric tensors up to a fixed order depending on the dimension of the algebra. On the other hand the representations of Clifford algebras are spinor representations of the same groups.

Since tensors and thus also Clifford algebras are genuine geometric objects one is tempted to use this in order to find a plausible geometric interpretation also for spinors. An obvious way to do that is to use representations of Clifford algebras by themselves. This can be done, as is prescribed by the theory of representations of algebras, by using onesided minimal ideals. This approach is not the most economical for Clifford algebras because of the complicated algebraic properties of the elements of such ideals. These elements are multiples of idempotents of the algebra.

We propose a different and in some sense complementary way of representing a Clifford algebra by itself resulting in spinor spaces related to a subalgebra of the original Clifford algebra, called the spinor algebra. This is inspired by the work of D. Hestenes[1)]. For the first time he introduced a representation of the real Dirac algebra in our sense and succeeded to give by means of it a geometric interpretation of the imaginary unit present in the Dirac equation. According to this the imaginary unit is a spacelike generator of the Lorentz group, related to spin.

J. S. R. Chisholm and A. K. Common (eds.), Clifford Algebras and Their Applications in Mathematical Physics, 49–60.

As a first step towards this representation we classify[2] all real Clifford algebras by means of three sequences of integers one of which is the Radon-Hurwitz sequence. Next we find the maximal number of independent commuting elements of a Clifford algebra whose square equals $+1$[3]. Independence here means the following: to each such element we associate a generating orthonormal basis of the original Clifford algebra, so that every such positive commuting element is odd with respect to its basis but even with respect to the associated bases of all other elements. By means of the grading involutions of all these bases we decompose the Clifford algebra into even and odd parts which are the spinor spaces. The part which is even with respect to all involutions is the spinor algebra. The elements of the spinor algebra, which commute with all positive commuting elements found above, generate a field of numbers, which is the base field of the spinor representation. The action of the Clifford algebra on the spinor spaces is then introduced by means of the positive commuting elements and grading involutions of their associated bases.

We call this construction geometric representation of Clifford algebras. By means of it spinors attain a geometric meaning. As is the case by the onesided minimal ideals here also spinors can be identified with parts or direct sums of tensors. But here because of the simple algebraic properties of spinors one can also identify them with elements of subgroups of some extensions of the groups associated to the original Clifford algebras. This result has not been proved in general. Nevertheless it holds for all low dimensional Clifford algebras of physical interest.

2. REAL CLIFFORD ALGEBRAS AND THEIR CLASSIFICATION

Given a real associative algebra A with unit 1 we define like I. Porteous[4] an <u>orthonormal subset</u> (ons) of signature (p,q) to be a subset $Q = e_1,\dots,e_n$, $n = p+q$, of A whose elements satisfy the relations

$$e_i e_j + e_j e_i = 0 \quad \text{for} \quad i \neq j\,(=1,\dots,n)$$

and

$$e_i^2 = +1 \text{ for } i=1,\dots,p \text{ and } e_i^2 = -1 \text{ for } i = p+1,\dots,n.$$

If Q generates A as an algebra over $\mathbb{R}$ then A is a real Clifford algebra for the orthogonal space $Q_{\mathbb{R}} \cong \mathbb{R}^{p,q}$. Q is then called an <u>orthonormal base</u> (onb) of A. $Q_{\mathbb{R}}$ denotes the real linear span of Q in A[4].

The classification of real Clifford algebras is bases on the following facts:

Let Q be an orthonormal subset of A of signature (p,q).

(i) If $\dim A = 2^n$ and $e_1 \cdots e_n = 1$ then A is a <u>universal Clifford algebra</u> of $\mathbb{R}^{p,q}$. and is denoted by $\mathbb{R}_{p,q}$.

(ii) If $\dim A = 2^{n-1}$ then A is a <u>non-universal Clifford</u> algebra of

$\mathbb{R}^{p,q}$, this is only possible if $p-q-1 = 0 \bmod 4$.

(iii) If $\dim A > 2^n$ and $e_1 \dots e_n \neq \pm 1$ then there exists a monomorphism

$\mathbb{R}_{p,q} \longrightarrow A$.

(i), (ii) and (iii) allow us to find all properties of Clifford algebras from the study of their ons. With their help we can prove the following simple results, which are the base of the classification of Clifford algebras.

$$\mathbb{R}_{0,0} \cong \mathbb{R} \; ; \; \mathbb{R}_{0,1} \cong \mathbb{C} \; ; \; \mathbb{R}_{1,0} \cong {}^2\mathbb{R} \; ; \; \mathbb{R}_{0,2} \cong \mathbb{H} \; ; \; \mathbb{R}_{1,1} \cong \mathbb{R}(2)^{4)} .$$

Having an ons of some signature we can make elementary transformations to change its signature or decompose it into two independent (commuting) parts. For this purpose we have two propositions:

<u>Prop. 1</u> If Q is an ons of signature $(p+1,q)$ and $a \in Q$ with $a^2 = 1$, then

$$Q' = \{ba : b \in Q - \{a\}\} \cup \{a\}$$

is an ons of signature $(q+1,p)$.
An immediate consequence of this proposition is

$$\mathbb{R}_{p+1,q} \cong \mathbb{R}_{q+1,p} .$$

<u>Prop. 2</u> Let A be an associative algebra with unit, $S = \{f_1, f_2\}$ an ons of A of signature (i) $(1,1)$, (ii) $(2,0)$ and (iii) $(0,2)$. Let Q be an ons of A of signature (p,q) such that $S \cup Q$ is an ons of A of signature (i) $(p+1,q+1)$, (ii) $(p+2,q)$ and (iii) $(p,q+2)$. Then $Q' = \{b f_1 f_2 : b \in Q\}$ is an ons of A of signature (i) (p,q), (ii) and (iii) (q,p) such that every element of Q' commutes with f_1 and f_2. Conversely the existence of Q' implies that of Q.

As consequences of this proposition we obtain

$$\mathbb{R}_{p+1,q+1} \cong \mathbb{R}_{p,q} \otimes \mathbb{R}_{1,1} \; ; \; \mathbb{R}_{p+2,q} \cong \mathbb{R}_{q,p} \otimes \mathbb{R}_{2,0} \; ; \; \mathbb{R}_{p,q+2} \cong \mathbb{R}_{q,p} \otimes \mathbb{R}_{0,2} .$$

From these relations we find also

$$\mathbb{R}_{p+m,q+m} \cong \mathbb{R}_{p,q} \otimes \mathbb{R}_{m,m} , \; \mathbb{R}_{p+4,q} \cong \mathbb{R}_{p,q} \otimes \mathbb{R}_{4,0} \cong \mathbb{R}_{p,q} \otimes \mathbb{R}_{0,4} \cong \mathbb{R}_{p,q+4}$$

The eight-periodicity theorem follows now immediately:
For $\mathbb{B} = \mathbb{K}$ or ${}^2\mathbb{K}$ and $\mathbb{K} = \mathbb{R}$, $\mathbb{C}$ or $\mathbb{H}$ if $\mathbb{R}_{p,q} \cong \mathbb{B}(N)$, N positive integer, then

$$\mathbb{R}_{p+8,q} \cong \mathbb{R}_{p+4,q+4} \cong \mathbb{R}_{p,q+8} \cong \mathbb{B}(16N).$$

We introduce now three sequences of integers:
For $n \in Z, n = 8k+\ell$, $k, \ell \in Z$ and $0 \leq \ell < 8$

$$s(n) = \begin{cases} 1 , & \text{if } \ell = 0,4 \\ 0 , & \text{otherwise} \end{cases}$$

$$\varphi(n) = \begin{cases} 0 , & \text{if } \ell = 0,1,2 \\ 1 , & \text{if } \ell = 3,7 \\ 2 , & \text{if } \ell = 4,5,6 \end{cases}$$

$$\chi(n) = \begin{cases} 4k , & \text{if } \ell = 0 \\ 4k+1 , & \text{if } \ell = 1 \\ 4k+2 , & \text{if } \ell = 2,3 \\ 4k+3 , & \text{if } \ell = 4,5,6,7 . \end{cases}$$

χ is the Radon-Hurwitz sequence extended to hold also for negative integers.
For $p,q \in Z$ with $p,q \geq 0$ we define

$$\zeta = \chi(p-q+2)+q-2 , \quad \eta = \varphi(p-q), \quad \sigma = s(p-q-1) , \quad \vartheta = 2^{\sigma} .$$

Using these numbers we can prove the following <u>classification formula</u> for all real Clifford algebras

$$\mathbb{R}_{p,q} \cong \mathbb{R}_{\zeta,\zeta+\eta} \otimes \mathbb{R}_{\sigma,0} \cong (\otimes^{\zeta} \mathbb{R}_{2,0}) \otimes \mathbb{R}_{0,\eta} \otimes \mathbb{R}_{\sigma,0} = {}^{\vartheta}\mathbb{K}_{\eta}(2^{\zeta}),$$

where ${}^1\mathbb{K} = \mathbb{K}$, $\mathbb{K}_0 = \mathbb{R}$, $\mathbb{K}_1 = \mathbb{C}$ and $\mathbb{K}_2 = \mathbb{H}$.

Equating the dimensions of $\mathbb{R}_{p,q}$ and ${}^{\vartheta}\mathbb{K}_{\eta}(2^{\zeta})$ gives a formula for the Radon-Hurwitz sequence

$$\chi(n) = 1 + \frac{1}{2}(n - \varphi(n-2) - s(n-3)) .$$

Inserting this in the defining equation for ζ we find

$$\zeta = \frac{1}{2}(n - \eta - \sigma) , \quad n = p+q .$$

3. GEOMETRIC REPRESENTATION ($\sigma = 0$)

As is seen from the classification formula in the case $\sigma = 0$ we have

$$\mathbb{R}_{p,q} = \mathbb{R}_{\zeta,\zeta+\eta} \ .$$

Thus we can find an onb Q_o of signature $(\zeta,\zeta+\eta)$. Let

$$Q_0 = \{e_1,\dots,e_\zeta \ ; \ e_{\zeta+1},\dots,e_{2\zeta+\eta}\}$$

be such a basis with $e_i^2 = +1$ for $i = 1,\dots,\zeta$ and $e_i^2 = -1$ for $i = \zeta+1,\dots,2\zeta+\eta$. We define the set H by

$$H = \{h_i := e_i e_{\zeta+i} \ : \ i = 1,\dots,\zeta\} \ .$$

H has the following properties:

(i) For all $h \in H$ $\quad h^2 = 1$.

(ii) For all $h,h' \in H$ $\quad h h' = h' h$.

(iii) For every $h_i \in H$ there exists an onb Q_i of $\mathbb{R}_{p,q}$, such that h_i is odd with respect to Q_i and even with respect to all Q_j with $j \neq i$.

(i) and (ii) follow immediately from the definition of H. To construct onbs as is prescribed in (iii) we set for some k, $0 \leq k \leq \zeta$,

$$Q_i = \{a\, e_{\zeta+i} \ : \ a \in Q_0 - \{e_{\zeta+i}\}\} \cup \{e_{\zeta+i}\}$$

of signature $(\zeta,\zeta+\eta)$ for $i = 1,\dots,k$ and

$$Q_i = \{a\, e_i \ : \ a \in Q_0 - \{e_i\}\} \cup \{e_i\}$$

of signature $(\zeta+\eta+1 \ , \zeta-1)$, for $i = k+1,\dots,\zeta$.
One can prove now the

<u>Theorem</u>. H is maximal with respect to the properties (i), (ii) and (iii)[5)]
In what follows we need some more notation. For $n \in Z$, $n > 0$, we introduce the set

$$I_n = \{A := (A_1,\dots,A_n) \ : \ A_i := 0,1 \ ; \ i = 1,\dots,n\} \ .$$

With the operations

$$A + B = C \quad , \quad C_i = (A_i + B_i) \bmod 2 \quad , \quad i = 1,\dots,n$$

and

$$AB = (A_1 B_1, \ldots, A_n B_n)$$

I_n becomes a ring. The zero element is $0 = (0,\ldots,0)$ and the unit is $\Delta = (1,1,\ldots,1)$. Furthermore, we define the length of A to be

$$|A| = \sum_{i=1}^{n} A_i \quad .$$

Using I_n we can write the elements of a linear basis of $\mathbb{R}_{p,q}$ in the form

$$e_A := \prod_{i=1}^{n} (e_i)^{A_i} \quad , \quad \text{for} \quad A = (A_1,\ldots,A_n) \in I_n \ ,$$

where $(e_i)^0 = 1$ and $(e_i)^1 = e_i$.

For every $A \in I_\zeta$ we set

$$h_A := \prod_{i=1}^{\zeta} (h_i)^{A_i}$$

and

$$\pi_A := \prod_{i=1}^{\zeta} \frac{1}{2} (1 + (-1)^{A_i} h_i) \quad .$$

The π_A's are idempotents with

$$\pi_A \pi_B = \pi_B \pi_A = \delta_{AB} \pi_A \ ,$$

$$\sum_{A \in I_\zeta} \pi_A = 1 \ ,$$

and

$$h_A \pi_B = \pi_B h_A = (-1)^{|AB|} \pi_B \, .$$

δ_{AB} is the Kronecker delta in I_ζ. The π_A's are primitive idempotents of $\mathbb{R}_{p,q}$[3].

To every onb Q_i there corresponds a grading involution ω_i of $\mathbb{R}_{p,q}$.

This is defined as follows:

for $a \in Q_{i\mathbb{R}}$, $a^{\omega_i} = -a$ and for $a,b \in \mathbb{R}_{p,q}$, $(ab)^{\omega_i} = a^{\omega_i} b^{\omega_i}$.

Since the grading involutions ω_i , $i = 1,\ldots,\zeta$, commute, we can define by composition the involutions

$$\omega_A = \prod_{i=1}^{\zeta} (\omega_i)^{A_i} \quad , \quad A \in I_\zeta \ .$$

Here $\prod$ denotes composition and $(\omega_i)^0 = id$, $(\omega_i)^1 = \omega_i$. One can prove now the relations

$$h_A^{\ \omega_B} = (-1)^{|AB|}\, h_A$$

$$\pi_A^{\ \omega_B} = \pi_{A+B} \ .$$

Using the above defined objects we introduce the <u>circle multiplication</u> in $\mathbb{R}_{p,q}$. For $a,b \in \mathbb{R}_{p,q}$ we set

$$a \circ b := \sum_{A \in I_\zeta} a^{\omega_A}\, b\, \pi_A \ .$$

Under this multiplication $\mathbb{R}_{p,q}$ becomes a left $\mathbb{R}_{p,q}$ module. We have

$$1 \circ b = b$$

and

$$a \circ (b \circ c) = (ab) \circ c \ .$$

We introduce now the spinor spaces. For $A = (A_1,\ldots,A_\zeta) \in I_\zeta$ we set

$$S_A = \{a \in \mathbb{R}_{p,q} : a^{\omega_i} = (-1)^{A_i} a \ , \ i = 1,\ldots,\zeta\} \ .$$

These are linear subspaces of $\mathbb{R}_{p,q}$ and have the following properties:

(i) For $a \in S_A$, $b \in S_B$ we have $ab \in S_{A+B}$,

(ii) $h_A \in S_A$.

From (i) we conclude that S_o is a subalgebra of $\mathbb{R}_{p,q}$ and

$$S_A = S_0 h_A \ , \quad A \in I_\zeta \ .$$

It turns out that S_0 is a Clifford algebra and has the onb

$$Q_S := \{e_1,\ldots,e_k \ ; \ e_{\zeta+k+1},\ldots,e_{2\zeta+\eta}\}$$

of signature $(k,\zeta-k+\eta)$. So

$$S_0 \cong \mathbb{R}_{k,\zeta-k+\eta} .$$

We set

$$s_i = e_i \text{ for } i=1,\dots,k \text{ , } s_i = e_{\zeta+i} \text{ for } i=k+1,\dots,\zeta \text{ and}$$

$$j_r = e_{2\zeta+r} \text{ for } r=1,\dots,\eta .$$

The subset $\mathbb{K}'_\eta := \{j_1,\dots,j_\eta\}$ of Q_S commutes with all $h_i \in H$ and is isomorphic to the field of numbers $\mathbb{K}_\eta$. So the sets $S_A, A \in I_\zeta$, can be understood as right linear spaces order $\mathbb{K}'_\eta$. They are also left $\mathbb{R}_{p,q}$-modules under the circle multiplication. We have for $A \in \mathbb{R}_{p,q}$, $b \in S_A$

$$(a \circ b)^{\omega_i} = a \circ (b^{\omega_i}) = (-1)^{A_i} a \circ b$$

and for $\lambda \in \mathbb{K}'_\eta$

$$a \circ (b\lambda) = (a \circ b)\lambda .$$

Hence S_A, $A \in I_\zeta$, are representation modules of $\mathbb{R}_{p,q}$ and so we call them <u>spinor spaces</u>. S_0 is called <u>spinor algebra</u>. One can prove also the

<u>Theorem</u>. The representation $\mathbb{R}_{p,q}$-modules $S_A, A \in I_\zeta$, over $\mathbb{K}'_\eta$ are faithful and irreducible[5)].
The resulting faithful and irreducible representation of $\mathbb{R}_{p,q}$ is called <u>geometric representation</u>.
A linear basis of S_0 is given by

$$s_A = \prod_{i=1}^{\zeta} (s_i)^{A_i} \quad \text{for} \quad A = (A_1,\dots,A_\zeta) \in I_\zeta .$$

An element $\psi \in S_0$ can be decomposed into

$$\psi = \sum_{A \in I_\zeta} s_A \psi^A , \quad \psi^A \in \mathbb{K}'_\eta .$$

The components ψ^A can be obtained by means of the formula

$$\psi^A = \pi_0 (s_A^{-1} \psi) .$$

Thus to every spinor $\psi \in S_0$ corresponds a column vector $(\psi^A) \in (\mathbb{K}_\eta)^{2^\zeta}$. For every element $a \in \mathbb{R}_{p,q}$ we obtain a matrix $(a^A{}_B) \in \mathbb{K}_\eta(2^\zeta)$ through

$$a \circ s_A = \sum_{B \in I_\zeta} s_B \, a^B{}_A .$$

Since

$$\mathbb{R}_{p,q} = \bigoplus_{A \in I_\zeta} S_A ,$$

we can decompose any element $a \in \mathbb{R}_{p,q}$ into components

$$a = \sum_{A \in I_\zeta} a_A \quad , \quad a_A \in S_A .$$

With this decomposition the circle multiplication takes the simpler form

$$a \circ b = \sum_{A \in I_\zeta} a_A \, b h_A .$$

4. GEOMETRIC REPRESENTATION ($\sigma = 1$).

As is seen from the classification formula in this case

$$\mathbb{R}_{p,q} = \mathbb{R}_{\zeta,\zeta+\eta} \otimes \mathbb{R}_{1,0} ,$$

where η can be either 0 or 2. We study the two cases separately.

A. ($\eta = 0$).

Here we have

$$\mathbb{R}_{p,q} \cong \mathbb{R}_{\zeta,\zeta} \otimes \mathbb{R}_{1,0} \cong \mathbb{R}_{\zeta+1,\zeta} .$$

Let

$$Q_0 = \{e_1, \ldots, e_{\zeta+1} \; ; \; e_{\zeta+2}, \ldots, e_{2\zeta+1}\}$$

be an onb of $\mathbb{R}_{p,q}$ of signature $(\zeta+1, \zeta)$. We construct the set H as in the last section

$$H = \{h_i = e_i\, e_{\zeta+1+i} : i = 1,\ldots,\zeta\} .$$

The associated onbs of the elements of H are given for $0 \le k \le \zeta$ by

$$Q_i = \{a\, e_{\zeta+1+i} : a \in Q_o - \{e_{\zeta+1+i}\}\} \cup \{e_{\zeta+1+i}\}$$

for $i = 1,\ldots,k$ and

$$Q_i = \{a\, e_i : a \in Q_o - \{e_i\}\} \cup \{e_i\}$$

for $i = k+1,\ldots,\zeta$. Both families have the signature $(\zeta+1, \zeta)$. It is easy to show that H is not maximal with respect to the three properties given in 3. Taking $h_o = e_{\zeta+1}$ with associated onb Q_o we obtain the maximal set $H \cup \{h_o\}$. Applying the construction of 3 to $H \cup \{h_o\}$ we obtain irreducible but not faithful representations.
Leaving h_o aside and applying this construction to H we find the following onb for the spinor algebra

$$Q_S = \{e_1,\ldots,e_k\, , e_{\zeta+1} ; e_{\zeta+k+2},\ldots e_{2\zeta+1}\}$$

of signature $(k+1, \zeta-k)$. The subset of Q_S whose elements commute with all h H is $\{e_{\zeta+1}\}$. We set

$$s_i = e_i \text{ for } i = 1,\ldots,h\ ,\ s_i = e_{\zeta+i+1} \text{ for } i = h+1,\ldots,\ \text{and } \alpha = e_{\zeta+1} .$$

$\{\alpha\}$ generates the double field ${}^2\mathbb{R}'$. Thus S_A, $A \in I_\zeta$, can be understood as 2^ζ-dimensional linear spaces over ${}^2\mathbb{R}'$.
The representations obtained in this way are faithful but non irreducible. S_A, $A \in I_\zeta$, possess two non trivial and irreducible submodules given by $S_A(1 \pm \alpha)$.

<u>B. $(\eta = 2)$.</u> In this case

$$\mathbb{R}_{p,q} \cong \mathbb{R}_{\zeta,\zeta+2} \otimes \mathbb{R}_{1,0} \cong \mathbb{R}_{\zeta,\zeta+3} .$$

Letting

$$Q_o = \{e_1,\ldots,e_\zeta ; e_{\zeta+1},\ldots,e_{2\zeta+3}\}$$

be an onb of signature $(\zeta, \zeta+3)$ we construct the set

$$H = \{h_i = e_i\, e_{\zeta+i} : i = 1,\ldots,\zeta\}$$

as usual with the associated onbs for some k, $0 \le k \le \zeta$. H is again not maximal. The element $e_{2\zeta+1}\, e_{2\zeta+2}\, e_{2\zeta+3}$ can be added to H together with the associated onb Q_o.

Applying the method of 3 to H we obtain the spinor spaces S_A, $A \in I_\zeta$, with the onb of S_o

$$Q_S = \{s_i = 1,\ldots,\zeta \ ; \ j_1, j_2 \ ; \ \alpha\, j_3\}$$

of signature $(k\ ,\ \zeta - k + 3)$. Here

$$s_i = e_i\ ,\ i = 1,\ldots,k\ ;\ e_i = e_{\zeta+i}\ ,\ i = k+1,\ldots,\zeta \ \text{ and}$$

$$j_1 = e_{2\zeta+1}\ ,\ j_2 = e_{2\zeta+2}\ ,\ j_3 = j_1 j_2 \ \text{ and } \alpha = e_{2\zeta+1}\, e_{2\zeta+2}\, e_{2\zeta+3}\ .$$

The set $\{j_1, j_2, j_3, \alpha\}$ commutes with H and generates the double field $2\mathbb{H}'$. The rest is as in case A.

5. INTERIOR PRODUCTS AND ISOMETRIES ON SPINOR SPACES

For two spinors $\psi, \varphi \in S_A$, $A \in I_\zeta$, and some fixed $Y \in I_\zeta$ we set

$$(\psi,\varphi)_Y := \pi_0 \circ (s_Y^{-1}\ \psi^\sigma \varphi)\ ,$$

where σ is an anti-involution of $\mathbb{R}_{p,q}$ commuting with all grading involutions involved in the circle multiplication. Using the identity $\pi_A \circ s_B = \delta_{AB}\, s_A$ we can show $(\psi,\varphi)_Y \in \mathbb{K}'_\eta$ and that $(\ ,\ )_Y$ defines an interior product on S_A. Setting $a^{s_Y} = s_Y^{-1}\ a\ s_Y$, for $a \in \mathbb{R}_{p,q}$, which defines an involution, $s_Y = \varepsilon_Y\, s_Y$, where $\varepsilon_Y = \pm 1$, we find

$$[(\psi,\varphi)_Y]^{\sigma s_Y} = \varepsilon_Y (\varphi,\psi)_Y\ .$$

Thus $(\ ,\ )_Y$ is a $\mathbb{K}'^{(\sigma s_Y)}_\eta$-symmetric or antisymmetric interior product,[5)] depending on whether $\varepsilon_Y = +1$ or -1.

An element $G \in \mathbb{R}_{p,q}$ is called an isometry of $(\ ,\ )_Y$ if for all $\psi, \varphi \in S_A$ $(G \circ \psi,\ G \circ \varphi)_Y = (\psi,\varphi)_Y\ .$

From this definition the simpler necessary and sufficient condition follows:

$$G^{\sigma\ \omega_{Y+\Sigma}}\ G = 1\ ,$$

where $\Sigma = (\Sigma_1, \ldots, \Sigma_\zeta)$ I_ζ is defined through $h_i^\sigma = (-1)^{\Sigma_i} h_i$, $i = 1, \ldots, \zeta$. Since there are 2^ζ basis elements s_A of S_0 we have defined 2^ζ interior products on S_B, $B \in I_\zeta$. These are in general not independent[4].

REFERENCES:

1) D. Hestenes, *J. Math. Phys.* **8**, 798 (1967); **14**, 893 (1973). See also Hestenes conference lecture, 'Clifford algebra and the Interpretation of Quantum Mechanics'

2) P. Lounesto, *Found. Phys.* **11**, 721 (1981)

3) There seems to be an overlapping between some of our results and those of P. Lounesto, 'On primitive idempotents of Clifford algebras', Report - HTKK-MAT-A 113, (Helsinki, 1977)

4) I.R. Porteous, *Topological geometry* (Van Nostrand, London, 1969)

5) Proofs for the results presented here are given in A. Dimakis, dissertation (Göttingen, 1983)

PRIMITIVE IDEMPOTENTS AND INDECOMPOSABLE LEFT IDEALS IN DEGENERATE CLIFFORD ALGEBRAS

Rafal Ablamowicz
Department of Mathematics
Gannon University
Erie PA 16541
U.S.A.

Pertti Lounesto
Institute of Mathematics
Helsinki University of Technology
02150 Espoo
Finland

ABSTRACT. Primitive idempotents of degenerate Clifford algebras are determined. A degenerate Clifford algebra A has a nilpotent Jacobson radical J(A) so that the factor algebra $\bar{A} = A/J(A)$ is a non-degenerate Clifford algebra isomorphic to a certain maximal Clifford subalgebra of A. Once primitive mutually annihilating idempotents of $\bar{A}$ are known, they can be lifted, modulo the radical, to primitive mutually annihilating idempotents of A. Moreover, each decomposition of $\bar{A}$ into a direct sum of principal indecomposable modules can be lifted to a corresponding decomposition of A. The resulting indecomposable summands of A need not be minimal. As an example, principal indecomposable modules of degenerate Clifford algebras with degeneracy in one dimension are found.

1. INTRODUCTION

The degenerate Clifford algebra A on an n-dimensional vector space V, $n = p+q+d$, $d \neq 0$, with quadratic form

$$Q(\underline{x}) = x_1^2 + \ldots x_p^2 - x_{p+1}^2 - \ldots - x_{p+q}^2$$

is a finite dimensional associative algebra over R containing a copy of R and a copy of V so that

i) the square of a vector $\underline{x}$ in V is equal to its quadratic form: $\underline{x}^2 = Q(\underline{x})$,

ii) the Clifford algebra A is generated as an algebra by the vector space V (but not by any proper subspace of V).

The vector space V is a direct sum of a non-degenerate orthogonal space V', where the quadratic form is Q, and a totally isotropic sub-

J. S. R. Chisholm and A. K. Common (eds.), Clifford Algebras and Their Applications in Mathematical Physics, 61–65.

space of dimension d annihilating all of V called the null space of V. The Clifford algebra generated by the null space, with null quadratic form, is a Grassmann algebra G. It is well known that the Clifford algebra A is a graded tensor product of the non-degenerate Clifford algebra A' on V' and the Grassmann algebra G [1], [2].

2. STRUCTURE OF DEGENERATE CLIFFORD ALGEBRAS

A degenerate Clifford algebra A contains non-trivial nilpotent ideals generated by the nilpotent elements of the Grassmann subalgebra of A. The largest such ideal, written as J(A), is called the Jacobson radical of A and it is usually defined as the intersection of all maximal right ideals of A [3].

Lemma 1. i) The algebra radical J(A) coincides with the ring radical of A.

ii) The radical J(A) contains every right and every left nilpotent ideal of A.

iii) J(A/J(A)) = (0).

iv) The radical J(A) is the set of all a in A such that 1 + ax is a unit for all x in A.

For example, the Jacobson radical of a Grassmann algebra G of dimension 2^d is the direct sum of non-scalar k-vector spaces, k = 1,2,...,d and, therefore, its dimension is 2^d-1. The radical of A is the tensor product of a copy of A' and the radical of G. Moreover, the degenerate Clifford algebra A is a direct A'-module sum of the nilpotent radical J(A) and a not necessarily unique copy of the non-degenerate Clifford algebra A' which is also isomorphic to A/J(A) [1], [4], [5].

Any finite dimensional algebra over a field is an Artinian ring. Therefore, the classical theory of such rings can be applied to non-degenerate as well as degenerate Clifford algebras. In particular, we would like to consider the problem of idempotent decompositions of unity involving primitive idempotents in such algebras. The importance of primitive idempotent decompositions is that they correspond to direct sum decompositions into principal indecomposable modules.

Let B be an arbitrary ring. Recall that an element $e \neq 0$ in B is called an idempotent if $e = e^2$. Two idempotents e_1 and e_2 are said to be mutually annihilating if $e_1e_2 = e_2e_1 = 0$, and an idempotent is called primitive if it is not the sum of two mutually annihilating idempotents. A B-module Be is indecomposable if and only if e is primitive. By an idempotent decomposition of 1 in B we mean a set of mutually annihilating idempotents $e_1,\ldots,e_r$ summing up to 1. An idempotent decomposition is called primitive if all the involved idempotents are primitive. It is well known that every right Artinian ring has primitive idempotent decompositions which are invariant in that the number of idempotents in each remains the same and that these sets of idempotents can be obtained from each other by an inner automorphism [6], [7].

Recall that the Radon-Hurwitz number r_i for i in Z is given by $r_{i+8} = r_i+4$, $r_0 = 0$, $r_1 = 1$, $r_2 = r_3 = 2$, $r_4 = r_5 = r_6 = r_7 = 3$. The primitive idempotent decompositions in non-degenerate Clifford algebras were determined in [8] where the proof of the following theorem can be found.

Theorem 2. In the canonical basis for A' there are always $k = q - r_{q-p}$ commuting elements $\underline{e}_{Ti}$, with $\underline{e}_{Ti}^2 = 1$, $i=1,\dots,k$ generating a group of order 2^k. The product e of the corresponding mutually non-annihilating idempotents $\frac{1}{2}(1+\underline{e}_{Ti})$ is a primitive idempotent in A' and A'e is a principal indecomposable module. Furthermore, letting the signs of the elements $\underline{e}_{Ti}$ in the product $\frac{1}{2}(1+\underline{e}_{T1})\dots\frac{1}{2}(1+\underline{e}_{Tk})$ vary independently, one obtains 2^k idempotents which provide a primitive idempotent decomposition of 1 in A'.

Corollary A' is a direct sum of 2^k principal indecomposable modules which are uniquely determined up to isomorphism.

Example Consider the Clifford algebra A' when the signature of Q is (3,0). Then, $k = -r_{-3} = -(r_5-4) = 1$, and a primitive idempotent has only one factor of the type $\frac{1}{2}(1\pm\underline{e}_1)$. Therefore, A' is a direct sum of two indecomposable (hence, minimal, since J(A') = (0)) left ideals $A'e_i$, $i = 1,2$ where $e_1 = \frac{1}{2}(1+\underline{e}_1)$, $e_2 = \frac{1}{2}(1-\underline{e}_1)$, $e_1e_2 = e_2e_1 = 0$ and $1 = e_1+e_2$.

Example Consider now the Clifford algebra A' when the signature of Q is (3,1). Then, $k = 1 - r_{-2} = 1 - (r_6-4) = 2$, and there are four primitive idempotents each having two commuting factors. For example, $e_1 = \frac{1}{2}(1+\underline{e}_1)\frac{1}{2}(1+\underline{e}_{24})$, $e_2 = \frac{1}{2}(1-\underline{e}_1)\frac{1}{2}(1+\underline{e}_{24})$, $e_3 = \frac{1}{2}(1+\underline{e}_1)\frac{1}{2}(1-\underline{e}_{24})$, $e_4 = \frac{1}{2}(1-\underline{e}_1)\frac{1}{2}(1-\underline{e}_{24})$. Then A' has a direct sum decomposition into four principal indecomposable modules $A'e_i$, i=1,2,3,4.

Let B be a ring with a non-zero Jacobson radical J(B) and N be an ideal of B contained in J(B). We say that idempotents modulo N can be lifted provided for every element f of B such that $f^2 - f$ is in N there exists an element $e = e^2$ in B such that $e - f$ is in N. It is now classical result, known as the Theorem on Lifting Idempotents, that idempotents modulo a nilpotent ideal can always be lifted in such a way that primitive idempotent decompositions of 1 in $\bar{B} = B/N$ are lifted to primitive idempotent decompositions of 1 in B with the same number of summands [3], [6], [7].

Therefore, since $A/J(A) \simeq A'$, any set of mutually annihilating primitive idempotents of a non-degenerate Clifford algebra A' summing up to 1 will be (lifted to) a set of mutually annihilating primitive idempotents summing up to 1 in a degenerate Clifford algebra A (here, A' is considered as a subalgebra of A). This observation together with the Theorem, when applied to degenerate Clifford algebras, yield the following result.

Theorem 3. i) If $1 = e_1+\dots+e_s$, $s = 2^k$, $k = q - r_{q-p}$ is a primitive idempotent decomposition of 1 in A', then it is also a primitive idempotent decomposition of 1 in A. If $1 = f_1+\dots+f_t$ is another such decomposition of 1 in A, then $t = s$ and there exists a unit u in A such that $e_iu = uf_{\pi(i)}$ for all i, where π is some permutation on $\{1,\dots,s\}$.

ii) If e is a primitive idempotent in A' then e is primitive in A. Thus, A'e is a minimal left ideal of A' and Ae is an indecomposable left ideal of A.

iii) The decomposition of A' into a direct sum of principal minimal ideals $A'e_1 \oplus \ldots \oplus A'e_s$ can be lifted to a direct sum $Ae_1 \oplus \ldots \oplus Ae_s$ of indecomposable ideals of A.

iv) Furthermore, for $i,j = 1,2,\ldots,s$ $A'e_i$ and $A'e_j$ are isomorphic as left A'-modules if and only if Ae_i and Ae_j are isomorphic as left A-modules.

Example The only primitive idempotent of a Grassmann algebra G is its unit element 1, and so the Grassmann algebra itself is an indecomposable left ideal. However, Grassmann algebra contains a descending chain of proper left ideals which ends with the 1-dimensional minimal left ideal of pseudoscalars generated by the d-vector $\underline{e}_1\underline{e}_2\ldots\underline{e}_d$.

Example Consider the degenerate Clifford algebra A when the signature of Q is (p,0,1). Thus, the even Clifford group of A generates rigid motions of the non-degenerate component V' in V [9]. Then, for p = 1,2,3,4 any primitive idempotent of A has only one factor $\frac{1}{2}(1+\underline{e}_1)$ and A has two principal indecomposable modules. If p = 5,6 then there are four primitive idempotents summing up to 1 in A, each having two factors of the type $\frac{1}{2}(1+\underline{e}_1)\frac{1}{2}(1+\underline{e}_{23})$, and which generate four principal indecomposable modules of A. If p = 7 then the number of factors is three and A decomposes into eight indecomposables. Finally, when p = 8, there are 16 primitive idempotents summing up to 1. Higher dimensions may be treated similarly. In the case when p = 3, the decomposition of A induces an analogous decomposition of its even subalgebra which is the algebra of biquaternions.

3. DISCUSSION

The technique of lifting idempotents, which is a fundamental tool in the classical theory of Artinian rings with a non-zero Jacobson radical, has been applied to degenerate Clifford algebras. It has been shown that the results presented in [8] can be extended to these algebras and their principal indecomposable module decompositions can then be established. In contrast to the non-degenerate case, these ideals need not be minimal. However, they remain indecomposable under regular representations of degenerate Clifford algebras and, as such, provide representation modules for spin groups associated with degenerate orthogonal spaces [4], [5].

REFERENCES

[1] A. Crumeyrolle: 'Algèbres de Clifford dégénérées et revêtements des groupes conformes affines orthogonaux et symplectiques.' Ann. Inst. H. Poincaré 33(1980), 235.

[2] T.Y. Lam: The Algebraic Theory of Quadratic Forms. Benjamin,

Reading, 1980.

[3] J. Lambek: Lectures on Rings and Modules. Blaisdell, Waltham, 1966.

[4] R. Ablamowicz: 'Structure of spin groups associated with degenerate Clifford algebras.' To appear in J. Math. Phys.(1986).

[5] J.A. Brooke: 'A Galileian formulation of spin. I. Clifford algebras and spin groups.' J. Math. Phys. **19**(1978), 52. 'II. Explicit realizations.' J. Math. Phys. **21**(1980), 617. 'Spin groups associated with degenerate orthogonal spaces.' NATO Advanced Research Workshop "Clifford Algebras and Their Applications in Mathematical Physics," Canterbury, 1985.

[6] P. Landrock: Finite Group Algebras and Their Modules. London Mathematical Society Lecture Note Series **84**, Cambridge University Press, Cambridge, 1983.

[7] Ch.W. Curtis and I. Reiner: Methods of Representation Theory With Applications to Finite Groups and Orders. Vol. I. Wiley Interscience, New York, 1981.

[8] P. Lounesto and G.P. Wene: 'Idempotent structure of Clifford algebras.' Submitted for publication.

[9] I.R. Porteous: Topological Geometry. Cambridge University Press, Cambridge, 1981.

GROUPES DE CLIFFORD ET GROUPES DES SPINEURS

Artibano MICALI
Université de Montpellier II
Institut de Mathématiques
Place Eugène Bataillon
34060 Montpellier
FRANCE

ABSTRACT. This is a review talk on Clifford groups and spin groups in the case of nondegenerate quadratic form over finite generated projective modules. If the quadratic form is singular, it is possible to construct such groups but with some technical complications.

1. PRELIMINAIRES

Soient K un anneau commutatif à élément unité, P un K-module, $f : P \to K$ une forme quadratique et $\varphi : P \times P \to K$, $(x,y) \mapsto f(x+y) - f(x) - f(y)$ la forme K-bilinéaire symétrique associée à f. On dira que f est <u>non dégénérée</u> si l'application K-linéaire $P \to P^*$ définie par $x \mapsto (y \mapsto \varphi(x,y))$ est un isomorphisme de K-modules, où $P^* = \mathrm{Hom}_K(P,K)$ est le K-module dual de P. Le couple (P,f) formé d'un K-module P et d'une forme quadratique f sur P est appelé un K-<u>module quadratique</u> (cf. [12], chapitre 1, pour la théorie des modules quadratiques). Etant donné un K-module quadratique (P,f), notons $C(P,f)$ son <u>algèbre de Clifford</u> (cf. [12], chapitre 2, pour la théorie des algèbres de Clifford). On sait qu'il s'agit d'une K-algèbre associative à élément, non nécessairement commutative, graduée sur $\mathbb{Z}/(2)$ dont les sous-K-modules des éléments homogènes de degré 0 et 1 sont notés, respectivement, par $C_0(P,f)$ et $C_1(P,f)$. On sait, de plus, que $C_0(P,f)$ est une sous-K-algèbre à élément unité de l'algèbre $C(P,f)$. Le résultat suivant nous montre le rapport entre formes quadratiques non dégénérées et algèbres séparables (cf. [9]) :

<u>Théorème</u> 1.1. <u>Soient</u> K <u>un anneau commutatif à élément unité et</u> (P,f) <u>un</u> K-<u>module quadratique</u>. <u>Les conditions suivantes sont équivalentes</u> : (i) <u>l'algèbre de Clifford</u> $C(P,f)$ <u>est séparable sur</u> K <u>au sens gradué</u> ; (ii) f <u>est une forme quadratique non dégénérée</u>.

On dira qu'une K-algèbre associative A à élément unité est <u>centrale</u> si son centre $Z_K(A) = \{x \mid x \in A,\ xy = yx,\ \forall\, y \in A\}$ est une K-algèbre

J. S. R. Chisholm and A. K. Common (eds.), Clifford Algebras and Their Applications in Mathematical Physics, 67–78.

isomorphe à K et on dira qu'une K-algèbre centrale A est une _algèbre d'Azumaya_ si A est séparable sur K et si A est une K-algèbre projective de type fini.

Théorème 1.2. _Soient_ K _un anneau commutatif à élément unité et_ (P,f) _un_ K-_module quadratique où_ P _est un_ K-_module projectif de type fini et_ f _est une forme quadratique non dégénérée_. _Si_ P _est de rang pair_, C(P,f) _est une algèbre d'Azumaya et_ $C_0(P,f)$ _est une algèbre séparable dont le centre est une extension quadratique de_ K. _Si_ P _est de rang impair_, $C_0(P,f)$ _est une algèbre d'Azumaya et_ C(P,f) _est une algèbre séparable dont le centre est une extension quadratique de_ K.

Pour la démonstration de ce théorème, qui nous sera utile par la suite, on renvoie à la bibliographie citée (cf. [9],[12]).

2. LE GROUPE DES RACINES CARREES ET QUESTIONS CONNNEXES

Soient K un anneau commutatif à élément unité et $\mu_2(K) = \{a \mid a \in K, a^2 = 1\}$ le groupe multiplicatif des racines carrées de l'unité. Le caractère fonctoriel de μ_2 est facile d'établir et si U(K) désigne le groupe multiplicatif des éléments inversibles de K, le morphisme de groupes abéliens $U(K) \xrightarrow{2} U(K)$ défini par $a \mapsto a^2$ nous fournit la suite exacte de groupes $1 \to \mu_2(K) \to U(K) \to U^2(K) \to 1$, où $U^2(K) = \operatorname{Im} U(K) \xrightarrow{2} U(K)$ est le sous-groupe de U(K) formé par les carrés des éléments inversibles de K. Si T désigne la topologie fppf sur Spec(K), on a la suite exacte de faisceaux $1 \to \mu_2 \to U \xrightarrow{2} U \to 1$ d'où la suite exacte de cohomologie $1 \longrightarrow \mu_2(K) \longrightarrow U(K) \xrightarrow{2} U(K) \longrightarrow H^1(T,K,\mu_2) \longrightarrow H^1(T,K,U) \xrightarrow{2} H^1(T,K,U)$. Le résultat suivant est bien connu (cf. [10]) :

Théorème 2.1. _Soient_ K _un anneau commutatif à élément unité et_ T _la topologie_ fppf _sur_ Spec(K). _Il existe alors un isomorphisme de groupes abéliens_ $H^1(T,K,U) \approx \mathrm{Pic}(K)$.

On rappelle ici que Pic(K) désigne le _groupe de Picard_ de K, c'est-à-dire, l'ensemble des classes d'isomorphismes [P] des K-modules P, projectifs de type fini et de rang 1, la structure de groupe étant définie par $[P] + [Q] = [P \otimes_K Q]$, quels que soient les éléments [P] et [Q] dans Pic(K).

Exemple 2.2. Si K est l'anneau $\mathbb{Z}$ des nombres entiers ou, plus généralement, si K est un anneau principal ou encore, si K est un anneau de polynômes à un nombre fini d'indéterminées à coefficients dans un corps commutatif k, alors Pic(K) = 0.

D'autre part, considérons l'ensemble $\mathcal{P}(K)$ formé des couples $(\mathcal{P},h)$ où P est un K-module projectif de type fini et de rang 1 et $h : P \otimes_K P \xrightarrow{\sim} K$ est un isomorphisme de K-modules. On dira encore que deux couples (P,h) et (P',h') sont <u>égaux</u> et on écrit $(P,h) = (P',h')$, s'il existe un isomorphisme de K-modules $g : P \xrightarrow{\sim} P'$ rendant commutatif le diagramme :

$$\begin{array}{ccc} P \otimes_K P & \xrightarrow[\sim]{g \otimes g} & P' \otimes_K P' \\ & {\scriptstyle h}\searrow\wr \qquad \wr\swarrow{\scriptstyle h'} & \\ & K & \end{array}$$

On fait de $\mathcal{P}(K)$ un groupe abélien, en posant $(P,h) + (P',h') = (P \otimes_K P', h * h')$ où $h*h' : (P \otimes_K P') \otimes_K (P \otimes_K P') \xrightarrow{\sim} (P \otimes_K P) \otimes_K (P' \otimes_K P') \xrightarrow[\sim]{h \otimes h'} K \otimes_K K \xrightarrow[\sim]{m} K$ est l'isomorphisme composé évident. L'élément neutre de $\mathcal{P}(K)$ pour cette structure de groupe est (K,m) où m est la multiplication de K et l'opposé de (P,h) est encore (P,h). On a le résultat suivant (cf. [10]) :

<u>Théorème</u> 2.3. <u>Soient</u> K <u>un anneau commutatif à élément unité et</u> T <u>la topologie</u> fppf <u>sur</u> Spec(K). <u>Il existe alors un isomorphisme de groupes abéliens</u> $H^1(T,K,\mu_2) \approx \mathcal{P}(K)$.

On peut montrer (cf. [1], section 2) que les éléments du groupe $\mathcal{P}(K)$ sont les <u>discriminants</u> des K-modules bilinéaires (P,h) où $h : P \times P \to K$ est une forme K-bilinéaire non dégénérée et P un K-module projectif de type fini et inversible. Ce groupe est aussi noté, dans la littérature, Discr(K) (cf. [1], section 2).

Pour tout anneau commutatif K à élément unité, on peut donc écrire la suite exacte de groupes abéliens

$$1 \to \mu_2(K) \to U(K) \xrightarrow{2} U(K) \to \mathcal{P}(K) \to \mathrm{Pic}(K) \xrightarrow{2} \mathrm{Pic}(K)$$

qui nous permettra de donner des exemples des <u>groupes des paires</u> $\mathcal{P}(K)$.

<u>Exemple</u> 2.4. Si K est un corps commutatif de caractéristique différente de 2, on a $\mathcal{P}(K) = K^*/K^{*2}$, où K^* est le groupe multiplicatif des éléments inversibles de K. D'autres exemples peuvent être facilement donnés à partir de la suite exacte ci-dessus.

Rappelons encore que si K est un anneau commutatif à élément unité et si $\mathrm{Ip}(K)$ désigne l'ensemble des idempotents de K, la loi de composition $e \mp e' = e + e' - 2ee'$ (somme booleienne), avec e et e' dans $\mathrm{Ip}(K)$, définit sur cet ensemble une structure de groupe abélien que nous appellerons le <u>groupe des idempotents</u> de l'anneau K. Le caractère fonctoriel de cette construction est facile d'établir et on a le résultat suivant (cf. [9]) :

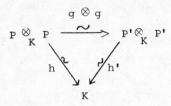

<u>Théorème</u> 2.5. <u>Soient</u> K <u>un anneau commutatif à élément unité</u>, T <u>la topologie</u> fppf <u>sur</u> Spec(K) <u>et</u> $\mathfrak{Q}(K)$ <u>le groupe des extensions quadratiques séparables de</u> K. <u>Il existe alors un isomorphisme de groupes abéliens</u> $H^1(T,K,Ip) \approx \mathfrak{Q}(K)$.

On note que pour tout anneau K commutatif à élément unité, l'application $Ip(K) \to \mu_2(K)$ définie par $e \mapsto 1-2e$ est un morphisme de groupes abéliens, ce qui nous permet de définir une transformation naturelle $Ip \to \mu_2$ donc un morphisme de groupes abéliens (en cohomologie) $H^1(T,K,Ip) \to H^1(T,K,\mu_2)$, pour toute topologie T sur Spec(K). En particulier, si T est la topologie fppf, ceci nous fournit, compte tenu des théorèmes 2.3. et 2.5., un morphisme de groupes abéliens $\mathfrak{Q}(K) \to \mathfrak{P}(K)$.

<u>Proposition</u> 2.6. <u>Si</u> K <u>est un anneau commutatif à élément unité et si</u> 2 <u>est inversible dans</u> K, <u>il existe un isomorphisme de groupes abéliens</u> $\mathfrak{Q}(K) \simeq \mathfrak{P}(K)$.

En effet, il suffit de voir que si 2 est inversible dans K, les morphismes de groupes abéliens $Ip(K) \to \mu_2(K)$, $e \mapsto 1-2e$ et $\mu_2(K) \to Ip(K)$, $a \mapsto \frac{1}{2}(1-a)$ sont des isomorphismes réciproques d'où l'isomorphisme en cohomologie $H^1(T,K,Ip) \simeq H^1(T,K,\mu_2)$. La proposition s'ensuit.

<u>Exemple</u> 2.7. Le fait que 2 soit inversible dans K est essentiel pour que l'on ait un isomorphisme de groupes abéliens $\mathfrak{Q}(K) \simeq \mathfrak{P}(K)$. En effet, on sait que si K est un anneau commutatif à élément unité dans lequel $2 = 0$, le morphisme $\mathfrak{Q}(K) \to \mathfrak{P}(K)$ est nul (cf. [12], proposition 2.4.15.). Et en vue d'un exemple, il suffit de voir que si K est un corps fini de caractéristique 2, on a $\mathfrak{Q}(K) \approx \mathbb{Z}/(2)$.

Soient K un anneau commutatif à élément unité et A une K-algèbre d'Azumaya. Si K est un corps, le <u>théorème de Skolem-Noether</u> nous dit que tout automorphisme d'une K-algèbre centrale simple est intérieur. Sinon, on sait que pour tout automorphisme σ de A, $J_\sigma = \{u \mid u \in A, \sigma(x)u = ux, \forall x \in A\}$ est un K-module projectif de type et de rang 1 et, de plus, J_σ est libre si et seulement si σ est un automorphisme intérieur (cf. [16]). On a encore, $J_\sigma \otimes_K J_{\sigma'} \approx J_{\sigma'\sigma}$, quels que soient σ et σ' dans le groupe $Aut_K(A)$. Ainsi, l'application $Aut_K(A) \to Pic(K)$ définie par $\sigma \mapsto [J_\sigma]$ est un morphisme de groupes dont le noyau est le groupe $Aut\ int_K(A)$ des automorphismes intérieurs de A, ce qui nous donne la suite exacte de groupes $1 \to Aut\ int_K(A) \to Aut_K(A) \to Pic(K)$.

Ainsi, pour que tout automorphisme de A soit intérieur, il suffira que le groupe Pic(K) soit trivial ou encore, que son sous-groupe noté $Pic_2(K) = \{[J_\sigma] \mid \sigma \in Aut_K(A)\}$ le soit. En effet, on a la suite exacte de groupes $1 \to Aut\ int_K(A) \to Aut_K(A) \to Pic_2(K) \to 1$. De plus, les éléments de $Pic_2(K)$ sont d'ordre 2 car pour tout σ dans $Aut_K(A)$, σ^2 est un automorphisme intérieur de A (cf. [11]).

3. LE GROUPE ORTHOGONAL

Soient K un anneau commutatif à élément unité et (P,f) un K-module quadratique où P est un K-module projectif de type fini et $f : P \to K$ est une forme quadratique non dégénérée. Le <u>groupe orthogonal</u> $\mathbb{O}(P,f)$ du K-module (P,f) est l'ensemble des K-isomorphismes linéaires $\sigma : P \xrightarrow{\sim} P$ tels que $f \circ \sigma = f$, muni de la structure de groupe définie par la composition d'applications linéaires. On sait que tout élément σ de $\mathbb{O}(P,f)$ s'étend à un automorphisme $C(\sigma)$ de l'algèbre de Clifford $C(P,f)$ qui laisse le K-module P fixe et réciproquement, tout automorphisme de $C(P,f)$ qui laisse fixe le K-module P donne, par restriction, un automorphisme orthogonal de P, c'est-à-dire, un élément du groupe $\mathbb{O}(P,f)$. Ainsi, si $Aut_K(C(P,f))$ désigne le groupe des automorphismes homogènes de degré zéro de $C(P,f)$, c'est-à-dire, ceux qui conservent la graduation de $C(P,f)$, l'application $\mathbb{O}(P,f) \to$ $\to Aut_K(C(P,f))$ définie par $\sigma \mapsto C(\sigma)$ est un morphisme de groupes. Etant donné que pour tout élément σ de $Aut_K(C(P,f))$ tel que $\sigma(P) \subset P$ on a $f(\sigma(x)) = \sigma(x)^2 = \sigma(x^2) = \sigma(f(x)) = f(x)$ pour tout x dans P, alors $\sigma_{|P}$ est un élément de $\mathbb{O}(P,f)$ et $C(\sigma_{|P}) = \sigma$. Ceci nous permet d'écrire le groupe orthogonal de (P,f) comme un sous-groupe du groupe $Aut_K(C(P,f))$ en posant $\mathbb{O}(P,f) = \{\sigma | \sigma \in Aut_K(C(P,f)),\ \sigma(P) \subset P\}$.

D'autre part, le <u>groupe orthogonal unimodulaire</u> $S\mathbb{O}(P,f)$ de l'espace quadratique (P,f) est le sous-groupe de $\mathbb{O}(P,f)$ formé des éléments σ tels que $C(\sigma)$ restreint au centre $Z_K(C_0(P,f))$ de $C_0(P,f)$ soit l'identité. Considérons le morphisme de groupes $Aut_K(C(P,f)) \xrightarrow{\gamma}$ $\xrightarrow{\gamma} Ip(K)$, $\sigma \mapsto e$, défini localement par $e = 0$ si $\sigma_{|Z_K(C_0(P,f))} = id$ et $e = 1$ sinon. Si $Aut_K^\circ(C(P,f))$ désigne son noyau, on a immédiatement que $S\mathbb{O}(P,f) = Aut_K^\circ(P,f) \cap \mathbb{O}(P,f)$. De plus, on a le résultat suivant (cf. [10], § 2.1, lemme 1) :

<u>Lemme</u> 3.1. <u>S'il existe un élément</u> σ <u>dans</u> $Aut_K(C(P,f))$ <u>tel que</u> $\gamma(\sigma) = 1$, <u>alors le morphisme</u> $\gamma : Aut_K(C(P,f)) \to Ip(K)$ <u>est surjectif</u>.

En effet, étant donné un idempotent e de K, $e\sigma + (1-e)id$ est une automorphisme de $C(P,f)$ et $\gamma(e\sigma + (1-e)id) = e$.

On cherche maintenant à calculer le déterminant d'un automorphisme orthogonal de (P,f), c'est-à-dire, d'un élément de $\mathbb{O}(P,f)$. On sait (cf. [10], § 2.2.) que si σ est un élément de $\mathbb{O}(P,f)$, alors $dét(\sigma)^2 =$ $= 1$ donc $dét(\sigma)$ est dans le groupe $\mu_2(K)$ et l'application $dét :$ $\mathbb{O}(P,f) \to \mu_2(K)$ définie par $\sigma \mapsto dét(\sigma)$ est un morphisme de groupes.

Le résultat suivant (cf. [10], § 2.2., théorème) nous permet de calculer le déterminant d'un élément de $\mathbb{O}(P,f)$:

<u>Théorème</u> 3.2. <u>Soient</u> K <u>un anneau commutatif à élément unité et</u> (P,f) <u>un</u> K-<u>module quadratique où</u> P <u>est projectif de type fini et</u> f <u>est non dégénérée</u>. <u>Si</u>, <u>pour</u> σ <u>dans</u> $\mathbb{O}(P,f)$, <u>on note</u> $e = \gamma(C(\sigma))$, <u>alors</u> $\text{dét}(\sigma) = 1-2e$.

4. GROUPES DE CLIFFORD

Voyons, tout d'abord, la définition classique du groupe de Clifford. Soient, à cet effet, K un corps commutatif, (P,f) un K-espace quadratique où P est un K-espace vectoriel de dimension finie et $f : P \to K$ est une forme quadratique non dégénérée, $\mathbb{O}(P,f)$ son groupe orthogonal et $C(P,f)$ son algèbre de Clifford. Le <u>groupe de Clifford</u> $\Gamma(P,f)$ du K-espace quadratique (P,f) est, par définition, l'ensemble des éléments inversibles u de $C(P,f)$ tels que $uPu^{-1} = P$ ou encore, tels uxu^{-1} est dans P pour tout x dans P. Si u est dans $\Gamma(P,f)$, l'application K-linéaire $\sigma(u) : P \to P$ définie par $x \mapsto uxu^{-1}$ vérifie $f(\sigma(u)(x)) = f(uxu^{-1}) = uxu^{-1}\,uxu^{-1} = ux^2u^{-1} = x^2uu^{-1} = x^2 = f(x)$ pour tout x dans P ou encore, $f\circ\sigma(u) = f$, c'est-à-dire, $\sigma(u)$ est dans $\mathbb{O}(P,f)$. On définit, de la sorte, un morphisme de groupes $\sigma : \Gamma(P,f) \to \mathbb{O}(P,f)$ donné par $u \mapsto \sigma(u)$. Cette définition est à l'origine de quelques difficultés et, en particuler, le fait que si la dimension de P est impaire, le morphisme $\sigma : \Gamma(P,f) \to \mathbb{O}(P,f)$ n'est pas surjectif, son image étant le groupe unimodulaire $S\mathbb{O}(P,f) = \{\alpha \mid \alpha \in \mathbb{O}(P,f), \text{dét}(\alpha) = 1\}$.

Pour éviter ces difficultés, on procède autrement. On définit $\Gamma(P,f)$ comme étant l'ensemble des éléments homogènes u dans $C(P,f)$, u inversible, tels que $(-1)^{d^\circ(u)}uxu^{-1}$ soit dans P pour tout x dans P, où $d^\circ(u)$ désigne le degré de l'élément u. Le théorème suivant, bien connu, nous dit que le morphisme de groupes $\sigma : \Gamma(P,f) \to \mathbb{O}(P,f)$ est surjectif :

<u>Théorème</u> 4.1. (théorème de E. Cartan et J. Dieudonné). <u>Le groupe orthogonal</u> $\mathbb{O}(P,f)$ <u>est engendré par les éléments</u> $\sigma(x)$ <u>avec</u> x <u>dans</u> P <u>et</u> $f(x) \neq 0$, <u>sauf si</u> $K = /(2)$ <u>et</u> P <u>est de dimension</u> 4.

Ainsi, il n'y a qu'une seule exception, à savoir, celle où $K = \mathbb{Z}/(2)$, P est de dimension 4 et $f : P \to K$ est la forme quadratique définie, sur une base $\{e_1,\dots,e_4\}$ de P sur K, par $f(\sum_{i=1}^{4} \lambda_i e_i) = \lambda_1\lambda_2 + \lambda_3\lambda_4$, quels que soient les scalaires $\lambda_1, \lambda_2, \lambda_3, \lambda_4$ dans K. Dans ce cas, les éléments $\sigma(x)$ avec x dans P et $f(x) \neq 0$ n'engendrent pas le groupe orthogonal $\mathbb{O}(P,f)$ mais le morphisme $\sigma : \Gamma(P,f) \to \mathbb{O}(P,f)$ est encore surjectif. En effet, il n'y a qu'un nombre fini d'éléments dans chaque groupe et il s'agit alors d'une vérification directe.

Par ailleurs, on montre aussi que le noyau du morphisme $\sigma : \Gamma(P,f) \to \mathbb{O}(P,f)$ est le groupe multiplicatif K^* des éléments inversibles de

K. Notons que ce résultat ne dépend pas de la caractéristique de K, étant vrai en caractéristique 2. Ceci nous permet d'énoncer le résultat suivant :

Proposition 4.2. *Soit K un corps commutatif. Pour tout K-espaces quadratique (P,f) où P est un K-espace vectoriel de dimension finie et f une forme quadratique non dégénérée sur P, il existe une suite exacte de groupes* $1 \to K^* \to \Gamma(P,f) \overset{\sigma}{\to} \mathbb{O}(P,f) \to 1$.

On passe maintenant à la construction du groupe de Clifford sur un anneau. Soient donc K un anneau commutatif à élément unité et (P,f) un K-module quadratique où P est un K-module projectif de type fini et f est une forme quadratique non dégénérée sur P. On dira qu'un élément u dans C(P,f) est *localement homogène* (cf. [1], § 1.1.) si pour tout idéal premier $\mathfrak{p}$ de K, l'image de u dans $C(P,f)_{\mathfrak{p}}$ ($\approx C(P_{\mathfrak{p}},f_{\mathfrak{p}})$, isomorphisme de $K_{\mathfrak{p}}$-algèbres, où $f_{\mathfrak{p}} : P_{\mathfrak{p}} \to K_{\mathfrak{p}}$ est la forme quadratique définie par $\frac{x}{s} \mapsto \frac{f(x)}{s^2}$) est homogène. Ainsi, pour tout élément u dans C(P,f) localement homogène et inversible, l'application $\sigma(u) : C(P,f) \to C(P,f)$, $x \mapsto (-1)^{d^\circ(x)d^\circ(u)} uxu^{-1}$, (x homogène) définit un automorphisme intérieure de C(P,f), i.e., σ(u) est dans $\mathrm{Aut\,int}_K C(P,f)$). Le *groupe de Clifford* Γ(P,f) du K-module quadratique (P,f) est défini comme étant l'ensemble des éléments inversibles et localement homogènes u tels que $(-1)^{d^\circ(u)} uxu^{-1}$ est dans P pour tout x dans P. Ainsi, si u est dans Γ(P,f), σ(u) est dans $\mathbb{O}$(P,f), ce qui définit un morphisme de groupes $\sigma : \Gamma(P,f) \to \mathbb{O}(P,f)$. Le *groupe de Clifford pair* est le sous-groupe $\Gamma_o(P,f)$ de Γ(P,f) défini par $\Gamma_o(P,f) = \Gamma(P,f) \cap C_o(P,f)$. On a ainsi la suite exacte de groupes $1 \to U(Z_K(C(P,f))) \to \Gamma(P,f) \overset{\sigma}{\to} \mathbb{O}(P,f)$ et comme l'algèbre de Clifford C(P,f) est d'Azumaya au sens gradué (cf. [1], § 1.2.), son centre gradué est K d'où la suite exacte de groupes $1 \to U(K) \to \Gamma(P,f) \overset{\sigma}{\to} \mathbb{O}(P,f)$. Ceci généralise le résultat précédemment obtenu, dans le cas où K est un corps.

5. GROUPES DES SPINEURS (cf. [1])

Soient K un anneau commutatif à élément unité et (P,f) un K-module quadratique où P est projectif de type fini et f est une forme quadratique non dégénérée sur P. Considérons l'antiautomorphisme $C(P,f) \to C(P,f)$, $u \mapsto \bar{u}$ obtenue en composant l'antiautomorphisme $C(P,f) \to C(P,f)$, $x_1 \ldots x_m \mapsto x_m \ldots x_1$ et l'automorphisme $C(-id_P) : C(P,f) \to C(P,f)$. Les propriétés suivantes sont immédiates : (i) pour tout x dans P, $\bar{x} = -x$; (ii) pour tout u dans C(P,f), $\bar{\bar{u}} = u$; (iii) quels que soient u et v dans C(P,f), $\overline{uv} = \bar{v}\,\bar{u}$.

L'ensemble $M = \{u \mid u \in C(P,f),\ u\bar{u} \in K\}$ muni de la structure multiplicative induite par celle de l'algèbre de Clifford $C(P,f)$ est un monoïde multiplicatif. En effet, quels que soient u et v dans M, on a $(uv)(\overline{uv}) = uv\bar{v}\bar{u} = (u\bar{u})(v\bar{v}) \in K$, donc uv est dans M. L'application $N : M \to K$ définie par $u \mapsto u\bar{u}$ est alors un morphisme de monoïdes que l'on appellera norme. On voit que pour tout x dans P on a $x\bar{x} = -x^2 = -f(x)$, c'est à dire, $P \subset M$ en tant que sous-monoïde et la restriction de N à P est égale à $-f$, i.e., $N_{|P} = -f$. De plus, si u est dans le groupe de Clifford $\Gamma(P,f)$, $\sigma(u)$ est dans $\mathbb{O}(P,f)$ donc pour tout x dans P, on a $\sigma(u)(x) = -\overline{\sigma(u)(x)} = (-1)^{d°(u)+1}\overline{uxu^{-1}} = (-1)^{d°(u)+2}\,\overline{u^{-1}}\,x\,\bar{u} = \overline{\sigma(u^{-1})}(x)$ soit $\sigma(u) = \overline{\sigma(u^{-1})}$. Ceci nous dit que $u\bar{u} \in \mathrm{Ker}(\sigma) = U(K)$ et la norme $N : \Gamma(P,f) \to U(K)$, $u \mapsto u\bar{u}$ est un morphisme de groupes. De plus, $N(a) = a^2$ pour tout a dans $U(K)$.

Ceci va nous permettre, par la suite de définir le groupe des spineurs. En effet, le groupe pin du K-module quadratique (P,f) est défini comme étant le noyau de la norme $N : \Gamma(P,f) \to U(K)$, que nous noterons $\mathrm{Pin}(P,f) = \mathrm{Ker}(N)$ et le groupe des spineurs du K-module quadratique (P,f) est le groupe noté $\mathrm{Spin}(P,f)$ et défini par $\mathrm{Spin}(P,f) = \mathrm{Ker}(\Gamma_o(P,f) \xrightarrow{N} U(K))$, c'est à dire, $\mathrm{Spin}(P,f) = \mathrm{Pin}(P,f) \cap \Gamma_o(P,f)$. On voit immédiatement que $\mathrm{Spin}(P,f) \cap U(K) = \mathrm{Pin}(P,f) \cap U(K) = \mu_2(K)$, le groupe des racines carrées de l'unité de K.

Pour ce qui est de la construction de la norme spinorielle $SN : \mathbb{O}(P,f) \to \mathcal{P}(K)$ on procède comme suit . Pour tout élément α de $\mathbb{O}(P,f)$, on pose $P_\alpha = \{u \mid u \in C(P,f),\ u$ homogène et $\alpha(x)u = (-1)^{d°(u)}ux,\ \forall x \in P\}$, celui-ci étant un K-module projectif de type fini et de rang 1 et on définit la norme spinorielle $SN : \mathbb{O}(P,f) \to \mathcal{P}(K)$, par $\alpha \mapsto (P_\alpha, h_\alpha)$ où $h_\alpha : P_\alpha \otimes_K P_\alpha \xrightarrow{\sim} K$ est l'isomorphisme de K-modules défini par $u \otimes v \mapsto u\bar{v}$. On montre facilement que SN est un morphisme de groupes et si l'on considère la suite exacte de groupes abéliens $1 \to U(K)/U^2(K) \to \mathcal{P}(K) \to {}_2\mathrm{Pic}(K) \to 1$ où ${}_2\mathrm{Pic}(K) = \mathrm{Ker}(\mathrm{Pic}(K) \xrightarrow{2} \mathrm{Pic}(K))$ (cf. §2), alors $U(K)/U^2(K)$ est isomorphe à un sous-groupe de $\mathcal{P}(K)$ et pour un α dans $\mathbb{O}(P,f)$, $SN(\alpha)$ est dans $U(K)/U^2(K)$ si et seulement si il existe un élément u dans $\Gamma(P,f)$ tel que $\sigma(u) = \alpha$ (cf. [1], proposition 3.3.2.), $\sigma : \Gamma(P,f) \to \mathbb{O}(P,f)$ étant le morphisme naturel du groupe de Clifford dans le groupe orthogonal.

Ces considérations nous permettent de donner un certain nombre de résultats. En effet, on a déjà les suites exactes de groupes (cf. [1], § 3.3.) $1 \to U(K) \to \Gamma(P,f) \to \mathbb{O}(P,f) \to {}_2\mathrm{Pic}(K)$ et $1 \to \mu_2(K) \to \mathrm{Pin}(P,f) \to \mathbb{O}(P,f) \to \mathcal{P}(K)$. On montre, de plus, que les suites de groupes $1 \to U(K) \to \Gamma_o(P,f) \to S\mathbb{O}(P,f) \to {}_2\mathrm{Pic}(K) \oplus \mathrm{Ip}(K)$ et $1 \to \mu_2(K) \to \mathrm{Spin}(P,f) \to S\mathbb{O}(P,f) \to \mathcal{P}(K) \oplus \mathrm{Ip}(K)$ sont exactes. Ceci nous montre la place du groupe des spineurs par rapport à des groupes déjà connus.

Supposons maintenant que $(P,f) = h(K^n)$, l'espace hyperbolique du K-module libre K^n et notons $Spin_n(K) = Spin(h(K^n))$ et $S\mathbb{O}_n(K) = S\mathbb{O}(h(K^n))$. La décomposition orthogonale $h(K^{n+1}) = h(K^n) \perp h(K)$ nous définit une inclusion $Spin_n(K) \to Spin_{n+1}(K)$ et $S\mathbb{O}_n(K) \to S\mathbb{O}_{n+1}(K)$ d'où, par passage à la limite inductive, on obtient les groupes notés $Spin(K)$ et $S\mathbb{O}(K)$. On a encore la suite exacte de groupes $1 \to \mu_2(K) \to Spin(K) \to S\mathbb{O}(K)$ et si T désigne la topologie fppf sur $Spec(K)$, on a la suite exacte de faisceaux $1 \to \mu_2 \to Spin \to S\mathbb{O} \to 1$. En passant en cohomologie, ceci nous permet de prolonger la suite exacte ci-dessus, à savoir, $1 \to \mu_2(K) \to Spin(K) \to S\mathbb{O}(K) \to \mathcal{P}(K) \to H^1(T,K,Spin) \to H^1(T,K,S\mathbb{O})$, compte-tenu de l'identification $H^1(T,K,\mu_2) \approx \mathcal{P}(K)$. De plus, le morphisme $S\mathbb{O}(K) \to \mathcal{P}(K)$ n'est autre que la norme spinorielle. On est ainsi en mesure de calculer des exemples concrets.

Exemple 5.1. Si K est un anneau commutatif dans lequel 2 est inversible, alors $Spin_2(K) \approx U(K)$, $Spin_3(K) \approx SL_2(K)$ et $Spin_4(K) \approx SL_2(K) \times SL_2(K)$, ceux-ci étant des isomorphisme de groupes. Ces calculs peuvent se faire aisément soit directement (cf. [1]) soit par voie cohomologique.

6. DEUX SUITES EXACTES (cf. [10])

On peut se demander ce qu'il advient du groupe des spineurs dans le cas où l'on ne tient pas compte de la structure graduée de l'algèbre de Clifford. Soient donc K un anneau commutatif à élément unité, (P,f) un K-module quadratique où P est projectif de type fini et f est une forme quadratique non dégénérée sur P et $\Gamma(P,f) = \{u \mid u \in C(P,f),$ u inversible, $uxu^{-1} \in P, \forall x \in P\}$ le groupe de Clifford. Il existe un morphisme de groupes $\sigma : \Gamma(P,f) \to \mathbb{O}(P,f)$ donc le noyau s'écrit $Ker(\sigma) = \{u \mid u \in \Gamma(P,f), ux = xu, \forall x \in P\} = U(Z_K(C(P,f)))$. On a ainsi une suite exacte de groupes $1 \to U(Z_K(C(P,f))) \to \Gamma(P,f) \overset{\sigma}{\to} \mathbb{O}(P,f)$.

Notons $C(P,f) \to C(P,f)$, $u \mapsto \bar{u}$ l'antiautomorphisme principal et $J_\alpha = \{u \mid u \in C(P,f), \alpha(x)u = ux, \forall x \in C(P,f)\}$ avec α dans $Aut_K(C(P,f))$. On sait que J_α est un K-module projectif de type fini et de rang 1 et l'application $N : \Gamma(P,f) \to U(Z_K(C(P,f)))$ définie par $u \mapsto u\bar{u}$ est un morphisme de groupes. En effet, $u\bar{u}x = u\alpha(x)\bar{u} = xu\bar{u}$ pour tout x dans P d'où $u\bar{u} \in Z_K(C(P,f))$. Si maintenant P est de rang pair, $C(P,f)$ est un K-algèbre d'Azumaya (au sens non gradué) donc, en particulier, $Z_K(C(P,f)) = K$. On a ainsi la suite exacte de groupes $1 \to U(K) \to \Gamma(P,f) \overset{\sigma}{\to} O(P,f)$ et $N : \Gamma(P,f) \to U(K)$. On définit le <u>groupe pin</u> en posant $Pin(P,f) = Ker(\Gamma(P,f) \overset{N}{\to} U(K))$ et le <u>groupe des spineurs</u>

par $\mathrm{Spin}(P,f) = \mathrm{Pin}(P,f) \cap C_0(P,f)$. Le morphisme de groupes $\mathbb{O}(P,f) \to$ $\to H^1(C/K,\mu_2)$ défini par $\alpha \mapsto (J_\alpha, h_\alpha)$, où $h_\alpha : J_\alpha \otimes_K J_\alpha \xrightarrow{\sim} K$ est l'isomorphisme de K-modules défini localement par $h(u \otimes \bar{u}) = u\bar{u} = N(u)$ avec u un générateur libre de J_α, nous fournit la suite exacte de groupes $1 \to \mu_2(K) \to \mathrm{Pin}(P,f) \to \mathbb{O}(P,f) \to H^1(C/K,\mu_2)$ avec $C = C(P,f)$. De même, on montre qu'il existe une suite exacte de groupes dans laquelle intervient le groupe des spineurs, à savoir, $1 \to \mu_2(K) \to \mathrm{Spin}(P,f) \to$ $S\mathbb{O}(P,f) \to H^1(C/A,\mu_2)$.

Note 6.1. Pour ce qui est des résultats concernant les topologies de Grothendieck ainsi que la construction des groupes de cohomologie correspondants, nous renvoyons au livre de M. Artin, Grothendieck Topologies, Havard University, 1962.

7. NOTE HISTORIQUE

La notion de spineur a été introduite par Eli Cartan en 1913 en connexion avec ses recherches portant sur les représentations linéaires des groupes simples (cf. E. Cartan, 'Les groupes projectifs qui ne laissent invariante aucune multiplicité plane', Bull. Soc. Math. France 41 (1913), 53-96). Mais le phénomène du spin d'une particule en Physique n'a été découvert, d'un point de vue expérimental, qu'en 1925. A propos de son équation de l'électron, P.A.M. Dirac (cf. The principle of Quantum Mechanics, Oxford University Press, 1930), introduit les spineurs dans un espace de Minkowski. Toutes ces constructions étaient relativement compliquées et d'une difficile compréhension. En 1935, R. Brauer et H. Weyl (cf. [3]) utilisent, pour la première fois, les algèbres de Clifford dans la construction des spineurs. Cette même construction a été reprise par C. Chevalley (cf. [6]) quelques années plus tard et rendue définitivement accessible. De ce point de vue on pourrait dire que les germes de la notion de spineur sont déjà dans les travaux de W.K. Clifford (cf. 'Applications of Grassmann's extensive algebra', Amer. J. Math. 1 (1878), 350-358).

L'extraordinaire développement de l'algèbre à partir des années 30 a permis de sortir du cadre des algèbres sur des corps et bâtir une théorie des spineurs pour des modules quadratiques sur un anneau quelconque. Parallèlement, la théorie des spineurs a été assortie de considérations cohomologiques qui n'apparaissaient pas dans le cas où l'anneau de base était un corps.

REFERENCES

[1] H. BASS, 'Clifford algebras and spinor norms over a commutative ring', Amer. J. Math. 96 (1974), 156-206.

[2] N. BOURBAKI, Formes sesquilinéaires et formes quadratiques, Hermann, Paris, 1959.

[3] R. BRAUER et H. WEYL, 'Spinors in n dimensions', Amer. J. Math. 57 (1935), 425-449.

[4] E. CARTAN, The theory of spinors, Dover Publications, Inc., New York, 1981.

[5] S.S. CHERN and C. CHEVALLEY, 'E. Cartan and his mathematical work', Bull. Amer. Math. Soc. 58 (1952), 217-250.

[6] C. CHEVALLEY, The algebraic theory of spinors, Columbia University Press, New York, 1954.

[7] Elie CARTAN 1869-1951, 'Hommage de l'Académie de la République Socialiste de Roumanie', Editura Academiei Republicii Socialiste România, Bucarest 1975.

[8] P. LOUNESTO, 'Clifford algebras and spinors', this volume.

[9] A. MICALI et O.E. VILLAMAYOR, 'Sur les algèbres de Clifford', Ann. Sc. Ecole Normale Sup. 4è série 1 (1968), 271-304.

[10] A. MICALI et O.E. VILLAMAYOR, 'Sur les algèbres de Clifford II', J. für die Reine und angewandte Math. 242 (1970), 61-90.

[11] A. MICALI et O.E. VILLAMAYOR, 'Algèbres de Clifford et groupe de Brauer', Ann. Sc. Ecole Normale Sup. 4è série 4 (1971), 285-310.

[12] A. MICALI et Ph. REVOY, 'Modules quadratiques', Bull. Soc. Math. France, Mémoire n° 63 (1979), 144 p.

[13] A. MICALI et A. PAQUES, 'Le groupe des algèbres quadratiques', Communications in Algebra 10 (16) (1982), 1765-1779.

[14] I.R. PORTEOUS, Topological Geometry, Second Edition, Cambridge University Press, Cambridge 1981.

[15] M. RIESZ, 'Clifford numbers and spinors', The Institute for Fluid Dynamics and Applied Mathematics, University of Maryland, 1958.

[16] A. ROSENBERG and D. ZELINSKY, 'Automorphisms of separable algebras', Pac. Journal of Math. 11 (1961), 1109-1117.

[17] C.T.W. WALL, 'Graded Brauer groups', J. für die Reine und angewandte Math. 213 (1964), 187-199.

[18] H. WEYL, The classical groups , Princeton University Press, Princeton, 1946.

ALGEBRES DE CLIFFORD $C^{+}_{r,s}$ DES ESPACES QUADRATIQUES PSEUDO-EUCLIDIENS STANDARDS $E_{r,s}$ ET STRUCTURES CORRESPONDANTES SUR LES ESPACES DE SPINEURS ASSOCIES. PLONGEMENTS NATURELS DES QUADRATIQUES PROJECTIVES REELLES $\widetilde{Q}(E_{r,s})$ ATTACHEES AUX ESPACES $E_{r,s}$.

Pierre ANGLES
U.E.R. de Mathématique
Université Paul Sabatier
118 route de Narbonne
31062 Toulouse Cedex - FRANCE.

ABSTRACT. Following the guiding method found out by R. Deheuvels for the case r-s = 4l+2 and r+s = 4k+2, [4], this paper systematically deals with the different structures given to the spaces S of Spinors for even Clifford algebras $C^{+}_{r,s}$ of quadratic standard spaces $E_{r,s}$ and the immersions of projective associated quadrics $\widetilde{Q}(E_{r,s})$.

RESUME. Les groupes d'automorphismes de produits scalaires sur les espaces de spineurs d'algèbres de Clifford réelles ont été étudiés dans [11], chapitre 13 et dans [9], chapitre 4 . R. Deheuvels a trouvé une méthode directrice, [4], dans le cas où r-s = 4l+2 et r+s = 4k+2 qui explique non seulement le plongement naturel du groupe spinoriel dans un groupe spécial pseudo-unitaire -bien connu et exploité par les physiciens dans un cas particulier, (cf. par exemple [10], [17])-, mais justifie le plongement, en toute généralité, des quadratiques $\widetilde{Q}(E_{r,s})$ associées dans un groupe unitaire. Ce travail a pour objet l'étude systématique des structures attachées aux espaces de spineurs des algèbres de Clifford paires $C^{+}_{r,s}$ des espaces standards pseudo-euclidiens $E_{r,s}$ et les plongements naturels des quadriques projective $\widetilde{Q}(E_{r,s})$ correspondantes qui en résultent. En raison de la place impartie, on n'étudiera brièvement que le cas quaternionien et on donnera seulement les résultats pour les autres cas. Les démonstrations complètes seront données dans un prochain article.

I. LES ALGEBRES DE CLIFFORD DES ESPACES PSEUDO-EUCLIDIENS REELS ET LES QUADRIQUES PROJECTIVES REELLES

I.1 - Les algèbres de Clifford $C_{r,s}$ et $C^{+}_{r,s}$

Soit $V = E_{r,s}$ l'espace pseudo-euclidien standard, de dimension m, de type (r,s), [4]. On note $(x|y) = x^1y^1+\ldots+x^ry^r-x^{r+1}y^{r+1}\ldots-x^{r+s}y^{r+s}$ son produit scalaire.

J. S. R. Chisholm and A. K. Common (eds.), Clifford Algebras and Their Applications in Mathematical Physics, 79–91.

$C(V) = C_{r,s}$ désigne l'algèbre de Clifford de V, quotient de $\otimes E_{r,s}$ par l'idéal bilatère engendré par les éléments $\{x \otimes x - (x|x).1,\ x \in V\}$. On utilise, ici, les notations de [3] et de [4]. C(V) est une algèbre associative unitaire de dimension 2^m sur $\mathbb{R}$. π désigne l'involution naturelle de C(V), τ l'unique antiautomorphisme de C(V) laissant invariants les éléments de V -τ est une anti-involution de l'algèbre C(V)- ; $\nu = \tau\circ\pi = \pi\circ\tau$ est la conjugaison de C(V). $C^+_{r,s} = C^+(V)$ est la sous algèbre des éléments pairs de C(V), de dimension 2^{m-1} sur $\mathbb{R}$. Comme dans [4], on convient d'appeler groupe de Clifford de V -ou encore, selon [3], groupe de Clifford régulier de V-, le groupe noté G, que forment avec l'unité $1_{C(V)}$ de l'algèbre C(V), pour la multiplication, les produits de vecteurs non isotropes de V. C'est aussi le groupe, que forment pour la multiplication, les éléments inversibles g de l'algèbre de Clifford C(V) qui satisfont à la condition : $\forall\ x\in V : (\pi.g)xg^{-1} = y\in V$. N, respectivement N', désigne la norme spinorielle ordinaire, respectivement norme spinorielle graduée, définies respectivement pour tout élément g du groupe de Clifford G (régulier) par : $\tau(g).g = g^{\tau}.g = N(g).1_{C(V)}$ et par $\nu(g).g = g^{\nu}.g = N'(g).1_{C(V)}$. N et N' sont des homomorphismes du groupe de Clifford (régulier) dans le groupe multiplicatif $\mathbb{R}^*$ qui appliquent le centre $\mathbb{R}^*.1_{C(V)}$ du groupe de Clifford régulier sur $(\mathbb{R}^*)^2$. Pour $g = x_1x_2\ldots x_p$, produit de vecteurs réguliers de V, $N'(g) = (-1)^p N(g)$ et $g^{-1} = \dfrac{g^{\tau}}{N(g)} = \dfrac{g^{\nu}}{N'(g)}$.

<u>On désigne par Spin V</u>, <u>le noyau de la restriction de N à $G^+ = C^+(V)\cap G$</u>, formé par les $g\in G$ produits d'un nombre pair de vecteurs réguliers de V.

Pour $m > 2$, ce groupe Spin V, le groupe spinoriel de V, est connexe et contenu dans G^{++}, groupe que forment les $g\in G$ qui sont produits d'un nombre pair de "vecteurs > 0" et d'un nombre pair de "vecteurs < 0".

Le groupe Spin V engendre linéairement $C^+(V) = C^+_{r,s}$, sous algèbre des éléments pairs, dans laquelle il est plongé. Nous allons maintenant étudier de façon précise la structure des algèbres $C_{r,s}$ et celle des $C^+_{r,s}$.

I.2. <u>Classification des algèbres de Clifford</u> $C^+_{r,s} = C^+(V)$

Comme il est noté dans [3], si u est un vecteur de V tel que $(u|u)=\varepsilon=\pm 1$, l'application φ de $u^\perp$ dans $C^+(V)$: $y \in u^\perp \to uy = \varphi(y)$ est telle que $(\varphi(y))^2 = -\varepsilon(y|y)$ et représente $C^+(V)$ comme l'algèbre de Clifford de de l'espace vectoriel $u^\perp$ muni de la forme quadratique induite de celle de V multipliée par $(-\varepsilon)$, donc de signature $(r,s-1)$ si $\varepsilon = -1$ ou $(s,r-1)$ si $\varepsilon = 1$. Toutes ces structures d'algèbres de Clifford sur $C^+(V)$ correspondant aux différents choix de u, définissent la même anti-involution de conjugaison qui coïncide avec la restriction de τ à $C^+(V)$.

On peut établir la table fondamentale qui donne explicitement la classification des algèbres de Clifford $C_{r,s}$ et $C^+_{r,s}$ suivant la congruence modulo 8 de r-s. On note $m(n,F)$ l'algèbre réelle des matrices nxn sur le corps F où F = $\mathbb{R}$, $\mathbb{C}$ ou $\mathbb{H}$, le corps usuel des quaternions. ([k] désigne la partie entière du réel k).

Table fondamentale :

r+s (modulo 2)	r-s (modulo 8)	$C^+_{r,s}$	$C_{r,s}$
0	0	$m(2^{[\frac{m-1}{2}]},\mathbb{R}) \oplus m(2^{[\frac{m-1}{2}]},\mathbb{R})$	$m(2^{\frac{m}{2}},\mathbb{R})$
1	1	$m(2^{\frac{m-1}{2}},\mathbb{R})$	$m(2^{[\frac{m}{2}]},\mathbb{R}) \oplus m(2^{[\frac{m}{2}]},\mathbb{R})$
0	2	$m(2^{[\frac{m-1}{2}]},\mathbb{C})$	$m(2^{\frac{m}{2}},\mathbb{R})$
1	3	$m(2^{\frac{m-1}{2}-1},\mathbb{H})$	$m(2^{[\frac{m}{2}]},\mathbb{C})$
0	4	$m(2^{[\frac{m-1}{2}]-1},\mathbb{H}) \oplus m(2^{[\frac{m-1}{2}]-1},\mathbb{H})$	$m(2^{\frac{m}{2}-1},\mathbb{H})$
1	5	$m(2^{\frac{m-1}{2}-1},\mathbb{H})$	$m(2^{[\frac{m}{2}]-1},\mathbb{H}) \oplus m(2^{[\frac{m}{2}]-1},\mathbb{H})$
0	6	$m(2^{[\frac{m-1}{2}]},\mathbb{C})$	$m(2^{\frac{m}{2}-1},\mathbb{H})$
1	7	$m(2^{\frac{m-1}{2}},\mathbb{R})$	$m(2^{[\frac{m}{2}]},\mathbb{C})$

Par exemple, la structure de $C^+_{4,7}$ est celle de $m(2^4,\mathbb{H})$, car $4-7 \equiv -3 \equiv 5$ (modulo 8) et celle de $C_{4,7}$: $m(2^4,\mathbb{H}) \oplus m(2^4,\mathbb{H})$

Remarque : On retrouve ainsi, simplement, le résultat général faisant l'objet de la proposition 3-20 de [8] chapitre 5, page 123, et celui de [12], page 280.

I.3 - Les quadratiques projectives réelles

Considérons $V = E_{r,s}$, l'espace pseudo-euclidien standard régulier réel de type (r,s) et de dimension $m = r+s$ dont le produit scalaire est $(x|y) = x^1y^1+...+x^ry^r-x^{r+1}y^{r+1}-...-x^{r+s}y^{r+s}$. Soit $e = \{e_1,...,e_r,e_{r+1},...,e_{r+s}\}$ la base orthogonale "canonique" de V pour ce produit scalaire. Le cône isotrope Q de $V = E_{r,s}$, épointé est une sous variété différentielle singulière de $E_{r,s}$. Si P désigne la projection de $V-\{0\}$ sur son espace projectif associé $P(V)$, $\tilde{Q} = \tilde{Q}(E_{r,s}) = P(Q-\{0\})$ est naturellement muni d'une structure pseudo-riemannienne conforme de type (r-1,s-1). Rappelons brièvement une construction classique.

On introduit $F = V \oplus H$ où H est le plan hyperbolique réel rapporté à une base isotrope (ε,η) telle que $2(\varepsilon|\eta) = 1$. F est ainsi un espace pseudo-euclidien réel standard régulier de type (r+1,s+1). Q(F), de dimension m+1, désigne son cône isotrope. $M = P(Q(F)- 0)$, image dans P(F) du cône isotrope, épointé de F, est de dimension m ; est appelé le compactifié conforme de $V = E_{r,s}$ et s'identifie à l'espace homogène $\frac{PO(F)}{S(V)}$ quotient du groupe conforme $PO(F) = \frac{O(r+1,s+1)}{\mathbb{Z}_2}$ par le groupe des similitudes $S(V)$ de V. On a une application naturelle de $S^r x S^s$ sur la quadratique projective $M = \tilde{Q}(F)$ qui identifie M au quotient de la variété $S^r x S^s$ par la relation d'équivalence $(a,b)\sim(-a,-b)$ et fait de $S^r x S^s$ un revêtement à deux feuillets de M, connexe si r et s sont tous deux différents de zéro. Si r et s sont tous deux $\geqslant 2$, $S^r x S^s$ est simplement connexe et constitue le revêtement universel de M, dont le groupe fondamental est $\mathbb{Z}_2$. Si r ou $s = 1$, $S^r x S^s$ n'est plus simplement connexe et le groupe fondamental de M est infini. [4]

I.4 - Plongement naturel des quadriques projectives des espaces pseudo-euclidiens $E_{r,s}$,

On désigne par $\mathcal{S}$ l'espace vectoriel des spineurs de l'algèbre de Clifford

$C_{r,s}$ et par $(x|y)$ le produit scalaire de $V = E_{r,s}$. Tout vecteur x de $E_{r,s}$ est représenté par un opérateur linéaire de $\mathcal{S}$. Si x est isotrope, comme $x^2 = 0$ dans $C_{r,s}$, $\mathrm{Im}\, x \subset \mathrm{Ker} x$.

Si x est isotrope non nul, on peut trouver y vecteur isotrope tel que $(x|y) = \frac{1}{2}$, ce qui équivaut dans $C_{r,s}$ à : $xy+yx = 1$. Les deux idempotents xy et yx appartiennent à $C^+_{r,s}$ et sont représentés dans tout espace où opère $C^+_{r,s}$ par exemple dans $\mathcal{S}$ par deux projecteurs supplémentaires.

Introduisons, alors, S espace de spineurs de $C^+_{r,s}$, stable par $C^+_{r,s}$, sous espace (propre ou non) de $\mathcal{S}$. Soit $[\,|\,]$ un produit scalaire sur S associé à l'anti-involution τ ; autrement dit, si $a \in C^+_{r,s}$, a et a^τ, opérateurs linéaires de S sont adjoints l'un de l'autre par rapport à $[\,|\,]$.

L'application injective : {droite isotrope $\{\lambda x\}$ de V} $\mapsto$
sous espace totalement isotrope maximal
$S(x) = \mathrm{Im}(xy)_S = \mathrm{Ker}(yx)_S$

\- où $(xy)_S$ et $(yx)_S$ sont les projecteurs de S définis par les éléments xy et yx de $C^+_{r,s}$ - détermine un plongement naturel de la quadrique projective $\tilde{Q}(E_{r,s})$ dans la grassmannienne des sous espaces de dimension moitié $G(S, \frac{1}{2} \dim S)$. [4].

II - STRUCTURE PSEUDO-QUATERNIONIENNE SUR LES ESPACES DE SPINEURS S DES $C^+_{r,s}$, $r+s = m = 2k+1$, $r-s \equiv \pm 3$ (modulo 8) ET PLONGEMENT NATUREL DES QUADRIQUES PROJECTIVES $\tilde{Q}(E_{r,s})$ ASSOCIEES.

II.1. Structures quaternioniennes sur les espaces vectoriels à droite sur $\mathbb{H}$.

a) Pour X et Y espaces vectoriels à droite de dimension finie sur $\mathbb{H}$, $L(X,Y)$ espace vectoriel des applications linéaires, (à droite), de X dans Y est un espace vectoriel sur le centre $\mathbb{R}$ de $\mathbb{H}$ et possède une structure d'algèbre réelle. On notera ν la conjugaison qui au quaternion $q = \alpha+i\beta+j\gamma+k\delta \in \mathbb{H}$, où 1, i, j, k sont les quatre "unités" de $\mathbb{H}$ associe $q^\nu = \alpha-i\beta-j\gamma-k\delta \in \mathbb{H}$. $\mathbb{C}$ peut être plongé dans $\mathbb{H}$ et opère par multiplication à droite dans $\mathbb{H}$ qui est un $\mathbb{C}$-espace vectoriel de base $\{1,j\}$. Soit S un espace vectoriel quaternionien à droite sur $\mathbb{H}$ de dimension n, dont $\varepsilon = \{\varepsilon_1,\ldots,\varepsilon_n\}$ est une base sur $\mathbb{H}$. Si b est une forme sesquilinéaire sur S, $b(x,y) = ({}^tX^\nu)BY$ où X, B, Y sont les matrices respectives dans ε de x, b, g. En restreignant le corps $\mathbb{H}$ à $\mathbb{C}$, S est un $\mathbb{C}$-espace vectoriel de base $\varepsilon = \{\varepsilon_1,\ldots,\varepsilon_n,\varepsilon_1 j,\ldots,\varepsilon_n j\}$ et les composantes de b relativement à la structure complexe de S sont les formes à valeurs complexes h et a définies par $b(x,y) = h(x,y) + ja(x,y)$, où a est bilinéaire complexe et h est sesquilinéaire, (linéaire en le second argument et antilinéaire en le premier). Soit $\{|\}$ un produit scalaire quaternionien sur S, [3]. Les composantes de $\{|\}$ sont, sur l'espace vectoriel complexe S, un produit scalaire hermitien noté $\langle|\rangle$ et un produit scalaire symplectique noté $[|]$: $\{x|y\} = \langle x|y\rangle + j[x|y]$. Ainsi le groupe symplectique unitaire $SpU(S) = U(S) \cap Sp(S)$. Si l'on munit S d'un produit scalaire pseudo-quaternionien de type (p,q) dont les composantes sur l'espace vectoriel complexe sont un produit scalaire pseudo-hermitien de type (2p,2q) et un produit scalaire symplectique, on obtient la décomposition : $SpU(p,q) = U(2p,2q) \cap Sp(2(p+q),\mathbb{C})$. Comme $U(p,q) = SO(2p,2q) \cap Sp(2(p+q),\mathbb{R})$, il en résulte que $SpU(p,q) = SO(4p,4q) \cap Sp(2(p+q),\mathbb{C}) \cap Sp(4(p+q),\mathbb{R})$. (S est, aussi, un $\mathbb{R}$-espace vectoriel de dimension $4n = 4(p+q)$ dont une base sur $\mathbb{R}$ est $\{\varepsilon_1,\ldots,\varepsilon_n,\varepsilon_1 i,\ldots,\varepsilon_n i,\varepsilon_1 j,\ldots,\varepsilon_n j,\varepsilon_1 k,\ldots,\varepsilon_n k\}$).

b) Si b est une forme sesquilinéaire antisymétrique pour la structure quaternionienne (ou encore antihermitienne), il est telle que pour tous x, y $\in$ S, $b(y,x) = -(b(x,y))^{\vee}$, $b(x,y) = h(x,y) + j\,a(x,y)$ où h est antihermitienne et a bilinéaire symétrique complexe. Supposons b non dégénérée. On trouve ainsi que le groupe unitaire des automorphismes de S qui conservent b, U(S,b) que l'on convient d'appeler groupe symplecto-quaternionien est l'intersection du groupe unitaire $U_{2n}(\mathbb{C},h)$ pour la forme antihermitienne h sur l'espace vectoriel complexe S et du groupe orthogonal complexe $O(2n,\mathbb{C})$; (toutes les formes sesquilinéaires b antihermitiennes sur S non dégénérées sont équivalentes et d'indice maximal $[\frac{n}{2}]$, [5,a] p. 383). On écrira $\underline{U_n(S,b) = U_{2n}(\mathbb{C},h) \cap O(2n,\mathbb{C})}$. Il en résulte que le groupe spécial unitaire $SU_n(S,b)$ -des automorphismes de S conservant b et de déterminant égal à 1- est l'intersection du groupe $SU_{2n}(\mathbb{C},h)$ et de $SO(2n,\mathbb{C})$ et par suite s'identifie à $SO^*(2n)$ par définition même ([6], p. 340 ; [17], p. 13].

$$\underline{SO^*(2n) = SU_n(S,b).}$$

II.2. <u>Anti-involutions sur l'algèbre réelle $\mathcal{L}_{\mathbb{H}}(S)$, où S est un espace vectoriel quaternionien à droite de dimension n sur $\mathbb{H}$.</u>

a) <u>Le théorème suivant ([1] page 154 ; [4]) est fondamental</u> :

<u>Théorème</u> :

L désigne un corps commutatif muni d'un automorphisme involutif : $\lambda \to \bar{\lambda}$. K est le sous corps des invariants.
Soit $\mathcal{A}$ une algèbre simple centrale ayant pour centre le corps L.
Soient α et β deux anti-involutions sur $\mathcal{A}$ associées au même automorphisme involutif de L. L'une d'elle, par exemple β, est la composée de l'autre α et d'un automorphisme intérieur par un élément u qui est soit α-symétrique soit α-antisymétrique : $a^\beta = (a^\alpha)^{j(u)} = u^{-1}.a^\alpha.u$, $(a\in\mathcal{A})$, avec $u^\alpha = u$ ou $u^\alpha = -u$. u est déterminé à un facteur scalaire non nul près.

Si $u^\alpha = u$ on a $u^\alpha = u^\beta$. Si $u^\alpha = -u$, on a $u^\beta = u^\alpha = -u$.

Soit S un espace vectoriel quaternionien à droite de dimension finie n sur $\mathbb{H}$. A un produit scalaire pseudo-quaternionien sur S est associé dans l'algèbre réelle $\mathcal{A} = \mathcal{L}_{\mathbb{H}}(S)$ des opérateurs linéaires (à droite) de S l'opérateur d'adjonction $a \to a^*$ tel que : $(a+b)^* = a^*+b^*$, $(a^*)^* = a$, $(ab)^* = b^*a^*$, $\forall \lambda \in \mathbb{R}$, $(\lambda.1)^* = \lambda.1$. L'algèbre réelle $m(n,\mathbb{H})$, isomorphe à $\mathcal{L}_{\mathbb{H}}(S)$, des matrices nxn sur le corps $\mathbb{H}$ est munie d'une structure d'espace vectoriel quaternionien à droite de dimension n^2 sur $\mathbb{H}$, dont une base est formée des n^2 matrices ε_{ij}, $1 \leq i,j \leq n$ telles que la matrice ε_{ij} a pour seul terme non nul celui de la $i^{\text{ième}}$ ligne et de la $j^{\text{ième}}$ colonne qui est égal à 1. L'adjointe de la matrice A est alors $A^* = {}^tA^\nu$ où A^ν est la matrice conjuguée de A.

Pour S, espace vectoriel à droite de dimension n sur $\mathbb{H}$, nous allons montrer que toute anti-involution α de l'algèbre réelle $\mathcal{L}_{\mathbb{H}}(S)$, peut être considérée comme l'opération d'adjonction relativement à une forme sesquilinéraire non dégénérée, sur S.

Si $\mathcal{A}$ est une algèbre centrale simple sur un corps K commutatif, d'après le théorème de Wedderburn ([3], p. 340 ; [13]), $\mathcal{A}$ est isomorphe à l'algèbre sur K, $\mathcal{L}_\Gamma(S)$ des endomorphismes d'un espace vectoriel à droite S sur un corps Γ ayant pour centre K. Comme $K = \mathbb{R}$, le cas qui nous intéresse ici, comme l'algèbre est centrale, est celui où $\Gamma = \mathbb{H}$. $\mathcal{A}$ possède un $\mathcal{A}$-module simple à un isomorphisme près, [3], chapitre VIII, et $\mathcal{A}$ s'identifie à l'algèbre réelle des opérateurs linéaires de l'espace vectoriel S quaternionien. Soit $\varepsilon = \{\varepsilon_1,\ldots,\varepsilon_n\}$ une base arbitrairement choisie de S. Cette base détermine sur S un produit scalaire quaternionien pour lequel ε est une base orthogonale. Un élément a de $\mathcal{A}$ est représenté par sa matrice A dans ε et l'opérateur d'adjonction est $A^* = {}^tA^\nu$.

Si α est une anti-involution de $\mathcal{A}$, d'après le théorème rappelé ci-dessus, elle s'écrit : $A^\alpha = U^{-1}({}^tA^\nu)U$ avec ${}^tU^\nu = U$ ou ${}^tU^\nu = -U$.

Si ${}^tU^\nu = U$, -(U est symmétrique pour la structure quaternionienne)-, U est la matrice dans ε d'une forme sesquilinéaire non dégénérée sur S qui détermine sur S un produit scalaire pseudo-quaternionien dont α est précisément l'opération d'adjonction.

Si ${}^tU^\nu = -U$, -(U est antisymétrique pour la structure quaternionienne)-, U est la matrice dans ε d'une forme sesquilinéaire non dégénérée antihermitienne sur S d'indice maximal $[\frac{n}{2}]$ (cf. II.1) dont α est l'opération d'adjonction.

Comme dans [4], le problème qui se pose alors est :

Si l'anti-involution α sur l'algèbre centrale simple réelle $\mathcal{A} = \mathcal{L}_{\mathbb{H}}(S)$, est associée à un produit scalaire pseudo-quaternionien, déterminer la signature (p,q) du produit scalaire pseudo-quaternionien sur S dont elle est l'opération d'adjonction.

b) Forme associée à une anti-involution α sur $\mathcal{A} = \mathcal{L}_{\mathbb{H}}(S)$:

$\mathcal{A} = \mathcal{L}_{\mathbb{H}}(S)$ est une algèbre centrale simple réelle. Si $\dim_{\mathbb{H}} S = n$, comme $\dim_{\mathbb{R}} \mathbb{H} = 4$, il en résulte que $\dim_{\mathbb{R}} \mathcal{A} = 4n^2$ (cf. théorème de Wedderburn). Comme dans [15], page 601, si $l(a)$ est l'endomorphisme de l'espace vectoriel réel sous-jacent à $\mathcal{A}$: $x \mapsto ax$, alors $Tr(l(a))$ est bien défini ; de plus $Tr(l(a))$ est invariant par tous les automorphismes de $\mathcal{A}$ et comme $\mathcal{A}$ est simple $Tr(l(a))$ est aussi invariant par tous les anti-automorphismes de $\mathcal{A}$, [15] page 601. Si λ est un anti-automorphisme de $\mathcal{A}$ alors $Tr(l(x^\lambda y))$ est une forme bilinéaire non dégénérée sur $\mathcal{A} \times \mathcal{A}$. Comme $(x^\lambda y)^\lambda = y^\lambda x^{\lambda^2}$ et donc $Tr(l(x^\lambda y)) = Tr(l(y^\lambda x^{\lambda^2}))$, la forme bilinéaire $(x,y) \mapsto Tr(l(x^\lambda y))$ est symétrique si et seulement si $\lambda^2 = 1$, idest si et seulement si λ est une anti-involution de $\mathcal{A}$.

Si α est une anti-involution de $\mathcal{A} = \mathcal{L}_{\mathbb{H}}(S)$, $(x,y) \mapsto Tr(l(x^\alpha y))$ est une forme bilinéaire symétrique non dégénérée sur $\mathcal{A}$, associée à l'anti-involution α de $\mathcal{A}$.

$m(n,\mathbb{H})$ est un espace vectoriel à droite de dimension n^2 sur $\mathbb{H}$, de base les $n^2 \varepsilon_{ul}$ $(1 \leq u,l \leq n)$. $m(n,\mathbb{H})$ est, ainsi, un $\mathbb{R}$-espace vectoriel de base les $4n^2$ éléments : $\mathcal{E}' = \{\varepsilon_{ul}, \varepsilon_{ul} i, \varepsilon_{ul} j, \varepsilon_{ul} k\}$. Supposons qu'il existe sur S un produit scalaire pseudo-quaternionien de type (p,q), dont α est l'opération d'adjonction. Proposons-nous de déterminer alors la signature de la forme quadratique réelle définie sur $\mathcal{A} = \mathcal{L}_{\mathbb{H}}(S)$ par $x \mapsto Tr(l(x^\alpha x))$. Nous allons utiliser $m(n,\mathbb{H})$, muni de sa structure d'anneau et d'espace vectoriel quaternionien à droite de dimension n^2 sur $\mathbb{H}$. Soit $l(A) : B \in m(n,\mathbb{H}) \mapsto AB$. $l(A)$ est une application linéaire de $m(n,\mathbb{H})$ dans $m(n,\mathbb{H})$. On peut définir $Tr(l(A)) = nTrA) \in \mathbb{H}$. (Cf. [5],b). Or $Tr({}^tA^\nu) = (TrA)^\nu$. Il en résulte que $Tr(l({}^tA^\nu)) = (Tr(l(A)))^\nu$. Soit α l'anti-involution de $\mathcal{A}$ qui vérifie donc, [4], $\forall$ a, b $\in \mathcal{A}$, $\forall \lambda \in \mathbb{R}$, $(ab)^\alpha = b^\alpha a^\alpha$, $(a^\alpha)^\alpha = a$, $(\lambda 1)^\alpha = \lambda.1$. La traduction matricielle dans $m(n,\mathbb{H})$ est : $(AB)^\alpha = B^\alpha A^\alpha$, $(A^\alpha)^\alpha = A$, $(\lambda Id)^\alpha = \lambda Id$, $\forall \lambda \in \mathbb{R}$. Autrement dit α est $\mathbb{R}$-linéaire de $m(n,\mathbb{H})$ dans $m(n,\mathbb{H})$ et α est un antiautomorphisme pour la structure d'anneau de $m(n,\mathbb{H})$. On vérifie, alors, que $Tr(l\,{}^t(A^\alpha)^\nu) = Tr(l(A))$ [cf. [15] pages 601-603], (car $A \mapsto {}^t(A^\alpha)^\nu$ est un anti-automorphisme), et donc $Tr(l(A^\alpha)) = (Tr(l(A)))^\nu$.

Ainsi l'application f de $m(n,\mathbb{H}) \times m(n,\mathbb{H})$ dans $\mathbb{H}$ définie par $f(A,B) = Tr(l(A^\alpha B))$ vérifie : $f(B,A) = (f(A,B))^\nu$, $f(A+A',B) = f(A,B)+f(A',B)$, $f(A,B+B') = f(A,B) + f(A,B')$ et $\forall \lambda \in \mathbb{R}$ $f(\lambda A,B) = \lambda f(A,B) = f(A,\lambda B)$. Mais f n'est pas sesquilinéaire sur $m(n,\mathbb{H})$.

Choisissons pour ε une base orthogonale pour le produit scalaire pseudo-quaternionien. $\mathcal{E}'$ est alors orthogonale pour la forme bilinéaire symétrique réelle $(x,y) \mapsto Tr(l(x^\alpha y)$, comme on le vérifie. Si $H = \mathrm{diag}(\lambda_1,\ldots,\lambda_n)$ est la matrice dans ε de ce produit scalaire pseudo-quaternionien avec $\lambda_1,\ldots,\lambda_p > 0$ et $\lambda_{p+1},\ldots,\lambda_n < 0$, $A^\alpha = H^{-1}({}^tA^\nu)H$. Si

$A = \sum_{i,j} \varepsilon_{ij} a_{ij}$ on trouve que $Tr(l(A^{\alpha}A)) = f(A,A) = n \sum_{ij} \frac{\lambda i}{\lambda j} |a_{ij}|^2$ où $|a_{ij}| = (a_{ij}\, {}^{\nu}a_{ij})^{\frac{1}{2}}$ est la valeur absolue classique du quaternion a_{ij}. Ainsi, $f(A,A)$ comporte p^2+q^2 signes positifs et $2pq$ signes négatifs.

Il en résulte immédiatement que la signature de la forme quadratique réelle définie sur $\mathcal{L}_{\mathbb{H}}(S)$ par $x \to Tr(l(x^{\alpha}x))$ est : $(4(p^2+q^2), 8pq)$. (On vérifie que $4(p^2+q^2) + 8pq = 4n^2 = \dim_{\mathbb{R}}(\mathcal{L}_{\mathbb{H}}(S))$).

II.3. Structures pseudo-quaternioniennes sur les espaces de spineurs S des $C^+_{r,s}$, $r+s = m = 2k+1$, $r-s \equiv \pm 3$ (modulo 8).

On a déjà noté comment réaliser $C^+_{r,s}$ en choisissant u : tel que $(u|u) = \varepsilon = \pm 1$, et en représentant $C^+_{r,s}$ comme l'algèbre de Clifford $C(u^{\perp})$, $u^{\perp}$ étant de type $(r,s-1)$ si $(u|u) = -1$ ou de type $(s,r-1)$ si $(u|u) = 1$. De plus, en raison du théorème rappelé au II.2, toute anti-involution α de $\mathcal{A} = C^+_{r,s}$ s'écrit $A^{\alpha} = U^{-1}\,({}^t A^{\nu})U$ avec ${}^tU^{\nu} = U$ ou ${}^tU^{\nu} = -U$. Si ${}^tU^{\nu} = U$, alors $U^{\alpha} = {}^tU^{\nu} = U$ et si ${}^tU^{\nu} = -U$, alors $U^{\alpha} = {}^tU^{\nu} = -U$. Il en résulte que l'élément U qui "dirige" l'automorphisme intérieur peut être précisé. Si l'on réalise, par exemple $C^+_{r,s}$ en choisissant pour u : e_{2k+1} tel que $(e_{2k+1}, e_{2k+1}) = -1$, on peut prendre : $J_1 = e_1 \dots e_{2k} \in C^+_{r,s}$ et U est proportionnel à J_1. Si l'on réalise, par contre, $C^+_{r,s}$ en choisissant pour U : e_1 tel que $(e_1, e_1) = 1$, on peut prendre $J'_1 = e_2 \dots e_{2k+1}$ et U est proportionnel à J'_1.

La dichotomie : k pair et k impair s'introduit alors naturellement.

Choisissons en effet pour anti-involution α : τ l'antiautomorphisme principal de $C^+_{r,s}$; $J_1^{\tau} = J_1$ (respectivement $J_1'^{\tau} = J'_1$) si et seulement si k est pair et $J_1^{\tau} = -J_1$ (respectivement $J_1'^{\tau} = -J'_1$) si et seulement si k est impair.

a) Etude du cas où $k \equiv 0$ (2). On obtient, alors, le théorème qui suit donné sans démonstration, faute de place.

Théorème :

L'espace des spineurs S attaché aux algèbres de Clifford $C^+_{r,s}$ des espaces pseudo-euclidiens $V = E_{r,s}$ de type (r,s) avec $r+s = 2k+1$, $k \equiv 0$ (2) et $r-s \equiv \pm 3$ (modulo 8), possède une structure naturelle pseudo-quaternionienne et un produit scalaire pseudo-quaternionien neutre, déterminé à un facteur scalaire près, invariant par le groupe spinoriel Spin V. Pour $m \geqslant 5$, on a une inclusion naturelle :

$$\mathrm{Spin}\, V \subseteq SpU\left(2^{\frac{m-1}{2}-2}, 2^{\frac{m-1}{2}-2}\right)$$

b) Etude du cas où $k \equiv 1\ (2)$. On établit le théorème suivant :

Théorème :

L'espace des Spineurs S attaché aux algèbres de Clifford $C^{+}_{r,s}$ des espaces pseudo-euclidiens $V = E_{r,s}$ de type (r,s) avec $r+s = 2k+1$, $k\equiv 1\ (2)$ et $r-s \equiv \pm 3\ (8)$ est muni d'une forme sesquilinéaire non dégénérée antisymétrique pour la structure quaternionienne.

De plus, $\mathrm{Spin}(E_{r,s}) \subseteq SO^{*}(2^{\frac{m-1}{2}})$. $(m \geqslant 3)$.

II.4 - Plongement naturel des quadriques projectives des espaces pseudo-euclidiens $E_{r,s}$, $r+s = 2k+1$, $r-s \equiv \pm 3$ (modulo 8)

a) Supposons $k \equiv 0\ (2)$

Ce qui a pour conséquence l'existence sur S d'un produit scalaire $\{|\}$ pseudo-quaternionien neutre de type (p,p) avec

$p = \frac{1}{2} \dim_{\mathbb{H}} S = 2^{k-2} = 2^{\frac{m-1}{2} - 2}$, associé à τ et $C^{+}_{r,s} = \mathcal{L}_{\mathbb{H}}(S)$. On obtient le

Théorème :

La quadrique projective $\widetilde{Q}(E_{r,s})$ associée à l'espace pseudo-euclidien $E_{r,s}$ où $m = r+s = 2k+1$, $k\equiv 0\ (2)$, $r-s\equiv\pm 3$ (modulo 8) est naturellement $m \geqslant 5$
plongée dans le groupe symplectique unitaire $SpU(2^{\frac{m-1}{2}-2})$. En particulier, le compactifié conforme $\widetilde{Q}(E_{4,1})$ de l'espace euclidien $E_{3,0}$ s'identifie naturellement au groupe symplectique unitaire SpU(1).

L'ensemble des sous espaces de S strictement positifs maximaux, donc de dimension $\frac{1}{2} \dim_{\mathbb{H}} S = 2^{k-2}$, forme un ouvert de la grassmannienne $G(S,\frac{1}{2}\dim S)$, appelé par convention, la semi-grassmannienne de $(S,\{|\})$, et noté $G_{+}(S)$.

$G_{+}(S)$ est l'espace symétrique classique -simplement connexe de type CII dans la classification d'Elie Cartan (cf. [6], [16] par exemple) :

$\dfrac{SpU(2^{k-2},2^{k-2})}{SpU(2^{k-2})\times SpU(2^{k-2})}$ et $\widetilde{Q}(E_{r,s})$ est plongé dans la "frontière" de $G_{+}(S)$ dans $G(S,2^{k-2})$.

$G_{+}(S)$ s'identifie à l'espace symétrique des anti-involutions de $\mathcal{L}_{\mathbb{H}}(S) = C^{+}_{r,s}$ qui commutent avec τ : $\alpha\tau = \tau\alpha$ et qui sont strictement positives, ce qui signifie ([4], [12], pages 273, 274) que la forme quadratique réelle sur l'algèbre réelle $C^{+}_{r,s} = \mathrm{Tr}(l(x^{\alpha}x))$ est définie positive.

b) Supposons $k \equiv 1$ (2), ce qui a pour conséquence l'existence sur S d'une forme sesquilinéaire b antisymétrique pour la structure quaternionienne non dégénérée d'indice $2^{\frac{m-1}{2}-2}$. On trouve le

Théorème :

La quadrique projective $\tilde{Q}(E_{r,s})$ associée à l'espace pseudo-euclidien $R_{r,s}$ où $m = r+s = 2k+1$, $k \equiv 1$ (2), $r-s \equiv \pm 3$ (modulo 8) est naturellement plongée dans le groupe $O(2^{\frac{m-1}{2}})$. ($m \geqslant 3$)

III. STRUCTURES REELLES SUR LES ESPACES DE SPINEURS S DES $C^+_{r,s}$ OÙ $m = 2k+1$, $r-s \equiv \pm 1$ (modulo 8) ET PLONGEMENT NATUREL DES QUADRIQUES PROJECTIVES ASSOCIEES.

III.1. Structure pseudo-euclidienne ou symplectique réelle sur les espaces de spineurs S des algèbres $C^+_{r,s}$ où $r+s = 2k+1$ et $r-s \equiv \pm 1$ (modulo 8).

On établit le théorème suivant :

Théorème :

L'espace des spineurs S attaché aux algèbres de Clifford $C^+_{r,s}$ des espaces pseudo-euclidiens $V = E_{r,s}$ de type (r,s) avec $r+s = 2k+1$ et $r-s \equiv \pm 1$ (modulo 8) possède une structure naturelle réelle pseudo-euclidienne neutre ou symplectique suivant que k est pair ou impair dont le produit scalaire est invariant par le groupe spinoriel $\mathrm{Spin}(E_{r,s})$. Ce groupe $\mathrm{Spin}(E_{r,s})$ est plongé dans $SO(p,p)$ ou $Sp(2p,\mathbb{R})$ avec $p = 2^{\frac{m-1}{2}-1} = 2^{k-1}$, suivant que k est pair ou impair.

III.2. Plongement naturel des quadriques projectives $\tilde{Q}(E_{r,s})$ des espaces $E_{r,s}$, $r+s = 2k+1$, $r-s \equiv \pm 1$ (modulo 8).

On obtient le théorème suivant :

Théorème :

La quadrique projective $\tilde{Q}(E_{r,s})$ pour $r+s = 2k+1$, $r-s = \pm 1+8l$ est naturellement plongée dans le groupe $O(2^{k-1})$ si k est pair et dans le groupe $\frac{U(2^{k-1})}{O(2^{k-1})}$ si k est impair. ($m \geqslant 3$).

Si k est pair, le même raisonnement qu'au paragraphe II.4 montre que l'ensemble des sous espaces de S strictement positifs maximaux, donc de dimension 2^{k-1} forme un ouvert de la grassmannienne $G(S,2^{k-1})$ appelé la semi-grassmannienne de $(S,(|))$ et noté $G_+(S)$. $G_+(S)$ est l'espace symétrique classique $\frac{SO_+(2^{k-1},2^{k-1})}{SO(2^{k-1})\times SO(2^{k-1})}$ de type BDI dans la classification d'Elie Cartan, ([6], page 349, par exemple). Comme au II.4, $G_+(S)$ s'identifie canoniquement à l'espace symétrique des anti-involutions α de $C^+_{r,s}$ qui commutent avec τ et qui sont strictement positives.

IV. ETUDE DES CAS OU $r-s \equiv 0$ (modulo 8) ET $r-s \equiv 4$ (modulo 8)

Si $r-s \equiv 0$ (modulo 8), $C^+_{r,s}$ est semi-simple et isomorphe à la composée directe de deux algèbres isomorphes à $m(2^{[\frac{m-1}{2}]},\mathbb{R})$ et si $r-s \equiv 4$ (modulo 8) $C^+_{r,s}$ est isomorphe à la composée directe de deux algèbres isomorphes à $m(2^{[\frac{m-1}{2}]-1},\mathbb{H})$. On obtient le

Théorème :

Si $r-s \equiv 0$ (modulo 8), $r+s = m = 2k$, et si k est pair, $\mathrm{Spin}(E_{r,s})$ est alors plongé dans $Sp(2p,\mathbb{R})$ avec $p = 2^{[\frac{m-1}{2}]-1} = 2^{k-2}$ et la quadratique projective $\tilde{Q}(E_{r,s})$ est naturellement plongée dans $\frac{U(2^{k-2})}{O(2^{k-2})}$, $(k \geqslant 2)$; si k est impair, $\mathrm{Spin}(E_{r,s})$ est plongée dans $SO(p,p)$ où $p = 2^{[\frac{m-1}{2}]-1} = 2^{k-2}$ et $\tilde{Q}(E_{r,s})$ dans $O(2^{k-2})$, $k \geqslant 2$. Si $r-s \equiv 4$ (modulo 8), $r+s = m = 2k$ et si k est pair, $\mathrm{Spin}\,(E_{r,s})$ est plongé dans $SO^*(2p)$, $p = 2^{k-3}$ et $\tilde{Q}(E_{r,s})$ dans $O(2^{k-3})$, $k \geqslant 3$; si k est impair, $\mathrm{Spin}(E_{r,s})$ est plongé dans $SpU(p,p)$, $p = 2^{k-3}$ et $\tilde{Q}(E_{r,s})$ dans $SpU(2^{k-3})$, $k \geqslant 3$.

V. ETUDE DES CAS $r-s \equiv \pm 2$ (modulo 8)

On établit le théorème suivant qui permet de retrouver les résultats donnés dans [4] pour $r+s \equiv 2$ (modulo 4) et $r-s \equiv 2$ (modulo 4).

Théorème :

Si $r-s \equiv \pm 2$ modulo 8, $m = r+s = 2k$, $\mathrm{Spin}\, E_{r,s}$ est alors plongé dans $SU(p,p)$, où $p = 2^{\frac{m}{2}-2}$, si k est impair et dans $SO^*(2p)$ avec $p = 2^{\frac{m}{2}-2}$, si k est pair ; $k \geqslant 2$. La quadrique projective $\tilde{Q}(E_{r,s})$ est naturellement plongée dans le groupe $U(p)$ où $p = 2^{\frac{m}{2}-2} = 2^{k-2}$ $(k \geqslant 2)$, quelle que soit la parité de k.

REMERCIEMENTS

L'auteur tient à exprimer sa profonde gratitude au Professeur René Deheuvels de l'Université de Paris VI qui lui a proposé ce sujet de recherche, a examiné le manuscrit et dont les remarques judicieuses lui ont été précieuses.

BIBLIOGRAPHIE

[1] - Albert (A.A.), 'Structures of Algebras', American Math. Soc., **Vol. XXIV**, New York, 1939.
[2] - Atiyah (M.F.), Bott (R.), and Shapiro (A.), 'Clifford modules', Topology, **Vol. 3**, 1964, p. 3-38.
[3] - Deheuvels (R.), Formes quadratiques et groupes classiques, Presses Universitaires de France, Paris, 1980.
[4] - Deheuvels (R.), 'Groupes conformes et algèbres de Clifford', Publication du Séminaire de Mathématique de l'Université de Turin, Turin, 1985.
[5] - Dieudonné (J.), a) 'On the structure of unitary groups', Trans. Amer. Math. Soc., **72**, p. 367-385, 1952.
b) 'Les déterminants sur un corps non commutatif', Bull. Soc. Math., France, **71**, p. 27-45, 1943.
[6] - Helgason (S.), Differential geometry and symmetric spaces, Academic Press, New York and London, 1962.
[7] - Husemoller (D.), Fibre bundles, Mc Graw Hill Inc., 1966.
[8] - Lam (T.Y.), The Algebraic Theory of quadratic forms, W.A. Benjamen Inc., 1973.
[9] - Lounesto (P.), 'Spinor valued regular functions in hypercomplex Analysis', Thesis, Report H.T.K.K. Mat. A. **154**, 1979, Helsinki University of Technology, p. 1-79.
[10]- Maia (M.D.), 'Conformal spinors in general relativity', Journal of Math. Physics, **Vol. 15**, n° 4, 1974, p. 420-425.
[11]- Porteous (I.R.), Topological Geometry, Cambridge University Press, 2ème édition, 1981.
[12]- Satake (I.), Algebraic structures of symmetric domains, Iwanami Shoten Publishers and Princeton Univsersity Press, 1980.
[13]- Serre (J.P.), 'Applications algébriques de la cohomogie des groupes. II - Théorème des algèbres simples', Séminaire H. Cartan, E.N.S. 1950-1951, 2 exposés, 6-10, 6-09 ; 7-01, 7-11.
[14]- Wall (C.T.C), 'Graded algebras anti-involutions, simple groups and symmetric spaces', Bull. American Math. Soc., **74**, 1968, P. 198-202.
[15]- Weil (A.), 'Algebras with involutions and the classical groups', Collected papers, **Vol. II**, (1951-1964), p. 413-447, reprinted by permission of the editors Journal of Ind. Math. Soc., Springer-Verlag, New York, 1980.
[16]- Wolf (J.A.), Spaces of constant curvature, Publish or Perish Inc., Boston, 1974.
[17]- Wybourne (B.G.), Classisal groups for Physicists, John Wiley and sons, Inc. New York, 1974.

SPIN GROUPS ASSOCIATED WITH DEGENERATE ORTHOGONAL SPACES

J.A. Brooke
Department of Mathematics
University of Saskatchewan
Saskatoon, Saskatchewan
Canada S7N OWO

ABSTRACT. A review of Clifford algebras over degenerate real bilinear forms is outlined and a discussion of the associated spin groups given. Attention is paid particularly to forms with nullspace of dimensions one and two, and explicit computations illustrate the connection to representations of inhomogeneous orthogonal groups of interest in physics. This provides a unified picture of relativistic and non-relativistic covariant wave equations for spinning particles.

1. INTRODUCTION AND REVIEW

Although the relationship between the real Clifford algebra and associated spin group of a non-degenerate real orthogonal space has been very well understood mathematically and exploited in a multitude of physical contexts, such is not the case when the underlying orthogonal space is degenerate. In fact there still remain a number of basic, unanswered mathematical questions in the degenerate situation concerning both the detailed structure of the Clifford algebra, and of its Clifford and spin groups.

For example, while some of the structural aspects of the Clifford algebra of an arbitrary degenerate real orthogonal space are known, there is still no algorithm to construct and classify minimal faithful realizations of the Clifford algebra as a matrix algebra over the reals, complexes or quaternions analogous to the well-known realizations in the general non-degenerate situation. As to the question of spin groups, a number of general results have been obtained in the degenerate case:

(a) when the underlying bilinear form has a one-dimensional nullspace, the spin group is the semi-direct product of a (pseudo-) Euclidean space with the spin group of the non-degenerate part; a similar result holds for the pin group.

(b) when the underlying bilinear form has a two-dimensional nullspace, the spin group has the structure of a semi-direct product of a Heisenberg group and the spin group of the non-degenerate part.

The <u>detailed</u> structure of the spin group is not known for higher degeneracies.

J. S. R. Chisholm and A. K. Common (eds.), Clifford Algebras and Their Applications in Mathematical Physics, 93–102.

Applications to the homogeneous and inhomogeneous Galilei groups are numerous. Spin is incorporated into non-relativistic physics in a manner directly analogous to its appearance in relativistic kinematics. A notion of Clifford algebra contraction can be seen to be a pre-cursor of the Inönü-Wigner contraction of the Poincaré to the Galilei group. The extended Galilei group of quantum mechanics is related to a doubly-degenerate spin group and the entire formalism provides a manifestly covariant framework for the treatment of non-relativistic wave equations for spinning particles. These and other issues are discussed.

By $\mathbb{R}^{r,p,q}$ we mean the real orthogonal space $\mathbb{R}^{r+p+q}$ together with the symmetric bilinear form:

$$B = \mathrm{diag}(0, \ldots, 0, -1, \ldots, -1, 1, \ldots, 1)$$

with r zero, p negative, and q positive entries.

Let $\mathbb{R}_{r,p,q}$ denote the universal Clifford algebra associated with the orthogonal space $\mathbb{R}^{r,p,q}$ [see Porteous (1981)].

An <u>orthonormal subset</u> of $\mathbb{R}^{r,p,q}$ is a subset $\{\gamma^i\}$ of $\mathbb{R}^{r,p,q}$ satisfying $\gamma^i\gamma^j + \gamma^j\gamma^i = -2B^{ij}$ with B^{ij} defined above.

Let $\hat{}$ denote the <u>main involution</u> of $\mathbb{R}_{r,p,q}$, i.e. the unique automorphism of $\mathbb{R}_{r,p,q}$ induced by $-\mathrm{id}$ on $\mathbb{R}^{r,p,q}$; and let $\bar{}$ denote the <u>conjugation anti-involution</u> of $\mathbb{R}_{r,p,q}$, i.e. the unique anti-automorphism of $\mathbb{R}_{r,p,q}$ induced by $-\mathrm{id}$ on $\mathbb{R}^{r,p,q}$.

The <u>even</u> elements of $\mathbb{R}_{r,p,q}$ - denoted $\mathbb{R}^+_{r,p,q}$ - are those members $a \in \mathbb{R}_{r,p,q}$ for which $\hat{a} = a$. The <u>odd</u> elements $\mathbb{R}^-_{r,p,q}$ are those members a of $\mathbb{R}_{r,p,q}$ for which $\hat{a} = -a$.

Following Atiyah, Bott, Shapiro (1964) and Porteous (1981) we define the <u>Clifford group</u> $\Gamma(\mathbb{R}^{r,p,q})$ [also written $\Gamma(r,p,q)$] as follows:

$$\Gamma(\mathbb{R}^{r,p,q}) = \{g \in \mathbb{R}_{r,p,q} : g^{-1} \text{ exists and } g\, x\, \hat{g}^{-1} \in \mathbb{R}^{r,p,q} \text{ for all } x \in \mathbb{R}^{r,p,q}\} .$$

This is also called the <u>Clifford group with twist</u> to distinguish it from the definition [Chevalley (1954)] which does not employ the main involution $\hat{}$.

The <u>even Clifford group</u> $\Gamma(\mathbb{R}^{r,p,q})^+$ is, by definition, the intersection of $\Gamma(\mathbb{R}^{r,p,q})$ with $\mathbb{R}^+_{r,p,q}$.

The <u>norm</u> $N(g)$ of an element g of $\mathbb{R}_{r,p,q}$ is, by definition, $\bar{g}g$.

Motivated by results in the non-degenerate case, when $r = 0$ [see

Porteous (1981)] we define:

$$\text{Pin}\ (\mathbb{R}^{r,p,q}) = \{g \in \Gamma(\mathbb{R}^{r,p,q}) : N(g) = \pm 1\}$$

$$\text{Spin}\ (\mathbb{R}^{r,p,q}) = \text{Pin}\ (\mathbb{R}^{r,p,q}) \cap \mathbb{R}^{+}_{r,p,q}$$

and also write for simplicity Pin (r,p,q) and Spin (r,p,q) respectively.

2. PIN AND SPIN GROUPS IN THE DEGENERATE CASE

While the detailed structures of Pin (r,p,q) and Spin (r,p,q) are well understood in the non-degenerate case, $r = 0$, the degenerate situation $r > 0$ has been little-studied. Of closely related interest is the work of Helmstetter (1977) on Clifford groups for $r > 0$. A number of general results have appeared in Brooke (1978, 1980b), Crumeyrolle (1980) for $r = 1$ and in a slightly different context, and in Abłamowicz (1985a, 1985b) also for $r = 1$. In addition, see also Abłamowicz, Lounesto (1986). It should be pointed out that the special case $r = 1$, $p = 0$, $q = 3$ appears implicitly in the representations of the group of Euclidean motions of Study (1890, 1891) who uses the biquaternions of Clifford (1873).

The following results are proved in Brooke (1978, 1980b).

For simplicity, we use ρ to denote any of the maps Pin $(r,p,q) \to 0\ (r,p,q)$, Spin $(r,p,q) \to SO\ (r,p,q)$ where $\rho(g)x = g\, x\, \hat{g}^{-1}$, $x \in \mathbb{R}^{r,p,q}$.

<u>Theorem 1</u>: Suppose that $n = p + q \geq 1$. Then

(a) Spin $(1,p,q) \cong \mathbb{R}^{n}$ Ⓢ Spin (p,q) where Spin (p,q), the abbreviation for Spin $(0,p,q)$, acts on $\mathbb{R}^{n}$ via the homomorphism ρ : Spin $(p,q) \to SO\ (p,q)$.

If Spin^{+} (r,p,q) denotes the set of those $g \in$ Spin (r,p,q) for which $N(g) = 1$, then

(b) Spin^{+} $(1,p,q) \cong \mathbb{R}^{n}$ Ⓢ Spin^{+} (p,q) and, except when $(p,q) \in \{(0,1), (1,0), (1,1)\}$, the surjective mapping ρ of Spin^{+} $(1,p,q)$ to SO^{+} $(1,p,q)$ is a double covering with kernel $\{1, -1\}$. □

<u>Theorem 2</u>: Suppose that $n = p + q \geq 1$. Then

(a) Pin $(1,p,q) \cong \mathbb{R}^{n+1}$ Ⓢ Pin (p,q) where Pin (p,q) acts on $\mathbb{R}^{n+1}$ as follows: $a \cdot (\lambda,x) = (\det \rho(a)\lambda,\ \rho(a)x)$ with ρ : Pin $(p,q) \to 0\ (p,q)$ the canonical homomorphism.

(b) The canonical mapping ρ of Pin $(1,p,q)$ to $0\ (1,p,q)$ is not surjective and in fact has kernel isomorphic to $\mathbb{R} \times \{-1, 1\}$. □

Turning now to orthogonal spaces $\mathbb{R}^{2,p,q}$ possessing a higher degeneracy one has:

Theorem 3: Let $K(p,q)$ denote the group of all triples $(\theta, x_1, x_2) \in \mathbb{R} \times \mathbb{R}^n \times \mathbb{R}^n$, with $n = p + q \geq 1$, and product:

$$(\theta, x_1, x_2)(\theta', x_1', x_2') = (\theta + \theta' + B(x_1, x_2') - B(x_2, x_1'), x_1 + x_1', x_2 + x_2').$$

Then:

$$\text{Spin}\,(2,p,q) \cong K(p,q) \ⓢ \text{Spin}\,(p,q)$$

with the action of $a \in \text{Spin}\,(p,q)$ on $(\theta, x_1, x_2) \in K(p,q)$ defined by:

$$a \cdot (\theta, x_1, x_2) = (\theta, \rho(a)x_1, \rho(a)x_2).$$

Moreover, $\text{Spin}\,(2,p,q)/\{-1,1\} \cong K(p,q) \ⓢ \text{Spin}\,(p,q)/\{-1,1\}$ and the canonical mapping $\rho : \text{Spin}\,(2,p,q) \to SO\,(2,p,q)$ has kernel isomorphic to $\mathbb{R} \times \{-1,1\}$ and is not surjective. □

The group $K(p,q)$ is one of the Heisenberg groups associated to the abelian group $\mathbb{R}^n \times \mathbb{R}^n$. Since we have assumed $p + q \geq 1$, $K(p,q)$ is a non-trivial central extension of $\mathbb{R}^n \times \mathbb{R}^n$.

Recently, the structure of Pin (2,p,q) has been determined.

Theorem 4: $\text{Pin}\,(2,p,q) \cong \tilde{K}(p,q) \ⓢ \text{Pin}\,(p,q)$ where

$$\tilde{K}(p,q) = \{(\theta, \lambda_1, x_1, \lambda_2, x_2)\} = \mathbb{R} \times \mathbb{R} \times \mathbb{R}^n \times \mathbb{R} \times \mathbb{R}^n$$

and $(\theta, \lambda_1, x_1, \lambda_2, x_2)(\theta', \lambda_1', x_1', \lambda_2', x_2')$

$$= (\theta + \theta' + B(x_1, x_2') - B(x_2, x_1'), \lambda_1 + \lambda_1', x_1 + x_1', \lambda_2 + \lambda_2', x_2 + x_2')$$

and the action of $a \in \text{Pin}\,(p,q)$ on $(\theta, \lambda_1, x_1, \lambda_2, x_2) \in \tilde{K}(p,q)$ is:

$$a \cdot (\theta, \lambda_1, x_1, \lambda_2, x_2) = (\theta, \det \rho(a)\lambda_1, \rho(a)x_1, \det \rho(a)\lambda_2, \rho(a)x_2).$$

The canonical mapping $\rho: \text{Pin}\,(2,p,q) \to O(2,p,q)$ has kernel isomorphic to $\mathbb{R}^3 \times \{-1,1\}$.

Proof (sketch): The argument proceeds along the lines of the proof of Theorem 3 [see Brooke (1980b)]. If $\{\gamma_1^\circ, \gamma_2^\circ, \gamma^1, \ldots, \gamma^n\}$ denotes an orthonormal basis of $\mathbb{R}^{2,p,q}$ which moreover generates

$\mathbb{R}_{2,p,q}$, and in which $(\gamma_1^\circ)^2 = 0 = (\gamma_2^\circ)^2$ and $\{\gamma^1, \ldots, \gamma^n\}$ generate $\mathbb{R}_{p,q}$, then $\tilde{K}(p,q)$ is isomorphic to the following subgroups of Pin (2,p,q) :

$$\tilde{K}(p,q) \cong \left\{1 + (\lambda_1 + x_1)\gamma_1^\circ + (\lambda_2 + x_2)\gamma_2^\circ + (\theta + (\lambda_2 x_1 - \lambda_1 x_2) - \frac{1}{2}[x_1, x_2])\gamma_1^\circ\gamma_2^\circ\right\}$$

where $\lambda_1, \lambda_2 \in \mathbb{R}$, $x_1, x_2 \in \mathbb{R}^{p,q}$, $\theta \in \mathbb{R}$. It can be shown also that Pin (2,p,q) $\cong \tilde{K}(p,q) \cdot$ Pin (p,q) , that $\tilde{K}(p,q)$ is normal in Pin (2,p,q) , and intersects Pin (p,q) in the identity. Thus Pin (2,p,q) has the claimed semi-direct product structure. □

The following recently obtained result illustrates the relationship between Pin (r,p,q) and subgroups Pin (r-1,p,q) , $\tilde{K}(r-1,p,q)$, and thereby yields a recursive method to compute Pin (r,p,q) . Before stating the theorem let us set up the notation. As usual we suppose $n = p + q \geq 1$. Let $\{\gamma_1^\circ, \ldots, \gamma_r^\circ, \gamma^1, \ldots, \gamma^n\}$ denote an orthonormal basis of $\mathbb{R}^{r,p,q}$ generating $\mathbb{R}_{r,p,q}$. Moreover let $\mathbb{R}^{r-1,p,q}$ denote the span of $\{\gamma_2^\circ, \ldots, \gamma_r^\circ, \gamma^1, \ldots, \gamma^n\}$ and let $\mathbb{R}_{r-1,p,q}$ denote the (universal) Clifford algebra generated by $\{\gamma_2^\circ, \ldots, \gamma_r^\circ, \gamma^1, \ldots, \gamma^n\}$. Thus, obviously we have $\mathbb{R}^{r-1,p,q} \subset \mathbb{R}^{r,p,q}$ and $\mathbb{R}_{r-1,p,q} \subset \mathbb{R}_{r,p,q}$.

Thus also, we may regard Pin (r-1,p,q) as a subgroup of Pin (r,p,q) .

Let $\tilde{K}(r-1,p,q)$ denote the subset of $\mathbb{R}_{r,p,q}$:

$$\tilde{K}(r-1,p,q) = \{1 + c\gamma_1^\circ : c \in \mathbb{R}_{r-1,p,q},\ \hat{c} = \bar{c}\ ,\ x\hat{c} - cx \in \mathbb{R}^{r-1,p,q} \text{ for all } x \in \mathbb{R}^{r-1,p,q}\} .$$

It can be shown that the conditions on c , that imply $1 + c\gamma_1^\circ \in \tilde{K}(r-1,p,q)$, are such as to guarantee that $1 + c\gamma_1^\circ$ belongs to $\Gamma(r,p,q)$, the Clifford group of $\mathbb{R}_{r,p,q}$. Moreover, not only is $\tilde{K}(r-1,p,q)$ a subgroup of Pin (r,p,q) , but also it is normal. In addition, Pin (r-1,p,q) and $\tilde{K}(r-1,p,q)$ intersect in the identity, and every member of Pin (r,p,q) can be written as a product of an element of $\tilde{K}(r-1,p,q)$ with an element of Pin (r-1,p,q) . We have therefore:

Theorem 5: Pin $(r,p,q) = \tilde{K}(r-1,p,q) \circledS \mathrm{Pin}\ (r-1,p,q)$ with the action of $a \in \mathrm{Pin}\ (r-1,p,q)$ on a member $1 + c\gamma^\circ_1$ of $\tilde{K}(r-1,p,q)$ by conjugation:

$$a \cdot (1 + c\gamma^\circ_1) = a(1 + c\gamma^\circ_1)a^{-1} = 1 + (\rho(a)c)\gamma^\circ_1$$

where, as usual, $\rho(a)c = ac\hat{a}^{-1}$ is the canonical map. □

The proof of Theorem 5 is computational and will be omitted. We shall be content with an explicit description of $\tilde{K}(r-1,p,q)$. Notationally, if I denotes a subset of $\{2, \ldots, r\}$ and γ°_I the ordered product $\gamma^\circ_{i_1}\gamma^\circ_{i_2} \ldots \gamma^\circ_{i_k}$ where

$$I = \{i_1, i_2, \ldots, i_k\}, \quad 2 \leq i_1 < i_2 < \ldots < i_k \leq r,$$

then we have the following:

Lemma:

$$\tilde{K}(r-1,p,q) = \{1 + c\gamma^\circ_1 : c = \sum_{I \subset \{2,\ldots,r\}} c_I \gamma^\circ_I + \sum_{i=1} x^i\gamma^i ; \quad c_I, x^i \in \mathbb{R}\}$$

where in the summation over I's we consider only those I of length $|I| = \mathrm{card}\ (I)$ congruent to 0 or 1 modulo 4. Thus, if $1 + c\gamma^\circ_1$ is a member of $\tilde{K}(r-1,p,q)$, then c is expressible as:

$$c = c_\phi + c_2\gamma^\circ_2 + \ldots + c_r\gamma^\circ_r + x^1\gamma^1 + \ldots + x^n\gamma^n +$$
$$+ c_{2345}\gamma^\circ_2\gamma^\circ_3\gamma^\circ_4\gamma^\circ_5 + \ldots$$ □

It is easy to see the corresponding result for Spin (r,p,q) by replacing "Pin" by "Spin", and $\tilde{K}(r-1,p,q)$ by the subset defined by c's of the form:

$$c = \sum_{I \subset \{2,\ldots,r\}} c_I\gamma^\circ_I + \sum_{i=1}^{n} x^i\gamma^i$$

in which $|I| \equiv 1 \pmod 4$.

The next lemma, concerning semi-direct products generally, is useful in determining the precise structure of Pin (r,p,q) as suggested by Theorem 5.

Lemma: Suppose $G = K \circledS H$ and $H = L \circledS J$, then $G = (K \circledS L) \circledS J$ where the action of J on $K \circledS L$ is determined by the action of H on K and J on L. □

By a straightforward extension of the lemma, we have by virtue of Theorem 5,

Theorem 6:

$$\mathrm{Pin}\,(r,p,q) = \tilde{K}_r(p,q) \circledS \mathrm{Pin}\,(p,q)$$

where $\tilde{K}_r(p,q)$ denotes the nested semi-direct product

$$\tilde{K}_r(p,q) = \tilde{K}(r-1,p,q) \circledS (\tilde{K}(r-2,p,q) \circledS (\ldots \circledS \tilde{K}(0,p,q) \ldots))$$

with the obvious actions resulting from the inclusions

$$\mathrm{Pin}(p,q) \subset \mathrm{Pin}(1,p,q) \subset \ldots \subset \mathrm{Pin}(r-1,p,q)$$

$$\tilde{K}(k-1,p,q) \subset \mathrm{Pin}(k,p,q)\,, \quad 1 \leq k \leq r\,.$$

Moreover, $g \in \mathrm{Pin}\,(r,p,q)$ has a unique decomposition:

$$g = (1 + c_1\gamma_1^\circ)\,(1 + c_2\gamma_2^\circ)\,\ldots\,(1 + c_r\gamma_r^\circ)a$$

with $a \in \mathrm{Pin}(p,q)$, $c_1 \in \mathbb{R}_{r-1,p,q}, \ldots, c_r \in \mathbb{R}_{0,p,q}$ such that $1 + c_i\gamma_i^\circ \in \tilde{K}(r-i,p,q)$, $1 \leq i \leq r$. □

Unfortunately the detailed structure of $\tilde{K}_r(p,q)$ is unknown except for small values of r. For example: $\tilde{K}_1(p,q) \cong \mathbb{R}^{p+q+1}$ (Theorem 2), $\tilde{K}_2(p,q) \cong \tilde{K}(p,q)$ of Theorem 4.

3. APPLICATIONS TO PHYSICS - BRIEF SUMMARY

3.1 Kinematical Groups

The homogeneous Galilei group is readily seen to be precisely $SO^+(1,0,3)$ and consequently the spin version is $\mathrm{Spin}\,(1,0,3)$. Thus, at least in a purely mathematical sense, the notion of spin in quantum theory (and therefore of the spin group) is no more a relativistic notion than it is a non-relativistic concept. This point has been made strenuously by Lévy-Leblond (1967).

The Poincaré group for a spinning particle is easily seen to be $\mathrm{Spin}\,(1,1,3) \cong \mathbb{R}^4 \circledS \mathrm{Spin}\,(1,3) \cong \mathbb{R}^4 \circledS SL(2,C)$.

The de Sitter group for a spinning particle, $\mathrm{Spin}\,(1,4)$, is the smallest "classical" spin group containing $\mathrm{Spin}\,(1,0,3)$ as a subgroup.

This follows since $\mathbb{R}^{1,4}$ is the smallest space containing $\mathbb{R}^{1,0,3}$ as an orthogonal subspace. Low-dimensional complex representations of Spin (1,0,3) can be obtained thereby [Brooke (1980b)].

The extended Galilei group for a spinning particle of mass $m > 0$, Spin $(G_{11}(m))$, can be realized as a subgroup of Spin (1,1,4), whereas perhaps surprisingly the ordinary inhomogeneous Galilei group Spin (G_{10}) is not a subgroup of Spin (1,1,4). From this it follows that Spin (2,5) is the smallest "classical" spin group containing Spin $(G_{11}(m))$ as a subgroup. Further, one is led to the obscure 6×6 representation of $G_{11}(m)$ found in Lévy-Leblond (1974) and called by Duval, Burdet, Künzle, Perrin (1985) the Bargmann representation [see Brooke (1980b)].

3.2 Contractions of Clifford Algebras and Spin Groups

A notion of Clifford algebra contraction, analogous to Inönü-Wigner-Saletan contraction of Lie algebras has been introduced [Brooke (1978, 1980b)] and can be seen to induce the usual Inönü-Wigner contraction of the Lorentz group to the homogeneous Galilei group. A very similar notion of contraction plays a prominent role in Abłamowicz (1985b).

Whether every Lie algebra contraction of kinematical groups has an underlying Clifford algebra contraction remains to be explored.

3.3 Galilei-Covariant Spin - 1/2 Wave Equations

The apparatus of Galilei spin group and Clifford algebra permits an intrinsic and manifestly covariant treatment of spin - 1/2 wave equations [Brooke (1980b)], first discussed in Lévy-Leblond (1967).

Very recently the situation has been extended by Künzle, Duval (1984) to general curved Newtonian space-times by making heavy use of the Galilei spin group formalism and particularly the embedding of Spin $(G_{11}(m))$ in Spin (1,1,4). In this regard and related to Newtonian gravity see also Duval et al. (1985).

Finally, from the standpoint of irreducible unitary representations of the de Sitter group Spin (1,4), the covariant spin - 1/2 wave equation serves to define an irreducible Hilbert space of wave functions on $\mathbb{R}^3$ with values in $\mathbb{C}^4$ which figures in the manifestly covariant realizations of Moylan (1984).

4. REFERENCES

Abłamowicz, R. (1985a): 'Structure of spin group associated with degenerate Clifford algebra', Gannon University, preprint.

Abłamowicz, R. (1985b): 'Deformation and contraction in Clifford algebras', Gannon University, preprint.

Abłamowicz, R., P. Lounesto (1986): 'Idempotents in degenerate Clifford algebras and Pfaffian chart', these proceedings.

Atiyah, M., R. Bott, A. Shapiro (1964): 'Clifford modules', Topology **3**, Suppl. 1, 3-38.

Brooke, J.A. (1978): 'A Galileian formulation of spin I: Clifford algebras and spin groups', J. Math. Phys. **19**, 952-959.

Brooke, J.A. (1980a): 'A Galileian formulation of spin II: Explicit realizations', J. Math. Phys, **21**, 617-621.

Brooke, J.A. (1980b): Clifford Algebras, Spin Groups and Galilei Invariance - New Persectives, thesis, U. of Alberta, unpublished.

Chevalley, C. (1954): Algebraic Theory of Spinors, Columbia University Press, New York.

Clifford, W.K. (1873): 'Preliminary sketch of biquaternions', Proc. Lond. Math. Soc. 4, 381-395.

Crumeyrolle, A. (1980): 'Algèbres de Clifford dégénérées et revêtements des groupes conformes affines orthogonaux et symplectiques', Ann. Inst. H. Poincaré, **33A**, 235-249.

Duval, C., G. Burdet, H.P. Künzle, M. Perrin (1985): 'Bargmann structures and Newton-Cartan theory', Phys. Rev. **D31**, 1841-1853.

Helmstetter, J. (1977): 'Groupe de Clifford pour des formes quadratiques de rang quelconque', C.R. Acad. Sc. Paris, **285A**, 175-177.

Künzle, H.P., C. Duval (1984): 'Diracfield on Newtonian space-time', Ann. Inst. H. Poincaré, **41A**, 363-384.

Lévy-Leblond, J.-M. (1967): 'Nonrelativistic particles and wave mechanics', Commun. Math. Phys., **6**, 286-311.

Lévy-Leblond, J.-M. (1974): 'The pedagogical role and epistemological significance of group theory in quantum mechanics', Riv. Nuovo Cimento, **4**, 99-143.

Moylan, P. (1984): 'Invariant equations and manifestly covariant group representations for SO(4, 1)', Max-Planck-Institut/Werner-Heisenberg-Institut, preprint.

Porteous, I. (1981): Topological Geometry, Cambridge University Press, Cambridge.

Study, E. (1890): 'Über systeme complexen Zahlen and ihre Anwendung in der Theorie der Transformationsgruppen', Monatshefte für Math. u. Physik **1**, 283-354.

Study, E. (1891): 'Von den Bewegungen and Umlegungen', Math. Annalen **39**, 441-566.

ALGEBRES DE CLIFFORD SEPARABLES II

Artibano MICALI et Cristian MALLOL
Université de Montpellier II
Institut de Mathématiques
Place Eugène Bataillon
34060 Montpellier
France

ABSTRACT. The aim of this note is to give a characterization of associative algebras with unit that can be written as quotients of Clifford algebras of free quadratic modules. In particular, we are able to say when an associative algebra with unit is the Clifford algebra of a free quadratic module.

1. INTRODUCTION

Cette note est, dans une certaine mesure, la suite de [1]. Nous donnons ici une caractérisation des algèbres qui s'écrivent comme quotients d'algèbres de Clifford de modules quadratiques libres et, en particulier, nous caractérisons les algèbres de Clifford elles-mêmes.

Soient K un anneau commutatif à élément unité, P un K-module, $f : P \to K$ une forme quadratique sur P et $\varphi : P \times P \to K$, $(x,y) \mapsto f(x+y)-f(x)-f(y)$ la forme K-bilinéaire symétrique associée à f. On dira que f est <u>non dégénérée</u> si l'application K-linéaire $P \to P^*$ définie par $x \mapsto (y \mapsto \varphi(x,y))$ est un isomorphisme de K-modules, où $P^* = \mathrm{Hom}_K(P,K)$ est le K-module dual de P. Pour les notions et résultats concernant les formes quadratiques utilisées dans cette note, on renvoie à [2], chapitre 1. De plus, si (P,f) est un K-<u>module quadratique</u>, i.e., un K-module P muni d'une forme quadratique $f : P \to K$, on notera $C(P,f)$ son <u>algèbre de Clifford</u>. Pour les propriétés concernant les algèbres de Clifford, on renvoie à [2], chapitre 2.

2. UNE CARACTERISATION DES ALGEBRES DE CLIFFORD

Nous commençons par donner une description des algèbres qui s'écrivent comme quotients d'algèbres de Clifford, notre caractérisation découlant de cette description.

<u>THEOREME</u> 2.1. <u>Soient</u> K <u>un anneau commutatif à élément unité et</u> A <u>une</u> K-<u>algèbre associative à élément unité</u>. <u>Les conditions suivantes sont équivalentes</u> :

(i) <u>L'algèbre</u> A <u>est isomorphe au quotient d'une algèbre de Clifford</u>

J. S. R. Chisholm and A. K. Common (eds.), Clifford Algebras and Their Applications in Mathematical Physics, 103–107.

$C(P,f)$ par un idéal bilatère J, où P est un K-module libre de rang n et $f : P \to K$ est une forme quadratique non dégénérée.
(ii) Il existe des applications K-linéaires $u_i : A \to A$ $(i = 1,\dots,n)$ vérifiant les conditions suivantes : (1) $u_i(xy) = u_i(x)y$ $(i = 1,\dots,n)$ quels que soient x, y dans A ; (2) $u_i u_j + u_j u_i = b_{ij}\, id_A$ $(i,j = 1,\dots,n\ ;\ i \neq j)$, où les b_{ij} sont dans K et $b_{ij} = b_{ji}$ $(i,j = 1,\dots,n\ ;\ i \neq j)$; (3) $u_i^2 = a_i\, id_A$ $(i = 1,\dots,n)$ où les a_i sont dans K ; (4) si l'on pose $b_{ii} = 2a_i$ $(i = 1,\dots,n)$, le déterminant de la matrice (b_{ij}) est inversible dans K ; (5) l'ensemble $\{1, u_1(1),\dots,u_n(1)\}$ est un système de générateurs de A en tant que K-algèbre.

En effet, supposons que la condition (i) soit vérifiée, notons $\{e_1,\dots,e_n\}$ une base du K-module libre P et définissons des applications K-linéaires $u_i : C(P,f) \to C(P,f)$ $(i = 1,\dots,n)$ par $u_i(x) = e_i x$ $(i = 1,\dots,n)$, pour tout x dans $C(P,f)$. Si l'on pose $b_{ij} = \varphi(e_i,e_j)$ $(i,j = 1,\dots,n\ ;\ i \neq j)$ et $a_i = f(e_i)$ $(i = 1,\dots,n)$, il est facile de voir que les applications $u_1,\dots,u_n$ vérifient les conditions (1) à (5) de (ii) pour l'algèbre $C(P,f)$, où $\varphi : P\times P \to K$ est la forme K-bilinéaire symétrique associée à f. Comme J est un idéal bilatère de $C(P,f)$, on a $u_i(J) \subset J$ $(i = 1,\dots,n)$ donc par passage aux quotients, il existe des applications K-linéaires notées encore $u_i : A \to A$ $(i = 1,\dots,n)$ rendant commutatif les diagrammes

$$\begin{array}{ccc} C(P,f) & \xrightarrow{\ u_i\ } & C(P,f) \\ \downarrow & & \downarrow \\ A & \xrightarrow{\ u_i\ } & A \end{array} \qquad (i = 1,\dots,n),$$

les flèches verticales étant les surjections canoniques. Et il est évident que ces nouvelles applications K-linéaires $u_i : A \to A$ $(i = 1,\dots,n)$ vérifient les conditions (1) à (5) de (ii).

Réciproquement, supposons que la condition (ii) soit vérifiée et soient $e_i = u_i(1)$ $(i = 1,\dots,n)$ et P le K-module engendré par $\{e_1,\dots,e_n\}$. Alors P est un K-module libre de base $\{e_1,\dots,e_n\}$ car si $\sum_{i=1}^n c_i e_i = 0$ où les c_i sont dans K, on a $\sum_{i=1}^n c_i(e_i e_j + e_j e_i) = 0$ $(j = 1,\dots,n)$ d'où $\sum_{i=1}^n b_{ij} c_i = 0$ $(j = 1,\dots,n)$. Etant donné que $\det(b_{ij})$ est inversible dans K, on déduit que $c_i = 0$ $(i = 1,\dots,n)$. On peut maintenant définir une forme quadratique $f : P \to K$ en posant, pour tout élément $\sum_{i=1}^n c_i e_i$ de P, $f(\sum_{i=1}^n c_i e_i) = \sum_{i=1}^n c_i^2 a_i +$

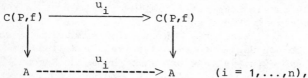

$+ \sum_{\{i,j\}} c_i c_j b_{ij}$, où $\sum_{\{i,j\}}$ désigne la somme étendue aux sous-ensembles à deux éléments de l'ensemble $\{1,\ldots,n\}$. En particulier, $f(e_i) = a_i$ $(i = 1,\ldots,n)$ et si $\varphi : P \times P \to K$ est la forme K-bilinéaire symétrique associée à f, alors $\varphi(e_i,e_j) = b_{ij}$ $(i,j = 1,\ldots,n\ ;\ i \neq j)$. De plus, la matrice de f par rapport à la base $\{e_1,\ldots,e_n\}$ étant (b_{ij}), il résulte que f est une forme quadratique non dégénérée. Comme $x^2 = f(x)$ pour tout x dans P, l'injection canonique $P \to A$ se prolonge en un unique morphisme de K-algèbre $C(P,f) \to A$ rendant commutatif le diagramme

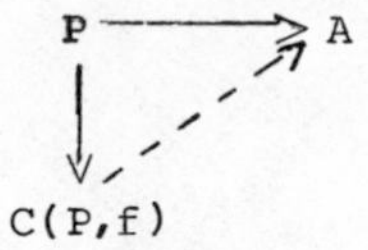

où la flèche verticale est canonique. Puisque les algèbres A et $C(P,f)$ ont le même système de générateurs, la flèche $C(P,f) \to A$ est surjective, d'où le résultat voulu.

COROLLAIRE 2.2. *Soient* K *un anneau commutatif à élément unité et* A *une* K-*algèbre associative à élément unité vérifiant les conditions équivalentes du théorème* 2.1. *Si* n *est pair, il existe une extension* $K \to K'$ *telle que* A *est une* K'-*algèbre de Clifford et si* n *est impair, il existe une extension* $K \to K'$ *telle que* A *est la composante homogène de degré zéro d'une* K'-*algèbre de Clifford*.

En effet, si n est pair, $C(P,f)$ est une K-algèbre centrale séparable (cf. [2], théorème 2.6.7) et on sait alors que tout idéal bilatère de $C(P,f)$ est de la forme $IC(P,f)$ où I est un idéal de K donc $A \approx C(P,f)/IC(P,f) \approx C(P,f) \otimes_K K' \approx C(P \otimes_K K', f \otimes_K K')$ où $K' = K/I$. Supposons maintenant que n soit impair. On sait alors que $C(P,f)$ est une K-algèbre séparable et son centre L est une extension quadratique séparable de K et, de plus, $C(P,f) \approx C_o(P,f) \otimes_K L$, isomorphisme de K-algèbres (cf. [2], théorème 2.6.9). Comme $C(P,f)$ est une L-algèbre centrale séparable, il existe un idéal I de L tel que $J = IC(P,f)$ d'où $A \approx C(P,f)/IC(P,f) \approx C(P,f) \otimes_L (L/I) \approx (C_o(P,f) \otimes_K L) \otimes_L (L/I) \approx$ $\approx C_o(P,f) \otimes_K K' \approx C_o(P \otimes_K K', f \otimes_K K')$, où $K' = L/I$. Ceci achève la démonstration du corollaire.

Si l'on ajoute, à la condition (ii) du théorème 2.1. le fait que $\{1,u_1(1),\ldots,u_n(1)\}$ soit un système libre de générateurs de A, on obtient le résultat suivant :

COROLLAIRE 2.3. *Soient* K *un anneau commutatif à élément unité et* A *une* K-*algèbre associative à élément unité. Les conditions suivantes sont équivalentes* : (i) *Il existe un* K-*module quadratique* (P,f), *où* P *est un* K-*module libre de rang* n *et* $f : P \to K$ *est une forme quadratique* non dégénérée, tel que $A \approx C(P,f)$, isomorphisme de K-algèbres. (ii) Il existe des applications K-linéaires $u_i : A \to A$ $(i = 1,\dots,n)$ vérifiant les conditions (1), (2), (3) et (4) du théorème 2.1. ainsi que la condition suivante : (5) l'ensemble $\{1,u_1(1),\dots,u_n(1)\}$ est un système libre de générateurs de A en tant que K-algèbre.

EXEMPLE 2.4. Soient K un corps commutatif de caractéristique différente de 2 et A la K-algèbre des quaternions dont la table de multiplication relativement à une base $\{1,e_1,e_2,e_1e_2\}$ s'écrit :

	1	e_1	e_2	e_1e_2
1	1	e_1	e_2	e_1e_2
e_1	e_1	a_1	e_1e_2	a_1e_2
e_2	e_2	$-e_1e_2$	a_2	$-a_2e_1$
e_1e_2	e_1e_2	$-a_1e_2$	a_2a_1	$-a_1a_2$

Notons P le K-module libre de base $\{e_1,e_2,e_3\}$ et $f : P \to K$ la forme quadratique définie par $f(e_1) = a_1$, $f(e_2) = a_2$, $f(e_3) = -a_1a_2$, $\varphi(e_1,e_2) = 0$, $\varphi(e_1,e_3) = 0$ et $\varphi(e_2,e_3) = 0$, $\varphi : P\times P \to K$ étant la forme K-bilinéaire symétrique associée à f. Le noyau du morphisme surjectif de K-algèbres $\psi : C(P,f) \to A$ est l'idéal bilatère de $C(P,f)$ engendré par les vecteurs $e_1e_2-e_3$, $e_1e_3-a_1e_2$, $e_2e_3+a_2e_1$ et $e_1e_2e_3+$ $+a_1a_2$. Etant donné que $(e_1e_2e_3+a_1a_2)e_3 = -a_1a_2(e_1e_2-e_3)$, $(e_1e_2e_3+a_1a_2)e_2 = -a_2(e_1e_3-a_1e_2)$ et que $(e_1e_2e_3+a_1a_2)e_1 = a_1(e_2e_3+$ $+a_2e_1)$, si l'on suppose que f soit non dégénérée, i.e., que $a_1 \neq 0$ et $a_2 \neq 0$, on conclut que $J = \mathrm{Ker}(\psi)$ est l'idéal bilatère de $C(P,f)$ engendré par le vecteur $e_1e_2e_3+ a_1a_2$ et il est clair que $A \approx C(P,f)/J$, isomorphisme de K-algèbres. Donc, si nous désignons par L le centre de l'algèbre de Clifford $C(P,f)$, il existe un idéal I de L tel que $J = IC(P,f)$ d'où, d'après le corollaire 2.2., $A \approx C_o(P,f) \otimes_K K' \approx$ $\approx C_o(P \otimes_K K', f \otimes_K K')$ avec $K' = L/I$. Notons que, dans notre exemple, l'algèbre des quaternions A est déjà au départ, une algèbre de Clifford.

EXEMPLE 2.5. Nous venons de voir qu'il faut des conditions très restrictives pour que une algèbre associative à élément unité soit une algèbre de Clifford (cf. théorème 2.1). En effet, même dans des cas simples comme, par exemple, le cas des extensions quadratiques, il n'est pas toujours vrai qu'une extension quadratique séparable A d'un anneau commutatif à élément unité K soit une algèbre de Clifford. En effet, supposons que $A = K[x]$, avec $x^2 = \alpha x + \beta$ soit une extension quadratique libre de K, où α, β sont dans K et supposons qu'il existe un isomorphisme de K-algèbres $\psi : C(P,f) \cong A$, où P est un K-module libre de base $\{e\}$ et $f : P \to K$ est une forme quadratique sur P. Etant donnée que $\psi(1) = 1$ et que $\psi(e) = \lambda x + \mu$ avec λ, μ dans K et λ inversible dans K, alors $e^2 = f(e)$ entraîne $f(e) = \psi(e)^2 = (\lambda x + \mu)^2$ d'où $\lambda^2\alpha + 2\lambda\mu = 0$ soit $\lambda\alpha + 2\mu = 0$ et $\lambda^2\beta + \mu^2 = f(e)$. L'équation $(\alpha^2 + 4\beta)\lambda^2 = 4f(e)$ nous montre que si A est une K-algèbre séparable isomorphe à $C(P,f)$, alors $f : P \to K$ est une forme quadratique non dégénérée et 2 est inversible dans K. Il est alors facile de donner des exemples d'algèbres associatives qui ne sont pas des algèbres de Clifford.

References

[1] A. MICALI et Ph. REVOY, 'Algèbres de Clifford séparables', Publication n° 64, Secrétariat des Mathématiques, Université de Montpellier, Montpellier 1970, 18 p. ; Math. Rev. 46, 1973, # 198 b.

[2] A. MICALI et Ph. REVOY, 'Modules quadratiques', Bull. Soc. Math. France, Mémoire n° 63, 1979, 144 p.

SUR UNE QUESTION DE MICALI-VILLAMAYOR

Gérard KIENTEGA
Institut de Mathématiques
et de Physique
UNIVERSITE DE OUAGADOUGOU
B.P. 7021 - OUAGADOUGOU
BURKINA FASO.

ABSTRACT : Let K be a field characteristic $\neq 2$, Br(K) its Brauer group and ${}_2Br(K) = \{x \in Br(K) \mid 2x = 0\}$. A generating set of ${}_2Br(K)$ using quaternion division algebras is given. When $K = \mathbb{Q}$, we compute ${}_2Br(\mathbb{Q})$ by elementary methods avoiding class field theory. From these computations a set of generators of ${}_2Br(\mathbb{Q})$ is deduced sharpening in this particular case the general result found in the first part of this work.

RESUME : Soit K un corps de caractéristique différente de 2, Br(K) son groupe de Brauer et ${}_2Br(K) = \{x \in Br(K) \mid 2x = 0\}$. Nous déterminons un système générateur de ${}_2Br(K)$ formé de corps de quaternions. Lorsque $K=\mathbb{Q}$, nous calculons ${}_2Br(\mathbb{Q})$ par une méthode élémentaire qui n'utilise pas la théorie du corps de classes. De ce calcul nous déduisons un système générateur de ${}_2Br(\mathbb{Q})$ qui affine dans ce cas particulier le résultat général trouvé au début de ce travail.

On se propose de calculer le sous-groupe du groupe de Brauer de $\mathbb{Q}$ formé des éléments dont l'ordre divise 2. Ce résultat sera possible grâce à l'utilisation de méthodes empruntées à la théorie des nombres et aux idées de [4] . On confirmera ainsi l'assertion de [2] disant : "D'autre part, le calcul de $W(\mathbb{Q})$ fait par W. Scharlau et le théorème 4.1 nous permettent de calculer $B_2(\mathbb{Q})$".

Les méthodes de théorie des nombres permettent de retrouver certains résultats de [4] et d'étudier d'autres exemples.

J. S. R. Chisholm and A. K. Common (eds.), Clifford Algebras and Their Applications in Mathematical Physics, 109–114.

1. Préliminaires. Soient K un corps de caractéristique différente de 2, L une extension galoisienne de K et $G = \mathrm{Gal}(L/K)$ le groupe de Galois de L sur K. Si p est un nombre premier et Br(K) le groupe de Brauer de K, ${}_p\mathrm{Br}(K)$ désigne le noyau du morphisme de multiplication par p : $\mathrm{Br}(K) \to \mathrm{Br}(K)$ et $\mathrm{Br}_p(K)$ désigne la partie p-primaire de Br(K). Rappelons que Br(L/K) est le noyau de $\mathrm{Br}(K) \to \mathrm{Br}(L)$ obtenu par extension des scalaires. Notons N l'application norme de L dans K. Le lemme suivant est bien connu :

Lemme 1.1 Si G est un groupe cyclique on a $\mathrm{Br}(L/K) \simeq K^*/N(L^*)$.

Nous nous proposons de décrire un ensemble de générateurs de ${}_2\mathrm{Br}(K)$ que nous noterons Quat(K). On sait, d'après Merkurjev (cf. [3]), que ${}_2\mathrm{Br}(K)$ est engendré par les classes de corps de quaternions lorsque la caractéristique de K est différente de 2. Nous notons Quat(K) l'ensemble de ces classes.

Lemme 1.2. Si K est un corps de caractéristique différente de 2, on a

$$\mathrm{Quat}(K) = \bigcup_{d \in K^* \setminus K^{*2}} \mathrm{Ker}'(\mathrm{Br}(K) \to \mathrm{Br}(K(\sqrt{d})))$$

où Ker' indique qu'on exclue l'élément neutre.

En effet, soit $A \in \mathrm{Quat}(K)$. Alors A est la classe d'un corps de quaternions D engendré par deux éléments x et y satisfaisant aux relations $x^2 = \alpha$, $y^2 = \beta$, $xy = -yx$, où α et β sont des éléments non nuls de K. En effet, si par exemple α est un carré, soit $\sqrt{\alpha}$ une détermination quelconque de la racine carrée de α, alors les quatre matrices

$$\begin{pmatrix} 1 & 0 \\ 0 & 1 \end{pmatrix} \quad \begin{pmatrix} \sqrt{\alpha} & 0 \\ 0 & -\sqrt{\alpha} \end{pmatrix} \quad \begin{pmatrix} 0 & \sqrt{x} \\ \sqrt{x} & 0 \end{pmatrix} \quad \begin{pmatrix} 0 & -\alpha \\ \alpha & 0 \end{pmatrix}$$

nous donnent l'isomorphisme $D \simeq M_2(K)$. On peut donc supposer que α et β ne soient pas des carrés de K^*. Alors $K(\sqrt{\alpha})$ est un corps neutralisant pour D. En effet, $D \otimes_K K(\sqrt{\alpha})$ possède la base suivante : $1 \otimes 1$, $x \otimes 1$, $y \otimes 1$, $xy \otimes 1$. Si l'on pose $u = x \otimes 1$, $v = y \otimes 1$, il vient que $uv = -vu$. Comme x est un carré de $K(\sqrt{\alpha})$ le calcul ci-dessus montre que $D \otimes_K K(\sqrt{\alpha})$ est isomorphe à $M_2(K(\sqrt{\alpha}))$. Réciproquement soit A un élément non nul de

$\mathrm{Ker}(\mathrm{Br}(K) \longrightarrow \mathrm{Br}(K(\sqrt{d})))$ où $d \in K^* \setminus K^{*2}$. Alors $K(\sqrt{d})$ est un sous-corps commutatif maximal d'une algèbre D semblable à A et D est alors de dimension 4 sur K : c'est une algèbre de quaternions (cf.[1]).

2. Groupe de Brauer relatif d'un corps quadratique. On suppose à présent, que $K = \mathbb{Q}$ et $L = \mathbb{Q}(\sqrt{d})$, où d est égal à 2 ou à -1 ou un nombre premier congru à 1 modulo 8 ou de la forme -q ou q est un nombre premier congru à -1 modulo 8. Pour a et b dans $\mathbb{Q}^*$ on note $(a,b)_v$ le symbole de Hilbert de a et b calculé dans le corps local $\mathbb{Q}_v$. Soit P l'ensemble des nombres premiers. Notons P^- (resp. P^+) les éléments de P tels que $(d,p)_p = -1$ (resp. $(d,p)_p = +1$).

Théorème 2.1. Si $d < 0$, $\mathrm{Ker}(\mathrm{Br}(\mathbb{Q}) \to \mathrm{Br}(\mathbb{Q}(\sqrt{d}))$ est isomorphe à $\mathbb{Z}/(2) \oplus (\bigoplus_{p \in P^-} B_p)$ et si $d > 0$,

$\mathrm{Ker}(\mathrm{Br}(\mathbb{Q}) \to \mathrm{Br}(\mathbb{Q}(\sqrt{d})))$ est isomorphe à $\bigoplus_{p \in P^-} B_p$, où $B_p \simeq Z/(2)$ pour tout $p \in P^-$.

Pour démontrer ce théorème nous avons besoin du lemme suivant :

Lemme 2.2. Si $d < 0$, $N(\mathbb{Q}(\sqrt{d})^*)$ est le sous-groupe abélien libre du groupe multiplicatif $\mathbb{Q}^*$ engendré par les nombres premiers p tels que $(d,p)_p = +1$ et les carrés des nombres premiers p tels que $(d,p)_p = -1$. Si $d > 0$, $N(\mathbb{Q}(\sqrt{d})^*)$ est le sous-groupe abélien libre du groupe multiplicatif $\mathbb{Q}^*$ engendré par -1, les nombres premiers p tels que $(d,p)_p = +1$ et les carrés des nombres premiers p tels que $(d,p)_p = -1$.

Démontrons le lemme d'abord dans le cas où $d = q > 0$ est un nombre premier tel que $q \equiv 1 (\mathrm{mod}\ 8)$. Soit $a \in \mathbb{Q}^*$ appartenant au sous-groupe abélien libre de $\mathbb{Q}^*$ ci-dessus cité. Pour démontrer que a est une norme il suffit de montrer que -1 est une norme et que tout nombre premier p tel que $(q,p)_p = +1$ est une norme puisque d'une part les carrés sont des normes et d'autre part la norme est multiplicative. Comme $\mathbb{Q}(\sqrt{q})$ est un corps quadratique réel, il possède une unité de norme -1. Soit à présent p un nombre premier tel que $(q,p)_p = +1$. Comme $f = x^2 - qy^2$ est une forme quadratique, on peut appliquer le théorème de Hasse-Minkowski : c'est une forme de rang 2, de discriminant -q et d'invariant $\varepsilon(f) = +1$ dans chaque $\mathbb{Q}_v$. Il faut vérifier que $(q, p)_v = +1$ pour tout v (cf. [7]). On a $(q,p)_\infty = +1$ car $p > 0$, $(q,p)_v = +1$ si $v \neq 2,q,p$. Or $(q,p)_p = +1$ par hypothèse. Si $v = 2$ on a $(q,2)_2 = +1$ car $q \equiv 1 \pmod 8$ et $(q,p)_2 = +1$ si $p \neq 2$. Pour $v = q$ on a alors $(q,p)_q = +1$ grâce à la formule du produit.

Réciproquement soit a une norme. Il existe $(x,y) \in \mathbb{Q}^* \times \mathbb{Q}^*$ tel que $a = x^2 - qy^2$. On peut clairement supposer que a, x et y sont des entiers. Posons $a = u \prod_{p \in P} p^{n(p)}$ la décomposition en facteurs premiers de a, où $u = +1$ ou -1. Il suffit de montrer que lorsque $n(p)$ est impair alors $(q,p)_p = +1$. Supposons d'abord p impair. En divisant x^2 et y^2 par une puissance convenable de p on peut supposer que $x^2 \not\equiv 0 \pmod p$, $y^2 \not\equiv 0 \pmod p$. Mais on a $x^2 - qy^2 \equiv 0 \pmod p$, donc q est carré modulo p, c'est-à-dire, $\left(\frac{q}{p}\right) = +1$. En particulier $q \neq p$. Or $\left(\frac{q}{p}\right) = (q,p)_p$ pour p impair. Si, à présent, $p = 2$ alors $(q,2)_2 = +1$ car $q \equiv 1 \pmod 8$. À quelques modifications près les autres cas sont semblables.

Démontrons à présent le théorème dans le cas où $d < 0$. D'après le lemme 1.1., on a $Br(\mathbb{Q}(\sqrt{d})/\mathbb{Q}) \simeq \mathbb{Q}^*/N(\mathbb{Q}(\sqrt{d})^*)$. D'après le lemme 2.2., $\mathbb{Q}^*/N(\mathbb{Q}(\sqrt{d})^*)$ est le $\mathbb{Z}/(2)$-espace vectoriel engendré par -1 et les nombres premiers p tels que $(d,p)_p = -1$. Le résultat est donc immédiat.

Avant de pouvoir appliquer ce théorème nous avons besoin de deux résultats préliminaires que nous donnerons sans démonstration.

Lemme 2.3. *Si* L/K *est degré 2, on a* ${}_2Br(L/K) = Br(L/K)$.

Lemme 2.4. *Soient* H_1, H_2, H_3 *des groupes et* $\varphi : H_1 \times H_2 \to H_1 \times H_3$ *un isomorphisme de groupes tel que* $\varphi(H_1) = H_1$. *Alors il existe un isomorphisme de groupes de* H_2 *sur* H_3.

3. Applications. Si $d = -1$, on a $(-1,p)_p = \left(\frac{-1}{p}\right) = (-1)^{\frac{1}{2}(p-1)}$ si $p \neq 2$ et $(-1,2)_2 = 1$. Ainsi $(-1,p)_p = -1$ si et seulement si $p \equiv 3 \pmod 4$ et d'après le théorème 2.1, on a

$$\mathrm{Br}(\mathbb{Q}(i)/\mathbb{Q}) \simeq \mathbb{Z}/(2) \oplus \Big(\bigoplus_{p \equiv 3 (\mathrm{mod} 4)} \mathbb{Z}/(2)\Big).$$

Si $W(\mathbb{Q})$ désigne le groupe de Witt de $\mathbb{Q}$, d'après [6] on a

$$W(\mathbb{Q}) \simeq \mathbb{Z} \oplus \mathbb{Z}/(2) \oplus \Big(\bigoplus_{p \equiv 3 (\mathrm{mod} 4)} \mathbb{Z}/(4)\Big) \oplus \Big(\bigoplus_{p \equiv 1 (\mathrm{mod} 4)} \mathbb{Z}/(2) \times \mathbb{Z}/(2)\Big)$$

et d'autre part, d'après [2], on a

$$W(\mathbb{Q}) \simeq \mathbb{Z} \oplus G_4 \oplus G_2 \oplus {}_2\mathrm{Br}(\mathbb{Q})/{}_2\mathrm{Br}(\mathbb{Q}(i)/\mathbb{Q}), \text{ où}$$

$$G_2 \simeq \mathbb{Z}/(2) \oplus \Big(\bigoplus_{p \equiv 1 (\mathrm{mod}\ 4)} \mathbb{Z}/(2)\Big) \text{ et } G_4 \simeq \bigoplus_{p \equiv 3 (\mathrm{mod} 4)} \mathbb{Z}/(4).$$

Les éléments d'ordre infini étant dans Z et les éléments d'ordre 4 dans G_4, le lemme 2.4 nous donne

$$\mathbb{Z}/(2) \oplus \Big(\bigoplus_{p \equiv 1 (\mathrm{mod} 4)} \mathbb{Z}/(2)\Big) \oplus \big({}_2\mathrm{Br}(\mathbb{Q})/{}_2\mathrm{Br}(\mathbb{Q}(i)/\mathbb{Q})\big) \quad \text{isomorphe à}$$

$$\mathbb{Z}/(2) \oplus \Big(\bigoplus_{p \equiv 1 (\mathrm{mod} 4)} \mathbb{Z}/(2) \times \mathbb{Z}/(2)\Big).$$

Cet isomorphisme de $\mathbb{Z}/(2)$-espaces vectoriels équivaut à

$${}_2\mathrm{Br}(\mathbb{Q})/{}_2\mathrm{Br}(\mathbb{Q}(i)/\mathbb{Q}) \simeq \bigoplus_{p \equiv 1 (\mathrm{mod} 4)} Z/(2)$$

Utilisant les résultats précédents on a finalement la suite exacte scindée de $Z/(2)$-espaces vectoriels

$$0 \to \mathbb{Z}/(2) \oplus \Big(\bigoplus_{p \equiv 3 (\mathrm{mod} 4)} \mathbb{Z}/(2)\Big) \to {}_2\mathrm{Br}(\mathbb{Q}) \to \bigoplus_{p \equiv 1 (\mathrm{mod} 4)} \mathbb{Z}/(2) \to 0$$

soit $${}_2\mathrm{Br}(\mathbb{Q}) \simeq \bigoplus_{p \in P} \mathbb{Z}/(2).$$

Il n'est pas difficile de déterminer les quaternions qui engendrent ${}_2\mathrm{Br}(\mathbb{Q})$. D'abord un lemme :

Lemme 3.1. *Soit* p *un nombre premier*, $p \equiv 1 \pmod 8$, *il existe un nombre premier* q *tel que* $2 < q < p$ *et* $(\frac{p}{q}) = -1$.

Pour α et β des éléments non nuls d'un corps K de caractéristique différente de 2, notons (α, β) la classe dans Br(K) de l'algèbre de quaternions engendrée par x et y satisfaisant aux relations:

$x^2 = \alpha$, $y^2 = \beta$, $xy = -yx$. D'après la démonstration du théorème 2.1, $(-1,-1)$ et les quaternions de la forme $(-1,p)$ avec p premier, $p \equiv 3 \pmod 4$, engendrent $Br(\mathbb{Q}(i)/\mathbb{Q})$. Lorsque p est un nombre premier, $p \equiv 1 \pmod 4$, distinguons deux cas. Si $p \equiv 5 \pmod 8$ alors $(2,p)$ est un corps de quaternions. Si $p \equiv 1 \pmod 8$, le lemme 3.1 nous permet de trouver un nombre premier q, $2 < q < p$ tel que (q,p) soit un corps de quaternions. L'ensemble de ces quaternions engendre de façon évidente ${}_2Br(\mathbb{Q})$.

Exemple 3.2 . En appliquant le théorème 2.1, il n'est pas difficile de voir que $Br(\mathbb{Q}(\sqrt{2})/\mathbb{Q}) \simeq (\bigoplus_{p \equiv -5 (mod 8)} \mathbb{Z}/(2)) \oplus (\bigoplus_{p \equiv 5 (mod 8)} \mathbb{Z}/(2))$.

Note 3.3. Cas de la caractéristique 2. Dans ce cas, avec une modification convenable de la notion de quaternion (cf.[1], définition I.8), les résultats de ce papier peuvent s'étendre à la caractéristique 2.

BIBLIOGRAPHIE

[1] Blanchard A., Les corps non commutatifs, PUF, Paris 1972.

[2] Larotonda A., Micali A., Villamayor O.E.,'Sur le groupe de Witt,' Symposia Mathematica 11 (1973), 211-219.

[3] Merkurjev A.S.,'Brauer group of fields', Communications in Algebra 11 (22) (1983), 2611-2624.

[4] Micali A., Villamayor O.E.,'Algèbres de Clifford et groupe de Brauer', Ann. Sc. Ec. Normale Sup. 4 (1971) 285-310.

[5] Milnor J., Husemoller D., Symmetric bilinear forms, Springer Verlag, Berlin 1973.

[6] Scharlau W., 'Quadratic reciprocity laws', Journal of Number Theory 4 (1972), 78-97.

[7] Serre J.P., Cours d'Arithmétique, PUF, Paris 1970.

[8] Serre J.P., Corps locaux, Hermann, Paris 1968.

SPINGROUPS AND SPHERICAL MONOGENICS

R. Delanghe and F. Sommen
Seminar of Algebra and Functional Analysis
State University of Gent
Galglaan 2
B-9000 Gent, Belgium

ABSTRACT. In this paper we give a survey on some results concerning the theory of spherical harmonics and its interaction with representations of the spingroup Spin(m). This leads to the spherical monogenics as a refinement of the spherical harmonics. We also develop a scheme to construct solutions to partial differential equations with constant coefficients which are Spin(m)-invariant.

INTRODUCTION. Spherical harmonics are classically introduced as the polynomial solutions to the Laplace equation $\Delta f=0$.
The application of spherical harmonics however goes far beyond the theory of the Laplacian. Indeed, when PD(f)=0 is a linear partial differential equation which is SO(m) invariant, its solutions are to be expressed as linear super-positions of solutions of the form

$$f(\vec{x})=g(r)S_k(\omega)$$

where $r=|\vec{x}|$, $\vec{x}\in R^m$, $\vec{x}=r\omega$ and where $S_k(\omega)$ is spherical harmonic of degree k.
Typical examples are e.g. the Helmholtz operator $\Delta+k^2$, the heat operator $\Delta+\frac{\partial}{\partial t}$, the Schrödinger operator $\frac{\hbar^2}{2m}\Delta+E-V(r)$, the elasticity operator and so on. This means that spherical harmonics come from the representation of SO(m) (see [11], [16], [20], [24], [25]).
For m>2 the spaces H_k of spherical harmonics are irreducible under SO(m), whereas for m=2, H_k splits into two one-dimensional irreducible pieces. In order to extend this splitting to several dimensions, one has to consider the double covering group Spin(m) of SO(m). This group arises naturally in quantum mechanics from the description of spin

J. S. R. Chisholm and A. K. Common (eds.), Clifford Algebras and Their Applications in Mathematical Physics, 115–132.

as well as from the splitting of the Klein-Gordon operator (see [26]).
In partial differential equations there is the following motivation to consider spingroups rather than rotation groups. When one studies SO(m)-invariant differential operators with constant coefficients (Euclidean operators) one obtains for m=1 all operators

$$\sum_{(j)} a_j \left(\frac{d}{dx}\right)^j, \quad a_j \in C,$$

whereas for m>1, one only obtains operators of even order of the form $\sum_{(j)} a_j \Delta^j$. In order to make the analogy between m=1 and m>1 more complete, we have to replace the group SO(m) by Spin(m). It will be proved that we obtain as spin-invariant operators (Spin-Euclidean operators) all operators of the form $\sum_{(j)} A_j D^j$, D being the Dirac operator and $A_j = a_{0j} + e_{1\ldots m} a_{mj}$; $a_{0j}, a_{mj} \in C$.

As $D^2 = -\Delta$, all Euclidean operators are Spin-Euclidean.
In order to construct concrete solutions to Spin-Euclidean differential equations, we use so called spherical monogenics, which arise automatically in the theory of the Dirac operator (see [14],[22]). They correspond to the eigenfunctions of the Casimir operator of the representations of Spin(m).
Furthermore every spherical harmonic may be split into two spherical monogenics such that for m=1 one reobtains the splitting of H_k into its irreducible pieces.
Although the modules of spherical monogenics are themselves not irreducible, they are quite useful tools both in physics and in the theory of partial differential equations. They can be made irreducible as vector spaces in a purely algebraic way by making use of spinors (see [14],[24]).

1. SPHERICAL HARMONICS

Let $\Delta = \sum_{j=1}^{m} \frac{\partial^2}{\partial x_j^2}$ be the Laplacian in R^m and let $\Omega \subseteq R^m$ be open.

Then a function $f \in C_2(\Omega)$ satisfying the equation $\Delta f = 0$ is called harmonic in Ω.

Definition 1. A homogeneous harmonic polynomial H_k of degree k is called a spherical harmonic of degree k, i.e.

$$\Delta H_k(\vec{x})=0 \text{ and } H_k(\vec{x})=r^k H_k(\omega).$$

The restrictions $S_k(\omega)=H_k(\vec{x})|S^{m-1}$ are also called spherical harmonics. The space of spherical harmonics of degree k on S^{m-1} will be denoted by $\mathcal{H}_k$.

Spherical harmonics are used to construct an orthonormal basis for the Hilbert space $L_2(S^{m-1})$, provided with the standard inner product and norm

$$(f,g)=\frac{1}{\omega_m}\int_{S^{m-1}} \overline{f}(\omega)g(\omega)dS_\omega, \quad \|f\|_2^2=(f,f),$$

dS_ω being the Lebesgue measure on S^{m-1} and ω_m the area of S^{m-1}.

Consider the Poisson kernel

$$P(\omega,\vec{x})=\frac{1-r^2}{(1-2<\vec{x},\omega>+r^2)^{\frac{m}{2}}},$$

where $<\vec{x},\omega>=\sum_{j=1}^{m} x_j\omega_j$, and put

$$h(\vec{x})=\frac{1}{\omega_m}\int_{S^{m-1}} f(\omega)P(\omega,\vec{x})dS_\omega=Pf(\vec{x}).$$

Then h is harmonic in $\overset{\circ}{B}(0,1)$ and

$$\lim_{r\uparrow 1} h(r\omega)=f(\omega) \text{ in } L_2(S^{m-1})\text{-sense}.$$

Next, we decompose the Poisson kernel in spherical harmonics

$$P(\omega,\vec{x})=\sum_{k=1}^{\infty} r^k N(m,k)P_{k,m}(<\omega,\frac{\vec{x}}{r}>),$$

where $N(m,k)=\dim \mathcal{H}_k$ and where

$$P_{k,m}(t)=(-\frac{1}{2})^k \frac{\Gamma(\frac{m-1}{2})}{\Gamma(k+\frac{m-1}{2})}(1-t^2)^{\frac{3-m}{2}} D_t^k(1-t^2)^{k+\frac{m-3}{2}}$$

are the Legendre polynomials in m dimensions (see [11], [16],[25]). We then may decompose f in spherical harmonics by means of the formula

$$f(\omega) = \sum_{k=0}^{\infty} S_k f(\omega),$$

where

$$S_k f(\omega) = \frac{N(m,k)}{\omega_m} \int_{S^{m-1}} f(\omega') P_{k,m}(<\omega,\omega'>) dS_{\omega}.$$

Futhermore the operators $S_k : f \to S_k f$ are orthogonal projection operators on $L_2(S^{m-1})$ satisfying $1 = \sum_{k=1}^{\infty} S_k$. Hence $f \in L_2(S^{m-1})$ if and only if the associated sequence $(\|S_k f\|_2)_{k\in N}$ belongs to l_2.

Similar expansions in spherical harmonics are obtained for other spaces of functions on the sphere.

<u>Examples</u>. (i) Let $f \in E(S^{m-1}) = C_\infty(S^{m-1})$; then $f(\omega) = \lim_{r\uparrow 1} h(r\omega)$ in C_∞-sense and the sequence $(S_k f)_{k\in N}$ of spherical harmonics satisfies

$$|S_k f(\omega)| \leqslant C_s (1+k)^{-s}, \text{ for all } s \in N \quad \text{(see [20])}.$$

(ii) In the case of the space $\mathfrak{a}(S^{m-1})$ of analytic functions on S^{m-1}, $f(\omega) = \lim_{r\uparrow 1} h(r\omega)$ in $\mathfrak{a}(S^{m-1})$-sense if and only if the sequence $(S_k f)_{k\in N}$ satisfies

$$|S_k f(\omega)| \leqslant C(1-\delta)^k, \text{ for some } C>0,\ 0<\delta<1$$

(see [21],[6]). Furthermore h is extendable to a harmonic function in a neighbourhood of $\overline{B}(0,1)$.
(iii) Let $T \in E'(S^{m-1})$, $E'(S^{m-1})$ being the space of Schwartz distributions. Then we may also decompose T in spherical harmonics $T = \sum_{k=0}^{\infty} S_k T$. Furthermore the space of distributions on S^{m-1} is characterized by the estimates

$$|S_k T(\omega)| \leqslant C(1+k)^s, \text{ for some } C>0,\ s \in N$$

and $h(\vec{x})$ admits a distributional boundary value on S^{m-1} if and only if

$$|h(r\omega)| < C|1-r|^{-s}, \text{ for some } C>0,\ s \in N \quad \text{(see [15])}.$$

(iv) The largest space of boundary values of harmonic functions is the space $\mathcal{A}'(S^{m-1})$ of hyperfunctions or analytic functionals on S^{m-1}.
Every harmonic function has a hyperfunction boundary value. Furthermore the sequences $(S_k T)_{k\in\mathbb{N}}$ associated to analytic functionals are characterized by the estimates (see [6], [7]),

$$|S_k f(\omega)| \leqslant C_\varepsilon (1+\varepsilon)^k, \text{ for all } \varepsilon>0.$$

Another way of introducing spherical harmonics is by means of the representation H_0 of $SO(m)$:

$$H_0 : T \to H_0(T) f(\omega) = f(T^{-1}\omega).$$

To that end, let Δ_S be the Laplace-Beltrami operator arising in the expression of Δ in spherical coordinates :

$$\Delta = \frac{\partial^2}{\partial r^2} + \frac{m-1}{r} \frac{\partial}{\partial r} + \frac{1}{r^2}\Delta_S .$$

Then Δ_S is the Casimir operator of H_0 and hence its eigenspaces, which are the spaces $\mathcal{H}_k$ with eigenvalue $-k(k+m-2)$, are invariant under H_0.
We now have that for $m>2$, H_0 is irreducible on each space $\mathcal{H}_k$ (see [25]). For $m=2$ however,

$SO(2)=U(1)$, and H_0 is given by

$$H_0(e^{i\psi}) f(e^{i\theta}) = f(e^{i(\theta-\psi)}).$$

Moreover the Fourier series expansion

$$f(e^{i\theta}) = \sum_{k=0}^{\infty} (c_k e^{ik\theta} + d_k e^{-ik\theta})$$

corresponds to the splitting $\mathcal{H}_k = \mathcal{H}_k^+ \oplus \mathcal{H}_k^-$ of $\mathcal{H}_k$ into the irreducible subspaces $\mathcal{H}_k^{\pm} = \{c e^{\pm ik\theta} : c \in \mathbb{C}\}$.

The generalization of this splitting to several dimensions will be realised by means of so called spherical monogenics, which arise only in the representation of the double covering group Spin(m) of SO(m).

2. THE GROUP Spin(m)

For basic literature on the group Spin(m) we refer to [14], [17], [18]. First we introduce some notions concerning Clifford algebras.

The complex Clifford algebra C_m consists of elements of the form

$$a=\sum_{A\subset N} a_A e_A,\quad a_A\in C,\ N=\{1,\ldots,m\},$$

where $e_A=e_{\alpha_1}\ldots e_{\alpha_k}$ for $A=\{\alpha_1,\ldots,\alpha_k\}$

with $1\leq\alpha_1<\ldots<\alpha_k\leq m$ and $e_\phi=e_0=1$.

The multiplication in C_m is determined by the relations

$$e_ie_j+e_je_i=-2\delta_{ij};\ i,j=1,\ldots,m.$$

Moreover an involution on C_m is given by $\bar{a}=\sum_{A\subset N}\bar{a}_A\bar{e}_A$, where $\bar{a}_A$ is complex conjugation and $\bar{e}_A=\bar{e}_{\alpha_h}\ldots\bar{e}_{\alpha_1}$, $\bar{e}_j=-e_j$ for $j=1,\ldots,m$.

Notice that for $m=2\ell$, C_m is the matrix algebra $C(2^\ell)$,

whereas for $m=2\ell+1$,

$$C_m=C(2^\ell)\oplus C(2^\ell) \qquad \text{(see [17])}.$$

The space of complex (resp. real) k-vectors $\sum_{|A|=k} a_A e_A$ will be denoted by $C_{m,k}$ (resp. $R_{m,k}$).

The most natural way of defining the group Spin(m) is by means of Hamilton's theory of reflections.

Let $\vec{\varepsilon}\in S^{m-1}$; then by $R(\vec{\varepsilon})$ we denote the reflection $R(\vec{\varepsilon}):\vec{x}\to-\vec{x}_{/\!/}+\vec{x}_\perp$, where $\vec{x}_{/\!/}=\langle\vec{x},\vec{\varepsilon}\rangle\vec{\varepsilon}$ and $\vec{x}_\perp\perp\vec{\varepsilon}$. Hence, as every $T\in SO(m)$ is representable as a composition of an even number of reflections $T=R(\vec{\varepsilon}_1)\ldots R(\vec{\varepsilon}_{2k})$, we have $T\vec{x}=s\vec{x}\bar{s}$, where $s=\vec{\varepsilon}_1\ldots\vec{\varepsilon}_{2k}$. This leads to

<u>Definition 2</u>. The spingroup is given by

$$\text{Spin}(m)=\{s=\vec{\varepsilon}_1\ldots\vec{\varepsilon}_{2k}:k\in N,\vec{\varepsilon}_j\in S^{m-1}\}.$$

Notice that the inverse of $s\in\text{Spin}(m)$ is its conjugate $\bar{s}$. By Hamilton's theory we obtain the representation $\nu:\text{Spin}(m)\to SO(m)$, given by $\nu(s)\vec{x}=s\vec{x}\bar{s}$, $s\in\text{Spin}(m)$, $\vec{x}\in R^m$. Furthermore, as $\ker\nu=Z_2$, we obtain the exact sequence

$$0\to Z_2\overset{i}{\to}\text{Spin}(m)\overset{\nu}{\to}SO(m)\to 0.$$

The Lie algebra of Spin(m) is the space $R_{m,2}$ of real

bivectors, provided with the induced bracket $[v,w]=vw-wv$. Left and right invariant vectorfields on Spin(m) are given by

$$X_l(w)\varphi(s)=\lim_{h\to 0}\frac{1}{h}(\varphi(s(1+hw))-\varphi(s)),$$

$$X_r(w)\varphi(s)=\lim_{h\to 0}\frac{1}{h}(\varphi((1+hw)s)-\varphi(s)).$$

3. BASIC REPRESENTATIONS OF Spin(m)

The basic representation $\nu:\text{Spin}(m)\to SO(m)$, may be extended to a representation acting on the whole $\mathcal{C}_m$, where it may be decomposed as $\nu=\lambda\circ\rho$, where

$$\lambda: s\to\lambda(s)a=sa$$

$$\rho: s\to\rho(s)a=a\bar{s},$$

for all $a\in\mathcal{C}_m$, $s\in\text{Spin}(m)$.

From these representations,we may construct several other representations of Spin(m) on spaces of functions defined on subsets of $\mathbb{R}^m$, with values in $\mathcal{C}_m$, namely

$$\begin{aligned}\mathcal{H}_0&: s\to\mathcal{H}_0(s)f(\vec{x})=f(\bar{s}\vec{x}s)\\ L&: s\to L(s)f(\vec{x})=sf(\bar{s}\vec{x}s)\\ \mathcal{H}_1&: s\to\mathcal{H}_1(s)f(\vec{x})=sf(\bar{s}\vec{x}s)\bar{s}\end{aligned}$$

Notice that $\mathcal{H}_0$ corresponds to the representation H_0 introduced before, whereas $\mathcal{H}_1$ may also be regarded as a representation of SO(m), by considering $\mathcal{C}_{m,k}$-valued functions. Both $\mathcal{H}_0$ and L are unitary representations on the right Hilbert module $L_2(S^{m-1},C_m)$ of C_m-valued L_2-functions on S^{m-1}, provided with the inner product

$$(f,g)=\frac{1}{\omega_m}\int_{S^{m-1}}\bar{f}g\,dS_\omega.$$

Let R(s) be any representation of Spin(m). Then the infinitesimal representation associated to R(s) is constructed as follows. First we construct the exterior derivative dR(s), given by $dR(s)(X)=\overline{XR(s)}$.
For the left and right invariant vector fields we have

$$dR(s)(X_\ell(w))=\lim_{h\to 0}\frac{1}{h}(R(s(1+hw))-R(s))$$

$$dR(s)(X_r(w))=\lim_{h\to 0} \frac{1}{h}(R((1+hw)s)-R(s)),$$

for all $w\in R_{m,2}$. Hence it is natural to define the infinitesimal representation by

$$dR(w)=dR(X_{\{\ell_r\}}(w))(1)=\lim_{h\to 0} \frac{1}{h}(R(1+hw)-1)$$

A straightforward computation yields

$$d\lambda(e_{ij})a=e_{ij}a,$$

$$d\nu(e_{ij})a=[e_{ij},a],\ a\in C_m,$$

$$dH_0(e_{ij})f(\vec{x})=-2L_{ij}f(\vec{x}),$$

$$dL(e_{ij})f(\vec{x})=(-2L_{ij}+e_{ij})f(\vec{x}),$$

where $L_{ij}=x_i \frac{\partial}{\partial x_j}-x_j\frac{\partial}{\partial x_i}$.

In order to construct the Casimir operator of a representation R of Spin(m), we start from the bilinear form on $R_{m,2}$,

$$B(w,w')=4[\overline{w}w']_0 ,$$

$[a]_0$ being the scalar part of $a\in C_m$. It is clear that B is a non-degenerate, symmetric, associative bilinear form on the Lie algebra $R_{m,2}$, for which

$\{\frac{1}{2}e_{ij}:i<j\}$ is an orthonormal basis (see [12],[17]). Hence for a representation R of Spin(m), the Casimir operator is given by

$$C_{B,R}=\sum_{i<j} (dR(\tfrac{1}{2}e_{ij}))^2$$

and is invariant under R, whence it commutes with the infinitesimal operators dR(w).
It is easy to see that the Casimir operators of $\lambda,\mathcal{H}_0$

and L are given by

$$C(\lambda)=\frac{1}{4}\sum_{i<j} e_{ij}^2=-\frac{1}{4}\binom{m}{2},$$

$$C(H_0)=\sum_{i<j} L_{ij}^2=\Delta_S,$$

$$C(L)=\sum_{i<j}(L_{ij}-\frac{1}{2}e_{ij})^2=\Delta_S+\Gamma-\frac{1}{4}\binom{m}{2},$$

where Δ_S is the Laplace-Beltrami operator and

$\Gamma=-\sum_{i<j} e_{ij}L_{ij}$ is a Dirac type operator on S^{m-1} (see [14], [22]).

In physics Δ_S, $-\frac{1}{4}\binom{m}{2}$ and Γ correspond respectively to "total orbital angular momentum", "total spin" and "spin-orbit coupling" whereas C(L) corresponds to "total angular momentum" of a spin-½ particle.

$C(H_0)$ and C(L) are unbounded, densely defined Hermitean operators on $L_2(S^{m-1},C_m)$.

The eigenfunctions of $C(H_0)$ are the spherical harmonics, with eigenvalues $-k(k+m-2)$.
As $C(H_0)$ and C(L) commute , we may consider the simultaneous eigenfunctions, which leads to one way to introduce spherical monogenics.

4. THE LAPLACE-BELTRAMI OPERATOR ON Spin(m)

We now express the Casimir operator of a representation R of Spin(m) in terms of the Laplace-Beltrami operator on Spin(m), which is classically given by (see e.g.[8])

$$\Delta_{Spin}=\sum_{i<j} X_r(e_{ij})^2.$$

In fact the procedure is the same for all representations of Spin(m), so that we restrict ourselves to the representation L.
First we consider

$$f(\vec{x},s)=sf(\bar{s}\vec{x}s)=L(s)f(\vec{x}),$$

as a function depending on $(\vec{x},s)\in\Omega\times Spin(m)$.

Then we have , ε being infinitesimal,

$$L(1+\varepsilon w)f(\vec{x},s)=(1+\varepsilon w)f((1-\varepsilon w)\vec{x}(1+\varepsilon w),s)$$
$$=f(\vec{x},(1+\varepsilon w)s),$$

so that

$$dL(w)f(\vec{x},s)=X_r(w)f(\vec{x},s).$$

Hence we obtain

$$C(L)f(\vec{x},s)=\frac{1}{4}\sum_{i<j}X_r(e_{ij})^2f(\vec{x},s)$$
$$=\frac{1}{4}\Delta_{spin}f(\vec{x},s).$$

We also can express the spherical Dirac operator Γ in terms of a Dirac type operator on Spin(m) as follows. First notice that

$$\frac{1}{2}\sum_{i<j}e_{ij}dL(e_{ij})=\Gamma-\frac{1}{2}\binom{m}{2}$$

and that $dL(e_{ij})f(\vec{x},s)=X_r(e_{ij})f(\vec{x},s)$.
Hence we have

$$\Gamma f(\vec{x},s)=\frac{1}{2}(D_{spin}+\binom{m}{2})f(\vec{x},s),$$

where $D_{spin}=\sum_{i<j}e_{ij}X_r(e_{ij})$ is a Dirac type operator on Spin(m).
As Δ_{spin} is an elliptic operator on the compact group Spin(m), we can decompose $f(\vec{x},s)$ into eigenfunctions of Δ_{spin},

$$f(\vec{x},s)=\sum_{k=0}^{\infty}f_k(\vec{x},s).$$

The decomposition of $f(\vec{x})$ into the eigenfunctions of C(L) is then simply given by

$$f(\vec{x})=\sum_{k=0}^{\infty}f_k(\vec{x},1).$$

The operator D_{spin} is more difficult to handle, since it need not be elliptic. At the identity D_{spin} behaves like the operator $D_{m,2} = \sum_{i<j} e_{ij} \frac{\partial}{\partial x_{ij}}$ on $R_{m,2}$, where in general,

$D_{m,k} = \sum_{|A|=k} e_A \frac{\partial}{\partial x_A}$ on $R_{m,k}$.

For m=4 already, we have

$$D_{4,2} = e_{12}\left(\frac{\partial}{\partial x_{12}} - \omega\frac{\partial}{\partial x_{34}}\right) + \text{cycl}, \quad \omega = e_{1234},$$

the characteristic variety being given by

$$y_{12} = \pm y_{34}, y_{23} = \pm y_{14}, y_{31} = \pm y_{24}.$$

Example. For m=3 we put $\varepsilon_1 = e_{23}$, $\varepsilon_2 = e_{31}$, $\varepsilon_3 = e_{12}$ and we then have $\mathrm{Spin}(3) = \{q_0 + \sum_1^3 q_j\varepsilon_j : q_0^2 + \sum_1^3 q_j^2 = 1\}$, which is the quaternion unit sphere.

For a function on $H\setminus\{0\}$, $X_r(\varepsilon_j)$ is defined by

$$X_r(\varepsilon_1) = \lim_{h\to 0} \frac{1}{h}\left(f((1+h\varepsilon_1)(q_0 + \textstyle\sum_1^3 q_j\varepsilon_j)) - f(q)\right)$$

$$= \left(q_0\frac{\partial}{\partial q_1} - q_1\frac{\partial}{\partial q_0} + q_2\frac{\partial}{\partial q_3} - q_3\frac{\partial}{\partial q_2}\right)f(q),$$

and so on. Hence $\Delta_{spin} = \Delta_{S^3}$ and $D_{spin} = \Gamma$, where

$$\bar{q}\sum_0^4 \varepsilon_j \frac{\partial}{\partial q_j} = |q|\frac{\partial}{\partial |q|} + \Gamma,$$

whence Γ is an elliptic operator satisfying $\Delta_{S^3} = \Gamma(2-\Gamma)$.

5. SPHERICAL MONOGENICS

We shall introduce spherical monogenics by means of the Dirac operator $D = \sum_{j=1}^m e_j \frac{\partial}{\partial x_j}$ in R^m. The solutions in $\Omega \subseteq R^m$ of Df=0 are called left monogenic functions. The system Df=0 has been studied in various contexts; from the geometric point of view we refer to [1], [4], [10] and from the function theoretic point of view to [2], [3] [5], [9], [13], [14], [19], [22], [23], [24].

In spherical coordinates,

$$D=\vec{e}_r\left(\frac{\partial}{\partial r}+\frac{1}{r}\Gamma\right),\ \vec{e}_r=\frac{\vec{x}}{r},\ r=|\vec{x}|$$

and as $D^2=-\Delta_m$ we find that

$$\Delta_S=\Gamma(m-2-\Gamma).$$

Furthermore the representation L leaves the equation $Df=0$ invariant, since D commutes with the infinitesimal representations $dL(w)$. Spherical monogenics may be introduced as follows .

Definition 3. (i) A homogeneous monogenic function of degree k, $k\in N$, $P_k(\vec{x})=r^kP_k(\omega)$, is called inner spherical monogenic of degree k.
(ii) A homogeneous monogenic function of degree $-(k+m-1)$, $k\in N$, $Q_k(\vec{x})=r^{-(k+m-1)}Q_k(\omega)$, is called outer spherical monogenic of degree k.

Notice that for $m=2$, spherical monogenics correspond to the holomorphic monomials $z\to z^k$ and $z\to z^{-(k+1)}$, $k\in N$.
From now on we consider spherical monogenics only as functions defined on the sphere S^{m-1}.
It follows immediately from the previous definition that

$$\Gamma P_k(\omega)=-kP_k(\omega)$$

$$\Gamma Q_k(\omega)=(k+m-1)Q_k(\omega)$$

so that the eigenvalues of Γ are $-k$ and $k+m-1$, $k\in N$.

As $C(L)=\Gamma(m-\Gamma-1)-\frac{1}{4}\binom{m}{2}$, the eigenvalues of $C(L)$ are $-k(k+m-1)-\frac{1}{4}\binom{m}{2}$.

Notice that the spherical monogenics are simultaneous eigenfunctions of $C(L)$ and $C(H_0)$. Conversely we have

Theorem 1. Every spherical harmonic S_k admits the canonical decomposition into spherical monogenics : $S_k=P_k+Q_{k-1}$, where

$$P_k=\frac{k-m-2-\Gamma}{2k+m-2}S_k,$$

$$Q_{k-1}=\frac{\Gamma+k}{2k+m-2}S_k.$$

Theorem 2. Every eigenfunction of $C(L)$ admits the canonical decomposition into spherical monogenics :

$E_k=P_k+Q_k$, where

$$P_k=\frac{k-m-1-\Gamma}{2k+m-1}E_k,$$

$$Q_k=\frac{\Gamma+k}{2k+m-1}E_k.$$

As C(L) is Hermitean, it admits a spectral decomposition

$$C(L)=\sum_{k=0}^{\infty}\left(-k(k+m-1)-\frac{1}{4}\binom{m}{2}\right)\Pi_k,$$

where $1=\sum_{k=0}^{\infty}\Pi_k$, Π_k being the projections onto the eigen-spaces in $L_2(S^{m-1},\mathcal{C}_m)$ of C(L).
Hence the eigenfunctions satisfy the orthogonality relations : $(E_k,E_\ell)=0$, $k\neq\ell$. As the same is true for $C(H_0)=\Delta_S$, we reobtain some basic results of [22].

Corollary. Spherical monogenics satisfy the orthogonality relations in $L_2(S^{m-1},\mathcal{C}_m)$:

$$(P_k,P_\ell)=(Q_k,Q_\ell)=0,\ k\neq\ell,$$

$$(P_k,Q_\ell)=0,\ k,\ell\in\mathbb{N}.$$

It is clear that all functions or distributions on the sphere may be decomposed into spherical monogenics. Let $f\in L_2(S^{m-1};\mathcal{C}_m)$ admit the expansion

$$f(\vec{x})=\sum_{k=0}^{\infty}(P_kf(\omega)+Q_kf(\omega));$$

then the Cauchy transform of f,which is defined by

$$\hat{f}(\vec{x})=\frac{1}{\omega_m}\int_{S^{m-1}}\frac{\vec{x}-\vec{u}}{|\vec{x}-\vec{u}|^m}\vec{u}f(\vec{u})dS_{\vec{u}},$$

is given by

$$\hat{f}(\vec{x})=\begin{cases}\sum_{k=0}^{\infty} r^kP_kf(\omega), & \text{if } \vec{x}\in\mathring{B}(0,1),\\ \sum_{k=0}^{\infty} r^{-(k+m-1)}Q_kf(\omega), & \text{if } \vec{x}\in\mathbb{R}^m\setminus\overline{B}(0,1)\end{cases}$$

Furthermore

$$f(\omega)=\lim_{r\uparrow 1}(\hat{f}(\tfrac{1}{r}\omega)-\hat{f}(r\omega))$$

in $L_2(S^{m-1};\mathcal{C}_m)$. Similar results may be established for various spaces of functions or distributions on S^{m-1}, such as $\mathcal{E}(S^{m-1}),\mathcal{E}'(S^{m-1}),\mathcal{Q}(S^{m-1}),\mathcal{Q}'(S^{m-1})$. The estimates for the expansions in spherical monogenics and the boundary value results are similar to the harmonic case.

6. SPIN-EUCLIDEAN OPERATORS

Expansions in spherical monogenics can be applied usefully to solutions of partial differential equations $P(D)f=0$ which are spin-invariant. This is the case if $P(D)$ commutes with the infinitesimal operators $dL(w)$, $w\in R_{m,2}$.

First we consider the group of Euclidean motions

$$M(m)=\{(\vec{y},T):\vec{y}\in R^m,\ T\in SO(m)\},$$

acting on functions as follows

$$H_0(\vec{y},T)f(\vec{x})=f(T^{-1}\vec{x}-\vec{y}).$$

Then an $M(m)$-invariant operator $P(D)$ will be called a Euclidean operator.
It is easy to see that, for $m=1$, the Euclidean operators are all operators of the form $\sum_{(j)} a_j(\frac{d}{dx})^j$, whereas for $m>1$, one only obtains the operators of even degrees : $\sum_{(j)} a_j\Delta^j$.

Furthermore it is natural to look for null-solutions $f(r)S_k(\omega)$, S_k being a spherical harmonic, since the equation $P(D)f=0$ then leads to a one dimensional differential equation.
However, in order to obtain also operators of odd degrees, leading to a complete analogy with the case $m=1$, it is more convenient to replace the group $M(m)$ by the larger group

$$\mathcal{M}(m)=\{(\vec{y},s):\vec{y}\in R^m,\ s\in Spin(m)\}$$

acting on functions as follows,

$$L(\vec{y},s)f(\vec{x})=sf(\bar{s}\vec{x}s-\vec{y}).$$

An operator $P(D)$ with values in $\mathcal{C}_m$ is $\mathcal{M}(m)$-invariant if $P(D)$ has constant coefficients and commutes with the operators $dL(w)$, $w\in R_{m,2}$. This leads to

Definition 4. A differential operator P(D) which is $M(m)$-invariant, is called a Spin-Euclidean operator.

Notice that all operators of the form $\sum_{(k)} A_k D^k$, D being the Dirac operator and $A_k = a_{0,k} + a_{m,k} e_{1\ldots m}$; $a_{0,k}, a_{m,k} \in C$, are Spin-Euclidean. Hence all Euclidean operators are also Spin-Euclidean.
In order to study nullsolutions of Spin-Euclidean operators, it will be useless to expand the solutions in spherical harmonics. However, as Spin-Euclidean operators commute with the Casimir operator C(L), it will be useful to expand them in eigenfunctions of C(L) and hence in spherical monogenics.
Hence it is natural to look for nullsolutions of the form

$$(A(r)+\vec{e}_r B(r))P_k(\omega),$$

where $P_k(\omega)$ is inner spherical monogenic of degree k and where A(r) and B(r) are C-valued radial functions. Indeed, as $\vec{e}_r P_k(\omega)$ is outer spherical monogenic of degree k (see [22]),

$$D((A(r)+\vec{e}_r B(r))P_k(\omega))$$
$$=[-B'(r)-\frac{k+m-1}{r}B(r)+\vec{e}_r(A'(r)-\frac{k}{r}A(r))]P_k(\omega),$$

which enables us to reduce the equation $\sum_{(j)} A_j D^j f=0$ to a one-dimensional one, by using the special form $f=(A(r)+\vec{e}_r B(r))P_k(\omega)$.

The question still remains whether all Spin-Euclidean operators are of the form $\sum_{(j)} A_j D^j$. This will be proved in the following characterization theorem.

Theorem 3. Every Spin-Euclidean operator P(D) is of the form $\sum_{(j)} A_j D^j$, with $A_j = a_{0j} + a_{mj} e_{1\ldots m}$, where $a_{0j}, a_{mj} \in C$.
Proof. Let P(D) be a differential operator with constant coefficients which commutes with the operators $L_{ij} - \frac{1}{2}e_{ij}$.

Then we use the Fourier transform of the identity

$$[L_{ij} - \frac{1}{2} e_{ij}, P(D)]\varphi = 0.$$

As $F((L_{ij}-\frac{1}{2}e_{ij})\varphi) = (L_{ij}-\frac{1}{2}e_{ij})F\varphi$, we have

$$\mathcal{F}([L_{ij}-\tfrac{1}{2}e_{ij},P(D)]\varphi)=[L_{ij}-\tfrac{1}{2}e_{ij},P(-i\vec{x})]\,\mathcal{F}\varphi\ .$$

Now

$$[L_{ij}-\tfrac{1}{2}e_{ij},P(-i\vec{x})]=L_{ij}(P(-i\vec{x}))-\tfrac{1}{2}[e_{ij},P(-i\vec{x})]\,,$$

so that, putting $Q(\vec{x})=P(-i\vec{x})$, we then have $Q(\vec{x})=sQ(\overline{s}\vec{x}s)\overline{s}$, for all $\vec{x}\in R^m\setminus\{0\}$ and $s\in \underline{\mathrm{Spin}}(m)$. Hence we also find that for all $\ell=0,\dots,m$, $sQ_\ell(\overline{s}\vec{x}s)\overline{s}=Q_\ell(\vec{x})$, Q_ℓ being the ℓ-vector part of $Q(\vec{x})$. Clearly $Q_0(\vec{x})$ and $e_{1\dots m}Q_m(\vec{x})$ are of the form $\sum_{(j)} c_j r^{2j}$, $c_j\in C$.

For $m>2$, the above relation means that for every $\vec{x}\in R^m\setminus\{0\}$ fixed, $Q_\ell(\vec{x})\in R_{m,\ell}$ is stable under the subgroup of SO(m), leaving $\vec{x}$ invariant.
Hence we can write $Q_\ell(\vec{x})=\vec{e}_r R_{\ell-1}+R_\ell$, $R_{\ell-1}\in R_{m-1,\ell-1}$, $R_\ell\in R_{m-1,\ell}$ being SO(m-1)-invariant. But then $R_{\ell-1}$ and R_ℓ have to be of the form $c_0+c_0'\vec{e}_r e_{1\dots m}$; $c_0,c_0'\in C$.

For $1<\ell<m-1$, this is only possible if $c_0=c_0'=0$, whereas for $\ell=1$, $R_\ell=0$ and $R_{\ell-1}=c_0$ and for $\ell=m-1$, $R_{\ell-1}=0$ and $R_\ell=c_0'\vec{e}_r e_{1\dots m}$.

Hence, as $\vec{x}Q_\ell(\vec{x})$ and $\vec{x}Q_{m-1}(\vec{x})$ are SO(m)-invariant polynomials, with values in $R_{m,0}$ and $R_{m,m}$ respectively,

$$Q_\ell(\vec{x})+Q_{m-1}(\vec{x})=\vec{x}\sum_{(j)}(c_{0j}+c_{mj}e_{1\dots m})r^{2j}\,,$$

for some c_{0j}, $c_{mj}\in C$. The case m=2 is left to the reader.■

ACKNOWLEDGEMENT. The second author (*) is a Senior Research Assistant, supported by the Belgian National Science Foundation.

REFERENCES.

[1] M.F. Atiyah, 'Classical groups and classical differential operators on manifolds', *Course given at Varenna, C.I.M.E.*, (1975).

[2] F. Brackx, R. Delanghe, F. Sommen, 'Clifford Analysis', *Research Notes in Math.*, 76 (Pitman, London, 1982).

[3] J. Bureš, V. Souček, 'Generalized hypercomplex analysis and its integral formulas', to appear.

[4] T. Friedrich, 'Der erste Eigenwert des Dirac-operators einer kompakten Riemannschen Mannigfaltigkeit nightnegativer Skalarkrümming', *Math. Nachr.* 97 (1980), 117-146.

[5] R. Fueter, 'Analytische Funktionen einer Quaternionenvariabelen', *Comentarii Matematici Helvetici* 4 (1932), 9-20.

[6] M. Hashizume, K. Minemura, K. Okamoto, 'Harmonic functions on Hermitean hyperbolic spaces', *Hiroshima Math. J.* 3 (1973), 81-108.

[7] S. Helgason, 'Eigenspaces of the Laplacian; Integral representations and irreducibility', *J. Functional Analysis* 17 (1974), 328-353.

[8] ———, 'Groups and Geometric Analysis; Integral Geometry, Invariant Differential Operators and Spherical Functions', *Pure and Applied Math.* (Academic Press, Orlando, London, 1984).

[9] D. Hestenes, 'Multivector Calculus', *J. Math. Anal. Appl.* 24 no. 2 (1968), 313-325.

[10] N. Hitchin, 'Harmonic Spinors', *Advances in Math.* 14 (1974), 1-55.

[11] H. Hochstadt, 'The functions of mathematical physics', *Pure and Applied Math.*, 23 (Wiley-Interscience, New York, 1971).

[12] J.E. Humphreys, 'Introduction to Lie Algebras and Representation Theory', *Graduate texts in Math.*, 9 (Springer-Verlag, New York, 1972).

[13] K. Imaeda, 'A New Formulation of Classical Electrodynamics', *Nuovo Cimento* 32 (1976), 138-162.

[14] P. Lounesto, 'Spinor Valued Regular Functions in Hypercomplex Analysis' (Thesis, Helsinki, 1979).

[15] M. Morimoto, 'Analytic functionals on the sphere and their Fourier-Borel transformations', to appear in a volume of the *Banach Center Publications*.

[16] C. Müller, 'Spherical Harmonics', *Lecture Notes in Math.* 17 (Springer, 1966).

[17] I. Porteous, 'Topological Geometry', (Van Nostrand Reinhold Cy., London, 1979).

[18] M. Riesz, 'Clifford numbers and spinors', *Lecture series* 38 (Institute for Physical Sciences and Technology, Maryland, 1958).

[19] J. Ryan, 'Complexified Clifford Analysis', *Complex Variables : Theory and Application* 1 (1982), 119-149.

[20] R.T. Seeley, 'Spherical Harmonics', *A.M.S. Monthly* 73 part II, no.4 (1966), 115-121.

[21] ———————, 'Eigenfunction expansions of analytic functions', *Proc. Amer. Math. Soc.* 21 (1969), 734-738.

[22] F. Sommen, 'Spherical Monogenic Functions and Analytic Functionals on the Unit Sphere', *Tokyo J. Math.* 4 (1981), 427-456.

[23] V. Souček, 'Complex-quaternionic Analysis applied to Spin - $\frac{1}{2}$ Massless Fields', *Complex Variables : Theory and Application* 1 (1983), 327-346.

[24] E. M. Stein, G. Weiss, 'Generalization of the Cauchy-Riemann equations and representation of the rotation group', *Amer. J. Math.* 90 (1968), 163-196.

[25] N.J. Vilenkin, 'Special functions and the theory of group representations', *Transl. Math. Monographs, Amer. Math. Soc.*, 22 (1968).

[26] H. Weyl, 'The theory of groups and quantum mechanics', (Dover Publ., New York, 1950).

LEFT REGULAR POLYNOMIALS IN EVEN DIMENSIONS, AND TENSOR PRODUCTS OF CLIFFORD ALGEBRAS

John Ryan
School of Mathematics
University of Bristol
Bristol BS8 1TW
Britain

ABSTRACT: In [4] Fueter describes a method for characterizing homogeneous polynomial solutions to the quaternionic analogue of the Cauchy-Riemann equations. In [3] Delanghe demonstrates that this method may be generalized to characterize homogeneous polynomial solutions to a generalized Cauchy-Riemann equation defined over a Clifford algebra. The function theory associated with this particular equation has been extensively pursued in recent years by a number of authors (eg [2,5,7,8,9,10]). However, the author shows in [7] that the methods used in [3,4] may be extended to characterize homogeneous polynomial solutions to homogeneous, first order, constant coefficient differential equations defined over arbitrary, finite dimensional, associative algebras with identity. For this reason it would appear desirable to find another method to characterize these solutions, to generalized Cauchy-Riemann equations over Clifford algebras, which is more closely aligned to the properties of Clifford algebras.

In this paper another approach is adopted to characterize homogeneous polynomial solutions, in even dimensions, of the equations considered by Delanghe in [3]. It is demonstrated that these polynomials form an orthogonal basis with respect to the inner product introduced, by Sommen [8] for solutions to the homogeneous Dirac equation in R^n. Moreover, the method used here to introduce the inner product makes use of a natural automorphism within the Clifford algebra and bypasses the inversion transform employed in [8]. We also show that these basis elements contain all homogeneous polynomial solutions to the split Dirac equations described by Sommen in [9]. We conclude by giving a local characterization of the solutions to these split Dirac equations in terms of canonical isomorphisms between Clifford algebras and tensor products of lower dimensional Clifford algebras.

1. PRELIMINARIES:

In this section we develop the necessary algebraic and analytic background required in this paper.

In [1] and [6, chapter 13] it is shown that from the space R^n, orthonormal basis $\{e_j\}_{j=1}^{n}$, it is possible to construct a real 2^n dimensional associative algebra A_n. This algebra has an identity $1(=e_o)$ and $R^n \subseteq A_n$, and the elements $\{e_j\}_{j=1}^{n}$ satisfy the relation

J. S. R. Chisholm and A. K. Common (eds.), Clifford Algebras and Their Applications in Mathematical Physics, 133–147.

$e_ie_j+e_je_i=-2\delta_{ij}$ where δ_{ij} is the Kroneker delta. This algebra is an example of a Clifford algebra, and it is spanned by the vectors $1, e_1, \ldots, e_n, e_1e_2, \ldots, e_{j_1} \ldots e_{j_r}, \ldots, e_1 \ldots e_n$, where $j_1 < \ldots < j_r$ and $1 \leq r \leq n$. This algebra possesses the following natural automorphism $-: A_{n,-} \to A_{n,-} : e_{j_1} \ldots e_{j_r} \to (-1)^r e_{j_r} \ldots e_{j_1}$, and for an element $Z \in A_{n,-}$ we denote $-(Z)$ by $\bar{Z}$. It may be deduced that for elements $Z_1, Z_2 \in A_{n,-}$ we have $\overline{Z_1Z_2} = \bar{Z}_2\bar{Z}_1$. By considering the identity component of $Z\bar{Z}$ it may be observed that $A_{n,-}$ is a trace algebra.

It is also shown in [1] and [6, chapter 13] that from the space R^n, with orthonormal basis $\{f_j\}_{j=1}^n$, it is possible to construct a real 2^n dimensional, associative algebra $A_{n,+}$. This algebra has an identity 1, and $R^n \subseteq A_{n,+}$, and the elements $\{f_j\}_{j=1}^n$ satisfy the relation $f_if_j+f_jf_i = 2\delta_{ij}$. This is another example of a Clifford algebra, and it is spanned by the vectors $1, f_1 \ldots, f_n, \ldots, f_{j_1} \ldots f_{j_r}, \ldots, f_1 \ldots f_n$. The two types of Clifford algebras introduced here are related to each other by the following isomorphisms:

<u>Proposition 1</u> [1] The algebra $A_{n+2,-}$ is canonically isomorphic to the algebra $A_{n,+} \otimes A_{2,-}$, and the algebra $A_{n+2,+}$ is canonically isomorphic to the algebra $A_{n,-} \otimes A_{2,+}$.

<u>Outline proof</u>: The canonical isomorphisms may be deduced by considering the maps

$$A_{n+2,-} \to A_{n,+} \otimes A_{2,-} : e_1 \to 1 \otimes e_1, e_2 \to 1 \otimes e_2 \text{ and}$$
$$e_i \to f_{i-2} \otimes e_1e_2 \text{ for } 3 \leq i \leq n+2$$

and

$$A_{n+2,+} \to A_{n,-} \otimes A_{2,+} : f_1 \to 1 \otimes f_1, f_2 \to 1 \otimes f_2 \text{ and}$$
$$f_i \to e_{i-2} \otimes f_1f_2 \text{ for } 3 \leq i \leq n+2 \quad \square$$

By induction we have:

<u>Corollary</u>: For each positive integer m the algebra $A_{n+2m,+}$ is canonically isomorphic to the algebra $A_{n,\mathrm{sgn}(-1)^m} \otimes A_{2m,-}$, and the algebra $A_{n+2m,-}$ is canonically isomorphic to the algebra $A_{n,\mathrm{sgn}(-1)} \otimes A_{2m,+}$, where $\mathrm{sgn}(-1)^m$ denotes the sign of $(-1)^m$. $\square$

Definition 1: Suppose that U is a domain in R^n and $f:U \to A_{n,-}$ is a smooth function which satisfies the equation

$$\sum_{j=1}^{n} e_j \frac{\partial}{\partial x_j} f(x) = 0$$

for all $x \in U$, then f is called an $A_{n,-}$ left regular function. If $g: U \to A_{n,+}$ is a smooth function which satisfies the equation $\sum_{j=1}^{n} f_j \frac{\partial g}{\partial x_j}(x) = 0$ for all $x \in U$, then g is called an $A_{n,+}$ left regular function.

In [3] it is shown that if $P_r: R^n \to A_{n,+}$ is an $A_{n,-}$ left regular polynomial, which is homogeneous of degree r with respect to the origin then $P_r(x) = \sum_{(\ell_1,\ldots,\ell_r)}^{n} P_{\ell_1\ldots\ell_r}(x) a_{\ell_1\ldots\ell_r}$, where $(\ell_1,\ldots,\ell_r) \varepsilon \{2,\ldots,n\}^r, P_{\ell_1\ldots\ell_r}(x) = \Sigma\frac{1}{r!}(x_{\ell_1} - e_1^{-1} e_{\ell_1} x_1)\ldots$ $\ldots(x_{\ell_r} - e_1^{-1} e_{\ell_r} x_1)$, where summation is taken over all permutations of the elements $\ell_1,\ldots,\ell_r$, and $a_{\ell_1\ldots\ell_r} \varepsilon A_{n,-}$.

It is also shown in [3] that each polynomial $P_{\ell_1\ldots\ell_r}(x)$ is an $A_{n,-}$ left regular function. By restricting the subspace spanned by the vectors $e_2,\ldots,e_n$ it may be observed that the elements of the set of $A_{n,-}$ left regular polynomials $P_{\ell_1\ldots\ell_r}$ is in one to one correspondence with the monomials, homogeneous of degree r, in the variables $x_2,\ldots,x_n$. Consequently, it may be observed that the polynomials $P_{\ell_1\ldots\ell_r}$ are linearly independant of each other. Similar results hold for homogeneous $A_{n,+}$ left regular polynomials.

In [7] we show that for A an arbitrary, real, associative algebra with an identity, then for each polynomial $P'_r:V \to A$ homogeneous of degree r, where V is the subspace of A spanned by the vectors $k_1,\ldots,k_p$ with k_1 invertible in A, and satisfying the equation $\sum_{j=1}^{p} k_j \frac{\partial P'_r}{\partial y_j}(y) = 0$, for each $y \in V$, then

$$P'_r(y) = \sum_{(\ell_1,\ldots,\ell_r)} P'_{\ell_1\ldots\ell_r}(y) b_{\ell_1\ldots\ell_r}, \qquad (1)$$

where $(\ell_1,\ldots,\ell_r)\varepsilon\{2,\ldots,p\}^r$, $P'_{\ell_1\ldots\ell_r}(y) = \Sigma\frac{1}{r!}(y_{\ell_1}-k_1^{-1}k_{\ell_1}y_1)\ldots$

$\ldots(y_{\ell_r}-k_1^{-1}k\ y_1)$, and summation is taken ovr all permutations of the elements $\ell_1,\ldots,\ell_r$ and $b_{\ell_1\ldots\ell_r}\in A$. Moreover, we have from [17] that

$$\sum_{k=1}^{p} k_j\frac{\partial P'_{\ell_1\ldots\ell_r}(y)}{\partial y} = 0 \ . \tag{2}$$

2. $A_{n,-}$ and $A_{n,-}$ LEFT REGULAR POLYNOMIALS: In view of the generality of equations (1) and (2) it would appear desirable to obtain linearly independent bases for the homogeneous $A_{n,+}$ and $A_{n,-}$ left regular polynomials, which are more closely related to the algebras $A_{n,+}$ and $A_{n,-}$. We begin by introducing such a basis for the special case $n = 4$. Here we have that

$$e_1\frac{\partial}{\partial x_1} + e_2\frac{\partial}{\partial x_2} + e_3\frac{\partial}{\partial x_3} + e_4\frac{\partial}{\partial x_4} = e_1(\frac{\partial}{\partial x_1} + e_1^{-1}e_2\frac{\partial}{\partial x_2}) + e_3(\frac{\partial}{\partial x_3}+e_3^{-1}e_4\frac{\partial}{\partial x_4}). \tag{3}$$

It may be observed from the right hand side of expression (3) that the set of $A_{n,-}$ left regular polynomials homogeneous of degree r include the polynomials $z_1^{r-p}z_2^p$, where

$$z_1 = x_1 + e_1^{-1}e_2x_2 \tag{4}$$

$$z_2 = x_3 + e_3^{-1}e_4x_4 \tag{5}$$

and $0 \leq p \leq r$. Moreover, the polynomial $z_1^{r-p}z_2^p(= z_2^pz_1^{r-p})$ satisfies the equations

$$(e_1\frac{\partial}{\partial x_1} + e_2\frac{\partial}{\partial x_2})z_1^{r-p}z_2^p = 0 = (e_3\frac{\partial}{\partial x_3} + e_4\frac{\partial}{\partial x_4})z_1^{r-p}z_2^p \ . \tag{6}$$

Equation (6) is a special case of a system of equations studied in [9]. On placing $e_1 = \sqrt{-1}f_1$ and $e_2 = \sqrt{-1}f_2$ it follows from expression (4) that the polynomial z_1^{r-p} is an $A_{2,+}$ left regular function. It now follows from the outline proof of proposition 1 that the polynomial $z_1^{r-p}z_2^p$ is the tensor product of an $A_{2,+}$ left regular polynomial with an $A_{2,-}$ left regular polynomial.

It may be deduced that for each pair of positive integers r_1 and r_2 and each pair of integers p_1 and p_2, with $0 \le p_1 \le r_1$ and $0 \le p_2 \le r_2$, the integral

$$\int_{S^3} z_1^{r_1-p_1} z_2^{p_2} \bar{z}_2^{p_2} \bar{z}_1^{r_2-p_2} \, ds^3 \qquad (7)$$

is identically equal to zero, unless $r_1 = r_2$ and $p_1 = p_2$, where S^3 is the unit sphere in R^4 and ds^3 is the Lebesgue measure on S^3. In the cases where $r_1 = r_2$ and $p_1 = p_2$ the integral (7) evaluates to a positive real number. For the polynomials $z_1^{r-p} z_2^p$ the integral (7) is identical to the inner product for $A_{n,-}$ left regular functions introduced in [8]. It follows that the polynomials $z_1^{r-p} z_2^p$ form part of an orthogonal basis for $A_{4,-}$ left regular functions with respect to this inner product. However, the polynomials $\{z_1^{r-p} z_2^p\}$ are not members of the set of linearly independent basis elements $\{P_{\ell_1 \ldots \ell_r}\}$ of $A_{4,-}$ left regular polynomials described in the previous section.

It may be observed from expressions (3), (4) and (5) that the vectors 1, $e_1^{-1}e_2$ and $e_3^{-1}e_4$ play a central role in the construction of the polynomials $\{z_1^{r-p} z_2^p\}$. These vectors are elements of the finite group $\{1, e_1e_2, -e_1e_2, e_1e_3, -e_1e_3, e_1e_4, -e_1e_4, e_2e_3, -e_2e_3, e_2e_4, -e_2e_4, e_3e_4, -e_3e_4, e_1e_2e_3e_4\}$.
This finite group is a subgroup of the spin group, Spin (4), described in [1] and elsewhere.

In view of the preceding remarks on the properties of the polynomials $\{z_1^{r-p} z_2^p\}$ it would appear desirable to construct a basis for the homogeneous $A_{4,-}$ left regular polynomials, which at the least included real scalar multiples of the polynomials $\{z_1^{r-p} z_2^p\}$, and were orthogonal to each other with respect to the inner product (7).

On placing $$P^m_{\ell,k}(x) = \sum_{r=0}^{\min(\ell,k)} (-1)^r z_1^{[m-k-\ell+r]} \bar{z}_1^{[r]} \bar{z}_2^{[\ell-r]} z_2^{[k-r]}$$

where $0 \le \ell, k \le m \in N$, and $m - \ell - k \ge 0$, and $z_1^{[r]}$ means $\frac{1}{r!} z^r$, and $P^m_{-1,k}(x) = 0$, we have

Theorem 1: For each positive integer m the set of homogeneous polynomials $Q^m_{\ell,m} = P^m_{\ell,m} - e_1^{-1}e_3P^m_{\ell-1,m}$ is a linearly independent basis for the set of $A_{4,-}$ left regular polynomials homogeneous of degree m, and are orthogonal with respect to the inner product

$$\int_{s^3} Q^m_{\ell_1,k_1}(x)\, \overline{Q^m_{\ell_2,k_2}}(x)\, ds^3.$$

(Quaternionic analogues of these polynomials are introduced by Sudbery in [10] in a development of harmonic analysis over the group s^3. For this reason a proof of theorem 1 is omitted.) □

On placing $e_i = \sqrt{-1}f_i$ it may be noted that the polynomials $Q^m_{\ell,k}$ are also $A_{4,+}$ left regular.

To give a generalization of theorem 1 for the case where $n = 6$ we first note that

$$\sum_{j=1}^{6} e_j\frac{\partial}{\partial x_j} = e_1(\frac{\partial}{\partial x_1} + e_1^{-1}e_2\frac{\partial}{\partial x_2}) + e_3(\frac{\partial}{\partial x_3} + e_3^{-1}e_4\frac{\partial}{\partial}) + e_5(\frac{\partial}{\partial x_5} + e_5^{-1}e_6\frac{\partial}{\partial x_6}). \tag{8}$$

Using the right hand side of expression (8), and the preceding arguments, it is straightforward to deduce that the polynomials $Q^m_{\ell,k}(x')_{z_3}$ are $A_{6,-}$ left regular polynomials, where $x' = x_1e_1+\ldots+x_4e_4$, and $z_3 = x_5 + e_5^{-1}e_6x_6$. Again, it may be observed that the polynomial $Q^m_{\ell,k}(x')z_3^p (= z_3^pQ^m_{\ell,k}(x'))$ satisfies the equations

$$\sum_{j=1}^{4} e_j\frac{\partial Q^m}{\partial x_j}\ell,k(x')z_3^p = 0 = \sum_{q=5}^{6} e_q\frac{\partial Q^m}{\partial x_q}\ell,k(x')z_3^p .$$

As a consequence of proposition 1 it may now be observed that the polynomial $Q^m_{\ell,k}(x')z_3^p$ is the tensor product of an $A_{4,+}$ left regular polynomial with an $A_{2,-}$ left regular polynomial.

On placing

$$P^m_{\ell_1,\ell_2,k_1+k_2+\ell_2,k_2}(x) = P'_{\ell_1,\ell_2,k_1',k_2}(x)$$

where

$$P'^m_{\ell_1,\ell_2,k_2',k_2}(x) = \sum_{r_1=0}^{\min(\ell_1,k_1)} \sum_{r_2=0}^{\min(\ell_2,k_2)} (-1)^{r_1+r_2}2z_1[m-k_1'-\ell_1+r_1]_{\bar{z}}[r_1]\times$$

$$\times z_2^{[k_1'-k_2-\ell_2-r_1+r_2]}\bar{z}_2^{[\ell_1-r_1]} z_2^{[r_2]} z_3^{[k_2-r_2]}\bar{z}_3^{[\ell_2-r_2]} + B(\ell_1)(x)$$

and $B(\ell_1)(x) = 0$ if $\ell_1 \leq k_1$, otherwise

$$B(\ell_1)(x) = \sum_{r_2=0}^{\min(\ell_2,k_2)} \sum_{p=1}^{\min(\ell_1-k_1,r_2)} (-1)^{r_2+p} z_1^{[m-k_1'-\ell_1+k_1+p]}\bar{z}_1^{[k_1+p]}$$

where $0 \leq \ell_1,\ell_2,k_1,k_2 \leq m \in \mathbb{N}$, and $m-\ell_1-\ell_2-k_1-k_2 \geq 0$, and

$$P'^m_{-1,\ell_2,k_1',k_2}(x) = P'^m_{\ell_1,-1,k_1',k_2}(x) = P'^m_{-1,-1,k_1',k_2}(x) = 0 ,$$

we have

Theorem 2: For each positive integer m the set of homogeneous polynomials

$$\{Q^m_{\ell_1,\ell_2,k_1,k_2}(x) = P^m_{\ell_1,\ell_2,k_1+k_2+\ell_2,k_2}(x) - e_1^{-1}e_3P^m_{\ell_1-1,\ell_2,k+\ell_2+k_2,k_2}(x)$$
$$-e_3^{-1}e_5P^m_{\ell_1,\ell_2-1,k_1+\ell_2+k_2,k_2}(x)\}$$

is a linearly independent basis for the set of $A_{6,-}$ left regular polynomials homogeneous of degree m. They are orthogonal with respect to the inner product

$$\int_{s^5} Q^m_{\ell_1,\ell_2,k_1,k_2}(x)\,\overline{Q}^m_{\ell_3,\ell_4,k_3,k_4}(x)\,ds^5,$$

where s^5 is the unit sphere in R^5.

(A proof of this result is contained in the proof of theorem 3.) □

Note that when $\ell_2 = 0$ we find that $Q^m_{\ell_1,0,k_1,k_2}(x) = Q^m_{\ell_1,k_1}(x')z_3^{[k_2]}$ and when $\ell_1 = 0$ we have

$$Q_{0,\ell_2,k_1,k_2}(x) = z_1^{m-k_1'}Q^{k_1'}_{\ell_2,k_2}(x'') \qquad (9)$$

where $x'' = x_3e_3+\ldots+x_6e_6$. It may be observed from the right hand side of equation (9) that Q^m_{0,ℓ_2,k_1,k_2} is the tensor product of an $A_{2,+}$ left regular polynomial with an $A_{4,-}$ left regular polynomial.

In general, on placing $k'_{n-j}=k_{n-j}+k_{n-j+1}+\dots+k_{n-1}+\ell_{n-j+1}+\dots+\ell_{n-1}$, we have

$$P'^{m}_{\ell_1,\dots,\ell_{n-1},k'_1,k'_2,\dots,k'_{n-2},k_{n-1}}(x)$$

$$= P'^{m}_{\ell_1,\dots,\ell_{n-1},K,}(x)$$

where

$$P'^{m}_{\ell_1,\dots,\ell_{n-1},K}(x) = \sum_{i=1}^{n-1}\sum_{r_i=0}^{\min(\ell_i,k_i)}(-1)^{r_1+\dots+r_{n-1}} z_1^{[m-k'_1-\ell_1+r_1]}\dots\times$$

$$\times z_{i+1}^{[k'_i-k'_{i+1}-\ell_{i+1}-r_i+r_{i+1}]}\bar{z}_{i+1}^{[\ell_i-r_i]}\bar{z}_{i+1}^{[r_{i+1}]}\dots z_n^{[k_{n-1}-r_{n-1}]}+$$

$$+B(\ell_1,\dots,\ell_{n-2})(x), \tag{10}$$

with

$$B(\ell_1,\dots,\ell_{n-2}) = \sum_{i=1}^{n-1}\sum_{r_{i+1}=0}^{\min(\ell_{i+1},k_{i+1})}\sum_{p_i=1}^{\min(r_{i+1},\ell_i-k_i)}(-1)^{r_1+\dots+r_{n-1}}\times$$

$$\times z_1^{[m-k'_1-\ell_1+k_1+p_1]}\dots z_{i+1}^{[k'_i-k_{i+1}-\ell_{i+1}-k_i+r_{i+1}-p_i]}\times$$

$$\times\bar{z}_{i+1}^{[\ell_i-k_i-p_i]}\bar{z}_{i+1}^{[r_{i+1}]}\dots z_n^{[k_{n-1}-r_{n-1}]},$$

where $0\le\ell_1,\dots,\ell_{n-1},k_1,\dots,k_{n-1}\le m\in\mathbb{N}$ with $m-\ell_1-\dots-\ell_{n-1}-k_1-\dots-k_{n-1}\le 0$. Also $P'^{m}_{\ell_1,\dots,\ell_{i-1},-1,\ell_{i+1},\dots,\ell_{n-1},K}=0$

<u>Theorem 3</u>: For each positive integer m the set of homogeneous polynomials $\{Q^m_{\ell_1\dots\ell_{n-1},k_1,\dots,k_{n-1}}\}$, where

$$Q^m_{\ell_1,\dots,\ell_{n-1},k_1,\dots,k_{n-1}} = P'^{m}_{\ell_1\dots k_{n-1},K} + \sum_{r=1}^{\{n/2\}}\sum_{j(1)=1}^{n-1}\dots\sum_{j(r)=j(r-1)+2}^{n-1}\times$$

$$\times(-1)^r e^{-1}_{2j(1)-1} e_{2j(1)+1} \cdots e^{-1}_{2j(r)-1} e_{2j(r)+1} \times$$

$$\times P'^m_{\ell_1,\ldots,\ell_{j(1)}-1,\ldots,\ell_{j(r)}-1,\ldots,\ell_{n-1},K} \qquad (11)$$

and $\{n/2\}$ denotes the integer part of $n/2$, is a linearly independent basis for the set of $A_{2n,-}$ left regular polynomials homogeneous of degree m. They are orthogonal with respect to the inner product

$$\int_{S^{2n-1}} Q^m_{\ell_1\ldots\ell_{n-1}k_1\ldots k_{n-1}}(x)\, \overline{Q}^m_{\ell'_1\ldots\ell'_{n-1}k'_1\ldots k'_{n-1}}(x)\, dS^{2n-1}, \qquad (12)$$

where S^{2n-1} is the unit sphere in R^{2n}.

Proof: To prove that the function (11) is $A_{2n,-}$ left regular we place

$$\sum_{k=1}^{2n} e_k \frac{\partial}{\partial x_k} = \sum_{j=1}^{n} e_{2j-1}\left(\frac{\partial}{\partial x_{2j-1}} + e^{-1}_{2j-1} e_{2j} \frac{\partial}{\partial x_{2j}}\right) .$$

It is now sufficient to observe from expression (10) that

$$e_{2j(s)+1}\left(\frac{\partial}{\partial x_{2j(s)+1}} + e^{-1}_{2j(s)+1} e_{2j(x)} + \frac{\partial}{\partial x_{2j(s)+2}}\right) \sum_{r=1}^{n/2} \sum_{j(1)=1}^{n-1} \cdots$$

$$\sum_{j(r)=j(r-1)+2}^{n-1} (-1)^r e^{-1}_{2j(1)-1} \cdots e_A e^{-1}_{2j(s)-3} e_{2j(s)-1} e^{-1}_{2j(s)+5} e_B \cdots$$

$$\cdots e_{2j(r)+1} P'^m_{\ell_1\ldots\ell_{j(s)-1}-1,\ell_{j(s)},\ell^{-1}_{j(s)+1},\ldots,\ell_{n-1},K}$$

$$= e_{2j(s)-3}\left(\frac{\partial}{\partial x_{2j(s)-3}} + e^{-1}_{2j(s)-3} e_{2j(s)-2} \frac{\partial}{\partial x_{2j(s)-2}}\right) \sum_{r=1}^{\{n/2\}} \sum_{j(1)=1}^{n-1} \cdots$$

$$\cdots \sum_{j(r)=j(r-1)+2}^{n-1} (-1)^r e^{-1}_{2j(1)-1} \cdots$$

$$\cdots e_A e^{-1}_{2j(s)-1} e_{2j(s)+1} e^{-1}_{2j(s)+3} e_{2j(s)+5} e_B \cdots$$

$$\times\, e_{2j(r)+1}\, P'^{m}_{\ell_1\ldots,\ell_{j(s)-1},\ell_{j(s)}-1,\ell_{j(s)+1}-1,\ldots,\ell_{n-1},K}$$

$$+\, e_{2j(s)+5}\left(\frac{\partial}{\partial x_{2j(s)+5}} + e^{-1}_{2j(s)+5}e_{2j(s)+6}\frac{\partial}{\partial x_{2j(s)+6}}\right)\sum_{r=1}^{\{n/2\}}\sum_{j(1)=1}^{n-1}\cdots$$

$$\times \sum_{j(r)=j(r-1)+2}^{n-1}(-1)^{r}e^{-1}_{2j(1)-1}\cdots e_A e_{2j(s)-3}e_{2j(s)-1}e_{2j(s)+1}e_{2j(s)+3}e_B\cdots$$

$$\times\, e_{2j(r)+1}\, P'^{m}_{\ell_1\ldots,\ell_{j(s)-1}-1,\ell_{j(s)}-1,\ell_{j(s)+1},\ldots,\ell_{n-1},K}$$

where $0 < A < 2j(s)-5$ and $B = 0$ if $2j(s)+3 = 2n-1$ and $B > 2j(s)+7$ otherwise.

$$e_{2j(s)+1}\left(\frac{\partial}{\partial x_{2j(s)+1}} + e^{-1}_{2j(s)+1}e_{2j(s)+2}\frac{\partial}{\partial x_{2j(s)+2}}\right)\sum_{r=1}^{\{n/2\}}\sum_{j(1)=1}^{n-1}\sum_{j(r)=j(r-1)+2}^{n-1}$$

$$\times (-1)^{r}e^{-1}_{2j(s)-1}\cdots e_{A'}e^{-1}_{2j(s)-3}e_{2j(s)-1}e_{B'}\cdots$$

$$\times\, e_{2j(r)+1}P'^{m}_{\ell_1\ldots,\ell_{j(s)-1}-1,\ell_{j(s)},\ldots,\ell_{n-1},K}$$

$$= e_{2j(s)-3}\left(\frac{\partial}{\partial x_{2j(s)-3}} + e^{-1}_{2j(s)-3}e_{2j(s)-2}\frac{\partial}{\partial x_{2j(s)-2}}\right)\sum_{r=1}^{\{n/2\}}\sum_{j(1)=1}^{n-1}$$

$$\times \sum_{j(r)=j(r-1)+2}^{n-1}(-1)^{r}e_{2j(1)-1}\cdots e_{A'}e_{2j(s)-1}e_{2j(s)-1}e_{2j(s)+1}e_{B'}\cdots$$

$$\times\, e_{2j(r)+1}P'^{m}_{\ell_1\ldots,\ell_{j(s)-1},\ell_{j(s)}-1,\ldots,\ell_{n-1},K}.$$

$$-e_{2j(s)+3}\left(\frac{\partial}{\partial x_{2j(s)+3}} + e^{-1}_{2j(s)+3}e_{2j(s)+4}\frac{\partial}{\partial x_{2j(s)+4}}\right)\sum_{r=1}^{\{n/2\}}\sum_{j(1)=1}^{n-1}$$

$$\times \sum_{j(r)=j(r-1)+2}^{n-1}(-1)^{r}e_{2j(1)-1}\cdots e_{A'}e_{2j(s)-3}e_{2j(s)-1}e_{2j(s)+1}e_{2j(s)+3}e_{B'}$$

$$\times \ldots e_{2j(r)+1} P'^{m}_{\ell_1 \ldots \ell_{j(s)-1}-1, \ell_{j(s)}, \ell_{j(s)+1}-1, \ldots, \ell_{n-1}, K}$$

where $0 \leq A' \leq 2j(s)-5$ and $B' = 0$ if $2j(s)+3 = 2n-1$, otherwise $B' \geq 2j(s) + 5$.

$$e_{2j(s)+1}\left(\frac{\partial}{\partial x_{2j(s)+1}} + e^{-1}_{2j(s)+1} e_{2j(s)+2} \frac{\partial}{\partial x_{2j(s)+2}}\right) \sum_{j(1)=1}^{n/2} \ldots \sum_{j(r)=j(r-1)+2}^{n-1}$$

$$(-1)^r e^{-1}_{2j(1)-1} \ldots e_{A''} e^{-1}_{2j(s)+3} e_{2j(s)+5} e_{B''} \ldots e_{2j(r)+1}$$

$$P'^{m}_{\ell_1, \ldots, \ell_{j(s)}, \ell_{j(s)+1}-1, \ldots, \ell_{n-1}, K}$$

$$= e_{2j(s)+5}\left(\frac{\partial}{\partial x_{2j(s)+5}} + e^{-1}_{2j(s)+5} e_{2j(s)+6} \frac{\partial}{\partial x_{2j(s)+6}}\right) \sum_{r=1}^{n/2} \sum_{j(1)=1}^{n-1}$$

$$\times \sum_{j(r)=j(r-1)+2}^{n-1} (-1)^r e^{-1}_{2j(1)-1} \ldots e_{A''} e^{-1}_{2j(s)+3} e_{2j(s)+5} e_{B''} \ldots$$

$$\times e_{2j(r)+1} P'^{m}_{\ell_1, \ldots, \ell_{j(s)}, \ell_{j(s)+1}-1, \ldots, \ell_{n-1}, K}$$

$$- e_{2j(s)-1}\left(\frac{\partial}{\partial x_{2j(s)-1}} + e^{-1}_{2j(s)-1} e_{2j(s)} \frac{\partial}{\partial x_{2j(s)}}\right) \sum_{r=1}^{\{n/2\}} \sum_{j(1)=1}^{n-1}$$

$$\times \sum_{j(r)=j(r-1)+2}^{n-1} (-1)^r e_{2j(1)-1} \ldots e_{A''} e_{2j(s)-1} e_{2j(s)+1} e_{2j(s)+s} e_{2j(s)+5} e_{B''}$$

$$\times \ldots e_{2j(r)+1} P'^{m}_{\ell_1 \ldots \ell_{j(s)-1}-1, \ell_{j(s)}, \ell_{j(s)+1}-1, \ell_{n-1}, K}$$

where $0 \leq A'' \leq 2j(s)-3$ and $B'' = 0$ if $2j(s)+s = n-1$ otherwise $B'' \geq 2j(s)+7$.

$$e_{2j(s)-1}\left(\frac{\partial}{\partial x_{2j(s)-1}} + e^{-1}_{2j(s)-1}e_{2j(s)}\frac{\partial}{\partial x_{2j(s)}}\right)\sum_{r=1}^{n/2}\sum_{j(1)=1}^{n-1}$$

$$\times \sum_{j(r-1)+2}^{n-1} (-1)^r e^{-1}_{2j(1)-1} \cdots$$

$$\times e_{A'''}e_{B'''}\cdots e_{2j(r)+1}P'^{m}_{\ell_1\ldots\ell_{j(s)-1},\ell_{j(s)},\ell_{j(s)+1},\ldots,\ell_{n-1},K}$$

$$= -e_{2j(s)-3}\left(\frac{\partial}{\partial x_{2j(s)-3}} + e^{-1}_{2j(s)-3}e_{2j(s)-1}\frac{\partial}{\partial x_{2j(s)-2}}\right)H_1(s)\sum_{r=1}^{\{n/2\}}\sum_{j(1)=1}^{n-1}$$

$$\times \sum_{j(r)=j(r-1)+2}^{n-1} (-1)^r e_{2j(1)-1}\cdots e_{A'''}e_{2j(s)-3}e_{2j(s)-1}e_{B'''}\cdots$$

$$\times e_{2j(r)+1}P'^{m}_{\ell_1\ldots\ell_{j(s)-1}-1,\ell_{j(s)+1},\ldots,\ell_{n-1},K}$$

$$-e_{2j(s)+1}\left(\frac{\partial}{\partial x_{2j(s)+1}} + e^{-1}_{2j(s)+1}e_{2j(s)+2}\frac{\partial}{\partial x_{2j(s)+2}}\right)H_2(s)\sum_{r=1}^{n/2}\sum_{j(1)=1}^{n-1}$$

$$\times \sum_{j(r)=j(r-1)+2}^{n-1} (-1)^r e_{2j(1)-1}\cdots e_{A'''}e^{-1}_{2j(s)-1}e_{2j(s)+1}e_{B'''}\cdots e_{2j(r)+1}$$

$$\times P'^{m}_{\ell_1,\ldots,\ell_{j(s)-1},\ell_{j(s)},\ell_{j(s)+1}-1,\ldots,\ell_{n-1},K},$$

where $H_1(s) = 0$ if $s = 1$ otherwise $H_1(s) = 1$, and $H_2(s) = 0$ if $s = n$ otherwise $H_2(s) = 1$, and $0 \le A''' \le 2j(s)-3$ and $B''' = 0$ if $2j(s) + 1 = n - 1$ otherwise $B''' \ge 2j(s) + 3$.

It is straightforward to deduce that the polynomials (11) are mutually orthogonal with respect to the inner product (12). Consequently, these polynomials are linearly independent of each other with respect to multiplication on the right by elements of the algebra $A_{2n,-}$. To deduce that these polynomials form a basis for left $A_{2n,-}$ left regular polynomials homogeneous of degree m it is sufficient to observe that the set $\{Q^m_{\ell_1}\ldots,\ell_{n-1},k_1,\ldots,k_{n-1} : 0 \le \ell_1 \ldots \ldots\ell_{n-1},k_1,\ldots,k_{n-1} \le m,$ with $m - \ell_1-\ldots-\ell_{n-1}-k_1\ldots-k_{n-1} \ge 0\}$

is in one to one correspondence with the set of monomials defined over R^{2n-1}, and homogeneous of degree m. It now follows from observations made in the preliminary section that this set of left regular polynomials is a basis set.

Observation 1: On placing $e_j = \sqrt{-1} f_j$ for $1 \leq j \leq 2n$ it may be observed that the polynomial (11) is also an $A_{2n,+}$ left regular polynomial.

Proposition 2: Suppose that for the $A_{2n,-}$ left regular polynomial $Q^m_{\ell_1 \ldots \ell_{n-1},K}$ we know that $\ell_j = 0$ for some j with $1 \leq j \leq n-1$ then the polynomial $Q^m_{\ell_1 \ldots \ell_{n-1} k_1 \ldots k_{n-1}}$ also satisfies the equations

$$\sum_{k=1}^{2j} e_k \frac{\partial}{\partial x} Q^m_{\ell_1 \ldots \ell_{n-1} K}(x) = \sum_{k=2j+1}^{2n} e_k \frac{\partial}{\partial x_k} Q^m_{\ell_1 \ldots \ell_{n-1},K}(x) = 0$$

Moreover,

$$Q^m_{\ell_1 \ldots \ell_{n-1},K}(x) = Q^{m-\ell_{j+1}-\ldots-\ell_{n-1}-k_j-\ldots-k_{n-1}}_{\ell_1 \ldots \ell_{j+1} k_1 \ldots, k_{j-1}}(x_1)$$

$$\otimes\, Q^{mk_j+\ldots+k_{n-1}+\ell_{j+1}+\ldots+\ell_{n-1}}_{\ell_{j+1},\ldots,\ell_{n-1},k_{j+1},\ldots,k_{n-1}}(x_2)$$

$$= Q^{k_j+\ldots+k_{n-1}+\ell_{j+1}+\ldots+\ell_{n-1}}_{\ell_{j+1} \ldots \ell_{n-1},k_{j+1},\ldots k_{n-1}}(x_2) Q^{m-\ell_{j+1}-\ldots-\ell_{n-1}-k_j-\ldots-k_{n-1}}_{\ell_1 \ldots \ell_{j-1} k_1 \ldots k_{j-1}}(x_1) \tag{13}$$

where $x_1 = y_1 e_1 + \ldots + y_{2j} e_{2j}$ and $x_2 = y_{2j+1} e_{2j+1} + \ldots + y_{2n} e_{2n}$.

It follows from expression (13) and the corollary to proposition 1 that in the cases where $\ell_j = 0$ the polynomial (11) is the tensor product of an $A_{2j,\mathrm{sgn}(-1)^{m-j}}$ left regular basis polynomial, and an $A_{2m-2j,-}$ left regular basis polynomial.

As a consequence of the previous remark, and proposition 2 we have:

Proposition 3: Suppose that $a \in A_{2m-2j,-}$ and $b \in A_{2j,sgm(-1)^{m-j}}$ then the function

$$F(x) = Q^{m_1}_{\ell_1 ..., \ell_{j-1} k_1 ... k_{j-1}}(x_1) \, Q^{m_2}_{\ell_{j+1} ... \ell_{n-1} k_{j+1} ... k_{n-1}}(x_2) a \otimes b$$

$$= Q^{m_2}_{\ell_1 ... \ell_{j-1} k_1 ... k_{j-1}}(x_1) b \otimes Q^{m_2}_{\ell_{j+1} ... \ell_{n-1} k_{j+1} ... k_{n-1}}(x_2) a,$$

where m_1 and $m_2 \in \mathbb{N}$, with $m_1 - \ell_1 - ... - \ell_{j-1} - k_1 - ... - k_{j-1} \geq 0$, and $m_2 - \ell_{j+1} - ... - \ell_{n-1} - k_{j+1} - ... - k_{n-1} \geq 0$, satisfies the equations

$$\sum_{p=1}^{2n} e_p \frac{\partial}{\partial x_p} F(x) = \sum_{p=1}^{2j} e_p \frac{\partial}{\partial x_p} F(x) = \sum_{p=2j+1}^{2n} e_p \frac{\partial F(x)}{\partial x_p} = 0.$$

As a consequence of the expression of the Taylor expansion of a left regular function given in [3, theorem 10] it now follows from theorem 3 and proposition 3 that:

Theorem 4: Suppose that $F' : U \subseteq R^{2n} \to A_{2n,-}$ is an $A_{2n,-}$ left regular function which satisfies the additional equations

$$\sum_{p=1}^{2j} e_p \frac{\partial}{\partial x_p} F'(x) = \sum_{p=2j+1}^{2n} e_p \frac{\partial F'(x)}{\partial x_p} = 0.$$

for each $x \in U$. Then for each $x \in U$ there is an open neighbourhood, $U_x \subseteq U$, of x on which we have

$$F'(x) = \sum_{m=0}^{\infty} A_m(\underline{x}_1) \otimes B_m(\underline{x}_2)$$

where $\underline{x}_1 = y_1 e_1 + ... + y_{2j} e_{2j}$ and $\underline{x}_2 = y_{2j+1} e_{2j+1} + ... + y_{2n} e_{2n}$ with $x = \underline{x}_1 + \underline{x}_2$, and $A_m(\underline{x}_1)$ is an $A_{2j,sgn(-1)^{n-j}}$ left regular function, and $B_m(\underline{x}_2)$ is an $A_{2n-2j,-}$ left regular function.

REFERENCES:

[1] M F Atiyah, R Bott, A Shapiro 'Clifford modules' Topology 3, 1965, 3-38.

[2] F Brackx, R Delanghe, F Sommen 'Clifford Analysis' Research Notes in Mathematics, No 76, Pitman, 1982

[3] R Delanghe 'On regular-analytic functions with values in a Clifford algebra' Mathematische Annalen 185, 1970, 91-111

[4] R Fueter 'Functions of a Hypercomplex Variable' Lecture Notes, University of Zurich, 1948, written and supplied by E Bareiss.

[5] M A M Murray 'The Cauchy integral, Calderón commutators and conjugations of singular integrals in R^n', Transactions of the Amer Math Soc, 289, 1985, 497-518

[6] I Porteous Topological Geometry Van Nostrand Company, 1969

[7] J Ryan 'Extensions of Clifford analysis to complex, finite dimensional, associative algebras with identity' Proceedings of the Royal Irish Academy 84A, 1984, 37-50

[8] F Sommen 'Spherical monogenics and analytic functionals on the unit sphere' Tokyo Journal of Mathematics 4, 1981, 427-456

[9] F Sommen 'Plane elliptic systems and monogenic functions in symmetric domains' Supplemento ai Rendiconti del Circulo Matematico di Palermo 11, 1984, 259-269

[10] A Sudbery 'Quaternionic analysis' Mathematical Proceedings of the Cambridge Phil Soc 85, 1979, 199-225.

SPINGROUPS AND SPHERICAL MEANS

Francıscus Sommen (*)
Seminar of Algebra and Functional Analysis
State University of Ghent
Galglaan 2
B-9000 Gent, Belgium

ABSTRACT. In this paper we generalize the notion of spherical mean of a function by making use of the representations of Spin(m). This leads to a factorisation as well as a generalization of the classical Euler Poisson-Darboux equation. Furthermore the boundary values of the generalized spherical means give rise to a set of fundamental biregular polynomials.

INTRODUCTION. The notion of spherical means of a function is known to be very useful in partial differential equations, as is clearly shown by F. John in [5]. It is especially designed for operators of the form $\Sigma c_j \Delta^j$, Δ being the m-dimensional Laplacian, since it satisfies the Euler-Poisson-Darboux equation

$$\Delta_x f(\vec{x},r)=\frac{\partial^2}{\partial r^2}f(\vec{x},r)+\frac{m-1}{r}\frac{\partial}{\partial r}f(\vec{x},r).$$

For the theory of special functions it is useful also to consider operators of the form $\Sigma c_j D^j$, D being the Dirac operator. This requires a refinement of the notion of spherical mean, which may be described by a first order system, replacing the second order Darboux equation. In order to obtain this, we make use of the representations of Spin(m) and the theory of spherical monogenics (see [7], [8], [6], [9]).
These ideas fit completely in the general setting of group representations and integral geometry, which were discussed by S. Helgason in [2].

(*) Senior Research Assistant supported by N.F.W.O. Belgium

J. S. R. Chisholm and A. K. Common (eds.), Clifford Algebras and Their Applications in Mathematical Physics, 149–158.

Furthermore we will use the full machinery of Clifford analysis (see [1]) and we will obtain, as a result of the theory of generalized spherical means, a new class of biregular homogeneous polynomials. They satisfy a **p**lane elliptic system, called special biregular plane wave system, which generalizes the classical Gegenbauer equations (see [2],[4],[12]).

1. BASIC TOOLS

Let C_m be the complex Clifford algebra, $\bar{a}$ the main involution on C_m and $s \in \mathrm{Spin}(m) \subseteq C_m$.

Then we consider the basic representations on $L_2(S^{m-1}; C_m)$

$$H_0 : s \to H_0(s),\ H_0(s)f(\vec{x}) = f(\bar{s}\vec{x}s),$$

$$L : s \to L(s),\quad L(s)f(\vec{x}) = sf(\bar{s}\vec{x}s).$$

The infinitesimal operators dH_0 and dL are given by

$$dH_0(e_{ij}) = -2L_{ij},$$

$$dL(e_{ij}) = -2L_{ij} + e_{ij},$$

where $R_{m,2} = \{\sum_{i<j} x_{ij} e_{ij} : x_{ij} \in R\}$ is the space of bivectors and $L_{ij} = x_i \frac{\partial}{\partial x_j} - x_j \frac{\partial}{\partial x_i}$.

The Casimir operators of L and H are given by

$$C(H_0) = \Delta_S,\ C(L) = \Delta_S + \Gamma - \frac{1}{4}\binom{m}{2},$$

where Δ_S is the Laplace-Beltrami operator on S^{m-1} and where $\Gamma = -\sum_{i<j} e_{ij} L_{ij}$ (see [6],[9]).

The eigenspace of $C(H_0)$ with eigenvalue $-k(k+m-2)$ is the space H_k of spherical harmonics of degree k, whereas the eigenspace of $C(L)$ with eigenvalue $-k(k+m-1) - \frac{1}{4}\binom{m}{2}$ is called the space M_k of spherical monogenics of degree k.

As $\Delta_S = \Gamma(m-2-\Gamma)$, both H_k and M_k admit decompositions of the form

$$H_k = M_{+,k} + M_{-,k-1},\ M_k = M_{+,k} + M_{-,k},$$

where $M_{\pm,k}$ are the eigenspaces of Γ corresponding to the eigenvalues -k and k+m-1, the elements of which are called inner and outer spherical monogenics of degree k.

Let $\vec{x}\in R^m$, $\vec{x}=r\vec{w}$, $\vec{w}\in S^{m-1}$, $|\vec{x}|=r$ and let $f\in L_2(S^{m-1};A)$. Then the projections of f on $H_k, M_k, M_{+,k}$ and $M_{-,k}$ are denoted by $S_k(f)$, $\Pi_k(f)$, $P_k(f)$ and $Q_k(f)$ and we have (see [9]).

$$P_k(f)(\vec{w})=\frac{(-1)^{k+1}}{k!\omega_m}\int_{S^{m-1}}\langle\vec{w},\nabla_u\rangle^k\left(\frac{\vec{u}}{|\vec{u}|^m}\right)\vec{u}f(\vec{u})\,dS_u,$$

$$Q_k(f)(\vec{w})=-\vec{w}P_k(\vec{w}f).$$

Let $D_x=\sum_{j=1}^{m} e_j\frac{\partial}{\partial x_j}$ be the Dirac operator. Then in spherical coordinates $D_x=\vec{w}(\frac{\partial}{\partial r}+\frac{1}{r}\ )$. Hence, if $P_k(\vec{w})$ and $Q_k(\vec{w})$ are inner and outer spherical monogenics of degree k, $r^kP_k(\vec{w})$ and $r^{-(k+m-1)}Q_k(\vec{w})$ are left monogenic ($D_xf=0$) in $R^{m+1}\setminus\{0\}$. Finally, as D_x is L(Spin(m))-invariant, $[D_x,\Pi_k]=0$, where $\Pi_k=P_k+Q_k$, $k\in N$.

2. GENERALIZED SPHERICAL MEANS

Let $\Omega\subseteq R^m$ be open and f be continuous in Ω. Then the function

$$P(f)(\vec{x},r)=\frac{1}{\omega_m}\int_{S^{m-1}}f(\vec{x}+r\vec{w})\,dS_w$$

is defined in $\hat{\Omega}=\{(\vec{x},r):0<r<d(\vec{x},\partial\Omega)\}$ and is called the spherical mean of f (see [5]).

$P(\vec{x},r)$ is an even function of r and satisfies the Euler-Poisson-Darboux equation

$$\frac{\partial^2}{\partial r^2}P(f)+\frac{m-1}{r}\frac{\partial}{\partial r}P(f)=\Delta P(f).$$

Furthermore $f(\vec{x})=\lim_{r\downarrow 0}P(f)(\vec{x},r)$.

In order to refine this notion, we consider the "oriented spherical mean"

$$Q(f)(\vec{x},r)=\frac{1}{\omega_m}\int_{S^{m-1}}\vec{w}f(x+r\vec{w})\,dS_w$$

and we have the refined Darboux system

$$D_x P(f)(\vec{x},r)=\frac{\partial}{\partial r}Q(f)(\vec{x},r)+\frac{m-1}{r}Q(f)(\vec{x},r)$$

$$D_x Q(f)(\vec{x},r)=-\frac{\partial}{\partial r}P(f)(\vec{x},r).$$

Furthermore $D_x P(f)=P(D_x f)$ and $D_x Q(f)=Q(D_x f)$ and for $f\in C_1(\Omega)$ (see [3]),

$$\lim_{r\downarrow 0}\frac{1}{r}Q(f)(\vec{x},r)=\frac{1}{m}D_x f.$$

Next, consider $f(\vec{x}+\vec{y})$, $\vec{x}$ fixed. Then

$$\Pi_0 f(\vec{x}+\vec{y})=P(f)\ (\vec{x},|\vec{y}|)-\frac{\vec{y}}{|\vec{y}|}Q(f)(\vec{x},|\vec{y}|),$$

since 1 and $\vec{w}$ form an orthonormal basis of M_0. Hence the Darboux equations follow immediately from

$$\Pi_0 D_x f(\vec{x}+\vec{y})=D_y\Pi_0 f(\vec{x}+\vec{y})$$

$$=\frac{\vec{y}}{|\vec{y}|}\ \frac{\partial}{\partial r}P(f)+\frac{\partial}{\partial r}Q(f)+\frac{m-1}{r}Q(f),\ r=|\vec{y}|.$$

Also in general, if we put $f_{\vec{x}}(\vec{y})=f(\vec{x}+\vec{y})$,

$$\Pi_k f_{\vec{x}}(\vec{y})=P_k(f_{\vec{x}})(\vec{y})-\frac{\vec{y}}{|\vec{y}|}P_k(-\frac{\vec{y}}{|\vec{y}|}f_{\vec{x}})(\vec{y}),$$

which we write in

$$\Pi_k(\vec{x},r)=P_{+,k}(f)(\vec{x},r)-\vec{w}P_{-,k}(f)(\vec{x},r),$$

$\vec{y}=r\vec{w}$, where $\Pi_k f$, $P_{\pm,k}(f)$ are functions in $\hat{\Omega}$ with values in M_k and $M_{\pm,k}$ respectively.

Definition 1. $P_{+,k}(f)(\vec{x},r)=P_k(f_{\vec{x}})(\vec{y})$ and $P_{-,k}(f)(\vec{x},r)=P_k(\vec{w}f_{\vec{x}})(\vec{y})$, $r\vec{w}=\vec{y}$, are called the k-th inner and outer spherical means of f. $\Pi_k f(\vec{x},r)$ is called the k-th spherical means of f.

We have

$$P_{+,k}(f)(\vec{x},r)(\vec{w})=\frac{(-1)^{k+1}}{k!\omega_m}\int_{S^{m-1}}\langle\vec{w},\nabla_u\rangle^k\left(\frac{\vec{u}}{|\vec{u}|^m}\right)\vec{u}f(r\vec{u}+\vec{x})\,dS_{\vec{u}}\ .$$

Furthermore it follows from $\langle D,\Pi_k\rangle=0$ that

$$P_{+,k}(D_x f)=\frac{\partial}{\partial r}P_{-,k}(f)+\frac{k+m-1}{r}P_{-,k}(f)$$

$$P_{-,k}(D_x f)=-\frac{\partial}{\partial r}P_{+,k}(f)+\frac{k}{r}P_{+,k}(f),$$

which may be considered as a generalized Darboux system. It indeed follows from $D_x^2=-\Delta_x$ that $P_{+,k}(f)$ and $P_{-,k-1}(f)$ satisfy

$$\Delta_x g(\vec{x},r)=\left(\frac{\partial^2}{\partial r^2}+\frac{m-1}{r}\frac{\partial}{\partial r}-\frac{k(k+m-2)}{r^2}\right)g(\vec{x},r).$$

Using the property that every homogeneous polynomial of degree k may be expressed in terms of spherical harmonics of degree k, one easily shows the following boundary behaviour of $P_{\pm,k}(f)(\vec{x},r)$.

Lemma 1. Let $f\in C_k(\Omega)$ (resp. $f\in C_{k+1}(\Omega)$). Then for $r\to 0$, $P_{+,k}(f)=O(r^k)$. (resp. $P_{-,k}(f)=O(r^{k+1})$).

In view of this lemma, the following definition of boundary value of $P_{\pm,k}(f)$ makes sense.

Definition 2. Let $f\in C_k(\Omega)$ (resp. $f\in C_{k+1}(\Omega)$). Then for every $\vec{w}\in S^{m-1}$ we define operators $D_{\pm,k}(\vec{w})$ by

$$D_{+,k}(\vec{w})f(\vec{x})=\lim_{r\downarrow 0}\frac{1}{r^k}P_{+,k}(f)(\vec{x},r)(\vec{w}),$$

$$D_{-,k}(\vec{w})f(\vec{x})=\lim_{r\downarrow 0}\frac{1}{r^{k+1}}P_{-,k}(f)(\vec{x},r)(\vec{w}).$$

Using distributional techniques one easily shows

Proposition 1. The operators $D_{\pm,k}(\vec{w})$ are homogeneous differential operators of the form

$$D_{+,k}(\vec{w})=\sum_{|\alpha|=k}R_{\alpha,+}(\vec{w})\,\partial^\alpha$$

$$D_{-,k}(\vec{w})=\sum_{|\alpha|=k+1} R_{\alpha,-}(\vec{w})\partial^{\alpha},$$

where $R_{\alpha,\pm}(\vec{w})$ are inner spherical monogenics of degree k, given by

$$R_{\alpha,\pm}(\vec{w})=P_{\pm,k}\left(\frac{x_1^{\alpha_1}\ldots x_m^{\alpha_m}}{\alpha_1!\ldots\alpha_m!}\Big|S^{m-1}\right) .$$

For example $D_{+,0}=1$, $D_{-,0}=\frac{1}{m}D$,

$$D_{+,1}(\vec{w})=\langle\vec{w},\nabla\rangle+\frac{1}{m}\vec{w}D.$$

Putting $r\vec{w}=\vec{x}$, the symbol of $r^k D_{\pm,k}(\vec{w})$ is denoted by $D_{\pm,k}(\vec{x},\vec{t})$ and using the formulae for $R_{\alpha,\pm}(\vec{w})$ we arrive at

$$D_{+,k}(\vec{x},\vec{t})=\frac{(-1)^{k+1}}{k!^2\omega_m}\int_{S^{m-1}}\langle\vec{x},\nabla_u\rangle^k\left(\frac{\vec{u}}{|\vec{u}|^m}\right)\vec{u}\langle\vec{u},\vec{t}\rangle^k dS_u,$$

$$D_{-,k}(\vec{x},\vec{t})=\frac{(-1)^{k}}{k!(k+1)!\omega_m}\int_{S^{m-1}}\langle\vec{x},\nabla_u\rangle^k\left(\frac{\vec{u}}{|\vec{u}|^m}\right)\langle\vec{u},\vec{t}\rangle^{k+1} dS_u.$$

Using the expansion of homogeneous polynomials in spherical harmonics, one can show that $D_{+,k}(\vec{x},\vec{t})$ is biregular with $D_x D_{+,k}(\vec{x},\vec{t})=D_{+,k}(\vec{x},\vec{t})D_t=0$, and that $D_{-,k}(\vec{x},\vec{t})D_t=-D_{+,k}(\vec{x},\vec{t})$.

Next, assume that f admits a radial expansion about the point $\vec{x}$: $f(\vec{x}+r\vec{w})=\sum_{l=0}^{\infty} r^l f_l(\vec{x},\vec{w})$.

Then we have

$$P_{+,k}(f)(\vec{x},r)=\sum_{l=k}^{\infty} r^l P_k(f_l(\vec{x},\vec{w})),$$

$$P_{-,k}(f)(\vec{x},r)=\sum_{l=k+1}^{\infty} r^l P_k(\vec{w}\, f_l(\vec{x},\vec{w})).$$

Using the definition of $D_{\pm,k}(\vec{w})$, it follows easily from the Darboux equation that $P_k(f_l(\vec{x},\vec{w}))=0$ for k+l odd, $P_k(\vec{w}f_l(\vec{x},\vec{w}))=0$ for k+l even and that

$$P_k(f_{k+2s}(\vec{x},\vec{w}))=\frac{\Delta_x^s}{\prod_{j=1}^{s}(4j^2+2j(2k+m-2))}D_{+,k}(\vec{w})f(\vec{x}),$$

$$P_k(\vec{w}f_{k+2s+1}(\vec{x},\vec{w}))=\frac{\Delta_x^s}{\prod_{j=1}^{s}(4j^2+2j(2k+m))}D_{-,k}(\vec{w})f(\vec{x}).$$

Hence, putting $D_{\pm,k}(\vec{y},\nabla_k)=r^kD_{\pm,k}(\vec{w})$, $\vec{y}=r\vec{w}$ we can write

the k th spherical means as infinite order differential operators of the form

$$P_{+,k}(f)(\vec{x},r)=\sum_{s=0}^{\infty}\frac{\Gamma(k+\frac{m}{2})r^{2s}\Delta_x^s}{4^s s!\Gamma(s+k+\frac{m}{2})}D_{+,k}(\vec{y},\nabla_x)f(\vec{x}),$$

$$P_{-,k}(f)(\vec{x},r)=\sum_{s=0}^{\infty}\frac{\Gamma(k+\frac{m}{2}+1)r^{2s+1}\Delta_x^s}{4^s s!\Gamma(s+k+\frac{m}{2}+1)}D_{-,k}(\vec{y},\nabla_x)f(\vec{x}).$$

Furthermore, using the Bessel functions (see [4]) we obtain a generalization of Fritz John's formula :

$$P_{+,k}(f)(\vec{x},r)=\Gamma(k+\tfrac{m}{2})\left(\frac{ir\sqrt{\Delta}}{2}\right)^{-\nu}J_\nu(ir\sqrt{\Delta})D_{+,k}(\vec{y},\nabla_x)f(\vec{x}),$$

where $\nu=k+\frac{m}{2}-1$, and

$$P_{-,k}(f)(\vec{x},r)=\Gamma(k+\tfrac{m}{2}+1)\left(\frac{ir\sqrt{\Delta}}{2}\right)^{-\nu}J_\nu(ir\sqrt{\Delta})D_{-,k}(y,\nabla_x)f(\vec{x}),$$

where $\nu=k+\frac{m}{2}$.

3. THE POLYNOMIALS $D_{+,k}(\vec{y},\vec{t})$

We already obtained the biregular polynomials $D_{+,k}(\vec{y},\vec{t})$ as integrals over the sphere. The explicit calculation of $D_{+,k}(\vec{y},\vec{t})$ may be done in terms of Gegenbauer polynomials, as follows.

From section 2 it follows that, for $f(\vec{x})=e^{i<\vec{t},\vec{x}>}$,

$$P_k(e^{i<\vec{t},\vec{y}>})=2^\nu\Gamma(k+\tfrac{m}{2})i^k(r|\vec{t}|)^{-\nu}J_\nu(r|\vec{t}|)D_{+,k}(\vec{y},\vec{t}),$$

where $\nu=k+\frac{m}{2}-1, r=|\vec{y}|, \vec{y}=r\vec{w}$. On the other hand we have the classical formula (see [4])

$$S_k(e^{i<\vec{t},\vec{y}>})=2^{\frac{m}{2}-1}\Gamma(\tfrac{m}{2}-1)(k+\tfrac{m}{2}-1)i^k(r|\vec{t}|)^{k-\nu}J_\nu(r|\vec{t}|)C_k^{\frac{m}{2}-1}(\theta),$$

where $\theta=\frac{<\vec{x},\vec{t}>}{|\vec{x}||\vec{t}|}$, $\nu=k+\frac{m}{2}-1$ and $C_k^\lambda(\theta)$ are the Gegenbauer polynomials. But in view of section one, we have

$$P_k=\frac{k+m-2-\Gamma}{2k+m-2}S_k,$$

$$Q_k=-\vec{w}P_k\vec{w}=\frac{k+1+\Gamma}{2k+m}S_{k+1}.$$

Hence, by comparing the formulae for $P_k(e^{i<\vec{t},\vec{y}>})$ and $P_k(\vec{w}e^{i<\vec{t},\vec{y}>})$ and using the identity

$$\Gamma_y C_k^\lambda\left(\frac{<\vec{y},\vec{t}>}{|\vec{y}||\vec{t}|}\right)=2\lambda\ \frac{\vec{t}\wedge\vec{y}}{|\vec{t}||\vec{y}|}C_{k-1}^{\lambda+1}\left(\frac{<\vec{y},\vec{t}>}{|\vec{y}||\vec{t}|}\right),$$

we obtain the formulae

$$D_{+,k}(\vec{y},\vec{t})=\frac{\Gamma(\frac{m}{2}-1)(|\vec{y}||\vec{t}|)^k}{2^{k+1}\Gamma(k+\frac{m}{2})}\ [(k+m-2)C_k^{\frac{m}{2}}(\theta)+(m-2)\frac{\vec{y}\wedge\vec{t}}{|\vec{y}||\vec{t}|}C_{k-1}^{\frac{m}{2}}(\theta)],$$

$$-\vec{y}D_{-,k}(\vec{y},\vec{t})=\frac{\Gamma(\frac{m}{2}-1)(|\vec{y}||\vec{t}|)^{k+1}}{2^{k+2}\Gamma(k+\frac{m}{2}+1)}\ \times$$

$$\times\,[(k+1)C_{k+1}^{\frac{m}{2}-1}(\theta)-(m-2)\ \frac{\vec{y}\wedge\vec{t}}{|\vec{y}||\vec{t}|}\ C_k^{\frac{m}{2}}(\theta)].$$

Furthermore, it follows from the identities

$$\frac{1}{|\vec{x}-\vec{y}|^{m-2}}=\sum_{k=0}^{\infty}\ \frac{|\vec{x}|^k}{|\vec{y}|^{m+k-2}}\ C_k^{\frac{m}{2}-1}(\theta)=\sum_{k=0}^{\infty}\ \frac{(-1)^{k+1}}{k!}<\vec{x},\nabla_y>^k\ \frac{\vec{y}}{|\vec{y}|^m}$$

and

$$\frac{\vec{x}-\vec{y}}{|\vec{x}-\vec{y}|^m}=\frac{D_x}{2-m}\ \frac{1}{|\vec{x}-\vec{y}|^{m-2}}=\frac{D_y}{m-2}\ \frac{1}{|\vec{x}-\vec{y}|^{m-2}}$$

that

$$D_{-,k}(\vec{x},\vec{y})=\frac{1}{2k+m}\ \vec{y}D_{+,k}(\vec{y},\vec{x})$$

and that

$$D_{+,k}(\vec{x},\vec{y})=\frac{\Gamma(\frac{m}{2})(-1)^{k+1}}{2^k k!\,\Gamma(k+\frac{m}{2})}\vec{x}\,|\vec{x}|^{2k+m-2}<\vec{y},\nabla_x>^k\left(\frac{\vec{x}}{|\vec{x}|^m}\right).$$

The main property of $D_{+,k}(\vec{x},\vec{y})$ is that they are biregular homogeneous polynomials of the form

$$A(|\vec{x}||\vec{y}|,\frac{\langle\vec{x},\vec{y}\rangle}{|\vec{x}||\vec{y}|})+\frac{\vec{x}\wedge\vec{y}}{|\vec{x}||\vec{y}|}B(|\vec{x}||\vec{y}|,\frac{\langle\vec{x},\vec{y}\rangle}{|\vec{x}||\vec{y}|}).$$

A left monogenic function of this simple form is automatically biregular and will be called a special biregular plane wave. The monogenicity conditions will be called special biregular plane wave equations (SBPW).

Let us put $\rho=|\vec{x}||\vec{y}|$, $\theta=\frac{\langle\vec{x},\vec{y}\rangle}{|\vec{x}||\vec{y}|}$; then the SBPW-equations are defined in $R_+\times]-1,1[$ and given by

$$\rho\frac{\partial A}{\partial\rho}+(1-\theta^2)\frac{\partial B}{\partial\theta}-(m-1)\theta B=0,$$

$$\rho\frac{\partial B}{\partial\rho}-\frac{\partial A}{\partial\theta}+(m-2)B=0.$$

We shall construct the global solutions. Let $A=\sum_0^\infty\rho^l A_l(\theta)$, $B=\sum_0^\infty\rho^l B_l(\theta)$. Then we obtain the relations

$$lA_l+(1-\theta^2)\frac{\partial B_l}{\partial\theta}-(m-1)\theta B_l=0$$

$$lB_l-\frac{\partial A_l}{\partial\theta}+(m-2)B_l=0,$$

which lead to the Gegenbauer equation,

$$(1-\theta^2)\frac{\partial^2 B_l}{\partial\theta^2}-(m+1)\theta\frac{\partial B_l}{\partial\theta}+(l-1)(l+m-1)B_l=0,$$

admitting the polynomial solution $C_{l-1}^{\frac{m}{2}}(\theta)=B_l$.
A_l then follows from the recursion formulae for $C_k(\theta)$:

$$lA_l+(1-\theta^2)mC_{l-2}^{\frac{m}{2}+1}(\theta)-(m-1)\theta\, C_{l-1}^{\frac{m}{2}}(\theta)=0$$

or

$$A_l=C_l^{\frac{m}{2}}(\theta)-\theta C_{l-1}^{\frac{m}{2}}(\theta)=\frac{l+m-2}{m-2}C_l^{\frac{m}{2}-1}(\theta).$$

Hence the global solutions of the SBPW-equations are all of the form

$$A+\frac{\vec{x}\wedge\vec{y}}{|\vec{x}||\vec{y}|}B=\sum_{k=0}^{\infty}a_k D_{+,k}(\vec{x},\vec{y}).$$

Furthermore the SBPW-equations may be regarded as a generalization to the strip $R_+\times]-1,1[$ of the class of Gegenbauer equations.

References

[1] F. Brackx, R. Delanghe, F. Sommen, 'Clifford Analysis', *Research Notes in Math.*, 76 (Pitman, London, 1982).
[2] S. Helgason, 'Groups and Geometric Analysis', *Pure and Applied Math.* (Academic Press, Orlando, London, 1984)
[3] D.Hestenes, 'Multivector Calculus', *J. Math.Anal.Appl.* 24 no.2 (1968), 313-325.
[4] H. Hochstadt, 'The functions of mathematical physics', *Pure and Applied Math.*, 23 (Wiley-Interscience, New York, 1971).
[5] F. John, *Plane Waves and Spherical Means*, (Springer, 1955).
[6] P. Lounesto, 'Spinor valued regular functions in hypercomplex analysis' (Thesis, Helsinki, 1979).
[7] I. Porteous, *Topological Geometry*, (Van Nostrand Reinhold Cy., London, 1979).
[8] M. Riesz, 'Clifford numbers and spinors', *Lecture series* 38 (Institute for Physical Sciences and Technology, Maryland, 1958).
[9] F. Sommen, 'Spherical monogenic functions and analytic functionals on the unit sphere', *Tokyo J. Math.* 4 (1981), 427-456.
[10] ———, 'Some connections between Clifford analysis and complex analysis', *Complex Variables : Theory and Application* 1 (1982), 97-118.
[11] ———, 'Plane waves, biregular functions and hypercomplex Fourier analysis', to appear in *Rend.Circ. Mat. Palermo*
[12] N.J. Vilenkin, 'Special functions and the theory of group representations', *Transl. Math. Monographs, Amer. Math. Soc.*, 22 (1968).

THE BIREGULAR FUNCTIONS OF CLIFFORD ANALYSIS: SOME SPECIAL TOPICS

F. Brackx and W. Pincket
State University of Gent
Seminarie voor Wiskundige Analyse
Sint-Pietersnieuwstraat 39
B - 9000 Gent, Belgium

ABSTRACT. In this paper we intend to show that certain features encountered in the theory of holomorphic functions of several complex variables, also occur in the theory of biregular functions, which are regular functions of two variables of arbitrary dimension with values in a Clifford algebra. For these functions there exist integral representations with regular and non-regular kernels, just as in the complex case. We will give a representation formula of the second kind using differential forms. Another topic is the study of the domains of biregularity which are the analogues of the domains of holomorphy of complex analysis. We will make it clear that multi-valued functions cannot be avoided in this study. Some examples of domains of biregularity will be given, by constructing biregular functions in the domains which cannot be extended to any larger domain.

INTRODUCTION

In [2,3,4,5] the biregular functions were intensively studied;they are functions of two variables of arbitrary dimension, which are left monogenic in one variable and right monogenic in the other. The holomorphic functions of two complex variables may be imbedded in the class of biregular functions. It is quite interesting to note that properties of those holomorphic functions extend to all biregular functions; this is the case for a.o. the Hartogs theorem on the global real-analyticity, the non-existence of pointwise singularities and the Hartogs extension theorem on analytical continuation.

Also in the field of integral representations of biregular functions, features similar to the complex case are encountered. In both cases there exist representations with regular kernels, but only valid in special regions - such as the Cauchy integral representation - and representations with non-regular kernels, but however valid in arbitrary domains - such as the Bochner-Martinelli formula. In part one of this paper we intend to give a representation formula of the second kind using special differential forms. The problem of representing biregular functions in arbitrary domains with biregular kernels is still an open problem.

Another special topic in the theory of several complex variables

J. S. R. Chisholm and A. K. Common (eds.), Clifford Algebras and Their Applications in Mathematical Physics, 159–166.

is the study of the domains of holomorphy. Although this notion is trivial in the plane, it is a hard problem to characterize these domains in higher dimensions. Again we notice the analogy between the holomorphic and the biregular functions. Indeed, the notion of a domain of biregularity becomes trivial in the monogenic case (see [6]). In the second part of this paper we will give a short introduction into the study of domains of biregularity, emphasizing the fact that multi-valued functions will appear ; some examples will be discussed in detail.

PART I

1. Let Ω be an open set in $\mathbb{R}^{n+1}\times\mathbb{R}^{n+1}$ $(n>1)$ and let A be the universal real 2^n-dimensional Clifford algebra over $\mathbb{R}^n$ with basis $(e_A : A=(h_1,\ldots,h_r)\in P\{1,\ldots,n\},\ 1\leqslant h_1<\ldots<h_r\leqslant n)$. We denote by A_1 the space of one-vectors $sp_{\mathbb{R}}(e_1,\ldots,e_n)$, which can be identified with $\mathbb{R}^n$.

The functions considered are of the form $f:\Omega\to A$, $(x,y) = (x_o,x_1,\ldots,x_n,y_o,y_1,\ldots,y_n) \mapsto f(x,y) = \sum_A e_A f_A(x,y)$, where the components f_A are real-valued.

Next we introduce the generalized Cauchy-Riemann operators $D_x = \sum_{i=0}^{n} e_i \partial_{x_i}$ and $D_y = \sum_{i=0}^{n} e_i \partial_{y_i}$.

We also have the following notations for the intersections of Ω parallel to the x- and y-space :

$$U_y = \{x\in\mathbb{R}^{n+1} : (x,y)\in\Omega\},\ y \text{ fixed in } \mathbb{R}^{n+1};$$
$$V_x = \{y\in\mathbb{R}^{n+1} : (x,y)\in\Omega\},\ x \text{ fixed in } \mathbb{R}^{n+1}.$$

DEFINITION 1 : A function $f:\Omega\to A$ is said to be biregular in Ω iff for each y fixed in $\mathbb{R}^{n+1}$, f is C^1 and $D_x f = 0$ in U_y, while for each x fixed in $\mathbb{R}^{n+1}$, f is C^1 and $fD_y = 0$ in V_x.

Notice that by the Hartogs Theorem (see [2]) a biregular function in Ω becomes a real-analytic function in all the variables $(x_o,\ldots,x_n,y_o,\ldots,y_n)$ together.

2. Let M be a (2n+2)-dimensional, compact, differentiable, oriented manifold-with-boundary contained in Ω, and let $u = \sum_{i=0}^{n} e_i u_i$ be a C^1-map from M to $\mathbb{R}\oplus A_1 \cong \mathbb{R}^{n+1}$. Then we define the following continuous differential forms :

$$dU = du_o \wedge du_1 \wedge \ldots \wedge du_n$$
$$du = \sum_{i=0}^{n} e_i du_i$$

and

$$d\sigma_u = \sum_{i=0}^{n} (-1)^i e_i \, du_o \wedge \ldots \wedge d\not{u}_i \wedge \ldots \wedge du_n .$$

If v is such a second C^1-map from M to $\mathbb{R} \oplus A_1$ then we define the 2n-form : $\omega(u,v) = d\sigma_v \wedge d\sigma_u$ and the 1-form : $\omega'(u,v) = u\, dv - du\, v$, whereby it has to be stipulated that exterior multiplication involving ω' is carried out as follows :

$$\{\omega' \wedge \omega'' g\} = u\, dv \wedge \omega'' g - g\omega'' \wedge du\, v$$
$$\{\{\omega' \wedge \omega''\} \wedge \omega'''\} = u\, dv \wedge \omega'' \wedge \omega''' - \omega''' \wedge \omega'' \wedge du\, v,$$

where g is a differentiable function and ω'', ω''' are differential forms. By a direct computation we obtain

PROPOSITION: The above differential forms satisfy the following properties :

(i) $d\omega(u,v) = 0$;

(ii) $(n+1)dU = d\bar{u} \wedge d\sigma_u = (-1)^n d\sigma_u \wedge d\bar{u}$;

(iii) for every real-valued C^1-function ϕ on M holds :

$$\omega'(\phi u, \phi v) = \phi^2 \omega'(u,v);$$

(iv) if $(u_o, \ldots, u_n, v_o, \ldots, v_n)$ are the co-ordinates of $(u,v) \in \mathbb{R}^{n+1} \times \mathbb{R}^{n+1}$, then for every fixed $(x,y) \in \mathbb{R}^{n+1} \times \mathbb{R}^{n+1}$,

$$d_{(u,v)}(\omega'(\overline{u-x}, \overline{v-y}) \wedge \omega(u,v)) = 2(n+1)^2 \, dU \wedge dV.$$

3. If $|(u,v)|$ denotes the Euclidean norm in $\mathbb{R}^{n+1} \times \mathbb{R}^{n+1}$, then for $(u,v) \neq (0,0)$ we put $\tilde{u} = \dfrac{u}{|(u,v)|^{2n+2}}$ and $\tilde{v} = \dfrac{v}{|(u,v)|^{2n+2}}$. For fixed $(x,y) \in \mathbb{R}^{n+1} \times \mathbb{R}^{n+1}$, the differential form

$$\{\omega'_{(u,v)}((\overline{u-x})^{\sim}, (\overline{v-y})^{\sim}) \wedge \omega(u,v)\} = \frac{\{\omega'_{(u,v)}(\overline{u-x}, \overline{v-y}) \wedge \omega(u,v)\}}{|(u-x,v-y)|^{2n+2}}$$

is smooth in $(u,v) \in (\mathbb{R}^{n+1} \times \mathbb{R}^{n+1}) \setminus \{(x,y)\}$, and has an integrable singularity at $(u,v) = (x,y)$. So we can define the following two integral operators.

If g is a bounded measurable function on ∂M then we put

$$(I_{\partial M} g)(x,y) = \frac{1}{(n+1)\omega_{2n+2}} \int_{\partial M} \{\omega'_{(u,v)}((\overline{u-x})^{\sim}, (\overline{v-y})^{\sim}) \wedge \omega(u,v) g(u,v)\},$$

while for a bounded one-form h on M we put

$$(I_M h)(x,y) = \frac{1}{(n+1)\omega_{2n+2}} \int_M \{\{\omega'_{(u,v)}((\overline{u-x})^{\sim}, (\overline{v-y})^{\sim}) \wedge \omega(u,v)\} \wedge h\},$$

ω_{2n+2} denoting the surface area of the unit sphere S^{2n+1} in R^{2n+2}.

4. For a C^1-function f in Ω we can introduce the one-form:

$$\bar{\partial} f = (fD_v)d\bar{v} - d\bar{u}(D_u f).$$

Clearly $\bar{\partial} f = 0$ iff f is biregular in Ω.
The Bochner-Martinelli representation theorem, already obtained in [3], may now be stated in the following elegant form, which is quite similar to the one of the several complex variables case.

THEOREM : Let Ω and M be as before. For a C^1-function f in Ω the formula

$$f = I_{\partial M} f - I_M \bar{\partial} f$$

holds in $\overset{\circ}{M}$.

COROLLARY : Let Ω and M be as before, and let f be biregular in Ω. Then in $\overset{\circ}{M}$:

$$f = I_{\partial M} f.$$

PART II

5. Intuitively one could say that a domain of biregularity is a domain in which a biregular function can be found, which cannot be extended biregularly to any larger domain. However in [5] we introduced the following

DEFINITION 2 : A domain $\Omega \subset R^{2n+2}$ is called a domain of biregularity iff it is impossible to find two domains U_1, U_2 in R^{2n+2} satisfying the following properties :

(i) $\emptyset \neq U_2 \subset \Omega \cap U_1 \subsetneq U_1$;

(ii) for each biregular function f in Ω, there exists a function f_1, biregular in U_1, such that $f \equiv f_1$ in U_2.

Let us illustrate the need for such a definition by means of an example in the monogenic setting. Consider R^3 (n=2) and the hypercomplex variables $z_1 = x_1 e_o - x_o e_1$, $z_2 = x_2 e_o - x_o e_2$. Then arg z_1 and arg z_2 are defined as the polar angles in respectively the (x_o, x_1)- and the (x_o, x_2)-plane w.r.t. the positive x_o-axis and taking values in between $-\pi$ and π(see figure 1).

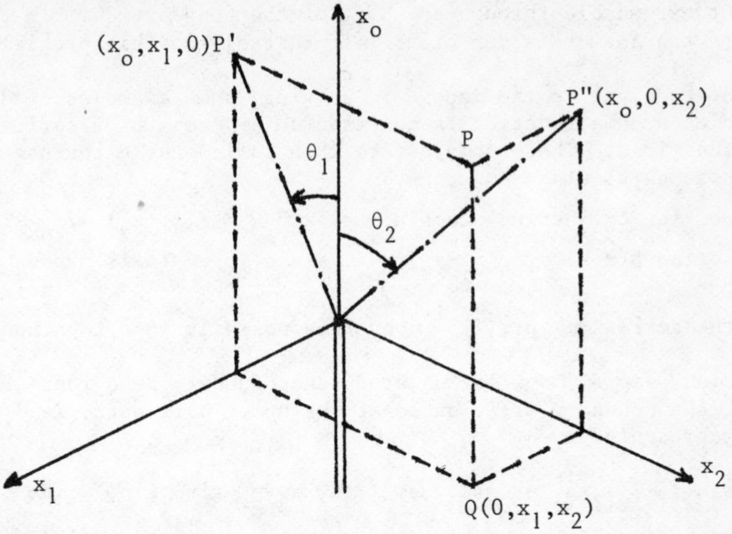

Figure 1. Definition of the functions arg z_1 and arg z_2 as polar angles w.r.t. the positive x_o-axis.

Clearly the function f given by

$$f(x_o,x_1,x_2) = e_o \ln \sqrt{(x_o^2+x_1^2)(x_o^2+x_2^2)}+e_1 \arg z_1+e_2 \arg z_2$$

is monogenic in $\Omega = R^3 \setminus S$, the set of singularities of f being the union of two half-planes: $S = \{x_2 = 0,\ x_o \leq 0\} \cup \{x_1 = 0,\ x_o \leq 0\}$. However Ω does not satisfy the conditions of Definition 2 for that particular function f. Take a point $P(x_o^\star,0,x_2^\star)$ with $x_o^\star < 0$, $x_2^\star > 0$, and consider the ball U_1 centred at P and with radius $r < \min(-x_o^\star,x_2^\star)$. For U_2 we take any connected subset of U_1 in the half-space $\{x_1 < 0\}$. Then, if arg z_1 is allowed to take values less than $-\pi$, one can find a function $f^\star$, monogenic in U_1, which coincides with f on U_2.

Of course it is seen at once that by in- or decreasing arg z_1 and arg z_2, f can be extended monogenically in four different ways. But then we end up with a multi-valued function with branches, such that from each branch it is possible to attain the four others, all being copies of Ω glued together along the cuts in R^3. So the analogue of the Riemann surfaces of complex analysis arises here.

6. There are several characterizations of the domains of biregularity both of an analytical and of a geometrical nature; details may be found in [5].

Now suppose that Ω fulfills the conditions of Definition 2; then the question arises whether a biregular function f in Ω may be found which is not extendable through any part of the boundary $\partial\Omega$. In the complex case the answer is yes ; in Clifford analysis this problem is still open.

We want to give, in the monogenic setting, some examples of domains in which such a function is constructed by means of a series of monogenic functions. These domains are then called the existence domains of those particular functions.

Consider the cylindrical domain $\Omega = \{x \in R^{n+1} ; (x_1+x_2)^2+2x_o^2 < 1\}$ and the function $p(x,\vec{a}) = e_o(x_1+x_2)-x_o\vec{a}$, $\vec{a} = e_1+e_2$. It is proven in [1] that the series $\sum_{\alpha=0}^{\infty} p(x,\vec{a})^{\alpha!}$ converges normally in Ω to a monogenic function, say f. Now let us study the boundary behaviour of f. In figure 2 the boundary of Ω, at least the upper half part, is drawn. It can be represented by

$$x_o = -\frac{1}{\sqrt{2}}\sin(\phi_1+\phi_2),\ x_1 = \cos\phi_1\cos\phi_2 + k,\ x_2 = -\sin\phi_1\sin\phi_2 - k,$$

$k \in Z$, (ϕ_1,ϕ_2) belonging to the rectangular domain shown in figure 3. If k runs through Z, $\partial\Omega$ is cut into "elementary parts", each of which is then represented by the rectangle of figure 3.

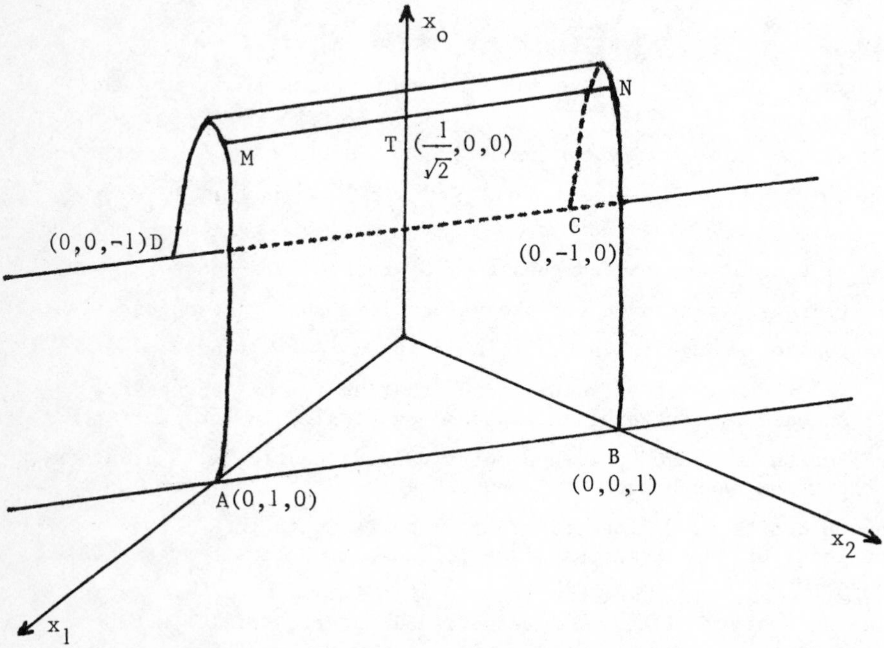

Figure 2. An "elementary part" of the upper half of the boundary of the cylindrical domain Ω.

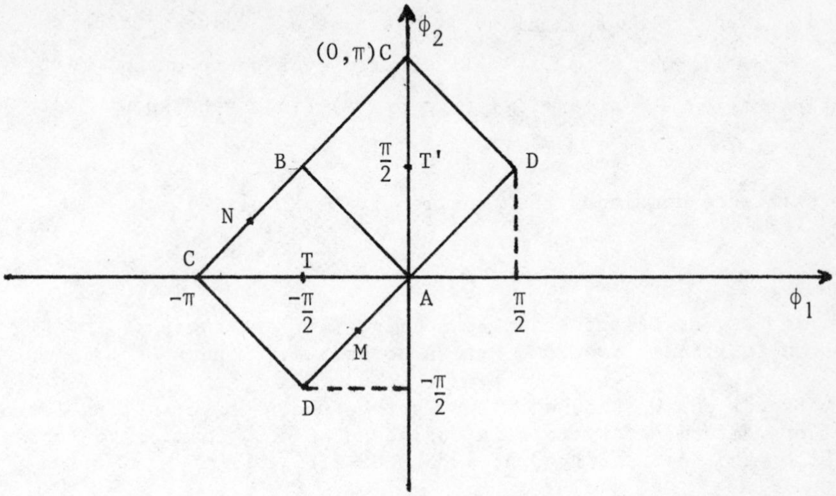

Figure 3. The rectangular domain of (ϕ_1,ϕ_2) corresponding to an "elementary part" of the boundary of Ω.

Now on $\partial\Omega$:

$$p(x,\vec{a}) = e_o \cos(\phi_1+\phi_2)+(e_1+e_2)\frac{1}{\sqrt{2}}\sin(\phi_1+\phi_2).$$

It may be shown that

$$p(x,\vec{a})^{\alpha!} = e_o \cos\alpha!(\phi_1+\phi_2)+(e_1+e_2)\frac{1}{\sqrt{2}}\sin\alpha!(\phi_1+\phi_2).$$

Take $\phi_1 = \frac{p_1}{q_1}\pi$, $\phi_2 = \frac{p_2}{q_2}\pi$, $p_1,p_2 \in N$, $q_1,q_2 \in N_o$.

The point with co-ordinates $x_o = -\frac{1}{\sqrt{2}} r \sin(\phi_1+\phi_2)$, $x_1=r(\cos\phi_1\cos\phi_2+k)$

$x_2 = -r(\sin\phi_1\sin\phi_2 + k)$, $r < 1$, belongs to Ω, so f is defined there. As $\alpha!(\phi_1+\phi_2) = 2\ell\pi$, $\ell \in Z$, for $\alpha \geq \beta$, β chosen sufficiently large, we can obtain

$$\left|\sum_{\alpha=0}^{\infty} p(x,\vec{a})^{\alpha!}\right|_o \geq \left|\sum_{\alpha=\beta}^{\infty} p(x,\vec{a})^{\alpha!}\right|_o - \left|\sum_{\alpha=0}^{\beta-1} p(x,\vec{a})^{\alpha!}\right|_o$$

$$\geq \sum_{\alpha=\beta}^{\infty} 2^{n/2} r^{\alpha!} - \left|\sum_{\alpha=0}^{\beta-1} p(x,\vec{a})^{\alpha!}\right|_o$$

If $r \gtrless$ then the right hand side diverges, whence $|f(x)|_o \to +\infty$ if

$x \in \partial\Omega$ is given by the special ϕ_1,ϕ_2. But as these values are dense in the set of allowed ϕ_1,ϕ_2, f will diverge everywhere on $\partial\Omega$. More generally, putting $\vec{a} = e_1+e_2+\ldots+e_j$ in $p(x,\vec{a})$, it may be shown that

$$\Omega_j = \{x \in R^{n+1} : (x_1+\ldots+x_j)^2 + j\, x_o^2 < 1\}$$

is the existence domain of the function $f_j = \sum_{\alpha=0}^{\infty} p(x,\vec{a})^{\alpha!}$

REFERENCES

[1] F. Brackx, R. Delanghe and F. Sommen, *Clifford Analysis, Research Notes in Mathematics* 76 , Pitman Books Ltd., London

[2] F. Brackx and W. Pincket, 'Two Hartogs theorems for null-solutions of overdetermined systems in Euclidean space', *Complex Variables: Theory and Applications*, 5, 1985, 205-222

[3] ________________________, 'A Bochner-Martinelli formula for the biregular functions of Clifford Analysis' (to appear in *Complex Variables : Theory and Applications*)

[4] ________________________, 'Series expansions for the biregular functions of Clifford Analysis' (to appear in *Simon Stevin*)

[5] ________________________, 'On domains of biregularity in Clifford Analysis' (to appear in *Rend. Circ. Mat. Palermo*)

[6] R. Delanghe, F. Brackx and W. Pincket, 'On domains of monogenicity in Clifford Analysis' (to appear in *Rend. Circ. Mat. Palermo*)

CLIFFORD NUMBERS AND MÖBIUS TRANSFORMATIONS IN R^n

Lars V. Ahlfors*
Harvard University
Department of Mathematics
Science Center, 1 Oxford Street
Cambridge, MA 02138

ABSTRACT. Möbius transformations in any dimension can be expressed through 2×2 matrices with Clifford numbers as entries. This technique is relatively unknown in spite of having been introduced as early as 1902. The present paper should be viewed as a strong endorsement for the use of Clifford algebras in this particular context. In addition to an expository introduction to Clifford numbers it features a discussion of the fixed points and classification of Möbius transformations.

What is known today as Clifford analysis had its origin in R. Fueter's theory of quaternion-valued left-regular and right-regular functions. It was preceded by his recognition that Möbius transformations in the complex plane can be extended to the upper halfspace by means of the simple device of replacing the complex variable by a special quaternion.

Fueter was not aware that this had been done before, even in the more general setting of Clifford numbers, in an obscure paper by K.Th.Vahlen published in 1902. It is hardly surprising that he did not know the paper, for it seems to have been completely ignored, except for rather restrained comments in E. Cartan's and E. Study's encyclopedia article on complex numbers. It was rediscovered by H. Maass who in 1949 recognized its value and showed that the main results could be proved according to modern standards.

1. CLIFFORD NUMBERS

For this audience there is no need to define Clifford numbers or Clifford algebras, but there are many notations, and I must specify my choice. I denote by $\underline{\underline{A}}$ the associative algebra over the reals generated by an infinite sequence of elements $e_1,\ldots,e_n,\ldots$ subject to the relations $e_h^2 = -1$, $e_h e_k = -e_k e_h$ for $h \neq k$, and no others. Each $a \in \underline{\underline{A}}$ can be written uniquely in the form

*Research supported by National Science Foundation and Forschungs-Institut für Mathematik

J. S. R. Chisholm and A. K. Common (eds.), Clifford Algebras and Their Applications in Mathematical Physics, 167–175.

$$a = \sum a_\nu E_\nu$$

where ν runs over all ordered, finite multi-indices $\nu = \nu_1 \nu_2 \ldots \nu_p$, $0 < \nu_1 < \ldots < \nu_p$, the a_ν are real numbers, only finitely many $\neq 0$, and $E = e_{\nu_1} \ldots e_{\nu_p}$. The empty index $\nu = 0$ corresponding to $p = 0$ and $E_0 = 1$ (or e_0) is included, and the coefficient a_0 is regarded as the real part of a.

The subalgebra generated by $e_1, \ldots, e_n$ is denoted by A_n. A and A_n are vector spaces, the latter of dimension 2^n. The square norm $|a|$ is defined by $|a|^2 = \Sigma a_\nu^2$. We note that $A_0 = R$, $A_1 \simeq C$ and $A_2 \simeq H$ (the quaternions).

There are three involutions: the main involution, denoted $a \to a'$, the reverse involution $a \to a^*$, and the complex involution $a \to \bar{a} = a'^* = a^{*\prime}$ (the notations a' and $\bar{a}$ are rather commonly adopted, but $a^\dagger$ is more usual than a^*).

2. VECTORS

Although A and A_n are vector spaces in their own right there is good reason to reserve the name vector for the elements of a smaller subspace. The most common choice is to single out the pure vectors $x = x_1 e_1 + \ldots + x_n e_n$ with the characteristic property $x^2 = -|x|^2$. For our purposes we prefer to focus on all vectors $x = x_0 + x_1 e_1 + \ldots + x_{n-1} e_{n-1}$. For fixed n the set of these vectors is denoted by V^n, the superscript referring to the dimension. The space V^n will be identified with R^n; the point of the notation is that $x \in V^n$ plays a double role as a Clifford number and as an n-tuple $(x_0, \ldots, x_{n-1})$. The union of the V^n is denoted by V.

All vectors $x \in V$ satisfy $x^* = x$, $\bar{x} = x'$ and $x\bar{x} = |x|^2$. The last relation implies that every vector $x \neq 0$ is invertible with $x^{-1} = \bar{x}/|x|^2$. We shall also need the formula $x\bar{y} + y\bar{x} = 2(x,y)$ where $x,y \in V$ and (x,y) is the inner product.

An interesting feature is that every power of a vector is again a vector. The same is true of any polynomial in a vector x. The property carries over even to fractional powers. For instance,

$$\sqrt{x} = \sqrt{\frac{|x|+x_0}{2}} + \frac{x-x_0}{|x-x_0|} \sqrt{\frac{|x|-x_0}{2}}$$

is a vector.

3. THE CLIFFORD GROUP

The multiplicative structure of A gives rise to an important group known as the Clifford group. We shall denote it by Γ and the subgroup formed by elements in A_n by Γ_n. It is usually defined as the multiplicative group of all elements in A with the following property:

Definition. An element $a \in A$ is in Γ if and only if a is invertible and $ax = xa'$ for all $x \in V$.

We list a number of well known and easy to prove consequences of the definition:

(i) Γ is a multiplicative group,

(ii) every $a \in \Gamma$ satisfies $a\bar{a} = |a|^2$, $a^{-1} = \bar{a}/|a|^2$,

(iii) the norm $|\ |$ is multiplicative on Γ, i.e., $|ab| = |a|\cdot|b|$ if $a,b \in \Gamma$.

(iv) every non-zero vector is in Γ, and every $a \in \Gamma$ is a product of non-zero vectors.

(v) the mapping $x \to axa'^{-1}$ with $a \in \Gamma_{n-1}$ is a sense-preserving isometry of V^n. In the standard matrix notation this means that there exists a matrix $\rho(a) \in SO(n)$ such that $axa'^{-1} = \rho(a)x$ for all $x \in V^n$.

(vi) for every $k \in SO(n)$ there exists an $a \in \Gamma_{n-1}$ with $\rho(a) = k$; a is unique up to a real factor.

For any $a \in \Gamma$ the axis of a is the set of all $x \in V$ with $axa'^{-1} = x$. The axis will be denoted by ax (a) and its intersection with V^n by $ax_n(a)$. The dimension of $ax_n(a)$ is odd or even together with n.

The following observation will be used repeatedly: If $a,b \in \Gamma$, then ab^{-1} and b^*a are simultaneously in V. This follows from $b'^{-1}(ab^{-1})b = |b|^{-2}b^*a$. In particular, $a^*b = b^*a$.

4. CLIFFORD MATRICES

The leading idea in Vahlen's paper was to imitate the role of $GL_2(C)$ in the theory of Möbius transformations in the complex plane. Rather than increase the size of the matrices he chose to stay with two by two matrices and replace the elements by Clifford numbers, subject to certain restrictions. His aim was to let a matrix $g = \begin{pmatrix} a & b \\ c & d \end{pmatrix}$ act on a vector $x \in V$ according to the rule $gx = (ax+b)(cx+d)^{-1}$, and he needed to impose conditions under which this makes good sense.

With little or no historical justification I have introduced the name Clifford matrix, and I shall continue to use it. The following definition can be traced to Vahlen and was made explicit by Maass.

Definition. A matrix $g = \begin{pmatrix} a & b \\ c & d \end{pmatrix}$ is a Clifford matrix if the following is true:

(i) $a,b,c,d \in \Gamma \cup \{0\}$

(ii) $ad^*-bc^* = 1$

(iii) ac^{-1} and $c^{-1}d \in \underset{\sim}{V}$ if $c \neq 0$.

In case $c = 0$ (iii) is replaced by

(iii)' $b^{-1}a$ and $db^{-1} \in \underset{\sim}{V}$ if $b \neq 0$.

If b and c are both zero (iii) is omitted.

Without motivation this definition makes little sense. Condition (i) is needed already for the inverses. The quantity ad^*-bc^* is known as the pseudo-determinant of g. It is essential that it be real, and (ii) is a normalization. As in the complex case $\underset{\sim}{V}$ must be replaced by $\overline{\underset{\sim}{V}} = \underset{\sim}{V} \cup \{\infty\}$. Condition (iii) is chosen so that $x \in \overline{\underset{\sim}{V}}$ implies $gx = (ax+b)(cx+d)^{-1} \in \overline{\underset{\sim}{V}}$ and that the mapping $x \to gx$ is a Möbius transformation.

I have omitted to specify the dimensions because they are essentially arbitrary. Condition (i) implies in any case that a,b,c,d are in some $\Gamma_{n-1} \cup \{0\}$, and by (iii) ac^{-1} and $c^{-1}d$ are then in $\overline{V}^n$. If it has been proved that g induces a Möbius transformation in $\overline{V}^n$, then the same proof shows that g is also a Möbius transformation in $\overline{V}^{n+1}$ which is an extension of the one in $\overline{V}^n$. One recognizes that the introduction of Clifford numbers does exactly the same for n dimensions as, in Fueter's case, the quaternions did for two dimensions. This, by itself, is a strong indication that Clifford numbers are a natural tool for the study of Möbius transformations in arbitrary dimension.

5. THE THEOREM

The definition of a Clifford matrix implies much more than it states. In the first place, the observation at the end of Section 3 shows that condition (iii) can be replaced by $a^*c = c^*a \in \underset{\sim}{V}$ and $cd^* = dc^* \in \underset{\sim}{V}$; similarly, (iii)' is the same as $ab^* = ba^* \in \underset{\sim}{V}$ and $b^*d = d^*b \in \underset{\sim}{V}$. When this is combined with (ii) one obtains, for instance, $(c^*a)^{-1} = b^{-1}a - c^{-1}d$ and $(d^*b)^{-1} = b^{-1}a - d^{-1}c$ from which it can be deduced that (iii) implies (iii)' and vice versa. Finally, it is also true that $d^*a - b^*c = 1$.

Matrices of Clifford numbers can be multiplied just as ordinary matrices, and multiplication is transitive. The identities satisfied by Clifford matrices lead to

$$\begin{pmatrix} a & b \\ c & d \end{pmatrix}\begin{pmatrix} d^* & -b^* \\ -c^* & a^* \end{pmatrix} = \begin{pmatrix} d^* & -b^* \\ -c^* & a^* \end{pmatrix}\begin{pmatrix} a & b \\ c & d \end{pmatrix} = \begin{pmatrix} 1 & 0 \\ 0 & 1 \end{pmatrix} .$$

In other words, the matrices $\begin{pmatrix} a & b \\ c & d \end{pmatrix}$ and $\begin{pmatrix} d^* & -b^* \\ -c^* & a^* \end{pmatrix}$ are inverse to each other.

The definition of Clifford matrices is inseparable from the end product, a theorem which I regard as having been proved jointly by Vahlen and Maass. For its formulation we recall that $M(R^n)$ denotes the Möbius group of $\overline{\underset{\sim}{R}}^n = \underset{\sim}{R}^n \cup \{\infty\}$ which is generated by all similarities of $\underset{\sim}{R}^n$ together with the inversion $x \to x/|x|^2$. The subgroup of sense-preserving mappings is denoted by $M(\underset{\sim}{R}^n)^+$.

<u>Theorem</u>. (Vahlen and Maass). The Clifford matrices form a group $SL_2(\Gamma)$ under matrix multiplication with subgroups $SL_2(\Gamma_n)$ formed by the matrices with elements in Γ_n. Each $g = \begin{pmatrix} a & b \\ c & d \end{pmatrix} \in SL_2(\Gamma_{n-1})$ induces a well defined mapping $\overline{\underset{\sim}{V}}^n \to \overline{\underset{\sim}{V}}^n$ given by $x \to gx = (ax+b)(cx+d)^{-1}$. When $\overline{\underset{\sim}{V}}^n$ is identified with $\overline{\underset{\sim}{R}}^n$ the mapping becomes a member of $M(\underset{\sim}{R}^n)^+$. This sets up a homomorphism $SL_2(\Gamma_{n-1}) \to M(\underset{\sim}{R}^n)^+$ with matrix multiplication corresponding to composition of mappings. The kernel consists of $\begin{pmatrix} 1 & 0 \\ 0 & 1 \end{pmatrix}$ and $\begin{pmatrix} -1 & 0 \\ 0 & -1 \end{pmatrix}$.

This is not the place to reproduce the proof in any detail. We have already noticed the existence of an inverse. Condition (ii) remains true for a product of Clifford matrices as seen by checking that the "pseudo-determinant" is multiplicative when it is real. All other assertions follow rather directly from the factorization

$$\begin{pmatrix} a & b \\ c & d \end{pmatrix} = \begin{pmatrix} 1 & ac^{-1} \\ 0 & 1 \end{pmatrix}\begin{pmatrix} c^{*-1} & 0 \\ 0 & 1 \end{pmatrix}\begin{pmatrix} 0 & -1 \\ 1 & 0 \end{pmatrix}\begin{pmatrix} 1 & c^{-1}d \\ 0 & 1 \end{pmatrix}$$

which is valid when $c \neq 0$. Indeed, all the factors represent simple Möbius transformations, and all Möbius transformations can be put together from such factors.

It is more important to treat a few special cases. Given $g = \begin{pmatrix} a & b \\ c & d \end{pmatrix}$ we pick out $g^{-1}0 = -a^{-1}b$, $g^{-1}\infty = -c^{-1}d$ and denote them by u and v. Then $g = \begin{pmatrix} a & -au \\ c & -cv \end{pmatrix}$ and condition (ii) yields $a(u-v)c^* = 1$ or $c = a^{*-1}(u-v)^{-1}$. The special case $a = 1$ leads to the Clifford matrix

$$g_{u,v} = \begin{pmatrix} 1 & -u \\ (u-v)^{-1} & -(u-v)^{-1}v \end{pmatrix}. \tag{5.1}$$

This matrix induces the mapping $x \to (x-u)(x-v)^{-1}(u-v)$. The factor $u-v$ is indispensable, for $(x-u)(x-v)^{-1}$ is not a vector. The most general transformation which takes u to 0 and v to ∞ is given by $\begin{pmatrix} \lambda & 0 \\ 0 & \lambda^{*-1} \end{pmatrix} g_{u,v}$, $\lambda \in \Gamma$.

We emphasize once more that the action of any $g \in SL_2(\Gamma_{n-1})$ on $\overline{V}^n$ extends automatically to $\overline{V}^{n+1}$ and even to any $\overline{V}^m$ with $m > n$. Because all the mappings are sense-preserving it is clear from the start that the Poincaré upper halfspace $H^{n+1} \subset \tilde{V}^{n+1}$ is mapped on itself. This can also be verified from the explicit formulas.

6. FIXED POINTS AND CLASSIFICATION

Because Clifford numbers permit explicit computations it is perhaps not too optimisitic to hope that these methods could lead to simpler proofs of known results, especially in connection with discrete groups, but at present I have no significant results in this direction. Instead I have studied the question of fixed points of Möbius transformations and their classification in terms of the fixed points.

In the complex case a Möbius transformation has one or two fixed points, and they can be found by solving a quadratic equation. The transformations with a single fixed point are called <u>parabolic</u>, and the ones with two fixed points are classified as <u>elliptic</u> or <u>loxodromic</u>, with <u>hyperbolic</u> a special case of loxodromic.

In higher dimensions things are not as easy because of the lack of commutativity. The problem is to find all fixed points of the transformation induced by an explicitly given Clifford matrix g acting on $\overline{V}^n$. If h is another Clifford matrix, then hgh^{-1} is a conjugate of g. If g has the fixed point v, then hgh^{-1} has the fixed point hv. Therefore, if the problem can be solved for g it can be solved for the whole conjugacy class of g.

A diagonal Clifford matrix is of the form $\begin{pmatrix} \lambda & 0 \\ 0 & \lambda^{*-1} \end{pmatrix}$, $\lambda \in \Gamma$. It has always the fixed points 0 and ∞. If $|\lambda| \neq 1$ these are the only fixed points, and the transformation is <u>loxodromic</u>; this includes the hyperbolic case of real $\lambda \neq \pm 1$. If $|\lambda| = 1$ the transformation is elliptic, and the set of fixed points consists of the axis $ax_n(\lambda)$ together with ∞. The axis can reduce to 0 or it can be a linear subspace of $\underset{\sim}{V}^n$. It is customary to exclude the identity mapping from the classification.

A Clifford matrix with a single fixed point is said to be parabolic. By conjugation the fixed point can be brought to ∞, in which case g has the form $\begin{pmatrix} a & b \\ 0 & a^{*-1} \end{pmatrix}$. It can be shown that ∞ is the only fixed point if and only if $|a| = 1$ and $b = au = ua'$, $u \in \underset{\sim}{V}^n$. It can also happen that there is no fixed point in $\overline{\underset{\sim}{V}}^n$, but for topological reasons there is then a fixed point in $\underset{\sim}{V}^{n+1}$.

We shall say that a Clifford matrix is diagonalizable if it is conjugate to a diagonal matrix. It is loxodromic or elliptic together with the diagonal matrix. Thus a loxodromic transformation has exactly two fixed points, while in the elliptic case the fixed point set is a Möbius image of a linear subspace. As such it is either a sphere or an affine subspace of $\underset{=}{V}^n$.

A parabolic matrix is never diagonalizable, but one with two fixed points always is. For if u and v are fixed points of g and $h_{u,v}$ is given by (5.1), then $h_{u,v}\, g\, h_{u,v}^{-1}$ has fixed points 0 and ∞; it is therefore of the form $\begin{pmatrix} \lambda & 0 \\ 0 & \lambda^{*-1} \end{pmatrix}$. Calculation yields

$$\lambda = (u-v)(cv+d)(u-v)^{-1}, \quad \lambda^{*-1} = cu+d . \tag{6.1}$$

It follows that g is loxodromic if $|cu+d|$ and $|cv+d| \neq 1$, hyperbolic if $cu+d$ and $cv+d$ are real and $\neq \pm 1$, elliptic if $|cu+d| = |cv+d| = 1$. This information is of course not sufficient to find the fixed points.

7. NORMALIZATION

The search for fixed points can be simplified by conjugation with a parallel translation, i.e., with a matrix $h = \begin{pmatrix} 1 & \beta \\ 0 & 1 \end{pmatrix}$, $\beta \in \underset{\sim}{V}$. With this choice $hgh^{-1} = \begin{pmatrix} a+\beta c & * \\ c & -c\beta+d \end{pmatrix}$ where * will not be needed. One observes that c has not changed, and neither has the quantity $\sigma = \frac{1}{2}(ac^{-1}+c^{-1}d)$; if $c \neq 0$, σ is a well defined vector. The special choice $\beta = \frac{1}{2}(c^{-1}d-ac^{-1})$ leads to

$$hgh^{-1} = \begin{pmatrix} \sigma c & \sigma c \sigma - c^{*-1} \\ c & c\sigma \end{pmatrix} \tag{7.1}$$

where the fourth element has been determined from $ad^*-bc^* = 1$. A Clifford matrix of the form (7.1) is said to be normalized. Clearly, $g = \begin{pmatrix} a & b \\ c & d \end{pmatrix}$ is normalized if $c \neq 0$ and $ac^{-1} = c^{-1}d$, or $g\infty = -g^{-1}\infty$. One sees at once that v is a fixed point of (7.1) if and only if

$$c(v+\sigma)c^*(v-\sigma) = -1 . \tag{7.2}$$

For matrices with $c \neq 0$ the fixed point problem is solved if one can solve it for normalized matrices.

If $\sigma \neq 0$ the normalization can be pushed one step further. As a vector σ has a square root $\sigma^{1/2}$. Conjugation with $\begin{pmatrix} \sigma^{1/2} & 0 \\ 0 & \sigma^{-1/2} \end{pmatrix}$ brings (7.1) to the form

(7.3) $$\begin{pmatrix} c & c-c^{*-1} \\ c & c \end{pmatrix}$$

where the new c is $\sigma^{1/2}c\sigma^{1/2}$ in the notation of (7.1) and the new σ is 1.

A Clifford matrix of type (7.3) might be called *strongly normalized*. The only matrices that cannot be strongly normalized are those of the form $\begin{pmatrix} a & 0 \\ 0 & a^{*-1} \end{pmatrix}$ and $\begin{pmatrix} 0 & -c^{*-1} \\ c & 0 \end{pmatrix}$. For these the fixed point problem is relatively easy.

As a simple application we shall state the condition for a Clifford matrix to be hyperbolic.

A matrix of type (7.3) *is hyperbolic if and only if* c *is real* and $|c| > 1$.

If (7.3) is hyperbolic with fixed point v, then $c(v+1)$ is real by (6.1). Because of (7.2), $v+1$ is a non-zero vector. Therefore $c = c^*$ and, again by (7.2), $c(v-1)$ is real. It follows that c and v are both real, and (7.2) has the obvious solutions $v = \pm (1 - c^{-2})^{1/2}$. The converse is proved the same way.

For (7.1) the condition becomes σc real with absolute value >1, and $\begin{pmatrix} a & b \\ c & d \end{pmatrix}$ is hyperbolic if and only if $a+d^*$ is real and $(a+d^*)^2 > 4$. In each case the fixed points can be found explicitly.

8. THE ELLIPTIC CASE

Geometrically there is little difference between the normalized and strongly normalized case, but to expedite the discussion we shall stay with the form (7.3).

Assume that g is elliptic of type (7.3), acting on $\overline{V}^n$. Let v be any fixed point of g. It follows from (6.1) with $|\lambda| = 1$ together with (7.2) with $\sigma = 1$ that $|v+1| = |v-1| = |c|^{-1}$. In other words, v lies on the intersection of the spheres of radius $|c|^{-1}$ centered at -1 and 1. Because there are more than one fixed point $|c| < 1$, and the intersection is an $S^{n-2}(r)$ with radius $r = (|c|^{-2}-1)^{1/2}$; the notation $S^{n-2}(r)$ will refer to this particular sphere which lies in the hyperplane orthogonal to the reals.

Because of the orthogonality $v = v_1e_1+\dots+v_{n-1}e_{n-1}$ and $v = -\bar{v}$. Condition (7.2) becomes $c(v+1)c^* = -(v-1)^{-1} = (v+1)|c|^2$ or, equivalently, $c(v+1)c'^{-1} = v+1$. This means that $v+1 \in ax_n(c)$. Conversely, if this is so, then v is a fixed point.

The matrix (7.3) *is elliptic if the parallel to the axis of* c *through the point* -1 *intersects* $S^{n-2}(r)$ *in more than one point. If it does, this intersection is the set of fixed points.*

The condition is fulfilled if the axis of c *has dimension at least two and if* $|c| < \cos\varphi$ *where* φ *is the acute angle between the real axis and the axis of* c.

The second part is evident by elementary geometry. If $c = c'$ the axis of c contains the real axis and the condition reduces to $|c| < 1$. We point out that the criterion is constructive, for at least in principle it is always possible to determine the axis.

Recall that c in (7.3) is not the same as in (7.1). To obtain the condition for (7.1) the point -1 has to be replaced by $-\sigma$, σc by c, and φ by the angle between σ and $ax_n(\sigma c)$.

The parabolic case is so similar to the elliptic that we shall deal with it forthwith. Assume that $g = \begin{pmatrix} a & b \\ c & d \end{pmatrix}$ has a single fixed point v; we are still excluding the case $c = 0$ so that $v \neq \infty$. Conjugation with $h = \begin{pmatrix} 0 & 1 \\ -1 & v \end{pmatrix}$ leads to $hgh^{-1} = \begin{pmatrix} cv+d & -c \\ 0 & a-vc \end{pmatrix}$. If g is parabolic this matrix has only the fixed point ∞. In Section 6 it was pointed out that this is possible only if $|cv+d| = |a-vc| = 1$. If g is of type (7.3) it follows that $|v+1| = |v-1| = |c|^{-1}$. This means that we are in exactly the same situation as in the elliptic case, except that the parallel to the axis of c is required to meet $S^{n-2}(r)$ in a single point.

The matrix (7.3) is parabolic if the parallel to the axis of c through the point -1 intersects the sphere $S^{n-2}(r)$ in a single point. This point is the fixed point. Geometrically, this means that the axis of c does not reduce to a point, and that $|c| = \cos \varphi$ where φ is the acute angle between the real axis and the axis of c.

9. SUMMARY

We have found the conditions under which $g \in SL_2(\Gamma_{n-1})$ is elliptic or parabolic in its action on $\overline{V}^n$. If neither criterion is fulfilled g is either loxodromic or acts freely on $\overline{V}^n$. The theorem below will show which alternative occurs.

Theorem. Suppose that $g \in SL_2(\Gamma_{n-1})$ with $c \neq 0$ and $\sigma \neq 0$. If $\dim ax_n(\sigma c) > 0$, let φ denote the acute angle between σ and $ax_n(\sigma c)$. Then the following is true:

(i) If $\dim ax_n(\sigma c) \geqq 2$, then g is elliptic if $|\sigma c| < \cos \varphi$, parabolic if $|\sigma c| = \cos \varphi$, loxodromic if $|\sigma c| > \cos \varphi$.

(ii) If $\dim ax_n(\sigma c) = 1$, then g is parabolic if $|\sigma c| = \cos \varphi$, it acts freely on $\overline{V}^n$ and is elliptic on $\overline{V}^{n+1}$ if $|\sigma c| < \cos \varphi$, and it is loxodromic if $|\sigma c| > \cos \varphi$.

(iii) If $\dim ax_n(\sigma c) = 0$, then g is always loxodromic.

This is seen by examining what happens when g acts freely on $\overline{V}^n$. We know that g has then a pair of fixed points in V^{n+1}, symmetrically placed with respect to V^n. It is clearly elliptic on $\overline{V}^{n+1}$. By our criterion $ax_{n+1}(\sigma c)$ has dimension $\geqq 2$, and since it contains e_n which is orthogonal to $ax_n(\sigma c)$ it follows that $\dim ax_n(\sigma c) \geqq 1$. In particular, φ is defined for $ax_n(\sigma c)$ and is the same as for $ax_{n+1}(\sigma c)$. If $\dim ax_n(\sigma c)$ were > 1, g would be elliptic and not act freely on $\overline{V}^n$. The contradiction shows that $ax_n(\sigma c)$ has dimension 1. The statements (i)-(iii) are direct consequences of these observations together with the facts proved in Section 8.

REFERENCES:

L. Ahlfors: 'Möbius transformations and Clifford numbers', Differential Geometry and Complex Analysis - in memory of H.E. Rauch, Springer Verlag 1985.

L. Ahlfors: 'On the fixed points of Möbius transformations in R^n', Ann. Acad. Sci. Fenn. Ser. AI, Vol. 10, 1984.

R. Fueter: 'Sur les groupes improprement discontinus', C.R.Acad. Sci. Paris 182, 432-434, 1926.

H. Maass: 'Automorphe Funktionen von mehreren Veränderlichen und Dirichletsche Reihen', Abh. Math. Sem. Univ. Hamburg 16, 53-104, 1949.

K.Th. Vahlen: 'Über Bewegungen und complexe Zahlen', Math. Annalen 55, 585-593, 1902.

A CLIFFORD CALCULUS FOR PHYSICAL FIELD THEORIES

R. W. Tucker
University of Lancaster,
Department of Physics,
Lancaster
LA1 4YB

ABSTRACT. A Clifford calculus on sections of a Clifford bundle associated with a (pseudo-) Riemannian metric is reviewed. Its use is illustrated by reference to the Einstein - Yang - Mills equations. The formalism highlights the difference between the Kähler and Dirac equations and their separability in a curved space-time is discussed. Some aspects of supersymmetric models are outlined.

INTRODUCTION

Many physicists have discovered and rediscovered properties of a Clifford algebra associated with some scalar product underlying their theories. The vector space isomorphism between the exterior and Clifford algebras|1| offers a powerful means of unifying the two algebras. Until recently most developments have given rise to a calculus based on a parallelisable space; often Euclidean 3-space |2| or the flat space-time described by the Minkowski metric|3|. For theories that neglect Einsteinian gravitation (or its refinements) this often suffices although such formulations rarely contemplate covariant derivatives in a non-parallel basis. The recent upsurge of interest in higher-dimensional non-flat geometries to describe the fundamental interaction between fields has focussed attention on higher-dimensional Clifford algebras, a curved space Clifford calculus, as well as the description of spinors and spin groups in this context. Physical models which exhibit some of these aspects are to be found in refs.|4,6,7|. They exemplify the richness of structure to be found in Clifford algebras and Clifford groups |25|.

However, to incorporate gravitational interactions in terms of the geometry of a manifold, the calculus must be generalised to non-parallelisable spaces. The description of spinors in terms of minimal left ideals of a Clifford algebra can also be accommodated in this more general context. The following will attempt to summarise one way of doing this using the language of Clifford bundles. In particular the introduction of a spinor covariant derivative on the space of spinor fields will hopefully clarify the distinction between the Kähler and

J. S. R. Chisholm and A. K. Common (eds.), Clifford Algebras and Their Applications in Mathematical Physics, 177–199.

Dirac operators that feature prominently in certain physical theories.

1. Preliminaries

A Clifford calculus on an n-dimensional smooth real manifold M will be defined in terms of a differential calculus on local sections $\Gamma C(T^*M,g)$ of a Clifford algebra bundle $C(T^*M,g)$ associated with the tangent bundle TM of M and its pseudo-Riemannian metric g.

If $\Gamma\Lambda(T^*M)$ denotes the space of local sections of the exterior bundle $\Lambda(T^*M)$ we have the standard exterior and interior operators:

$$d : \Gamma\Lambda(T^*M) \to \Gamma\Lambda(T^*M) \qquad 1.1$$

$$i_X : \Gamma\Lambda(T^*M) \to \Gamma\Lambda(T^*M), \quad \forall\, X \in \Gamma TM \qquad 1.2$$

where $i_X\alpha = \alpha(X)$, with properties

$$i_X(\alpha\wedge\beta) = i_X\alpha\wedge\beta - \alpha\wedge i_X\beta \qquad 1.3$$

$$d(\alpha\wedge\beta) = d\alpha\wedge\beta - \alpha\wedge d\beta, \qquad \alpha,\beta \in \Gamma\Lambda^1(T^*M) \qquad 1.4$$

extended by linearity and associativity to all sections.
If we write the metric-dual 1-form $\tilde{X}$ for any vector field X

$$\tilde{X}(Y) = g(X,Y) \qquad \forall\, Y \in \Gamma TM$$

then we define the Hodge map $*$ by:

$$* : \Gamma\Lambda(T^*M) \to \Gamma\Lambda(T^*M)$$

$$*(\alpha_1\wedge\alpha_2\wedge\ldots\wedge\alpha_r) = i_{\tilde{\alpha}_r}\, i_{\tilde{\alpha}_{r-1}} \ldots i_{\tilde{\alpha}_1} z, \quad \alpha_j \in \Gamma\Lambda^1(T^*M)\,. \qquad 1.5$$

The n-form z is an orientation fixing n-form. [If $\{e^k\}$ denotes an ordered set of g-orthonormal 1-forms - a local orthonormal co-frame - we may adopt $z \equiv *\,1 = \prod_k{}_{\wedge}(e^k)$]. It follows that on a pseudo-Riemannian manifold $**$ acting on p-forms satisfies

$$** = (-1)^{p(n-p)}t \qquad 1.6$$

where t (the signature) is the product of eigenvalues of the metric tensor. Furthermore:

$$*(\phi\wedge A) = i_{\tilde{A}}*\phi \qquad 1.7$$

for any $\phi \in \Gamma\Lambda(T^*M)$, $A \in \Gamma\Lambda^1(T^*M)$. We may call sections of $\Lambda(T^*M)$ real inhomogeneous differential forms. They may be expressed as

$\alpha = \sum_{p=o}^{n} S_p(\alpha)$ in terms of p-form projectors satisfying:

$$\sum_{p=o}^{n} S_p = 1$$

$$S_k S_\ell = \delta_{k\ell} S_k \qquad \text{(no sum) } k,\ 1 = 0,1,\ \ldots .\ n$$

$$d\, S_k = S_{k+1}\, d$$

$$* S_k = S_{n-k} * .$$

The exterior algebra of sections $\Gamma\Lambda(T^*M)$ has two involutory operations ξ, η defined by

$$\xi(\alpha \wedge \beta) = \xi\beta \wedge \xi\alpha \qquad 1.9$$

$$\eta(\alpha \wedge \beta) = \eta\alpha \wedge \eta\beta \qquad 1.10$$

where

$$\eta\alpha = \sum_p \eta S_p(\alpha) = \sum_p (-1)^p S_p(\alpha) \qquad 1.11$$

$$\xi\alpha = \sum_p \xi S_p(\alpha) = \sum_p (-1)^{[\frac{p}{2}]} S_p(\alpha) \qquad 1.12$$

and [r] is the integer part of r.

Since (T^*M, g) is a pseudo-Riemannian structure on M it is possible to give $\Lambda(T^*M)$ a Clifford structure by exploiting a vector space isomorphism between exterior and Clifford algebras. For every $\alpha \in \Gamma\Lambda^1(T^*M)$, $\Phi \in \Gamma\Lambda(T^*M)$ we define $C(T^*M,g)$ to be the bundle $\Lambda(T^*M)$ on which a Clifford product $_\vee$ between sections is induced from

$$\alpha_\vee \Phi = \alpha \wedge \Phi + i_{\tilde{\alpha}} \Phi \qquad 1.13$$

by linearity and associativity of the Clifford algebra product. Local sections $\Gamma C(T^*M,g)$ define Clifford forms on M via a pull-back. The correspondence between $_\vee$ and $\wedge$ enables us to induce from (9,10) the Clifford algebra involutions

$$\eta(\alpha_\vee \beta) = \eta\alpha\ _\vee\ \eta\beta \qquad 1.14$$

$$\xi(\alpha_\vee \beta) = \xi\beta\ _\vee\ \xi\alpha \qquad 1.15$$

$\alpha, \beta \in \Gamma C(T^*M,g)$ and no confusion arises in assigning to them the same notation.

If $\{e^a\},\{X_b\}$ are any naturally dual local bases for $\Gamma(T^*M)$ and ΓTM respectively; $e^a(X_b) = \delta^a{}_b$, then repeated use of (1.13) gives

$$\alpha_\vee\beta = \sum_{p=0}^{n} \frac{(-1)^{[\frac{p}{2}]}}{p!} (\eta^p i_{X_{a_1}} \ldots i_{X_{a_p}} \alpha)\wedge(i_{\tilde{e}^{a_1}} \ldots i_{\tilde{e}^{a_p}} \beta) \quad 1.16$$

$$\alpha\wedge\beta = \sum_{p=0}^{n} \frac{(-1)^{[\frac{p}{2}]}}{p!} (i_{X_{a_1}} \ldots i_{X_{a_p}} \eta^p\alpha)_\vee(i_{\tilde{e}^{a_1}} \ldots i_{\tilde{e}^{a_p}} \beta) . \quad 1.17$$

From these basic ideas it will be noted that the underlying structures for our calculus depend on the existence of differentiable manifolds with a (pseudo) Riemannian structure. It is not necessary to consider such manifolds as embeddings in a larger parallelisable space. Traditional methods of differential geometry will be exploited so that established and powerful concepts remain accessible.

In particular let us denote by ∇ the metric compatible (pseudo) Riemannian connection used in the construction of covariant derivatives of tensor fields. Thus $\nabla_X:\Gamma C(T^*M,g) \to \Gamma C(T^*M,g)$ $\forall X \in \Gamma TM$. In terms of ∇_X we introduce the frame independent differential operator

$$\not{d} : \Gamma C(T^*M,g) \to \Gamma C(T^*M,g) . \quad 1.18$$

In terms of the arbitrary naturally dual frames above

$$\not{d} = e^a{}_\vee\nabla_{X_a} , \quad 1.19$$

or exploiting (1.13)

$$\not{d} = e^a\wedge\nabla_{X_a} + i_{\tilde{e}^a}\nabla_{X_a} . \quad 1.20$$

If ∇ is chosen to be torsion free then $e^a\wedge\nabla_{X_a} = d$ and $i_{\tilde{e}^a}\nabla_{X_a} = -\tilde{\delta}$ where, in terms of the Hodge map $\tilde{\delta} = *^{-1}d*\eta$. For space-time (n=4) with a Lorentzian signature (t = -1)

$$\not{d} = d - \delta \quad 1.21$$

and $\delta \equiv *d*.\not{d}$ is often referred to as the Hodge-de-Rham operator. Since $d^2 = \delta^2 = 0$, it has the property

$$\not{d}^2 = -(d\delta + \delta d)$$

which is the Laplace-Beltrami operator on M. Since ∇_X is metric compatible, for any Clifford forms α, β

$$\nabla_X(\alpha_\vee\beta) = \nabla_X\alpha_\vee\beta + \alpha_\vee\nabla_X\beta \ . \qquad 1.22$$

From this is follows that :

$$\not{d}(\alpha_\vee\beta) = \not{d}\alpha_\vee\beta + \eta\alpha_\vee\not{d}\beta + 2i_{\tilde{e}^a}\alpha_\vee\nabla_{X_a}\beta \qquad 1.23$$

$$\not{d}(\alpha\wedge\beta) = \not{d}\alpha\wedge\beta + \eta\alpha\wedge\not{d}\beta + \eta\nabla_{X_a}\alpha\wedge i_{\tilde{e}^a}\beta + i_{\tilde{e}^a}\alpha\wedge\nabla X_a\beta \qquad 1.24$$

$$d(\alpha_\vee\beta) = d\alpha_\vee\beta + \eta\alpha_\vee d\beta + i_{\tilde{e}^a}\alpha_\vee\nabla_{X_a}\beta - \eta\nabla_{X_a}\alpha_\vee i_{\tilde{e}^a}\beta \ . \qquad 1.25$$

2. Classical Field Theories

It has long been recognised that the calculus of exterior forms on a manifold affords a natural language for many classical field theories in physics. Not only does this language simplify the formulation of such theories, it provides an efficient calculus for analysing their properties and offers a means for globalising local descriptions to accommodate the topological effects induced by the underlying manifold. Most dynamical field theories involve the space-time metric in an essential way. As such they are amenable to a formulation in terms of a Clifford calculus. This calculus can be constructed to retain many of the nice features of the exterior calculus but is often richer in structure and open to considerable generalisation. One of the prime motivations of our research programme has been to treat both tensor and spinor field theories in the same general framework, adopting a unified language for both their description and analysis.

Let us briefly summarise the Yang-Mills (including Maxwell) and Einstein classical field theories in the language of Clifford forms.

3. The Yang-Mills System

If $\mathbb{A} = A_i T^i$ is a Lie algebra valued 1-form, $A_i \varepsilon \Gamma\Lambda^1(T^*M)$ in a basis $\{T^i\}$ for some Lie algebra, the Yang-Mills field strength is $\mathbb{F} = d\,\mathbb{A} + [\,\mathbb{A}, \mathbb{A}]$, where the bracket between Lie algebra valued elements is $[\,\mathbb{H}, \mathbb{B}] = H_{i\wedge} B_j[T^i, T^j]$ in terms of the Lie bracket. Similarly define $\mathbb{H}_\vee\,\mathbb{B} = H_{i\vee}B_j[T^i, T^j]$. A Yang-Mills field system satisfies the field equation

$$\not{D}\,\mathbb{F} = 0 \qquad 3.1$$

where $\not{D} \equiv \not{d} + \mathbb{A}_\vee$. In a torsion-free space-time this equation is equivalent to the 3-form equation

$$d\,\mathbb{F} + [\,\mathbb{A}, \mathbb{F}] = 0 \qquad 3.2$$

and the 1-form equation

$$- \tilde{\delta}\, \mathbb{F} + i_{\tilde{A}_j} F_k[T^j,T^k] = 0 \ . \qquad 3.3$$

The first will be recognised as a Bianchi identity for $\mathbb{F}$ and the second may be written

$$d * \mathbb{F} + [\, \mathbb{A} \,, * \mathbb{F}] = 0 \qquad 3.4$$

using (1.6) and (1.7). The source-free Maxwell equations arise when the Yang-Mills group is chosen to be U(1).

4. Theories of Gravitation

It is convenient to introduce on sections of the bundle $\Lambda^p C \equiv \Lambda^p T^*M \otimes C(T^*M,g)$ an algebra $(\wedge, \vee)$ by

$$\alpha_\wedge : \Gamma\Lambda^p C \to \Gamma\Lambda^p C \ , \ \gamma \otimes \rho \to (\alpha_\wedge \gamma) \otimes \rho \qquad 4.1$$

$$\beta_\vee : \Gamma\Lambda^p C \to \Gamma\Lambda^p C \ , \ \gamma \otimes \rho \to \gamma \otimes (\beta\ \rho) \qquad 4.2$$

$$(\alpha \otimes \beta)(\gamma \otimes \rho) = (\alpha_\wedge \gamma) \otimes (\beta_\vee \rho) \qquad 4.3$$

for $\alpha \varepsilon \Gamma\Lambda^p(T^*M)$, $\beta \varepsilon \Gamma C(T^*M,g)$. The operators d, *, i_X are naturally extended to act on the first factor:

$$d : \Lambda^p C \to \Lambda^{p+1} C \qquad 4.4$$

$$* : \Lambda^p C \to \Lambda^{n-p} C \qquad 4.5$$

$$i_X : \Lambda^p C \to \Lambda^{p-1} C \ . \qquad 4.6$$

Classical theories of gravity may be written in terms of local sections $\{e^a\}$ of the orthonormal frame bundle and for each local frame a set of connection 1-forms $\{\omega^a{}_b\}$ on M. The metric tensor of space-time in such a local frame is

$$g = - e^a \otimes e^a + \sum_{k=1}^{3} e^k \otimes e^k \equiv \eta_{ab} e^a \otimes e^b \qquad 4.7$$

and the connection forms $\omega_b{}^c$ of ∇ are defined by 4.8

$$\nabla_X e^c = \omega_b{}^c(X) e^b \quad \forall X \ \varepsilon\ \Gamma TM \ . \qquad 4.9$$

These give rise to an element $\Omega = \omega_{ab} \otimes (e^a{}_\vee e^b)/4 \ \varepsilon\ \Gamma\Lambda^1 C$

$(\omega_{ab} = \eta_{ac}\omega^c{}_b)$. Define a Clifford curvature of ∇

$$\hat{R} = d\Omega + \Omega\Omega \qquad \varepsilon\ \Gamma\Lambda^2 C \ . \qquad 4.10$$

In terms of the curvature 2-forms $R_{ab} = d\omega_{ab} + \omega_a{}^c \wedge \omega_{cb}$

$$\hat{R} = R_{ab} \otimes \frac{1}{4}(e^a{}_v e^b) \ . \qquad 4.11$$

If we introduce $e = e^b \otimes e_b \in \Gamma\Lambda^1 C$ the vacuum Einstein equations are simply

$$e\hat{R} = 0 \ . \qquad 4.12$$

It is an interesting exercise in the "mixed" algebra introduced here to show that all solutions to this equation satisfy the double duality condition

$$*\hat{R} = z_v \hat{R} \ . \qquad 4.13$$

Given an action 4-form Λ on M for "matter" fields, the stress forms may be computed as coefficients of metric variations that deform the Clifford algebra structure of ΛT^*M. For example, for the Yang-Mills system above we may write

$$\Lambda = \mathrm{Tr} S_o(\mathbb{F}_v \mathbb{F}) z \ . \qquad 4.14$$

The orthonormal components of the stress tensor are then computed from this action to be

$$T_{ab} = \frac{1}{2}\mathrm{Tr} S_o\{e_{av}\, \mathbb{F}_v e_{bv} \mathbb{F} + e_{bv}\, \mathbb{F}_v e_{av}\, \mathbb{F}\} \ . \qquad 4.15$$

5. Space-Time Algebraic Spinors

For g with signature (- ,+,+,+), $C(T^*M,g)$ is isomorphic to a bundle of real 4 x 4 matrix algebras. Since M is a real manifold we shall concentrate on the real field but may complexify to the complex algebra $C_{3,1}(\mathbb{R}) \otimes \mathbb{C}(\mathbb{R})$ and denote by $\mathbb{C}(T^*M,g)$ the induced bundle of algebras with fibre $C_{3,1}(\mathbb{R}) \otimes \mathbb{C}(\mathbb{R})$. $\Gamma\mathbb{C}(T^*M,g)$ may be decomposed into 4 minimal left ideals characterised by a complete set of 4 minimal rank (primitive) idempotents (P_j) such that $P_{iv} P_j = P_i \delta_{ij}$, (no sum) i.e. they are pairwise 'orthogonal' under Clifford multiplication. The set is maximal in the sense that it is not possible to find a set of more than 4 idempotents having these properties. The primitivity of each P_j implies

$$P_{jv}\Phi_v P_j = 4S_o(\Phi_v P_j)P_j \ , \qquad \Phi \in \Gamma\mathbb{C}(T^*M,g) \ . \qquad 5.1$$

Since $C_{3,1}(\mathbb{R}) \otimes \mathbb{C}(\mathbb{R})$ is a total matrix algebra it is always possible to construct a basis $\{\underline{\varepsilon}_{ij}\} \in \Gamma\mathbb{C}(T^*M,g)$ i,j = 1,2,3,4, for it satisfying

$$\underline{\varepsilon}_{ijv}\underline{\varepsilon}_{jk} = \underline{\varepsilon}_{ik} \qquad \text{(no sum)} \qquad 5.2$$

$$\underline{\varepsilon}_{ijv}\underline{\varepsilon}_{pk} = 0 \qquad j \neq p \qquad 5.3$$

Clearly there exists a whole class of such bases related by an inner automorphism i.e.

$$\underline{\varepsilon}_{ij} \to S \vee \underline{\varepsilon}_{ij} \vee S^{-1} \tag{5.4}$$

where S is any invertible element in $\Gamma\mathbb{C}(T^*M,g)$. Elements in this class constitute a matrix basis for $\Gamma\mathbb{C}(T^*M,g)$ since if we write $\phi = \sum_{i,j} \Phi_{ij}\,\underline{\varepsilon}_{ij}$, $\psi = \sum_{i,j} \psi_{ij}\,\underline{\varepsilon}_{ij}$ $\varepsilon \Gamma C(T^*M,g)$ where Φ_{ij},ψ_{ij} are $\mathbb{C}$ valued functions on M then

$$\Phi \vee \Psi = \sum_{i,j} \rho_{ij}\,\underline{\varepsilon}_{ij} , \tag{5.5}$$

where

$$\rho_{ij} = \sum_{k=1}^{4} \Phi_{ik}\psi_{kj} . \tag{5.5}$$

In terms of $\{P_j\}$ the basis elements $\{\underline{\varepsilon}_{ij}\}$ satisfy:

$$\underline{\varepsilon}_{ij} = P_i \vee c_{ij} \vee P_j = P_i \vee c_{ij} = c_{ij} \vee P_j \qquad \text{(no sum)} \tag{5.6}$$

for some elements $c_{ij}\ \varepsilon\ \Gamma\mathbb{C}(T^*M,g)$.

A particular basis $\{\underline{\varepsilon}_{ij}\}$ constructed from the 4 primitive idempotents $P_{\varepsilon\sigma} = \frac{1}{4}(1 + \varepsilon e^0 \vee e^1)\vee(1 + i\sigma z), \varepsilon,\sigma = \pm 1$, is given in the following table:

$\underline{\varepsilon}_{ij}$ i \ j	1	2	3	4
1	P_1	$e^2 \vee P_2$	$e^0 \vee e^2 \vee P_3$	$e^0 \vee P_4$
2	$e^2 \vee P_1$	P_2	$- e^0 \vee P_3$	$e^2 \vee e^0 \vee P_4$
3	$e^0 \vee e^2 \vee P_1$	$e^0 \vee P_2$	P_3	$e^2 \vee P_4$
4	$- e^0 \vee P_1$	$e^2 \vee e^0 \vee P_2$	$e^2 \vee P_3$	P_4

where $P_1 = P_{++}, P_2 = P_{+-}$, $P_3 = P_{-+}$, $P_4 = P_{--}$. We have the relations

$$e^0 \vee P_{\varepsilon\sigma} = P_{-\varepsilon-\sigma} \vee e^0 \quad , \quad e^2 \vee P_{\varepsilon\sigma} = P_{\varepsilon-\sigma} \vee e^2$$

$$e^1 \vee P_{\varepsilon\sigma} = P_{-\varepsilon-\sigma} \vee e^1 \quad , \quad e^3 \vee P_{\varepsilon\sigma} = P_{\varepsilon-\sigma} \vee e^3 \ .$$

An element $\psi\ \varepsilon\ \Gamma\mathbb{C}(T^*M,g)$ belongs to a minimal left ideal $I_L(P)$ represented by a primitive idempotent $P\ \varepsilon\ \Gamma\mathbb{C}(T^*M,g)$ if $\psi = \psi \vee P$. We refer to such elements as <u>algebraic spinor fields</u>, although it should be noted that as a Clifford module $I_L(P)$ carries a representation of the full Clifford algebra. Pin (3,1) representations are included as subgroup representations. A space-time algebraic spinor field may be written in a matrix basis as $\psi = \sum_{k=1}^{4} \psi_k f^k$ where for each J $\{f^k\} = \{\underline{\varepsilon}_{kJ}\}$ is a basis for $I_L(P_J)$, J = 1, 2, 3, 4 and $\{\psi_k\}$ denotes a set of $\mathbb{C}$ valued components in this basis.

6. The Kähler Equation

In 1928 Darwin was experimenting with tensor equations in order to |8| understand the properties of electrons. He eventually made contact with Dirac's wave equation but considered his method uneconomical. Apparently Landau and Ivanenko |8| around the same time had similar intentions. In 1961 E. Kähler |9| introduced the equation

$$\not{d}\Phi = \mu\Phi - iA_{\vee}\Phi \qquad 6.1$$

for a complex inhomogeneous differential form Φ on a pseudo-Riemannian manifold and recovered Dirac's solution for the wave mechanics of a relativistic electron of mass μ in a Hydrogen atom when he analysed (6.1) in flat Minkowski space-time. Kähler's equation was formulated in terms of a torsion-free $\not{d}$ and may be written down on any manifold with a pseudo-Riemannian structure as

$$d\Phi - \tilde{\delta}\Phi = \mu\Phi - i\, i_{\tilde{A}}\Phi - iA\wedge\Phi \quad , \quad A \in \Gamma T^*M \, . \qquad 6.2$$

Apart from a penetrating paper by Graf |10| in 1978 this equation remained dormant until around 1982 when it was simultaneously taken up by a number of groups. Many of these were interested in using it to describe fermions on a lattice |11|. In Lancaster its properties as a quantum field theory have been examined |12,13,14|. It also proved to be a catalyst for a reformulation of spinor wave equations |15| on a spin-manifold in terms of local sections of an exterior bundle equipped with a Clifford structure and the development of a Clifford calculus that has proved of great value in a number of investigations - |16|, |17|,|18|.

Let us attempt to find real solutions to Kähler's equation with A = 0, that lie in some real minimal left ideal characterised by a real primitive idempotent P_j, i.e. algebraic spinor solutions. First, right multiply by P_j:

$$\not{d}\Phi_{\vee}P_j = \mu\Phi_{\vee}P_j \qquad 6.3$$

and use:

$$\not{d}\Phi_{\vee}P_j = e^a{}_{\vee}\nabla_X\, \Phi_{\vee}P_j = e^a{}_{\vee}\nabla_{X_a}(\Phi_{\vee}P_j) - e^a{}_{\vee}\Phi_{\vee}\nabla_{X_a}P_j \, .$$

Consequently $\psi_{(j)} \equiv \Phi_V P_j \in I_L(P_j)$ is an algebraic spinor solution

$$d\!\!/\psi_{(j)} = \mu\psi_{(j)} \qquad 6.4$$

provided $e^a{}_V \Phi_V \nabla_{X_a} P_j = 0$. In flat Minkowski space we can always choose inertial co-ordinates in which $e^a = dx^a$, $a = 0,1,2,3$, and a primitive idempotent P, satisfying $\nabla_X P = 0$, $\forall X \in \Gamma TM$. Furthermore $d\!\!/ = dx^a{}_V \nabla_{\partial/\partial x^a}$ so Kähler's equation reduces to Dirac's equation for the components $\psi_{(j)k}$ of a Majorana spinor $\psi_{(j)} \in \Gamma I_L(P_j)$,

$$(\gamma^a \frac{\partial}{\partial x^a} - \mu)_{ik}\psi_{(j)k} = 0 \qquad i,j,k = 1,2,3,4 \ . \qquad 6.5$$

In this reduction the 4 x 4 matrix γ^a has as its elements the components of the 1-form e^a in the matrix basis $\{\underline{\varepsilon}_{jk}\}$ associated with the set $\{P_j\}$:

$$e^a = \sum_{j,k} (\gamma^a)_{jk}\, \underline{\varepsilon}_{jk} \ . \qquad 6.6$$

Since we have 4 different primitives P_j we see that Kähler's equation in Minkowski space can be decoupled into 4 identical copies of one Dirac equation for the components of a spinor. There is no difficulty in analysing the Kähler equation including an electromagnetic interaction with the Kähler field and recovering 4 minimally coupled Dirac equations in Minkowski space. Clearly the Kähler field carries more dynamical degrees of freedom than a single Majorana spinor and these are not distinguished by generalising the theory to include minimal electromagnetic interactions. In a general curved space, however, we do not expect to be able to decouple the Kähler equation into equations for an algebraic spinor.

As an illustration of the utility of the Clifford calculus of sections of C(T*M,g), the equation $d\!\!/\Phi = \mu\Phi$ has been recently analysed |18| in spherically-symmetric space-times with metrics of the form

$$\begin{aligned} g &= - H_o{}^2(r)dt \otimes dt + e^{2\lambda(t)}\{H_1{}^2(r)dr \otimes dr \\ &\qquad + H_2{}^2(r)[d\theta \otimes d\theta + \sin^2\theta d\phi \otimes d\phi]\} \\ &= - e^0 \otimes e^0 + \sum_{k=1}^{3} e^k \otimes e^k \end{aligned} \qquad 6.7$$

in a local chart with co-ordinates (t,r,θ,ϕ), $0 \leqslant r < \infty$, $0 < \theta < \frac{\pi}{2}$, $0 \leqslant \phi < 2\pi$. The angular dependence of Φ may be separated into the forms $Y_\ell^m(\theta,\phi)$ and $dY_\ell^m(\theta,\phi)$ where $Y_\ell^m(\theta,\phi)$ are the standard spherical harmonics. Writing

$$\Phi = \sum_{\ell,m} (U_\ell dY^m_\ell + W_\ell Y^m_\ell) \qquad 6.8$$

where U_ℓ and W_ℓ are Clifford forms independent of θ, ϕ and both annihilated by $i_{\tilde{e}^2}$ and $i_{\tilde{e}^3}$, one obtains the coupled equations:

$$\not{d}W_\ell - \ell(\ell + 1)\frac{e^{-2\lambda}}{H_2^2}\eta U_\ell = \mu W_\ell \; ; \qquad 6.9$$

$$\not{d}U_\ell + \frac{2\dot{\lambda}}{H_o^2}\, i\frac{\partial}{\partial t}U_\ell - \frac{2e^{-2\lambda}}{H_1^2 H_2}(\frac{\partial}{\partial r}H_2)\, i\frac{\partial}{\partial r}U_\ell + \eta W_\ell = \mu W_\ell \; . \qquad 6.10$$

In Schwarzschild (or Reisner-Nordstrom) backgrounds, for example, a complete set of solutions exist in terms of

$$U_\ell = f_\ell(r) + g_\ell(r)e^1 \qquad 6.11$$

$$W_\ell = F_\ell(r) + G_\ell(r)e^1 \; , \qquad 6.12$$

where the real functions f_ℓ, g_ℓ, F_ℓ, G_ℓ satisfy the ordinary differential equations that are trivial to decouple:

$$\partial_r(H_1^{-1}\partial_r f_\ell) + \rho\partial_r f_\ell - (\ell(\ell + 1)H_2^{-2} + \mu^2)H_1 f_\ell$$
$$= - \mu\gamma H_1 g_\ell \qquad 6.13$$

$$\partial_r(H_1^{-1}\partial_r g_\ell) - \partial_r(\sigma g_\ell) - (\ell(\ell + 1)H_2^{-2} + \mu^2)H_1 g = 0 \qquad 6.14$$

$$\partial_r(\frac{H_2^2}{\ell(\ell+1)H_1}\partial_r G_\ell) + \partial_r(\frac{\rho H_2^2}{\ell(\ell+1)}G_\ell) -$$
$$- (\frac{H_2^2}{\ell(\ell+1)} + \mu^2)\frac{H_1 H_2^2 G_\ell}{\ell(\ell+1)} = \frac{-\mu\gamma}{\ell(\ell+1)}H_1 H_2^2 F_\ell \qquad 6.15$$

$$\partial_r(\frac{H_1 H_2^2}{\ell(\ell+1)}\partial_r F_\ell) - \frac{H_2^2\sigma}{\ell(\ell+1)}\,\partial_r F_\ell -$$
$$- (\frac{\ell(\ell+1)}{H_2^2} + \mu^2)\frac{H_1 H_2^2}{\ell(\ell+1)}\,F_\ell = 0 \qquad 6.16$$

where $\sigma = - \frac{\partial_r H_o}{H_o H_1}$, $\gamma = - \frac{2\partial_r H_2}{H_1 H_2}$, $\rho = - (\sigma + \gamma)$.

In flat space-time we may write

$$\Phi = e^{i\omega t} \sum_{\ell,m} R_\ell \vee r^{1-\ell} d(r^\ell Y_\ell^m) \vee \tfrac{1}{2}(1 + idt) \tag{6.17}$$

where $R_\ell \equiv A_\ell(r) + B_\ell(r)dr$, $A_\ell, B_\ell \in \Gamma\Lambda^0(T^*M)$.

Then

$$d\!\!/\Phi = \{d\!\!/R_\ell + \xi\eta R_\ell \vee dr \frac{(1-\ell)}{r}\} \vee r^{1-\ell} d(r^\ell Y_\ell^m) \tag{6.18}$$

and A_ℓ and B_ℓ decouple to satisfy spherical Bessel equations

$$\partial_r^2 A_\ell + \frac{2}{r}\partial_r A_\ell + \{\omega^2 - \mu^2 + \frac{(-\ell)(-\ell+1)}{r^2}\} A_\ell = 0 \tag{6.19}$$

$$\partial_r^2 B_\ell + \frac{2}{r}\partial_r B_\ell + \{\omega^2 - \mu^2 + \frac{\ell(\ell+1)}{r^2}\} B_\ell = 0 \,. \tag{6.20}$$

It is possible to multiply (6.17) by any ∇-parallel primitive idempotent and construct an algebraic spinor Kähler solution (in a polar chart) belonging to a minimal ideal.

7. Action and Conserved Currents for Kähler Solutions

If N, ϕ, $\psi \in \Gamma\mathbb{C}(T^*M,g)$ we may define an inner product $\{\psi,\phi\} = \mathrm{ReS}_0(\xi\psi^*_\vee\phi)$ with the properties $\{\psi,\phi\} = \{\phi,\psi\}$, $\{\phi,N_\vee\psi\} = \{\psi,\xi N^*{}_\vee\phi\}$. For a general metric-compatible ∇ we have the identity

$$\{\phi,d\!\!/\psi\}\star + \{d\!\!/\phi,\psi\}\star = dJ(\phi,\psi) + W(\phi,\psi) \,, \tag{7.1}$$

where $J(\phi,\psi) \equiv -J(\psi,\phi) \equiv *\mathrm{ReS}_1(\psi_\vee\xi\phi^*)$

$W(\phi,\psi) \equiv \mathrm{ReS}_0(\xi\phi^*{}_\vee i_{\tilde{e}^b} T^b{}_\vee\psi)\star$

in terms of the torsion 2-forms associated with ∇ and the orthonormal co-frame $\{e^b\}$. Since $\nabla\star = 0$ and $d\!\!/(\star_\vee\psi) = (-1)^{n-1}\star_\vee d\!\!/\psi$ this implies

$$\{d\!\!/\phi,\star_\vee\rho\}\star - (-1)^n\{\phi,\star_\vee d\!\!/\rho\} = dJ(\phi,\star_\vee\rho) + W(\phi,\star_\vee\rho) \tag{7.2}$$

by setting $\psi = \star_\vee\rho$. In a Minkowski-signatured space-time with n = 4 and zero torsion it follows that if α and β are any solutions of Kähler's equation $d\!\!/\alpha = \mu\alpha$ then $dJ(\alpha,\beta) = 0$ where $J(\alpha,\beta) = -J(\beta,\alpha) = \mathrm{ReS}_3(\beta^*{}_\vee\xi\alpha)$. For a local Killing vector field K, $\mathcal{L}_K * = * \mathcal{L}_K$ where $\mathcal{L}_K$ denotes the Lie derivative with respect to K. Thus, if α is a Kähler solution so is $\beta = \mathcal{L}_K\alpha$ and the Killing current $J(\alpha, \mathcal{L}_K\alpha)$ is conserved. Furthermore if we restrict to a space-time that admits a ∇-parallel element $\sigma \in \Gamma\mathbb{C}(T^*M,g)$ $(\nabla\sigma = 0)$, then if α is a Kähler solution so is $\alpha_\vee\sigma$ and $J(\alpha,\alpha_\vee\sigma)$ is another conserved current. In particular, in flat Minkowski space, $d\tilde{K}$ is ∇-parallel for a Killing vector field generating SO(3) and the current $J(\alpha,\mathcal{L}_k\alpha + \tfrac{1}{4}\alpha_\vee d\tilde{K})$ may be shown |12| to give rise to angular momentum "spin" currents with half-integer eigenvalues for "rest" solutions.

The inner product { , } is also useful for constructing an action principle for the U(1) coupled Kähler equation. If A is the real Maxwell 1-form with F = dA, the action 4-form is

$$\Lambda = \{\eta\phi, (\not{d} - iA_v - \mu)\phi\}z + \{F,F\}z \ . \qquad 7.3$$

8. Algebraic Spinor Solutions to the Kähler Equation

We have defined an algebraic spinor field to be an inhomogeneous differential ϕ satisfying $\phi = \phi_v P$ for some primitive idempotent P in $\Gamma\mathbb{C}(T^*M,g)$. As such it is independent of any choice of local frame in the bundle of linear frames. The question as to the nature of the manifold that admits such forms as solutions to the Kähler equation is an intriguing one. Sections of the Clifford bundle belonging to any minimal ideal of the Clifford algebra are not in general preserved under the action of ∇ or $\not{d}$. The existence of a ∇ - parallel primitive idempotent places severe restrictions on the base manifold |19|. Furthermore the existence of a single ∇ - parallel primitive does not in general guarantee the existence of a complete set.

On any Euclidean-signatured 4-dimensional manifold, $\frac{1}{2}(1 \pm z)$ comprise a complete set of ∇-parallel primitive idempotents and may be used to decouple the Kähler equation into two equations for a pair of algebraic spinors.

If Ω is the 2-form associated with a Kähler structure on a Euclidean-signatured 4 dimensional space-time ($\Omega = J^a{}_b g_{ac} e^b \wedge e^c$, $\nabla J = 0$, $\nabla g = 0$) then $\Omega_v\Omega$ lies in a minimal left ideal that is ∇-parallel. It is interesting to note that on manifolds such as CP^2 that possess no spin structure it is possible |20| to decouple the Kähler equation into equations whose solutions lie in minimal left ideals.

From a physical point of view the relevance of a spin-structure in formulating physical theories is far from clear. Although each algebraic spinor contains the same number of degrees of freedom as a spinor field there are important differences to note. A quantisation in the absence of gravitation usually requires a consideration of a complete basis of solutions. As such a quantised Kähler field can create more quanta than a quantised Dirac field. In the presence of gravitation no simple relation exists between Dirac and Kähler field solutions. This has led to speculation relating the observed structure of fermion families to an underlying dynamics based on a Kähler equation {18},{21}.

9. Spinor Fields

In this and the following section let C stand for the Clifford algebra $C_{p,q}(\mathbb{R})$ associated with a metric of signature (p,q). A manifold M is a spin manifold if TM admits a spin structure. TM admits a spin structure if $OM(M,SO^+(p,q))$, the orthonormal frame bundle of TM and $P = P(M,SPIN^+(p,q))$, a principal bundle with structure group $SPIN^+(p,q)$

are linked by a principal bundle homormorphism f, such that the diagram

$$
\begin{array}{ccc}
P \times SPIN^{+}(p,q) & \xrightarrow{\;\tilde{h}\;} & P \searrow \\
\downarrow{\scriptstyle f \times \varrho} & & \downarrow{f} \qquad M \\
OM \times SO^{+}(p,q) & \xrightarrow{\;h\;} & OM \nearrow
\end{array}
$$

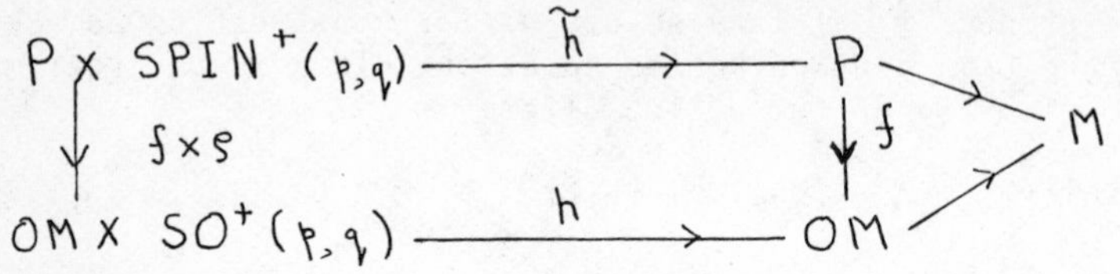

commutes. In this diagram $\rho : SPIN^{+}(p,q) \to SO^{+}(p,q)$ is a covering group homomorphism and $\tilde{h}$, h denote the group actions on P and OM respectively. For each such f (if it exists) the spin structure will be denoted (P,f). The bundle $PF = P \times_{spin^{+}(p,q)} \hat{F}$ whose standard fibre $\hat{F} = I_{L}(\hat{P})$, ($\hat{P}$ primitive in $C_{p,q}(R^{n}) \times C$), carries the $SPIN^{+}(p,q)$ representation: $(\xi,Q) \varepsilon \hat{F} \times SPIN_{(p,q)} \to \xi_{v}Q^{-1} \varepsilon \hat{F}$, is the associated spinor bundle. A spinor field is a section ΓPF of PF.

Given a local orthonormal co-frame $\{X_{a}\} \varepsilon \Gamma OM$ one may construct a complete set of primitive idempotents $\{P_{i}\}$ and a local basis $\{f^{J}\}$ $\varepsilon \Gamma C(T*M,g)$ for a minimal left ideal $I_{L}(\hat{P})$ for each P in the set. Such a basis will be said to be e-related to the choice of co-frame. This enables us to establish a correspondence between a standard basis for $I_{L}(\hat{P})$ and $\{f^{J}\}$.

We shall define a spinor field in $C(T*M,g)$ as an assignment of an algebraic spinor belonging to an e-related minimal ideal $\math{J}$, to an element ξ_{Q} of the standard fibre $\hat{F}$ of the associated bundle $P \times_{SPIN(p,q)} \hat{F}$, such that for every Q, Q_{o}, $Q_{1} \varepsilon SPIN^{+}(p,q)$, the linear map

$$
\psi_{Q_{i}} : \hat{F} \to \math{J}
$$

satisfies

$$
\psi_{Q_{1}}(\hat{f}_{Q_{1}}) = \psi_{Q_{o}}(\hat{f}_{Q_{o}})_{v}Q^{-1}
$$

where

$$
Q_{1} = Q_{v}Q_{o} .
$$

A spinor frame in $C(T*M,g)$ is an ordered basis of spinor fields for $\math{J}$. Once a local section in OM is selected and the corresponding local basis $\{f^{J}\}$ established we shall write $\psi_{Q}(\hat{f}_{Q})$ as ψ_{Q} or simply ψ with components ψ^{J} :

$$
\psi = \sum_{j} \psi^{j} f^{j} \quad , \quad \psi = \psi_{v}P . \tag{9.1}
$$

If the transposition map is denoted T

$$
T : \underline{\varepsilon}_{jk} \to (\underline{\varepsilon}_{jk})^{T} = \underline{\varepsilon}_{kj}
$$

then since

$$S_o(f^j{}_v(f^k)^T) = S_o(\underline{\varepsilon}_{ji}{}_v\underline{\varepsilon}_{ik}) = \delta_{jk}S_o(P)$$

we have

$$\psi^j = S_o(\psi_v(f^j)^T)/S_o(P) \; . \qquad 9.2$$

The notion of a spinor field in C(T*M,g) is useful for the formulation of spinor wave equations on a spin-manifold. Although the Kähler equation admits algebraic spinor solutions under certain conditions it will not in general. Although ∇ is compatible with the Clifford structure on Λ(T*M) it will not in general leave $I_L(P)$ invariant for any P.

10. Spinor Covariant Derivatives on the Clifford bundle

We shall define a spinor covariant derivative

$$S_X^{Q_o} : \mathcal{J} \rightarrow \mathcal{J} \qquad \forall X \in \Gamma TM \qquad 10.1$$

to be a derivation on $\mathcal{J}$, the space of sections lying in a minimal left ideal of a simple component of the algebra of sections ΓC(T*M,g), satisfying:

$$S_X^{Q_o}(\phi_v\psi_{Q_o}) = \nabla_X\phi_v\psi_{Q_o} + \phi_v S_X^{Q_o}\psi_{Q_o} \qquad 10.2$$

$$S_{fX}^{Q_o}\psi_{Q_o} = fS_X^{Q_o}\psi_{Q_o} \qquad 10.3$$

$\psi_{Q_o} \in \mathcal{J}$, $\phi \in \Gamma$C(T*M,g) , f any function M; such that under

$Q_o \rightarrow Q_1 = Q_vQ_o$

$$S_X^{Q_1}\psi_{Q_1} = (S_X^{Q_o}\psi_{Q_o})_vQ^{-1} \; . \qquad 10.4$$

These conditions do not uniquely determine a spinor covariant derivative.

If α lies in a minimal left ideal (represented by a primitive P) of a simple component of ΓC(T*M,g) then the element $P_v\alpha_vP$ lies in a division algebra $\mathcal{D}_P$ with unit element P. Writing $P_v\alpha_vP = (\alpha)_vP$, then ($\alpha$) lies in one of the division algebras $\mathcal{D}_1 \equiv \mathbb{R}$, $\mathbb{C}$ or $\mathbb{H}$. Let J : ΓC(T*M,g) $\rightarrow$ ΓC(T*M,g), $\phi \rightarrow \phi^J$ denote an involution and for each P denote by the same letter the element satisying

$$P = J^{-1}{}_vP^J{}_vJ \; . \qquad 10.5$$

(Such a J is unique up to multiplication by an element in $\mathcal{D}_P$.) The Levi-Civita connection will be taken to be compatible with this involution in the sense that:

$$\nabla_X\phi^J = (\nabla_X\phi)^J \qquad \forall\phi \in \Gamma C(T^*M,g) \; . \qquad 10.6$$

The involution J induces an involution j on $\mathcal{D}_p$ given by $\rho \to \rho^j = J^{-1}{}_\vee\rho_\vee{}^J J$.

We shall call $\alpha, \beta \in \mathcal{J} \times \mathcal{J} \to \{\alpha,\beta\} = J^{-1}{}_\vee\alpha^J{}_\vee\beta \in \mathcal{D}_p$ an inner product on $\mathcal{J}_Q$ and introduce for each choice of primitive P^Q and associated J^Q the adjoint spinor $\tilde{\psi}_Q = (J_Q)^{-1}\psi_Q^J$. Such an inner product is $SPIN^+(p,q)$ invariant:

$$\{\alpha_{Q_o},\beta_{Q_o}\} = \{Q_\vee\alpha_{Q_o},Q_\vee\beta_{Q_o}\} \qquad \forall\alpha_{Q_o},\beta_{Q_o} \in \mathcal{J} \qquad 10.7$$

$Q \in SPIN^+(p,q)$ (covering $SO^+(p,q) \in \Gamma C(T^*M,g)$).
Furthermore since it is possible to choose J such that $J^J = \pm J$ we may arrange:

$$\{\alpha,\beta\}^j = \pm \{\beta,\alpha\} \ . \qquad 10.8$$

For each choice of involution and associated J there exists a matrix basis $\{\underline{\varepsilon}_{ij}^{Q_o}\}$ for $\Gamma C(T^*M,g)$ such that

$$\underline{\varepsilon}_{ij}^{Q_o} = J^{Q_o}{}_\vee\underline{\varepsilon}_{ji}^{Q_o}{}_\vee(J^{Q_o})^{-1} \ . \qquad 10.9$$

In such a basis J effects transposition on matrix frames. If a new basis $\{\underline{\varepsilon}_{ij}^{Q_1}\}$ is induced by the inner-automorphism

$$P^{Q_o} \to P^{Q_1} = Q_\vee P^{Q_o}{}_\vee Q^{-1} \qquad 10.10$$

then a new element

$$J^{Q_1} = (Q^{-1})^J{}_\vee J^{Q_o}{}_\vee Q^{-1} \qquad 10.11$$

will effect transposition with respect to $\underline{\varepsilon}_{ij}^{Q_1} = Q_\vee\underline{\varepsilon}_{ij}^{Q_o}{}_\vee Q^{-1}$ and in particular

$$P^{Q_1} = (J^{Q_1})^{-1}{}_\vee(P^{Q_1})^J{}_\vee J^{Q_1} \ . \qquad 10.12$$

A spinor covariant derivative $S_X^{Q_o}$ will be said to be compatible with the inner product { , } if

$$\{S_X^{Q_o}\alpha_{Q_o},\beta_{Q_o}\} + \{\alpha_{Q_o},S_X^{Q_o}\beta_{Q_o}\} = P_\vee\nabla_X\{\alpha_{Q_o},\beta_{Q_o}\}_\vee P \ . \qquad 10.13$$

The axioms 10.2, 10.3, 10.4, 10.13 uniquely fix S_X in all cases provided j leaves the centre of (the simple component of) $\Gamma C(T^*M,g)$ invariant, i.e. j is the identity involution. Then $S_X^{Q_o}$ may be expressed in terms of ∇_X according to

$$S_X^{Q_o}\alpha_{Q_o} = \nabla_X\alpha_{Q_o\,v}P^{Q_o} + $$

$$+ \frac{1}{2}\alpha_{Q_o\,v}P^{Q_o}{}_v\nabla_X(J^{Q_o})^{-1}{}_vJ^{Q_o}{}_vP^{Q_o} \; . \qquad 10.14$$

If $\mathfrak{D}_1 = \mathbb{R}$, both ξ and $\xi\eta$ provide permissible involutions J and (suppressing Q_o)

$$S_X\alpha = \nabla_X\alpha_vP + 2S_o(\nabla_XJ^{-1}{}_vJ_v\rho)\alpha \; . \qquad 10.15$$

If $\mathfrak{D}_1 = \mathbb{C}$ then J is fixed by demanding $z^J = z$. Since $\xi z = (-1)^{[\frac{n}{2}]}z$, J is either ξ or $\xi\eta$ and

$$S_X\alpha = \nabla_X\alpha_vP + S_{\mathfrak{D}_1}(\nabla_XJ^{-1}{}_vJ_vP)_v\alpha \qquad 10.16$$

where

$$S_{\mathfrak{D}_1} \equiv S_o + S_n .$$

If $\mathfrak{D}_1 = \mathbb{H}$ then j must be chosen to define quaternionic conjugation to obtain a unique covariant derivative from the above axioms.
It may be noted that if we write $\{\alpha,\beta\} = (\alpha,\beta)_vP$ where $(\alpha,\beta) \in \mathfrak{D}_1$ then it follows from the fact that $P_v\nabla_XP_vP \equiv 0 \quad \forall X \in \Gamma TM$ that when $\mathfrak{D}_1 = \mathbb{R}$ or $\mathbb{C}$, compatibility of S_X with $\{\alpha,\beta\}$ implies

$$(S_X\alpha,\beta) + (\alpha,S_X\beta) = \nabla_X(\alpha,\beta) \; . \qquad 10.17$$

Furthermore in all cases:

$$\nabla_X[\alpha_v\tilde{\beta}] = S_X\alpha_v\tilde{\beta} + \alpha_v\widetilde{S_X\beta} \qquad \alpha_v\beta \in \mathcal{J} \qquad 10.18$$

If $\{e^a\}$ and $\{X_a\}$ are any naturally dual co-frame and frame (not necessarily orthonormal) we define

$$\$ = e^a{}_vS_{X_a} \qquad 10.19$$

and for $\psi = \psi_vP$ call

$$\$\psi = m\psi \qquad 10.20$$

a generalised Dirac equation.

To make contact with the familiar space-time spinor covariant derivative we note that it is possible to find elements P and J such that

$$\nabla_X P = [\overset{o}{\Sigma}_X, P] \qquad 10.21$$

$$\nabla_X J = [\overset{o}{\Sigma}_X, J] \qquad 10.22$$

in terms of Clifford commutators, where

$$\overset{o}{\Sigma}_X = \frac{1}{4} \nabla_X e^b \wedge e_b \qquad 10.23$$

and $g(e^a, e^b) = \eta^{ab}$. In an ideal represented by such a P

$$S_X \psi = \nabla_X \psi + \psi \vee \overset{o}{\Sigma}_X \ . \qquad 10.24$$

Under a change of left ideal generated by (10.10), (10.11), $S_X^{Q_o}\psi^{Q_o} \to S_X^{Q_1}\psi^{Q_1} = \nabla_X \psi^{Q_1} + \psi^{Q_1} \vee \Sigma_X$ where $\psi^{Q_1} = \psi^{Q_o} \vee Q$ and $\Sigma_X = Q \vee \overset{o}{\Sigma}_X \vee Q^{-1} + \nabla_X Q \vee Q^{-1}$. For the choice of connection 10.23, the spinor components of equation 10.24 may be identified with the usual local Dirac equation in a space with Levi-Civita connection ∇.

In view of the historical motivation for the Dirac equation it is interesting to note that $\$$ does not square to the Laplace-Beltrami operator. In fact:

$$\$^2 \psi = (S_{X^a} - \tilde{\delta} e^a) S_{X_a} \psi - \frac{1}{4} \mathcal{R} \psi \qquad 10.25$$

whereas for the Kähler Hodge-de-Rham operator:

$$\not{d}^2 \phi = (\nabla_{X^a} - \tilde{\delta} e^a) \nabla_{X_a} \phi - \frac{1}{4} \mathcal{R} \psi - \frac{1}{4} R_{cd} \vee \phi \vee e^c \vee e^d \ . \qquad 10.26$$

In these expressions the curvature scalar is defined by $\mathcal{R} = - R_{cd} \vee e^c \vee e^d$.

In spherically-symmetric space-time metrics of the type (6.7) it is possible |18| to find separable solutions to (10.20) of the form $\psi_j^m = W_j(r) \vee \mathcal{S}_j^m(r,\theta,\phi) \vee T(t)$ where the $\mathcal{S}_j^m$ are related to spinor fields on S^2.

11. Lie Derivatives on Spinor Fields

For any $\phi \in \Gamma C(T^*M, g)$, $\mathcal{L}_X \phi$ is well defined for any vector field X where $\mathcal{L}_X$ is the usual Lie derivative. It will only act as a derivation on a Clifford product of forms, however, if X is a Killing vector K of g. In that case we have, for a torsion-free ∇, the identity:

$$\mathcal{L}_K \phi = \nabla_K \phi + [\frac{1}{4} d\tilde{K}, \phi] \qquad 11.1$$

in terms of a Clifford commutator. $\mathcal{C}$ is said to be a conformal Killing vector if $\mathcal{L}_{\mathcal{C}}\, g = 2\lambda g$ where $n\lambda = - \tilde{\delta} \tilde{\mathcal{C}}$.

For $\psi \in \mathcal{J}$ and $\mathcal{C}$ any conformal Killing vector field we define a Lie derivative $\mathcal{L}_{\mathcal{C}}$ on spinors by :

$$\mathcal{L}_{\mathcal{C}}\,\psi = S_{\mathcal{C}}\psi + \tfrac{1}{4}d\tilde{\mathcal{C}}_{\vee}\psi \; . \qquad 11.2$$

This definition implies for any conformal Killing vector fields $\mathcal{C},\mathcal{C}_1,\mathcal{C}_2$

$$[\mathcal{L}_{\mathcal{C}_1},\mathcal{L}_{\mathcal{C}_2}] = \mathcal{L}_{[\mathcal{C}_1,\mathcal{C}_2]} \qquad 11.3$$

and

$$\mathcal{L}_{\mathcal{C}}(\alpha_{\vee}\tilde{\beta}) = \mathcal{L}_{\mathcal{C}}\alpha_{\vee}\tilde{\beta} + \alpha_{\vee}\widetilde{\mathcal{L}_{\mathcal{C}}\beta} \; . \qquad 11.4$$

Unlike $\mathcal{L}_{\mathcal{C}}$, the spinor Lie derivative is type-preserving on $\Gamma I_L(P)$.

12. Conserved Currents and Spinor Actions

If $(\phi,\psi) \equiv S_o(\tilde{\phi}_{\vee}\psi)$ is the real valued inner product on $\Gamma I_L(P)$ with $J^J = -\lambda J$, $\lambda^o = \pm 1$, then for any $N \in \Gamma C(T^*M,g)$

$$(\phi,N_{\vee}\psi) = -\lambda(\psi,\xi N_{\vee}\phi) = (\xi N_{\vee}\phi,\psi) \; . \qquad 12.1$$

From the compatability of S_X with this inner product we obtain the identity:

$$(\$\phi,\psi)z + (\phi,\$\psi)z = d\mathcal{J}(\phi,\psi) + (\phi,i_{\tilde{e}^a}T^a{}_{\vee}\psi)z \qquad 12.2$$

where $\mathcal{J}(\phi,\psi) = *S_1(\psi_{\vee}\tilde{\phi}) = -\lambda S_{n-1}(\phi_{\vee}\tilde{\psi}_{\vee}z)$

$$= -\lambda\,(-)^{[\frac{n-1}{2}]}\,\mathcal{J}(\psi,\phi) \quad , \qquad n = \dim M \; . \qquad 12.3$$

Since $\$(z_{\vee}\psi) = -z_{\vee}(-1)^n\$\psi$, setting ψ to $z_{\vee}\psi$ gives the alternative identity:

$$(\$\phi,z_{\vee}\psi)z - (-1)^n(\phi,z_{\vee}\$\psi) = dU(\phi,\psi) + (\phi,i_{\tilde{e}^a}T^a{}_{\vee}z_{\vee}\psi) \qquad 12.4$$

where $U(\phi,\psi) = \mathcal{J}(\phi,z_{\vee}\psi) = -\lambda S_{n-1}(\phi_{\vee}\widetilde{z_{\vee}\psi}_{\vee}z)$

$$= -\lambda(-)^{[\frac{n-1}{2}]}\,U(\psi,\phi) \; . \qquad 12.5$$

Thus for even n and a torsion-free ∇, if ϕ and ψ are solutions of $\$\alpha = m\alpha$, $m \in \mathbb{R}$, $U(\phi,\psi)$ is a closed current :

$$dU(\phi,\psi) = 0 \; . \qquad 12.6$$

In Minkowski-signatured space-time, $n = 4$, $z_{\vee}z = -1$, $U(\phi,\psi) = S_3(\phi_{\vee}\tilde{\psi})$. If the space-time admits conformal isometries generated by a conformal Killing vector $\mathcal{C}$ then we may show that :

$$[\hat{\mathcal{L}}_e, \$] = -\lambda\$ \qquad 12.7$$

where $\hat{\mathcal{L}} = \mathcal{L} + \frac{3}{2}\lambda$. Thus if α is a solution to $\$\,\alpha = 0$ so is $\hat{\mathcal{L}}_e\alpha$. This conformal symmetry of the massless Dirac equation in a curved space with conformal isometries gives rise to a conserved current U

$$dU_e = 0 \qquad 12.8$$

where $U_e = S_3(\alpha_\vee \widetilde{\hat{\mathcal{L}}_e\alpha})$. A similar argument may be used to construct conserved currents associated with every Killing vector of a space-time and the massive solutions to (10.20).

13. Pure Spinors on C(T*M,g)

If we choose a basis {x,y,a,b} for ΓT*M in which the space-time metric takes the form

$$g = 2(x \otimes y + y \otimes x) + a \otimes a + b \otimes b\ , \qquad 13.1$$

then
$$x_\vee x = y_\vee y = 0\ ,\ a_\vee x + x_\vee a = 0\ ,\ a_\vee y + y_\vee a = 0$$

$$a_\vee a = b_\vee b = 1\ ,\ b_\vee x + x_\vee b = 0\ ,\ b_\vee y + y_\vee b = 0$$

$$x_\vee y + y_\vee x = 1 \qquad 13.2$$

and a basis for $\Gamma I_L(\frac{1}{2}y_\vee x_\vee(1 + a))$ is $\{1, b, y, y_\vee b\}\frac{x}{\sqrt{2}}{}_\vee(1 + a)$. Any Majorana spinor in space-time can be written in the basis:

$$\psi = (\psi_1 + \psi_2 b + \psi_3 y + \psi_4 y_\vee b)\frac{x}{\sqrt{2}}{}_\vee(1 + a) \qquad 13.3$$

and it is known as a pure spinor basis. Since the 2 form $x \wedge a = x_\vee a$ defines an isotropic 2-plane $(x_\vee x = 0)$ the pure spinor $\frac{1}{2}x_\vee(1 + a)$ readily lends itself to an elegant geometrical interpretation. In general in our signatured space-time one may always find 'pure-spinors' of the form $\psi = (1 + a')_\vee x'_\vee \phi_\vee (1 + a)_\vee x$ for $\phi \in \Gamma C(T^*M,g)$ and the primed elements have an algebra isomorphic to the unprimed ones. Such a representation correlates the two isotropic 2-planes $a' \wedge x'$ and $a \wedge x$. The notion of a pure spinor can be generalised to higher dimensions in different ways. If the algebra is kept real, pure spinors of C(V, g), where the metric g on V has signature $(\underbrace{-----}_{r},\underbrace{+++++++}_{r+2})$, have been studied in |22|.

Not all spinors in higher dimensions are necessarily pure. This has led a number of authors |7| to examine the constraints on pure spinors as a key to understanding some of the regularities observed among the fundamental fermions in nature. Budinich has stressed that a single wave equation together with a pure spinor constraint in higher dimensions induces interesting correlations when dimensionally reduced.

Other applications of the use of pure spinors in Lorentzian-signatured space-time will be found in the article by I.M. Benn in these proceedings.

14. Curved Space Supersymmetry without Gravitinos

As a non-trivial application of the methods of Clifford analysis we have studied supersymmetric models in curved space-time |23|. The bundle of algebras is extended from $C(T^*M,g)$ to $C(T^*M,g) \otimes G$ where G is an (infinite dimensional) Grassman algebra. If $\psi \in \Gamma(I_L(P) \otimes G^-) \equiv \mathfrak{J}^-, \phi \in \Gamma((CT^*M,g) \otimes G^+)$ where $G^{\pm}$ are the even and odd graded parts of G; the action 4-form

$$\Lambda = S_o(\hat{\psi}\not{\$}\psi + \bar{\phi}\not{d}(\not{d} + \not{\delta})\phi)\text{z} \qquad 14.1$$

is supersymmetric, $Q(\alpha)\Lambda = 0$ mod d, under the action

$$Q(\alpha)\psi = (\not{d} + \not{\delta})\phi_{\pm\vee}\alpha \qquad 14.2$$

$$Q(\alpha)\phi_{\pm} = \frac{1}{2}(1 \pm \eta)(\psi_{\vee}\hat{\alpha}) \qquad 14.3$$

provided $\alpha \in \mathfrak{J}^-$ is S-parallel; $S_X\alpha = 0\ \forall X \in \Gamma TM$. In (14.1) $\bar{\phi} \equiv \xi\eta\phi$, $\hat{\psi} \equiv B^{-1}\psi$ where B satisfies $P = B^x\bar{P}B^{-1}$ and $\not{\delta}\bar{\phi} = \overline{\not{d}\phi}$. ψ and $\phi_{\pm}$ together comprise a Bose-Fermi supermultiplet where $\phi_{\pm} \equiv \frac{1}{2}(1 \pm \eta)\phi$ and Λ describes a pair of supersymmetric models. The supersymmetry algebra is, for any S-parallel spinor "parameters" $\alpha, \beta \in \mathfrak{J}^-$:

$$[Q(\alpha),Q(\beta)]\psi = \mathcal{L}_K\psi \qquad 14.4$$

$$[Q(\alpha),Q(\beta)]\phi_{\pm} = \mathcal{L}_K\phi - d(\lambda_{\pm})_+ + \delta(\lambda_{\pm})_- \qquad 14.5$$

$$[\mathcal{L}_{K_j},Q(\alpha)]\phi = Q(\tfrac{1}{4}d\tilde{K}_{j\vee}\alpha)\phi \qquad 14.6$$

$$[\mathcal{L}_{K_j},Q(\alpha)]\psi = Q(\tfrac{1}{4}d\tilde{K}_{j\vee}\alpha)\psi \qquad 14.7$$

where the space-time admits the set of Killing vectors $\{K_j\}$ and $K \equiv 4S_1(\alpha_\vee\hat{\beta})$. Since α and β are S-parallel, $\forall X \in \Gamma TM$,

$$\nabla_X\tilde{K} = 4S_1(S_X\alpha_\vee\hat{\beta} + \beta_\vee S_X\hat{\alpha}) = 0 \qquad 14.8$$

and K is ∇-parallel, therefore a Killing vector that satisfies $d\tilde{K} = 0$. This model illustrates the general feature of all so called 'rigid' supersymmetric models that exist in a background metric that exhibits Killing isometries generated by a set $\{K_i\}$. The essential ingredient is the existence of at least a pair of S-parallel spinors. The traditional arena for such models is flat Minkowski space with its 10-parameter Poincaré isometry group. However a plane-fronted gravitational wave has a 5-parameter isometry group with a 2-dimensional space of S-parallel pure spinors and will consequently admit super-symmetric matter interactions.

The methods of Clifford analysis introduced thus far are still being developed. Higher-spin wave equations have been formulated in terms of $\Gamma(\mathcal{S} \otimes \Lambda^p(T^*M))$ for general dimensions. The techniques have been applied |24| to study the zero-mode structure of the spin $\frac{3}{2}$ wave operator in the context of Kaluza-Klein theories.

Acknowledgements

I wish to thank I.M. Benn and M. Panahi for helpful discussions and advice in preparing this review.

References

1. C. Chevalley, 'The Algebraic Theory of Spinors', Columbia University Press, New York, 1954.

2. N. Salingaros, Y. Ilamed, Found. of Phys. **14** 777 (1984).

3. N. Salingaros, M. Dresden, Adv. in Appl. Maths. **4** 1 (1983); Phys. Rev. Letts. **43** 1 (1970).

4. D. Hestenes, 'Space-time Algebra' (Gordon and Breach, New York, 1966); Found. in Phys. **12**, 153, 1982.

5. J.S.R. Chisholm, R.S. Farwell, Proc. R. Soc. London, Ser. A, **377** 1 (1981).
Il. Nuovo. Cim. **82** 145, 185, 210 (1984).

6. T.T. Truong, H.J. Vega, Phys. Letts. **151B** 135 (1985).

7. P. Budinich, Proc. 8th Inst. Nathiagali, Pakistan (1983).
P. Budinich, K. Bugajska, J. Math. Phys. **26** 588 (1985).
P. Budinich, L. Dabrowski, Lett. in Math. Physics **10** (1985) 7.

8. C.G. Darwin, Proc. R. Soc. **118** 654 (1928).
D. Ivaneko and L. Landau, Zeits f. Phys. **48** (1928) 340.

9. E. Kähler, Rend. Mat. (3-4) **21** 425 (1962).

10. W. Graf, Ann. Inst. Henri Poincaré, XXIX **85** (1978).

11. P. Becher, H. Joos, Zeit. Phys. C. Particles and Fields **15**, 343, (1982).
J.M. Rabin, Nucl. Phys. **B201**, 315 (1982).
T. Banks, Y. Dothan and D. Horn, Phys. Lett. **117B** (1982) 413.
A.K. Common, 'Reduction of Dirac-Kähler Equation to Dirac Equation', University of Kent Preprint.

12. I.M. Benn, R.W. Tucker, Comm. Math. Phys. **89** 341 (1983).

13. P. Basarab-Horwath, R.W. Tucker, Comptes Renduc Acad. Sc. Paris, t, 299, Series I No. 20, 1984.

14. P. Basarab-Horwath, R.W. Tucker, 'A Quantisation for Kähler Fields in Static Space-Time. Lancaster University Preprint (1985).

15. I.M. Benn, R.W. Tucker, Comm. Math. Phys. **98** 53, (1985), Phys. Letts. B130 177 (1983).

16. I.M. Benn, R.W. Tucker, Phys. Lett. B125 47 (1983) J. Phys. A 16 4147 (1983).

17. I.M. Benn, R.W. Tucker, Phys. Lett. B132 325 (1984). Il Nuovo. Cim. 88a 273 (1984). Proc. of Colloquium on Differential Geometry, Debrecen, Hungary 1984. J. Phys. A (Math) **16** 4123 (1983).

18. T. Dereli, M. Onder, R.W. Tucker, Clas. Quantum Grav. 1 **L67** (1984). M. Panahi, R.W. Tucker, 'Separation of Dirac and Kähler Equations in Spherically Symmetric Space-Times'. Lancaster University Preprint 1985.

19. A. Al-Saad, I.M. Benn, 'Ideal Preserving Lorentzian Connections', Lancaster University Preprint (1985).

20. I.M. Benn, B. Dolan, R.W. Tucker, Phys. Letts. **150B** 100 (1985).

21. I.M. Benn, R.W. Tucker, Phys. Letts. **B119** 348 (1982).

22. I.M. Benn, R.W. Tucker, 'Pure Spinors and Real Clifford Algebras', Lancaster University Preprint 1984.

23. I.M. Benn, M. Panahi, R.W. Tucker, Clas. Quantum Grav. **2** L71 (1985).

24. I.M. Benn, M. Panahi, R.W. Tucker, 'A Note on the Zero Mode Structure of the Rarita-Schwinger Operator', Clas. Quantum Grav. **2** L109 (1985).

25. I.M. Benn, R.W. Tucker, 'A Modern Introduction to Spinors and Geometry with Applications in Physics', Adam Hilger Ltd., Techno House, Redcliffe Way, Bristol, BS1 6NX, England. (To be published.)

GENERALIZED C-R EQUATIONS ON MANIFOLDS

Vladimír Souček
Dep. Mathematical Analysis
Charles University
Sokolovská 83, 186 00 Praha
Czechoslovakia

ABSTRACT. The paper presents a generalization of the classical ∂ and $\overline{\partial}$ operators from complex analysis on Riemann surfaces to vector-valued differential forms on conformal n-dimensional manifolds. An abstract scheme for such generalization is based on the splitting of the vector-valued de Rham sequence. The possible generalizations are classified by couples of irreducible $CO(n)$-modules and by a choice of a connection on the associated vector bundle. Various generalizations of C-R equations, studied by different authors during last 50 years, are discussed and it is shown how they fit into the scheme. A special attention is paid to the most interesting case of dimension 4 and to the connection of the described systems of equations with equations in mathematical physics.

1. INTRODUCTION

The classical complex analysis (in the plane and even more on manifolds) is so rich and beautiful part of mathematics that there were many attempts to look for a similar theory in higher dimensions. The generalizations went to many directions. They usually consist of a system of first order linear PDE with constant coefficients ([2] , [14] , [15] , [20] , [22] , [34] , [37] , [38] , [39]), sometimes more general elliptic systems (with variable coefficients or with nonlinear 0-order terms) are considered ([13] and references therein), or even a fully nonlinear system of self-dual Yang-Mills field equations was suggested as a generalization of C-R equations to higher dimensions ([23]). Some generalizations to maps defined on manifolds were also presented ([1] , [15] , [21] , [32]).

Any of these generalizations has its own merit and it is difficult to decide what a proper generalization of C-R equations should be. It depends clearly on the point of view,

J. S. R. Chisholm and A. K. Common (eds.), Clifford Algebras and Their Applications in Mathematical Physics, 201–217.

on the choice what aspect of the classical complex analysis wants to be preserved under the generalization.

Not all generalizations, mentioned above, are included in the scheme presented in the paper. The guiding principles for the generalizations studied here can be stated as follows.

Firstly, to open possibilities to look for generalizations of the fascinating field of complex analysis on manifolds, the wanted scheme for the generalization should be necessarily formulated not for functions on domains in R_n, but for maps defined on an appropriate type of n-dimensional manifolds.

Secondly, the classical C-R equations in the complex plane are substituted by ∂ operator, acting on complex valued differential forms. The holomorphic functions form the kernel of $\bar{\partial}$ operator on 0-forms and the whole standard split de Rham sequence

$$\begin{array}{ccccc} & & \mathcal{E}^{1,0} & & \\ & \overset{\partial}{\nearrow} & & \overset{\bar{\partial}}{\searrow} & \\ \mathcal{E}^{0} & & \oplus & & \mathcal{E}^{2} \\ & \underset{\bar{\partial}}{\searrow} & & \underset{\partial}{\nearrow} & \\ & & \mathcal{E}^{0,1} & & \end{array} \qquad (1)$$

forms the inseparable picture. It would be highly desirable to find a generalization of the diagram (1) to higher dimensions.

Thirdly, 'a higher dimensional generalization' should mean that the equations reduce (at least in flat cases) back to the classical Cauchy-Riemann equations for dimension $n=2$.

The general scheme, satisfying the principles stated above, is described in the paper (following [7]). Basic idea is a certain kind of invariance under an appropriate group. The generalization of the domains of definitions of our maps is based on the fact that complex 1-dimensional manifolds (Riemann surfaces) coincide with (real) 2-dimensional manifolds with conformal structure. In higher dimensions, our maps will be defined on n-dimensional manifolds with the structure group $\tilde{G} = CO(n)$ (resp. the universal covering group G of $\tilde{G}$). Target spaces of maps will be irreducible finite dimensional G-modules over R (note that for $n=2$ such modules look like $R_2 = C$).

The basic idea of the scheme was inspired by the paper by Stein and Weiss ([38]) on generalized C-R equations and by the paper on conformally invariant first order operators by Fegan ([11]). The essence of the approach is simple, we shall describe it now in the flat case.

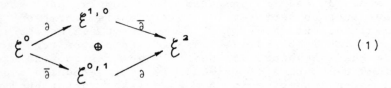

Let us denote by $\mathcal{E}^{\infty}(V)$ the space of smooth maps of a domain $\Omega \subset R_n$ to the vector space V and let us denote $\Lambda^j = \Lambda^j(R_n^*)$. For $n=2$ we take the de Rham sequence

$$\mathcal{E}^\infty(\Lambda^0) \xrightarrow{d} \mathcal{E}^\infty(\Lambda^1) \xrightarrow{d} \mathcal{E}^\infty(\Lambda^2) \tag{2}$$

and we shall consider its complexification (i.e. maps into $\Lambda_c^j = \Lambda^j \otimes_R C$). The splitting

$$\Lambda_c^1 = \Lambda^{1,0} \oplus \Lambda^{0,1} \tag{3}$$

leads directly to the split de Rham sequence (1). Now, any (nontrivial) SO(2)-module V looks like $R_2 \cong C$ and the tensor product $\Lambda^1 \otimes_R V$ splits (as SO(2)-module) into two pieces $F_1 \oplus F_2$, both isomorphic to R_n. The equations coming from this splitting are equivalent to C-R equations.

Let us take a G-module V (over R) and let us tensor the de Rham sequence

$$\mathcal{E}^\infty(\Lambda^0) \xrightarrow{d} \dots \xrightarrow{d} \mathcal{E}^\infty(\Lambda^n) \tag{4}$$

with V (over R). The products $\Lambda^j \otimes_R V$ can be decomposed into irreducible pieces (as G-modules), say

$$\Lambda_V^j = \Lambda^j \otimes_R V = F_1^j \oplus \dots \oplus F_{m_j}^j \tag{5}$$

It induces the splitting of V-valued de Rham sequence

$$\begin{array}{ccccccc} & \overset{\partial_1}{\nearrow} & \mathcal{E}^\infty(F_1^1) & \longrightarrow \dots \longrightarrow & \mathcal{E}^\infty(F_1^{n-1}) & \searrow & \\ & & \oplus & & \oplus & & \\ \mathcal{E}^\infty(V) & \longrightarrow & \vdots & & \vdots & \longrightarrow & \mathcal{E}^\infty(\Lambda^n \otimes_R V) \\ & & \oplus & & \oplus & & \\ & \underset{\partial_{m_1}}{\searrow} & \mathcal{E}^\infty(F_{m_1}^1) & \longrightarrow \dots \longrightarrow & \mathcal{E}^\infty(F_{m_n-1}^{n-1}) & \nearrow & \end{array} \tag{6}$$

which is proposed to be the proper generalization of (1) to higher dimensions. Every individual operator in (6) for every V is in such a way a generalization of ∂ or $\bar{\partial}$ operator from (1). Note that the structure of the diagram (6) can be much more complicated than that of (1) and that it depends generally on the choice of V. Moreover, the operators ∂ and $\bar{\partial}$ in (1) were quite similar one to another, while the individual operators in (6) will be in general quite different in character and as to properties of their solutions.

The best way to classify the individual operators

in (6) is by the highest weights of the corresponding spaces F_1^j and F_1^{j+1}, (note that the same operator can appear in the diagram (6) for different modules V and on different places and that every one can appear in the first column of operators).

The aim of this review is to discuss how various generalizations of C-R equations, studied by different authors during last 50 years, fit into the general scheme. So the general definition of the split de Rham sequence for vector-valued forms is introduced first (§.2.) and then various generalizations of C-R equations and their relations to the suggested scheme are discussed (§.3. and §.4.).

No attempt was made to make the list of examples, results and references complete (in fact, it is almost impossible, for example, for spinor fields on space-times). They were chosen with respect to the knowledge and interests of the author.

2. VECTOR-VALUED DIFFERENTIAL FORMS

Let us consider the conformal group $\tilde{G} = CO(n)$ and its universal covering group G. Let V be an irreducible, finite dimensional G-module over R. The fundamental representation of CO(n) on R_n induces the structure of G-module on $\Lambda^j = \Lambda^j(R_n^*)$ and the tensor product $\Lambda^j \otimes_R V$ splits (as G-module) into irreducible pieces:

$$\Lambda^j \otimes_R V = F_1^j \oplus \dots \oplus F_{m_j}^j , \quad j=1,\dots,n-1 \tag{7}$$

Let M be a (real) oriented n-dimensional manifold with conformal structure, i.e. we have the corresponding principal CO(n)-bundle $\tilde{P}$. Suppose that $\tilde{P}$ lifts to a principal G-bundle P, i.e. that we have a (fibre bundle) homomorphism $f : P \to \tilde{P}$. We shall denote by $\underline{V}$, $\underline{F}_1^j$ and $\underline{\Lambda}_j$ the vector bundles associated to the G-modules V, F_1^j and $\Lambda^j(R_n^*)$.

Finally, let us choose a covariant derivative $\nabla : \Gamma(\underline{V}) \longrightarrow \Gamma(\underline{\Lambda}^1 \otimes \underline{V})$ and extend it in the usual way to

$$\nabla : \Gamma(\underline{\Lambda}^j \otimes \underline{V}) \longrightarrow \Gamma(\underline{\Lambda}^{j+1} \otimes \underline{V}), j=1,\dots,n-1.$$

Then the sequence $\Gamma(\underline{V}) \xrightarrow{\nabla} \dots \xrightarrow{\nabla} \Gamma(\underline{\Lambda}^n \otimes \underline{V})$ for V-valued differential forms splits as

$$\Gamma(\underline{V}) \begin{array}{c} \xrightarrow{\partial_1} \Gamma(\underline{F}^1_1) \\ \oplus \\ \vdots \\ \oplus \\ \xrightarrow{\partial_{m_1}} \Gamma(\underline{F}^1_{m_1}) \end{array} \longrightarrow \cdots \longrightarrow \begin{array}{c}\Gamma(\underline{F}^{n-1}_1)\\ \oplus\\ \vdots\\ \oplus\\ \Gamma(\underline{F}^{n-1}_{m_{n-1}})\end{array} \longrightarrow \Gamma(\underline{\Lambda}^n \otimes \underline{V}) \qquad (8)$$

Remarks. 1. In flat spaces the definition simplifies a lot. It is not necessary to introduce the principle fibre bundles $\tilde{P}$ and P, all bundles being trivial. There is the distinguished connection $\nabla = d$ on $R_n \times V$. The sections in $\Gamma(\underline{\Lambda}^j \otimes \underline{V})$ are simply smooth maps in $\mathcal{C}^\infty(\Lambda^j \otimes_R V)$ and the diagram (8) coincide with (6).

To give a G-module V, it is necessary to choose a Spin(n)-module V and to specify the conformal weight. But the conformal weight does not influence the splitting (7), so the scheme (8) depends in the flat cases only on the choice of the Spin(n)-module V.

2. It is possible (and useful) to use the same procedure also for the case of the group $G = CO(p,q)$. The corresponding operators cannot be, of course, considered as generalizations of C-R equations. But, the representations of Spin(p,q) being the same as those for Spin(p+q), the corresponding operators are 'analytic continuation' of those for CO(p+q) case. This is especially interesting in dimension 4, where the operators in CO(4) case are analytic continuations to Euclidean spacetime of operators used in mathematical physics. It is also possible to relate properties of solutions of corresponding systems of equations (see e.g. 3.7).

3. Another variation of the described scheme can be often found, which works with complexifications of all vector bundles. The only changes needed are to consider complex valued forms and for a G-module V over C to consider the splitting of the tensor product $\Lambda^j_c \otimes_C V$. Basically, there is only a little difference between these two versions (for a more careful discussion see [8]), but the analogy with the classical case $n=2$ is closer for the version given here.

3. EXAMPLES (FLAT SPACES)

In all examples, presented in this paragraph, we shall consider only maps on R_n (flat space), so (with respect to Remark 1) we shall use the simpler diagram (6). Note that in these examples the scheme (8) gives the generalization of the corresponding equations to conformal manifolds.

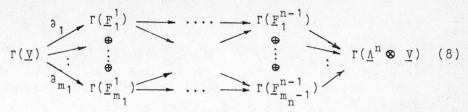

3.1 Regular spinor fields

Let us consider the case, when the module V is one of the basic spinor modules (over R). We shall include the more general case of the group $\mathrm{Spin}(p,q)$. The Clifford algebra $\mathcal{C}_{p,q}$ (corresponding to the quadratic form with p pluses and q minuses) is very useful for the description of these modules. For $q \geq 1$ we have the inclusion $\mathrm{Spin}(p,q) \subset \subset \mathcal{C}^{+}_{p,q} \cong \mathcal{C}_{p,q-1}$ and any minimal left ideal in $\mathcal{C}_{p,q-1}$ is the basic $\mathrm{Spin}(p,q)$-module (for more details see [20])

The tensor product splits in this case into two pieces

$$\Lambda^1 \otimes_R V = F_1 \oplus F_2 ,$$

where one summand on the right hand side (say F_1) is again a basic spinor module ([8]). So the first column in the diagram (6) looks like

$$\mathcal{C}^\infty(\Lambda^1 \otimes_R V) \begin{array}{l} \xrightarrow{\partial_1} \mathcal{C}^\infty(F_1) \\ \xrightarrow[\partial_2]{} \mathcal{C}^\infty(F_2) \end{array}$$

Choosing an orthonormal basis $e_1,\dots,e_{p+q}$ in R_{p+q} and denoting $a_1=\dots=a_p=-a_{p+1}=\dots=-a_{p+q}=1$, we can write the coordinate description of these operators:

$$\partial_1 : f \longmapsto \sum_k dx_k \otimes e_k \Big(\sum_j a_j e_j \frac{\partial f}{\partial x_j} \Big) \frac{1}{n} \tag{9}$$

$$\partial_2 : f \longmapsto \sum_k dx_k \otimes \left\{ \frac{\partial f}{\partial x_k} - e_k \Big(\sum_j a_j e_j \frac{\partial f}{\partial x_j} \Big) \frac{1}{n} \right\} ,$$

where the multiplication $e_k e_j v$ means the action of $e_k e_j$ on $v \in V$ ([8]). In the Riemannian case all a_j disappear.

The equation $\partial_1 f = 0$ is (after a simple redefinition $e_i \to -e_i$, $i=p+1,\dots,p+q$) just the condition for regular spinor field presented by Lounesto ([20]) in general $\mathrm{Spin}(p,q)$ case. The case $p=o$ (or $q=0$) is the most interesting and the most common from all generalization of C-R equations. The operator is usually called Dirac operator. There are many more results known for the Dirac operator than for any other considered operator. In the paper [20] the generalized Cauchy integral formula is proved (in the elliptic case) for the Dirac operator. We shall discuss it again in Examples 3.2,3.4,3.7,3.8,4.1,4.2,4.3,4.5, where further results will be discussed.

3.2 Clifford analysis

Let us consider the Clifford algebra $\mathcal{C}_n = \mathcal{C}_{o,n}$ with the standard grading $\mathcal{C}_n = \mathcal{C}_n^+ + \mathcal{C}_n^-$. The Clifford algebra $\mathcal{C}_n$ is the (reducible) Spin(n)-module due to the inclusion Spin(n) $\subset \mathcal{C}_n^+$ and, say, left multiplication. We can decompose it into irreducible pieces $\mathcal{C}_n = V_1 \oplus .. \oplus V_k$, where all V_j are basic Spin(n)-modules. As in 3.1, all tensor products $V_j \otimes_R \Lambda^1$ decomposes into $F_{j,1} \oplus F_{j,2}$, where all $F_{j,1}$ are again basic spinor modules. Adding all pieces together, we shall obtain two operators for maps $\Psi : R_n \longrightarrow \mathcal{C}_n$

$$\mathcal{C}^\infty(V_1 \oplus \ldots \oplus V_k) \cong \mathcal{C}^\infty(\mathcal{C}_n) \begin{array}{l} \xrightarrow{\partial_1} \mathcal{C}^\infty(F_{1,1} \oplus .. \oplus F_{k,1}) \cong \mathcal{C}^\infty(\mathcal{C}_n) \\ \xrightarrow[\partial_2]{} \mathcal{C}^\infty(F_{1,2} \oplus .. \oplus F_{k,2}) \end{array}$$

To describe the coordinate form of the operator ∂_1, let us choose an orthonormal basis $e_1, \ldots, e_n$ of $\mathcal{C}_n$. Applying the results of the section 3.1 to every piece V_j, the equation $\partial_1 \Psi = 0$ for $\mathcal{C}_n$-valued maps looks like

$$\sum_1^n e_j \frac{\partial \Psi}{\partial x_j} = 0 \quad . \tag{10}$$

This is the equation, studied by Delanghe ([9]). Note that multiplication by e_j need not preserve the individual pieces V_j, but preserves, of course, the whole algebra $\mathcal{C}_n$.

Multiplying the equation (10) by $(-e_1)$ we shall obtain the equivalent equation

$$\sum_1^n f_j \frac{\partial \Psi}{\partial x_j} = 0 \tag{11}$$

where $f_1 = 1$ and $f_j = -e_1 e_j$, $j=2,\ldots,n$ are generators of the algebra $\mathcal{C}_n^+ \cong \mathcal{C}_{n-1}$. Restricting our attention to maps $\Psi : R_n \longrightarrow \mathcal{C}_{n-1} = \mathcal{C}_n^+$, we shall get the equation for monogenic functions in Clifford analysis.

The study of solutions of the equation (11) is well advanced. A remarkable amount of results is known already for monogenic functions (see [2]), they will be described in more details in the lecture by Prof. Delanghe. Note that even if the system of equations studied in Clifford analysis is reducible (it consists of several copies of the equation for regular spinor maps), there are definitive advantages in notation (use of Clifford numbers) and concepts, which makes this setting of the problem valuable.

3.3 Fueter's regular functions

Fueter and his coworkers started to study quaternionic analysis 50 years ago. Their regular functions are special cases of monogenic functions in Clifford analysis for the case $n=3$, where maps $f : R_4 = H \longrightarrow \mathcal{C}_3 = H \oplus H$ are considered. To obtain Fueter's equation for regular functions, it is sufficient to split the value $f = (f_1, f_2)$ and consider only, say, f_1. A nice and modern account of basic results in quaternionic analysis can be found in [39].

3.4 Generalized C-R equations of Stein and Weiss

In this case we shall use the complexified form of the procedure, described in Remark 3 (section 2). For any Spin(n)-module V (over C) we shall split the tensor product $V \otimes_C \Lambda^1_c$ into two pieces. There is the exceptional irreducible piece in the product, called Cartan (or Jung) product of V and Λ^1_c. It is characterized by the fact that its highest weight is the sum of the highest weights of V and Λ^1_c. Let us denote it by F_1, so $V \otimes_C \Lambda^1_c = F_1 \oplus F_2$, where F_2 is (not necessarily irreducible) Spin(n)-module. Denoting π_2 the projection to F_2, we can write the equations of Stein and Weiss as

$$(\pi_2 \circ d)\, \Psi = 0 .$$

It was proved in [38] that solutions of such equations are (componentwise) harmonic functions and that the modulus of the maps to the power p is subharmonic function for $p \geq (n-2)/(n-1)$. The most interesting special cases are:

i) V is the basic spinor module (see 3.1), the case $n=3$ was studied by Moisil and Theodoresco ([22])

ii) $V = \Lambda^r_c$, then the tensor product $\Lambda^r_c \otimes_C \Lambda^1_c$ splits into 3 pieces (see [38]): the Cartan product F_1, $F_2 = \Lambda^{r+1}_c$, and $F_3 = \Lambda^{r-1}_c$; the equations defined using the projections F_2, resp. F_3 are just the operators d, resp. δ, the equation coming from the projection onto $F_2 \oplus F_3$ being the Hodge operator $d + \delta$. Both these cases are well studied on manifolds.

3.5 Massless fields

More complete discussion can be given for dimension $n=4$. At the same time it is the most interesting case because of its close connection to mathematical physics. All bundles and tensor products considered in this section will be complex. There is the isomorphism $Spin(4) \cong Sp(1)xSp(1)$. The two basic spinor modules V_+ and V_- can be realized e.g. by left multiplication by one of the factors on $V = H$. All irreducible Spin(4)-modules can be realized as submodules of tensor products of the two basic modules V^+ and V^- and can be classified by a couple of nonnegative integers (j,k):

$$V^{j,k} = S^j V_+ \otimes S^k V_- ,$$

where S^j denotes symmetrized tensor product.

It can be shown ([11],[17],[8]) that

$$V^{j,k} \otimes_C \Lambda^1_c = V^{j+1,k+1} \oplus V^{j+1,k-1} \oplus V^{j-1,k+1} \oplus V^{j-1,k-1}$$

(if j=0 or k=0, then there are only two pieces in the decomposition).

The four (resp. two) differential operators obtained in such a way are 'analytic continuation' of operators for massless fields, described by Gårding ([12]). Many of these equations are used often in mathematical physics, sometimes they are considered on complexified Minkowski space. The most important cases, when j=0 (or k=0) leads to the equations, which are usually called massless field equations and twistor (or Killing) equations. The description of these operators on manifolds is given in 4.2. A lot of results are known for them (see e.g. [17],[24],[27],[28]).

3.6 Quaternionic valued differential forms on R_4

Here we shall describe the whole split de Rham sequence (6) in the special case n=4 and $V = V_+$ (see 3.5). Only real modules will be considered here.

The spaces $V^{j,k}$, considered as modules over R, are irreducible only for j+k odd. For j+k even there is a real subspace $rV^{j,k} \subset V^{j,k}$ such that $rV^{j,k} \otimes_R C = V^{j,k}$ (see [18]) The exterior powers Λ^j have the following description in the classification:

$$\Lambda^1 = rV^{1,1} , \Lambda^2 = rV^{2,0} \oplus rV^{0,2} , \Lambda^3 = \Lambda^1 ,$$

which then leads to the splittings

$$V^{1,0} \otimes_R \Lambda^1 = V^{0,1} \oplus V^{1,2} , V^{1,0} \otimes_R \Lambda^2 = V^{3,0} \oplus V^{1,0} \oplus V^{1,2}$$

(the splitting for 3-forms being the same as for 1-forms). The diagram (6) hence has the form

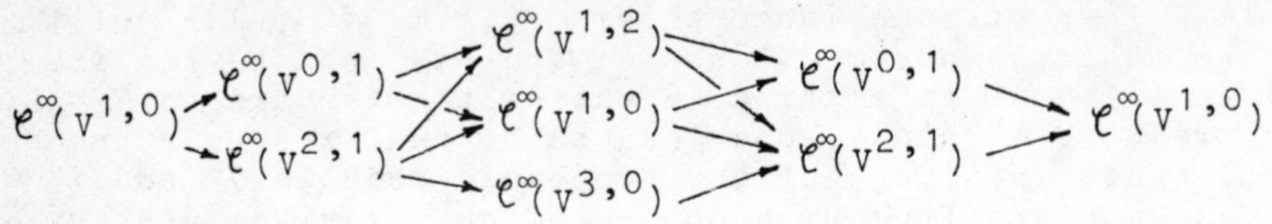

It can be compared with the splitting of quaternionic valued forms, described in [34] using quaternionic coordinates. The similar splitting can be written for other spinor modules, too.

3.7 Integral formulae and Leray residue

If the equations coming from the diagram (6) are reasonable generalizations of C-R equations, it should be possible to write a generalized Cauchy integral formula for their solutions. It is possible to do it in many cases (e.g. if another first order operator can be found such that the composition of both gives the Laplace operator, see [6]). Generalized Cauchy integral formulae were discussed in many papers (e.g. [2],[5],[6],[9],[10],[13],[14],[15],[20],[35],[37]). All these integral formulae are of elliptic type, i.e. the value at the point P is expressed using values on a sphere around P.

It was mentioned already that the equations coming from the diagram (6) have both elliptic and hyperbolic versions. The integral formulae for hyperbolic equations have quite different character, but even if these two types of integral formulae are very different indeed, there is very simple and nice principle how to deduce one from another using the Leray residue ([19],[29]). Leray's extension of the classical residue theorem can be described as follows.

Let X be a complex manifold of a (complex) dimension n, let S be a submanifold of X of codimension 1. For every $(p-1)$-cycle γ in S we shall denote by $\delta\gamma$ the Leray cobord of γ. It is, roughly speaking, the boundary of a tubular neighborhood around γ, so $\delta\gamma$ is a p-cycle in $X \setminus S$.

<u>Theorem (Leray)</u>:

Let τ be a smooth, closed p-form on $X \setminus S$, then there is a $(p-1)$-form $\operatorname{Res}\tau$ closed on S, such that

$$\int_{\delta\gamma} \tau = 2\pi i \int_{\gamma} \operatorname{Res}\tau$$

holds for every $(p-1)$-cycle γ in S.

The theorem holds for vector valued forms, too, and can be used in the study of integral formulae in the following way.

Suppose that the map $f : R_n \to V$ satisfies equations for which Cauchy integral formula holds. So suppose that there is a $(n-1)$-form ω_P, depending on P such that

$$f(P) = \int_{S_{n-1}} \omega_P \tag{12}$$

where the point P lies inside $S_{n-1} \subset R_n$.

Suppose further that the form ω_P is the restriction of a closed form on $C_n \setminus N$, where N is the complex null cone in C_n with the vertex in P.

Then, defining the index of the point P with respect to $(n-1)$-dimensional cycle $\Sigma \subset C_n \setminus N$ properly, the formula

$$\operatorname{Ind}_\Sigma P.f(P) = \int_\Sigma \omega_P \tag{13}$$

holds for every cycle $\Sigma \subset C_n \setminus N$ ([6]).

But then the Leray residue theorem tells us that

$$\operatorname{Ind}_{\delta\gamma} P.f(P) = 2\pi i \int_\gamma \operatorname{Res} \omega_P \tag{14}$$

for every $(n-2)$-cycle γ in N. Taking the cycle γ especially inside the intersection of N with the Minkowski slice through P gives then the corresponding integral formula for solutions of the hyperbolic system. In C_n we have both possibilities (either to express $f(P)$ using $\Sigma \subset C_n \setminus N$ or using $\gamma \subset N$).

Using the procedure it is possible e.g. to deduce Riesz's integral formula for solutions of the wave equation ([30]) from the standard integral formula for harmonic functions ([10]) or to deduce integral formulae, due to Penrose for spin n/2 massless fields on Minkowski space ([26],[27]) from the Cauchy integral formulae in hypercomplex analysis ([36]). A new integral formula for the Dirac operator on Minkowski space M_n was deduces by Bureš from the standard Cauchy integral formula in Clifford analysis ([5]).

3.8 Vector valued forms and cohomology

To illustrate the usefulness of the diagram (6) let us consider the following well known fact from the classical complex analysis. The number of holes in a domain $\Omega \subset C$ can be found using only properties of holomorphic functions on Ω. Holomorphic functions without primitives can exist on Ω and the dimension of the vector space

$$H(\Omega) \;/\; \frac{d}{dz} H(\Omega)$$

is equal to the number of holes in Ω.

Let us express it in more modern language.

Let us consider a part of the diagram (1)

$$0 \longrightarrow C \longrightarrow \mathcal{E}^{\infty}(\Omega,\Lambda^{0}) \xrightarrow{\ \partial\ } \mathcal{E}^{\infty}(\Omega,\Lambda^{1,0})$$

and its subcomplex $0 \longrightarrow C \longrightarrow M^{0} \xrightarrow{\ \partial\ } M^{1}$,

where $M^{0} = \{f \in \mathcal{E}^{\infty}(\Omega,\Lambda^{0}) \mid \bar{\partial}f=0\}$, $M^{1} = \{f \in \mathcal{E}^{\infty}(\Omega,\Lambda^{1,0}) \mid \bar{\partial}f=0\}$.

Then $H^{1}(\Omega,C) = M^{1} / \partial(M^{0})$.

All this can be generalized to R_n using the diagram (6). To give an example, let V be the basic spinor module (over R). Then we shall consider the spaces

$$E^{j} = \mathcal{E}^{\infty}(\Omega,\Lambda^{j} \otimes_{R} V)$$

of V-valued differential forms on $\Omega \subset R_n$. It may be proved by induction that one of irreducible pieces in the decomposition of $\Lambda^{j} \otimes_{R} V$, which we denote by F_1^j, is isomorphic to a basic Spin(n)-module. Let us denote by F_2^j the complementary invariant subspace, then we have the splitting $E^j = E_1^j \oplus E_2^j$, where $E_1^j = \mathcal{E}^{\infty}(\Omega, F_1^j)$, $E_2^j = \mathcal{E}^{\infty}(\Omega,F_2^j)$. So we shall obtain the diagram

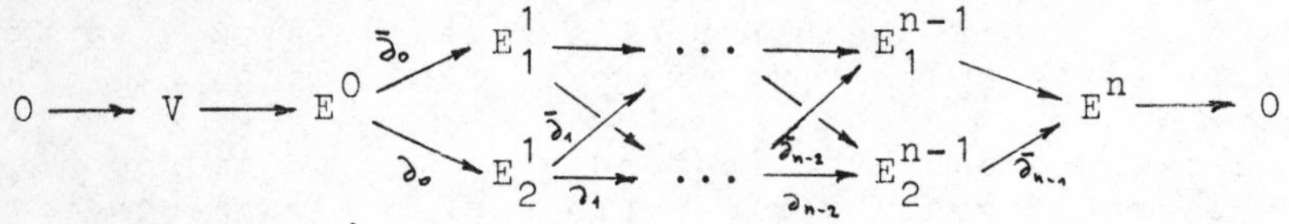

Let us denote $M^j = \operatorname{Ker} \bar{\partial}_j$, then the homology of the sequence

$$0 \longrightarrow V \longrightarrow M^{0} \xrightarrow{\partial_0} M_1^{1} \xrightarrow{\partial_1} \cdots \xrightarrow{\partial_{n-2}} M^{n-1} \longrightarrow 0$$

coincide with $H^j(\Omega, V)$ for j=o,...,n-1 (see [31],[33]).

4. EXAMPLES (MANIFOLDS)

4.1 Dirac operator

It is the only case where important and deep global results are already known. The Dirac operator plays, for example, the important role in the Atiyah-Singer index theorem ([25]).

To describe the Dirac operator using the general procedure of the section 2 (for complex bundles, see Remark 3), three choices have to be made - a Spin(n)-module V, the conformal weight w and a connection ∇ . We shall consider only the case of even dimension, n=2m.

Let us take w = (n-1)/2. Consider a minimal left ideal V in the complex Clifford algebra $\mathcal{C}_n^c$ with generators $e_1,\ldots,e_n$. Let us choose a Riemannian metric in the given conformal class. It will induce the connection ∇ on the Spin bundle V.

The tensor product $\Lambda_c^1 \otimes_C V$ again splits into two pieces F_1, F_2; F_1 being the basic spinor module. The space F_1 is often identified with V using the map, given in coordinates by

$$v \in V \longmapsto \frac{1}{n}\left\{\sum dx_k \otimes e_k . v\right\} \in \Lambda_c^1 \otimes_C V .$$

Then the projection of $\Lambda_c^1 \otimes_C V$ onto $F_1 = V$ is given by (compare with (9))

$$m : \quad \sum dx_j \otimes v_j \longrightarrow \sum e_j . v_j \in V .$$

Under such identification the Dirac operator looks like

$$\Gamma(V) \xrightarrow{\nabla} \Gamma(\Lambda_c^1 \otimes_C V) \xrightarrow{m} \Gamma(V)$$

and its symbol is given by Clifford multiplication.

Solutions of the Dirac equation are called harmonic spinors by Hitchin ([16]). The module V splits into two irreducible spinor modules V^+ and V^-, which leads to operators D^+, D^- and to spaces of positive (negative) harmonic spinors. The special value chosen for w is quite important, because in this case the Dirac operator does not depend on the choice of the Riemannian connection (see 4.2). It was proved in [16] that the dimension of the space of harmonic spinors is conformally invariant, but that it cannot be expressed in terms of topological invariants of the manifold.

In the case of conformally flat manifolds it is possible to give a nice coordinate description of the spinor bundle and of the Dirac operator. It was shown by Ahlfors ([0]) that conformal maps in higher dimensions can be expressed compactly using Clifford numbers in a very close analogy with the complex case. This notation can be used to give a simple formula for transition functions of the spinor bundle. The condition that solutions transform into solutions under conformal transformations (with a weight) picks out again the conformal weight $w = (n-1)/2$ as the only possibility (for details see [4],[21]).

4.2 Conformally invariant operators

Let us consider again complex bundles. Let us take any irreducible Spin(n)-module V and any conformal weight w. Let us consider further any irreducible piece W in $\Lambda_c^1 \otimes_C V$.

A natural possibility for the choice of ∇ is to consider a Riemannian metric inside the given conformal structure and to take the associated connection ∇ on V. It will give us the corresponding operator ∂ in the first column of the diagram (8). The operators ∂ depend generally on the choice of the Riemannian metric, but there is an exceptional, unique case, when all operators ∂ for all possible Riemannian metrics in the given conformal class coincide. It was proved by Fegan ([11]) (and by algebraic methods by Hitchin ([17])) that for every choice of V , W there is exactly one conformal weight w such that the operator ∂ is conformally invariant (in the sense described above).

Another very interesting subject is discussed in the paper by Hitchin ([17]). The twistor theory, created by R. Penrose, is nowadays very rich and extended theory. The theory is firmly rooted in physics (namely in general relativity) and its evolution has led to a lot of deep and important mathematical results for nonlinear equations. It was a part of the Penrose's twistor programme to study massless fields using the correspondence between Minkowski and twistor spaces. This transformation is studied in [17] for massless fields (i.e. $V = V^{j,o}$ in 3.5) and for the Laplace equation on self-dual 4-dimensional manifolds.

4.3 Generalized spherical C-R operator

Let M be n-dimensional Riemannian manifold with an exterior structure, given by the Weingarten map. Take the Clifford algebra $\mathcal{C}_n$ for the G-module V . A special connection was defined on the associated vector bundle using the Riemannian connection and the Weingarten map in [32]. The corresponding piece in the decomposition was used there as the generalization of the spherical C-R operator.

4.4 Kähler equation

Let us take the space Λ^r for the G-module V . Then (see the section 3.4) the tensor product $\Lambda^r \otimes_R \Lambda^1$ splits into 3 pieces F_1, F_2, F_3 and $F_2 \cong \Lambda^{r+1}$, $F_3 \cong \Lambda^{r-1}$. The corresponding operators ∂_2 and ∂_3 are just d and δ .

Now, if we take $V = \Lambda^* = \Lambda^o \oplus \ldots \oplus \Lambda^n$, then we can apply the procedure piece by piece and we shall end with the operators d and δ on Λ^* . The values of these operators are, strictly speaking, in $\Lambda^* \otimes_R \Lambda^1$, but we can identify them with Λ^* . We can even consider their sum d + δ , but then the result differs in both cases. The operator d + δ with values in Λ splits (locally) in a quite different way and it coincides (after the usual identification of Λ^* with $\mathcal{C}_n$) with the basic operator in Clifford analysis (so it splits into the sum of the Dirac operators for spinor valued fields).

The operator d+δ in the second sense is usually called the Dirac operator and its physical interpretation was studied in [1].

ACKNOWLEDGEMENT

In the paper the common work with J. Bureš is reported ([7]).

REFERENCES

[0] L.V.Ahlfors: 'Möbius transformations and Clifford numbers'. See elsewhere in this volume.

[1] I.Benn, R.Tucker: 'Fermions without spinors', *Com.Math.Physics*,89,3,1983, 341-362

[2] F.Brackx, R.Delanghe, F.Sommen: *Clifford analysis*, Res. Notes in Math. 76, Pitman,1982

[3] R.Delanghe, F.Brackx: 'Hypercomplex function theory and Hilbert modules with reproducing kernel', *Proc.Lond.Math.Soc.*,37, 1978, 545-576

[4] J.Bureš: 'Hypercomplex analysis on conformally flat manifolds', to be published

[5] J.Bureš: 'Integral formulas for left regular spinor-valued functions in Clifford analysis', to appear

[6] J.Bureš, V.Souček: 'Generalized hypercomplex analysis and its integral formulas', to appear in *Complex Variables*

[7] J.Bureš, V.Souček: 'On generalized Cauchy-Riemann equations on manifolds', Proc. of the 12th Winter School, Srni *Suppl.Rend.Circolo Mat.Palermo*,II,6,1984, 31-42

[8] J.Bureš, V.Souček: 'Regular spinor valued mappings', to appear in *Seminarii di geometria*, Bologna

[9] R.Delanghe: 'On regular-analytic functions with values in a Clifford algebra', *Math.Ann.*,185,1970,91-111

[10] M.Dodson, V.Souček: 'Leray residue applied to solutions of the Laplace and wave equations', to appear in *Seminarii di geometria*, Bologna

[11] H.D.Fegan: 'Conformally invariant first order differential operators', *Q.Jour.Math.*,27,1976,371-378

[12] L.Gårding: 'Relativistic wave equations for zero restmass', *Proc.Camb.Phil.Soc.*,41,1945,49-56

[13] R.P.Gilbert, J.L.Buchanan: *First order elliptic systems,* Math. in Science and Eng. 163, Academic Press, 1983

[14] H.Haefeli: 'Hyperkomplexe Differentiale', *Comment.Math.Helv.*,20,1947,382-420

[15] D.Hestenes, G.Sobczyk: *Clifford algebra to geometric calculus,* D.Reidel Publ. Comp., Dordrecht, 1984

[16] N.J.Hitchin: 'Harmonic spinors' *Adv.Math.*,14,1974,1-55

[17] N.J.Hitchin: 'Linear field equations on self-dual spaces, *Proc.R.Soc.London* A 370,1980,173-191

[18] D.Husemoller: *Fibre bundles,* McGraw Hill, 1966

[19] J.Leray: 'Le calcul differentiel et integral sur une variete analytique complex, Probleme de CauchyIII *Bull.Soc.math.Fr.*,87,1959,81-180

[20] P.Lounesto: 'Spinor-valued regular functions in hypercomplex analysis', thesis, Helsinky University of Techn. Report-HTKK-MAT-A154, 1979

[21] M.Markl: 'Regular functions over conformal quaternionic manifolds', *Com.Math.Univ.Carolinae,*22,1981,579-583

[22] G.C.Moisil, N.Theodoresco: Fonctions holomorphes dans l'espace" *Mathematica V,*1931,142-159

[23] W.Nahm: Self-dual magnetic monopoles and generalizations of holomorphic functions', Proc. of the 12th Winter School, Srni *Suppl.Rend.Circ.Mat.Palermo,*II,6,1984,233-242

[24] P. van Nieuwenhuizen, N.P.Warner: 'Integrability conditions for Killing spinors', *Com.Math.Physics,*1984, 277-284

[25] R.S.Palais: *Seminar on the Atiyah-Singer index theorem,* Princeton Univ. Press, 1965

[26] R.Penrose: 'Null hypersurface initial data for classical fields of arbitrary spin and for general relativity', *Gen.Rel.Grav.*,12,1980225-264

[27] R.Penrose, W.Rindler: *Spinors and space-time I,* Camb.Univ.Press, 1984

[28] R.Penrose, M.Walker: 'On quadratic first integrals of the geodesic equations for type 22 spacetimes', *Com.Math.Physics,*18,1970,265-274

[29] F.Pham: *Introduction a l etude topologique des singularites de Landau,*
Paris, Gauthier.Villars, 1967

[30] M.Riesz: 'A geometric solution of the wave equation in space-time of even dimension',
*Comm.Pure Appl.Math.,*XIII,1960,329-351

[31] F.Sommen: 'Monogenic differential forms and homology theory',
*Proc.Royal Irish Acad.,*84A,2,1984,87-109

[32] F.Sommen: 'Monogenic functions on surfaces'
to be published

[33] F.Sommen, V.Souček: 'Hypercomplex differential forms applied to the de Rham and the Dolbeault complex',
to appear in *Seminarii di geometria,*Bologna

[34] V.Souček: 'H-valued differential forms on H',
Proc. of the 11th Winter School, Zelezna Ruda
*Suppl.Rend.Circ.Mat.Palermo,*II,3,1984,293-299

[35] V.Souček: 'Complex-quaternionic analysis applied to spin-$\frac{1}{2}$ massless fields',
*Complex Variables,*1,1983,327-346

[36] V.Souček: 'Boundary value type and initial value type integral formulae for massless fields',
*Twistor Newsletters,*14,1982,

[37] V.Souček: 'Holomorphicity in quaternionic analysis'
Seminarii di geometria, 1982-83,
Istituto di Geometria, Universita Bologna, 147-171

[38] E.M.Stein, G.Weiss: 'Generalization of the Cauchy-Riemann equations and representations of the rotation group',
*Amer.J.Math.,*90,1968,163-196

[39] A.Sudbery: 'Quaternionic analysis',
Math.Proc.Camb.Phil.Soc., 85,1979,199-225

INTEGRAL FORMULAE IN COMPLEX CLIFFORD ANALYSIS.

Jarolím Bureš
Dep.of Mathematics
Charles University
Sokolovská 83,18600 Praha
Czechoslovakia

ABSTRACT. In this note some integral formulae for complex left regular mappings and for solutions of the complex Laplace equation are presented. These integral formulae are of two types (elliptic and hyperbolic) and there is a general procedure (described by V.Souček and M.Dodson for the Laplace equation in [3]) to transform one type into another, namely the Leray residue formula. In this way it is possible to derive from integral formulae in Clifford analysis e.g. Riesz' integral formula for the solution of the wave equation in the Minkowski space and Penrose's integral formula for a spin-$\frac{1}{2}$ massless field. There are good hopes that the new integral formula of hyperbolic type can be used to give further information on left regular mappings and spinor fields.

1. COMPLEX CLIFFORD ANALYSIS.

Let n be an odd number, say $n=2h-1$, $\mathbb{C}_n$ the complex Clifford algebra generated by $\{e_j, j=1,\dots,n\}$ with multiplication rules

$$e_i e_j + e_j e_i = -2\delta_{ij} e_0 \ , \ e_0 = 1$$

and denote

$$\mathbb{C} = \mathbb{C}\, e_0 \ , \ \mathbb{C}^{n+1} = \mathrm{span}_{\mathbb{C}}\{e_0,\dots,e_n\} .$$

For $z \in \mathbb{C}^{n+1}$, $z = \sum_{\alpha=0}^{n} z_\alpha e_\alpha$, put $z^+ = z_0 e_0 - \sum_{j=1}^{n} z_j e_j$, $|z|^2 = z^+ z$ and $E^{n+1}(z) = z + \mathbb{R}^{n+1} \subset \mathbb{C}^{n+1}$ the Euclidean slice at z, $M^{n+1} = z + \mathbb{R}\, e_0 + \sum_{j=1}^{n} i\mathbb{R}\, e_j$ the Minkowski slice at z. Further put $N^{\mathbb{C}}(z) = \{u \in \mathbb{C}^{n+1}, |u-z|^2 = 0\}$ and $N(z) = N^{\mathbb{C}}(z) \cap M^{n+1}(z)$; they represent resp. the complex and the real singularity cones at z.

J. S. R. Chisholm and A. K. Common (eds.), Clifford Algebras and Their Applications in Mathematical Physics, 219–226.

Consider on $\mathbb{C}^{n+1}$ the differential operators

$$\partial = \sum_{\alpha=0}^{n} e_\alpha \frac{\partial}{\partial z_\alpha}, \quad \partial^+ = e_0 \frac{\partial}{\partial z_0} - \sum_{j=1}^{n} e_j \frac{\partial}{\partial z_j}, \quad \square^{\mathbb{C}} = \partial\partial^+.$$

The following two types of holomorphic mappings will be studied in this paper :

(I) left regular mappings $\Phi : \Omega \subset \mathbb{C}^{n+1} \to \mathbb{C}_n$ satisfying $\partial\Phi = 0$;

(II) complex harmonic functions $\Psi : \Omega \subset \mathbb{C}^{n+1} \to \mathbb{C} \subset \mathbb{C}_n$, satisfying $\square^{\mathbb{C}}\Psi = 0$,

Ω being a domain in $\mathbb{C}^{n+1}$.

2. INTEGRAL FORMULAE OF ELLIPTIC TYPE.

It is a well-known fact (see for example [5]) that the value of a left regular mapping Φ at a point u can be expressed using the integral formula :

$$(1) \quad \Phi(u) = \varkappa_n^{-1} \int_{\Sigma^n} G(z-u)\, Dz\, \Phi(z)$$

where $\varkappa_n$ is the volume of S^n, $Dz = \sum_{j=0}^{n} (-1)^j e_j d\hat{z}_j$ with $d\hat{z}_j = dz_0 \wedge \dots \wedge dz_{j-1} \wedge dz_{j+1} \wedge \dots \wedge dz_n$, $G(w) = |w|^{-h} w^+$, Σ^n is a cycle in Ω homological in Ω -N(u) to the sphere $S^n(u)$ in the Euclidean slice $E^{n+1}(u)$ (with definite restriction of the fundamental complex quadratic form).

Similarly the value of a complex harmonic function at a point u is given by

$$(2) \quad \Psi(u) = \varkappa_n^{-1} \int_{\Sigma^n} (G(z-u)Dz\,\Psi(z) + g(z-u)D^+z\, \partial\Psi(z)),$$

where

$$g(w) = -2^{-1}(h-1)^{-1}|w|^{1-h}, \quad D^+z = e_0 d\hat{z}_0 + \sum_{j=1}^{n}(-1)^{j+1} e_j d\hat{z}_j : e_j d\hat{z}_j .$$

This formula is a Complex Clifford algebra version of a well-known formula for complex harmonic functions(for details see [2]).

In summary,the value of a mapping at u can be expressed by its values on some cycle homological to the sphere in the Euclidean slice at u.

3. INTEGRAL FORMULAE OF HYPERBOLIC TYPE.

For both types of mappings (I) and (II) integral formulae of another type can be written down too.Instead of the Euclidean slice at u we use the Minkowski slice at u ,and the value of a mapping at u depends on the values on some (n-2)-

dimensional submanifold s(u) lying in the singularity cone N(u) with vertex u ,its derivatives in characteristic directions at the points of s(u),and geometric properties of s(u).

The formulae have the following form:

in case (I)

$$(1^+)\qquad \Phi(u) = (-\pi)^{1-h}\int_{s(u)} (b^+ \cdot c)\ \frac{\partial^{h-1}}{\partial\tau^{h-1}}\Big|_{\tau=0} (F\cdot\Phi)\ ds\ ,$$

and in the case (II):

$$(2^+)\qquad \Psi(u) = (-\pi)^{1-h}\int_{s(u)} \frac{\partial^{h-2}}{\partial\tau^{h-2}}\Big|_{\tau=0} \left(\tfrac{1}{2}\Psi\frac{\partial F}{\partial\tau} + \frac{\partial\Psi}{\partial\tau}\cdot F\right) ds$$

where b is the vector function describing s(u), $(b=z-u, z\in s(u))$ c(z) is the null vector which is normal to the tangent space to s(u) at z normalized with respect to b,that is with respect to the induced Lorentzian scalar product in $M^{n+1}(u)$

$$(c,c) = (c,db) = 0$$

$$(c,b) = \tfrac{1}{2}$$

and F is a function which depends only on geometric properties of s(u) (see the Appendix).Let τ be the parameter in the direction c. See Fig.1.

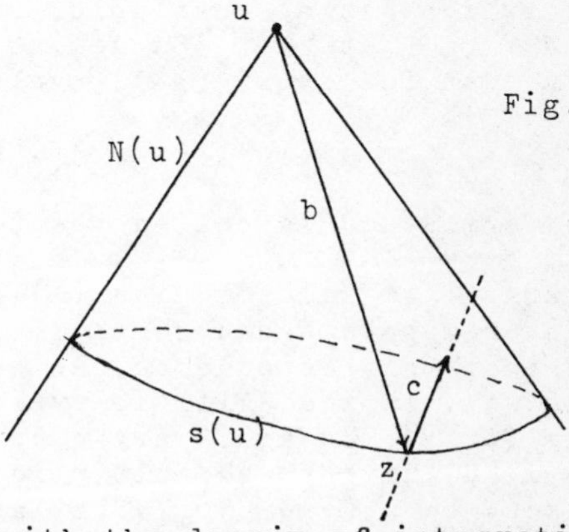

Fig.1

The singularity cone N(u) with the domain of integration.

Remark 1.The formula (2^+) is in fact the Riesz formula presented in [6].

Remark 2.The corresponding data for mappings of type (I) or (II) can be given on some n-dimensional space-like surface S in M^{n+1} . The surface s(u) is then the intersection of S with N(u). For complex harmonic functions (a system of second order) two data functions on S are needed as initial data representing values and derivatives in characteristic directions.For left regular functions one data function is enough since the derivative in the characteristic direction can be derived from derivatives on S using the equation for regularity.

4. LERAY RESIDUE FORMULA.

There is a procedure for obtaining formulae (1^+) and (2^+)

from the corresponding formulae (1) and (2). This procedure uses besides a complex residue formula a certain deformation in $\mathbb{C}^{n+1}-N(u)$ of the domain of integration from the sphere $S^n \subset E^{n+1}(u)$ to the surface $s(u) \times S^1$, where $s(u) \subset N(u)$ in $M^{n+1}(u)$ and S^1 lies in the complex direction out of $M^{n+1}(u)$. This procedure was used first by K. Imaeda in [7] and then by V. Souček in [8]. The equivalence of the formulae is a special case of a general formula called the Leray residue formula. We shall describe it briefly (for details see for example [3]).

Let X be a complex manifold of complex dimension n and $S \subset X$ a submanifold of codimension 1. For any (p-1)-cycle γ in S it is possible to define a p-cycle $\delta\gamma$ in X-S, called the Leray cobord of γ. There is also a generalized theorem of Leray for a V-valued form, (V being a vector space).

Theorem (Leray residue theorem): Let $\omega \in \mathcal{E}^p_V(X-S)$ be a closed form which has at most polar singularity at S. Then there is a form $\mathrm{Res}\,\omega \in \mathcal{E}^{p-1}_V(S)$ defined modulo an exact form on S such that for every (p-1)-cycle γ in S

$$\int_{\delta\gamma} \omega = 2\pi i \int_{\gamma} \mathrm{Res}\,\omega .$$

The form Res ω is called the Leray residue of ω.

Remark. Integral formulae (1), (2), (1^+) and (2^+) for left regular mappings and complex harmonic functions in $\mathbb{C}^{n+1}$, and their correspondence, can be generalized in the following way. In the elliptic type formulae (1) and (2) we can take for Σ^n an arbitrary n-cycle in $\mathbb{C}^{n+1}-N^{\mathbb{C}}(u)$, but $\Phi(u)$ and $\Psi(u)$ have to be multiplied by the index of Σ^n with respect to u (see [5]). Similarly, for s(u) in (1^+) and (2^+) we can take an arbitrary (n-1)-cycle on $N^C(u)$ not containing u and again we have to use a properly defined index in the corresponding generalized formula. The correspondence of the generalized formulae is again given by a Leray residue formula, but the co-ordinate expressions need not be so simple in general.

5. SPECIAL CASES FOR n=3.

For n = 3 we have the following more geometrical formulae:

$$(3) \qquad \Psi(u) = \frac{1}{4\pi}\int_{s(u)} \Psi\cdot K\cdot ds + \frac{1}{2\pi}\int_{s(u)} \frac{\partial\Psi}{\partial\tau}\Big|_{\tau=0}\, ds ,$$

$$(4) \qquad \Phi(u) = \frac{1}{2\pi}\int_{s(u)} (b^+\cdot c)\,\Phi\, K ds + \frac{1}{\pi}\int_{s(u)} (b^+\cdot c)\,\frac{\partial\Phi}{\partial\tau}\Big|_{\tau=0}\, ds ,$$

where K is the Gaussian curvature of s(u) and b and c are presented in $\mathbb{C}_n$-form.

The formula (3) is the Riesz integral formula presented by Hormander in [9] .

If Φ is a regular spinor mapping on $\mathbb{C}^4$,it can be identified with a spin-½ massless field on complex Minkowski space.The formula (4) is then a Clifford algebra version of Penrose's integral formula given in [10]. A simpler version of this result was proved in [2].Now another description of this equivalence will be given.

Let $\mathbb{C}^4 \subset \mathbb{C}_3$ be spanned by $\{e_0 = 1, e_1, e_2, e_3\}$,let

$$z = \sum_{j=0}^{3} z_j e_j \quad \text{and denote } z^+ = z_0 e_0 - \sum_{j=1}^{3} z_j e_j \ , \ \bar{z} = \sum_{j=0}^{3} \bar{z}_j e_j$$

and $z^* = \bar{z}^+ = \overline{z^+}$.

Let Minkowski space $M \subset \mathbb{C}^4$ be given by

$$w = (w_0, w_1, w_2, w_3) \in R^4 \longrightarrow \quad w = w_0 e_0 + \sum_{j=1}^{3} i w_j e_j \in \mathbb{C}^4 \subset \mathbb{C}_3$$

and complex Minkowski space $\mathbb{C}M$ by the same formula but with complex numbers instead of reals,so that

$$|w|^2 = w_0^2 - \sum_{j=1}^{3} w_j^2 .$$

Spinor spaces can be considered as minimal left or right ideals in C_3.We shall consider the following spinor spaces and their equivalents in the spinor formalism due to Penrose.

$$S_A = S = \{a = a_0 \tfrac{1}{2}(ie_1 - e_2) + a_1 \tfrac{1}{2}(e_0 + ie_3)\}$$
$$S^A = S^+ = \{a^+ = a^0 \tfrac{1}{2}(e_0 - ie_3) + a^1 \tfrac{1}{2}(ie_1 - e_2)\}$$
$$S_{A'} = S^* = \{a^* = \bar{a}_0 \tfrac{1}{2}(ie_1 + e_2) + \bar{a}_1 \tfrac{1}{2}(e_0 + ie_3)\}$$
$$S^{A'} = \bar{S} = \{\bar{a} = \bar{a}^0 \tfrac{1}{2}(e_0 - ie_3) + \bar{a}^1 \tfrac{1}{2}(ie_1 + e_2)\}$$

Moreover we have the identifications

$$\alpha_A \bar{\beta}_{A'} = \beta^* \alpha \ ; \ \alpha_A \beta^A = \alpha \beta^+ \in S \cap S^+ \cong \mathbb{C} \ ; \ \alpha^{A'} \beta_{A'} = \alpha^* \bar{\beta} \in S^* \cap \bar{S} \cong \mathbb{C}$$

and (in the integral formula (4))

$$b = r\, \zeta_A \bar{\zeta}_{A'} \to b = r\, \beta^* \cdot \beta \ ; \ c = r^{-1}\, \xi_A \bar{\xi}_{A'} \to c = r^{-1}\, \gamma^* \cdot \gamma \ ;$$

$$\xi_A \xi^A = 1 \to \beta \cdot \gamma^+ = 1 \ ; \ \phi^A \to \phi^+ .$$

Hence (4) for u = 0 can be put into the form

$$\Phi^+(0) = (2\pi)^{-1} \int_{s(0)} \beta^+ r^{-1} \, (\alpha \cdot D\Phi^+ - 2\, \alpha \cdot \Phi^+ \cdot \rho) \ ds ,$$

which is equivalent with the Penrose's formula

$$\hat{\Phi}(0) = (2\pi)^{-1} \int_{s(0)} \xi_A r^{-1} ((D\Phi^B)\xi_B - 2\rho(\Phi^B \xi_B))\, ds.$$

6. APPENDIX.

Keeping the notation from above we now give a proof of the integral formula (2^+). Let us suppose for simplicity that $u = 0$ (the origin of the co-ordinates) and suppose that Φ is a holomorphic left regular mapping from $\mathbb{C}^{n+1}$ to $\mathbb{C}_n$. Starting from the formula (2) we get by Stokes' theorem after deformation of the domain of integration (see [3]):

$$(5) \quad \Phi(0) = \varkappa_n^{-1} \int_{s\times S^1} |z|^{-h}\, z^+\, Dz\, \Phi(z)\, .$$

Let $\lambda = (\lambda_1, \ldots, \lambda_{n-1})$ be (real) co-ordinates on s and let

$$s\times S^1 = \{ (\lambda, \tau)\, ,\ \tau \in \mathbb{C}\, ,\ |\tau| = \varepsilon \} \quad \text{for some } \varepsilon > 0$$

We have :

$$z^+ Dz = (z_0 e_0 - \sum_{j=1}^n z_j e_j)(d\hat{z}_0 e_0 + \sum_{j=1}^n (-1)^j d\hat{z}_j e_j)$$

$$= (z_0 d\hat{z}_0 + \sum_{j=1}^n (-1)^j z_j d\hat{z}_j) e_0 - \sum_{j=1}^n ((-1)^j z_0 d\hat{z}_j + z_j d\hat{z}_0) e_j -$$

$$- \sum_{1\le j<k\le n} ((-1)^k z_j d\hat{z}_k - (-1)^j z_k d\hat{z}_j) e_j e_k\, .$$

Let $z_0 = w_0, z_j = iw_j$, then

$$z^+ Dz = i^n \{(w_0 d\hat{w}_0 + \sum_{j=1}^n w_j d\hat{w}_j) e_0 - i \sum_{j=1}^n ((-1)^j w_0 d\hat{w}_j + w_j d\hat{w}_0) e_j -$$
$$- \sum_{1\le j<k\le n} ((-1)^k w_j d\hat{w}_k - (-1)^j w_k d\hat{w}_j)\}\, e_j e_k .$$

According to part 3 we have $w_j = b_j + \tau c_j$, $dw_j = db_j + c_j d\tau + \tau dc_j$ for $j = 0, 1, \ldots, n$, $|w| = \tau$ and

$$F(\tau,\lambda) = D(0,\lambda)^{-1}\, D(\tau,\lambda)\ ,\quad D(\tau,\lambda) = \left| b.c, \frac{\partial b}{\partial \lambda_1} + \tau \frac{\partial c}{\partial \lambda_1}, \ldots, \frac{\partial b}{\partial \lambda_{n-1}} + \tau \frac{\partial c}{\partial \lambda_{n-1}} \right|$$

(see [6] for details). It is easy to show that

$$w_0 d\hat{w}_0 + \sum_{j=1}^n (-1)^j w_j d\hat{w}_j = D\, d\tau \wedge d\lambda$$

$$\sum_{j=1}^n ((-1)^j w_0 d\hat{w}_j + w_j d\hat{w}_0) e_j = 2D \sum_{j=1}^n (b_j c_0 - b_0 c_j) e_j\, d\tau \wedge d\lambda$$

$$\sum_{1\le j<k\le n} (-1)^k (w_j d\hat{w}_k - (-1)^j w_k d\hat{w}_j) e_j e_k = 2D \sum_{1\le j<k\le n} (c_j b_k - c_k b_j)\cdot$$

$\cdot e_j e_k \cdot d\tau \wedge d\lambda$, and we obtain from the residue theorem

$$\Phi(0) = i^n \varkappa_n^{-1} \int_s \int_{|\tau|=\varepsilon} \frac{2}{\tau}h \quad (b^+ \cdot c \cdot D \cdot \Phi) \quad d\tau \wedge d\lambda =$$

$$= (-\pi)^{1-h} \int_s \quad (b^+ \cdot c) \quad \frac{d}{d\tau}\Big|_{\tau=0} (F\Phi) \quad ds$$

($ds = 2D(0,\lambda)\, d\lambda$ is the volume element of s), which is the desired result.

ACKNOWLEDGEMENT.

I am very indebted to V.Souček for valuable suggestions and discussions on the subject.

REFERENCES

[1] J.Bureš :'Some integral formulas in complex Clifford analysis', Proc.11th Winter School ,Srni Suppl.Rend.Circolo Mat.Palermo,II, 3,1984,81-87

[2] J.Bureš :'A comparison of integral formulae of hyperbolic and elliptic type', Proc.Conf.Diff.Geom. and its Appl. Nové Město 1983,II,35-40.

[3] M.Dodson,V.Souček :'Leray residue applied to solutions of the Laplace and wave equations ', to appear in Seminarii di geometria,Bologna.

[4] J.Ryan : 'Complexified Clifford analysis', Complex variables,1,1983,115-149.

[5] J.Bureš,V.Souček : 'Generalized hypercomplex analysis and its integral formulae', Complex variables,5, 1985,53-70

[6] M.Riesz : 'A geometric solution of the wave equation in space-time of even dimension', Comm.Pure Appl.Math. XIII,19,1960,329-351

[7] K.Imaeda :'A new formulation of electromagnetism', Nuovo Cimento 32B,1976,138-162

[8] V.Souček :'Complex-quaternionic analysis applied to spin-½ massless field', Complex variables,1,1983,327-346

[9] L.Hörmander : The analysis of linear partial differential operators I,II,GMW 257,Springer Verlag,Berlin 1983

[10] R.Penrose : 'Null hypersurface initial data for classical fields of arbitrary spin and for general relativity', Gen.Rel.and Grav. **12** No.3,1980,225-264.

KILLING VECTORS AND EMBEDDING OF EXACT SOLUTIONS IN GENERAL RELATIVITY

G. E. Sobczyk
Department of Mathematics
Spring Hill College
4000 Dauphin Avenue
Mobile, AL 36608

ABSTRACT. Two ways in which exact solutions of Einstein's field equations can be classified are by the existence of preferred vector fields, such as Killing vectors, and by its embedding class in a higher dimensional pseudoeuclidean space. The present paper shows how the notion of Killing and conformal Killing vectors find simple expression in the geometric calculus on vector manifolds, and considers a number of different isometric embeddings of exact solutions in pseudoeuclidean space.

1. INTRODUCTION

In the theory of vector manifolds, set down in [HS], the notion of a geometric algebra $\mathcal{G}$ of directions is taken as primitive. In order to familiarize the reader with these ideas and the notation used, we will highlight the parts of the theory which we will most refer to in this paper.

The geometric algebra $\mathcal{G}$ is Clifford algebra of unspecified dimension and signature. We write

$$\mathcal{G} = \mathcal{R}\text{eals} + \mathcal{V}\text{ectors} + \mathcal{B}\text{ivectors} + \mathcal{T}\text{rivectors} + \ldots . \tag{1.1}$$

By a vector manifold $(\mathcal{M}, P)$ of $\mathcal{G}$, we mean a set of points $x \in \mathcal{M}$ and projection operators $P = P(x)$, where

$$\mathcal{M} = \mathcal{M} \cap \mathcal{G} \subset \mathcal{V}, \text{ and } P(\mathcal{G}) = \mathcal{D}. \tag{1.2}$$

The $\mathcal{D} = \mathcal{D}(x)$ in (1.2) is called the tangent algebra to the manifold at the point x, and is a subalgebra of $\mathcal{G}$. Let $I = I(x)$ be the unit pseudoscalar field of the tangent algebra at each point of x of $\mathcal{M}$. The unit pseudoscalar field I is a simple n-vector at each point x of an n-dimensional orientable manifold. Since I is a simple n-vector, there are n orthonormal vector fields e_i, such that

$$I = e_1 e_2 \cdots e_n = e_1 \wedge e_2 \wedge \cdots \wedge e_n . \tag{1.3}$$

J. S. R. Chisholm and A. K. Common (eds.), Clifford Algebras and Their Applications in Mathematical Physics, 227–244.

The unit vectors e_i can be taken to be the generators of the 2^n-dimensional tangent algebra $\mathcal{D}$ at the point x. If p of the vectors e_i have positive square, and n-p have negative square, we say the manifold $\mathcal{M}$ has signature (p,n-p).

So far, we have defined a manifold only in terms of the geometric algebra $\mathcal{G}$. We will also make use of the Geometric Calculus of functions defined on $\mathcal{G}$, which has been set down in Chapter 2 of [HS]. In particular, the manifold $\mathcal{M}$ inherits a unique Riemannian connection $\nabla = \nabla_x$. This connection can be defined in terms of the vector derivative $\partial = \partial_x$ of the tangent algebra $\mathcal{D}(x)$, by the operator equation

$$\nabla = P\partial P . \tag{1.4}$$

This equation shows that the Riemmannian connection of a manifold is intrinsic in character, meaning that its range and domain are restricted ted by the projection P to multivector fields on the tangent algebra. Exclusive use of the connection ∇ precludes the study of the extrinsic geometry which a vector manifold has in addition to the intrinsic geometry of Riemannian manifolds. In this sense, the study of vector manifolds is more general than the study of Riemannian manifolds. As will be seen below, this is an unfortunate and unnecessary restriction imposed by the more standard approaches, because many of the intrinsic properties of a manifold can be more simply expressed by going outside of the tangent algebra.

Let $A_1 = A_1(x)$, $A_2 = A_2(x)$, ... , $A_k = A_k(x)$, be functions defined on the manifold $\mathcal{M}$ and with range in the geometric algebra $\mathcal{G}$. By an extensor T on $\mathcal{M}$, we mean a function

$$T(A_1,A_2,\ldots,A_k) \equiv T(x,A_1(x),A_2(x),\ldots,A_k(x)), \tag{1.5}$$

which in addition to being a function of $x\in\mathcal{M}$, is linear in each of the arguments $A_1,A_2,\ldots,A_k$. If, in addition, T satisfies the condition that

$$T(A_1,A_2,\ldots,A_k) = PT(P(A_1),P(A_2),\ldots,P(A_k)), \tag{1.6}$$

we say that T is a tensor field on $\mathcal{M}$. Now let a be a tangent vector in the tangent algebra $\mathcal{D}(x)$. The differential T_a of the extensor T is defined by

$$\begin{aligned} T_a(A_1,\ldots,A_k) &\equiv a\cdot\dot{\partial}\dot{T}(A_1,\ldots,A_k) \qquad (1.7)\\ &= a\cdot\partial T(A_1,\ldots,A_k) - T(a\cdot\partial A_1,A_2,\ldots,A_k) - T(A_1,a\cdot\partial A_2,\ldots,A_k)\\ &\quad - \ldots - T(A_1,A_2,\ldots,a\cdot\partial A_k), \end{aligned}$$

and is itself an extensor. By the coderivative $T_{,a}$ of the extensor T, we mean the differential of T restricted to the tangent algebra. Thus,

$$T_{,a}(A_1,\ldots,A_k) \equiv PT_a(PA_1,PA_2,\ldots,PA_k). \tag{1.8}$$

Clearly, the coderivative of an extensor field is a tensor field.

For more details and properties of the differential of an extensor, see [HS, p.142].

The projection $P(A) \equiv P(x,A)$ is the most fundamental tensor defined on the manifold $\mathcal{M}$. In terms of the differentials P_a, and higher order differentials P_{ab}, ..., $P_{ab\cdots d}$, all the local properties of both the intrinsic and extrinsic curvature of the vector manifold can be expressed. We gather together here some of the most important definitions and properties that will be used in this paper.

The differential $P_b(a)$ is a generalization of the normal to a hypersurface, and from it we construct the <u>shape operator</u> $S(a)$, where a is any vector in the tangent algebra $\mathcal{D}(x)$.

$$S(a) \equiv \partial_v \wedge P_v(a) = \dot{\partial}\dot{P}(a). \tag{1.9}$$

Definition (1.9) shows that the shape operator $S(a)$ is a bivector-valued extensor. In terms of the shape extensor, and its differential $S_b(a)$, we can express the appropriate generalization of the well-known Codazzi-Mainardi equation:

$$S_b(a) - S_a(b) = 2S(a)xS(b). \tag{1.10}$$

The cross product on the right hand side of (1.10) denotes the commutator product in geometric algebra. A more general definition of the shape operator (1.9), and a discussion of (1.10) can be found in [HS, p149,198].

The shape operator can also be expressed in terms of the differential of the unit pseudoscalar (1.3) of the manifold ([HS,p163]):

$$S(a) = I^{-1}\, a\cdot\partial I = I^{-1}\, P_a(I). \tag{1.11}$$

Equation (1.11) provides the key to the geometric interpretation of the shape operator; it is a measure of the angular velocity of the pseudoscalar I as it moves in the direction of a.

The <u>Riemann curvature</u> tensor $R(B)$ is a bivector-valued tensor defined for bivectors B in the tangent algebra $\mathcal{D}(x)$. We give below several equivalent definitions of $R(B)$:

$$R(a\wedge b)\cdot c \equiv [b\cdot\nabla, a\cdot\nabla]c - [b,a]\cdot\nabla c = c_{,ab} - c_{,ba}, \tag{1.12}$$

where

$$[b\cdot\nabla, a\cdot\nabla] \equiv b\cdot\nabla a - a\cdot\nabla b \quad , \text{ and } [a,b] \equiv a\cdot\nabla b - b\cdot\nabla a.$$

Definition (1.12) is equivalent to the usual definition of curvature in terms of the connection ∇. We can also define $R(a\wedge b)$ directly in terms of the differentials P_a, or in terms of the shape tensors $S(a)$:

$$R(a\wedge b) = \partial_u\wedge\partial_v P_u(a)\cdot P_v(b) = P(S(a)xS(b)). \tag{1.13}$$

The curvature tensor is symmetric,

$$R(A)\cdot B = R(B)\cdot A, \tag{1.14}$$

and satisfies the Ricci identity

$$a \cdot R(b \wedge c) + b \cdot R(c \wedge a) + c \cdot R(a \wedge b) = 0. \tag{1.15}$$

In addition, the curvature tensor satisfies the famous Bianchi identity, which can be cast in the form

$$\dot{\nabla} \wedge \dot{R}(a \wedge b) = 0. \tag{1.16}$$

Note in the definition of $R(a \wedge b)$, on the right side of (1.13), a projection operator is involved. This motivates defining the total curvature of the manifold to be $S(a) \times S(b)$. Breaking $S(a) \times S(b)$ into intrinsic and extrinsic parts gives,

$$S(a) \times S(b) = R(a \wedge b) + E(a \wedge b), \tag{1.17}$$

where the extrinsic curvature $E(a \wedge b)$ is defined by

$$E(a \wedge b) \equiv \partial_u \cdot \partial_v P_u(b) \wedge P_v(a) = P_b(\partial_v) \wedge P_v(a).$$

For a detailed discussion of properties of the curvature tensor $R(a \wedge b)$, and the extrinsic curvature of a manifold, see [HS,p189].

An important relationship between the second coderivative and the second differential of a vector field b is

$$(a \cdot \nabla)^2 b - P[(a \cdot \partial)^2 b] = P_a^2(b) = -(1/2) P P_{aa}(b). \tag{1.18}$$

This relationship can be proved by taking the first and second differentials of the fundamental property $P^2 = P$ of the projection operator, and remembering the basic relationship (1.8) between differentials and coderivatives.

Important for the next section is the notion of a geodesic. We close this section by giving the most important property of a vector field which is everywhere tangent to a bundle of geodesics, [HS,p206].

<u>Property</u>: A vector field $u = u(x)$ is everywhere tangent to a bundle of geodesics if and only if there exists a scalar field $\lambda = \lambda(x)$ such that $v = \lambda u$ satisfies the identity

$$v \cdot \nabla v = v \cdot (\nabla \wedge v) + (1/2) \nabla v^2 = 0. \tag{1.19}$$

Note, in definition (1.19), that the first equality is true for every vector field.

2. KILLING VECTOR FIELDS

We begin this section with a definition.

<u>Definition</u>: A vector field k on $\mathcal{M}$ is said to be a <u>Killing vector</u>, if there exists a bivector field Ω with the property that

$$a\cdot\nabla k = a\cdot\Omega \iff \dot{\nabla}\dot{k}\cdot a = \Omega\cdot a \tag{2.1}$$

for each tangent vector $a \in \mathcal{T}$.

Taking the vector derivative ∂_a on the right of both sides of either of the equivalent equalities in (2.1), gives

$$\nabla k = 2\Omega \iff \left\{ \begin{array}{l} \nabla\cdot k = 0 \\ \Omega = (1/2)\nabla\wedge k \end{array} \right\}. \tag{2.2}$$

Equation (2.2) shows that a Killing vector is divergence free, and that the bivector Ω of a Killing vector is completely determined by its curl.

A simple, but important consequence of the fact that a Killing vector is divergence free is that

$$\mathcal{L}_k(I) \equiv -[I,k] = (I\cdot\nabla)\wedge k - k\cdot\nabla I = 0, \tag{2.3}$$

where the generalized Lie bracket $[A,B]$ has been defined in [HS,p159]. To prove (2.3), note that coderivative $a\cdot\nabla I$ of the unit pseudoscalar is always zero, and $(I\cdot\nabla)\wedge k = I\,\nabla\cdot k = 0$.

Dotting both sides of (2.1) with the tangent vector variable a, gives

$$(a\cdot\nabla k)\cdot a = a\cdot(a\cdot\Omega) = (a\wedge a)\cdot\Omega = 0. \tag{2.4}$$

If we now differentiate (2.4) by $b\cdot\partial_v$, we get

$$(a\cdot\nabla k)\cdot b + (b\cdot\nabla k)\cdot a = 0. \tag{2.5}$$

Equation (2.5) is recognized as being equivalent to the well-known equations of Killing, when the tangent vectors a and b are chosen to be basis vectors of a frame $\{e_i\}$ of vector fields on the manifold. Thus, we see that (2.1) implies (2.5) which is Killing's equation. Conversely, suppose that a vector field k satisfies (2.5). By reversing the steps, we can see that (2.5) implies (2.4), which in turn implies (2.1).

Killing vectors enjoy very many nice properties, which reflect their geometric significance as the generators of the transformation group of isometries on a manifold. The classical reference on the subject is Eisenhart's book [E]. To show how simply the properties of a Killing vector follow from our definition (2.1), a number of them will be reproduced here.

Suppose that a vector field $u = u(x)$ is tangent to a bundle of geodesics on $\mathcal{M}$. Then, by property (1.19), there exists a vector field $v = \lambda u$ such that $v\cdot\nabla v = 0$. Since

$$v\cdot\nabla(v\cdot k) = (v\cdot\nabla v)\cdot k + v\cdot(v\cdot\nabla k) = 0 + (v\wedge v)\cdot\Omega = 0, \tag{2.6}$$

it follows that v makes a constant angle with k along the geodesics of v.

We say that the metric $a\cdot b$ is <u>compatible</u> with the connection ∇ , because it satisfies the Liebnitz product rule

$$c\cdot\nabla a\cdot b = (c\cdot\nabla a)\cdot b + a\cdot(c\cdot\nabla b). \tag{2.7}$$

(2.7) is just an expression of the fact that the coderivative of the metric tensor of a Riemannian manifold is zero. We will now show that the metric $a\cdot b$ is compatible with the Lie derivative with respect to a Killing vector, i.e.,

$$\mathcal{L}_k a\cdot b \equiv k\cdot\nabla a\cdot b = (\mathcal{L}_k a)\cdot b + a\cdot(\mathcal{L}_k b), \tag{2.8}$$

where $\mathcal{L}_k a \equiv [k,a] = k\cdot\nabla a - a\cdot\nabla k$ is the familiar Lie bracket of the vector fields k and a. To prove (2.8), it is only necessary to note the identity

$$\mathcal{L}_k a = k\cdot\nabla a - a\cdot\nabla k = k\cdot\nabla a - a\cdot\Omega ,$$

from which it follows, by using definition (2.1), that

$$(\mathcal{L}_k a)\cdot b = (k\cdot\nabla a)\cdot b - (a\cdot\Omega)\cdot b = (\mathcal{L}_k a)\cdot b + (a\wedge b)\cdot\Omega. \tag{2.9}$$

The proof is completed by adding (2.9) to a copy of (2.9) with a and b interchanged, and noting that $a\wedge b = - b\wedge a$.

The property (2.8) of a Killing vector has many important consequences. For example, a coordinate frame of vector fields $\{e_i\}$ can always be chosen on $\mathcal{M}$ (at least locally), in such a way that

$$k = \partial_1 x \equiv (\partial/\partial x^1)x(x^1,x^2,\ldots,x^n), \tag{2.10}$$

i.e., so that k is the tangent vector to each of the coordinate curves of x^1, see [E,p209]. For such a choice of coordinates, it follows from (2.8), by noting that $[e_i,e_j] = 0$, that

$$k\cdot\nabla e_i\cdot e_j = \partial_1 g_{ij} = 0, \tag{2.11}$$

where $g_{ij} \equiv e_i\cdot e_j$ is the metric tensor of the coordinate system.

Actually, we can prove a more general result, and without reference to a coordinate frame (2.10). Let $\mathcal{S}$ be the family of geodesically parallel hypersurfaces which make a given constant angle with k. This family can be uniquely parameterized by a function $s = \varphi(x)$ by marking off equal distances of arc length s along the geodesic through the point x in the direction of k. If we define the vector field $\bar{k} = \nabla\varphi$, then, by (2.6),

$$k\cdot\bar{k} = k\cdot\nabla\varphi = \text{constant}. \tag{2.12}$$

The tangent pseudoscalar to these hypersurfaces is the (n−1)-vector $I\bar{k}$. The integrability of these hypersurfaces with respect to the Killing vector is expressed by the basic relationship

$$\mathcal{L}_k(I\bar{k}) = -[I\bar{k},k] = 0. \tag{2.13}$$

The proof of (2.13) follows from the steps

$$[I\bar{k},k] = [(I\bar{k})\cdot\partial]\wedge k - k\cdot\nabla I\bar{k} = I[(\bar{k}\wedge\nabla)\cdot k - k\cdot\nabla\bar{k}]$$
$$= -I[\dot{\nabla}\dot{k}\cdot\bar{k} + k\cdot\nabla\bar{k}] = 0.$$

The first equality above is just the definition of the generalized Lie bracket. The second two equalities are consequences of (2.2) and algebraic identities relating the duality of the inner and outer products by multiplication by I. The last equality is a consequence of the fact that

$$\nabla\wedge\bar{k} = 0 \quad \Longrightarrow \quad k\cdot\nabla\bar{k} = \dot{\nabla}\,\dot{\bar{k}}\cdot k = -\dot{\nabla}\,\dot{k}\cdot\bar{k}. \tag{2.14}$$

A direct consequence of (2.14) and the second half of (2.1) is

$$k\cdot\nabla\bar{k} = -\dot{\nabla}\dot{k}\cdot\bar{k} = k\cdot\Omega. \tag{2.15}$$

If the trajectory of each point in the motion generated by a Killing vector is a geodesic, i.e., if

$$k\cdot\nabla k = 0, \tag{2.16}$$

then the motion is said to be a translation, [E,p211]. It follows from the identity

$$a\cdot\nabla k^2 = 2k\cdot(a\cdot\nabla k) = 2k\cdot(a\cdot\Omega) = -2a\cdot(k\cdot\nabla k), \tag{2.17}$$

that the Killing vector k generates a translation iff k^2 is constant. From (2.17) and (2.6), we can conclude that if k_1 and k_2 are translations then, the angle between k_1 and k_2 is constant, so that

$$\nabla k_1\cdot k_2 = 0. \tag{2.18}$$

Using (2.17) and (2.18), we find that

$$\nabla(s_1k_1 + s_2k_2)^2 = \nabla(s_1^2k_1^2 + 2s_1s_2k_1\cdot k_2 + s_2^2k_2^2) = 0, \tag{2.19}$$

from which it follows that $s_1k_1 + s_2k_2$ generates a translation, [E,p212].

Let us examine how the existence of a Killing vector k restricts the degrees of freedom of the curvature tensor $R(a\wedge b)$. Recalling (2.1), we take the first and second coderivaties of k, with the results

$$k_{,a} = a\cdot\nabla k = a\cdot\Omega, \text{ and } k_{,ab} = a\cdot\Omega_{,b}. \tag{2.20}$$

Then, using (1.12), we find

$$R(a\wedge b)\cdot k = k_{,ab} - k_{,ba} = a\cdot\Omega_b - b\cdot\Omega_a. \tag{2.21}$$

Dotting both sides of (2.21) by c, and with the help of (1.14), we get

$$R(a\wedge b)\cdot(k\wedge c) = (a\wedge b)\cdot R(k\wedge c) = a\cdot(\Omega_b\cdot c) - b\cdot(\Omega_a\cdot c). \tag{2.22}$$

Taking successive vector derivatives of (2.22), first with respect to a, and then with respect to b, and using the equivalence relationship

$$\nabla\wedge\nabla\wedge k = 0 \iff \Omega_{,a} + \partial_b\wedge(\Omega_{,b}\cdot a) = 0,$$

we find that

$$R(k\wedge c) = \partial_b\wedge(\Omega_{,b}\cdot c) = (\partial_b\wedge\Omega_{,b})\cdot c - \Omega_{,c} = -\Omega_{,c}. \tag{2.23}$$

From (2.23), and the basic identity

$$k_a\cdot b + k_b\cdot a = 0 \iff k_{,ac}\cdot b + k_{,bc}\cdot a = 0$$

we find that

$$R(k\wedge c)\cdot a = -(1/2)[\partial_v\wedge k_{,vc}]\cdot a = (1/2)(k_{,ac} - \partial_v k_{,vc}\cdot a) \tag{2.24}$$

$$= k_{,ac}\ .$$

With (2.24) at our disposal, and the aid of the Ricci identity (1.15), we can now show that the Lie bracket $[k_1, k_2]$ of two Killing vectors k_1 and k_2, is itself a Killing vector. We find that

$$[k_1, k_2] = k_1\cdot\Omega_2 - k_2\cdot\Omega_1, \tag{2.25}$$

and

$$a\cdot\nabla[k_1, k_2] = a\cdot(\Omega_1\times\Omega_2) + a\cdot R(k_1\wedge k_2), \tag{2.26}$$

see [E,p216]. Thus, the set of all motions form a subgroup of the Lie group of all continuous transformations on a manifold $\mathcal{M}$, with the corresponding Lie subalgebra of Killing vectors.

All of the above formulas can be generalized to apply to conformal Killing vectors. Generalizing (2.1), we say that a vector c is a <u>conformal Killing vector</u>, if

$$a\cdot\nabla c = (1/2)\lambda(x)a + a\cdot\Omega, \tag{2.27}$$

where

$$\lambda(x) \equiv (2/n)\nabla\cdot c, \quad \text{and} \quad \Omega = (1/2)\nabla\wedge c.$$

Clearly, a conformal vector c is a Killing vector when $\lambda = 0$.

Rather than carry out the somewhat more involved calculations for the conformal vectors c, we will just give some of the results. For the generalization of (2.8), the Lie derivative of the metric tensor g(a,b) $\equiv a\cdot b$, we find

$$(\mathcal{L}_c g)(a,b) = \mathcal{L}_c a\cdot b - (\mathcal{L}_c a)\cdot b - a\cdot(\mathcal{L}_c b)$$

$$= c\cdot\nabla a\cdot b - [c,a]\cdot b - a\cdot[c,b] = \lambda(x)a\cdot b. \tag{2.28}$$

The formulas (2.23) and (2.24) generalize to

$$R(a\wedge c) = \tfrac{1}{2}a\wedge(\nabla\lambda) + \Omega_{,a} \tag{2.29}$$

and

$$R(a\wedge c)\cdot b = \tfrac{1}{2}(a\wedge(\nabla\lambda))\cdot b + \lambda_a b - c_{,ba}. \qquad (2.30)$$

The Lie bracket of two conformal Killing vectors is

$$[c_1, c_2] = \lambda_2 c_1 - \lambda_1 c_2 + c_1\cdot\Omega_2 - c_2\cdot\Omega_1 , \qquad (2.31)$$

and the coderivative of the Lie bracket is

$$[c_1, c_2]_{,a} = (\lambda_{21} - \lambda_{12})a + a\cdot(c_2\wedge\nabla\lambda_1 - c_1\wedge\nabla\lambda_2) + \qquad (2.32)$$
$$+ a\cdot(\Omega_1\times\Omega_2) + a\cdot R(c_1\wedge c_2),$$

where $\lambda_{jk} \equiv c_k\cdot\nabla\lambda_j$. Comparing (2.32) with (2.27), we can conclude that the Lie bracket of conformal Killing vectors is itself a conformal Killing vector.

As an example, we give the conformal Killing vectors which generate the conformal group for the linear space $\mathcal{A}(p,n-p)$:

translations	<--->	a (constant vectors)	
rotations	<--->	$x\cdot B$ (B constant bivector)	
			(2.33)
dilations	<--->	sx (constant scalar s)	
special conformal	<--->	xax (constant vector a),	

[CWD,p336].

We have shown how the notion of Killing and conformal Killing vector fields can be efficiently reformulated on a vector manifold. It is not surprizing, therefore, that the theory of Lie groups and Lie Algebras is itself amenable to an efficient reformulation in terms of the geometric Calculus on a vector manifold. The foundation for such a reformulation has already been worked out in [HS, Chp8].

The decomposition (2.27) can be generalized to apply to the coderivative of any vector field $v = v(x)$, by writing

$$a\cdot\nabla v = (1/n)(\nabla\cdot v)a + (1/2)a\cdot(\nabla\wedge v) + r(a), \qquad (2.34)$$

where $r = r(x,a)$ is a vector-valued tensor linear in a. By taking the vector derivative ∂_a of both sides of (2.34), we find that

$$\nabla v = \nabla\cdot v + \nabla\wedge v + \partial_a r(a),$$

from which it follows that

$$\partial_a r(a) = \partial_a\cdot r(a) + \partial_a\wedge r(a) = 0, \qquad (2.35)$$

meaning that r(a) is a trace-free, symmetric vector operator. A decomposition related to (2.34) plays a central role in the classification of exact solutions of Einstein's field equations, [KSMH,p77]. The algebraic classification of r(a), and the other parts of (2.34), has been worked out for space-time in [S1-S3]; so there is no doubt that the whole classification scheme can be carried out in geometric algebra. If we perform the same calculations as we did to get (2.29), we find that

$$R(a\wedge v) = \tfrac{1}{2}a\wedge(\nabla\lambda) + \Omega_{,a} - \dot{\nabla}\wedge\dot{r}(a), \tag{2.36}$$

a formula much simpler than the related formula [KSMH,p79,(6.25)].

3. Embeddings of Exact Solutions.

Exact solutions of Einstein's field equations can also be classified according to its embedding class. The embedding class is defined to be the least number of extra dimensions k that are necessary in order that the solution $\mathcal{M}_n$ can be locally embedded in the pseudoeuclidean space E_{n+k}, [KSMH,p354]. A number of different authors have studied embeddings, among them [Fn],[Go],[Fl],[R], and [Gr]. The framework of vector manifolds is ideally suited for this problem, since a vector manifold by its very definition is a subspace of an unspecified higher dimensional pseudoeuclidean space. In this section, we will study a number of different embeddings, and carry out explicit calculations in the geometric calculus of vector manifolds.

Following Hestenes in [STA], we shall identify Minkowskian flat space-time with the real Dirac algebra $\mathcal{D}$ having signature (+---). The Dirac algebra algebra can be written as the sum of the linear subspaces

$$\mathcal{D} = \mathcal{R} + \mathcal{D}_1 + \mathcal{D}_2 + \mathcal{D}_3 + \mathcal{P}, \tag{3.1}$$

of, respectively, scalars, vectors, bivectors, trivectors, and pseudoscalars. A vector $x \in \mathcal{D}_1$ is interpreted as being a point or "event" in spacetime, and can be split, with respect to a unit time-like vector, into the components

$$x = x\gamma_0\gamma_0 = x\cdot\gamma_0\gamma_0 + x\wedge\gamma_0\gamma_0 = ct\gamma_0 + \underset{\sim}{x}\gamma_0, \tag{3.2}$$

where ct is the time and $\underset{\sim}{x}$ is the position vector of the event as measured by an observer in the inertial frame of γ_0.

The Einstein model, as given in [R,p204], is a transformation mapping Minkowskian spacetime into a hypersurface of a pseudoeuclidean space of one extra dimension. The transformation f has the form

$$x' = f(x) = x + \beta n \tag{3.3}$$

for $\beta = \beta(r)$, where $r = |\underset{\sim}{x}|$, and where n is a constant unit space-like vector orthogonal to $\mathcal{D}$. We can express the conditions on n by

$$n^2 = -1, \quad \text{and} \quad n\cdot i = 0,$$

where $i \in \mathcal{P}$ is the unit pseudoscalar of the Dirac algebra $\mathcal{D}$. The Dirac algebra $\mathcal{D}$ is, of course, a subalgebra of the universal geometric algebra $\mathcal{G}$, and n is a vector orthogonal to $\mathcal{D}$ in $\mathcal{G}$.

General transformations of the form (3.3) have been studied in [HS, p176]. The differential outermorphism of f(x) is given by

$$\underline{f}(A_k) = A_k + A_k\cdot\nabla\beta\, n, \tag{3.4}$$

where A_k is an k-vector in $\mathcal{D}$. Letting $A_4 = i$ in (3.4), we find that the pseudoscalar i' of the Einstein space is given by

$$i' = J_f^{-1}\,\underline{f}(i) = J_f^{-1}\, i(1+\partial\beta\, n) \tag{3.5}$$

where $J_f^2 = (1-(\partial\beta)^2)$, is the Jacobian of the transformation f(x). If we define $I = ni = n'i'$, then I is the unit pseudoscalar of the enveloping pseudoeuclidean space $\mathcal{E}_{1,4}$, and n and n' are the respective normal vectors to the Minkowskian space and the hypersurface at the points x and x' = f(x), respectively. Solving (3.5) explicitly for n', we find

$$n' = J_f^{-1}(n + \partial\beta). \tag{3.6}$$

We have assumed in (3.3) that the scalar function β is only a function of the distance $r = |\underset{\sim}{x}|$ as measured in the inertial system of a given time-like vector γ_0. To see what this implies, we note the identity $a\cdot\partial x = a$, and calculate

$$a\cdot\partial r^2 = a\cdot\partial(x\wedge\gamma_0)\cdot(x\wedge\gamma_0) = 2(a\wedge\gamma_0)\cdot(x\wedge\gamma_0) = 2\underset{\sim}{a}\cdot\underset{\sim}{x}, \tag{3.7}$$

from which it follows that

$$\partial r^2 = 2\,\partial_a\langle a\wedge\gamma_0\, x\wedge\gamma_0\rangle_0 = 2\,\partial_a a\cdot(\gamma_0\cdot(x\wedge\gamma_0)) = -2\underset{\sim}{x}\,\gamma_0. \tag{3.8}$$

With the help of (3.7) and (3.8), we then calculate

$$a\cdot\partial\beta = (1/r)\beta'\underset{\sim}{a}\cdot\underset{\sim}{x}, \quad\text{and}\quad \partial\beta = -(1/r)\beta'\underset{\sim}{x}\gamma_0, \tag{3.9}$$

where $\beta' \equiv (d/dr)\beta(r)$. Employing (3.9), (3.4) and (3.6) can be put in the form

$$\underline{f}(A_k) = A_k - (\beta'/r)A_k\cdot(\underset{\sim}{x}\gamma_0)n \tag{3.10}$$

and

$$n' = J_f^{-1}(n - (1/r)\beta' x), \tag{3.11}$$

where $J = (1 + \beta'^2)^{\frac{1}{2}}$.

From (1.11) and (1.13), it follows that the curvature bivector for the hypersurface defined by the transformation (3.3) can be expressed in the form

$$R'(\underset{\sim}{a}'\wedge\underset{\sim}{b}') = \langle n'_a\, n'_b\rangle_2 = \underset{\sim}{n}'_b\wedge\underset{\sim}{n}'_a, \ (*) \tag{3.12}$$

so we need to calculate the differentials n'_a of the normal vector n'

(*) for a further explanation of this point, see [HS,p196].

given in (3.11). After a fairly involved, but straight forward computation, with the help of (3.9)-(3.11), we find that

$$R'(\underset{\sim}{a}'\wedge\underset{\sim}{b}') = (\beta'/r^2 J_F^2)[s(\underset{\sim}{x}\cdot(\underset{\sim}{a}'\wedge\underset{\sim}{b}'))\wedge P'(\underset{\sim}{x}) - \beta'\underset{\sim}{a}'\wedge\underset{\sim}{b}'] , \qquad (3.13)$$

where $s = (1/r^2)(\beta' + (\beta')^3 - r\beta'')$, and P' is the projection operator onto the tangent space of the hypersurface defined by (3.3).

The curvature tensor (3.13) of the hypersurface defined by (3.3) is a complicated function of the parameter $\beta = \beta(r)$, where r is the distance to the origin of the inertial system of the time-like vector γ_0. For the Einsteinian space refered to in [R], β is given by

$$\beta(r) = R^2 - r^2; \qquad (3.14)$$

for which case s = 0, and (3.13) simplifies to

$$R'(\underset{\sim}{a}'\wedge\underset{\sim}{b}') = (1/R^2)\underset{\sim}{a}'\wedge\underset{\sim}{b}', \qquad (3.15)$$

the curvature tensor of a space with constant curvature.

The basic relationship (3.15) of an Einstein space can be derived directly by noting that (3.14) and (3.3) together imply that

$$ct' = x'\cdot\gamma_0 = x\cdot\gamma_0 + \beta n\cdot\gamma_0 = ct,$$

and

$$\underset{\sim}{x}'^2 = (\underset{\sim}{x} + \beta n)^2 = \underset{\sim}{x}^2 + \beta^2\underset{\sim}{n}^2 = r^2 + \beta^2 = R^2. \qquad (3.16)$$

From (3.16) it immediately follows that

$$\underset{\sim}{n}' = \underset{\sim}{x}'/R, \quad \text{and} \quad \underset{\sim}{n}'_{\underset{\sim}{a}} = -\underset{\sim}{a}'/R, \qquad (3.17)$$

which, taken with (3.12), implies (3.15). For a discussion of the geometry of an Euclidean n-sphere, see [HS,p199].

Next, consider transformations of Minkowski space-time of the form

$$x' = f(x) = \underset{\sim}{x}\gamma_0 + e^{\Omega}\gamma_4 , \qquad (3.18)$$

where $\Omega = \Phi(r) + \omega(t)\gamma_{40}$, γ_4 is a unit spacelike vector orthogonal to the standard basis of Minkowski space $\gamma_0, \gamma_1, \gamma_2, \gamma_3$, and $\gamma_{40} \equiv \gamma_4\gamma_0$. This transformation is a generalization of the embedding of the "de Sitter model", considered by [R, p205]. Calculating the differential of (3.18), we find that

$$a' = \underline{f}(a) = \underset{\sim}{a}\gamma_0 + e^{\Omega}\Omega_a\gamma_4, \qquad (3.19)$$

where $\Omega_a = \Phi_a + \omega_a\gamma_{40} = (1/r)\Phi'\underset{\sim}{a}\cdot\underset{\sim}{x} + \dot{\omega}a\cdot\gamma_0\gamma_{40}$. The apostrophe and dot indicate differentiation with respect to r and t, respectively, of the scalar-valued functions $\Phi(r)$ and $\omega(t)$. By defining

$$\Psi = e^{\Omega}\gamma_{40}, \quad \text{and} \quad \Psi_a = e^{\Omega}\Omega_a\gamma_{40},$$

we can put (3.18) and (3.19) into the 'spinor forms'

$$x' = (\underset{\sim}{x} + \Psi)\gamma_0, \quad \text{and} \quad a' = (\underset{\sim}{a} + \Psi_a)\gamma_0 . \tag{3.20}$$

From (3.19) and (3.20), we calculate the metric form of the transformation (3.18), getting

$$\begin{aligned} a'^2 &= a'a' = (\underset{\sim}{a} + \Psi_a)\gamma_0(\underset{\sim}{a} + \Psi_a)\gamma_0 = (\underset{\sim}{a} + \Psi_a)(-\underset{\sim}{a} + \overline{\Psi}_a) = \\ &= -\underset{\sim}{a}^2 + \langle \Psi_a \overline{\Psi}_a \rangle_0 = -\underset{\sim}{a}^2 - e^{2\Phi}[\dot{\omega}^2(a\cdot\gamma_0) - (1/r^2)\Phi'^2(\underset{\sim}{a}\cdot\underset{\sim}{x})^2] . \end{aligned} \tag{3.21}$$

Comparison of (3.21) with the metric of the de Sitter model given in [R], shows that (3.21) is equivalent to it for

$$\omega(t) = t/R, \quad \Phi' = -r/(R^2-r^2), \text{ and } e^{2\Phi} = R^2 - r^2 . \tag{3.22}$$

We will return to this choice of the parametric functions $\omega(t)$ and $\Phi(r)$, after we have carried out calculations for the more general transformation (3.18).

Evaluating the differential (3.19) for the orthonormal frame of vectors γ_0, $\hat{\underset{\sim}{x}}\gamma_0$, $\hat{\underset{\sim}{\theta}}\gamma_0$, and $\hat{\underset{\sim}{\phi}}\gamma_0$, where $\hat{\underset{\sim}{x}} = \underset{\sim}{x}/r$, and $\hat{\underset{\sim}{\theta}}$ and $\hat{\underset{\sim}{\phi}}$ are unit vectors chosen orthogonal to $\hat{\underset{\sim}{x}}$, such that $\hat{\underset{\sim}{x}}\hat{\underset{\sim}{\theta}}\hat{\underset{\sim}{\phi}} = i$, we get

$$\begin{aligned} &\underline{f}(\gamma_0) = e^{\Lambda}\dot{\omega}\gamma_0, \quad \underline{f}(\hat{\underset{\sim}{x}}\gamma_0) = \hat{\underset{\sim}{x}}\gamma_0 + e^{\Lambda}\Phi'\gamma_4, \\ &\underline{f}(\hat{\underset{\sim}{\theta}}\gamma_0) = \hat{\underset{\sim}{\theta}}\gamma_0, \text{ and } \underline{f}(\hat{\underset{\sim}{\phi}}\gamma_0) = \hat{\underset{\sim}{\phi}}\gamma_0 . \end{aligned} \tag{3.23}$$

From (3.23), we can calculate $i' = J_f^{-1}\,\underline{f}(i) = \underline{f}(\gamma_0)\wedge f(\hat{\underset{\sim}{x}}\gamma_0)\wedge f(\hat{\underset{\sim}{\theta}}\gamma_0)\wedge f(\hat{\underset{\sim}{\phi}}\gamma_0)$. We find that

$$i' = J_f^{-1}[e^{\Lambda} + e^{2\Phi}\Phi'\gamma_{40}\hat{\underset{\sim}{x}}]\dot{\omega}i . \tag{3.24}$$

If we define the normal vector n' to the hypersurface of the transformation (3.18) by $n' = -i'I = -i'i\gamma_4$, then using (3.24),

$$n' = \dot{\omega}J_f^{-1}(e^{\Lambda}\gamma_4 - (1/r)e^{2\Phi}\Phi'\underset{\sim}{x}\gamma_0). \tag{3.25}$$

Now, calculating the differential n'_a of n', and introducing the new parameters $\alpha = (1/r)e^{2\Phi}\Phi'$ and $\alpha_a = a\cdot\partial\alpha$, we find, upon simplification

$$n'_a = \dot{\omega}J_f^{-1}\; P'[(1+\alpha)\Psi_a\gamma_0 - \alpha_a\underset{\sim}{x}\gamma_0 - \alpha a'] , \tag{3.26}$$

which with the help of (3.12) can be used to calculate $R'(\underset{\sim}{a}'\wedge\underset{\sim}{b}')$.

For the choice of ω and Φ given in (3.22), $\alpha = -1$, $\alpha_a = 0$, and $J_f = 1$, in which case (3.26) drastically simplifies to

$$n'_a = a'/R, \tag{3.27}$$

from which it follows, with the help of (3.12), that the curvature tensor is

$$R'(a'\wedge b') = n'_a \wedge n'_b = (1/R\ )a'\wedge b', \tag{3.28}$$

showing that the de Sitter model of the universe has constant curvature.

For the de Sitter universe, the expression for the normal vector (3.25) can be immediately written down by noting, from (3.20) and (3.22), that

$$x'^2 = [(\underset{\sim}{x} + \psi)\gamma_0]^2 = -\underset{\sim}{x}^2 - e^{2\Phi} = -R^2. \tag{3.29}$$

The relationship (3.29) shows that the de Sitter universe is a hypersphere in $\mathcal{E}_{1,4}$. It follows that the normal vector (3.25) at a point x' is given by

$$n' = \hat{x}' = x'/R, \tag{3.30}$$

from which (3.27) and (3.28) can be easily derived.

The relationships (3.29) and (3.30) show clearly the advantage in working with the hypersurface directly defined by (3.29), rather than through the mapping f(x) specified by (3.18). We will follow this procedure as much possible in the case of the Schwarzschild universe discussed below.

The embedding of the Schwarzschild model is specified by

$$x' = f(x) = \underset{\sim}{x}\gamma_0 + 2e^{\nu}\gamma_4 + g\gamma_5, \tag{3.31}$$

where,

$$e^{2\nu} = (1-r^{-1})e^{t\gamma_{40}}, \text{ and } g'^2 = [g'(r)]^2 = (r^{-3})(r^2+r+1),$$

see [F1]. The vectors $\gamma_0, \gamma_1, \ldots, \gamma_5$, are chosen to be an orthonormal basis for the pseudoeuclidean space $\mathcal{E}_{1,5}$, and such that the Dirac algebra (3.1) is a subspace. The form of the embedding (3.31) is a generalization of (3.3) and (3.18). Introducing the multivectors

$$S = \gamma_1\wedge\gamma_2\wedge\gamma_3, \quad \text{and} \quad T = \gamma_4\wedge\gamma_0, \tag{3.32}$$

it is easy to check that $S^2 = T^2 = 1$, and $S\cdot T = 0$. Decomposing (3.31) into components with respect to the subspaces of S and T, we can write (3.31) in the form

$$x' = f(x) = x'_S + x'_T + g\gamma_5, \tag{3.33}$$

where,

$$x'_S \equiv (x'\cdot S)S \quad \text{and} \quad x'_T = (x'\cdot T)T.$$

The differential of the mapping (3.31) is

$$a' = \underline{f}(a) = \underset{\sim}{a}\gamma_0 + 2e^{\nu}\nu_a\gamma_4 + g'\underset{\sim}{a}\cdot\hat{\underset{\sim}{x}} \equiv a'_S + \nu_a x'_T + \underset{\sim}{a}\cdot\hat{\underset{\sim}{x}}g'\gamma_5. \tag{3.34}$$

Evaluating the differential (3.34) for the orthonormal frame of vectors used in (3.23), we obtain

$$\begin{aligned} a'_0 &\equiv \underline{f}(\gamma_0) = e^{\nu}\gamma_0 = -(1/2)x'_T\,T, \quad a'_\theta = f(\hat{\underset{\sim}{\theta}}\gamma_0) = \hat{\underset{\sim}{\theta}}\gamma_0, \\ a'_r &= \underline{f}(\hat{\underset{\sim}{x}}\gamma_0) = (1/r)x'_S + \Phi' x'_T + g'\gamma_5, \quad a'_\phi = f(\hat{\underset{\sim}{\phi}}\gamma_0) = \hat{\underset{\sim}{\phi}}\gamma_0, \end{aligned} \tag{3.35}$$

where $e^{2\Phi} = (1-1/r)$.

Consider now the families of hypersurfaces defined by the scalar-valued functions

$$F(x') = x_S'^2 + h^2 , \quad \text{and} \quad G(x') = x_T'^2 + 4(1-1/h), \tag{3.36}$$

where $h \equiv g^{-1}$, is the inverse of the monotonic function $s = g(r)$ defined in (3.31). The embedding of the Schwarzschild solution (3.31) can be considered to be the intersection of the two 5-dimensional hypersurfaces defined by the equations

$$F(x') = 0 \quad \text{and} \quad G(x') = 0. \tag{3.37}$$

The gradients $m' = \partial'F$ and $n' = \partial'G$, are normal vectors to the respective hypersurfaces passing through the point x', so the bivector defined by $M' = m'\wedge n'$ is normal to the Schwarzschild manifold at a point x' satisfying (3.37). Calculating the gradients of the functions (3.36), we find

$$\begin{aligned} m' &= \partial'F = 2x_S' + 2h\partial'h = 2(x_S' - r(g')^{-1}\gamma_5), \\ \text{and} \quad n' &= \partial'G = 2x_T' + 4h^{-2}\partial'h = 2(x_T' - [2/r^2g']\gamma_5), \end{aligned} \tag{3.38}$$

where we are using the fact that $\partial'h = (1/g')\dot{\partial}(-x'\cdot\gamma_5) = -(1/g')\gamma_5$. It follows from (3.38) that

$$M' = m'\wedge n' = 4x_S' \, x_T' - (8/r^2g')x_S'\gamma_5 + 4(r/g')x_T'\gamma_5 . \tag{3.39}$$

Recalling the definitions (3.31) and (3.33), we see that

$$x_S'^2 = -\underset{\sim}{x}^2 = -r^2, \text{ and } x_T'^2 = -4(1-1/r),$$

which we can use with (3.39) to calculate

$$M'^2 = -(8r/g')^2 . \tag{3.40}$$

If we normalize M' by defining

$$N' \equiv (g'/8r)M' = (g'/2r)x_S'x_T' - (1/r\ )x_S'\gamma_5 + (1/2)x_T'\gamma_5, \tag{3.41}$$

then N' is a unit bivector ($N'^2 = -1$) normal to the Schwarzschild manifold at all points x' satisfying (3.37). The normal bivector N' to an 4-dimensional surface embedded in a 6-dimensional space is the natural generalization of the normal vector to a hypersurface, and, just as the curvature tensor of a hypersurface can be expressed in terms of differentials of the unit normal vector (recall (3.12)), we have with the help of (1.11) and (1.13),

$$R'(a'\wedge b') = P'(S'(a')\times S'(b')) = P'(N_{a'}'\times N_{b'}'). \tag{3.42}$$

Writing (3.41) in the form,

$$N' = \alpha x_S' x_T' + \beta x_S' \gamma_5 + (1/2) x_T' \gamma_5 ,$$

where $\alpha = r'/2r$, $\beta = -1/r$, and taking the differential of N' with respect to a', we get

$$\begin{aligned} N'_{a'} &= \alpha_a x_S' x_T' + \alpha a_S' x_T' + \alpha x_S' a_T' + \beta_a x_S' \gamma_5 + \beta a_S' \gamma_5 + (1/2) a_T' \gamma_5 \\ &= a_S'(\alpha x_T' + \beta \gamma_5) + (\alpha x_S' - (1/2)\gamma_5)\mathcal{N}_a x_T' + x_S'(\alpha_a x_T' + \beta_a \gamma_5). \end{aligned} \tag{3.43}$$

Evaluating the differential (3.43) for the vectors of the frame specified in (3.35), we get

$$\begin{aligned} N_o' &\equiv N'_{a_o'} = (1/2)(\alpha x_S' - (1/2)\gamma_5) T x_T' , \\ N_r' &\equiv N'_{a_r'} = (1/r) x_S'(\alpha x_T' + \beta \gamma_5) + (\alpha x_S' - (1/2)\gamma_5)\Phi' x_T' + \\ &\quad + x_S'(\alpha_r x_T' + \beta_r \gamma_5), \end{aligned} \tag{3.44}$$

$$N_\theta' \equiv N'_{a_\theta'} = \hat{\underset{\sim}{\theta}} \gamma_o (\alpha x_T' + \beta \gamma_5),$$

and

$$N_\phi' \equiv N'_{a_\phi'} = \hat{\underset{\sim}{\phi}} \gamma_o (\alpha x_T' + \beta \gamma_5),$$

where $\alpha_o = \beta_o = 0$, $\mathcal{N}_o = T/2$, $\alpha_r = -(3r^2+4r+5)/(4g' r^5)$, and $\beta_r = -3\beta/r$.

Defining a frame of bivectors in terms of the outer products of the frame of vectors in (3.35), we have

$$\begin{aligned} A'_{ro} &\equiv a_r' \wedge a_o' = a_r' a_o' = -(1/2r^2)[(r x_S' + r^2 g' \gamma_5) x' T - 2T], \\ A'_{\theta o} &\equiv a_\theta' \wedge a_o' = a_\theta' a_o' = -(1/2)\hat{\underset{\sim}{\theta}} \gamma_o x_T' T = \hat{\underset{\sim}{\theta}} e^{\overline{\mathcal{N}}} , \\ A'_{\phi o} &\equiv a_\phi' \wedge a_o' = a_\phi' a_o' = -(1/2)\hat{\underset{\sim}{\phi}} \gamma_o x_T' T = \hat{\underset{\sim}{\phi}} e^{\overline{\mathcal{N}}} , \\ A'_{r\theta} &\equiv a_r' \wedge a_\theta' = a_r' a_\theta' = (x_S' / r + \Phi' x_T' + g' \gamma_5)\hat{\underset{\sim}{\theta}} \gamma_o = \\ &= -i\hat{\underset{\sim}{\phi}} + (\Phi' x_T' + g' \gamma_5)\hat{\underset{\sim}{\theta}} \gamma_o , \end{aligned} \tag{3.45}$$

and

$$\begin{aligned} A'_{r\phi} &\equiv a_r' \wedge a_\phi' = a_r' a_\phi' = (x_S' / r + \Phi' x_T' + g' \gamma_5)\hat{\underset{\sim}{\phi}} \gamma_o = \\ &= -i\hat{\underset{\sim}{\theta}} + (\Phi' x_T' + g' \gamma_5)\hat{\underset{\sim}{\phi}} \gamma_o . \end{aligned}$$

Using (3.42) in conjunction with (3.44) to evaluate the curvature tensor $R'(a' \wedge b')$ on the frame of bivectors given in (3.45), we find, after some straight forward but tedious computation, that

$$R'(A'_{\theta o}) = N_\theta' x N_o' = -(1/2r^3) A'_{\theta o}, \quad R'(A'_{\phi o}) = N_\phi' x N_o' = -(1/2r^3) A'_{\phi o} ,$$

$$R'(A'_{r\theta}) = N_r' x N_\theta' = -(1/2r^3) A'_{r\theta}, \quad R'(A'_{r\phi}) = N_r' x N_\phi' = -(1/2r^3) A'_{r\phi} ,$$

and

$$R'(A'_{\theta\phi}) = N_\theta' x N_\phi' = (1/2r^3) A'_{\theta\phi}, \quad R'(A'_{r\pi}) = N_r' x N_\pi' = (1/2r^3) A'_{r\pi} . \tag{3.46}$$

It follows from (3.46) that the bivectors (3.45) are eigen-bivectors of the curvature tensor $R'(A')$ of the Schwarzschild manifold, and that the Schwarzschild manifold is not of constant curvature. We can also conclude from (3.46) that $R'(A')$ is dual symmetric, which means that

$$R'(i'A') = i'R'(A'), \tag{3.47}$$

for all tangent bivectors A', where i' is the unit pseudoscalar of the Schwarzschild manifold at the point x'. Finally, we conclude from (3.46) that the extrinsic curvature $E'(A')$, defined in (1.17), of the embedding (3.31) of the Schwarzschild manifold, is identically zero.

Generally, there has been opposition by mathematicians and physicists to the study of vector manifolds embedded in a higher dimensional pseudoeuclidean space, preferring instead a coordinate based "intrinsic" approach. The great advantage of geometric algebra is that it is a language in which the extrinsic properties of a vector manifold can be as efficiently expressed as its intrinsic properties. Even when the extrinsic properties are not of interest, the unique features of the study the vector manifolds using geometric calculus make computations easier than the corresponding computations carried out using differential forms or tensor analysis. Thus, for example, the intrinsic properties (3.46) and (3.47) of the Schwarzschild manifold can be efficiently calculated without reference to the embedding (3.31), see [H].

Two other embeddings of exact solutions are given by

$$x' = f(x) = (1+r^2/a^2)N_{tz} + r(1+r^2/a^2)^{-1} e_\phi + g(r)e_7, \tag{3.48}$$

where

$$N_{tz} \equiv e^{te_{12}}e_2 + e^{ze_{56}}e_6, \quad e_\phi = e^{\phi e_{43}}e_4,$$

and

$$g(r) = \int \left\{ \left[(1+r^2/a^2)^6 + (1-r^2/a^2)^2\right]^{\frac{1}{2}} dr/(1+r^2/a^2)^2 \right\},$$

for a Static cylindrically symmetric magnetic or electric geon, and

$$x' = f(x) = r\exp(-\Psi)N_\phi + \exp(\Psi)N_t + \exp(\gamma-\Psi)N_{rz}, \tag{3.49}$$

where

$$N_\phi = e_1 + e^{\phi e_{98}}e_8, \quad N_t = e_{10} + e^{te_{23}}e_2, \text{ and } N_{rz} = e^{ze_{45}}e_4 + e^{re_{76}}e_6$$

for Weyl's static rotationally symmetric solution, [R,p208]. Many questions pertaining to the value and possible physical significance of the embedding of exact solutions have been addressed in [Go]. The geometric calculus of vector manifolds, and, in particular, the extrinsic curvature tensor $E'(A')$ defined in (1.11), offers new tools in terms of which some of these questions might be fruitfully re-examined. Other questions regarding the extrinsic curvature tensor also arise. For example, are there embeddings of the Schwarzschild manifold for which the extrinsic curvature does not vanish, and if so, to what extent is the extrinsic curvature of a manifold independent from its intrinsic curvature?

ACKNOWLEDGEMENTS

The author wishes to thank Spring Hill College and the organizers of the Workshop for making possible my participation in the Workshop.

REFERENCES

[HS] D. Hestenes and G. Sobczyk, Clifford Algebra to Geometric Calculus, Reidel, Dordrecht/Boston (1984).

[E] L. Eisenhart, Continuous Groups of Transformations, Dover, NY (1961).

[CWD] Y. Choquet-Bruhat, C. De Witt-Morette and M. Dillard-Bleick, Analysis, Manifolds and Physics, North-Holland, NY (1977).

[KSMH] D. Kramer, H. Stephani, M. MacCallum and E. Herlt, Exact Solutions of Einstein's Field Equations, Cambridge Univ. Press (1980).

[S1] G. Sobczyk, 'Plebanski Classification of the Tensor of Matter,' Acta Physica Polonica, Vol. **B11**, 579 (1980).

[S2] G. Sobczyk, 'Conjugations and Hermitian Operators in Space Time,' Acta Physica Polonica, Vol. **B12**, 509 (1981).

[S3] G. Sobczyk, 'Algebraic Classification of Important Tensors in Physics,' Phys. Letters **84A**, 49 (1981).

[Fn] A. Friedman, 'Isometric Embedding of Riemannian Manifolds into Euclidean Spaces,' Rev. of Mod. Phys. Vol. **37**, No 1 (1965).

[Go] H. Goenner, 'Local Isometric Embedding of Riemannian Manifolds and Einstein's Theory of Gravitation', in General Relativity and Gravitation, edited by Held, Plenum (1980).

[Fl] C. Fronsdal, 'Completion and Embedding of the Schwarzschild Solution,' Physical Review, Vol. **116**, No. 3, 778 (1959).

[R] J. Rosen, 'Embedding of Various Relativistic Riemannian Spaces in Pseudo-Euclidean Spaces,' Rev. of Mod. Phys., Vol. **37**, No. 1, 204 (1965).

[Gr] R. Greene, 'Isometric Embedding of Riemannian and Pseudo-Riemannian Manifolds,' Memoirs Am. Math. Soc., No. **97** (1970).

[STA] D. Hestenes, Space-Time Algebra, Gordon and Breach, N.Y. (1966).

[H] D. Hestenes, 'Curvature Calculations with Spacetime Algebra,' preprint - Arizona State University, to be published in Int. J. for Theo. Phys.

FROM GRASSMANN TO CLIFFORD

Zbigniew Oziewicz
Institute of Theoretical Physics
University of Wrocław
Cybulskiego 36, 50-205 Wrocław
Poland

ABSTRACT

The aims of this note are to convince the readers in as elementary as possible way that:
1. Clifford algebra is the *particular* case of the Grassmann algebra which is the most fundamental from the point of view of the physics, geometry and analysis. Grassmann algebra with the (pseudo-) Riemannian structure (including degenerate case), or with the (pre-) symplectic structure, or with the Hermitian (Hilbert) structure, contain the corresponding Riemannian or symplectic or Hermitian Clifford algebra.
2. Grassmann algebra is the tautology of the formalism of the (multi-) fermion and antifermion creation and annihilation operators. This relation does not need at all to fix any kind of the Riemannian (Euclidean) or Hilbertian, Hermitian, etc, structures.
3. Clifford algebra is the tautology of the formalism of the quasi-particles in nuclear physics and therefore can be viewed as the Bogoliubov transformation of the Grassmann algebra.
4. The last two statements open wide possibilities of the applications of the Grassmann and Hermitian Clifford algebras particularly to nucleons and quarks in the nuclear shell theory.
5. The Clifford product can always be reexpressed in terms of the fermion creation and annihilation operators.
6. It is interesting that symplectic Clifford algebras has not yet been explored in the classical Lagrangian and Hamiltonian mechanics.

The pair of the mutually dual linear spaces and modules. Whenever we consider or deal with some linear space L, over say field $\mathbb{R}$ or $\mathbb{C}$, with elements having this or another physical interpretation then we have on hand automatically, independent of our wishes, another linear space $L^*=\mathrm{Hom}(L,\mathbb{R})$, dual to L, of all linear mappings

$$L^* \ni \alpha: L \ni v \to \alpha v \in \mathbb{R} . \tag{1}$$

Whenever L has a physical interpretation, then L^* has also a physical interpretation. Roughly speaking L^* is the set of the measurement apparatuses and αv in the above (1) is a result of an orientation-dependent measurement of v by means of the instrument α (unfortunately we do not

J. S. R. Chisholm and A. K. Common (eds.), Clifford Algebras and Their Applications in Mathematical Physics, 245–255.

take into account that any physical measurement has finite accuracy). In the finite dimensional case (which for simplicity we will consider here) $\dim L = \dim L^* < \infty$. Also both spaces L and L^* are mutually dual to each other, $L^{**} = L$, and within the algebra are not distinguished, and can *never* can be identified! As everybody knows, one member of this pair $\{L, L^*\}$, say L, naturally gives rise to the family of contravariant tensors, and another is generating the family of covariant tensors, the names of course having only historical meaning. The modelling of the physical quantities by the elements of L or L^* is not simply a convention but comes from experimental observations as first was realized by James Clerk Maxwell in his Treatise in 1864. Introduction of the Euclidean structure or any other bilinear form $\phi: L \times L \to \mathbb{R}$, greatly obscures discrimination between L and L^* forcing identification of these spaces. To say that we do not need to consider "any" dual space is meaningless because, independently of our wishes, dealing with L we have to do at least with the pair $\{L, L^*\}$. In fact all tensor products are involved. However, to say that we do not need to fix any particular Euclidean structure is meaningful. If we consider the particular linear space L, then of course we have on hand naturally the space of all bilinear mappings $L \times L \to \mathbb{R}$, i.e. the space of the second degree covariant tensors. But of course this is not the same as fixing the particular member of this space. For what follows it is important to understand the geometrical representation of the covector $\alpha \in L^*$ as the codimension one hyperplane $\mathrm{Ker}\alpha \equiv \{v \in L, \alpha v = 0\} \subset L$. In fact the covector α is completely and uniquely determined by $\mathrm{Ker}\alpha$ and any vector $v \in L$ such that $\alpha v = 1$. The wave fronts for instance are described by covectors (forms).

The above without change is also true for the pair of the mutually dual modules $\{\mathcal{V}^1, \Lambda^1\}$ generated by the commutative algebra $\mathcal{F}$. The commutative algebras, for instance the algebra of the real-valued smooth functions on some smooth manifold M, serve as the model of the set of the classical (exact) measurements. In what follows, however, we do not need to pick up any of the particular functional realizations of the given algebra $\mathcal{F}$ as the algebra of the functions on some manifold. In fact we do not need the notion of a manifold, because analysis on the manifolds is and ought to be fully contained in a set $\mathcal{V}^1$ of derivations on commutative algebras. This set of derivations $\mathcal{V}^1$ is just the $\mathcal{F}$-module of the vector fields. Dual to $\mathcal{V}^1$, is the $\mathcal{F}$-module Λ^1 of the differential forms, so $\alpha \in \Lambda^1$ is an $\mathcal{F}$-linear mapping

$$\alpha : \mathcal{V}^1 \ni x \to \alpha x \in \mathcal{F}. \tag{2}$$

Also here within the algebra the modules $\mathcal{V}^1$ and Λ^1 are not distinguished; however, analysis is clearly discriminating them because only vector fields are generating the flows (evolutions) on algebra $\mathcal{F}$ and only differential forms could be integrated, and so on.

Therefore there is no advantage or simplification if one does not need to talk about the pair of the mutually dual linear spaces or modules and instead is using only one linear space of vectors or one module of fields. In most cases the price is the obscure identification of L^* with L.

Grassmann multiplications. The geometrical and physical meaning of the Grassmann multiplications (exterior and interior) can *not* be understood if we don't wish to consider the dual pair $\{L,L^*\}$ or $\{\Lambda^1, V^1\}$. If $\alpha^i \in \Lambda^1$ and $x_i \in V^1$ for $i \in \{1,2,\ldots\}$, then both the exterior Grassmann multiplication of the forms and the exterior Grassmann multiplication of the vectors (the wedge products $\wedge$) are *defined* through the determinants, having the clear geometrical meaning as oriented volumes, viz

$$(\alpha^1 \wedge \ldots \wedge \alpha^k)(x_1 \wedge \ldots \wedge x_k) \equiv \det\{\alpha^i x_j\} . \tag{3}$$

Evidently (3) generalizes the $\mathbb{R}$- or F-linear mappings (1-2). Here $\alpha^1 \wedge \ldots \wedge \alpha^k \in \Lambda^k$ is called a decomposable (differential) form of degree k, etc. The decomposable multiforms and decomposable multivectors are generating linearly *two* copies of the Grassmann algebras V and Λ of the multivector fields and of the differential multiforms. We put $V^0 = \Lambda^0 \equiv F$. All algebraic properties of the Grassmann products $\wedge$ are equivalent to and follow from the theory of determinants [1]. In particular the Grassmann exterior products are associative (theorem!).

It is important to realize that the formula (3) has nothing to do with the scalar product, because it is independent of any Euclidean structure. Also determinant is the *value* of the form and not the form itself. With this respect cf. with [5] (formula (3.17)) and with ([6], Ch.1-4) where the definitions are essentially different.

The Grassmann exterior product

$$\wedge : V \times V \ni x,y \to x \wedge y \in V \tag{4}$$

(and similiar for the multiforms) where $\deg(x \wedge y) = \deg x + \deg y$, is much better represented by the equivalent tensor operator e of the left multiplication (also known as the left adjoint representation)

$$V \ni x \overset{e}{\to} e_x \in \operatorname{End} V ,$$

where $$e_x y \equiv x \wedge y . \tag{5}$$

Then: $$e_x \circ e_y = (-)^{(\deg x)(\deg y)} e_y \circ e_x \tag{6}$$

$$(\Rightarrow (e_x)^2 = 0 \quad \text{for } \deg x = \text{odd}).$$

The associativity of the exterior Grassmann products $\wedge$, means that

$$e_x \circ e_y = e_{e_x y} \tag{7}$$

and has nothing to do with the obvious associativity in End V: $e_x \circ \circ (e_y \circ e_z) = (e_x \circ e_y) \circ e_z$.

The pair of the Grassmann algebras $\{V,\Lambda\}$ is interrelated through the interior Grassmann products "i", dual, by definition, to the exterior Grassmann products e, $i \equiv$ pull back of e, all of which are F-linear mappings

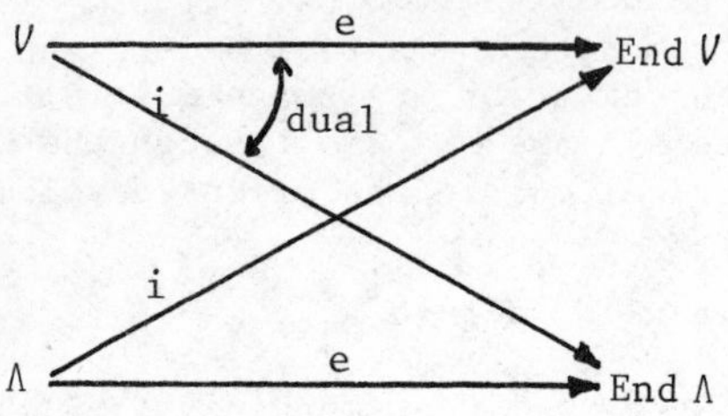

(8)

Here

$$(i_x\omega)y \equiv \omega(e_x y) \in F$$

$$\text{for } \deg\omega = \deg x + \deg y,$$

and

$$\beta(i_\alpha x) \equiv (e_\alpha\beta)x \in F$$

$$\text{for } \deg x = \deg\alpha + \deg\beta . \tag{9}$$

($\alpha, \beta, \omega \in \Lambda$ and $x, y \in V$).

Also $\deg(i_x\omega) = \deg\omega - \deg x$

and

$$i_x\omega = 0 \qquad \text{for } \deg\omega < \deg x \text{, etc.} \tag{10}$$

(One can consider instead of (9) also the right as well as the left Grassmann exterior products and their duals). From (9) it follows that

$$i_x \circ i_y = (-)^{(\deg x)(\deg y)}\, i_y \circ i_x \tag{11}$$

$$(\Rightarrow (i_x)^2 = 0 \qquad \text{for } \deg x = \text{odd}),$$

and

$$i_x \circ i_y = i_{e_y x} . \tag{12}$$

The above discussion shows clearly that the names exterior and interior should in fact be interchanged. For example it is meaningless to ask about the associativity of the interior product.

In the mathematical literature the interior Grassmann product usually is introduced as the antiderivation of the (abstract) Grassmann algebra with respect to the involution $\alpha \to (-)^{\deg\alpha}\alpha$ (see e.g. [4] Ch.V § 4). However, here this property of the interior products is simply a consequence of the definitions (3) and (9). In fact one can easily show that for instance for $\deg x = 1$

$$i_x \circ e_\alpha = e_{i_x \alpha} + (-)^{\deg\alpha} e_\alpha \circ i_x . \tag{13}$$

Here i_x and $e_\alpha \in \text{End}\,\Lambda$. We prefer, however, to present relations similar to (13) (including arbitrary degx) in another form more suitable for the proper physical interpretation. Namely, let's define the *superbracket*

$$\{A,B\} \equiv A \circ B - (-)^{(\deg A)(\deg B)} B \circ A \tag{14}$$

for $A,B \in \text{End}\,\Lambda$ or End V. Then for instance in End Λ we have the following relations (cf with (6) and (11))

$$\{e_\alpha, e_\beta\} = 0$$

$$\{i_x, i_y\} = 0$$

$$\{i_x, e_\alpha\} = e_{i_x \alpha} + (\ldots) \tag{15}$$

In particular, if the degrees of all multiforms and multivectors in (15) are equal to one, then the superbrackets are simply the anticommutators; and because

$$e_{i_x \alpha} = \alpha x ,$$

then

$$\{i_x, e_\alpha\} = \alpha x \in F \tag{16}$$

(here $\deg x = \deg\alpha = 1$). In this way we recover the familiar anticommutation relations for the fermion creation and anihilation operators (15-16). We see also that what is called in mathematics an antiderivation of the Grassmann algebra is the same as an fermion anihilation operation in particle physics. However, in order to have the proper physical interpretation of the Grassmann products e and i in (8) we must identify the pair of the mutually dual Grassmann algebras $\{V,\Lambda\}$ with the Fock spaces of the fermions and anti-fermions. If e is the multi-fermion creation operator then $i=e^*$ is the multi-*anti*-fermion anihilation operator and vice-versa.

The terms on the right hand side of equation (15) denoted by dots indicate the violation of the ideal fermions and bosons (anti-) commutation relations if these particles are built up from fermions. The boson-fermion duality and construction of bosons from fermions can be achieved only in ∞-dimensional spaces. For example if degx=degy=1 and degα=2 then the *commutator* of the creation and anihilation operators of the pairs of (anti-) fermions has the explicit form

$$[i_{x\wedge y}, e_\alpha] = \alpha(x\wedge y) + e_{i_y \alpha} \circ i_x - e_{i_x \alpha} \circ i_y , \tag{17}$$

which shows that pair of fermions is not an ideal boson.

In what follows we will understood the Grassmann algebra as the pair $\{V,\Lambda\}$ with the full set of the exterior and interior (right and left) Grassmann products (8).

Quasi-particles, super-operators and Dirac relation. Because e_α and $i_x \in \mathrm{End}\,\Lambda$ (see (8)) for all α in Λ and all x in V, then we have within this pure Grassmann algebra also the quasi-(anti-) particle operators (nuclear physicist's terminology)

$$e_\alpha + i_x \in \mathrm{End}\,\Lambda ,$$

and their pull-back's (duals),

$$(e_\alpha + i_x)^* = e_x + i_\alpha \in \mathrm{End}\,V. \tag{18}$$

Therefore having the Grassmann algebra $\{V,\Lambda\}$, (8), we can consider the new algebra $V\times\Lambda$ with the multiplication taken as the quasi-particle operator and denoted by γ,

$$V \times \Lambda \ni \{x,\alpha\} \xrightarrow{\ \gamma\ } \gamma_{x,\alpha} \in \mathrm{End}(V \times \Lambda). \tag{19}$$

Here

$$\gamma_{x,\alpha}\{y,\beta\} \equiv \{(e_x + i_\alpha)y, (e_\alpha + i_x)\beta\}. \tag{20}$$

This new algebra $\{V\times\Lambda,\gamma\}$ is the clear generalization of the Kähler-Atiyah algebra [7],[2] and [3]. Extremly important is here that we do *not* fix yet any kind of the correlation between V and Λ. Choosing particular Riemannian structure g (see below) we get from (20) exactly the original Kähler-Atiyah multiplication. However, our multiplication (20) does *not* depend on either Riemannian or symplectic structures, etc.

Consider only *odd* degree elements in Grassmann algebra (8). The Grassmann multiplications e and i of *odd* degrees are called *super*-operators or *super*-charges; here $\deg e_\alpha = \deg\alpha$ and $\deg i_x = -\deg x$. Then the anti-commutator of the Kähler-Atiyah super-operators (20) (restricted say to V) obey

$$\{\gamma_{x,\alpha}, \gamma_{y,\beta}\} = \{i_\alpha, e_y\} + \{i_\beta, e_x\} . \tag{21}$$

This looks like the celebrated relations in say supersymmetric quantum mechanics [8] and super-quasi-particles operators γ are analogous to the super-charges, except that we do not used any kind of the correlation between Λ and V, and relation (21) is independent of either Hilbert or Riemannian structures.

Also it should be clear that (21) generalize the Dirac relations for the familiar Dirac matrices. However, the γ-"matrix" now represents a *pair* of a, say, vector x and a covector α, which are independent. For example γ now represents (in a set $\mathrm{End}(V\times\Lambda)$) a space-time splitting on space $\{x\}$ and time $\{\alpha\}$ *without* pseudo-Riemannian (Minkowskian) structure.

In particular, from (21) it follows that

$$(\gamma_{x,\alpha})^2 = \{i_\alpha, e_x\} , \tag{22}$$

which means that the super-operators are the square-roots of the anti-commutators. Combining (22) with eq.(16) for degx=degα=1, we have

$$(\gamma_{x,\alpha})^2 = \alpha x \in \mathbb{R} \quad \text{or } F \quad .$$

The point is that the multiplication (20) has the unpleasant feature of non-associativity like the original Kähler-Atiyah Riemannian multiplication. Namely for the associator we get (cf. with eq.(8)),

$$\begin{aligned}\gamma_{x,\alpha}\circ\gamma_{y,\beta} - \gamma_{\gamma_{x,\alpha}\{y,\beta\}} &= i_\alpha \circ i_\beta - i_{e_\alpha \beta} \\ &+ i_\alpha \circ e_y - e_{i_\alpha y} \\ &+ e_x \circ i_\beta - i_{i_x \beta} \;.\end{aligned}$$

This hold true for arbitrary degrees.

Clifford algebras. Suppose ϕ is any (semi-) F-linear mapping

$$\phi : V \to \Lambda , \qquad ,$$

induced for example from the Poincaré isomorphism ([4], Ch.VI § 2) or from the (pseudo-) Riemannian, (pre-) symplectic or Hermitian structures (not necessarily invertible: similarly one can consider the mapping $\Lambda \to V$). For instance $\phi = g + \omega$, where g is a (pseudo-) Riemannian structure and ω is a bi-form. Both cases of the Riemannian *and* symplectic Clifford algebras can be considered jointly.

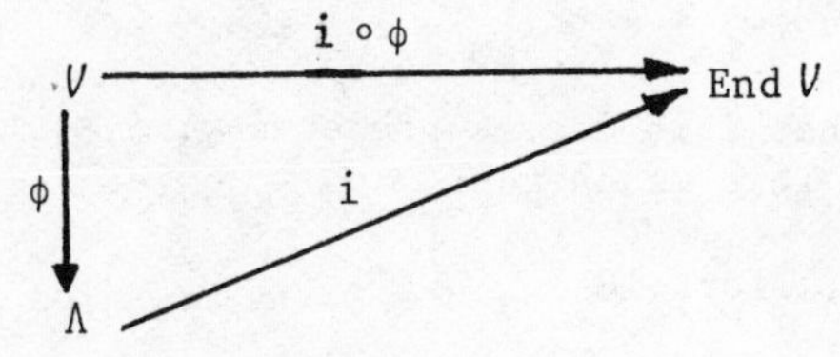

Then we can introduce the *Clifford multiplication* in the set V as F-linear mapping,

$$V \ni v \overset{\gamma}{\to} \gamma_v \in \text{End}\, V$$

which we *define* as follows

1. $\gamma_v = e_v$ for degv = 0,

2. $\gamma_v = (e + i \circ \phi)_v$ for degv = 1,

3. $\gamma_{\gamma_v w} = \gamma_v \circ \gamma_w$ (associativity). (23)

The above conditions determine γ uniquely and the set of the multivector fields V with the multiplication γ (23) is the Clifford algebra. Both Riemannian and symplectic Clifford algebras defined in this way have the *same* dimension (=dim V).

We like to show that the Clifford product (23) can be reexpressed in terms of the fermion creation and annihilation operators. This one can show through the induction on the decomposable elements. If degv=1 then

$$\gamma_v w = v \wedge w + i_{\phi v} w ,$$

and the associativity condition gives

$$\gamma_{\gamma_v w} = \gamma_{v \wedge w} + \gamma_{i_{\phi v} w} = \gamma_v \circ \gamma_w .$$

This shows that

$$\gamma_{v \wedge w} = \gamma_v \circ \gamma_w - \gamma_{i_{\phi v} w} \qquad \text{(degv=1)}. \tag{24}$$

The above means that the associativity of the Clifford product imply that each homogeneous Clifford operator can be expressed by means of the Clifford operators of smaller degrees. For example for degw=1 in (24) we get

$$\gamma_{v \wedge w} = \gamma_v \circ \gamma_w - \phi(v,w) . \tag{25}$$

This can be evaluated further:

$$\gamma_{v \wedge w} = (e - i \circ \phi)(v \wedge w) + e_v \circ i_{\phi w} - e_w \circ i_{\phi v} .$$

In (25) we tacitly assumed that ϕ is degree preserving, i.e. that degϕw = degw for all w in V. This means that

$$\phi = g + \omega \quad , \text{ for } \omega \in \Lambda^2 . \tag{26}$$

Eq.(25) one can present through the commutator,

$$\gamma_{v \wedge w} = \frac{1}{2}\left[\gamma_v , \gamma_w\right] - \omega(v \wedge w) ,$$

because anticommutator is independent of the (pre-) symplectic part of ϕ (26),

$$\{\gamma_v, \gamma_w\} = 2g(v,w) \quad ,\text{for } \deg v=\deg w=1.$$

These formulae indicate clearly once more why the Clifford multiplication γ should *not* be considered as the primitive concept (along the lines developed in [5] and [6]). Most natural is the definition of the Clifford multiplication γ (23) in terms of the primitive Grassmann multiplications e and e^* with some structure ϕ. In fact Clifford algebra is a pair: Grassmann algebra *and* some ϕ. Putting $\phi=0$ is *not* the same as not choosing any structure ϕ at all. We consider the definition of the Grassmann multiplication e in terms of the Clifford multiplication as completely inappropriate.

Let k,l,m,p,q,r,s be vectors in V^1. Then for the degree preserving mapping ϕ (26) we have

$$\gamma_{k\wedge l\wedge m} = \gamma_k \circ \gamma_l \circ \gamma_m$$

$$+ \phi(k,m)\gamma_l - \phi(l,m)\gamma_k - \phi(k,l)\gamma_m .$$

$$\gamma_{p\wedge q\wedge r\wedge s} = \gamma_p \circ \gamma_q \circ \gamma_r \circ \gamma_s$$

$$+ \phi(q,s)\gamma_p \circ \gamma_r - \phi(r,s)\gamma_p \circ \gamma_q - \phi(q,r)\gamma_p \circ \gamma_s$$

$$- \phi(p,q)\gamma_r \circ \gamma_s - \phi(p,r)\gamma_q \circ \gamma_s - \phi(p,s)\gamma_q \circ \gamma_r$$

$$+ \phi(p,q)\phi(r,s)-\phi(p,r)\phi(q,s)+\phi(p,s)\phi(q,r). \qquad (27)$$

The above formulae are sufficient for the effective applications of the Clifford calculus for the four dimentional spacetime. Eq.(26) clearly could be interpreted in framework of the Einstein nonsymmetric unified theory of the gravitational $\{g\}$ and electromagnetic $\{\omega\}$ fields.

One can ask: why we are postulating the associativity in (23)? The algebra can be defined by means of any other condition on associator. The most important notion related to associative Clifford algebras is a spinor space as minimal ideal in algebra. One can say that the spinors imply an associativity. From (23) it follows that

$$\gamma_p \circ \gamma_p = \gamma_p \Leftrightarrow \gamma_p\, p = p \in V \quad . \qquad (28)$$

Suppose that p is a primitive idempotent, i.e. $\gamma_p p=p$ and p is generating the left minimal ideal $I_p \subset V$ in Clifford algebra. Then the Dirac operator Γ_p in the spinor space ${}^pI_p$ is explicitly p-dependent,

$$V \ni v \xrightarrow{\Gamma_p} \delta_p \circ \gamma_v = \gamma_v \circ \delta_p \in \text{End } I_p$$

Here $\delta_a b \equiv \gamma_b a$, i.e. δ defines the multiplication opposite to γ[4] .

We have

$$\gamma_a \circ \gamma_b = \gamma_{\gamma_a b} \Leftrightarrow \delta_a \circ \delta_b = \delta_{\delta_a b} ,$$

Therefore (28) is equivalent to

$$\delta_p \circ \delta_p = \delta_p \Leftrightarrow \delta_p p = p . \tag{29}$$

The opposite product δ can also be expressed in terms of the fermion creation and anihilation operators e and e^*. Choosing the particular basis in I_p we get the particular matrix representation of the Dirac operator Γ^p. However, this matrix representation is completely irrelevant. What *is* essential is the p-dependence of the Dirac operator dependence which is mostly ignored or not noticed. One should stress that the idempotent p is a multifield on a manifold and not every manifold allows the global existence of such a multifield: this is the reason why spinors do not exist globally on arbitrary manifolds.

Suppose that the idempotent p is of the simple form

$$2p = \lambda + v + a \wedge b, \tag{30}$$

where degλ=0 and v,a and b are vectors in V^1.
Then eqs.(28-29) imply that (using (26) and for p≠1)

$$v \wedge a \wedge b = 0 ,$$

$$\lambda = 1 + \omega(a \wedge b),$$

$$v\omega(a \wedge b) + b\omega(v \wedge a) - a\omega(v \wedge b) = 0 ,$$

$$g(v,v) + [g(a,b)]^2 - g(a,a)g(b,b) = 1 . \tag{31}$$

One thing this shows is how important are a Riemannian structure and its signature (for g=0 there are no nontrivial idempotents, at least of the form (30)). From (30-31) we have two examples.

$$A = \frac{1}{2}(1 + v) \qquad \text{with } g(v,v) = +1 ,$$

and $$B = \frac{1}{2}(1 + \omega(a \wedge b) + a \wedge b) \quad \text{with}$$

$$[g(a,b)]^2 - g(a,a)g(b,b) = 1.$$

These idempotents A and B are not necessary primitive ones: we do not yet fix the dimension. If they commute,

$$\delta_A \circ \delta_B = \delta_B \circ \delta_A , \tag{32}$$

then $\delta_A B=\delta_B A$ is again an idempotent. Neglecting ω in (31) we get from (32) that $g(v,a)=g(v,b)=0$ (ω in (31) could be interpreted as the distortion of the spinor space through the electromagnetic field). For the case of the four dimensional space-time the idempotent $\delta_A B$ is primitive one and is determined, as we see, by the complete rèpere mobile (vierbein), because in this case $v \wedge a \wedge b \neq 0$.

If $\phi: \Lambda \to V$ we can then build the Clifford algebra of the differential multiforms. From superbrackets

$$\{d, e_\alpha\} = e_{d\alpha}$$

$$\{d, i_x\} = L_x$$

$$\{L_x, i_y\} = i_{L_x y}, \qquad \text{etc}$$

(for all α in Λ and all x,y in V) it follows that

$$\{d,\gamma_\alpha\} = e_{d\alpha} + L_{\phi\alpha} \qquad \text{for } \deg\alpha = 1.$$

Therefore if $\beta \in \Lambda$ is idempotent then the exterior derivative operator d does *not* commute with the spinor projection: $d \circ \delta_\beta \neq \delta_\beta \circ d$.

Acknowledgments

I am indebted to Professor Jan Rzewuski who created stimulating Thursday seminars at Wrocław University giving most inspirations to my research. Writing these notes I remember Professor Garret Sobczyk who introduced me to the Clifford algebras many years ago and our endless discussions at Pęgów with him and with Jerzy Kocik. I would like to thank many friends and colleagues, particularly to Professor Leopold Halpern and to all participants of Thursday seminars for their kind interest in this work.

REFERENCES

[1] V. Arnold *Les Méthodes Mathématiques de la Mécanique Classique* Mir: Moscou (Russian edition: 1974)§32-33,1974
[2] M.F. Atiyah, R. Bott and A.A. Shapiro *Clifford Modules*, *Topology* 3 (Supp.1)3-38,1964
[3] I.M. Benn and R.W. Tucker "The Dirac equation in exterior form" *Commun.Math.Phys.* 98, 53-63,1985
[4] W.H. Greub *Multilinear Algebra* Springer-Verlag: New York, 1967
[5] D. Hestenes *Space-Time Algebra* Gordon and Breach: New York 1966
[6] D. Hestenes and G. Sobczyk *Clifford Algebras to Geometric Calculus* D. Reidel Publishing Company: Dordrecht, 1984.
[7] E. Kähler *Rendiconti di Matematica* 3-4, 21,425,1962
[8] E. Witten "Supersymmetry and Morse Theory" *J. Differential Geometry* 17, 661-692,1982

LORENTZIAN APPLICATIONS OF PURE SPINORS

I.M. Benn
Department of Physics
University of Lancaster
Lancaster
U.K.

ABSTRACT. It was Cartan who first observed that certain spinors associated with a metric g on a vector space V can be put into correspondence with maximal isotropic subspaces of V. Such spinors are called pure. Chevalley later gave a thorough treatment of pure spinors. In Cartan's treatment the vector space was over the complex field. Chevalley allowed an arbitrary field but assumed the quadratic form to be of maximal index (i.e. the totally isotropic subspaces have the largest possible dimension). I will firstly show how the notion of pure spinors may be extended to the case of a real vector space whose metric has a signature with r + 2 (r) plus (minus) signs; the pure spinors being in correspondence with oriented isotropic (r + 1)-planes. For r = 1 all spinors are pure. As an example of how this may be used I shall show how to find Lorentzian manifolds for which the Kähler equation admits spinorial solutions. Similarly all Lorentzian manifolds admitting parallel spinors may be found, giving possible background geometries for supersymmetry.

The notion of a pure spinor is a very useful one, giving a geometrical interpretation to (certain) spinors. Certain elements in minimal left ideals of the Clifford algebra C(V,g) associated with the vector space V and symmetric non-degenerate bilinear form g may be related to maximal totally isotropic sub-spaces of V. Chevalley [1] gave a thorough treatment of pure spinors and it would be nice to employ these results in physics. However, there is an obstacle to the direct application of these results. Chevalley required the quadratic form on the vector space to have maximal index (so that the maximal totally isotropic sub-spaces have the largest possible dimension) whereas of paramount physical interest is the case of a real four-dimensional vector space with a Lorentzian metric. After briefly reviewing the case of maximal index I shall examine possible generalisations to cover the Lorentzian case by scrutinising possible ideas for pure spinors in the algebra $C_{r+2,r}(\mathbb{R})$ (associated with a metric whose signature has r+2 plus and r minus signs). Here one can usefully associate spinors with isotropic (but not totally isotropic) oriented (r+1)-planes. After briefly outlining the results I will give some

J. S. R. Chisholm and A. K. Common (eds.), Clifford Algebras and Their Applications in Mathematical Physics, 257–263.

examples of how these ideas may be employed in the case of $C_{3,1}(\mathbb{R})$.

Let V be a 2r-dimensional vector space over F with a non-degenerate bilinear form g of maximal index. The vector space will be identified with its image in the Clifford algebra C(V,g) and juxtapositioning of symbols will denote the Clifford product. If M is a maximal (r-dimensional) totally isotropic sub-space of V then Z_M will denote the product of some basis; that is, if $\{x^1,....,x^r\}$ is a basis then $Z_M = x^1 x^2 ... x^r$. We may associate with M the minimal left ideal $C(V,g)Z_M$. If N is an isotropic sub-space such that $V = M \oplus N$ then $C(V,g)Z_M \simeq C(N,g)Z_M \simeq \Lambda(N)Z_M$ and so this ideal is of dimension 2^r. Since for maximal index C(V,g) is isomorphic to the algebra of all matrices over F of order 2^r this ideal is indeed minimal. Of course we could equally well have defined a right ideal.

If M,T are maximal totally isotropic sub-spaces then any element of $\{C(V,g)Z_M\} \cap \{Z_T C(V,g)\}$ is a representative spinor for T (with respect to M). A spinor that represents some T is called pure. (1)

Since the intersection of a minimal left ideal and a minimal right ideal of a total matrix algebra is a one-dimensional space, representatives for T are determined up to a multiple of F. It is worth stressing that this definition is independent of the bases chosen for M and T. The definition assigns spinors to isotropic sub-spaces; the following enables such a sub-space to be reconstructed from a pure spinor.

$$u \in C(V,g)Z_M \text{ is a representative for T iff } ux=0 \ \forall \ x \in T. \quad (2)$$

Many important properties of pure spinors can be found in Chevalley's book. Here there is only space enough to note the following:-

If $r \leqslant 3$ then all spinors that are even or odd are pure. (3)

Here even and odd refer to behaviour under the involutory automorphism η which is induced on C(V,g) by minus the identity on V. By spinor is meant an element of some minimal left ideal. Thus the result should not be read as stating that all spinors are pure, since some minimal left ideals have no elements that are even or odd.

Now the case of a real vector space whose metric has a signature with r+2 (r) plus (minus) signs will be considered. To save notation let $A \equiv C_{r+2,r}(\mathbb{R})$. It is worth giving consideration to different possible generalisations of the concept of a pure spinor to this algebra. Five potential directions of departure may be put under the following rough headings.

(i) Use the isomorphism $C_{r+2,r}(\mathbb{R}) \simeq C_{r+1,r+1}(\mathbb{R})$.
(ii) Complexify, use $C_{r+2,r}(\mathbb{R}) \otimes \mathbb{C}(\mathbb{R})$.
(iii) Use non-minimal ideals.
(iv) Use the even sub-algebra.
(v) Use planes that are not totally isotropic.

Since for $C_{r+1,r+1}(\mathbb{R})$ one may use the usual results and $C_{r+2,r}(\mathbb{R}) \simeq C_{r+1,r+1}(\mathbb{R})$ one may wish to call spinors in $C_{r+2,r}(\mathbb{R})$ pure if they are the isomorphic images of pure spinors in $C_{r+1,r+1}(\mathbb{R})$. However, this obscures any geometrical significance that pure spinors may have since obviously the isomorphism does not preserve the underlying vector space V; neither is it canonical, that is, no particular such isomorphism is singled out. One could of course choose some (in some sense) standard isomorphism but I shall eschew this approach in favour of the fifth option which will be seen to lead to different conclusions.

Another way in which one might hope to employ the established results is by considering $A \otimes C(\mathbb{R})$. This real algebra has the involutory automorphism of complex conjugation which leaves A invariant. Regarded as a complex algebra this algebra is isomorphic to the Clifford algebra associated with a complex Euclidean space. Here the maximal isotropic sub-spaces are (r+1)-dimensional. Elements of minimal left ideals of $A \otimes C(\mathbb{R})$ may be associated with these (r+1)-dimensional complex isotropic planes. The operation of complex conjugation enables a real isotropic sub-space to be determined. However this will in general be less than r-dimensional. Only for the special case of $r = 1$ does every complex maximal isotropic sub-space contain a maximal real sub-space. This special case will be referred to later.

The maximal totally isotropic sub-spaces can be used to define 2^{r+2}-dimensional ideals in A; that is, of twice the minimal dimension. If M and T are maximal totally isotropic sub-spaces then we could take the space of representatives of T (with respect to M) to be $Z_T A Z_M$. Now there exists an s in the Clifford group such that $Z_T = sZ_M s^{-1}$ so $Z_T A Z_M = sZ_M s^{-1} AZ_M = sZ_M A Z_M$. We may choose a totally isotropic N and a two-dimensional R such that $V = M \oplus N \oplus R$ and $T \subset M \oplus N$. Then $Z_M A Z_M = Z_M Z_N C(R,g) Z_M = C(R,g) Z_M$. Now R is in the conjugate of M and $C(R,g) \simeq M_2(\mathbb{R})$ thus these ideals intersect in an algebra which is the tensor product of a total matrix algebra and a one-dimensional algebra. In general then if ϕ and ψ are in $Z_T A Z_M$ we are not guaranteed that $\phi = m\psi$ for some m. For the special case of $C_{3,1}(\mathbb{R})$ there is a connection between these non-minimal ideals and the minimal left ideals in the complexified algebra; this will be mentioned later.

Although the maximal totally isotropic sub-spaces are not large enough to define minimal left ideals in A we can use them to define minimal ideals in the even sub-algebra, A^+, simply by projecting out the even part of the non-minimal ideals of A defined above. Since $A^+ \simeq M_{2^r}(\mathbb{R}) \otimes C(\mathbb{R})$ the intersection of a minimal left ideal and a minimal right ideal is of one complex dimension and so in this way we could represent maximal totally isotropic sub-spaces by a class of spinors in A^+ determined up to a complex scaling. Rather than pursue this approach I shall proceed with the final option, and show how this enables one to recover even spinors associated with maximal totally isotropic sub-spaces.

If M is a maximal totally isotropic sub-space and a is a unit norm element of the conjugate of M then $A(1+a)Z_M$ is a minimal left ideal. Ostensibly this prescription requires not only an isotropic (but not totally isotropic) (r+1)-plane but also a specific choice of the element a. The situation is not quite as bad as this for if a' is any unit norm element in the plane containing M and a then $a' = \pm \alpha$, $\alpha \epsilon M$, and $a'Z_M = \pm aZ_M$. Thus the ideal is associated with an isotropic (r+1)-plane which has a definite orientation for non-null vectors. This is the price that has to be paid for using planes that are not totally isotropic, it is not just the vector sub-space but also this preferred orientation that determines a minimal left ideal. I shall use Penrose's terminology and refer to such a plane as a "flag". It is convenient to specify a flag by the pair (M,a). The following definition can now be made.

Any element of $\{(1+a')Z_{M'}A\} \cap \{A(1+a)Z_M\}$ is a representative spinor for the flag (M',a') (with respect to (M,a)). Any spinor that represents some flag will be called pure. (4)

The consequences of such a definition can be derived in close analogy to the case of maximal index. Full results are given in reference 2. It has already been anticipated that this definition allows an even spinor to be associated with a totally isotropic r-plane. It straightforwardly follows the if ϕ represents (M',a') with respect to (M,a) then the even part of ϕ lies in $Z_{M'}A^+Z_M$.

The following is the analogue of (2):

$u \varepsilon A(1+a)Z_M$ represents a flag (M',a') iff $xu = +(-)\sqrt{x^2}u$ $\forall$ positively (negatively) oriented x in the flag plane. (5)

The case of four dimensions is particularly simple.

If r = 1 all spinors are pure. (6)

This analogue of (3) is different in two respects. For the case of maximal index all even or odd spinors in six or less dimensions are pure, whereas the definition in (4) renders all spinors (with no caveats) pure in up to four dimensions.

We return now to the complexified algebra for the special case of four dimensions. To simplify notation let $B \equiv C_{3,1}(\mathbb{R}) \otimes \mathbb{C}(\mathbb{R})$. In this low dimensional case, and this case only, the maximal totally isotropic complex sub-spaces contain maximal real sub-spaces. That is, if M is a two-dimensional isotropic plane we can choose a basis $\{x,u\}$ where $x^* = x$. It is convenient to introduce as basis for the complexified vector space $\{x,y,u,u^*\}$ where

$$xy + yx = 1$$

$$uu^* + u^*u = 1 \qquad (7)$$

and all other pairs are mutually orthogonal. We may then choose the volume 4-form Z to be

$$Z = i[x,y][u,u^*] \qquad (8)$$

where the brackets denote the Clifford commutator. It will later prove convenient to be able to write the minimal left ideal associated with M in a different way. It readily follows from (7) and (8) that

$$xu = -\tfrac{1}{2}ux(1-iZ)$$

and $\quad x(1-iZ)=-2u^*xu.$ (9)

Thus $\quad Bxu = Bx(1-iZ).$ (10)

So if $\phi\varepsilon Ax$ then $\psi = \phi(-iZ)$ is in BZ_M for some totally isotropic complexified plane M. The arbitrariness in the orientation given by Z reflects the fact that a real null vector can be embedded in only two distinct complex isotropic 2-planes. It follows that ψ will be pure if ϕ is even or odd. Conversely if $\psi\varepsilon BZ_M$ and $\phi = \psi+\psi^*$ then ϕ is in Ax and will be even or odd if ψ is pure.

The motivation for considering the algebra associated with a real vector space equipped with a Lorentzian metric is that such a vector space is provided by the cotangent space of the space-time manifold. In this case the vector space of the Clifford algebra is naturally identified with the space of differential forms on space-time. There is a close analogy between pure spinors and decomposable differential forms. If ω is a decomposable p-form in n dimensions then a distribution of (n-p)-planes Σ may be associated with ω by defining

$$i_v\omega = 0 \qquad \forall v\varepsilon\Sigma. \tag{11}$$

This purely algebraic condition gives an (n-p)-plane in each tangent space. The condition that these planes form the leaves of a foliation is given by the well known Frobenius condition

$$d\omega = A\wedge\omega \quad \text{for some A.} \tag{12}$$

Thus a differential condition on a decomposable form is related to a geometrical property of the associated planes. The same is true for pure spinors. The relation between a pure spinor and a plane is algebraic; conditions on the spinorial derivatives impart information about the distribution of planes.

Of relevance to the study of the curved space Kähler equation [3] is a classification of space-times for which some minimal left ideal of the Clifford algebra is preserved by the Lorentzian connection ∇. If the ideal is characterised by a primitive idempotent P then Graf [4] gave the condition as $P\nabla_X P = 0 \;\; \forall X$. Since all spinors in $C_{3,1}(\mathbb{R})$ are pure, every minimal left ideal is associated with an isotropic 2-plane and it readily follows that ∇_X is an endomorphism on the ideal if and only if it is an endomorphism on this plane. An integrability condition shows that the curvature must be algebraically special with the preserved plane containing a principal null direction. A theorem due to Walker gives a parameterisation of the most general metric such that these conditions are satisfied. Details are to be found in reference (5). It is also of interest to know when the connection preserves some minimal left ideal of $C_{3,1}(\mathbb{R}) \otimes \mathbb{C}(\mathbb{R})$. Let P be some primitive idempotent in the complexified algebra B such that BP is a minimal left ideal preserved by the connection. Let

$$Q_\varepsilon = \tfrac{1}{2}(1 + i\varepsilon Z) \tag{13}$$

with $\varepsilon = \pm 1$. Then

$$(PQ_\varepsilon)^2 = \lambda_\varepsilon PQ_\varepsilon \quad \text{with} \quad \lambda_\varepsilon \in \mathbb{C} \tag{14}$$

since P is primitive. We cannot have $\lambda_\varepsilon = 0$ for both signs of ε and so if we drop the subscript in the case which corresponds to $\lambda_\varepsilon \neq 0$ then P' is idempotent where

$$P' = \lambda^{-1} PQ \ . \tag{15}$$

If S_o projects out the sub-space spanned by the identity then since

$$S_o(P') = \lambda^{-1} S_o(PQ) = S_o(P). \tag{16}$$

It thus follows that P' is primitive. Now if $\phi = \phi P'$ then $\eta\phi = -Z\phi Z$ and since $Q_\varepsilon Z = -i\varepsilon Q_\varepsilon$ we have $\eta\phi = i\varepsilon Z\phi$. Thus the ideal BP' contains even and odd elements. Since the complexified algebra can be regarded as being associated with a complex four-dimensional Euclidean space all ideals containing even and odd elements are associated with a complex isotropic 2-plane. It follows from (10) that

$$BP' = Bx(1 - iZ) \tag{17}$$

for some real null x and an appropriate choice of orientation. Now if ∇_X preserved BP it certainly preserved BP' since Q_ε is parallel. From (17) it then follows that x must be a recurrent null vector and such geometries can be classified as in reference [5].

In models of "rigid" supersymmetry the transformations are parameterised by spinors that are parallel with respect to the spinorial covariant derivative S_X. (See the talk given by R.W. Tucker.) If ψ represents a plane Σ then $S_X\psi = \lambda(X)\psi$ for some function $\lambda(X)$ if and only if ∇_X is an endomorphism of Σ [5]. If $S_X\psi = 0$ then ε contains a parallel (with respect to ∇) null vector. Thus the most general background geometry for "rigid" supersymmetry is a p-p gravitational wave [6]. Here a definition of pure spinor has been given that applies to the Lorentzian case, and its usefulness demonstrated. Although there are other dimensions and signatures where the ideas presented here are directly applicable it is not clear how the notion of pure spinor can be usefully developed to cover the general case. The current interest in theories of the Kaluza-Klein type motivates the desire for a clear geometrical interpretation of the spinors occurring in such theories.

References

(1) C. Chevalley, The algebraic theory of spinors, Columbia University Press, New York 1954.
(2) I.M. Benn, R.W. Tucker, 'Pure spinors and real Clifford algebras', Lancaster preprint.
(3) E. Kähler, Rend. Mat. (3-4) **21**, 425 (1962).
(4) W. Graf, Ann. Inst. Henri Poincaré XXIX, 85 (1978).

(5) A. Al Saad, I.M. Benn, 'Ideal Preserving Lorentzian Connections', Lancaster Preprint.

(6) I.M. Benn, M. Panahi, R.W. Tucker, Classical and Quantum Gravity 2 (1985) L71.

THE POINCARÉ GROUP

Erich Kähler
Mozartstraße 42
2000 Wedel (Holstein)
West Germany

ABSTRACT: The relativity of Einstein-Minkowski should be substituted by another relativity which postulates an absolute origin of time. This being at a great distance in the past, the new relativity will be a good approximation of the classical one.

1. THE INNER DIFFERENTIAL CALCULUS

An n-dimensional manifold V with differential structure of infinite order determines by its differential forms a ring $D_\wedge$ in which the multiplication of two forms u and v has the effect

$$u \wedge v$$

and the addition is realized by admitting the summation

$$u_o + u_1 + u_2 + \quad . \quad . \quad + u_n$$

of differential forms u_m of different degrees m. The elements of this Ring $D_\wedge$ shall be called simply the d i f f e r e n t i a l s of the Manifold V.

Given a Riemannian metric

$$g_{ik}.dx^i.dx^k$$

in V, one can define another multiplication designated

$$u \vee v$$

which supplies the totality $D_\wedge$ of the differentials of V in a second manner with the constitution of an associative ring $D_\vee$.

In order to describe the relation of the "o u t e r" multiplication $u \wedge v$ to the "i n n e r" multiplication $u \vee v$ it will be necessary to introduce the linear operators η, ζ,

J. S. R. Chisholm and A. K. Common (eds.), Clifford Algebras and Their Applications in Mathematical Physics, 265–272.

γ, e_i. Their effect is

$$\eta u = (-1)^p . u, \quad \zeta u = (-1)^{\binom{p}{2}} . u, \quad \gamma u = p.u$$

where u is an (outer) differential form of degree p, whereas e_i is defined by $e_i dx^k = \delta_i^k$ and the postulate that generally

$$e_i(u \wedge v) = e_i u \wedge v + \eta u \wedge e_i v .$$

The formula

$$u \vee v = \sum_{p=0}^{n} \frac{1}{p!} . (-1)^{\binom{p}{2}} \eta^p e_{i_1} .. e_{i_p} u \wedge e^{i_1} .. e^{i_p} v$$

expresses the inner multiplication as a sequence of outer multiplications, showing for instance that

$$dx^i \vee dx^k = dx^i \wedge dx^k + g^{ik}$$

and therefore $D_\vee$ is a C l i f f o r d - ring-bundle.

Covariant differentiation d_i of a differential u can be defined in such a manner that the $d_i u$ become the components of a covariant vector, and therefore

$$du = dx^i \wedge d_i u, \qquad \delta u = dx^i \vee d_i u$$

are definitions of two invariant linear operators d and δ. The operators d is known as the o u t e r d i f f e r e n t i a l, and δ as the i n n e r d i f f e r e n t i a l ; they have proved their importance in the D i r a c - T h e o r y of the electron.

The formula

$$\delta(u \vee v) = \delta u \vee v + \eta u \vee \delta v + 2.e^i u \vee d_i v$$

corresponds to the classical rule

$$d(u \wedge v) = du \wedge v + \eta u \wedge dv .$$

The covariant derivatives of the volume-differential

$$\tau = \sqrt{|g_{ij}|} . dx^1 \wedge . . \wedge dx^n$$

being zero, one finds

$$\delta(u \vee \tau) = \delta u \vee \tau .$$

For physical applications it is important to introduce the s c a l a r p r o d u c t s (u,v), $(u,v)_1$ defined by

$$(u,v) = (\zeta u \vee v) \wedge \tau , \qquad (u,v)_1 = e_i(dx^i \vee u, v),$$

related by the G r e e n 's f o r m u l a

$$d(u,v)_1 = (u,\delta v) + (v,\delta u) .$$

2. D I R A C - E Q U A T I O N S

Let $a = (a_{ij})$ be a $(m\times m)$-matrix and $u = (u_i)$ a $(m\times 1)$-matrix, the elements of which are differentials in V.

If these matrices are related by the equation

2.1 $$\delta u = a\vee u ,$$

we shall say that u is a solution of a D i r a c - e q u a - t i o n with the p o t e n t i a l a and the f i e l d o f f o r c e

$$Tr(da+\eta a\wedge a) .$$

The a d j o i n t D i r a c - e q u a t i o n of 2.1 is

2.2 $$\delta v = - \zeta\, {}^t a \vee v ,$$

where ${}^t a$ means the transpose of the matrix a. Both equations are related by the fact, that from them follows the c o n - s e r v a t i o n - t h e o r e m

2.3 $$d(Tr({}^t u,v)_1) = 0$$

with

$$Tr({}^t u,v)_1 = \sum_{i=1}^{m} (u_i,v_i)_1 .$$

If the elements f_{ij} of the $(m\times m)$-matrix $f=(f_{ij})$ have degree 0, that is to say that f_{ij} are functions, the inner differentiation of f.u simplifies to

$$\delta(f\vee u) = \delta f\vee u + f\vee\delta u .$$

When u satisfies the Dirac-equation 2.1 and f is invertible the matrix $v = f\vee u$ proves to be a solution of the Dirac-equation

2.4 $$\delta v = b\vee v$$

with the potential

2.5 $$b = df\vee f^{-1} + f\vee a\vee f^{-1} ,$$

and the relation

2.6 $$db + \eta\, b \wedge b = f \vee (da + \eta\, a \wedge a) \vee f^{-1}$$

shows that 2.4 has the same field of force as the equation 2.1 .

3. THE POINCARÉ-GROUP

The notion of elementary particle has no philosophic fundament. A better way to start understanding nature is to conceive the unity behind all phenomena as a monad endowed with an infinity of symmetries, and to derive the manifold of physical phenomena by the mathematical tactics of "symmetry breaking".

This presupposes a mathematical monadology offered by a philosophic transposition of the local algebra, about which I have reported elsewhere. A philosophical mobilization of mathematics cannot ignore that geometry is only one aspect of this science and that immersion of reality in a space is not the finest use of mathematical art.

Dynamics of monads find their best representation in arithmetic and purely algebraic relations. Therefore it must be esteemed as a case of "prestabilized harmony" (in the sense of Leibniz), that Hamilton's invention of the quaternion leads one immediately to consider time and space in the algebraic alliance of a quaternion.

3.1 $$q = x + y.i + z.i.j + t.j \ ,$$

the real linear combination of 4 units 1, i, j, i.j bound by the relations $i^2 = j^2 = -1$, $i.j + j.i = 0$. The 4-dimensional totality H of the quaternions obtains a topology by the provisional introduction of a metric, which defines the norm $|q-r|$ of the difference of two quaternions as the distance of q and r.

Reproducing the classical passage from the Gaussian plane to the Riemannian sphere one can define a compact space $H \cup \infty$ containing the space H and the point at infinity ∞ .

The topology of $H \cup \infty$ is such, that each union of ∞ with the set of quaternions q satisfying a condition

$$|q| > \varrho \text{ (a fixed real number)}$$

is to be considered as a neighbourhood of infinity. Given four constant quaternions a,b,c,d satisfying only the condition that the right-column-rank of the matrix

$$M = \begin{pmatrix} a & b \\ c & d \end{pmatrix}$$

must be 2, one can define a topological map $H \cup \infty \rightarrow H \cup \infty$ changing q into

$$M(q) = (a.q + b).(c.q + d)^{-1} \ .$$

The totality of such maps is a group to be named: the symmetry of the cell $H \cup \infty$.

A matrix

$$T = \begin{pmatrix} \alpha & \bar{\beta} \\ \beta & \gamma \end{pmatrix}$$

formed by real numbers α, γ and an arbitrary quaternion β, determines a real function

$$T(q) = \alpha.q.\bar{q} + \beta.q + \bar{q}.\bar{\beta} + \gamma = (\bar{q},1).T.\begin{pmatrix} q \\ 1 \end{pmatrix} .$$

The quotient

3.2 $$\frac{dq.d\bar{q}}{T(q)^2}$$

changes by the action of an element $q \mapsto M(q) = \hat{q}$ of the symmetry of the cell into

3.3 $$\frac{d\hat{q}.d\bar{\hat{q}}}{T(\hat{q})^2} = |M|^2 . \frac{dq.d\bar{q}}{\hat{T}(q)^2}$$

with $\hat{T} = {}^t\bar{M}.T.M$ and $|M| = |a.c^{-1}.d.c - b.c|$ when $c \neq 0$, $|M| = |a.d|$ when $c = 0$. The relation 3.3 shows that only in the case

$$T = \begin{pmatrix} 0 & cj \\ -cj & 0 \end{pmatrix} \quad \text{with } c \in \mathbf{C}$$

are all substitutions $q \mapsto M(q)$ with $M \in SL_2(\mathbf{C})$ isometries of the metric 3.2. If the Universe is supposed to be measurable with such metric, the passage from the cell $H \cup \infty$ to the Universe will have the character of a c e l l - d i v i s i o n :

The symmetry of that cell is broken to become the subgroup of those maps, which map the two " p a r t i a l c e l l s "

$$T(q) > 0 \quad \text{and} \quad T(q) < 0$$

into themselves.

The fact that the proper homogeneous L o r e n t z g r o u p is isomorphic to $SL_2(\mathbf{C})$ and that a B i g B a n g at the origin of time is imaginable induces me to risk proposing the Hypothesis:

A g e n e r a l t h e o r y o f m a t t e r m u s t b e f o u n d e d o n a P o i n c a r é - g r o u p w h i c h i s t h e t o t a l i t y o f t h e i s o m e t r i e s o f t h e m e t r i c

3.4 $$\frac{4.dq.d\bar{q}}{(\bar{q}.j - j.q)^2} = \left(\frac{dx}{t}\right)^2 + \left(\frac{dy}{t}\right)^2 + \left(\frac{dz}{t}\right)^2 + \left(\frac{dt}{t}\right)^2$$

with

3.1 $$q = x + y.i + z.i.j + t.j$$

This new Poincaré-group contains the 10-dimensional Lie group of the substitutions

$$q = x + i.y + i.j.z + j.t \mapsto \hat{q} = M(q) = \hat{x} + i.\hat{y} + i.j.\hat{z} + j.\hat{t}$$

with

$$^t\bar{M}.\begin{pmatrix} 0 & j \\ -j & 0 \end{pmatrix}.M = \begin{pmatrix} 0 & j \\ -j & 0 \end{pmatrix}$$

As in classical relativity, time t and space (x,y,z) are relative. Time must be measured as purely imaginary number, and the relation

$$\hat{t} = t.|c.q + d|^{-2}, \qquad M = \begin{pmatrix} a & b \\ c & d \end{pmatrix},$$

shows this condition to be invariant against change of coordinates.

Let t = T be the time of any fixed historical event. Then for all normal physical observations the coordinate t will be such that

3.5 $$\frac{t}{T} = s$$

is a real number in the neighbourhood of s = 1, whereas $\frac{i}{T}$ is an extremely minute real constant. T being constant, the Poincaré group can also be conceived as the totality of the isometries of the metric

3.6 $$\left(\frac{dx}{s}\right)^2 + \left(\frac{dy}{s}\right)^2 + \left(\frac{dz}{s}\right)^2 + \left(\frac{dt}{s}\right)^2$$

which is nearly the same as

$$(dx)^2 + (dy)^2 + (dz)^2 + (dt)^2,$$

the metric responsible for comparison of theory and observation. Therefore 3.6 will be the metric destined to govern the inner differential calculus in the new relativity.

Each binary inner product of differentials dx, dy, dz, dt other than

$$dx \vee dx = dy \vee dy = dz \vee dz = dt \vee dt = s^2$$

equals then the corresponding outer product, and the volume differential

$$\tau = \frac{dx}{s} \vee \frac{dy}{s} \vee \frac{dz}{s} \vee \frac{dt}{s}$$

is idempotent: $\tau \vee \tau = 1$. If a differential u is *spatial* in the sense that it can be written in the form $u = a + b_1.dx + b_2.dy + b_3.dz + c_1.dy \wedge dz + c_2.dz \wedge dx + c_3.dx \wedge dy + h.dx \wedge dy \wedge dz$, it determines two scalars a and h and two vector fields $\mathfrak{b} = (b_1, b_2, b_3)$ and $\mathfrak{c} = (c_1, c_2, c_3)$. For every differential u one can write u and its differentials in the form

$$u = u^+ \vee e^+ + u^- \vee e^-, \ \delta u = v^+ \vee e^+ + v^- \vee e^-, \ \delta\delta u = w^+ \vee e^+ + w^- \vee e^-$$

with spatial differentials $u^{\pm}, v^{\pm}, w^{\pm}$ and $e^{\pm} = \frac{1}{2}.(1 \pm \tau)$.

If a, h, $\mathfrak{b}$, $\mathfrak{c}$ are the scalars and vector fields of $u^{\pm}$, the corresponding a*, h*, $\mathfrak{b}^*$ $\mathfrak{c}^*$ of $v^{\pm}$, a**, h**, $\mathfrak{b}^{**}$, $\mathfrak{c}^{**}$ of $w^{\pm}$ will be

$$a^* = s^2.\mathrm{div}\,\mathfrak{b} \mp s^4.\frac{\partial h}{\partial t}$$
$$h^* = \mathrm{div}\,\mathfrak{c} \mp s^{-2}.\frac{\partial a}{\partial t}$$
$$\mathfrak{b}^* = \mathrm{grad}\,a + s^2.(\pm.\frac{\partial \mathfrak{c}}{\partial t} - \mathrm{rot}\,\mathfrak{c})$$
$$\mathfrak{c}^* = s^2.\mathrm{grad}\,h \pm \frac{\partial \mathfrak{b}}{\partial t} + \mathrm{rot}\,\mathfrak{b}$$

3.7
$$a^{**} = s^2.\Delta a - \frac{2s}{T}.\frac{\partial a}{\partial t}$$
$$h^{**} = s^2.\Delta h + \frac{4s}{T}.\frac{\partial h}{\partial t} \mp \frac{2}{s.T}.\mathrm{div}\,\mathfrak{b}$$
$$\mathfrak{b}^{**} = s^2.\Delta \mathfrak{b} + \frac{2s^3}{T}.\mathrm{grad}\,h$$
$$\mathfrak{c}^{**} = s^2.\Delta \mathfrak{c} + \frac{2s}{T}.(\frac{\partial \mathfrak{c}}{\partial t} \mp \mathrm{rot}\,\mathfrak{c}) ,$$

where Δ means the euclidean L a p l a c e - O p e r a t o r

$$\Delta = \frac{\partial^2}{\partial x^2} + \frac{\partial^2}{\partial y^2} + \frac{\partial^2}{\partial z^2} + \frac{\partial^2}{\partial t^2} .$$

The Poincaré group contains a 10-dimensional Lie group generated by the operators

3.8
$$A_1 = y.\frac{\partial}{\partial z} - z.\frac{\partial}{\partial y},\quad A_2 = z.\frac{\partial}{\partial x} - x.\frac{\partial}{\partial z},\quad A_3 = x.\frac{\partial}{\partial y} - y.\frac{\partial}{\partial x} ,$$
$$A_o = x.\frac{\partial}{\partial x} + y.\frac{\partial}{\partial y} + z.\frac{\partial}{\partial z} + t.\frac{\partial}{\partial t} ,$$
$$B_1 = \frac{\partial}{\partial x},\quad B_2 = \frac{\partial}{\partial y},\quad B_3 = \frac{\partial}{\partial z} ,$$
$$C_1 = 2x.A_o - q.\bar{q}.B_1,\quad C_2 = 2y.A_o - q.\bar{q}.B_2,\quad C_3 = 2z.A_o - q.\bar{q}.B_3,$$

with
$$q.\bar{q} = x^2 + y^2 + z^2 + t^2$$
and the relations

3.9
$$(A_i,A_k) = - e_{ikj}.A_j ,$$
$$(A_i,B_k) = - e_{ikj}.B_j,\quad (A_i,C_k) = - e_{ikj}.C_j ,$$
$$(B_i,C_k) = 2.\delta_{ik}.A_o - 2.e_{ikj}.A_j ,$$

($i \neq 0$, $k \neq 0$, summation of j over 1,2,3)

$$(B_i,B_k) = 0 = (C_i,C_k) ,$$
$$(A_o,A_k) = 0,\quad (A_o,B_k) = - B_k,\quad (A_o,C_k) = C_k ,$$

(e_{ikj} = 1, -1, 0 according as ikj is an even, or odd, or no permutation of 1,2,3) .

The conditions $A_1u = A_2u = A_3u = 0$ are equivalent to

$$u = v_1 \vee e^+ + v_2 \vee e^-$$

where v_1 and v_2 have the form

3.10 $$R = g.rdr \vee dt + b.rdr + c.dt + h$$

and the functions g,b,c,h depend only on t and $r = \sqrt{x^2+y^2+z^2}$.

From 3.10 it follows that

3.11 $$\delta R = g^*.rdr + b^*.rdr + c^*.dt + h^*$$

with $s = t/T$ and

$$g^* = \frac{1}{r}.\frac{\partial c}{\partial r} - \frac{\partial b}{\partial t}$$

$$b^* = \frac{1}{r}.\frac{\partial h}{\partial r} - s^2.\frac{\partial g}{\partial t}$$

3.12 $$c^* = (r.\frac{\partial g}{\partial r} + 3g).s^2 + \frac{\partial h}{\partial t}$$

$$h^* = (r.\frac{\partial b}{\partial r} + 3b + \frac{\partial c}{\partial t}).s^2 - 2c.\frac{s}{T} .$$

Each homogeneous and harmonic polynomial $f = f(x,y,z)$ of degree k determines three differentials

3.13 $$Y = r^{-k}.f(x,y,z), \quad S = r^{1-k}.df(x,y,z), \quad S^* = S - (2k+1).Y.dr$$

satisfying the relations

$$dY = \frac{1}{r}.S - \frac{dr}{r}.Y = \frac{1}{r}.S^* + (k+1).\frac{dr}{r}.Y$$

$$\delta S = \frac{1-k}{r}.dr \vee S, \quad \delta S^* = \frac{2+k}{r}.dr \vee S^*$$

3.14 $$\delta(R \vee Y) = \delta R \vee Y + \eta R \vee dY = \frac{1}{r}.\eta R \vee S + (\delta R - \frac{k}{r}.\eta R \vee dr) \vee Y$$

$$\delta(R \vee S) = (\delta R + \eta R \vee \frac{1-k}{r}.dr + \frac{2s^2}{t}.e_t R) \vee S + 2k.s^2.e_r R \vee \frac{dt}{t} \vee Y$$

$$\delta(R \vee S^*) = (\delta R + \eta R \vee \frac{2+k}{r}.dr + \frac{2s^2}{t}.e_t R) \vee S^* - (2k+2).s^2.e_r R \vee \frac{dt}{t} \vee Y,$$

where R is defined by 3.10 and

3.15 $$e_t R = -g.rdr + c, \quad e_r R = g.r.dt + b.r .$$

These formulae prepare for the separation of variables in the physically important case of Dirac equations with Potential of the Form 3.10.

MINIMAL IDEALS AND CLIFFORD ALGEBRAS IN THE PHASE SPACE REPRESENTATION OF SPIN-1/2 FIELDS

P. R. Holland*
Laboratoire de Physique Théorique
Institut Henri Poincaré
11 rue Pierre et Marie Curie
75231 Paris Cedex 05 France

ABSTRACT. The Liouville superoperator representation of spin-1/2 wave equations (Dirac, Feynman-Gell-Mann) is studied following the phase space approach of Prigogine, Wigner-Moyal, Bohm and Schönberg. We give the conditions for the phase space tensor theory to reproduce the usual formalism in terms of a restriction of states to minimal ideals in the spacetime Dirac and Jordan-Wigner Clifford Algebras. Consequences of these results are discussed.

1. STATES, OPERATORS AND MINIMAL IDEALS

The notion of a "superoperator" has proved a useful one for emphasizing the distinction between a quantum mechanical operator in Hilbert space H and an operator in the product space H⊗H. Of particular interest are the non-factorizable superoperators which cannot be decomposed into a product of operators acting in the lower space. Taking the Liouville superoperator as the generator of motions, acting on the density matrix treated as a supervector, Prigogine et al have shown that one may introduce an entropy function having Lyapounov properties in the space of superoperators. Thus the algebraic structure of physics is widened and on the level of analysis where the thermodynamical properties of matter become relevant, pure states may evolve into mixed states(1).

Similar notions lay behind the Wigner-Moyal approach(2) which applies a non-local unitary mapping to the density matrix to yield a quasi-distribution function in phase space obeying a quantum Liouville equation. Again, this method goes outside of the usual local unitary transformations of quantum theory and the density matrix is treated as a state vector in the higher space.

What the Prigogine and Wigner-Moyal approaches bring out is that the description of different aspects of physical processes (e.g. dynamics or thermodynamics) may not be possible within one given vector space or algebraic structure, but that a hierarchy of spaces or algebras

*Royal Society/SERC European Fellow.

J. S. R. Chisholm and A. K. Common (eds.), Clifford Algebras and Their Applications in Mathematical Physics, 273–283.

are required for a complete account. Quantum mechanics has only investigated the first two levels of such a hierarchy and the usual formalism moreover only makes use of those operators in the second level which are factorizable. Generalizing, though, we may extend this idea and propose a sequence of levels, each corresponding to different facets of physical processes, which is non-reductionist in the sense that no level is more basic or fundamental than any other level. It will be a property of the algebras associated with a given level (i.e. that they possess non-factorizable elements) that the account they provide is not reducible to that provided by the other levels. Rather, the levels are complementary in that all are needed for a complete account (but not in the mutually exclusive sense of Bohr) and each level refers to an objectively real feature (e.g. irreversibility is not an illusion based on ignorance of initial conditions). The third or higher levels, for example, may deal with processes whose complexity transcends even that of thermodynamics.

At each level, then, we shall have states and operators associated with the relevant algebra. The states are the operators of the algebra of the next lower level, the operators the states of the next higher level. To make the relation between the levels more mathematically precise, we note that the observation of Riesz(3), that spinors may be treated as belonging to minimal left or right ideals in the Clifford algebra that operates on them as irreducible subspaces (so that if E is a primitive idempotent in a Clifford algebra, then Ψ is an algebraic left spinor iff $\Psi = \Psi E$) may be extended to a general characterization of the relation between states and operators. It is obvious, for example, that in a pure state density matrix $\rho(x, x') = \Psi(x)\Psi^*(x')$ the wave function appears as an element of a minimal left ideal in the algebra of observables (i.e. a column times a row). We are thus led to propose that the states relevant to a given level in the hierarchy of algebras are primitive elements(3) (i.e. ones which generate left or right minimal ideals of which the primitive idempotents are a subset) in the corresponding algebra. As such, states are particular kinds of factorizable superoperators and consequently, as stated above, operators at a lower level.

Following ref.4 we shall apply this scheme to the Wigner-Moyal representation of a massive Dirac or Feynman-Gell-Mann spin-1/2 particle in an external electromagnetic field (emphasizing in particular the Clifford algebraic aspects of the problem) by treating the density matrix as an operator, and then as a factor in a minimal ideal in a higher phase space algebra.

2. PRIMITIVE DIRAC NUMBERS

Consider a four-component spinor field $\xi^a(x\ )$ which satisfies at the spacetime point x_1 the Dirac equation

$$e^i\left(i\partial_{1i} - eA_i(x_1)\right)\xi(x_1) = m\xi(x_1),\ i = 0,1,2,3, \tag{2.1}$$

and a dual spinor field $\eta_a(x_2)$ satisfying the adjoint Dirac equation at the point x_2 (where x_1 and x_2 independently range over all of spacetime). Define the pure state characteristic matrix

$$\Psi^a{}_b(x_1, x_2) = \xi^a(x_1)\, \eta_b(x_2), \quad a,b = 1,2,3,4. \tag{2.2}$$

Then a general characteristic matrix $\rho^a{}_b\,(x_1,x_2)$ (sum of terms of the form (2.2)) satisfies the two Dirac equations

$$e^i\left[i\partial_{1i} - eA_i(x_1)\right]\rho(x_1\; x_2) = m\rho(x_1\; x_2) \tag{2.3a}$$

$$\rho(x_1\; x_2)\, e^i\left[i\overleftarrow{\partial}_{2i} + eA_i(x_2)\right] = -m\,\rho(x_1\; x_2). \tag{2.3b}$$

When ρ is of the form (2.2) one may deduce from these equations the original two dual Dirac equations, and it is from eqs. (2. 3) that we may derive phase space relations. Here the Dirac algebra C_4 is generated by the unity $1_{C_4} \in \mathbb{C}$ and the real 1-forms e^i, i = 0.1,2,3, spanning a vector space such that

$$e^i.e^j = \tfrac{1}{2}(e^i e^j + e^j e^i) = g^{ij}\, 1_{C_4} \tag{2.4}$$

where g^{ij} = diag (1, -1, -1, -1).

Now, we may treat a general matrix ρ as a Dirac number which can be developed as a sum of two-point antisymmetric tensors. Neglecting for the moment the spacetime dependence, any $\rho \in C_4$ may be expressed as a sum of forms

$$\rho = S + V + B + T + P = \sum_A c_A e^A, \tag{2.5}$$

where $V = V_i e^i$, $B = \frac{1}{2}B_{ij}e^{ij}$, $T = \frac{1}{3!}T_{ijk}e^{ijk}$, $P = P_5 e^5$ with complex tensor coefficients and $e^{i_1 \ldots i_r} = e^{i_1}\wedge \ldots \wedge e^{i_r} = e^A$. Conditions must evidently be imposed on the tensors in order that ρ factorizes in the form (2.2). It has been shown(5), generalizing a result of Riesz(3), that for a non-nilpotent element Ψ of a simple Clifford algebra C_n whose Wedderburn division algebra is equal to the centre Z, Ψ factorizes into itself pre- or post-multiplied by a primitive idempotent iff

$$\Psi C \Psi = z\Psi \;\forall C \in C_n, \text{ some } z \in Z. \tag{2.6}$$

From this we may deduce that an element of the form (2.5) with $S \neq 0$ is primitive iff

$$\Psi^2 = 4S\Psi. \tag{2.7}$$

(Given(2.7), the requirement $S \neq 0$ is equivalent to $\Psi^2 \neq 0$; the case of primitive nilpotents is discussed elsewhere(6)).

There are various equivalent ways of expressing (2.7) and one of those deduced by assuming all the p-forms to be non-zero is summarized below (for the definition of inner and outer products and their relation see ref.6):

$$\left.\begin{aligned} V.V &= S^2 - P.P \\ T.T &= S^2 - P.P \\ V \wedge T &= 0 \\ SB &= V.T + B.P \end{aligned}\right\} \qquad (2.8)$$

Seven degrees of freedom in Ψ remain arbitrary and we must have $B \neq 0$. Eq. (2.7) can be expressed entirely in terms of the exterior product and the Hodge dual and is thus conformally invariant.

Let $\Psi \varepsilon C_4$ be represented by a complex matrix $\Psi^a{}_b$, a,b = 1,...,4. A primitive matrix has rank 1 and this is true iff(3)

$$\Psi^a{}_b = \xi^a \eta_b. \qquad (2.9)$$

When $S \neq 0$, (2.7) or (2.9) may be written as(7)

$$\Psi^2 = (Tr\Psi)\Psi. \qquad (2.10)$$

Keeping η fixed and varying ξ, Ψ generates a minimal left ideal. That ξ and η are four-component Dirac spinors in dual spaces follows from the transformation law of an algebraic basis element. Thus under a Lorentz rotation $e^i \to Re^iR^{-1}$ with $R \varepsilon$ spin (1,3)(8) we have $\Psi \to R\Psi R^{-1}$ and

$$\xi \to R\xi, \; \eta \to \eta R^{-1} \qquad (2.11)$$

(note that R itself is not a primitive element since such elements have no inverse in the algebra). An algebraic spinor then, in its general form, is not directly related linearly to the components of a column spinor but is a quadratic combination of unrelated dual column spinor components which may be prescribed by certain combinations of tensors. The usefulness of these general conditions may be seen from the following applications:

(i) All <u>primitive idempotents</u> in C_4 follow from the requirement $\Psi^2 = 4S\Psi = \Psi$, which is true iff $S = 1/4$ in addition to the primitivity conditions. In general such numbers are composed of more than four components and cannot be decomposed into products of commuting <u>2-component</u> subelements of the kind given elsewhere(9). However since all primitive idempotents in a simple algebra are equivalent we can make a similarity transformation to such a canonical form, a result proved also by Eddington(7). An example of a non-standard 6-component primitive idempotent is

$$E = \tfrac{1}{4}\left(1 + (1/\sqrt{2})(e^{01} + ie^{23} + ie^{12} + e^{03}) + ie^5\right)$$

which decomposes into a product of commuting 3-component subelements.

(ii) The linear <u>spin space</u> theory is recovered by fixing one of the factors in (2.9), say letting $\eta = (0100)$, which reduces Ψ to a column ξ in a matrix. This process is equivalent to choosing a specific ideal basis. For example, the primitive idempotent $E = \tfrac{1}{4}(1-e^{03}+ie^{12}-ie^5)$ generates an ideal with basis $(E, -e^{01}E, e^0E, -e^1E)$. Representing the

spacetime basis e^i with respect to this basis yields a Weyl-type representation and Ψ has the form given by the above choice of η.
(iii) As another example of a spin space basis, one generated by a nilpotent bivector ($\Psi = B$, $B.B = B\wedge B = 0$) is appropriate to a Witt decomposition of the spacetime basis, and has been used to generate a basis for twistor space(11).
(iv) We can generate tensor representations of spinors by imposing relations between ξ and η which entail corresponding covariant restrictions on the primitivity conditions.

(a) Algebraically, Hermitian conjugation becomes for $C\epsilon C_4$

$$C^\dagger = e^0\tilde{C}^*e^0,\ (C_1C_2)^\dagger = C_2^\dagger C_1^\dagger \tag{2.12}$$

for a given timelike 1-form e^0 where $\tilde{C}$ is the reversion and C^* means take the complex conjugate of the tensor coefficients(10). Impose on Ψ the condition that Ψe^0 is Hermitian. As a matrix relation this is satisfied by $\eta = \xi^\dagger e^0$. The corresponding restrictions on the forms of Ψ are

$$S,\ V,\ P \text{ real; } B,\ T \text{ pure imaginary.} \tag{2.13}$$

In this way we recover the tensor theory of Dirac spinors derived by Takabayasi(12) in a different way.

(b) Spin space duality is defined through an antisymmetric spin metric and this arises through the nilpotent restriction. Thus $\Psi = \xi\eta$ with $\eta\xi = 0$ is satisfied by $\eta_a = \xi_a$ where $\xi_a = \Sigma_{ab}\xi^b$, $\Sigma_{ab} = -\Sigma_{ba}$. If now we decompose ξ into Weyl spinors and impose on Ψ the above Hermitian restriction we deduce the 'flag-plus-pole' model of 2-component relativistic spinors(13): $\Psi = V + iB$ where V and B are real with $V.V = 0$, $B = V\wedge U$, $V.U = 0$.
(v) The above are uadratic representations. A linear representation follows from (2.10) which may be written

$$(\Psi - \mathrm{Tr}\Psi)\xi = 0.$$

This is satisfied by

$$\xi = \Psi\xi_0,\ \Psi^2 = 4S\Psi \tag{2.14}$$

where ξ_0 is constant. Tensors are mixed states of spinors (e.g. $V^i = \Sigma_{ab}e^i{}_{ab}\bar{\xi}^a\xi^b$). Eq. (2.14) answers the question: what combination of tensors yields a spinor? (and generalizes a result based on (2.13)(14)).
(vi) The factorization of Dirac spinors into a product of momentum and spin projectors(15) is a particular case of primitivity. Thus (2.7) implies that

$$\Psi = (S + V + P)\left[1 + (S^2-P^2)^{-1}(S - P)\,T\right].$$

Putting here $S = 1/4$, $V = (1/4)p$, $P = 0$ yields the idempotent

$$\Psi = (1/4)\ (1 + p)\ (1 + T),$$

the standard expression if p is identified as the momentum and T is the dual of the spin vector.

Finally, generalizing to include the point dependence of Ψ, we obtain (assuming the integrals converge)

$$\int \Psi^a{}_b(x_1,x_2)d^4x_2\Psi^b{}_c(x_2,x_3) = \int \Psi^b{}_b(x_2,\ x_2)d^4x_2\Psi^a{}_c(x_1,x_3) \qquad (2.15)$$

as the necessary and sufficient conditions on a set of interconnected non-local antisymmetric tensors to describe a characteristic matrix (2.2)

3. SUBSIDIARY CONDITION IN PHASE SPACE CLIFFORD ALGEBRA

A Clifford algebra for relativistic phase space was shown by Schönberg(16) to be determined by a Jordan-Wigner algebra. Suppose that a basis of 1-vectors I_i in the tangent space T_4 and a basis of 1-forms I^i in the co-tangent space T^*_4 generate two dual Grassmann algebras:

$$\left.\begin{aligned} &I^iI^j + I^jI^i = I_iI_j + I_jI_i = 0, \\ &I_iI^j + I^jI_i = \delta^j{}_i\, 1_{G_4},\ i,\ j = 0,1,2,3. \end{aligned}\right\} \qquad (3.1)$$

We denote by G_4 the resulting linear associative algebra with unity 1_{G_4}. G_4 over $\mathbb{C}$ is evidently a Jordan-Wigner algebra of creation and annihilation operators for spacetime. A general element of G_4 may be written in the form

$$\Gamma = \sum_{p,q} \frac{1}{p!}\frac{1}{q!} c^{j_1\ldots j_p}_{k_1\ldots k_q} I^{k_1\ldots k_q} \wedge I_{j_p\ldots j_1} \qquad (3.2)$$

where $I_{j_p\ldots j_1} = I_{j_p} \wedge \ldots \wedge I_{j_1}$, $I^{k_1\ldots k_q} = I^{k_1} \wedge \ldots \wedge I^{k_q}$, $c^{j_1\ldots j_p}_{k_1\ldots k_q}$ is a tensor antisymmetric in its upper and lower indices respectively, and $\Lambda = I_{0123}I^{0123}$ is a scalar element which plays the role of a vacuum state since

$$\Lambda^2 = \Lambda,\ I_i\Lambda = 0,\ \Lambda I^i = 0. \qquad (3.3)$$

Moreover Λ is a primitive idempotent since $\Lambda\Gamma\Lambda = c\Lambda\ \forall\Gamma\in G_4$, where c is the coefficient of Λ in (3.2). G_4 is a total 16x16 matrix algebra and is the Clifford algebra of relativistic phase space $S_8 = T_4\ T^*_4$ having (independent of any spacetime metric) a quadratic form $(1/2)U_iV^i$ where (U_i, V^i) are the coordinates of a point in S_8.

When the spacetime metric is available, we may perform a linear transformation to define generators of two Clifford algebras corresponding to opposite signs of the metric:

$$e_{\pm i} = I_i \pm g_{ij}I^j \qquad (3.4)$$

with

$$e_{\pm}^{i}e_{\pm}^{j} + e_{\pm}^{j}e_{\pm}^{i} = \pm 2g^{ij}1_{G_4}, \; e_{+}^{i}e_{-}^{j} + e_{-}^{j}e_{+}^{i} = 0. \tag{3.5}$$

Since the e^i's form a linearly independent basis for S_8 and a set of generators for G_4 we see from (3.5) that G_4 is the Clifford algebra of an 8-dimensional space with quadratic form $g^{ij}(x_i x_j - x_{4+i}x_{4+j})$.

A similar construction is known in supersymmetry(17).

Define the element $\omega = e_+^5 e_-^5$ where $e_\pm^5 = e_\pm^0 e_\pm^1 e_\pm^2 e_\pm^3$ and the generators $e_1^i = e_+^i$, $e_2^i = \omega e_-^i$. Then by (3.5)

$$e_1^i e_1^j + e_1^j e_1^i = e_2^i e_2^j + e_2^j e_2^i = 2g^{ij}1_{G_4}, \; e_1^i e_2^j = e_2^j e_1^i \tag{3.6}$$

and so $G_4 = C_4 \otimes C_4$ i.e. the direct product of two Clifford algebras corresponding to the same metric. Starting from (3.6) we may derive (3.5) by writing $e_+^i = e_1^i$, $e_-^i = e^9 e_2^i$ where $e^9 = e_1^5 e_2^5$. Clearly $e^9 = \omega$ and $\omega^2 = 1_{G_4}$.

The identification of G_4 as a simple Clifford algebra implies that we may apply much of the theory of the previous section. Thus an element $\Gamma \in G_4$ of the form (3.2) (with $\mathrm{Tr}\Gamma \neq 0$) is primitive iff

$$\Gamma^2 = (\mathrm{Tr}\Gamma)\Gamma \tag{3.7}$$

where $\mathrm{Tr}\Gamma = \sum_p^{0,\ldots,4} \frac{1}{p!} c_{i_1,\ldots i_p}^{\;i_1,\ldots i_p}$ since $\mathrm{Tr}\Lambda = 1$. Eq.(3.7) means that $\Gamma = \Xi H$ where Ξ, H are 16-component dual column spinors in phase space.

An important special case of (3.7) relates to the minimal ideals generated by Λ, whereby the spacetime covariant antisymmetric tensors become contravariant spinors of S_8. Thus any $\Psi \in G_4$ which lies in the minimal left ideal generated by Λ, i.e. for which $\Psi = \Psi\Lambda$, has the form

$$\Psi = \sum_{p=0}^{4} \frac{1}{p!} c_{i_1 \ldots i_p} I^{i_1 \ldots i_p}\Lambda . \tag{3.8}$$

A basis for the ideal is provided by the 16 elements $I^A\Lambda$, $A = i_1 \ldots i_p$, and the covariant tensor components c_A are spinor coefficients. Similarly the contravariant antisymmetric tensors are spinor components in the minimal right ideal generated by Λ. Taking I^i as creation operators, (3.8) represents a state vector of the fermion field. In a matrix representation, the spinors Ξ, H are respectively themselves the spacetime covariant or contravariant antisymmetric tensors, depending on whether we have a left or right ideal.

For $\Psi \in G_4\Lambda$ of the form (3.8) and $\Phi = \sum_{p=0}^{4} \frac{1}{p!} d^{i_1 \ldots i_p}\Lambda I i_1 \ldots i_p \in \Lambda G_4$, the product $\Psi\Phi$ gives the outer product of the tensors and

$$\Phi\Psi = \sum_{p=0}^{4} \frac{1}{p!} c_{i_1 \ldots i_p} d^{i_1 \ldots i_p}\Lambda = \mathrm{Tr}(\Psi\Phi)\Lambda$$

gives the scalar product.

A general $\Gamma \in G_4$ is a sum of 16 primitive elements. In accordance with the general scheme set out in §1 we have in (3.7) the condition on the 256 antisymmetric spacetime tensor components to generate a minimal ideal and so determine a state in a lower space. In this way bosons may be represented by spinors - we shall call them "Schönberg" spinors.

4. PHASE SPACE FORMULATION OF SPIN-1/2 WAVE EQUATIONS

We now have the algebraic tools to give superoperator phase space relations equivalent to the relativistic spin-1/2 wave equations appropriate to a lower space.

Writing $\Psi = (1c_4\ \rho)\Lambda$ with $\rho(x_1,x_2) = \sum_A c_A(x_1,x_2)e^A$ and noting that from (3.3) and (3.4)

$$e_+^{i_1\ldots i_p}\Lambda = (-1)^p\, e_-^{i_1\ldots i_p}\Lambda = I^{i_1\ldots i_p}\Lambda \quad ,$$

we may express the Dirac equations (2.3) in terms of the Jordan-Wigner operators (which is the representation of the Kähler-Dirac equation(18) in the higher space):

$$(I^i + I^{+i})\left[i\partial_{1i} - eA_i(x_1)\right] \Psi(x_1,x_2) = m\, \Psi(x_1,x_2) \qquad (4.1a)$$

$$(I^i - I^{+i})\left[i\partial_{2i} + eA_i(x_2)\right] \Psi(x_1,x_2) = m\,\omega\Psi(x_1,x_2) \qquad (4.1b)$$

with $I^{+i} = g^{ij}I_j$. Here the characteristic Dirac matrix is represented by an algebraic spinor $\Psi \epsilon G_4$. The restriction (2.15) must be imposed on this spinor in the higher space in order that it describes a primitive Dirac matrix in the lower space.

Introducing c.m. and relative coordinates $X^i = (1/2)(x_1^i + x_2^i)$, $z^i = x_2^i - x_1^i$ a relativistic-spinor Wigner transformation is defined, in the fermionic representation, by

$$F(X,P) = \left(\frac{1}{2\pi}\right)^4 \int_{-\infty}^{\infty} \Psi(X,z)e^{-iP_jz^j}\, d^4z$$
$$= \sum_A f_A(X,P)I^A\Lambda,\ f_A = \left(\frac{1}{2\pi}\right)^4 \int_{-\infty}^{\infty} c_A(X,z)e^{-iP_jz^j} d^4z \qquad (4.2)$$

which is a 16-component algebraic spinor field in phase space, f_A being a set of complex antisymmetric tensor fields. Integrating $I_AF(X,P)$ over the momentum variables and summing over spin indices we find

$$\mathrm{Tr}\int_{-\infty}^{\infty} I_A F(X,P)d^4P = c_A(X). \qquad (4.3)$$

When Ψ represents a primitive Hermitian Dirac matrix Ψ, characterized by (2.13), then (4.3) gives as the spinor coefficients the set of bilinear covariants of ordinary Dirac theory. In particular, the (real) vector coefficient c_i is the probability density current $\xi^\dagger e^0 e_i \xi$. $\mathrm{Tr}\left[I_i F(X,P)\right]$ thus represents a phase space distribution function for the Dirac field.

Instead of applying (4.2) to (4.1) we shall instead transform the four-component Feynman-Gell-Mann equation(19)

$$\left[(i\partial_i - eA_i)\,(i\partial^i - eA^i) - \frac{ie}{2}\, F_{ij}e^{ij}\right]\xi \;= m^2\xi \ , \qquad (4.4)$$

where $F_{ij} = \partial_i A_j - \partial_j A_i$. We then find the following phase space relations for the Feynman-Gell-Mann equation in terms of the Jordan-Wigner algebra (where we have Taylor expanded the potentials):

$$\Big[(p^i - eA^i(X))\frac{\partial}{\partial X^i} + e(p^i - eA^i(X))\frac{\partial A_i}{\partial X^j}\frac{\partial}{\partial P_j} + \frac{e}{2}(I^i I^{+j} + I^{+i} I^j)F_{ij} + $$

$$+ \frac{ie}{8}(I^i I^j + I^{+i} I^{+j})\frac{\partial F_{ij}}{\partial X^k}\frac{\partial}{\partial P_k}\Big] F(X,P) = 0 \qquad (4.5a)$$

$$\Big[\frac{1}{4}\Box_X - (P_i - eA_i(X))(P_i - eA^i(X)) + \frac{e}{2}\frac{\partial A^i}{\partial X^j}\frac{\partial}{\partial X^i}\frac{\partial}{\partial P_j} + $$

$$+ \frac{ie}{2}(I^i I^j + I^{+i} I^{+j})F_{ij}(X) + \frac{e}{8}(I^i I^{+j} + I^{+i} I^j)\frac{\partial F_{ij}}{\partial X^k}\frac{\partial}{\partial P_k} + m^2\Big] F(X,P) = 0 \,. \qquad (4.5b)$$

Eq.(4.5a) is a generalized Liouville equation for the spin-1/2 field and may alternatively be obtained by squaring the phase space Dirac equation. Eq.(4.5b) is a generalized "mass-shell" equation in phase space involving quantum potential contributions.

We now construct a conserved phase space current for the second order spin-1/2 field. From (4.4) we can deduce the existence of a conserved convection current:

$$j^i(x) = (1/2m)\xi^\dagger e^o(i\overleftrightarrow{\partial}^i - 2eA^i)\xi, \; \partial_i j^i = 0. \qquad (4.6)$$

To define a distribution function in phase space it is appropriate in a relativistic theory to introduce a vector-valued function. To this end consider the Lorentz vector

$$W_i(X,P) = \left(\frac{1}{2\pi}\right)^4 \int_{\infty}^{\infty} \rho_i(X,Z) e^{-iP_j z^j} d^4 z$$

where

$$\rho_i(X,Z) = -(1/2m)\xi(X - \tfrac{Z}{2})\left(\frac{i\overleftrightarrow{\partial}}{\partial X^i} + 2eA_i(X)\right)\xi + (X + \tfrac{Z}{\Sigma})\; e^o \,.$$

From this we may define the Wigner-Moyal transformation of a vector-valued Schönberg spinor:

$$F_i(X,P) = \left[1_{c_4} \otimes W_i(X,P)\right]\Lambda = \left(\frac{1}{2\pi}\right)^4 \int_{-\infty}^{\infty} \Psi_i(X,z) e^{-iP_j z^j} d^4 z$$

where

$$\Psi_i(X,z) = \left[1c_4 \otimes \rho_i(X,z)\right]\Lambda.$$

We then treat $TrF^i(X,P)$ as a phase space current for a Feynman-Gell-Mann particle, since it is a real vector and we have

$$Tr \int_{-\infty}^{\infty} F^i(X,P) d^4 P = j^i(X),$$

that is the current (4.6), a relation analagous to (4.3) for the Dirac field. It may be shown that this definition is reasonable in that TrF^i satisfies a conservation law in phase space, and the appearance of negative densities (TrF^o may become negative in certain regions) may be interpreted in terms of antiparticles(20).

5. CONCLUSION

This analysis illustrates in a specific but typical example how states
and operators, traditionally conceived as lying in distinct spaces,
may be treated as part of the same structure.

The information contained in a characteristic matrix solution to
spin-1/2 wave equations has been represented in two distinct ways:
firstly as a Dirac matrix operator in spacetime and then in terms of
fermionic creation operators as an algebraic spinor in phase space i.e.
as a factor in a superoperator. The latter interpretation leads to
algebraic wave equations for tensor functions in phase space and these
tensors must be inter-related (i.e. the Dirac matrix must be restricted
to a minimal ideal) so that the phase space theory represents the
initial spin-1/2 wave equations. We may illustrate the relations
between spaces developed in this paper as follows:

State   Primitive operator ε C_4

$\xi \longrightarrow \rho = \Psi, \ \Psi = \Psi \, E, \ \Psi^2 = 4S\Psi$

Supervector   Primitive superoperator ε G_4

$\rho \longrightarrow \Gamma = \mp, \ \mp = \mp \Lambda, \ \mp^2 = s \mp$

$\Big\downarrow$ Wigner–Moyal

$F(X,P)$

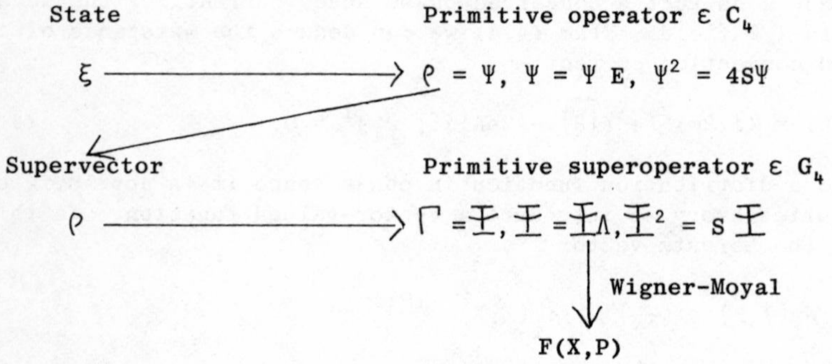

REFERENCES

1. I. Prigogine, C. George, F. Henin and L. Rosenfeld, Chem.Scripta 4, 5 (1973).
C. George, F. Henin, F. Mayné and I. Prigogine, Hadronic J. 1, 520 (1978).
C. George and I. Prigogine, Physica 99A, 369 (1979).
B. Misra, I. Prigogine and M. Courbage, Proc.Nat.Acad.Sci. USA 76, 4768 (1979).
I. Prigogine, From Being to Becoming (W.H. Freeman, San Francisco, 1980).
2. E. Wigner, Phys.Rev. 40, 749 (1932).
J.E. Moyal, Proc.Camb.Phil.Soc. 45, 99 (1949).
P.R. Holland, A. Kyprianidis, Z. Maric and J.P. Vigier, Phys.Rev.A (1986, in press).
3. M. Riesz, C.R. 10ème Cong.Math. Scand. (Copenhague, 1946) 123; C.R. 12ème Cong. Math. Scand. (Lund, 1953) 241.
4. D. Bohm and B.J. Hiley, in Old and New Questions in Physics, Cosmology, Philosophy and Theoretical Biology ed. A. van der Merwe (Plenum, London, 1983). p.67.
5. P.R. Holland, J.Phys.A: Math.Gen. 16, 2363 (1983).
6. P.R. Holland, Found.Phys. 16, (1986, in press).
7. A.S. Eddington, Fundamental Theory (University Press, Cambridge 1946).
8. I.R. Porteous, Topological Geometry (Van Nostrand, Lond. 1969).
9. S. Teitler, J.Math.Phys. 7. 1730 (1966).
P. Lounesto, Report - HTKK-MAT-A113 (1977); Found.Phys.11, 721 (1981).
10. D. Hestenes, Space-Time Algebra (Gordon & Breach, New York, 1966).
11. R. Ablamowicz, Z. Oziewicz and J. Rzewuski, J.Math.Phys.23, 231 (1982).
R. Ablamowicz and N. Salingaros, Lett.Math.Phys. 9, 149, (1985)
12. T. Takabayasi, Prog.Theor.Phys. Suppl. N°4 (1957).
13. R. Penrose and W. Rindler, Spinors and Space-Time, Vol 1. (University Press, Cambridge, 1984).
14. J.P. Crawford, J.Math.Phys. 26, 1439 (1985)
15. J.D. Bjorken and S.D. Drell, Relativistic Quantum Mechanics (McGraw-Hill, New York, 1964).
K.R. Greider, Found.Phys. 14, 467 (1984).
16. M. Schönberg, An.Acad.Bras.Ci. 28, 11 (1956); 29, 473 (1957); 30, 1, 117, 259, 429 (1958); Suppl.Nuovo Cimento (X) 6, 356 (1957); in Max-Planck-Festchrift 1958 (veb Deutscher Verlag der Wissenschaften, Berlin).
17. J. Wess and J. Bagger, Supersymmetry and Supergravity (Princeton, University Press, 1983).
18. E. Kähler, Rend.Mate.Roma, 21, 425 (1962)
P. Becher and H. Joos, Z.Phys. C15, 343 (1982).
19. R.P. Feynman and M. Gell-Mann, Phys.Rev. 109, 193 (1958).
Ph. Guéret, P.R. Holland, A. Kyprianidis and J.P. Vigier, Phys.Lett. 107A, 379 (1985).
20. P.R. Holland, A. Kyprianidis and J.P. Vigier, IHP preprint(1985)

SOME CONSEQUENCES OF THE CLIFFORD ALGEBRA APPROACH TO PHYSICS

Krystyna Bugajska
Institute for Theoretical Physics
Technical University Clausthal
D-3392 Clausthal-Zellerfeld
West Germany

ABSTRACT. We investigate the physical consequences of the realisation of the abstract Milnor-Lichnerowicz spinor bundle by a concrete bundle of algebraic spinors.

A. In our considerations we will use the Clifford algebra approach for the problem of spinor structure on a space-time manifold.

Let M be a Lorentzian 4 manifold and ξ_o the principal $\mathcal{L}_o$-bundle of oriented, orthonormal frames over M. According to the Milnor-Lichnerowicz definition the spinor structure on M is given by the $SL(2,\mathbb{C})$ - principal bundle ξ_s being a prolongation of the bundle ξ_s. It means that the following diagram

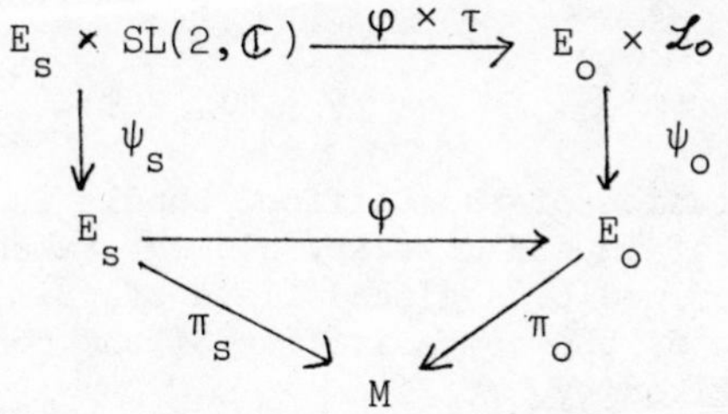

(A.1)

commutes. (Here E_s and E_o are total spaces of ξ_s and ξ_o respectively, τ is the covering map $\tau : SL(2,\mathbb{C}) \rightarrow \mathcal{L}_o$ and ψ_s, ψ_o denote the right actions of appropriate groups on the total spaces.) The spin representation of $SL(2,\mathbb{C})$ on $\mathbb{C}^4$ gives rise to the associated bundle $\xi_s[\mathbb{C}^4] =: \mathcal{S}$ whose sections can be considered as physical spinor fields.

However the bundle ξ_o allows us to construct the cotangential Clifford bundle $\mathcal{C}(T^*M)$ so we can try to introduce on M an algebraic spinor structure. It will be defined as follows: it is a bundle of minimal left ideals of the complexification $\mathcal{C}^{\mathbb{C}}(T^*M)$ of the Clifford bundle determined by a global field f of primitive idempotents. In a general case it is not true that the existence of a spinor-structure

J. S. R. Chisholm and A. K. Common (eds.), Clifford Algebras and Their Applications in Mathematical Physics, 285–292.

in the Milnor-Licherowicz sense implies the existence of an algebraic spinor structure and vice versa [1]. However for any Lorentzian space-time manifold these two approaches are completely equivalent.

Let f be some fixed global field of primitive idempotents on M. Let us denote the bundle of minimal left ideals determined by f as Ψ

$$\Psi := \mathcal{C}^{\mathbb{C}}(T^*M)f \quad . \tag{A.2}$$

Thus we can describe a physical spinor field either as a section $\tilde{\psi}$ of $\mathcal{S}$ or as a section ψ of Ψ. However although the bundles $\mathcal{S}$ and Ψ are isomorphic as vector bundles, the action of any vector field $X \in \Gamma(TM)$ on the sections of these bundles is different. Namely, the Levi-Civita connection ∇, on the one side induces (by the pullback) the unique connection ∇^s on $\mathcal{S}$:

$$\nabla^s_x : \Gamma(\mathcal{S}) \to \Gamma(\mathcal{S}) \qquad \forall\, X \in \Gamma(TM) , \tag{A.3}$$

and on the other passes to the Clifford bundle $\mathcal{C}^{\mathbb{C}}(T^*M)$ and determines the connection

$$\nabla_x : \Gamma(\mathcal{C}^{\mathbb{C}}(T^*M)) \to \Gamma(\mathcal{C}^{\mathbb{C}}(T^*M)) \qquad \forall\, X \in \Gamma(TM) \tag{A.4}$$

(it will be denoted by the same symbol ∇_x). Now we see that in a general case ∇_x transforms a section $\psi \in \Gamma(\Psi)$ into a section of $\mathcal{C}^{\mathbb{C}}(T^*M)$ which does not belong to $\Gamma(\Psi)$.

However we have the vector bundle isomorphism

$$\chi : \Psi \otimes \Psi \longrightarrow \mathcal{C}^{\mathbb{C}}(T^*M) \tag{A.5}$$

given by the formula

$$\chi(\psi \otimes \psi') = \psi\omega\beta(\psi') \qquad \forall\, \psi,\psi' \in \Gamma(\Psi) . \tag{A.5'}$$

Here β is the anti-involution of the Clifford bundle induced by the identity transformation of TM and ω is any element taken from the set of Clifford fields determined by a global field of primitive idempotents [2]. Now the requirement of the compatibility of the connection $\tilde{\nabla}$ on Ψ with the inner product given by

$$(\psi,\psi') = \omega\beta(\psi)\psi' \qquad \forall\, \psi,\psi' \in \Gamma(\Psi) \tag{A.5''}$$

determines a unique connection $\tilde{\nabla}$ on Ψ which produces ∇ on $\mathcal{C}^{\mathbb{C}}(T^*M)$. The easiest way to see this is to pass to a matrix realisation of the bundle $\mathcal{C}^{\mathbb{C}}(T^*M)$. In an appropriate basis the formulae (A.5') and (A.5'') have the same form as in the abstract, Penrose [2]. So we can use the known fact of the existence of the unique spinor-connection which is compatible with the spinor inner product and generates the standard four-vector connection [12].

Hence if we treat a field of algebraic spinors $\psi \in \Gamma(\Psi)$ as a section of the Clifford bundle (i.e. as given by some differential form), then the action of any vector field X on ψ is given by ∇_X and produces another section of the Clifford bundle which does not (in a general case) belong to $\Gamma(\Psi)$. But if we treat ψ as a Ψ-valued 0-form on M then the same vector field X acts by $\tilde{\nabla}_X$ and produces another field of algebraic spinors.

We have the following property of the isomorphism χ: the Clifford multiplication of Ψ by any element $s \in \mathrm{spin}_+ \subset \mathcal{C}^{\mathbb{C}}(T^*M)$ implies the following transformation of elements of the Clifford bundle

$$\chi(s\psi, s\psi') = s\chi(\psi,\psi')s^{-1} ,$$

i.e.

$$u \overset{s}{\rightsquigarrow} sus^{-1} \qquad \forall\, u \in \mathcal{C}^{\mathbb{C}}(T^*M) . \tag{A.6}$$

Since Ψ is contained in $\mathcal{C}^{\mathbb{C}}(T^*M)$, $\Psi \subset \chi(\Psi \otimes \Psi)$, we have to do with two kinds of transformations of algebraic spinors. Namely

$$\psi \overset{s}{\rightsquigarrow} s\psi \tag{A.7}$$

and

$$\psi \overset{s}{\rightsquigarrow} s\psi s^{-1} . \tag{A.8}$$

Now the parallel translation of ψ along any loop on M determined by ∇ corresponds to the transformation (A.8) whereas $\tilde{\nabla}$ is related with the transformation (A.7).

In the flat case when the holonomy group of g contains only the identity of both these actions (A.7) and (A.8) coincide and

$$\nabla_\mu = \tilde{\nabla}_\mu = \partial_\mu \qquad \left(\text{here} \quad x = \frac{\partial}{\partial x_\mu}\right). \tag{A.9}$$

If the holonomy group of the Levi-Civita connection contains elements $s \in \mathrm{spin}_+$ such that

$$sfs^{-1} = sf = f , \tag{A.10}$$

then we also have $\nabla_\mu = \tilde{\nabla}_\mu$. It was proved by Crumeyrolle [3] that the set of elements $s \in \mathrm{spin}_+$ which satisfy (A.10) is mapped by the covering map τ into the nilpotent subgroup $\mathcal{C}$ of $\mathcal{L}_0$ in its Iwasawa decomposition.

However in the general case we do not have such a situation and $\nabla_\mu \neq \tilde{\nabla}_\mu$. In this case we have to add some additional terms to $\tilde{\nabla}_\mu$ to compensate the action of ∇_μ i.e.

$$\nabla_\mu = \tilde{\nabla}_\mu + B_\mu . \tag{A.11}$$

Let us try to find the physical interpretation of these additional terms B_μ . For this let us notice that any global field of primitive idempotents has the following form:

$$f(x) = \frac{1}{4}(1 + \omega_1(x))(1 + \omega_2(x)) , \tag{A.12}$$

where $\omega_i^2 = 1$, $i = 1,2$; $\omega_1\omega_2 = \omega_2\omega_1$. They determine the rest of mutually anihilating primitive idempotents fields of the form

$$f^\alpha(x) = \frac{1}{4}(1 + \varepsilon_1^\alpha\omega_1(x))(1 + \varepsilon_2^\alpha\omega_2(x)); \qquad \varepsilon_i^\alpha = \pm 1 \quad . \tag{A.13}$$

They define the decomposition of the Clifford bundle into 4 mutually supplementing algebraic spinor bundles

$$\mathcal{C}^{\mathbb{C}}(T^*M) = \bigoplus_{\alpha/1}^{4} \Psi^\alpha \quad . \tag{A.14}$$

So we can introduce 4 different kinds of algebraic spinor fields $\psi^\alpha \in \Gamma(\Psi^\alpha)$, $\alpha = 1 . . 4$. Now any element $s \in spin_+$ operating according to (A.7) does not change the decomposition (A.14) but the action given by (A.8) mixes different spinor fields ψ^α between themselves. Hence for some gravitational fields for which their holonomy group is contained in $\mathcal{C}$ the decomposition (A.14) is irrelevant and cannot be observed. But in other cases we have to introduce some internal interaction B_μ which changes the just introduced internal coordinate α. Also the fields B_μ correspond to those generators of the holonomy group of g which do not belong to the Lie algebra of $\mathcal{C}$.

B. Let us consider the case when the holonomy group of our metric structure on M is given by $\mathcal{C}$. In this case there exists a global field of primitive idempotents f which is parallel translated with respect to both connections ∇_μ and $\tilde{\nabla}_\mu$(of course $f \in \Gamma(\Psi)$). Namely let us fix some orthonormal frame ε_x at $x \in M$. Let us take the holonomy bundle $\xi_{\mathcal{C}}$ through ε_x . Let $\varepsilon(x) = (e_1(x), e_2(x), e_3(x), e_4(x))$ be a global section of $\xi_{\mathcal{C}}$. Then the holonomy group can be generated by

$$A_1 = e_{14} - e_{31} , \qquad A_2 = e_{24} + e_{23} \tag{B.1}$$

and

$$f^1(x) = \frac{1}{4}(1 + e_1(x))(1 - e_{34}(x)) , \tag{B.2}$$

$$f^2(x) = \frac{1}{4}(1 + e_2(x))(1 - e_{34}(x)) \tag{B.2'}$$

are two parallel global fields of primitive idempotents which have the form $\frac{1}{4}(1 + e_I)(1 + e_J)$ in frames $\varepsilon(x)$ (I and J are multi-indexes) [1]. It is obvious that any global field f(x) corresponds to some nonvanishing section $\tilde{\psi}(x)$ of the standard 2-spinors

$$\xi_s[\mathbb{C}^2] =: \Sigma \ . \tag{B.3}$$

Now any element $\tilde{\psi}(x)$ of Σ determines a null vector, say $X \in T_x M$ and some space-like polarisation vector, say $K \in T_x M$. Vector K is tangent to a "celestial" sphere S^2 at the point fixed by X. The light vector X is determined by the spinor $\tilde{\psi}(x)$ up to a phase coefficient $e^{i\theta}$. When a phase of a spinor is altered by θ then the polarisation vector K turns through an angle 2θ [4]. Let the first field f^1 correspond to a field, say $\tilde{\psi}(x) \in \Gamma(\Sigma)$. One can see that it determines the light vector field $X(x) = (e_3(x) + e_4(x))/\sqrt{2}$ whose polarisation vector field can be given by $K^1(x) = e_1(x)$. The second field f^2 corresponds to the field $e^{i\pi/4}\tilde{\psi}(x)$ and determines the same light vector field $X(x)$ but with a polarisation $K^2(x) = e_2(x)$. In this way we have the following parallel distributions on M:

i) 1-dimensional spanned by $X(x)$

ii) 2-dimensional spanned by $\{X(x), K^1(x)\}$

iii) 2-dimensional spanned by $\{X(x), K^2(x)\}$

iv) 3-dimensional spanned by $\{X(x), K^1(x), K^2(x)\}$.

Since for torsionless connection we have

$$\nabla_X Y - \nabla_Y X = [X,Y] \qquad \forall\, X, Y \in \Gamma(TM)$$

these distributions are involutive. The codimension-one distribution (iv) will be called the Penrose distribution and denoted by P. So we obtain a codimension-one foliation F of our space-time manifold M. Moreover it is a transversally oriented and autoparallel foliation. But although any leaf of F is a totally geodesic submanifold of M it possesses a degenerate metric of signature $(1^0, 2^+, 0^-)$ (induced by a Lorentzian metric g on M). It implies, among others, that a parallel distribution complementary to P does not exist, and the notion of a totally geodesic foliation is not interchangeable with the notion of a bundle-like metric. Nevertheless we could try to introduce some Riemannian metric on M and investigate our codimension-one foliation F by means of it. It is known that for any concrete Lorentzian metric g on M any nonvanishing time-like vector field, say V, defines a Riemannian metric $\hat{g}$ on M given by the following formula:

$$\hat{g}(X,Y) = g(X,Y) - \frac{2g(X,V)g(Y,V)}{g(V,V)} \qquad \forall\, X,Y \in \Gamma(TM) .$$

Let us take a vector field $V(x) = 1/2\,(X(x) - Y(x)) = e_4(x)$ (here $Y(x)$ spans an isotropic line bundle complementary to P). We see immediately that the field of orthonormal frames $\varepsilon(x)$ introduced above is also orthonormal with respect to $\hat{g}$. So the principal $SO(4,R)$ bundle ξ_R of orthonormal frames related with $\hat{g}$ is formed by the set

$$\xi_R = \varepsilon(x)\ SO(4,R)$$

of linear frames. We can check [5] that the holonomy bundle of the Riemannian structure $(M,\hat{g})$ through ε_x is generated by $\hat{A}_1 = 1/2(\hat{e}_{14}-\hat{e}_{34})$, $\hat{A}_2 = 1/2(\hat{e}_{24} - \hat{e}_{23})$, $\hat{A}_3 = \hat{e}_{12}$ where $\hat{e}_i$ is the same element as e_i but with euclidean metric properties. It appears [5] that the integrable Penrose distribution P determines a totally geodesic foliation F with respect to both the g and $\hat{g}$ metric structures on M. It means that $\hat{g}$ is a bundle-like metric compatible with a foliation F. In a general case it is difficult to determine when a foliated manifold admits a bundle-like metric. However in a codimension-one case a necessary condition [6] is that the Pontrjagin ring of M vanishes in the top dimension. Because we have related the existence of a parallel spinor field with a Penrose codimension-one foliation we have a relationship between the existence of a nontrivial parallel spinor field on M and its topological invariant.

C. As we have said, we can always construct a global field of primitive idempotents f and realise the Milnor-Lichnerowicz spinor bundle $\mathcal{S}$ by an algebraic spinor bundle Ψ. Moreover when we choose some concrete field f(x) then automatically we fix some concrete codimension-one Penrose distribution P. If the holonomy group of our metric tensor is equal to $\mathcal{C}$ then the Penrose distribution is integrable (i.e. determines a foliation) and also the Kähler equation is equivalent to the Dirac equation. However in a general case the Penrose distribution is not involutive. The "breaking" of a Penrose foliation is manifested by the presence of an additional structure which is responsible for:

i) introducing a new degree of freedom α (A.14)

ii) introducing a new interaction B_μ (A.11) between spinor fields (whose origin is given by the kind of gravitational fields).

In this way we can look at the fields B_μ as some interaction which appears when we destroy a totally geodesic Penrose foliation of a space-time manifold M. So we can come to the following questions:

a) Is a Penrose distribution homotopic to a foliation?

b) Can a Penrose distribution P be homotopic to two nonhomotopic foliations?

c) How many concordance classes of codimension-one foliation can we construct i.e. how many "basic" states for B_μ interaction do exist?

The answer for (a) is given by the Phillips theorem [9]. It asserts that on any open manifold M every codimension-one plane field is homotopic to a smooth foliation. This is also true for a transversely orientable distribution on a compact manifold (which is the case for a Penrose distribution P).

The answer for (b) is negative. Namely using the Gromov-Phillips theorem and the fact that M is a spin manifold we find that P cannot be homotopic to nonhomotopic foliations.

To answer (c) we use the Haefliger classifying maps [10]. For an open manifold M the "basic" states for B_μ are given by the class of integrably homotopic foliation whereas for compact M (according to the Koschorke result) we can construct at most $2^{a(M)}-1$ "basic" states where $a(M) = \dim_{Z_2} H^1(M,Z_2)$. In both cases the "ground" states for interaction B_μ correspond to the concordant classes of the Lorentz structures on M.

D. The notion of inequivalent spinor structures (or exotic spinors) appears naturally in the Milnor-Lichnerowicz approach. In this section we introduce this notion for algebraic spinors.

Let f be some global field of primitive idempotents which determines the realisation of $\mathcal{S}$ by Ψ . Let f have a form

$$f(x) = \frac{1}{4}(1 + e_I(x))(1 + e_J(x)) \tag{E.1}$$

with respect to some global field $\varepsilon(x)$ of orthonormal frames on M. Let us take another field of primitive idempotents, say $f'(x)$, which has exactly the form (E.1) but in another field of frames, say $\varepsilon'(x)$. We have

$$\varepsilon'(x) = \varepsilon(x) \cdot \kappa(x) \tag{E.2}$$

for some function $\kappa\colon M \longrightarrow \mathcal{L}_o$. If κ induces a nontrivial homomorphism [11]

$$\kappa^*\colon \Pi_1(M) \longrightarrow \Pi_1(\mathcal{L}_o) \tag{E.3}$$

then $f'(x)$ determines an inequivalent algebraic spinor structure Ψ^χ which corresponds to the inequivalent Milnor-Lichnerowicz spinor structure $\mathcal{S}^\chi$. Here χ belongs to $H^1(m,Z_2)$ and is represented by our function κ. From the homotopical property of $\mathcal{L}_o$ it is enough to consider the functions

$$\kappa : M \longrightarrow SO(3,R) \subset \mathcal{L}_o .$$

Thus we see that we have to introduce an internal interaction B_μ for exotic algebraic spinor fields even when the holonomy bundle of Lorentzian structure is the Crumeyrolle principal bundle $\xi_{\mathcal{C}}$.

E. Although we have not confined ourselves to the Clifford algebra language, the results presented here have arisen from the Clifford algebra approach to the problem of a spinor structure. Maybe algebraic spinors have no physical meaning and all such investigations merely justify that the algebraic spinor fields are useless. But it may be also that physical fields describing fermions are given just by algebraic spinor fields; then our considerations should be useful to understand their nature.

REFERENCES

[1] K. Bugajska: Int.J.Th.Phys. 18 (2)(1979) 77-94
K. Bugajska:'Some problems of spinor and algebraic spinor structures' sub. for publ.
[2] P. Budinich, K. Bugajska: JMP 26 (1985) 588-592
K. Bugajska: JMP 27 (1986) 143-150
[3] A. Crumeyrolle: C.R.Acad.Sci. Paris 271 (1970) 1070
[4] R. Penrose, M. MacCallum: Phys. Rep. 6 (4) (1972) 241-316
[5] see [1](b)
[6] D. Jonson, A. Naveira: Geom. Dedicata 11 (1981) 347-352
[7] B. Lawson: Bull. Am. Math. Soc. 80 (3)(1974) 369-418
[8] K. Bugajska:'Another look at the gauge theories' sub.for pub.
[9] A. Phillips: Topology 6 (1967) 171-206 and [7]
[10] R. Bott:'Lectures on characteristic classes and foliations' Springer Lect. Notes, Springer N.Y. 1973 and [7]
[11] C. Isham: Proc. R. Soc. Lond. A364 (1978) 591-598
[12] C.P. Luehr, M. Rosenbaum: JMP 15 (1974) 1120-1137

ALGEBRAIC IDEAS IN FUNDAMENTAL PHYSICS FROM DIRAC-ALGEBRA TO SUPERSTRINGS

M. Dresden
Institute for Theoretical Physics
State University of New York at Stony Brook
Stony Brook, New York

ABSTRACT. It is suggested that algebraic methods are especially appropriate for the introduction of new physical concepts in physical theories. Examples are given to illustrate this procedure. It has been previously been used to introduce para statistics and supersymmetry. It is proposed here that a deep connection exists between Kac-Moody algebra's, Clifford algebras strings, and integrable systems.

OUTLINE

(1) Prologue - The relation between physics and mathematics in fundamental physics.
(2) Some aspects of the Dirac-Clifford algebra
(3) Comments on the Bose-Einstein, Fermi-Dirac dichotomy.
(4) Supersymmetry - A Clifford algebra of forms.
(5) From supersymmetry to superstrings
Moral - if any!

*This is a slightly expanded and somewhat altered version of a lecture delivered at the "Conference on Clifford algebra's", on Thursday, September at the University of Kent at Canterbury.

J. S. R. Chisholm and A. K. Common (eds.), Clifford Algebras and Their Applications in Mathematical Physics, 293–312.

SECTION 1 - PROLOGUE

The laws of classical physics are usually expressed in terms of differential equations. The entities occuring in those equations have themselves an immediate physical meaning, in principle they are measurable. The Hamiltonian equations of mechanics, or the Maxwell equations are typical examples. The mathematical language was that of analysis, algebraic notions, or algebraic structures occur only rarely in classical physics. (The theory of finite groups in crystallography is a notable exception). With quantum mechanics the mathematical character of physics changed radically. The quantum state is represented by a ray in a Hilbert space; the result of an individual measurement of a classical quantity $Q_{c\ell}$, is an eigenvalue of the operator Q_{op} corresponding to Q_{cl}. The physical super-position principle of states, is expressed mathematically by the linearity of the quantum space of states. There is an extraordinary "match" between the physical quantum concept and the mathematical notions, even to the extent that the mathematical notion "the spectrum of an operator", was introduced without ever referring to the physical notion of a spectrum. It is a little short of amazing that the observed physical spectrum of a Hydrogen atom, is indeed identical with the mathematical spectrum of the energy operator of that system.

Probably the most fundamental approach to physics (at least until the advent of the string theory), is that of quantum field theory, which is a straightforward adaptation of quantum notions to classical fields. The basic mathematical objects, space time dependent states $|\Psi\rangle$, are again rays in a Hilbert space, measured values of $\phi(x)$ are $\langle\Psi|\phi(x)|\Psi\rangle$, (the scalar product of ψ and $\phi(x)\ \Psi$). The space time dependence of the operators is determined by the equations of motion of the field operators. This is an analytic problem, similar to the classical problem of the solutions of the equation of motion. The operator character itself is determined by the prescribed (equal time) commutator of the field operators ϕ. This problem, which is partially algebraic, has no classical counterpart. The coupling between fields is again taken from, or transcribed from classical physics, the point nature of the classical particles, is expressed by the local nature of the interaction. If F(x) is a Fermi field, B(x), a Bose field, the coupling assumed is of the form $F(x)B(x)\overline{F}(x)$ ($\overline{F}$ is the adjoint of F).

This theoretical structure in spite of its well known divergence difficulties (renormalization theory), has been most effective in predicting and correlating phenomena in quantum electrodynamics and weak interactions. It's mathematical character is still predominantly analytic, although algebraic notions play an increasingly important role. Operator theory, functional analysis, invariance principles are the mathematical tools, which allow the formalization of the underlying physical ideas.

The particle features are usually described and not derived [but see the important exception in section (2)] by introducing multiple indexed wave functions or state vectors. It is at this juncture that algebraic notions especially group theory becomes important. Under a symmetry transfor-mation of the internal degrees of freedom, a state vector, in a representation space of that symmetry group, transforms according to an irreducible representation of the symmetry. Pictorially one studies vector or tensor analysis in the representation space of a symmetry group. The only unusual feature is that the vectors in that space possess quantum numbers. A suitable, possible, internal symmetry group is one whose dimensions of the irreducible representations coincide with the experimentally observed multiplicity of the states.

It is not all that simple to delineate the precise or the proper role of mathematics in physics.

Mathematics, codifies, organizes, describes physics. Certain physical ideas find a natural, sometimes almost miraculous description in terms of mathematical concepts. As such mathematics provides a formal framework, which allows, and facilitates the analysis of physics. However mathematics, all by itself does not solve physical problems. Sometimes the identification or association of physical notions with mathematicl concepts is helpful in understanding the logical structure of physics. But the essence in such a case is always the appropriate identification of the physical concepts with mathematical notions. It's meaning and uniqueness define the theory.

Sometimes an extension of the mathematical framework is suggestive and useful to obtain new physics, but the relationship between the new mathematical objects and the physics they describe must be analyzed with great care in each case. There exists a celebrated theorem by V. Neumann and Birkhoff,[1] to the effect that if one wishes to maintain the Hilbert space stucture of quantum mechanics a state vector $|\psi\rangle$ in the Hilbert space may be multiplied by a real number, a complex number or a quaternion. The last possibility leading to quanternion quantum mechanics has been frequently discussed as a possible formalism, incorporating physical, internal, degrees of freedom, such as isotopic spin. So far, however, no or at best very few new physical insights have resulted from this new identification. Presumably octonion quantum mechanics, which goes beyond the scope of the Hilbert space formulation of quantum mechanics, does lead to new physical predictions, but the physical meaning of the product of a state vector and an octonion, would be something new.

Thus octonion quantum mechanics, a mathematical extension, of the usual theory does contain new physics . Another attempt to introduce new physics, via an extension of the mathematical framework, was made quite a long time ago by Jordan and Wigner[2]. They proposed to describe the quantum theory of nuclear processes (as distinguished from atomic processes) by a non-associative algebra. (The usual quantum theory is of course associative, but not commutative). Although the attempt failed, its general pattern is clear, using "natural", or bold extensions of an existing mathematical formalism, one hopes to obtain a broader framework, which incorporates or in deed predicts new physics via a suitable physical interpretation of the new mathematical objects. As a final, quite current example of this procedure, it is pertinent to mention the compactification scheme of particle physics and string theory . It was first suggested by Kaluza and Klein, that the way to a genuine unification of electrodynamics and gravitation in 4 dimensions, was via a purely qravitational theory in five dimensions. In this case the extension from a four-dimensional to an n dimensional geometry is fairly obvious, but the reinterpretation, or the elimination of the extra-dimensions is far from obvious, it contains the real physics. But the basic idea: to use generalization of mathematical structures and give new physical interpretations of the new entities, is the same in all these cases.

It is quite striking that many of the resulting procedures and subsequent reinterpretations can be expressed in algebraic terms. Perhaps most striking is the recognition that quantum mechanics is a non-commutative extension of classical Hamiltonian mechanics. But there are many other instances also where the algebraic formulation appears the most direct and gives the deepest insight into the physical interrelations. This will be pursued in the sequel, but it is well to remember that the mere convenient mathematical formulation of a physical concept, is no guarantee that this particular identification is unique, or indeed the most appropriate.

2. Two aspects of the Dirac (Clifford algebra)

a. Charge conjugation

The simplest Clifford algebra is obtained from the generators e_1, $e_2, \ldots$, which satisfy

$$e_i e_j + e_j e_i = 2\delta_{ij} \quad , \; i,j = 1 \ldots \infty \tag{1}$$

An element a of the algebra is the sum

$$a = c_0 + \sum^{\infty} c_i e_i \tag{2}$$

The C_0, C_i are complex numbers.

The algebra first obtain by Dirac, has 4 generators, traditionally called γ_μ (μ=0,1,2,3). The behavior of particles of spin 1/2 is governed by the Dirac equation

$$(\partial_\mu \gamma^\mu + m)\psi = eA_\mu \bar{\psi}\gamma^\mu \tag{3}$$

ψ is the 4 component Direc spinor, A_μ is the external vector potential, $\overline{\psi}$ is the adjoint spinor.

Crucial for the interpretation of the theory, is the existence of an additional spinor ψ^c, which satisfies the equation.

$$\psi^c(\partial_\mu \gamma^\mu + m) = - eA_\mu \gamma^\mu \overline{\psi}^c \qquad (4)$$

It is this equation for ψ^c, with the characteristic (-) sign in the last term which allows the particle-antiparticle interpretation of the theory. The existence of ψ^c in turn depends on the mathematical existence of a maxtrix C, which is unitary, skew symmetric, Lorentz in variant. C must satisfy

$$C\gamma_\mu C^{-1} = \gamma_\mu{}^T \qquad (5a)$$

(γ_μ^T is γ_μ transposed)

In fact once C is known ψ^c is obtained as

$$\psi^c = C\psi \qquad (5b)$$

In the usual 4 dim. Dirac theory C is obtained by explicit construction. Since the existence of ψ^c, hence of C, is so important for the interpretation of the theory, it is important to inquire whether in n dimensions, the theory allows a similar interpretation. In n dimensions the Dirac Clifford algebra has n generators: $\gamma_0, \ldots \gamma_{n-1}$. The basic elements are collectively called γ_A

$$\begin{array}{l} 1 \\ \gamma_\mu \\ \gamma_\mu \gamma_\nu \\ \gamma_0 \cdots \gamma_{n-1} \end{array} \qquad (6)$$

A general element of this algebra , called χ is

$$\chi = \sum_A C_A \gamma_A \qquad (7)$$

C_A are again complex numbers.

This algebra is of order 2^n

possesses a unit

is not commutative

possesses divisors of zero

is associative

Using these properties one can show that indeed a C always exists, the interesting point is that C behavior quite differently in spaces of <u>even</u> dimensions (n=2v) and old dimension n=2v+1 .

The conditions C must satify are

$$C^T = (-1)^{\frac{\nu}{2}(\nu-1)} C \quad \text{even dimension } E^{2\nu} \tag{8a}$$

$$C^T = (-1)^{\frac{\nu}{2}(\nu+1)} C \quad \text{odd dimension } E^{2\nu+1} \tag{8b}$$

This purely algebraic result [3], shows that the physical relationship between particle and <u>anti</u> particle is very different in spaces of different dimension.

b. Dirac obtained the defining relations for γ matrices

$$\gamma_\mu\gamma_\nu + \gamma_\nu\gamma_\mu = 2\delta_{\mu\nu} \quad \mu,\nu=0,1,2,3 \tag{9}$$

by factoring a quadratic expression in χ:

$$Q = \sum_{i=0}^{3} \varepsilon_i \chi_i^2 \qquad \varepsilon_i = \pm 1 \tag{10}$$

into two linear factors in χ. Symbolically

$$Q = LL \tag{11}$$

The resulting Clifford algebra (9) has irreducible representations of dimension 4, yielding the well known 4 component Dirac spinors. One can also factor a quadratic form, in a product of a quadratic and linear form, symbolically

$$Q = QL \tag{12}$$

The coefficients β in the linear form, now satisfy an algebra (The Kemmin-Duffin algebra)

$$\beta_\mu\beta_\nu\beta_\rho + \beta_\rho\beta_\nu\beta_\mu = \delta_{\mu\nu}\beta_\rho + \delta_{\rho\nu}\beta_\mu \tag{13}$$

This is not a traditional Clifford algebra. The irreducible representation of (13) are of dimensions 10, 5, 1.

Correspondingly the equation

$$(\partial_\mu \beta^\mu + m)\ \psi = 0 \qquad (14)$$

describes particles of spin 1 (a 10 component wave function or spin 0 (a 5 component wave function).

Similarly a factorization Q = LLLL describes particles of spin 3/2 and spin 1/2 .

Unfortunately this factorization schemes looks more interesting than it is. It would seem that this would allow a simple unified treatment of supersymmetry by transformations of the type $\vec{\beta} \leftarrow \gamma$, but this does not appear to be the case.

3. The Bose-Fermi Dichotomy

Nowhere is the algebraic formulation of quantum ideas more important then in the characterization of states of systems of identical particles. In the analytical (Schrodinger) formulation, the wave function $\Psi\ (x_i \ldots x_n)$ is either symmetrical (for Bosons) or antisymmetrical (for Fermions). This same information is contained in the commutation relations for the operators.

$$[a_i, a_j^\dagger] = \delta_{ij} \qquad (15a)$$

$$[a_i, a_j] = [a_i^+, a_j^+] = 0 \qquad \text{for Bosons}$$

$$\{a_i, a_j^+\} = \delta_{ij} \qquad (15b)$$

$$\{a_i, a_j\} = \{a_i^+, a_j^+\} = 0 \qquad \text{for Fermions}$$

$\{\ \}$ is the anticommutator

The ground state $|0\rangle$ in either case is defined by

$$a_i|0\rangle=0 \qquad (16)$$

The physical states are obtained again in both cases by acting creation operators on $|0\rangle$. It is easy to check that the states so obtained are symmetric or antisymmetric according to the commutation relations (15a) or (15b) used for the operators.

In non relativistic quantum mechanics, the choice between symmetrical or antisymmetrical wave functions is made by experiment. In relativistic quantum field theory the consistency of the theory, imposes limitations in the choice. Particles with integer spin, must satisfy Bose- or Bose like statistics, the operators must satisfy, appropriate commutation relations, (see below) while particles of half integer spin satisfy Fermi, or Fermi like statistics and corresponding anti commutation rules. These statistics, and the corresponding particles are called para-statistics, and para-Fermions or para-Bosons. A parafermion may be qualitatively described in terms of the number of particles, p, which can be contained in a single, non-degenerate quantum state. For the usual Fermions this number is 1. One could describe the fact that 3 quarks can occupy the identical quantum level in a proton by individualizing the quarks by means of a new attribute: The color. This is the way it is most often done, quarks possess a color degree of freedom, capable of 3 values (red, white and blue). But it is also possible that quarks should be described as para fermions of degree 3, so that 3 quarks, could occupy the same state.

The commutations relations for parabosons and parafermions are instead of (5), trilinear quadrilinear etc. The algebraic structure of these relations becomes rapidly more complicated. For example for parabosons of rank (2), the commutation relations read:

$$[a_k,\{a_\ell^+,a_m\}] = 2\delta_{k\ell}a_m$$

$$[a_k,\{a_\ell^+,a_m^+\}] = 2\delta_{k\ell}a_m^+ + 2\delta_{km}a_\ell^+$$

$$[a_m, \{a_\ell, a_m\}] = 0 \tag{17}$$

These commutation relations, are similar, but not identical to the commutation rules for the β matrics (The Kemmer-Duffin matrics formula). The discussion of the object satisfying the commutation algebraic problem. If one assumes the existence of a vector $|0>$ (a state the M vacuum state, as far as physics is concerned) such that

$$a_k|0>=0$$

$$a_k a_\ell^+|0>=\delta_{k\ell} \tag{18}$$

The algebraic problem has a unique solution. It is possible but a little involved to find a matrix representation for both the para fermi and the parabose operators. The Clifford algebra's are not of the standard types, but more or less standards techniques provide a reasonable control of the mathematical problems. This however is not true for the physical interpretation of the formalism. Whether there is a compelling physical reason for the (so far) exclusive occurence of just Bosons and Fermions is really an unanswered question. If color is intrinsically, in principle unobservable, it would appear that quarks are indeed parafermions. In that case why would there not be parabosons? It should be stressed that all current thinking goes in the opposite directions, parafermions or bosons in spite of their natural mathematical appearance seem to have no place in scheme of the physical world.

4. Supersymmetry and a Clifford algebra of forms

It has always been somewhat of a miracle that Bosons and Fermions, "live such independent lives". The masses, the couplings, the life times of Fermions and Bosons appear totally unrelated. This is even stranger, since most if not all interactions are composites of a basic coupling, whose Hamiltonian is

$$H \sim F\bar{F}B. \quad (19)$$

The coupling may be visualized as a process where one fermion disappears, another appears while a boson is emitted or absorbed. In spite of this close connection, bosons and fermions by themselves are independent and unrelated. There is no obvious transformation or symmetry which relates them. It was first shown by Volkov and Akulov[4], that it is possible to consider transformations of fields which change Fermions into Bosons. Later Wess and Zumino[5] constructed a field theoretic model which exhibited a new invariance property: it was supersymmetric, the theory remained unchanged under an exchange of Fermion and Bosons. The supersymmetry transformations, which can physically transform a Fermion into a Boson, are mathematically described by an extension of the ordinary Lie algrebra of the generators of the symmetrics. Instead of a Lie algebra, the appropriate structure is a graded Lie algebra.

A graded Lie algebra, is characterized by a set of generators A_m (m=1...D), a set of graded generators Q_α (α=1...d) which satisfy the following system of commutation relations:

$$(20)$$

$$[A_m, A_n] = f_{mn}^{\ \ \ell} A_\ell$$

$$[A_m, Q_\alpha] = S_{m\alpha}^{\ \ \ \beta} Q_\beta$$

$$\{Q_\alpha, Q_\beta\} = g^m{}_{\alpha\beta} A_m$$

The structure coefficients f, s, g, possess and contain symmetry properties, which follow from the Jacobi identities of the generators.

Physically most important is that the physical fields (or physical field operators) in a supersymmetric theory depend not only on the space time variables x^μ, but also on a set of Grassmann variables θ_i, which satisfy $\{\theta_i, \theta_j\} = 0$.
The field $\phi = \phi(x, \theta)$ is the basic ingredient of a supersymmetric theory. Mathematically the supersymmetric theories have many attractive properties.

1. There are <u>no</u>, no-go theorems which prohibited the non trivial amalgamation of space time and internal symmetries.

2. They have better renormalizability properties.

3. They allow the natural combination of interactions.

Unfortunately the <u>physical</u> properties of supersymmetric theories, although striking, are not so successful. Supersymmetry predicts that all the masses of the members of a multiplet are the same, further the couplings should be the same as well. There is no experimental evidence of either. Even so it is widely believed that strict supersymmetry is a, in fact the, proper starting point for basic physical theories. Supersymmetry can be expressed very naturally in algebraic terms. Consider the one forms σ^μ on a flat Minkovski space. The metric used here is $g^{\mu\mu} = (-1,-1,-1,1) = (x,y,z,t)$. Construct the set of 16 forms $f_{[\alpha]}$ defined by the Wedge products

$$f_{[\alpha]} = 1, \sigma^\mu, \sigma^\mu \wedge \sigma^\nu, \sigma^\mu \wedge \sigma^\nu \wedge \sigma^\rho, \sigma^1 \wedge \sigma^2 \sigma^3 \wedge \sigma^4 = \omega \tag{22}$$

It is easy to check that the multiplication of the forms is associative, but they do not form a group. (In general there is no inverse). An inner (scalar) product between forms σ^m is always defined by

$$(\sigma^\mu, \sigma^\nu) = g^{\mu\nu} \tag{23}$$

Define a "new" round or $\vee$ multiplication by[6 7]

$$\sigma^\mu \frown \sigma^\nu = \sigma^\mu \wedge \sigma^\nu + g^{\mu\nu} \tag{24}$$

$$\sigma^\mu \frown \sigma^\nu \frown \sigma^\rho = \sigma^\mu \wedge \sigma^\nu \wedge \sigma^\rho + g^{\mu\nu}\sigma^\rho + \sigma^\mu g^{\nu\rho} - \sigma^\nu g^{\mu\sigma}$$

$$\sigma^\mu \frown \sigma^{\mu 2} \ldots \frown \sigma^{\mu\nu} = \sum_{k=0}^{n} \sum_{P} (-1)^P g^{\mu_1 \mu_2} g^{\mu_3 \mu_4} \ldots g^{\mu_{2k-1}, \mu_{2k}} \sigma^{\mu_{2k+1}} \wedge \ldots \wedge \sigma^{\mu n}$$

Using these formulae, the set $f_{\lfloor\alpha\rfloor}$ can be written in terms of the one forms

and the round multiplication. The results are

1. The multiplication is associative.
2. The 16 forms $f_{[\alpha]}$ form a multiplicative group g(1,3) with a as multiplication.
3. The ring R(1,3) = $\sum C_{[\alpha]} f_{[\alpha]}$ (with respect to addition and multiplication, $C_{[\alpha]}$ is complex) contain sums of differential forms of different rank.
4. Using the multiplication it is possible to construct an algebra
(1,3) of tensor fields. A typical element $\hat{\alpha}$ has the form

$$\hat{\alpha} = F_0 + \sum_\mu F_1 \sigma^\mu + \sum F_2{}^{\mu\nu}\sigma^\mu \wedge \sigma^\nu + \sum F_3{}^{\mu\nu\rho}\sigma^\mu \wedge \sigma^\nu \wedge \sigma^\rho + \omega F_4 \qquad (25)$$
$$+WF_4$$

In (25) the F_0 F_1 F_2 F_3 depend on x,y,z,t, they transform as a contravariant scalar, a vector, an antisymmetric tensor, a pseudoscalar. The expansion (25) of an element of the algebra (1,3) is identical with the expansion of the superfield Φ (in terms of the Grassman variables θ:

$$\Phi - \Phi_0 + A^\mu\theta^\mu + F^{\mu\nu}\theta^\mu\theta^\nu + F\ {}^{\mu\nu\rho}\theta^\mu\theta^\nu\theta^\rho \qquad (26)$$
$$+G\,\theta^1\theta^2\theta^3\theta^4$$

The advantage of the formulation (25), is that the mathematical character of $\hat{\alpha}$ is completely transparent, (1,3) is the Clifford algebra of spacetime forms, it is an associative algebra. It is neither a normed, nor a division algebra. Although several manipulations can be carried out more conveniently within this formalism (it allows an especially simple description of duals) no strikingly new results have been obtained. It would seem, that extensions to a Riemann space, in n dimensions, might provide a stringent test for the utility and power of this formalism. It was at one time believed that this formalism could exploit the formal similarities the factorization method in describing particles of spin 0, 1, 1/2... in a uniform manner but so far this suggestive idea has not led anywhere.

5. From supersymmetry to superstrings

a. Motivation and background

The theory of superstrings is, at this moment, certainly the most exciting, the most fashionable part of theoretical physics. It is interesting that the same interaction of physical and mathematical ideas, the same reinterpretation of mathematical entities as relevant physical object recurs in this development. In addition the remarkable organizational simplicity of algebraic methods also reappears.

The basic idea is that on the scale where quantum gravity becomes important ($\ell \simeq 10^{-33}$cm), the fundamental objects are not points but instead 1 dimensional strings. This by itself is not so remarkable, the idea of finite fundamental objects is neither unusual nor new. However in this new version, the physical objects are one-dimensional strings in a 10 or possibly 26 dimensional space, with either 6 or 22 dimensions curled up, so that they escape direct detections. Such an extended object possesses classical vibrational, rotational and possibly twisting motions. In the present philosophy it is assumed, or in any case believed, that the "usual" quantization of such modes leads to a set of states, whose energies or masses, correspond to the observed particles. In addition the joining or splitting of strings corresponds to the, in fact all possible interactions.

b. The old string

It was noted in 1971-1975 that the mass spectrum and the scattering of hadrons, pions, protons, and many others could be described in terms of a simple S matrix. There were many striking regularities: for example, particles, of the same quantum numbers, but different masses, had masses, so that they all would lie on a straight line in the spin-(mass) plane. It wasstressed by Nambu, that the family of the particles and their interactions could be described as the excitation spectrum of a one-dimensional string. Dynamically a string is described by variables x , which are functions of time and a string parameter σ. A very suggestive lagrangian to describe the dynamics of such strings would be

$$L = \int \delta A \tag{27}$$

Here δA is the element of area so that the equations of motion would be given by $\delta L=0$. This is in obvious analogy to the equations of point mechanics, which are obtained from the principle of minimum length. δA in (27) is the element of area on the world sheet swept out by the string in its motion through space time.

From the lagrangian, the theory can be constructed in the usual manner

Lagrangian

↓

Equation of Motion

↓

Quantization ← Hamiltonian → Conserved Quantities

↓

S matrix

Indeed the S matrix so obtained for a Bosonic string was very similar to the one first suggested by Veneziano for hadronic scattering.

This idea is of course very appealing, a hadron is made up from two quarks, held together by a string - the dynamical modes of the string represent, are, the states of the particle. One could even derive the relation (for bosons) that the excitation, satisfy

$$J = \alpha M^2 \tag{28}$$

The angular momentum J is proportional to the square of the mass.

c. This theory was suggestive but had lots of troubles (and a few good points).

1. It predicted massless states, there were (and are) no massless hadrons.
2. There were states of negative norm.
3. It appeared difficult to quantize the theory.

The bosonic strings could be quantized straight forwardly only in a space of 26 dimensions (25 space + 1 time dimension), while the fermionic string could only be quantized in 10 dimensions. This circumstance convinced many physicists to abandon the string theory approach.

4. These theories possess a very large invariance group. The generators of this symmetry satisfy an algebra, the Virasoro algebra. It plays a fundamental role in string theory. It also makes it possible to connect the string theory with two dimensional conformal theory. It is of some interest to record just where the Virasoro algebra originates in the elementary string theory.

In principle this is quite straightforward, it follows directly from the usual quantization procedure. Expand the string coordinate $x^\mu(\sigma, \;)$ in a Fourier series

$$x^\mu(\sigma,\tau) = x^\mu_0 + 2\alpha p^\mu \tau + \sum_n \sqrt{\frac{2\alpha}{n}} \cos n\sigma \left[a_n^\mu e^{-in\tau} + (a_n^\mu)^+ e^{in\tau}\right] \quad (29)$$

x^μ_0 is a constant, α a string parameter, p the momentum of the center of mass, the a^μ_n are the Fourier coefficients. The Hamiltonian is found to be

$$H = \alpha p^2 + \sum_n a^+_{n\mu}\, a_n{}^\mu \equiv L_0 \quad (30)$$

The quantization is now clear, the a's should satisfy the Bose commutation rules:

$$\left[a_m{}^\mu, (a_n{}^\nu)^+\right] = g^{\mu\nu}\delta_{mn} \quad (31)$$

The difficulty comes from the fact that $g^{\mu\nu}$, the metric of the Minkovski space, is necessarily indefinite. This is the source of the negative norm states. However by imposing constraints on the allowed states $|\phi\rangle$ of the string, it is possible to eliminate the states of negative norm.

$$L_0|\phi\rangle = \alpha|\phi\rangle \tag{32}$$

$$L_n|\phi\rangle = 0$$

$$\langle\phi|L^+_{-n}| = 0$$

Here the L_n are the constrained operators

$$L_n = -i\sqrt{2n\alpha}\, p_u a^\mu_n + \sum_{\mu>0} \sqrt{m(m+n)}\,(a^\mu_n)^+ a_{(m+n)\mu} \tag{33}$$

$$-\frac{1}{2}\sum_{m=1}^{n-1}\sqrt{(n-m)m}\, a^\mu_m a_{(n-m)\mu}$$

d. Much more important than the explicit form of the constraint operators, is the algebra these operators satisfy. It is not difficult to show that this algebra, the Virasoro algebra is given by

$$[L_m, L_n] = (m-n)L_{m+n} + \frac{1}{12}\, d(m^3-m)\,\delta_{m,-n} \tag{34}$$

d is the dimensionality of the space time.

It is remarkable that this infinite dimensional symmetry, occurs in many parts of physics.

1. It is a symmetry possessed by the string theory, because of the reparametrization invariance of the coordinates $(\overline{\sigma\phi})$ in the string coordinates $x^\mu(\sigma,\tau)$.
2. The self dual solution of the 4 dimensional Euclidan Yang Mills equations possess this same symmetry.
3. The dual models exhibit this symmetry.
4. The critical behavior of two dimensional lattice systems, again shows this same symmetry.

e. The advent of the modern super string theory started from a change in attitude. The string excitations were not considered as hadronic states, instead they were to describe quarks, gluons, leptons, $W^\pm$, gravitons. Then instead of considering the massless states of the string theory as a disadvantage, they became a success, incorporating, the massless Yang Mills quanta and the gravitons within one single scheme. The states of negative norm are eliminated via the Virasoro constraints. Further by generalizing the Virasoro algebra, to a super algebra, or a graded algebra (following the pattern of super symmetry) one can eliminate all further problems of the string theory (including the tachyons) by requiring invariance under the Neveu-Schwartz algebra

$$[L_m, L_n] = (n-m)L_{n+m} + \frac{1}{8} d(n^3-n)\delta_{n,-m} \qquad (35)$$

$$\{Q_\alpha, Q_\beta\} = 2L_{\alpha+\beta} + \frac{1}{2} d(\alpha^2-\frac{1}{4})\delta_{\alpha-\beta}$$

$$[L_n, Q_\alpha] = \frac{1}{2} (m-\alpha)Q_{\alpha+m}$$

m and n are integers, α, β, half integers. The detailed discussion of superstrings has led to algebraic problems. The equations (35) are a graded version of a general class of algebras, the Kac-Moody algebras. A general affine Kac Moody algebra relative to a simple Lie group G (with structure coefficients F^{ijk}) and generators $T^i{}_r$, is defined by

$$[T^i_m, T^j_n] = iF^{ijk}T^k_{m+n} + dm\delta^{ij}, \delta_{m,-n} \qquad (36)$$

This is for example the symmetry algebra of the self dual solution of the Yang-Mills field equation. $(T^i_n = (T^i_{-n})^+)$

The importance of (35) and (36) for string theory, shows once again that algebraic questions play a central role in many parts of physics. It is not clear, whether the Kac-Moody algebra have a particularly close relation to Clifford algebra's, although there might well exist a more comprehensive scheme that includes both. As a final comment it is important to stress that the algebraic structures emphasized in this lecture, still have a tantalizing connection to apparently totally unrelated mathematical structures. For example the generators of the Virasoro algerbra L_n (34), are strangely connected to the Korteweg equation[8] Define

$$U(\sigma) = \frac{6}{d} \sum_{n=-\alpha}^{+\infty} L_n e^{in\sigma} - \frac{1}{4} \tag{37}$$

Then the U operators satisfy

$$[U(\sigma),U(\sigma')] = \frac{6\pi}{d} [-\delta''(\sigma-\sigma')+4U(\sigma)\delta'(\sigma-\sigma')+2U'\delta(\sigma-\sigma')] \tag{38}$$

From (37) one may deduce that the commutator of $U(\sigma)$, with the field Hamiltonian H_f

$$H_f = \frac{1}{2} \int d\sigma U^2 \tag{39}$$

leads to the equation

$$\frac{d}{16\pi} \lfloor U(\sigma),H_f \rfloor = \dot{U} = -U +6UU' \tag{40}$$

(40) is the Korteweg-de Vries equations.

The precise meaning of this interrelation is quite unclear. Perhaps it indicates a deep relation between Yang Mills Fields, Kac Moody algebra strings, and integrable systems (de K de V equation is an example). W truth this conjecture might have there is no doubt that the continual interplay of mathematics and physics leads to striking and unexpected insights. It is a pleasure to thank Dr. A. Common and Professor J.S.R Chisholm for a beautifully organized, smoothly run conference. The co interaction between physicists and mathematicians was informative, instructive, congenial and extremely pleasant.

REFERENCES

1. J. V. Neumann and G.D. Birkhoff, Annals of Math 37. 023, 1936.
2. P. Jordan and E. Wigner and J.V. Neumann; Ann. of Math. 35, 29, 1934.
3. A. Pais, Journ. of Math. Phys. 6, 581, 1965.
4. Volkov and Akulov, Phys. Letters B46, 1973, 109.
5. J. Wess and B. Zumino Nucl. Physics B70, (1974) 39.
6. N. Salingaros, M. Dresden, Phys. Rev. Letters 43, 1, 1979. Actually the new "V" multiplication defined there wasn't so new. It only appeared that way because the authors were unaware of the next important reference. It is late, but at least one of the authors wishes to apologize for this oversight.
7. E. Kahler Rendiconti Matematica di Palermo 21, 425, 1962.
8. This was first observed by Gervais. (preprint)

ON TWO SUPERSYMMETRIC APPROACHES TO QUANTUM GRAVITY: CLIFFORD ALGEBRA DEGENERACY v EXTENDED OBJECTS

J.G. Taylor
Department of Mathematics
King's College London
Strand, London WC2R 2LS
England

ABSTRACT. We review the central charge and the superstring approach to quantum gravity. A complete classical theory of the former is described for maximal super Yang-Mills and supergravity, and problems in its quantisation. The construction of higher loop superstring amplitudes is sketched, as are their divergence properties.

1. Introduction

One of the main goals of theoretical physics over the last few decades has been to obtain a sensible theory of quantum gravity. The need to obtain such a theory has been made stronger by the recent discovery of the W and Z particles at CERN, so justifying unification of the electromagnetic and weak interactions. Grand unification is also a distinct possibility, though it has yet to receive any experimental justification. Yet gravity is still outside these unification schemes. The difficulty is that of the quantisation of gravity, as is well known. Recent work on this problem has led to two different avenues as possibly leading to the desired goal. Neither of these may ultimately succeed, but there is enough progress on each to feel it useful to report here on them. Moreover it is relevant to describe this work in a symposium on Clifford algebras. For both the approaches I wish to describe deal with the most sophisticated use of supersymmetry, which itself involves Clifford algebra concepts. One approach will lead inevitably to the investigation and use of degenerate irreps of extended supersymmetry algebras with central charges - I will explain these words shortly - and the other to considering field theories of extended objects (strings) embedded in a higher dimensional space-time. In fact the first approach also involves us in higher dimensions, as we will soon see, so the difference between the two approaches might be regarded as of degeneracy versus extension.

Both approaches use supersymmetry. That is a symmetry between bosons and fermions introduced in the early 70's[1]. Besides being an intriguing symmetry, relating as it does particles with integer spin to those with half odd integral spin, it also allows for reduction of the

J. S. R. Chisholm and A. K. Common (eds.), Clifford Algebras and Their Applications in Mathematical Physics, 313–320.

virulence of the old and well-known infinities of quantum field theory. This can be seen in the simplest models of the vacuum energy H_0 of a set of particles of mass m_s and spin $s(s=0,\frac{1}{2},---)$, with

$$H_o = \sum_s (-1)^{2s} \int d^3 p \, (p^2 + m_s^2)^{\frac{1}{2}} \ . \tag{1.1}$$

The expression on the r.h.s. of (1.1) is just the sum of the kinetic energies of the individual species, suitably weighted so as to give positive energy to the bosons (with s = 0,1,---) and negative energy to the fermions (with $s = \frac{1}{2},\frac{3}{2},---$). Each integral in (1.1) is divergent, so we need to introduce a cut-off, $|p| < \Lambda$, to make the integrals well-defined. The term depending on Λ^4, the most divergent term, has coefficient

$$(N_B - N_F) \tag{1.2}$$

where N_B, N_F are the total numbers of bosons and fermions respectively. Equality of N_B and N_F clearly annihilates the most divergent contribution in H_o.

Supersymmetry extends this relationship by imposing symmetry under the interchange

$$\text{boson} \leftrightarrow \text{fermion} \tag{1.3}$$

This is achieved mathematically by means of a spinor (Grassmann-valued) operator, denoted S_α, which changes a bosonic to a fermionic state and vice-versa:

$$S_\alpha|\text{boson}> \sim |\text{fermion}> ; | S_\alpha |\text{fermion}> \sim |\text{boson}> \ . \tag{1.4}$$

The labels α run from 1 to 4 (for four-dimensional space-time), α denoting the labels of a spinor representation of SO(1,3).

The appropriate algebra $\mathcal{S}_1$ to which the S_α's belong may be deduced by various considerations [1], and is an extension of the Poincare algebra of the inhomogeneous Lorentz algebra. The crucial relation is

$$[S_\alpha, S_\beta]_+ = -(\not{P}C^{-1})_{\alpha\beta} \tag{1.5}$$

where $\not{P} = P_\mu \gamma^\mu, \gamma^\mu$ are the four Dirac matrices, P_μ the translation generator, and C the charge conjugation matrix. We may extend $\mathcal{S}_1$ by adding internal symmetry labels to S_α, so obtaining $S_{\alpha i}$; the only modification to the r.h.s. of (1.5) is the extra factor δ_{ij}. Here $1 \leq i, j \leq N$, and the corresponding algebra is denoted by $\mathcal{S}_N$.

2. The N = 3 Barrier

Vigorous attempts have been made since the mid-70's to construct an extension of gravity which obeys extended supersymmetry[1]. These have been successful as far as correctly incorporating the physical component fields, but have not over determining the so-called auxiliary fields. These are needed to ensure supersymmetry when the equations of motion are not imposed. For if we consider N=1 supersymmetric Yang-Mills theory, there are the physical fields of the gauge vector A_μ and the associated 'gaugino' λ_α. For massless fields these each have 2 degrees of freedom; without requiring the equations of motion A_μ has 3 but λ_α 4 degrees of freedom. A further scalar field D is needed to satisfy (1.2) in general.

The problem of auxiliary fields becomes more severe as N increases. Irreps of $\mathcal{S}_N$ on states with a given p_μ are those of the associated Clifford algebra (1.5). There are 4N fermionic generators $S_{\alpha i}$ in $\mathcal{S}_N$, so the dimension of the unique irrep is $2^{2N} r_N$ (where r_N is an additional reality factor, with $r_N = \frac{1}{2}$ for N even, $r_N = 1$ for N odd). Thus the number of fermionic degrees of freedom in the irrep is $2^{N-1} r_N$. However for N-extended supersymmetric Yang-Mills (N-SYM) there are only 2N fermionic degrees of freedom for the physical fields (the gauginos $\lambda_{\alpha i}$). In order to build N-SYM one therefore only requires the factor $(N/2^{N-2} r_N)$ of an irrep of $\mathcal{S}_N$. This factor is 1 for N=1 and 2 but $\frac{1}{8}$ for N = 4. We conclude that there do not exist auxiliary fields for N-SYM beyond N = 3. The above argument can be made more precise, and also extended to N-supergravity, with the same result: there exists a barrier at N = 3 in the construction of N-SYM or N-supergravity[2].

3. Central Charges

Various methods have been attempted to avoid the N=3 barrier. This is particularly important to achieve since we wish to build versions of N-Supergravity with the largest value of N. This is N=8, since beyond that value there will be physical fields with spin greater than 2 for which it is very difficult (if not impossible) to construct consistent interacting theories (with N=4 being the maximal N for N-SYM). Moreover we wish to use what are called superfield techniques, in which the various components are collected together into a function $\Phi(x,\theta)$ of the space-time variable x and a Grassmann-valued spinor variable $\theta_{\alpha i}$. Expansion of $\Phi(x,\theta)$ in θ has only a limited number of powers, the coefficient functions being the various component fields for a set of irreps of $\mathcal{S}_N$. This allows super-actions and quantum supergraphs to be constructed, using Berezhin integration on the θ-variables with $\int d\theta = 0$, $\int \theta d\theta = 1$.

Those supergraph approaches have allowed proof of the finiteness of N=4 SYM to all orders in perturbation theory [3] (expressed in N=2 superfields, with $1 \le i \le 2$ in $\theta_{\alpha i}$). If we were able to build a maximal superfield version of N=8 supergravity we may be able to discover if it also has similar finiteness, due to fermi-bose cancellations of infinities. However we expect to need to use the full N=8 superfield version to make such cancellations manifest, since only then will we have the maximal symmetry made explicit.

In order to broach the N=3 barrier by superfield techniques it appears necessary to add an extra term to the r.h.s. of (1.5), so that now

$$[S_{\underline{\alpha}}, S_{\underline{\beta}}]_+ = -(\not{P}C^{-1})_{\underline{\alpha}\,\underline{\beta}} + C_{\alpha\beta}Z^{ij} \qquad (3.1)$$

where $\underline{\alpha}$ denotes the pair (α,i) and $Z^{ij} = -Z^{ji}$ and commutes with all the generators $S_\alpha, P_\mu, J_{\mu\nu}$ of $\mathcal{S}_N$ (where $J_{\mu\nu}$ is the generator of SO(1,3)).

The extended algebra $\mathcal{S}_N^{(z)} = \mathcal{S}_N \cup \{Z^{ij}\}$ is the central charge algebra, and the Z^{ij} are called central charges. As for $\mathcal{S}_N$, we may obtain representations on superfields for $\mathcal{S}_N^{(z)}$, now of the form $\Phi(x^\mu, z_{ij}, \theta_{\underline{\alpha}})$, $Z^{ij} = \partial/\partial z_{ij}$.

The irreps of $\mathcal{S}_N^{(z)}$ can be obtained by using the Casimirs of $\mathcal{S}_N$, suitably modified[4] to make account of the extra term on the r.h.s. of (3.1). The most crucial feature of this modification is in the superspin Casimir, the extension of the Pauli-Lubanski vector of the inhomogeneous Lorentz group. The details of this are given in ref. [4] but the crucial point is the term containing (in obvious matrix rotation)

$$(\not{P} - Z)S/(P^2 - Z^*Z) \qquad (3.2)$$

where $Z^*Z = Z^{*ij} Z^{ij}$. We may therefore classify the irreps into

$$\begin{aligned} &\text{(a) non-degenerate: } P^2 \neq Z^*\underline{Z} \\ &\text{(b) degenerate} \quad : \not{P}\cdot S = ZS \\ &\text{(c) indecomposable: } P^2 = Z^*Z,\ \not{P}S \neq ZS \quad . \end{aligned} \qquad (3.3)$$

The above classification agrees with that for irreps of classical superalgebras, except what we have called degenerate, (b), is termed atypical there.

Neither non-degenerate nor indecomposable irreps can help us broach the N=3 barrier, whilst degenerate can do so since they effectively reduce the number of spinor generators S_α by a factor of two by means of the Dirac equation (3.3). We must thus learn how to handle the degeneracy constraint (3.3), and especially how to incorporate it into a covariant construction of N=8 supergravity.

4. Boundary Control Theory of Central Charges

In order to construct field actions as full superspace integrals we must also include integration over the z_{ij} variables if we wish to have a positive dimensional measure. Thus for N=4 we need six real central charges z_{ij} ($1 \leq i, j \leq 4$) and can construct the measure $d^4 x\, d^6 z\, d^{16}\theta$, of dimension two (in length) appropriate to the construction of N=4 SYM. Similarly we need fourteen real central charges for N=8 with the same dimension measure $d^4 x\, d^{14} z\, d^{32}\theta$ appropriate for N=8 supergravity.

How are we to interpret integration over z's? We have developed[5] actions which involve superfields constrained by (3.3). The constraint dictates the dependence of the fields on the z-variables. Thus the action only depends on the values of the fields and their z-derivatives at a suitable boundary in z-space - hence the title of this section. Thus for N=2 supersymmetry the doublet superfield Φ_i has action

$$\int d^4 x \int_{\Gamma} d^2 z \int d^8 \theta\, \Phi_i^+ \Phi_i \quad , \tag{4.1}$$

with constraints

$$D_{\alpha(i}\Phi_{j)} = 0 \ , \tag{4.2}$$

where $D_{\underline{\alpha}}$ are covariant super-derivatives which anticommute with $S_{\underline{\alpha}}$. The region Γ of integration in (4.1) is over a cone in (z,x)-space, whose vertex is an $\underline{R}^4$. We may show[5] that the equations of motion for (4.1) and (4.2) are the correct physical ones for fields on the $\underline{R}^4$ at the vertex of Γ.

This approach has been extended to N=4 SYM and N=8 supergravity. In each case covariant constraints have been written down in superspace which imply a covariant form of (3.3), for both the above theories. Superfield actions have also been constructed which give the correct equations of motion at the classical level when the constraints are used[6].

5. Quantisation

Having constructed an apparently satisfactory superfield theory of N=4 SYM and N=8 supergravity, with full symmetry, we must now ask for their quantum properties. That is much more difficult, because (3.3), (4.2) or their covariant form, play a role not usual in dynamical constraints systems. It is possible to obtain a satisfactory quantum theory at the component level for N=4 SYM[7], but this does not possess the full symmetry nor does it appear finite. Quantisation of actions like (4.1), but with interaction, have proved very difficult. It may be there are insuperable difficulties in the program, so whilst further work needs to be done it is also valuable to explore another avenue to quantum gravity which has recently become very topical.

6. Superstring Theory

A resurgence of interest in string theory, very popular in the 70's, has occurred with the discovery of anomaly-free and finite theories of superstrings at one loop. The superstring itself is an extended object described by the embedding vector $X^{\mu}(\sigma,\tau)$ of the world sheet variables σ and τ of the string, σ being the spatial extent and τ the time. μ runs over d values in a d-dimensional space-time. There is an associated spinor (Grassmann) variable $S^{\alpha}_{A}(\sigma,\tau)$, where A = 1,2 are the world sheet spinor indices whilst α runs over the space-time spinor variables. There are both open and closed strings; we will concentrate on the closed case as being mathematically simpler. Thus we take $0 \leq \sigma \leq 2\pi$ to parametrise the string.

We would expect that an amplitude describing the initial fusion, interaction and subsequent splitting of various strings would be described by a suitable sum over possible parameters necessary to specify a world sheet. From the conformal equivalence of the theory we only expect to sum over the class of conformally inequivalent surface with a given number of handles corresponding to the loops. Thus we expect the amplitude to be of form[9]

$$\int_{\Sigma} d(\text{interaction times})\ d(\text{widths of internal loops})\,\mathcal{D}X\,\mathcal{D}S\ \times$$
$$\times \exp(\text{'area' of world sheet } \textstyle\sum) \times \delta(X - X_{\text{initial}})\ \delta(X - X_{\text{final}})$$
$$\times\ \delta(S - S_{\text{initial}})\ \delta(S - S_{\text{final}}) \tag{6.1}$$

where the area is quadratic in X and S, and the δ-functions fix X or S at some very early initial time and very late final time. The quadratic nature of the area allows (6.1) to be evaluated by Gaussian integration to give

$$\int d\ (\text{interaction times})\ d(\text{widths of internal loops}) \times$$
$$\times\ [\det{}_{\Sigma}\Delta]^{-M} \times \exp[\text{external sources} \times \text{Greens function of } \Sigma] \tag{6.2}$$

where M is a constant to be discussed shortly, and Δ is the Laplacian on Σ. Since $\det_{\Sigma}\Delta$ is infinite unless it be regularised[10], we see that (6.2) possesses a divergence unless M=10. This is the case for closed superstrings and very likely true for the heterotic string in one particular version (though Δ is replaced by $\partial/\partial z$ on Σ in (6.2) for the heterotic string).

7. Divergence of Superstring Theory

In order to analyse the divergences of (6.2) it is necessary to obtain it in a more explicit form. This has been done using conformal mapping techniques and a specially soluble case[11], and leads to an elegant expression for the amplitude in terms of the natural parameters (moduli) describing Σ originally introduced by Ahlfors, and the first and third Abelian differentials on Σ. In terms of these we find the amplitude (6.2) reduces, to within a simple kinematic factor, to

$$\int d(\text{moduli}) \int d(\text{external sources } z_i)\, f(\text{moduli}, z_i) \times$$

$$\times \exp [- p_r p_s G(z_r, z_s)] \quad , \qquad (6.3)$$

where p_r are the momenta of the external string states and f is a known function. We may now investigate where putative divergences may arise by determining the singularities of the integrand. These are only at possibly topology-change values for the moduli.

A careful analysis[12] indicates that topology change by degeneration of Σ into surfaces of lower genus, with no total loss of genus, can give factors of form

$$\int_{\underline{C}} d^2\rho \cdot |\rho|^{-r} \quad ,$$

where r = 2 for the closed superstring and r = 1 for the heterotic string. Our result, coupled with that at the end of the previous section, indicates that there may only be one version of the heterotic string which can be finite. Since open superstrings have much less chance to be finite one expects that there is the possibility of a unique finite quantum gravity ahead of us.

7. Acknowledgement

I would like to thank my numerous students over the last few years for their hard work on the central charge programme: G. Bufton, C. Card, P. Davis, D. Gorse, J. Hassoun, B. Rands, V. Rivelles and especially my research Colleague A. Restuccia, both in that programme and in our more recent work on superstrings.

References

[1] For reviews of supersymmetry see J. Wess and J. Bagger Supersymmetry and Supergravity, Princeton Univ. Press 1983; S.J. Gates, M. Grisaru, M. Rocek and W. Siegel Superspace Benjamin Cummings 1983; Supersymmetry and Supergravity '82, World Scientific Pub. Co., 1983, ed S. Ferrara P. van Nieuwenhuizen and J.G. Taylor; J.G. Taylor 'A Review of

Supersymmetry and Supergravity', Prog. in Particle and Nucl. Physics. 12, 1(1984).

[2] V.O. Rivelles and J.G. Taylor, Phys. Lett. 121B, 37 (1983), and references therein.

[3] P.S. Howe, K. Stelle and P. Townsend, Nucl. Phys. B. 214, 519 (1983); ibid B236, 125 (1984).

[4] B. Rands and J.G. Taylor, J. Phys. A16 1005, 3921 (1983).

[5] A. Restuccia and J.G. Taylor, J. Phys. A16, 4097 (1983).

[6] C.T. Card. P.R. Davis, A. Restuccia and J.G. Taylor, J. Class. and Quant. Grav. 2, 351 (1985).

[7] C.T. Card, P.R. Davis, A. Restuccia and J.G. Taylor, Phys. Lett 151B, 235 (1985).

[8] M.B. Green and J. Schwarz, Phys. Lett. 149B, 444 (1982); ibid 149B, 117 (1984); D.J. Gross, J. Harvey, E. Martinex and R. Rohm, Phys. Rev. Lett. 54, 54, (1985).

[9] S. Mandelstam, Phys. Rep. 13C, 259 (1974)

[10] A. Restuccia and J.G. Taylor, 'Another Reason Why Strings Must be Supersymmetric', Phys. Lett. 162B, 109 (1985).

[11] A. Restuccia and J.G. Taylor, 'On the Construction of Higher Loop Closed Superstring Amplitudes', KCL preprint, Aug. 1985.

[12] A. Restuccia and J.G. Taylor, 'Divergence Analysis of Closed Superstrings', KCL preprint, October 1985.

CLIFFORD ALGEBRA AND THE INTERPRETATION OF QUANTUM MECHANICS

David Hestenes
Physics Department
Arizona State University
Tempe, Arizona 85287
USA

ABSTRACT. The Dirac theory has a hidden geometric structure. This talk traces the conceptual steps taken to uncover that structure and points out significant implications for the interpretation of quantum mechanics. The unit imaginary in the Dirac equation is shown to represent the generator of rotations in a spacelike plane related to the spin. This implies a geometric interpretation for the generator of electromagnetic gauge transformations as well as for the entire electroweak gauge group of the Weinberg-Salam model. The geometric structure also helps to reveal closer connections to classical theory than hitherto suspected, including exact classical solutions of the Dirac equation.

1. INTRODUCTION

The interpretation of quantum mechanics has been vigorously and inconclusively debated since the inception of the theory. My purpose today is to call your attention to some crucial features of quantum mechanics which have been overlooked in the debate. I claim that the Pauli and Dirac algebras have a geometric interpretation which has been implicit in quantum mechanics all along. My aim will be to make that geometric interpretation explicit and show that it has nontrivial implications for the physical interpretation of quantum mechanics.

Before getting started, I would like to apologize for what may appear to be excessive self-reference in this talk. I have been pursuing the theme of this talk for 25 years, but the road has been a lonely one where I have not met anyone travelling very far in the same direction. So I will not be able to give much in the way of reference to the work of others, except on occasion when I found my road crossing theirs. I have reached a vantage point from which I can see where the road has been taking me pretty clearly. I will describe what I see so you can decide if you would like to join me on the trip.

Since the pursuit of my theme has been a personal Odyssey, I will supply a quasi-historical account of my travels to give you some sense of where the ideas came from and how they developed. I began a serious

J. S. R. Chisholm and A. K. Common (eds.), Clifford Algebras and Their Applications in Mathematical Physics, 321–346.

study of physics and mathematics only after a bachelors degree in philosophy and other meanderings in the humanities. Although that handicapped me in technical skills, which are best developed at an earlier age, it gave me a philosophical perspective which is unusual among American students of science. I had found in my studies of modern epistemology that the crucial arguments invariably hinged on some authoritative statement by the likes of Einstein, Bohr, Schroedinger and Heisenberg. So I concluded (along with Bertrand Russell) that these are the real philosophers, and I must scale the Olympus of physics to see what the world is really like for myself. I brought along from philosophy an acute sensitivity to the role of language in understanding, and this has been a decisive influence on the course of my studies and research.

Under the influence of Bertrand Russell, I initially believed that mathematics and theoretical physics should be grounded in Symbolic Logic. But, as I delved more deeply into physics, I soon saw that this is impractical, if not totally misguided. So I began to search more widely for a coherent view on the foundations of physics and mathematics. While I was a physics graduate student at UCLA, my father was chairman of the mathematics department there. This gave me easy access to the mathematics professors, students and courses. Consequently, I spent as much time on graduate studies in mathematics as in physics. I still regard myself as much mathematician as physicist.

I mention these personal details because I believe that influences from philosophy, physics and mathematics converged to produce a result that would have been impossible without any one of them. To be more specific about training that bears on my theme, from physics I became very skilful at tensor analysis because my mentor, Robert Finkelstein, was working on unified field theories, and I became familiar with the Pauli and Dirac algebras from courses in advanced quantum theory. In mathematics, I had one of the first courses in "modern" differential geometry from Barret O'Neill, and I studied exterior algebra and differential forms when the only good books on the subject were in French. Now this is what that background prepared me for:

Immediately after passing my graduate comprehensive examinations I was awarded a research assistantship with no strings attached, whereupon, I disappeared from the physics department for nearly a year. My father got me an isolated office on the fourth floor of the mathematics building where I concentrated intensively on my search for a coherent mathematical foundation for theoretical physics. One day, after about three months of this, I sauntered into a nearby math-engineering library and noticed on the "New Books shelf" a set of lecture notes entitled <u>Clifford Numbers and Spinors</u> by the mathematician Marcel Riesz [1]. After reading only a few pages, I was suddenly struck by the realization that the Dirac matrices could be regarded as vectors, and this gives the Dirac algebra a geometric meaning that has nothing to do with spin. The idea was strengthened as I eagerly devoured the rest of Reisz's lecture notes, but I saw that much would be required to implement it consistently throughout physics. That's what got me started.

About two months later, I discovered a geometrical meaning of the Pauli algebra which had been completely overlooked by physicists and mathematicians. I went excitedly to my father and gave him a lecture on what I had learned. The following is essentially what I told him, with a couple of minor additions which I have learned about since.

Physicists tacitly assign a geometric meaning to the Pauli matrices $\underset{\sim}{\sigma}_k$ by putting them in one-to-one correspondence with orthogonal directions in Euclidean 3-space. The $\underset{\sim}{\sigma}_k$ can be interpreted as unit vectors representing these directions, because their products have a geometric meaning. Thus, the orthogonality of $\underset{\sim}{\sigma}_1$ and $\underset{\sim}{\sigma}_2$ is expressed by the anticommutative product $\underset{\sim}{\sigma}_1\underset{\sim}{\sigma}_2 = -\underset{\sim}{\sigma}_2\underset{\sim}{\sigma}_1$, which can also be regarded as Grassmann's outer product $\underset{\sim}{\sigma}_1\wedge\underset{\sim}{\sigma}_2$, so the result can be interpreted geometrically as a directed area (a bivector). This implies a geometric meaning for the formula

$$\underset{\sim}{\sigma}_1\underset{\sim}{\sigma}_2\underset{\sim}{\sigma}_3 = i, \qquad (1.1)$$

which appears only as formal result in the textbooks on quantum mechanics. This formula tells us that i should be interpreted as the unit pseudoscalar for Euclidean 3-space, for it expresses i as a trivector formed from the outer product $\underset{\sim}{\sigma}_1\wedge\underset{\sim}{\sigma}_2\wedge\underset{\sim}{\sigma}_3 = \underset{\sim}{\sigma}_1\underset{\sim}{\sigma}_2\underset{\sim}{\sigma}_3$ of orthogonal vectors.

Equation (1.1) suggests that the unit imaginary $(-1)^{\frac{1}{2}}$ in quantum can be interpreted geometrically as the unit pseudoscalar i for physical 3-space, though, strictly speaking, i is related to $(-1)^{\frac{1}{2}}I$, where I is the identity matrix for the Pauli algebra. This idea turned out to be wrong, as we shall see. But the suggestion itself provided a major impetus to my research for several years. It demanded an analysis of the way the Pauli and Dirac algebras are used in physics.

Physicists generally regard the $\underset{\sim}{\sigma}_k$ as three components of a single vector, instead of an orthonormal frame of three vectors as I have suggested they should. Consequently, they write

$$\underset{\sim}{\sigma}\cdot\underset{\sim}{a} = \sum_k \underset{\sim}{\sigma}_k a_k \qquad (1.2)$$

for the inner product of a vector σ with a vector $\underset{\sim}{a}$ having ordinary scalar components a_k. To facilitate manipulations they employ the identity

$$\underset{\sim}{\sigma}\cdot\underset{\sim}{a}\underset{\sim}{\sigma}\cdot\underset{\sim}{b} = \underset{\sim}{a}\cdot\underset{\sim}{b} + i\underset{\sim}{\sigma}\cdot(\underset{\sim}{a}\times\underset{\sim}{b}) \qquad (1.3)$$

This is a good examples of a redundancy in the language of physics which complicates manipulations and obscures meanings unnecessarily.

Formula (1.3) is a relation between two distinct mathematical languages, the vector algebra of Gibbs and the Pauli matrix algebra. This relation expresses the fact that the two languages have overlapping "geometric content," and it enables one to translate from one language to the other. However, by interpreting the σ_k as vectors generating a geometric algebra we can eliminate all redundancy incorporating both languages into a single coherent language. Instead of (1.2) we write

$$a = \sum_k a_k \sigma_k \tag{1.4}$$

expressing the expansion of a vector in terms of an orthonormal basis. Then (1.3) takes the form

$$ab = a\cdot b + i(a\times b), \tag{1.5}$$

where
$$a\times b = -i(a\wedge b) \tag{1.6}$$

defines the vector cross product as the dual of the outer product. By the way, this shows that the conventional distinction between polar and axial vectors in physics is really the distinction between vectors and bivectors in disguise, for (1.6) exhibits the vector $a\times b$ as a representation of the bivector $a\wedge b$ by its dual.

Sometime later I realized that the conventional interpretation of (1.2) as an inner product of a with a spin vector σ is completely consistent with my interpretation of the σ_k as a frame of vectors, for the physicists always sandwich the σ between a pair of spinors before calculations are completed. Thus, if ϕ is a 2-component Pauli spinor, one gets a spin density

$$\phi^\dagger \sigma_k \phi = \frac{1}{2}\,\mathrm{Tr}(\phi\phi^\dagger \sigma_k) = \rho s\cdot\sigma_k. \tag{1.7}$$

The notation on the right indicates that the matrix trace is equivalent to projecting out the components of a vector ρs which is inherent in the matrix $\phi\phi^\dagger$ by dotting it with the basis vectors σ_k. Thus, (1.2) gives us

$$\phi^\dagger \sigma\cdot a\phi = \sum_k \phi^\dagger \sigma_k \phi a_k = \rho s\cdot a, \tag{1.8}$$

which is an ordinary inner product on the right.

These observations about the Pauli algebra reveal that is has a universal significance that physicists have overlooked. It is not just a "spinor algebra" as it is often called. It is a matrix representation for the geometric algebra $\mathcal{R}_3$, which, as was noted in my first lecture, is no more and no less than a system of directed numbers representing the geometrical properties of Euclidean 3-space. The fact that vectors in R_3 can be represented as hermitian matrices in the Pauli algebra has nothing whatever to do with their geometric interpretation. It is a consequence of the fact that multiplication in R_3 is associative and every associative algebra has a matrix represention. This suggests that we should henceforth regard the $\underset{\sim}{\sigma}_k$ only as vectors in R_3 and dispense with their matrix representations altogether, because they introduce extraneous artifacts like imaginary scalars.

I wondered aloud to my father how all this had escaped notice by Herman Weyl and John von Neumann, not to mention Pauli, Dirac and other great physicists who has scrutinized the Pauli algebra so carefully. When I finished my little talk, my father gave me a compliment which I remember word for word to this day, because he never gave such compliments lightly. He has always been generous with his encouragement and support, but I have never heard him extend genuine praise for any mathematics which did not measure up to his own high standards. He said to me, "you understand the difference between a mathematical concept and its representation by symbols. Many mathematicians never learn that."

My initial insights into the geometric meaning of the Pauli algebra left me with several difficult problems to solve. The first problem was to learn how to represent spinors in terms of geometric algebra without using matrices. Unfortunately, Marcel Riesz never published the chapter in his lectures which was supposed to be about spinors. From his other publications I learned that spinors can be regarded as elements of minimal left ideals in a Clifford algebra, and ideals are generated by primitive idempotents. But I had to find out for myself how to implement these ideas in quantum mechanics. So I spent much of the next three years intensively studying spinors and ideals in the Pauli algebra, the Dirac algebra and Clifford algebras in general. I was unaware that Professor Kähler was engaged in a similar study at about the same time. I could have profited from his publications [2,3], but I did not learn about them till more than a decade later.

The mathematical problem of constructing spinors and ideals for the Dirac and Pauli algebras is fairly simple. The real problem is to find a construction with a suitable geometrical and physical interpretation. Let me describe for you the solution which I developed in my doctoral dissertation [4]. As you know, the Dirac algebra is mathematically the algebra $C(4)$ of complex 4×4 matricies and this is isomorphic to the complex Clifford algebra C_4. A physico-geometrical interpretation is imposed on C_4 by choosing an orthonormal frame of vectors γ_μ (for $\mu = 0, 1, 2, 3$) to represent directions in spacetime. Of course, the γ_μ correspond exactly to the Dirac matrices in $C(4)$. In addition, I provided the unit imaginary in $\mathcal{C}_4$ with a physical interpretation by identifying it with the unit pseudoscalar of $\mathcal{R}_3$ as specified by (1.1). This entails a factorization of $\mathcal{C}_4$ into

$$\mathcal{C}_4 = \mathcal{R}_3 \otimes \mathcal{R}_2 \tag{1.9}$$

Of course, this factorization is to be done so that the σ_k in $\mathcal{R}_3$ correspond to the same physical directions as the γ_k (for k = 1, 2, 3) in $\mathcal{C}_4$.

The factorization (1.9) implicitly identifies $\mathcal{C}_4$ with a real geometric algebra $\mathcal{R}_{p,q}$. As Lounesto [5] and others have noted, every complex Clifford algebra $\mathcal{C}_{2n}$ can be identified with a real algebra $\mathcal{R}_{p,q}$ where $p + q = 2n + 1$ and $n(2n + 1) + q$ are odd integers. These conditions on p and q imply that a unit pseudoscalar of $\mathcal{R}_{p,q}$ commutes with all elements of the algebra and has negative square. Therefore, it has the algebraic properties of the unit imaginary scalar in $\mathcal{C}_{2n}$. In the case of physical interest $p + q = 4 + 1 = 5$, and we can choose an orthonormal basis e_0, e_1, e_2, e_3, e_4 for $\mathcal{R}^{p,q}$ so the unit pseudoscalar can be written $i = e_0e_1e_2e_3e_4$. From this we can see explicitly that

$$i^2 = e_0^2e_1^2e_2^2e_3^2e_4^2 = -1, \tag{1.10}$$

provided q = 1, 3 or 5. Thus, $\mathcal{C}_4$ is isomorphic to $\mathcal{R}_{4,1}$, $\mathcal{R}_{2,3}$ and $\mathcal{R}_{0,5}$. Among these alternatives the best choice is determined by geometrical considerations.

The simplest relation to the spacetime algebra $\mathcal{R}_{1,3}$ is determined by the projective conditions

$$\gamma_\mu = e_\mu e_4. \tag{1.11}$$

Imposing the spacetime metric

$$\gamma_0^2 = -\,e_0^2e_4^2 = 1,\ \gamma_k^2 = -\,e_k^2e_4^2 = -1,$$

we find that $e_4^2 = e_k^2 = -\,e_0^2 = 1$. Thus, we should identify $\mathcal{C}_4$ with $\mathcal{R}_{4,1}$. You will recognize (1.11) as defining a projective map, whose general importance was pointed out in my first lecture. It identifies $\mathcal{R}_{1,3}$ with the even subalgebra of $\mathcal{R}_{4,1}$, as expressed by

$$\mathcal{R}_{4,1}^{+} = \mathcal{R}_{1,3}. \tag{1.12}$$

To show that this is consistent with the factorization (1.9), we write

$$\underset{\sim}{\sigma}_k = \gamma_k e_0 = e_k e_{40}. \tag{1.13}$$

Then we find that the pseudoscalars for $\mathcal{R}_3$, $\mathcal{R}_{4,1}$ and $\mathcal{R}_{1,3}$ are related by

$$\underset{\sim}{\sigma}_1 \underset{\sim}{\sigma}_2 \underset{\sim}{\sigma}_3 = i = e_{01234} = \gamma_5 e_4, \tag{1.14}$$

where $\gamma_5 = \gamma_{0123} = e_{0123}$. Thus, all the requirements for a geometrical identification of i have been met. It may be of interest also to note that (1.14) and (1.11) can be solved for

$$e_4 = -i\gamma_5 \qquad \text{and} \qquad e_\mu = i\gamma_5\gamma_\mu, \tag{1.15}$$

which expresses the generating vectors of the abstract space $\mathcal{R}_{4,1}$ in terms of quantities with direct physical interpretations.

I did not employ the projective mapping (1.11) in my dissertation, because I did not appreciate its significance until later, but what I did was equivalent to it for practical purposes. With a definite physico-geometrical interpretation for C_4 in hand, I went on to the study of ideals and spinors in C_4. As most of you know, the columns of a matrix are minimal left ideals in a matrix algebra, because columns are not mixed by matrix multiplication from the left. The Dirac matrix algebra $C(4)$ has four linearly independent minimal left ideals, because each matrix has four columns. The Dirac spinor for an electron or some other fermion can be represented in $C(4)$ as a matrix with nonvanishing elements only in one column, like so

$$\begin{bmatrix} \psi_1 & 0 & 0 & 0 \\ \psi_2 & 0 & 0 & 0 \\ \psi_3 & 0 & 0 & 0 \\ \psi_4 & 0 & 0 & 0 \end{bmatrix}, \tag{1.16}$$

where the ψ_i are complex scalars. The question arises: Is there a physical basis for distinguishing between different columns? The question looks more promising when we replace $C(4)$ by the isomorphic geometric algebra $\mathcal{R}_{4,1}$ in which every element has a clear geometric meaning. Then the question becomes: Is there a physical basis for distinguishing between different ideals?

The Dirac theory clearly shows that a single ideal (or column if you will) provides a suitable representation for a single fermion. This suggests that each ideal should represent a different kind of fermion, so the space of ideals is seen as a kind of fermion isospace. I developed this idea at length in my dissertation, classifying leptons and baryons in families of four and investigating possible interactions with symmetries suggested by features of the geometric algebra, including SU(2) gauge invariant electroweak interactions.

I was not very convinced by my own dissertation, however, because there was too much guesswork in associating ideals with elementary particles, though my theory seemed no less satisfactory than other theories around at the time. I was bothered even more by the relation (1.14) between the pseudoscalar i for $\mathcal{R}_3$ and γ_5 for $\mathcal{R}_{1,3}$, because the factor e_4 does not seem to make any sense in terms of spacetime geometry. I should add that this problem is inherent in conventional applications of the Dirac theory, as can be seen by rewriting (1.13) in the form

$$\gamma_k = \underset{\sim}{\sigma}_k e_0. \tag{1.17}$$

In conventional matrix representations this is expressed as the decomposition of a 4×4 matrix into a Kroenecker product of 2×2 matrices $\underset{\sim}{\sigma}_k$ and e_0.

Before continuing with the story of my Odyssey, let me point out that I have been exploiting an elementary property of geometric algebras with general significance. Let $\mathcal{G}_n = \mathcal{R}_{p,q}$ be a geometric algebra generated by orthonormal vectors e_λ where $\lambda = 1, 2, \ldots, n$. Let $\mathcal{G}_{n-2}$ be the geometric algebra generated by

$$\sigma_k = e_k e_{n-1} e_n \tag{1.18}$$

where $k = 1, 2, \ldots, n - 2$, and let $\mathcal{G}_2$ be the geometric algebra generated by the vectors e_n and e_{n-1}. It is easy to see that the elements of $\mathcal{G}_{n-2}$ commute with those of $\mathcal{G}_2$. Therefore, $\mathcal{G}_n$ can be expressed as the Kroenecker product

$$\mathcal{G}_n = \mathcal{G}_{n-2} \otimes \mathcal{G}_2. \tag{1.19}$$

This factorization has been known since Clifford, but it seems to me that its geometric meaning has been overlooked. So I want to emphasize that the vectors of G_{n-2} are actually trivectors in G_n with a common bivector factor. Geometrically, they determine a family of 3-spaces which intersect in a common plane.

It is important to realize that the decomposition (1.19) does not involve the introduction of any new kind of multiplication aside from the geometric product. I have employed the notation and terminology of the Kroenecker product only to emphasize that the factors commute. I would like to add that in publications since my thesis I have avoided use of the Kroenecker factorization (1.19) in order to emphasize geometric interpretation. However, I believe that the factorization will prove to be very important in Geometric Function Theory, because it reduces geometric functions to commuting factors which can be differentiated independently.

The Kroenecker decomposition (1.19) should be compared with the projective decomposition of a geometric algebra which I emphasized in my first lecture. These are two basic ways of relating geometric algebras with different dimensions, and I believe they should be employed systematically in the classification of geometric algebras and their properties. Most work on the classification of Clifford algebras ignores geometric considerations and develops a classification in terms of matrix representations. Without denying the value of such work, I suggest that a classification without matrices is desirable. This is in accord with the viewpoint of my first lecture that geometric algebra is more fundamental than matrix algebra.

2. REAL SPINOR REPRESENTATIONS

About six months after completing my doctorate, I found a way to resolve the problem of geometric interpretation which bothered me in my thesis. I had worked out detailed representations for Lorentz transformations and the equations of electrodynamics in terms of the real spacetime algebra $R_{1,3}$ and separately in terms of the Pauli algebra before it dawned on me that the representations are identical if R_3 and $R_{1,3}$ are projectively related by

$$\underset{\sim}{\sigma}_k = \gamma_k\gamma_0, \tag{2.1}$$

so that $i = \underset{\sim}{\sigma}_1\underset{\sim}{\sigma}_2\underset{\sim}{\sigma}_3 = \gamma_0\gamma_1\gamma_2\gamma_3 = \gamma_5.$ (2.2)

As we saw in my first lecture, this simplifies computations and makes perfect sense geometrically. It also eliminates the need to supplement spacetime with additional degrees of freedom in order to interpret the Dirac algebra geometrically, as in the extension to $R_{4,1}$ which we just discussed. At first sight, however, this appears to be incompatible with quantum mechanics, because the imaginary scalar $i' = (-1)^{\frac{1}{2}}$ of the

Dirac algebra appears explicitly in the Dirac equation, and, unlike the pseudoscalar $i = \gamma_5$, it commutes with the γ_μ. To see how this difficulty can be resolved, we need to understand the representation of spinors in the real spacetime algebra $R_{1,3}$.

The number of linearly independent minimal left ideals in $R_{1,3}$ is two, half the number we found for the Dirac algebra. An orthogonal pair of such ideals is generated by the idempotents $\frac{1}{2}(1 \pm \sigma_3)$, where $\sigma_3 = \gamma_3\gamma_0$ is a unit timelike bivector. Any multivector M in $R_{1,3}$ can be written in the form

$$M = M_+ + M_-, \tag{2.3}$$

where
$$M_\pm = M\tfrac{1}{2}(1 \pm \sigma_3) \tag{2.4}$$

are its component in the two orthogonal ideals. Since the entire algebra has $2^4 = 16$ components, each of the ideals has 8 components, exactly the number of real components in a Dirac spinor. Therefore, we should be able to write the Dirac equation for a spinor $\Psi = \Psi_+$ in one of the ideals. The "imaginary problem" is handled by observing that the ideals are invariant under multiplication not only on the left by any element of the algebra but also on the right by the pseudoscalar $i = \gamma_5$, since i commutes with σ_3. To help clarify the correspondence with the Dirac theory, let me define right multiplication by i as a linear operator $\underline{i}$ by writing

$$\underline{i}\Psi = \Psi i. \tag{2.5}$$

This operator commutes with any operator multiplying Ψ from the left, for

$$\underline{i}(\gamma_\mu\Psi) = (\gamma_\mu\Psi)i = \gamma_\mu(\Psi i) = \gamma_\mu(\underline{i}\Psi), \tag{2.6}$$

which justifies the operator equation $\underline{i}\gamma_\mu = \gamma_\mu\underline{i}$. Note that this is a mapping of the associative law onto the commutative law.

Now the Dirac equation for a particle with mass m and charge e can be written in the form

$$\Box\Psi i - eA\Psi = m\Psi, \tag{2.7}$$

where $\mathbf{A} = A_\mu\gamma^\mu$ is the electromagnetic vector potential. This looks just like the conventional Dirac equation when it is written in the form

$$\gamma^{\mu}(\underline{i}\partial_{\mu} - eA_{\mu})\Psi = m\Psi, \tag{2.8}$$

but it employs the only real spacetime algebra $\mathcal{R}_{1,3}$ in which every element has a clear geometric-physical meaning. As soon as I understood that (2.7) really is equivalent to the Dirac equation, I salvaged what I could from my dissertation and quickly wrote a manuscript which was eventually published as the book Space-Time Algebra [6].

I learned much later that already in (1962) Kahler [3] had proposed a form of the Dirac equation with γ_5 operating on the right which is essentially equivalent to (2.7). He employed complex scalars, but the real and imaginary parts obey separate equations of the same form as long as the imaginary unit does not appear explicitly in the equation.

One thing bothered me about the Dirac equation in the form (2.7): What is the physical significance of the bivector $\underset{\sim}{\sigma}_3$ which determines the ideal of the spinor? It took me nearly two years to answer that question. As described in Ref. [7], I found the answer by factoring Ψ into

$$\Psi = \psi U, \tag{2.9}$$

where ψ is an element of the even subalgebra $\mathcal{R}^{+}_{1,3}$, and

$$U = \tfrac{1}{2}(1 + \gamma_0)(1 + \underset{\sim}{\sigma}_3). \tag{2.10}$$

Note that
$$\gamma_0 U = U \tag{2.11}$$

and
$$i\underset{\sim}{\sigma}_3 U = Ui. \tag{2.12}$$

Also, it can be proved that any ψ in $\mathcal{R}^{+}_{1,3}$ for which $\psi\tilde{\psi} \neq 0$ can be written in the canonical form

$$\psi = (\rho e^{i\beta})^{\frac{1}{2}}R, \tag{2.13}$$

where ρ and β are scalars and $R\tilde{R} = 1$. I should mention that the reverse $\tilde{M}$ of any multivector M in $\mathcal{R}_{1,3}$ with k-vector parts M_k can be defined by

$$\tilde{M} = M_0 + M_1 - M_2 - M_3 + M_4. \tag{2.14}$$

Note that (2.11) enables us to regard ψ as even in (2.9), because if ψ had an odd part, that part could be made even by multiplying it with γ_0 without altering (2.9).

The even multivector ψ is a new representation of a Dirac spinor, for which the Dirac equation takes the form

$$\Box\psi i\sigma_3 - eA\psi = m\psi\gamma_0, \tag{2.15}$$

as can be verified by multiplying it on the right by U to get back (2.7). At first this may look more complicated than the conventional form of the Dirac equation, but it gets much simpler as we learn to understand it. Ironically, I derived (2.15) in my book and promptly forgot about it for two years until I rederived it by a different method and discovered how ψ is related to the observables of the Dirac theory.

In spite of the similarity of (2.7) and (2.15) to the Dirac equation, some people doubt that they are equivalent, so I should provide a proof. I will employ the complex Clifford algebra C_4, which we know is isomorphic to the Dirac matrix algebra $C(4)$. A minimal ideal in C_4 is generated by

$$U' = \frac{1}{2}(1 + \gamma_0)(1 - i'i\sigma_3) = \frac{1}{2}(1 + \gamma_0)(1 - i'\gamma_5\gamma_3), \tag{2.16}$$

where, as before, i' denotes the imaginary scalar and $i = \gamma_5$ is the pseudoscalar. Note that U' has the idempotent property $(U')^2 = 2U'$. Also

$$\gamma_0 U' = U', \tag{2.17}$$

and

$$i\sigma_3 U' = \gamma_5\gamma_3 U' = i'U'. \tag{2.18}$$

Now it is obvious that every spinor Ψ' in the ideal generated by U' can be written in the form

$$\Psi' = \psi U', \tag{2.19}$$

where ψ is an even multivector exactly as before. For any odd part of ψ in (2.19) can be eliminated by using (2.17), and if i' appears explicity in any part of ψ, (2.18) allows us to replace it with $i\sigma_3$. It is easy to find a matrix representation for U' which puts Ψ' in the column matrix form (1.16), so Ψ' is clearly equivalent to a Dirac spinor. Thus, we have established that Ψ, ψ and Ψ' are equivalent representations of a Dirac spinor. (See Appendix A of Ref. [8] for more details and a slightly different proof). Now the proof that (2.15) is equivalent to the Dirac equation is trivial. We simply multiply it on the right by U' and use (2.17), (2.18) and (2.19) to get the conventional form of the Dirac equation

$$\gamma^\mu(i'\partial_\mu - eA_\mu)\Psi' = m\Psi'. \tag{2.20}$$

The main import of this proof is that complex scalars do not play an essential role in the Dirac theory. As (2.18) indicates, the unit imaginary i' is replaced by a spacelike bivector $i\underset{\sim}{\sigma}_3 = \gamma_2\gamma_1$, when we go from the complex to the real representation. This implies that a geometrical meaning for i' is implicit in the Dirac theory, for the bivector $i\underset{\sim}{\sigma}_3$ is manifestly a geometrical quantity. This presents us with an important question to investigate: What is the physical significance of the fact that the unit imaginary in the Dirac theory represents a spacelike bivector?

To establish a physical interpretation for the spinor "wave functions" Ψ, ψ and Ψ', we need to relate them to "observables". I will simply assert an interpretation at the beginning, but by the time we are finished it will be clear that my interpretation agrees with the conventional interpretation of the Dirac theory. I begin with the interpretation of ψ, because it is simplest and most direct. Using $\tilde{\psi} = (\rho e^{i\beta})^{\frac{1}{2}}\tilde{R}$, the Dirac probability current is given by

$$J = \psi\gamma_0\tilde{\psi} = \rho R\gamma_0\tilde{R} = \rho v. \tag{2.21}$$

As shown in Ref. [6], R is the "spin representation" of a Lorentz transformation, so $v = R\gamma_0\tilde{R}$ describes a Lorentz transformation of the timelike vector γ_0 into the "velocity vector" v. The factor ρ is therefore to be interpreted as a probablility density. Note that the factor $e^{i\beta}$ disappears from (2.21) because i anticommutes with γ_0.

The spin (or polarization) vector of a Dirac particle is given by

$$\rho s = \psi\gamma_3\tilde{\psi} = \rho R\gamma_3\tilde{R}. \tag{2.22}$$

Strictly speaking the spin S is bivector quantity, but it is related to the vector s by

$$S = R(i\underset{\sim}{\sigma}_3)\tilde{R} = iR\gamma_3\tilde{R}R\gamma_0\tilde{R} = isv, \tag{2.23}$$

so one determines the other.

The vectors γ_0 and γ_3 in (2.21) and (2.22) are not necessarily related to the reference frame of any observer. They are singled out because they appear as constants in the Dirac equation. These constants are not seen in the conventional form of the Dirac theory because they are buried in an idempotent. Having exhumed them, we may attend to their physical meaning. The constant $i\underset{\sim}{\sigma}_3 = i\gamma_3\gamma_0 = \gamma_2\gamma_1$ is especially important because of its role as the imaginary unit. Equation (2.23) shows that the wave function relates it to the spin.

Indeed, we could replace $i\sigma_3$ by the spin S in the Dirac equation (2.15) by employing the identity

$$\psi i\sigma_3 = S\psi . \tag{2.24}$$

This enables us to interpret a rotation generated by $i\sigma_3$ as a rotation in the S-plane. The bivector $i\sigma_3$ is also generator of electromagnetic gauge transformations. The Dirac equation (2.15) is invariant under a gauge transformation of the wave function replacing ψ by

$$\psi' = \psi e^{i\sigma_3 \chi} = e^{S\chi}\psi \tag{2.25}$$

while A is replaced by $A' = A - e^{-1}\Box\chi$. This transformation also leaves the spin and velocity vectors invariant, for $\psi'\gamma_0\tilde{\psi}' = \psi\gamma_0\tilde{\psi}$ and $\psi'\gamma_3\tilde{\psi}' = \psi\gamma_3\tilde{\psi}$. Indeed, we could define the electromagnetic gauge transformation as a rotation which leaves the spin and velocity vectors invariant.

Now we can give a detailed physical interpretation of the spinor wave function ψ in its canonical form (2.13). We can regard ψ as a function of 8 scalar paramenters. Six of the parameters are needed to determine the factor R which represents a Lorentz transformation. Five of these determine the velocity and spin directions in accordance with (2.21) and (2.22). The sixth determines the gauge or phase of the wave function (in the S-plane). We have already noted that ρ is to be interpreted as a probability density. The interpretation of the remaining parameter β presents problems which I do not want to discuss today, although I will make some observations about it later. As you know, this parameter is not even identified in the conventional formulation of the Dirac theory.

Having related ψ to observables, it is easy to do the same for Ψ and Ψ'. From (2.9) and (2.10), we find that $\Psi\tilde{\Psi} = 0$, but

$$\Psi\gamma_0\tilde{\Psi} = \psi(1 + \gamma_0)\tilde{\psi} = \rho e^{i\beta} + J, \tag{2.26}$$

and

$$\Psi i\gamma_0\tilde{\Psi} = \psi i\gamma_3(1 + \gamma_0)\tilde{\psi} = \rho is(1 + ve^{i\beta}). \tag{2.27}$$

On the other hand, if we introduce the definition $\tilde{i}' = -i'$ for the scalar imaginary, then (2.16) and (2.19) give us

$$\Psi'\gamma_0\tilde{\Psi}' = \Psi'\tilde{\Psi}' = \psi U^2\tilde{\psi} = \psi 2U\tilde{\psi} = \rho e^{i\beta} + J - i'\rho\gamma_5 s(1 + ve^{i\beta}). \tag{2.28}$$

Equating corresponding k-vector parts of (2.26), (2.27) and (2.28) and calculating their components, we get the following set of equivalent expressions for the so-called "bilinear covariants" of the Dirac theory:

$$\langle\tilde{\Psi}'\Psi'\rangle = \langle\Psi\gamma_0\tilde{\Psi}\rangle = \langle\psi\tilde{\psi}\rangle = \rho\cos\beta, \tag{2.29a}$$

$$\langle\tilde{\Psi}'\gamma_\mu\Psi'\rangle = \langle\gamma_\mu\Psi\gamma_0\tilde{\Psi}\rangle = \langle\gamma_\mu\psi\gamma_0\tilde{\psi}\rangle = J\cdot\gamma_\mu, \tag{2.29b}$$

$$i'\langle\tilde{\Psi}'\gamma_\mu\wedge\gamma_\nu\Psi'\rangle = \langle\gamma_\mu\wedge\gamma_\nu\Psi\gamma_0\tilde{\Psi}\rangle = (\gamma_\mu\wedge\gamma_\nu)\cdot(\rho S e^{i\beta}), \tag{2.29c}$$

$$i'\langle\tilde{\Psi}'\gamma_5\gamma_\mu\Psi'\rangle = \langle i\gamma_\mu\Psi\gamma_0\tilde{\Psi}\rangle = \langle\gamma_\mu\psi\gamma_3\tilde{\psi}\rangle = \rho s\cdot\gamma_\mu, \tag{2.29d}$$

$$\langle\tilde{\Psi}'\gamma_5\Psi'\rangle = \langle i\Psi\gamma_0\tilde{\Psi}\rangle = \langle i\psi\tilde{\psi}\rangle = -\rho\sin\beta. \tag{2.29e}$$

The angular brackets here mean scalar part, and I have used the theorem $\langle AB\rangle = \langle BA\rangle$ to put the terms on the left in the standard form of the Dirac theory. Equations (2.29b) and (2.29c) justify our earlier identifications of the Dirac current and the spin in (2.21) and (2.22).

We have completed the reformulation of the Dirac theory in terms of the real spacetime algebra $R_{1,3}$. It should be evident that of the three different representations for a Dirac spinor, the representation ψ is the easiest to interpret geometrically and physically. So I will work with ψ exclusively from here on, with full confidence that its equation of motion (2.15) is 100% equivalent to the conventional Dirac equation. I suggest that we refer to ψ as the operator representation of a Dirac spinor, because it produces "ideal representations" by operating on idempotents and it produces observables by operating on vectors as in (2.21) and (2.22).

The most important thing we have learned from the reformulation is that the imaginary i' in the Dirac equation has a definite geometrical and physical meaning. It represents the generator of rotations in a spacelike plane associated with the spin. Indeed, we saw that i' can be identified with the spin bivector S. I want to emphasize that this interpretation of i' is by no means a radical speculation; it is a fact! The interpretation has been implicit in the Dirac theory all the time. All we have done is make it explicit.

Clearly the identification of the imaginary i' with the spin bivector has far-reaching implications about the role of complex numbers in quantum mechanics. Note that it applies even to Schroedinger theory [7] when the Schroedinger equation is derived as an approximation to the Dirac equation. It implies that a degenerate representation of the spin direction by the unit imaginary has been implicit in the Schroedinger equation all along.

This is the kind of idea that can ruin a young scientist's career. It appears to be too important to keep quiet about. But if you try to explain it to most physicists, they are likely to dismiss you as some kind of crackpot. The more theoretical physics they know the harder it is to explain, because they already have fixed ideas about the mathematical formalism, and you can't understand this idea without reanalyzing such basic concepts as how to multiply vectors. They quickly become impatient with any discussion of elementary concepts, so they employ the ultimate putdown: "What are the new predictions of your theory." If you can't come up with a mass spectrum or branching ratio, the conversation is finished. I learned early that you must be very careful when and where and how you voice such a crackpot idea.

Since I first published the real spinor formulation of the Dirac theory in (1967), the only people (besides my students) who took it seriously enough to use it for something were the Frenchmen Casanova [9], Boudet [10] and Quilichini [11]. I suppose that is partly because they are not conventional physicists. They found out some interesting things about the solutions of the Dirac equation for the hydrogen atom. But most physicists would be more concerned about how the formalism applies to quantum electrodynamics (QED). There they would encounter difficulties at once, which I suppose has induced some to dismiss the entire formalism. Conventional QED calculations involve multiplications with a commuting imaginary all over the place, but the real formalism does not contain such an entity. It appears at first that one cannot even define the conventional commuting momentum and spin projection operators without it. I must admit that I was perplexed by this difficulty for a long time myself. However, it has a simple solution: One merely handles right multiplications by defining them as operators acting from the left, as I did in defining $\underline{i}$ by (2.5). Let me illustrate this by defining projection operators for the real spinor Ψ. The momentum projection operators are just the usual idempotents

$$P_{\pm} = \frac{1}{2m}(1 \pm p), \tag{2.30}$$

where $p^2 = m^2$. The spin projection operators are defined by

$$\Sigma_{\pm} = \frac{1}{2}(1 \pm \underline{i}\gamma_5 s), \tag{2.31}$$

where s is a spacelike unit vector satisfying $p \cdot s = 0$. Operating on Ψ this gives

$$\Sigma_{\pm}\Psi = \frac{1}{2}(1 \pm \underline{i}\gamma_5 s)\Psi = \frac{1}{2}(\Psi \pm is\Psi i). \tag{2.32}$$

Note that $\Sigma_{\pm}$ is not an idempotent of the algebra $R_{1,3}$; nevertheless, it has the idempotent property $(\Sigma_{\pm})^2 = \Sigma_{\pm}$ of a projection operator. Also, the operators $P_{\pm}$ and $\Sigma_{\pm}$ commute, like they do in the conventional theory.

One can apply the same trick to carry out conventional QED calculations with the spinor ψ. However, the use of ψ suggests new methods of calculation without projection operators which may prove to be superior to conventional methods. But that is a long story itself which I cannot get into today.

Now that we have settled the relation of the real spinor formalism to the conventional complex spinor formalism (I hope), we are prepared to discuss implications for the interpretation of quantum mechanics. First let me point out some negative implications. The complex probability amplitudes that appear in physics have led some to propose that quantum mechanics entails some new kind of "quantum logic" which is essentially different from the logic of classical thoery. However, the association of the unit imaginary with the spin shows that complex amplitudes arise from some physical reason, so the quantum logic idea appears to be on the wrong track.

Another negative implication of our reformulation concerns the interpretation of operators in quantum mechanics. It is widely believed by physicists that the Pauli and Dirac matrices have some special quantum mechanical significance, so their commutation relations have bearing on questions about quantum mechanical measurement. But the reformulation in terms of spacetime algebra shows that these commutation relations express geometrical relations which are no more quantum mechanical than classical. Therefore, we can dismiss most of that stuff as arrant nonsense. It has validity only to the extent that it is merely an expression of geometrical relations. For example, the γ_μ are often said to be velocity operators in the Dirac theory. But we see them merely as ordinary vectors, which are velocity operators only in the sense that they pick out velocity components from the wave function by the ordinary inner product, as, in fact, they do in (2.29b). To attribute more meaning than that to the γ_μ is to generate nonsense. There is, indeed, an extensive literature on such nonsense. We should not be surprised that this literature is muddled and barren.

Turning now to the positive implications of the real reformulation, we have seen that it reveals geometrical features of the Dirac theory which are hidden in the conventional matrix formulation. Of course, it cannot <u>per se</u> produce any new predictions, because the reformulated Dirac theory is isomorphic to the original formulation. We will get new physics only if the new geometrical insights guide us to significant modifications or extensions of the Dirac theory. I want to tell you next about some promising possibilities along this line.

3. CLASSICAL SOLUTIONS OF THE DIRAC EQUATION

The relation of the Dirac theory to classical relativistic electrodynamics is not well understood. I aim to show you that it is more intimate than generally suspected. The classical limit is ordinarily obtained as an "eikonal approximation" to the Dirac equation. To express that, in our language, the wave function is put in the form

$$\psi = \psi_0 e^{i\underset{\sim}{\sigma}_3\chi}, \tag{3.1}$$

and the "amplitude" ψ_0 is assumed to be slowly varying compared to the "phase" χ, so it is a good approximation to neglect derivatives of ψ_0 in the Dirac equation. Thus, inserting (3.1) into the Dirac equation

(2.15), multiplying by $\tilde{\psi}$ on the right, and using the canonical form (2.13) for ψ, we obtain

$$\Box\chi + eA = -mve^{i\beta} \tag{3.2}$$

This implies that $e^{i\beta} = \pm 1$, because the trivector part on the right must vanish. These two signs can be absorbed in the sign of the charge e on the left. That tells us that the parameter β distinguishes between particle and antiparticle solutions of the Dirac equation. Assuming $\beta = 0$, by squaring (3.2) we obtain

$$(\Box\chi + eA)^2 = m^2. \tag{3.3}$$

This is exactly the classical relativistic Hamilton-Jacobi equation for a charge particle. On the other hand, the curl of (3.2) gives us

$$-m\Box\wedge v = e\Box\wedge \mathbf{A} = eF, \tag{3.4}$$

where $F = \Box\wedge A$ is the electromagnetic field. Now we use the general identity $v\cdot(\Box\wedge v) = v\cdot\Box v - \frac{1}{2}\Box v^2$. Since $v^2 = 1$ here, this implies that $v\cdot(\Box\wedge v) = v\cdot\Box v = \dot{v}$, where the overdot indicates a derivative along the streamlines (integral curves) of the vector field $v = v(x)$. Let me remind you that the probability conservation law $\Box\cdot J = \Box\cdot(\rho v) = 0$ implies that these curves are well-defined. Thus, by dotting (3.4) with v, we obtain the equation of motion for any streamline of the Dirac current,

$$m\dot{v} = eF\cdot v. \tag{3.5}$$

You will recognize this as the classical equation of motion for a point charge.

All this is supposed to be an approximation to the Dirac theory. But I want to point out that it holds exactly for solutions of the Dirac equation when

$$\Box\psi_0 = 0, \tag{3.6}$$

in which case our assumption that derivatives of ψ_0 are negligible is unnecessary. Members of this audience will recognize (3.6) as a generalization of the Cauchy-Riemann equations to spacetime, so we can expect it to have a rich variety of solutions. The problem is to pick out those solutions with physical significance. To see how that can be done, we write $\psi_0 = \rho^{\frac{1}{2}}R$ with $v = R\gamma_0\tilde{R}$, from which we obtain $v\psi_0 = \psi_0\gamma_0$. Differentiating, we have

$$\Box(v\psi_0) = (\Box v)\psi_0 - v(\Box\psi_0) + 2v\cdot\Box\psi_0 = \Box\psi_0\gamma_0.$$

Then using (3.6) and $\square\cdot(\rho v) = \rho\,\square\cdot v + v\cdot\square\,\rho = 0$, we obtain an equation of motion for the spinor R along a streamline,

$$\dot{R} = -\tfrac{1}{2}(\square\wedge v)R. \tag{3.7}$$

When (3.4) is used to eliminate $\square\wedge v$, this becomes

$$\dot{R} = \frac{e}{2m}\,FR, \tag{3.8}$$

which, as we saw in my first lecture, can regarded as a classical equation of motion. And from the first lecture we know that besides implying the classical Lorentz force (3.5), it determines the precession of the spin vector along a streamline.

Our derivation shows that equations (3.6) and (3.8) are related by the probability conservation law. Any solution of the Dirac equation which satisfies these equations can fairly be called a classical solution. Let me tell you that exact classical solutions of the Dirac equation actually exist, though I do not have time to spell out the details here. The so-called Volkov solution for an electron in an electromagnetic plane wave field is one. It may be that there are solutions of this type for any electromagnetic field. However, the standard solution for the hydrogen atom is not one of them, for the parameter β in that case is a nonvanishing, nontrivial function of position. Nevertheless, it must have a definite relation to the known classical solution of (3.8) for the hydrogen atom. More work is needed along this line.

The intimate relation between streamlines of the Dirac theory and trajectories of the classical theory which we have just uncovered provides a much more detailed correspondence between the classical and quantum theories than the conventional approach using expectation values and Ehrenfest's theorem. Let me remark in passing that it also suggests a relativistic generalization of the Feynman path integral including spin, where one integrates the entire classical spinor R along each path instead of just the gauge factor exp $(i\underset{\sim}{\sigma}_3\chi)$, where χ is the classical action. But what I want to emphasize most is that the basic idea which we have been exploiting provides a general geometrical approach to the interpretation of the Dirac theory as follows: Any solution $\psi = \psi(x)$ of the Dirac equation (2.15) with the form (2.13) determines a field of orthonomal frames $e_\mu = e_\mu(x)$ defined by

$$\psi\gamma_\mu\tilde{\psi} = \rho e_\mu, \qquad \text{where} \qquad e_\mu = R\gamma_\mu\tilde{R}, \tag{3.9}$$

with $e_0 = v$ and $e_3 = s$ as before. Through each spacetime point there is a streamline $x = x(\tau)$ with tangent $v = v(x(\tau))$, and we can regard $e_\mu = e_\mu(x(\tau))$ as a "comoving frame" on the streamline with vectors e_1

and e_2 rotating about the "spin axis" $e_3 = s$. For the classical solutions discussed above, the general precession of the comoving frame is determined by (3.8), while an additional rotation of e_1 and e_2 is determined by the "gauge factor" in (3.1). It should be of genuine physical interest to identify and analyze any deviations from this classical rotation which quantum mechanics might imply.

The Dirac theory provides a beautiful mathematical theory of spinning frames on the spacetime manifold. But a spinning frame is not a spinning thing, and physicists want to know if the Dirac theory can be interpreted as a mathematical model for some physically spinning thing. I am afraid that question cannot be answered without becoming embroiled in speculations. But the question is too important to be avoided for that. Some physicists have attempted to model the electron as a small spinning ball. But that introduces all kinds of theoretical complications and, as far as I have been able to see, no significant insights. Along with Asim Barut and others, I think a much more promising possibility is the idea that the electron is a particle executing a minute helical motion, called the zitterbewegung, which is manifested in the electron spin and magnetic moment. As I have recently published a speculative article on that idea [12], I will not go into details here. I only want to mention a key idea of that article, namely, that the Coulomb field ordinarily attributed to an electron is actually the time average of a more basic periodic electromagnetic field oscillating with the de Broglie frequency $\omega = mc^2/\hbar \approx 10^{21}\ s^{-1}$ of the electron. This is a new version of wave-particle duality, where the electron is a particle to which this high frequency electromagnetic field (or wave) is permanently attached. As the article points out, this gives us a mechanism for explaining the most perplexing features of quantum mechanics from diffraction to the Pauli principle. What I want to add here is that this version of wave-particle duality may be viable even without the literal electron zitterbewegung. In particular, it may fit into the gauge theory which I turn to next.

4. GAUGE STRUCTURE OF THE DIRAC THEORY

One thing that bothered me for a long time after I discovered the underlying geometrical structure of the Dirac theory is the question: How does the concept of probability fit in with all this geometry? I now have an answer which I find fairly satisfying. To get a spacetime invariant generalization of the concept of a continuous probability density in 3-space, we need to introduce the concept of a probability current J obeying the conservation law $\Box\cdot J = 0$. Now, it should be evident from our earlier discussion that <u>every timelike vector field</u>

$J = J(x)$ can be written in the form

$$J = \psi\gamma_0\tilde{\psi}, \tag{4.1}$$

where $\psi = \psi(x)$ is an even spinor field. This theorem automatically relates the probability current to a spinor field and all the geometry that goes with it. You will notice that the bilinear relation between the spinor field ψ and the vector field J is perfectly natural in this language, and it is as applicable to classical theory as to quantum mechanics. This removes much of the mystery from the bilinear relation of the wave function to observables in quantum mechanics. It shows that the bilinearity has a geometrical origin, which certainly has nothing to do with any sort of quantum logic.

The relation of vector J to spinor ψ in (4.1) cannot be unique, because only four scalar parameters are required to specify J uniquely, while eight are required to specify ψ. Indeed, the same J results from (4.1) if ψ is replaced by

$$\psi' = \psi S \tag{4.2}$$

where $S\gamma_0\tilde{S} = \gamma_0$. (4.3)

This is a gauge transformation of the wave function ψ, so the set of all such transformations is the <u>gauge group of the Dirac probability current</u>, that is, the group of gauge transformations leaving the Dirac current invariant.

We can identify the structure of this group by decomposing S into

$$S = e^{i\alpha}U, \tag{4.4}$$

where α is a scalar parameter and $U\tilde{U} = 1$. The set of all $S = U$ satisfying (4.3) is the spin "little group" of Lorentz transformations leaving the timelike vector γ_0 invariant; it has the SU(2) group structure. It follows from (4.4), therefore, that the gauge group of the Dirac current has the SU(2) ⊗ U(1) group structure. Note that this is a 4-parameter group, so it accounts completely for the difference in the number of parameters needed to specify J and ψ. You will notice also that it has the same structure as electroweak gauge group in the Weinberg-Salam (W-S) model of weak and electromagnetic interactions. I want to make the stronger claim that the <u>electroweak gauge group should be identified with the gauge group of the Dirac current</u>. This, of course, is to claim that the electroweak gauge group has been inherent in the Dirac theory all the time, though one could not see it without the reformulation in terms of spacetime algebra. It is a strong claim, because it relates the electroweak gauge group to spacetime geometry of the wavefunction. That requires some justification.

The gauge group of the Dirac current has a subgroup which also leaves the spin vector $\rho s = \psi\gamma_3\tilde{\psi}$ invariant, namely, the 2-parameter group of elements with the form

$$S = e^{i\alpha}e^{i\sigma_3\chi}. \tag{4.5}$$

We recognize the last factor as an electromagnetic gauge transformation, which, as we have seen, leaves the Dirac equation as well as the spin and velocity vectors invariant. Now, the main idea of the W-S model is to generalize the electromagnetic gauge group of the Dirac theory to account for weak interactions. Since we have proved that the generator of electromagnetic gauge transformations is a bivector $i\sigma_3$ associated with the spin, we should not be satisfied with a generalization which does not supply an associated geometric interpretation for the generators of weak gauge transformations. But we have just what is needed in the gauge group of the Dirac current. This appears to be ample justification for identifying that group with the electroweak gauge group. Of course, that requires a modification of the Dirac equation to accommodate the larger gauge group. As I have shown eleswhere [13] how that can be done to conform perfectly to the W-S model, I will not go into such details here. But I want to add some observations about the general structure of the theory.

An important feature of the W-S model is the separation of "right-handed" and "left-handed" components of the wave functions. To show how the conventional expressions for these components should be translated into our geometrical language, it is convenient to introduce the imaginary operator $\underline{i}$ defined by

$$\underline{i}\psi = \psi i\sigma_3. \tag{4.6}$$

Now the projections onto left and right handed components ψ_+ and ψ_- are defined by

$$\tfrac{1}{2}(1 \pm \underline{i}\gamma_5)\psi = \tfrac{1}{2}(\psi \pm i\psi i\sigma_3) = \psi\tfrac{1}{2}(1 \mp \sigma_3) = \psi_{\mp}. \tag{4.7}$$

Thus, we are back to spinors in ideals again, but now we have a physical interpretation for the ideals.

In my formulation of the W-S model in terms of spacetime algebra [13], the two ideals are interpreted as electron and neutrino eigenstates of the lepton wave function. Equation (4.7) shows that there is a relation between the representation of different particles by the two ideals and the decomposition of the wave function for one particle into left and right handed components. This is a relation between two features introduced as independent assumptions in the original W-S model. Moreover, from (4.5) we see that the quantities

$$iQ_{\pm} = \tfrac{1}{2}(1 \pm \sigma_3)i \tag{4.8}$$

are generators of the "spin invariance group" (4.5). This relates the electromagnetic gauge group to the "chiral projection operators" in (4.7). All this suggests possibilities for a deeper geometrical justification of the phenomenological W-S model. I should mention that the full W-S model involves a spinor with both even and odd parts, whereas (4.7) is an even spinor only. That must be taken into account in relating (4.7) to the W-S model.

For the purpose of physical interpretation we need to relate the decomposition (4.7) to observables. Since $(1 + \underset{\sim}{\sigma}_3)\gamma_0(1 - \tilde{\underset{\sim}{\sigma}}_3) = 0$, the ideals are orthogonal in the sense that

$$\psi_+\gamma_0\tilde{\psi}_- = \psi_-\gamma_0\tilde{\psi}_+ = 0. \tag{4.9}$$

This implies that the Dirac current separates into uncoupled left and right handed currents $J_\pm$. Thus, using $\psi = \psi_+ + \psi_-$, we find

$$J = \psi\gamma_0\tilde{\psi} = J_+ + J_-, \tag{4.10}$$

where $$J_\pm = \psi_\pm\gamma_0\tilde{\psi}_\pm = \psi\tfrac{1}{2}(\gamma_0 \pm \gamma_3)\tilde{\psi}. \tag{4.11}$$

The expression on the right of (4.11) follows from $\frac{1}{2}(1 \pm \underset{\sim}{\sigma}_3)\gamma_0\,\frac{1}{2}(1 \mp \underset{\sim}{\sigma}_3) = \frac{1}{2}(1 \pm \gamma_3\gamma_0)\gamma_0 = \frac{1}{2}(\gamma_0 \pm \gamma_3)$. The corresponding decomposition of the spin vector is given by

$$\rho s = \psi\gamma_3\tilde{\psi} = \rho(s_+ + s_-), \tag{4.12}$$

where $$\rho s_\pm = \psi_\pm\gamma_3\tilde{\psi}_\pm = \psi\tfrac{1}{2}(\gamma_3 \pm \gamma_0)\tilde{\psi} = \pm J_\pm. \tag{4.13}$$

The separation (4.10) into uncoupled currents suggest that the two spin components $\psi_\pm$ might be identified with different particles, or coupled differently to gauge fields, as, in fact, they are in the W-S theory. Equation (4.11) shows that the currents $J_\pm$ are null vectors, as required for massless particles. And (4.12) shows that spin vectors $\rho s_\pm$ are, respectively, parallel or antiparallel to their associated currents $J_\pm$. I believe these elementary observations will be important for understanding and assessing the geometric structure of the W-S model.

5. CONCLUSION

My objective in this talk has been to explicate the geometric structure of the Dirac theory and its physical significance. My approach may seem radical at first sight, but I hope you have come to recognize it as ultimately conservative. It is conservative in the sense that, by restricting my mathematical language to the spacetime algebra, I allow nothing in my formulations of physical theory without an interpretation in terms of spacetime geometry. I am not opposed to investigating the possibilities for unifying physical theory by extending spacetime geometry to higher dimensions, and I believe geometric algebra is the ideal tool for that. But we still have a lot to learn about the physical implications of conventional spacetime structure, so I have focused my attention on that.

REFERENCES

[1] Marcel Riesz, 'Clifford Numbers and Spinors', Lecture series No. 38, The Institute for Fluid Dynamics and Applied Mathematics, University of Maryland (1958).
[2] E. Kähler, Abh. Dt. Akad. Wis. MPTJ 5 (1961).
[3] E. Kähler, Rend. Mate. (Roma) Ser. V, 21, 425 (1962).
[4] D. Hestenes, Geometric Calculus and Elementary Particles (doctoral dissertation) UCLA (1963).
[5] P. Lounesto, Found. Phys. 11, 721 (1981).
[6] D. Hestenes, Space-Time Algebra, Gordon and Breach, N.Y. (1966).
[7] D. Hestenes, J. Math. Phys. 8, 798 (1967).
[8] D. Hestenes, J. Math. Phys. 14, 893 (1973).
[9] G. Casanova, C.R. Acad. Sc. Paris, 266B, 1551 (1968; 271A, 817 (1970); 275B, 53, 121, 267, 399 (1972).
[10] R. Boudet, C.R. Acad. Sci. Paris, 272A, 767 (1971); 278A, 1063 (1974).
[11] P. Quilichini, C.R. Acad. Sc. Paris, 273B, 1063 (1971).
[12] D. Hestenes, Found. Phys. 15, 63 (1985).
[13] D. Hestenes, Found. Phys. 12, 153 (1982).

APPENDIX: The Kähler Algebra

Since the Dirac-Kähler equation has been a topic for much discussion at this conference, I will provide a dictionary here to show how easy it is to translate the Clifford algebra employed in this paper into Kähler's representation in terms of differential forms. We consider only algebras of spacetime; the generalization to algebras of higher dimension is trivial.

Let Ψ be any element of a Clifford algebra (real or complex) generated by an orthonormal frame of spacetime vectors γ_μ as in the text above. The expansion of Ψ into k-vector parts Ψ_k is given by

$$\Psi = \sum_{k=0}^{4} \Psi_k . \tag{A.1}$$

For $k > 0$, the expansion of Ψ_k with respect to a basis can be written

$$\Psi_k = \frac{1}{k!}\phi_{\mu_1 \ldots \mu_k}\gamma^{\mu_1}\wedge\ldots\wedge\gamma^{\mu_k}$$

$$= \frac{1}{k!}\phi^{\mu_1 \ldots \mu_k}\gamma_{\mu_1}\wedge\ldots\wedge\gamma_{\mu_k}, \qquad \text{(A.2)}$$

where the coefficients may be real of complex, and $\{\gamma^\mu\}$ is the frame reciprocal to $\{\gamma_\mu\}$, as defined by the equations $\gamma^\mu\cdot\gamma_\nu = \delta^\mu{}_\nu$. Let us write

$$\gamma^\mu\wedge\Psi = \sum_k \gamma^\mu\wedge\Psi_k, \qquad \gamma^\mu\cdot\Psi = \sum_k \gamma_\mu\cdot\Psi_k, \qquad \text{(A.3)}$$

so $\Box\wedge\Psi = \sum_k \Box\wedge\Psi_k, \qquad \Box\cdot\Psi = \sum_k \Box\cdot\Psi_k, \qquad$ (A.4)

and $\Box\Psi = \Box\cdot\Psi + \Box\wedge\Psi. \qquad$ (A.5)

We can map all this into differential forms by introducing a multivector differential of mixed grade defined by

$$D = \sum_{k=0} D_k, \qquad \text{(A.6)}$$

where $D_0 = 1$ and for $k > 0$,

$$D_k = dx_1\wedge\ldots\wedge dx_k = \frac{1}{k!}dx^{\mu_1}\ldots dx^{\mu_k}\gamma_{\mu_1}\wedge\ldots\wedge\gamma_{\mu_k}. \qquad \text{(A.7)}$$

To every k-vector Ψ_k there corresponds a differential k-form Φ_k given by

$$\Phi_k = \langle\tilde{D}\Psi_k\rangle = \tilde{D}_k\cdot\Psi_k = \frac{1}{k!}\phi_{\mu_1 \ldots \mu_k}dx^{\mu_1}\ldots dx^{\mu_k}. \qquad \text{(A.8)}$$

More generally, to every multivector Ψ there corresponds a differential "multiform" Φ given by

$$\Phi = \langle\tilde{D}\Psi\rangle = \langle\Psi\tilde{D}\rangle = \sum_{k=0}^{4}\Phi_k. \qquad \text{(A.9)}$$

This generalizes the mapping of multivectors into forms which I considered in my first lecture.

Now the <u>exterior product</u> of a 1-form $dx^\mu = \gamma^\mu\cdot dx$ with the form Φ can be defined by

$$dx^\mu\wedge\Phi = \langle(\gamma^\mu\wedge\Psi)\tilde{D}\rangle, \qquad \text{(A.10)}$$

and the <u>contraction</u> of Φ with the vector γ^{μ} can be defined by

$$\gamma^{\mu} \lrcorner \Phi = \langle (\gamma^{\mu} \cdot \Psi)\tilde{D} \rangle . \quad (A.11)$$

Kähler defined a "vee product" for differential forms by writing

$$dx^{\mu} \vee \Phi = dx^{\mu} \wedge \Phi + \gamma^{\mu} \lrcorner \Phi . \quad (A.12)$$

But (A.10) and (A.11) imply that this is equivalent to

$$dx^{\mu} \vee \Phi = \langle \gamma^{\mu} \Psi \tilde{D} \rangle , \quad (A.13)$$

which defines the vee product by a linear mapping of the geometric product into forms. Thus, we have established a one-to-one mapping of Clifford algebra onto differential forms. This representation of Clifford algebra by an algebra of differential forms is called the <u>Kähler algebra</u>.

The induced mapping of the curl into the differential forms gives us the <u>exterior derivative</u>, as expressed by

$$d\Phi = \langle \tilde{D} \Box \wedge \Psi \rangle . \quad (A.14)$$

This divergence maps to

$$-\delta \Phi = \langle \tilde{D} \Box \cdot \Psi \rangle . \quad (A.15)$$

Therefore, the vector derivative maps to

$$(d - \delta)\Phi = \langle \tilde{D} \Box \Psi \rangle = dx^{\mu} \vee \langle \tilde{D} \partial_{\mu} \Psi \rangle . \quad (A.16)$$

Now the Dirac equation in any of the forms we discussed above can easily be mapped into an equivalent Dirac-Kähler equation in the Kähler algebra. Of course, ideals in the Clifford algebra map into corresponding ideals in the Dirac algebra. I should mention, though, that for ideal spinors on curved manifolds the Kähler derivative defined by (A.16) differs from the derivative in the usual form of the Dirac equation. In particular, it couples minimal left ideals. This point is discussed by Benn and others in their lectures. I want to point out, however, that no such coupling occurs if one employs the operator representation of a Dirac spinor, and there is more than one possible way to define a covariant derivative for spinors. Unfortunately, there appears to be no hope of distinguishing between the various possibilities by any sort of experimental test.

REPRESENTATION-FREE CALCULATIONS IN RELATIVISTIC QUANTUM MECHANICS

M. F. Ross
University of the Pacific
Stockton, California 95211

ABSTRACT. It was shown by Pauli that expectation values in relativistic quantum mechanics for spin-1/2 particles are independent of the representation chosen for the Dirac matrices. In this paper we extend and generalize the works of Eddington, Sauter, Riesz, and Teitler to show that these expectation values may be calculated without choosing <u>any</u> representation for the Dirac matrices. This is done by replacing the column spinor wave function by a square matrix to obtain an equation which is similar but not identical to the Dirac equation. This new equation, invented by Sauter, provides a framework in which only the algebraic properties of the matrix operators and wave functions are needed for all calculations in relativistic quantum mechanics.

1. INTRODUCTION

When Dirac[1] invented the equation which now bears his name, he did so by postulating a first-order matrix differential equation which gives the relativistic energy-momentum relation when it is iterated. This requirement on the equation leads to algebraic conditions for the "Dirac" matrices in the equation. These constraints, however, do not uniquely specify these matrices. This ambiguity was shown not to be a problem by Pauli[2] who proved that any set of matrices with the correct algebraic properties would lead to the same bilinear densities and therefore to the same expectation values. Thus, the expectation values of physical quantities are independent of the representation chosen for the Dirac matrices. In other words, one can choose any set of Dirac matrices and calculate expectation values which are the same as those obtained by choosing any other set of Dirac matrices.

This paper goes one step further and shows that one can calculate these same expectation values using only the algebraic properties of the Dirac matrices and never choosing a matrix representation at all. Thus the expectation values of physical quantities are not merely representation <u>independent</u>, they are representation <u>free</u>.

J. S. R. Chisholm and A. K. Common (eds.), Clifford Algebras and Their Applications in Mathematical Physics, 347–352.

2. REPRESENTATION FREE EXPECTATION VALUES

The Dirac equation for an electron in an external field is

$$(-i\hbar\gamma_\mu\partial^\mu - \frac{e}{c}\gamma_\mu A^\mu + imc)\Psi = 0 \qquad (1)$$

where $-\gamma_0{}^2 = \gamma_1{}^2 = \gamma_2{}^2 = \gamma_3{}^2 = 1$ and $\gamma_\mu\gamma_\nu = -\gamma_\nu\gamma_\mu$ for $\mu \neq \nu$.

The γ_μ are 4 x 4 matrices and Ψ is a 4 x 1 column matrix. Consider an equation introduced by Sauter[3-5] which appears similar to Eq. 1 but is formally and fundamentally different:

$$(-i\hbar\gamma_\mu\partial^\mu - \frac{e}{c}\gamma_\mu A^\mu + imc)M = 0. \qquad (2)$$

The γ_μ are 4 x 4 matrices as above, but M is not a column matrix, it is also a 4 x 4 matrix.

For each solution M of Eq. 2 we can find a solution Ψ of Eq. 1 as follows: choose a matrix representation for M, multiply M by any constant column matrix v on the right that does not give zero, and then normalize Mv to obtain Ψ.

For a free particle, for example, where $A^\mu = 0$ for $\mu = 0, 1, 2, 3$, one solution to Eq. 2 is

$$M = \exp(ip_\mu x^\mu/\hbar)(mc+i\gamma_\mu p^\mu)/2mc$$

where $p_\mu p^\mu = -m^2c^2$. We can choose a representation for the γ_μ to find the matrix M, multiply M by any constant column v, and then normalize the result to find Ψ.

For each solution Ψ to Eq. 1 there are many solutions to Eq. 2. These solutions can be written in the form

$$M = (a\Psi\ b\Psi\ c\Psi\ d\Psi) \qquad (3)$$

where a,b,c,d are arbitrary complex constants.
Each of these entries in M is a 4 x 1 matrix, and the four columns constitute a 4 x 4 matrix when placed side-by-side. We can simplify our notation if we let

$$M = \Psi(a\ b\ c\ d) = \Psi v^\dagger \qquad (4)$$

where $v^\dagger = (a\ b\ c\ d)$ is a 1 x 4 row matrix.

Expectation values are calculated by integrating bilinear densities over all of space. For the column matrix Ψ from Eq. 1, the bilinear density associated with the matrix operator Q is $\Psi^\dagger Q\Psi$. For the square matrix M from Eq. 2, we consider instead the following trace:

$$\mathrm{Tr}(M^\dagger QM) = \mathrm{Tr}(v\Psi^\dagger Q\Psi v^\dagger) = \Psi^\dagger Q\Psi\ \mathrm{Tr}(vv^\dagger) = \Psi^\dagger Q\Psi v^\dagger v. \qquad (5)$$

If we now require that M be normalized such that

$$\mathrm{Tr}(M^{\dagger}M) = \Psi^{\dagger}\Psi, \qquad (6)$$

we find from Eq. 5 with Q = 1 that

$$v^{\dagger}v = |a|^2 + |b|^2 + |c|^2 + |d|^2 = 1. \qquad (7)$$

We therefore obtain the following relation:

$$\mathrm{Tr}(M^{\dagger}QM) = \Psi^{\dagger}Q\Psi. \qquad (8)$$

Except for some algebraic restrictions on M, our goal has been accomplished. Whenever we wish to calculate the bilinear density on the right of Eq. 8 we can instead calculate the trace on the left of the same equation.

Note that both sides of Eq. 8 are representation independent: the calculation of either side of this equation will give the same results in any representation. However, only the left side of the equation is also representation free: this trace can be calculated using only the algebraic properties of the γ_μ which constitute both M and Q so that no matrix representation is needed.

In standard calculations one solves Eq. 1 for Ψ in order to find $\Psi^{\dagger}Q\Psi$ and likewise one solves Eq. 2 for M in order to find $\mathrm{Tr}(M^{\dagger}QM)$. The difference between the two calculations is that one must choose a specific representation for the Dirac matrices in order to solve Eq. 1 while only the algebraic properties of the Dirac matrices are needed to solve Eq. 2.

3. ALGEBRAIC PROPERTIES OF THE WAVE FUNCTION M

Although not every matrix M which is a solution to Eq. 2 has the form $\Psi v^{\dagger}$, we can determine which solutions do have this form and therefore satisfy Eq. 8 without explicitly finding Ψ and $v^{\dagger}$. The algebraic properties of the solutions to Eq. 2 that have the form $\Psi v^{\dagger}$ can be seen from the following three theorems.

Let L be the set of all 4 x 4 square matrices of the form $\Psi v^{\dagger}$ where Ψ is any 4 x 1 column matrix and $v^{\dagger}$ is a particular 1 x 4 row matrix. Let A be the algebra formed by the set of all 4 x 4 square matrices.

3.1. Theorem 1. L is a left ideal of the algebra A.

Proof: Let $\Psi v^{\dagger}$ be any element of L and a be any element of A. From the rules of matrix multiplication, $a(\Psi v^{\dagger}) = (a\Psi)v^{\dagger} = \Psi' v^{\dagger}$ where $\Psi' = a\Psi$. Since Ψ' is a 4 x 1 column matrix, $\Psi' v^{\dagger}$ is also an element of L. This argument is true for all elements of L and A so L must be a left ideal of A.

3.2. Theorem 2. L is a minimal left ideal of the algebra A.

Proof: Let S be a nontrivial proper subideal of L. Then there exists at least one matrix $\Psi v^\dagger$ which is an element of both L and of S and another matrix $\Psi' v^\dagger$ which is an element of L but not S. There also exists an element of A, namely $a = \Psi'\Psi^\dagger/\Psi^\dagger\Psi$, such that $a(\Psi v^\dagger) = \Psi' v^\dagger$. Since $\Psi' v^\dagger$ is not an element of S we have contradicted the assumption that S is an ideal of A. There are no proper subideals of L other than the set which contains only the zero matrix so L must be a minimal left ideal of A.

Corollary. Let R be the set of all 4 x 4 square matrices of the form $v\Psi^\dagger$ where $\Psi^\dagger$ is any 1 x 4 row matrix and v is a particular 4 x 1 column matrix. R is a minimal right ideal of the algebra A.

3.3. Theorem 3. The elements of any minimal left ideal of the algebra A can be written in the form $\Psi v^\dagger$ where $v^\dagger$ is the same for all elements in the ideal.

Proof: Let R be the right minimal ideal of A defined above. Let Λ be any minimal left ideal of A. If r is an element of R and λ is an element of Λ, then the product $r\lambda$ is an element of both R and Λ. Thus at least one element of Λ, namely $r\lambda$, can be written in the form $\Psi v^\dagger$. By left multiplication on $r\lambda$ we can generate every other element of this form and obtain a new left ideal Λ' which is a subideal of Λ. Since Λ is a minimal left ideal in A, Λ and Λ' must be the same ideal. Thus each element in Λ can be written in the form $\Psi v^\dagger$.

We have found from the above theorems that we can make a one-to-one correspondence between the column matrix solutions of Eq. 1 and the square matrix solutions of Eq. 2 as long as these square matrix solutions are elements of the same minimal left ideal. We have also shown that every solution of Eq. 2 which is an element of a minimal left ideal has the correct algebraic properties such that Eq. 8 is satisfied. We can therefore abandon Eq. 1 with the confidence that we can calculate using only the algebraic properties of the solutions to Eq. 2.

The free particle solution that we found earlier was an element of a left ideal, but not a minimal left ideal. A solution that is an element of a minimal left ideal and will give the correct expectation values is

$$M = \exp(ip_\mu x^\mu/\hbar)(mc+i\gamma_\mu p^\mu)(1+\gamma_5\gamma_\mu s^\mu)/4mc$$

where $p_\mu p^\mu = -m^2c^2$, $s_\mu s^\mu = 1$, $p_\mu s^\mu = 0$, and $\gamma_5 = \gamma_0\gamma_1\gamma_2\gamma_3$. That this wave function can be written in the form $\Psi v^\dagger$ follows from the above theorems as well as a theorem by Eddington[6] since the last two factors form a primitive idempotent. The requirement that the solutions to Eq. 2 be elements of minimal left ideals was given first by Riesz[7-8], and later by Teitler[9-13], Greider[14-16], and Holland[17].

Bilinear densities of the form $\Psi_1{}^\dagger Q\Psi_2$ are also found in relativistic quantum mechanics whenever the wave function Ψ is expanded in a complete set of states:

$\Psi = \sum c_i \Psi_i \quad c_i$ are complex constants.

In order to expand the matrix M in a complete set of states we must choose all of the M_i to be in the same minimal left ideal as M:

$$M = \sum c_i M_i \qquad M_i = \Psi_i v^\dagger \qquad M = \Psi v^\dagger .$$

When this requirement is imposed on the M_i, we obtain a generalization of Eq. 8:

$$\mathrm{Tr}(M_1{}^\dagger Q M_2) = \mathrm{Tr}(v\Psi_1{}^\dagger Q\Psi_2 v^\dagger) = \Psi_1{}^\dagger Q\Psi_2 \mathrm{Tr}(vv^\dagger) = \Psi_1{}^\dagger Q\Psi_2 . \qquad (9)$$

A solution to Eq. 2 which is an element of a minimal left ideal which is independent of p and s is

$$M = \exp(ip_\mu x^\mu/\hbar)(mc+i\gamma_\mu p^\mu)(1+\gamma_5\gamma_\mu s^\mu)(1+i\gamma_0)(1+\gamma_5\gamma_3)/16mc.$$

4. CONCLUSION

We have seen that there are two equivalent methods to calculate expectation values in relativistic quantum mechanics, starting from Eq. 1 or from Eq. 2. Although each method gives us the same result, there is an advantage in using Eq. 2. If we find M from Eq. 2 and then calculate $\mathrm{Tr}(M^\dagger QM)$, we can do so using only the algebraic properties of the γ_μ. Since we never need to find the matrix elements of the γ_μ, M, or Q, we can change our view of these matrices to elements of an algebra which has the same properties as these matrices. This algebra is a Clifford algebra of four generators taken over the complex numbers and its properties have been studied in this connection by Riesz[7-8] and Teitler[9-13], and more recently by Greider[14-16] and Holland[17].

REFERENCES

1. P. Dirac, Roy. Soc. Proc. 117, 610 (1928).
2. W. Pauli, Ann. Inst. Henri Poincaré 6, 109 (1936).
3. F. Sauter, Z. Physik 63, 803 (1930).
4. F. Sauter, Z. Physik 64, 295 (1930).
5. F. Sauter, Z. Physik 97, 777 (1935).
6. A. Eddington, Fundamental Theory (Cambridge University Press, Cambridge, 1946), p. 143.
7. M. Riesz, Comptes Rendus du Dixième Congrès des Mathematiques des Pays Scandanaves, (Copenhagen, 1946), p. 123.
8. M. Riesz, Comptes Rendus du Douzième Congrès des Mathematiques des Pays Scandanaves, (Lund, 1953), p. 241.
9. S. Teitler, Nuovo Cimento Suppl. 3, 1 (1965).
10. S. Teitler, Nuovo Cimento Suppl. 3, 15 (1965).
11. S. Teitler, J. Math. Phys. 6, 1976 (1965).
12. S. Teitler, J. Math. Phys. 7, 1730 (1965).
13. S. Teitler, J. Math. Phys. 7, 1739 (1966).
14. K. Greider, Phys. Rev. Lett. 44, 1718 (1980).

15. K. Greider, Found. Phys. 14, 467 (1984).
16. K. Greider, Found. Phys. 15, 693 (1985).
17. P. Holland, J. Phys. A, 16, 2363 (1983).

DIRAC EQUATION FOR BISPINOR DENSITIES

J. P. Crawford
The Pennsylvania State University
Fayette Campus
Uniontown, PA 15401
USA

ABSTRACT. The relationship between the non-linear algebra satisfied by the real bispinor densities $\rho_i \equiv \overline{\Psi}\Gamma_i\Psi$ and the underlying Fierz and Clifford algebras is examined in Minkowskian spaces with one temporal and N-1 spacial coordinates. A general expansion for a spinor in terms of the bispinor densities is found and used to solve the inverse problem: Given a set of functions ρ_i which satisfy the bispinor algebra, construct the spinor Ψ to which they correspond. Finally, the Dirac equation for the spinor wavefunction in the presence of arbitrary potentials is transformed into a set of differential equations for the bispinor densities.

1. INTRODUCTION

In the context of quantum mechanics all physical observables are obtained from real quadratic functionals of the quantum wavefunction. For a Dirac particle in a spacetime of dimension N, represented by a $D=2^{[N/2]}$ component complex spinor wavefunction Ψ, there exist D^2 real quadratic forms $\rho_i \equiv \overline{\Psi}\Gamma_i\Psi$ which do not involve derivatives. These bispinor densities ρ_i are not independent since the underlying spinor, being composed of D complex functions, contains a total of 2D independent functions. Furthermore, as the overall phase of the spinor has no effect on the bispinor densities, only (2D-1) of the ρ_i may be considered independent, and therefore must satisfy $(D-1)^2$ equations. This system of equations is here referred to as the bispinor algebra and, not surprisingly, is related to the Clifford[1] and Fierz[2] algebraic systems.

In this article the problem of writing the Dirac equation in terms of the bispinor densities is addressed. This problem is reminiscent of writing the Schrodinger equation in "hydrodynamic form" and has been considered previously in various forms by Riordan[3], des Cloizeaux[4], and Mikhov[5]. Herein lies an alternate formulation.

The outline of the paper is as follows. In Section II, the relationships among the Clifford, Fierz, and bispinor algebras are discussed, with particular attention paid to projection operators. In Section III, a general method of writing a spinor in terms of the

J. S. R. Chisholm and A. K. Common (eds.), Clifford Algebras and Their Applications in Mathematical Physics, 353–361.

corresponding bispinor densities is presented. In Section IV, the Dirac equation written in terms of bispinor densities is obtained. Finally, Section V contains some comments regarding open questions pertaining to this remarkable algebraic system.

2. CLIFFORD, FIERZ, AND BISPINOR ALGEBRAS

To begin, some properties and definitions of the Clifford, Fierz, and bispinor algebras are presented. This will constitute a brief review and define the notation.

2.1. Clifford Algebra

The basic defining relation of the Clifford algebra is given by:

$$\{\gamma_\mu, \gamma_\nu\} = 2g_{\mu\nu}I \tag{2.1}$$

where here we take $g_{\mu\nu}$ to be the metric of $R^{1,N-1}$. The elements γ_μ and I are the generators of the Clifford algebra. There are various ways of completing the algebra (generating real or complex algebras, for example). Here the algebra is defined to be Dirac self adjoint. That is, all of the elements of the algebra Γ_i have the property of generating real functions when contracted between any spinor and its Dirac adjoint:

$$\rho_i \equiv \overline{\Psi}\Gamma_i \Psi \equiv \Psi^\dagger\gamma\Gamma_i \Psi \ , \ \rho_i^* = \overline{\Psi}\gamma\Gamma_i^\dagger\gamma\Psi = \rho_i \tag{2.2}$$

$$\overline{\Gamma}_i \equiv \gamma\Gamma_i^\dagger\gamma = \Gamma_i \tag{2.3}$$

where γ is any matrix which satisfies:

$$\gamma^\dagger = \gamma \ , \ \gamma^2 = I \ , \ \gamma\gamma_\mu^\dagger\gamma = \gamma_\mu \ . \tag{2.4}$$

(Note that in the familiar situation of N=4 one can choose $\gamma=\gamma_0$). To be specific, a possible choice for completing the algebra is to add the following matrices to the set:

$$\tilde{\gamma}_{\mu_1\mu_2\cdots\mu_{N-n}} \equiv (i)^{n(n-1)/2}(n!)^{-1} \times \times \varepsilon_{\mu_1\mu_2\cdots\mu_N}\gamma^{\mu_{N-n+1}}\gamma^{\mu_{N-n+2}}\cdots\gamma^{\mu_N} \tag{2.5}$$

where n = 2,3,...,N. Then the set of matrices forming the Clifford algebra is:

$$\{\Gamma_i\}=\{I,\gamma_\mu,\tilde{\gamma}_{\mu_1\cdots\mu_{N-2}},\ldots,\tilde{\gamma}_\mu,\tilde{\gamma}\} \tag{2.6}$$

where we have defined: $\tilde{\gamma} \equiv \tilde{\gamma}_{\mu_1\cdots\mu_N}$. As an illustrative example consider the familiar situation of N=4. In this case we find[6]:

$$n=2:\ \tilde{\gamma}_{\mu\nu} = (i)(2!)^{-1}\varepsilon_{\mu\nu\lambda\tau}\gamma^{\lambda}\gamma^{\tau} = \tfrac{1}{2}\,\varepsilon_{\mu\nu\lambda\tau}\sigma^{\lambda\tau}$$

$$n=3:\ \tilde{\gamma}_{\mu} = (i)^{3}(3!)^{-1}\varepsilon_{\mu\nu\lambda\tau}\gamma^{\nu}\gamma^{\lambda}\gamma^{\tau} = \gamma_5\gamma_{\mu}$$

$$n=4:\ \tilde{\gamma} = (i)^{6}(4!)^{-1}\varepsilon_{\mu\nu\lambda\tau}\gamma^{\mu}\gamma^{\nu}\gamma^{\lambda}\gamma^{\tau} = i\gamma_5 \quad .$$

Counting the number of elements in the algebra we find $d=2^N$, however if N is odd then each matrix appears twice in the set so that $d=2^N$ if N is even and $d=2^{N-1}$ if N is odd and, therefore, the matrices are $D\times D$ where D, the dimension of the corresponding spinor space, is given by:

$$D = 2^{[N/2]}, \tag{2.7}$$

where [N/2] is the integer part of N/2.

The metric on the Clifford algebra is defined as follows:

$$G_{ij} \equiv D^{-1}\,\mathrm{tr}(\Gamma_i\Gamma_j) \tag{2.8a}$$

$$G_{ij}G^{jk} = \delta_i^{\ k} \quad . \tag{2.8b}$$

For the choice of matrices given in equation (2.6) this metric is purely diagonal and has the signature of SO(d/2,d/2).

The structure factors for the algebra are defined as:

$$\Gamma_i\Gamma_j \equiv C_{ijk}\,\Gamma^k \tag{2.9a}$$

$$C_{ij\ell} = D^{-1}\mathrm{tr}(\Gamma_i\Gamma_j\Gamma_\ell) \quad , \tag{2.9b}$$

where the second equation follows from the first by multiplying by Γ_ℓ, taking the trace, and using equation (2.8a). Higher order structure factors can also be defined as follows:

$$\Gamma_i\Gamma_j\Gamma_k \equiv C_{ijk\ell}\,\Gamma^\ell \tag{2.10a}$$

$$C_{ijk\ell} = D^{-1}\mathrm{tr}(\Gamma_i\Gamma_j\Gamma_k\Gamma_\ell) \tag{2.10b}$$

$$C_{ijk}C^{k}_{\ \ell m} = C_{ij\ell m} \quad . \tag{2.10c}$$

Note that all of the structure factors are invariant under cyclic permutations of the indices. Also note that, given the C_{ijk}, equation (2.9a) can be considered the definition of the Clifford algebra, allowing for higher dimensional representations than the spinor representation. In particular, the d dimensional adjoint representation is given by the $d\times d$ matrices ${}^i[\Gamma_j]_k = C^i_{\ jk}$, i and k being the row and column indices.

2.2. Fierz Algebra

The Fierz algebra (or Fierz rearrangement theorem) arises when one considers the reordering of the spinors χ, Φ, ζ and ξ in the quartic form $(\bar{\chi}\Gamma_i \Phi)\cdot(\bar{\zeta}\Gamma_j \xi)$. Using α, β, γ and δ for spinor indices, this reordering can generally be written as:

$$ {}^{\alpha}\Gamma^{i}_{\beta}\,{}^{\gamma}\Gamma^{j}_{\delta} = F^{ij}{}_{k\ell}\,{}^{\alpha}\Gamma^{k}_{\delta}\,{}^{\gamma}\Gamma^{\ell}_{\beta} \tag{2.11}$$

where the $F_{ijk\ell}$ are the structure factors of the Fierz algebra. The relationships between the structure factors for the Fierz and Clifford algebras may be obtained as follows, using equations (2.8a), (2.10b), and (2.11), and contracting with ${}^{\beta}\Gamma^{m}_{\alpha}\,{}^{\delta}\Gamma^{n}_{\gamma}$:

$$\mathrm{tr}(\Gamma^m\Gamma^i)\,\mathrm{tr}(\Gamma^n\Gamma^j) = F^{ij}{}_{k\ell}\,\mathrm{tr}(\Gamma^m\Gamma^k\Gamma^n\Gamma^\ell)$$

$$F_{ijk\ell}\,C_m{}^k{}_n{}^\ell = DG_{im}G_{jn} \quad , \tag{2.12}$$

and similarly:

$$F_{ijmn} = D^{-1}C_{minj} \quad . \tag{2.13}$$

2.3. Bispinor Algebra

The bispinor algebra now follows easily from the Fierz algebra. Contracting equation (2.11) twice on a spinor Ψ and its Dirac adjoint and making use of equation (2.2) we find:

$$\rho_i\rho_j = F_{ijk\ell}\,\rho^k\rho^\ell \quad . \tag{2.14}$$

(Note that we have ignored the anticommutation property of the spinor Ψ. When dealing with quantum field theory equation (2.14) obtains a minus sign on the right hand side.) These equations constitute the bispinor algebra . Only $(D-1)^2$ of these d^2 equations are independent, leaving $(2D-1)$ independent functions in the set $\{\rho_i\}$.

2.4. Projection Operators

We now turn to a remarkable property of the bispinor algebra. Consider any set of real functions $\{R_i\}$ which satisfy the bispinor algebra. Then there exists some spinor such that $R_i = \bar{\Phi}\Gamma_i\Phi$. (The explicit construction of Φ occurs in Section 3.) For convenience we define $R_s \equiv \bar{\Phi}\Phi \equiv \sigma$, and also $\Gamma_s \equiv I$. Construct the Dirac self adjoint matrix:

$$R = R^i\Gamma_i \quad , \quad \bar{R} = \gamma R^\dagger\gamma = R \tag{2.15}$$

and consider:

$$R\Gamma_i R = R^j R^k \Gamma_j \Gamma_i \Gamma_k = F^{jk\ell m} C_{jikn} R_\ell R_m \Gamma^n$$
$$= DG^\ell{}_n G^m{}_i R_\ell R_m \Gamma^n = DR_i R \tag{2.16}$$

where we have used equations (2.10,), (2.14), and (2.12). Then the matrix $(D\sigma)^{-1}R$ is idempotent since:

$$RR = DR_s R = (D\sigma)R \quad . \tag{2.17}$$

Also note that:

$$\mathrm{tr}R = \mathrm{tr}(R^i \Gamma_i) = \sigma\,\mathrm{tr}(I) = D\sigma \tag{2.18}$$

and therefore R has one non-zero eigenvalue $(D\sigma)$, and the matrix $(D\sigma)^{-1}R$ is a projection operator.

Finally, consider the matrix:

$${}^i[P]_j \equiv {}^iP_j \equiv C^i{}_{jk}R^k \quad . \tag{2.19}$$

This matrix is Hermitian since:

$$C^{*i}{}_{jk}\Gamma^k = \overline{C^i{}_{jk}\Gamma^k} = \overline{\Gamma^i\Gamma_j} = \Gamma_j\Gamma^i = C_j{}^i{}_k\Gamma^k$$
$$C^i{}_{jk} = C^*{}_j{}^i{}_k \quad . \tag{2.20}$$

Using equations (2.14) and (2.13) we find:

$$\sigma R_i = D^{-1}C_{ks\ell i}R^k R^\ell = D^{-1}C_{k\ell i}R^k R^\ell = D^{-1}{}_iP_j R^j$$
$${}^iP_j R^j = (D\sigma)R^i \tag{2.21}$$

and therefore R^i is an eigenvector of ${}^iP_j$ with eigenvalue $(D\sigma)$. In fact $(D\sigma)^{-1}\,{}^iP_j$ is idempotent:

$${}^iP_j{}^jP_k = C^i{}_{jm}C^j{}_{kn}R^m R^n = C_m{}^i{}_{kn}R^m R^n$$
$$= C^i{}_{k\ell}C^\ell{}_{nm}R^n R^m = C^i{}_{k\ell}{}^\ell P_m R^m$$
$$= C^i{}_{k\ell}(D\sigma)R^\ell = (D\sigma)\,{}^iP_k \quad . \tag{2.22}$$

Now observe, using equations (2.12) and (2.13):

$$DG_{im}G_{jn} = D^{-1}C_{ki\ell j}C_m{}^k{}_n{}^\ell$$
$$DG_{is}G_{ss} = DG_{is} = D^{-1}C_{ki\ell}G^{k\ell} = D^{-1}C^k{}_{ki}$$
$$C^k{}_{ki} = D^2 G_{is} \tag{2.23}$$

so that:

$$\mathrm{tr}(P) = C^{k}{}_{k\ell} R^{\ell} = D^{2}\sigma \tag{2.24}$$

and therefore P has D non-zero eigenvalues $(D\sigma)$ and the matrix $(D\sigma)^{-1}P$ is a projection operator. This is not surprising since R (equation (2.15)) and P (equation (2.19)) are essentially the same operators written in different representations, R in spinor representation and in adjoint representation.

3. GENERAL SPINOR EXPANSION AND INVERSION THEOREM

In this section an assertion concerning the form in which any spinor may be written is presented[7]. In the course of demonstrating the veracity of this proposition the inversion theorem will also be proven[7,8].

Proposition: Any spinor Ψ may be written in the following form:

$$\Psi = e^{-i\phi} R^{j} \Gamma_{j}\, \eta = e^{-i\phi} R\eta \tag{3.1}$$

where η is an arbitrary constant spinor, ϕ and R_i are real functions, and furthermore the set $\{R_i\}$ constitutes a bispinor algebra. Note that the set $\{\phi, R_i\}$ comprise 2D real independent functions.

To demonstrate this result we begin by constructing the bispinor densities generated by the spinor:

$$\rho_i = \bar{\Psi}\Gamma_i \Psi = \bar{\eta} R \Gamma_i R\eta = DR_i \bar{\eta} R\eta \ , \tag{3.2}$$

where we have used equation (2.16) in the last step. At this point, we note that instead of beginning with a known spinor Ψ (to be cast in the form of equation (3.1)) we could begin with known bispinor densities ρ_i, in which case the explicit construction of the spinor Ψ in the form of equation (3.1) constitutes the inversion theorem. Now observe that $\bar{\eta}R\eta$ is a real scalar function so we have:

$$R_i = \Omega \rho_i \tag{3.3}$$

for some function Ω. To determine Ω we substitute equation (3.3) into equation (3.2) to obtain:

$$\Omega^{2} = [D\bar{\eta}\rho^{i}\Gamma_{i}\,\eta]^{-1} \ . \tag{3.4}$$

Therefore, equations (3.4) and (3.3) determine the functions R_i. If the spinor Ψ is known, the phase function ϕ may be obtained via:

$$\bar{\eta}\Psi = e^{-i\phi}\bar{\eta}R\eta = e^{-i\phi}\Omega\bar{\eta}\rho\eta \ . \tag{3.5}$$

This construction completes the proof of the proposition (equation (3.1)). Note that if we begin only with the knowledge of the physical observables it is not possible to find the phase function ϕ.

4. DIRAC EQUATION FOR BISPINOR DENSITIES

We are now in position to find the differential equations satisfied by the functions ϕ and R_i. Consider the Dirac equation for a spinor particle in the presence of arbitrary potentials:

$$(i\partial\!\!\!/ - \theta^i\Gamma_i)\Psi = 0 \quad . \tag{4.1}$$

The potentials θ_i are arbitrary real functions. (In the familiar case of N=4 this set is comprised of scalar, pseudoscalar, vector, axial vector, and antisymmetric tensor functions.) Substitution of the general spinor expansion (equation (3.1)), into equation (4.1) yields:

$$\begin{aligned} 0 &= [i\partial\!\!\!/ - (\theta^i\Gamma_i - \partial^\mu\phi\gamma_\mu)]R^j\Gamma_j\eta \\ &\equiv Q^k\Gamma_k\eta = Q\eta \quad , \end{aligned} \tag{4.2}$$

where Q^i are D^2 complex functionals of the real functions ϕ and R_i. Next we contract equation (4.2) with $\bar{\eta}\Gamma^i$ to obtain:

$$\begin{aligned} 0 &= \bar{\eta}\Gamma^i\Gamma_j\eta Q^j = C^i{}_{jk}\bar{\eta}\Gamma^k\eta Q^j \\ &\equiv C^i{}_{jk}h^kQ^j \equiv {}^iH_jQ^j \end{aligned} \tag{4.3}$$

where we have defined the constant bispinor algebra formed by the spinor η as h_i and the corresponding projection operator as ${}^iH_j$ (see equation (2.19)). Recall that the matrix ${}^iH_j$ has D non-zero eigenvalues and $(D^2\text{-}D)$ zero eigenvalues, so we define the set of eigenvectors:

$$\{X^i{}_j\} = \{Y^i{}_a, Z^i{}_A\} \tag{4.4a}$$

$${}^iH_jY^j{}_a = Dh_sY^i{}_a \qquad a = 1,2,\cdots,D \tag{4.4b}$$

$${}^iH_jZ^j{}_A = 0 \qquad A = D+1,\cdots,D^2 \tag{4.4c}$$

$$X^j{}_\ell X_{jm} = G_{\ell m} \quad . \tag{4.4d}$$

The general solution of equation (4.3) may be written as:

$$Q^i = Z^i{}_Aq^A \tag{4.5}$$

where the q^A are $(D^2\text{-}D)$ arbitrary complex functions, and therefore:

$$Q^jY_{ja} = 0 \tag{4.6}$$

follows from the normalization condition, equation (4.4d). This is the Dirac equation for the scaled bispinor densities R_i and the phase function ϕ.

There are (D^2+1) real functions in the set $\{\phi, R_i\}$. The bispinor algebra (equation (2.14)) comprises $(D-1)^2$ equations, and equation (4.6) contains a total of 2D equations (D complex equations) so there are a total of (D^2+1) equations for the (D^2+1) unknowns, and therefore the system is complete.

Inspection of equation (4.2) for the definition of the Q^i leads to the conclusion that the differential equations (equation (4.6)) are "almost" linear, the only non-linearity arising from the presence of the gradient of the phase function in the potential term. In a theory in which the vector potential is a gauge field, this non-linearity may be eliminated by a suitable choice of gauge. The reason we obtain linear differential equations is two-fold: the functions R_i must also satisfy the bispinor algebra which is a system of non-linear algebraic equations, and the functions R_i are themselves not the physical observables, as inspection of equation (3.2) reveals, but are scaled by the common factor $\bar{\eta}R\eta$ which itself depends on R.

5. CONCLUDING REMARKS

Clearly, this study of the bispinor algebra and the Dirac equation is incomplete. There are many more questions and problems to be addressed, including: (1) It is likely that there exists other spinor expansions aside from equation (3.1) which involve a bispinor algebra directly. One possible candidate is:

$$\Psi = \sigma^{1/2} e^{-i\phi} \eta \ , \ \phi = \phi^i \Gamma_i \qquad (5.1)$$

where σ is a real positive scalar function, η is an arbitrary constant spinor, and $\{\phi^i\}$ forms a real bispinor algebra. This form suggests generalizations of gauge transformations. (2) Aid in understanding the global structure of spinor solutions to non-linear Dirac equations may come from examining the geometric structure of the bispinor algebra. In particular, the bispinor densities ρ_i can be imagined to form a D^2 dimensional flat space with metric G_{ij} (equation (2.8)). Imposition of the non-linear bispinor algebra (equation (2.14)) induces an embedding in this flat space to a (2D-1) dimensional curved space. This geometric structure can also be used to generate non-linear sigma models. For the special case of two or three spacetime dimensions, the metric of the three dimensional curved bispinor space is Robertson-Walker.

These problems are currently under investigation.

6. REFERENCES

(1) W.K. Clifford, Am. J. Math. 1 (1878) 350; See also R.H. Good, Rev. Mod. Phys. 97 (1955) 810 for a review.
(2) M. Fierz, Z. Phys. 104 (1937) 553; W. Pauli, Ann. Inst. Henri Poincare 6 (1936) 109.
(3) F. Riordan, J. Phys. A 16 (1982) 71.
(4) J. des Cloizeaux, J. Physique 44 (1983) 885.
(5) S.G. Mikhov, Lett. Math. Phys. 9 (1985) 313.

(6) For the Clifford algebra in four dimensional spacetime we use the conventions of J.D. Bjorken and S.D. Drell, Relativistic Quantum Mechanics (McGraw-Hill, New York, 1964). In particular $\gamma_5 = i\gamma_0\gamma_1\gamma_2\gamma_3$ and $\sigma_{\mu\nu} = (i/2)[\gamma_\mu, \gamma_\nu]$.

(7) J.P. Crawford, J. Math. Phys. 26 (1985) 1439. The results of this reference are specific to the case N=4, whereas the material contained in this paper is valid for arbitrary N.

(8) Y. Takahashi, Prog. Theor. Phys. 69 Letters (1983) 369; Phys. Rev. D 26 (1982) 2169. See also: A.A. Campolataro, I.J.T.P. 19 (1980) 99, 127.

UNIFIED SPIN GAUGE THEORY MODELS

R. S. Farwell
St. Mary's College
Twickenham
England

J. S. R. Chisholm
University of Kent
Canterbury
Kent England

ABSTRACT When Gauge Theories of Particle Physics are formulated consistently within the framework of a Clifford Algebra, this formulation is referred to as Spin Gauge Theory. The basic principles and aims of spin gauge theories are presented. They are illustrated by an example: a spin gauge theory which generates exactly the Glashow-Salam-Weinberg helicity symmetric SU(2) interactions for the electron and its neutrino. Finally, two unified spin gauge theory models are briefly described; the first is a theory of the electrostrong interactions and the second unifies this theory with the weak interaction theory of the Glashow-Salam-Weinberg Standard Model.

1. Introduction

A gauge theory is a Lagrangian field theory; the form of the Lagrangian density is determined by requiring invariance under certain local (that is, space-time dependent) transformations. We shall describe the gauge principle with reference to a free Lagrangian density L_o for a massless Dirac bispinor, describing a fermion such as an electron. We are free to transform the spinor by an arbitrary phase factor without changing the physics described by the Lagrangian. It is this freedom which we shall gauge.

We represent the Dirac spinor by a 4 component column vector $\psi(x)$ and define a set of 4×4 matrices $\{\gamma_\mu,\ \mu=1,2,3,4\}$ to be generators of the complex Dirac algebra C_4. The 'bar' or hermitian conjugate of ψ is given by $\bar{\psi}=\psi^+\gamma$, where the conjugation matrix γ is defined by $\gamma_\mu{}^+=\gamma\gamma_\mu\gamma^{-1}$ and is such that $\gamma^+=\gamma$.

The free Lagrangian density is

$$L_o = [\bar{\psi}i\gamma_\mu(\partial^\mu\psi) - (\partial^\mu\bar{\psi})i\gamma_\mu\psi]/2 \ ; \qquad (1)$$

if θ is a scalar, this is invariant under the global phase transformations

$$\psi \to R\psi \ , \quad \bar{\psi} \to \bar{\psi}R^{-1} \qquad (2a)$$

J. S. R. Chisholm and A. K. Common (eds.), Clifford Algebras and Their Applications in Mathematical Physics, 363–370.

$$R = e^{-i\theta} , \tag{2b}$$

since R and γ_μ commute.

However, if R is x-dependent, that is, if we consider a local phase transformation (2), then

$$L_o \to L_o + \bar{\psi} i\gamma_\mu R^{-1}(\partial^\mu R)\psi$$

and the invariance of the Lagrangian is lost. Invariance can be restored by replacing the derivative ∂^μ in (1) by a covariant derivative

$$D^\mu = \partial^\mu - ieA^\mu(x) .$$

The resultant Lagrangian L_1 is given by

$$L_1 = L_o + e\bar{\psi}\gamma_\mu A^\mu \psi \tag{3}$$

and is invariant under x-dependent transformations (2) provided that

$$A^\mu \to RA^\mu R^{-1} + e^{-1}R(\partial^\mu R^{-1}) . \tag{4}$$

The new term in the Lagrangian (3) is interpreted as the interaction between the fermion and the electromagnetic field; the gauge field $A^\mu(x)$ is the photon field.

2. Spin Gauge Theories

Dirac spinors are elements of a minimal left ideal of the Dirac algebra. The set of Dirac matrices $\{\gamma_\mu\}$ is a representation of the set of generators of this same algebra. The gauge principle described in §1 involves a transformation (2) of the spinor ψ and its bar conjugate $\bar{\psi}$. Since the Dirac spinors and the Dirac matrices are both expressed in terms of a representation of the algebra, if we choose to change the representation of the spinor, then we should also change the representation of the matrices in a corresponding way. We suggest that the matrices should transform according to

$$\gamma_\mu \to R\gamma_\mu R^{-1} . \tag{5}$$

Since the phase transformation (2b) has only the scalar of the Dirac algebra in the exponent, it commutes with γ_μ. Hence, if the transformation (5) were incorporated into the example of §1, nothing would change. However, the gauge principle will be effected if we generalise the phase transformations to include, as generators in the exponent, other elements of the Dirac algebra which do not necessarily commute with γ_μ.

This is the basic principle of a spin gauge theory. We gauge the freedom to choose a different representation of spinors and spin γ-matrices at each point x in space-time. Local transformations of spinors have been studied in the past in order to define the spin connection and to define the parallel transport of spinors[1]; the purpose of this work was not to define spin gauge theories.

The aim of proposing such a gauge theory is that it could provide a framework for the unification of the strong, weak and electromagnetic interactions in which the three interactions would be associated with different aspects of the same symmetry principle. A single bispinor cannot be used, since, for instance, the weak interactions mix the spinors of the electron and its neutrino. A multispinor consisting of several spinors is formed and the Clifford algebra enlarged accordingly. If the multispinor has 2^n components then the Clifford Algebra C_{2n} is used. If, in addition, the generators of the symmetry transformations can be selected from amongst the bivectors of the Clifford algebra, then they are subsets of the generators of Spin(2n). It is therefore possible that gravitational interactions might then be generated in an analogous way to the other interactions.

We propose to use a particular representation of the algebra C_{2n}, with an appropriate value of n, to generate the interactions of the electron and its neutrino, say. Our second aim is to obtain the interactions of other particles, the up and down quarks, for instance, by globally transforming the representation of the algebra.

We now define the spin gauge principle, which has been used by several authors[2]:
The form of the Lagrangian density for a 2^n component multispinor ψ is determined by requiring local gauge invariance under the spin gauge transformations

$$\psi \to R\psi \equiv \psi' \ , \quad \Gamma_i \to R\Gamma_i R^{-1} \equiv \Gamma'_i \ , \tag{6a}$$

where $\{\Gamma_i,\ i=1,2,\ldots,2n\}$ is the set of generators of the Clifford algebra C_{2n} to which ψ belongs. The generators of the transformation R are elements of the algebra C_{2n}. We assume that the conjugation matrix Γ transforms to Γ', the form of which can be deduced by demanding that $\Gamma_i'^+ = \Gamma'\Gamma'_i\Gamma'^{-1}$. We find

$$\Gamma \to \Gamma' \equiv (R^{-1})^+\Gamma R^{-1} \tag{6b}$$

and hence

$$\bar{\psi} \to \bar{\psi}' \equiv \bar{\psi}R^{-1} \ . \tag{6c}$$

An advantage of spin gauge theories is that there are more allowable transformations leaving a given Lagrangian invariant. For instance, invariance can be retained even when R and Γ_i do not commute and also when R is not unitary. If some of the elements Γ_A of the Clifford algebra used as generators of R do not commute with Γ_i, then it is only necessary that they are anti-self conjugate in the sense that

$$\Gamma_A = -\bar{\Gamma}_A = -\Gamma^{-1}\Gamma_A^+\Gamma \ .$$

It should be noted that, in retaining a column vector form for the multispinor ψ, we are not treating ψ and the generators Γ_i equally within the algebra. Hence their transformations (6) have different forms. In a full Clifford algebra framework, the spinor ψ would be taken as a matrix of the same dimensions as the Γ_i and it would transform to $R\psi R^{-1}$. Since ψ is a minimal left ideal, right multiplication of ψ by R^{-1} leaves its rank invariant. This can probably be used to explain why right multiplication does not change physically measurable quantities.

3. Example

We shall illustrate the principles of a spin gauge theory by referring to an example: we shall reproduce, within the context of a spin gauge theory, the Glashow-Salam-Weinberg (G-S-W) helicity symmetric SU(2) interactions for the electron and its neutrino.

A 16 component multispinor

$$\psi = (\varepsilon_L \; \varepsilon_R \; \nu_L \; \nu_R)^T$$

is used, where ε and ν are respectively the 4 component electron and neutrino spinors and the subscripts L and R denote left and right handed helicities. For example, the left and right handed electron spinors are defined by

$$\varepsilon_L = (I + i\gamma_5)\varepsilon/2 \quad \text{and} \quad \varepsilon_R = (I - i\gamma_5)\varepsilon/2 \;,$$

where $\gamma_5 = \gamma_1\gamma_2\gamma_3\gamma_4$. Both the left and right spinors are 4 component column vectors, although in each spinor two of the four components are zero. Thus ψ has eight zero components and does not appear to be the most economical choice of multispinor for our purposes. However, we shall see later in this section that this form for ψ is most important.

The spin gauge theory is formulated using a 16×16 representation of C_8 acting on the spinor ψ. By reproducing the G-S-W interactions, we mean that on 'decomposition' or 'reduction' of the multispinor Lagrangian to 4 component spinors, we obtain the G-S-W helicity symmetric interactions, that is,

$$\bar{\varepsilon}\gamma_\mu W_3{}^\mu \varepsilon + \bar{\varepsilon}\gamma_\mu W_-{}^\mu \nu + \bar{\nu}\gamma_\mu W_+{}^\mu \varepsilon - \bar{\nu}\gamma_\mu W_3{}^\mu \nu \;, \tag{7}$$

where $W_a{}^\mu$ (a=1,2,3) are the standard gauge fields associated with the three SU(2) generators and $W_\pm{}^\mu = (W_1{}^\mu \pm iW_2{}^\mu)/\sqrt{2}$. Hence the 16×16 matrix representation of C_8 that we use is built up from the set of 4×4 γ-matrices together with two sets of Pauli matrices. The elements of the representation are all written in the form of direct products

$$\lambda_i \rho_j \gamma_A \quad (i,j = 1,2,3,4) \;,$$

with $\lambda_4 = \rho_4 = I_2$. The ρ-matrices mix the left and right handed electron or neutrino spinors, while the λ-matrices mix the electron and neutrino spinors.

Before we give our choice of vectors for the 'lepton representation' of C_8, we first define the 16×16 conjugation matrix Γ. Its form is determined from the requirement that the norm of ψ be equal to the sum of the standard electron and neutrino norms, that is,

$$\bar{\psi}\psi = \bar{\varepsilon}\varepsilon + \bar{\nu}\nu \;.$$

We deduce that Γ is given by $\lambda_4\rho_1\gamma$ so that

$$\bar{\psi} = \psi^+\lambda_4\rho_1\gamma = (\bar{\varepsilon}_L\ \bar{\varepsilon}_R\ \bar{\nu}_L\ \bar{\nu}_R) ,$$

where $\bar{\varepsilon}_L = \bar{\varepsilon}(I + i\gamma_5)/2$ and $\bar{\varepsilon}_R = \bar{\varepsilon}(I - i\gamma_5)/2$.

The anti-commuting set of vectors for the 'lepton representation' of C_8 is taken as

$$\begin{aligned} \Gamma_\mu &= \lambda_4\rho_1\gamma_\mu \qquad (\mu=1,2,3,4) \\ \Gamma_5 &= i\lambda_1\rho_2 I \\ \Gamma_6 &= i\lambda_2\rho_2 I \\ \Gamma_7 &= \lambda_4\rho_1\gamma_5 \\ \Gamma_8 &= \lambda_3\rho_2 I , \end{aligned}$$

with $\Gamma_i^2 = -1$, i=1,2,3,5,6,7 and $\Gamma_4^2 = \Gamma_8^2 = 1$. In general, the vector basis and signature are chosen so that:-
(a) all (or as many as possible) of the gauge transformations required are generated by bivectors in C_8;
(b) given (a), the resultant gauge interactions are not imaginary and hence are not zero in the helicity symmetric Lagrangian density;
(c) the generators of the gauge transformations which anti-commute with the 16×16 matrices in the multispinor kinetic energy term are anti-self conjugate.

The free Lagrangian density for ψ is given by

$$L_o = [\bar{\psi}i\Gamma_\mu(\partial^\mu\psi) - (\partial^\mu\psi)^+\Gamma i\Gamma_\mu\psi]/2 , \qquad (8)$$

with summation over μ from 1 to 4. We shall illustrate the decomposition of a 16 component term into familiar 4 component terms by considering the first term in (8):-

$$\bar{\psi}i\Gamma_\mu(\partial^\mu\psi) = (\bar{\varepsilon}_L\ \bar{\varepsilon}_R\ \bar{\nu}_L\ \bar{\nu}_R)i \begin{pmatrix} 0 & \gamma_\mu & 0 & 0 \\ \gamma_\mu & 0 & 0 & 0 \\ 0 & 0 & 0 & \gamma_\mu \\ 0 & 0 & \gamma_\mu & 0 \end{pmatrix} \begin{pmatrix} \partial^\mu\varepsilon_L \\ \partial^\mu\varepsilon_R \\ \partial^\mu\nu_L \\ \partial^\mu\nu_R \end{pmatrix}$$

$$= \bar{\varepsilon}_L i\gamma_\mu\partial^\mu\varepsilon_R + \bar{\varepsilon}_R i\gamma_\mu\partial^\mu\varepsilon_L + \bar{\nu}_L i\gamma_\mu\partial^\mu\nu_R + \bar{\nu}_R i\gamma_\mu\partial^\mu\nu_L$$

$$= [\bar{\varepsilon}(I+i\gamma_5)i\gamma_\mu\partial^\mu\varepsilon+\bar{\varepsilon}(I-i\gamma_5)i\gamma_\mu\partial^\mu\varepsilon+\bar{\nu}(I+i\gamma_5)i\gamma_\mu\partial^\mu\nu+\bar{\nu}(I-i\gamma_5)i\gamma_\mu\partial^\mu\nu]/2$$

$$= \bar{\varepsilon}i\gamma_\mu\partial^\mu\varepsilon + \bar{\nu}i\gamma_\mu\partial^\mu\nu .$$

The terms in the 16 component spinor kinetic energy thus decompose into the sum of the corresponding 4 component terms for the electron and its neutrino.

The bivectors in the lepton representation of C_8 include

$$i\Gamma_5\Gamma_6 = \lambda_3\rho_4 I,\ i\Gamma_5\Gamma_7 = i\lambda_1\rho_3\gamma_5,\ i\Gamma_6\Gamma_7 = i\lambda_2\rho_3\gamma_5\ ,$$

which generate rotations in the three-dimensional space with $\{\Gamma_5,\Gamma_6,\Gamma_7\}$ as basis, that is, they are the generators of an SU(2) group. We select these as the generators of the G-S-W helicity symmetric SU(2) spin gauge transformations. The local gauge transformation R(x) is

$$R(x) = \exp[-ig(i\lambda_j\rho_3\gamma_5\theta_j(x) + \lambda_3\rho_4 I\theta_3(x))/2]\ , \tag{9}$$

with j summed over 1 and 2.

The Lagrangian (8) is not invariant under the spin gauge transformations of the form (6). However, if we define an interaction Lagrangian L_1 by

$$L_1 = [\bar{\psi}i\Gamma_\mu(D^\mu\psi) - \bar{\psi}\overleftarrow{D}^\mu i\Gamma_\mu\psi]/2\ , \tag{10}$$

where

$$D^\mu\psi = \partial^\mu\psi - \Omega^\mu\psi,\ \bar{\psi}\overleftarrow{D}^\mu = (\partial^\mu\psi)^+\Gamma - \bar{\psi}\bar{\Omega}^\mu\ ,$$

$$\Omega^\mu = ig(i\lambda_j\rho_3\gamma_5 W_j{}^\mu(x) + \lambda_3\rho_4 I W_3{}^\mu(x))/2\ ,$$

then L_1 is invariant under (6) provided that Ω^μ undergoes a gauge transformation of the form

$$\Omega^\mu \to R\Omega^\mu R^{-1} + R(\partial^\mu R^{-1})\ .$$

The interaction terms in the Lagrangian (10) are

$$-\ \bar{\psi}i\lambda_4\rho_1\gamma_\mu\Omega^\mu\psi = [g\bar{\psi}(-\lambda_j\rho_2\gamma_5\gamma_\mu W_j{}^\mu + \lambda_3\rho_1\gamma_\mu W_3{}^\mu)\psi]/2. \tag{11}$$

We claim that these are the G-S-W helicity symmetric electron and neutrino SU(2) interactions (7), as required. If we decompose the interaction terms (11) following a similar method to that adopted above for the kinetic energy term, we obtain

$$\bar{\varepsilon}_L\gamma_\mu W_3{}^\mu\varepsilon_R + \bar{\varepsilon}_L i\gamma_5\gamma_\mu W_-{}^\mu\nu_R + \bar{\varepsilon}_R\gamma_\mu W_3{}^\mu\varepsilon_L - \bar{\varepsilon}_R i\gamma_5\gamma_\mu W_-{}^\mu\nu_L$$
$$-\ \bar{\nu}_L\gamma_\mu W_3{}^\mu\nu_R + \bar{\nu}_L i\gamma_5\gamma_\mu W_+{}^\mu\varepsilon_R - \bar{\nu}_R\gamma_\mu W_3{}^\mu\nu_L - \bar{\nu}_R i\gamma_5\gamma_\mu W_+{}^\mu\varepsilon_L. \tag{12}$$

However, using the identities

$$(I+i\gamma_5)i\gamma_5 = I+i\gamma_5 \quad \text{and} \quad (I-i\gamma_5)i\gamma_5 = -(I-i\gamma_5)\ ,$$

the terms (12) reduce to the required G-S-W terms (7). Hence we see that the presence of the γ_5 in the W_1 and W_2 interaction terms in (11) does not indicate that they have a pseudovector quality. Rather, the γ_5 is 'absorbed' into the left and right handed 4-component spinors, thus demonstrating the importance of our choice of the form of the multispinor

ψ.

We conclude this section by noting several points about spin gauge theories and this example in particular. Firstly, the kinetic energy for the gauge fields W_a^{μ} (a=1,2,3) can be obtained from the commutator of two covariant derivatives. The commutator gives the usual covariant gauge field strengths provided that we impose the condition that the vectors Γ_i are covariantly constant. This condition has no analogue in standard gauge theories, since the γ-matrices, for instance, are assumed to be x-independent. In the example,

$$[D^{\mu},D^{\nu}] = g(i\lambda_j\rho_3\gamma_5W_j^{\mu\nu} - \lambda_3\rho_4IW_3^{\mu\nu})$$

with

$$W_a^{\mu\nu} = \partial^{\mu}W_a^{\nu} - \partial^{\nu}W_a^{\mu} - g\, f_{abc}W_b^{\mu}W_c^{\mu}, \quad a=1,2,3,$$

and then the required kinetic energy terms are

$$- \mathrm{Tr}\{[D^{\mu},D^{\nu}][D_{\mu},D_{\nu}]\}/16 .$$

Secondly, the G-S-W U(1) interaction can easily be added into the example above by including a term $\lambda_4\rho_3\gamma_5\phi(x)$ into the gauge transformation (9); this term commutes with the generators in (9). In addition, the SU(2) interactions can be made helicity asymmetric by introducing the projection operator $(I_{16}+i\lambda_4\rho_4\gamma_5)/2$ into the transformation generators in (9).

4. Unified Spin Gauge Theory Models

(a) Model 1 of Electrostrong Interactions[3]

A 16 component multispinor, and hence a 16×16 representation of C_8, is used. The lepton multispinor ψ is defined by

$$\psi = (\varepsilon_L\ \nu_L'\ \varepsilon_R\ \nu_R')^T ,$$

where ν' is the charge conjugate neutrino spinor. Most of the generators of the spin gauge transformations are bivectors of C_8. They produce lepton multispinor interactions with two neutral fields and, in addition, new 'strong' interactions. The strong interactions are associated with an SU(1,1) symmetry group and are mediated by two pseudovector fields and a neutral field. The neutral fields mix to give the photon and a 'strong neutral field'. The form of the multispinor ψ ensures that, on decomposition to 4 component spinors, the electron and neutrino interactions with the strong neutral and pseudovector fields reduce to zero, leaving only the correct lepton electromagnetic interactions.

A global transformation of the lepton representation of C_8 defines the 'dion representation'. The dion spinor X resembles with the electron and anti-neutrino replaced by a 'down dion' and an 'anti-up dion' respectively. The dions have non-zero interactions with the strong neutral and pseudovector fields. Their electromagnetic interactions are those for particles of charges -e/3 and -2e/3. It is hoped that the dions might ultimately be identified with quarks and that the pseudovector character of the 'strong interactions' should account for confinement. Dions of different colours are defined by discrete global transformations of the dion representation.

(b) Model 2 of Electroweak and Strong Interactions [4]

The G-S-W weak interactions are incorporated into Model 1. The algebra is enlarged to C_{10} and 32 component spinors are used. The representation of the vectors $\{\Gamma_i, i=1,2,\ldots,10\}$ in the algebra is such that $\Gamma_i^2 = -1$, $i=1-3,5-9$ and $\Gamma_4^2 = \Gamma_{10}^2 = 1$. The basis $\{\Gamma_i, i=1,2,\ldots,10\}$ for the ten-dimensional vector space associated with the algebra C_{10} divides into three subsets: the set $\{\Gamma_\mu, \mu=1,2,3,4\}$ is a basis for Minkowski space-time; the generators of rotations in the space with $\{\Gamma_5,\Gamma_6,\Gamma_7\}$ as basis are the generators of the weak SU(2) interactions and the generators of hyperbolic rotations in the space with $\{\Gamma_8,\Gamma_9,\Gamma_{10}\}$ as basis are the generators of the 'strong' SU(1,1) symmetry group. The U(1) interaction is generated by the only element commuting with both the SU(2) and SU(1,1) generators, and gives the correct electroweak interactions for both leptons and quarks.

5. Conclusions

There is much still to be done in the area of spin gauge theories. Our models are certainly controversial and may not be correct. However, we believe that spin gauge theories are a logical consequence of the Clifford algebra formulation, and that they may provide a unified field theory.

References

[1] E. Schrodinger, Sitzber, preuss. Akad. Wiss. Physik-Math kl. **XI**, 105 (1932).
H. G. Loos, Ann. Phys. **25**, 91 (1963).
E. A. Lord, Proc. Camb. Soc. **69**, 423 (1971).

[2] J. S. R. Chisholm and R. S. Farwell, Lecture Notes in Physics **116**, 305 (1980) and Proc. Roy Soc. London **A377**, 1 (1981).
Z. Dongpei, Phys. Rev. D **22**, 2027 (1980).
W. J. Wilson, Phys. Lett. A **75**, 156 (1980).
A. O. Barut and J. McEwan, Phys. Lett. B **135**, 172 (1984) and Kent University preprint (1985).

[3] J. S. R. Chisholm and R. S. Farwell, Nuov. Cim. A **82**, 145 (1984).

[4] J. S. R. Chisholm and R. S. Farwell, Nuov. Cim. A **82**, 185 (1984).

U(2,2) Spin-Gauge Theory Simplification by use of the Dirac Algebra

J. McEwan
Mathematical Institute
University of Kent
Canterbury
Kent CT2 7NF
England

ABSTRACT. Exponentials of elements of the Lorentz covariant Dirac algebra with 16 parameters define local U(2,2) spin-gauge transformations of bispinor fields $\psi(x)$, $\bar{\psi}(x)$ over "flat" Minkowski space-time. From general algebraic identities, 12 vector gauge fields are shown to be sufficient to define local U(2,2) covariant derivatives $\gamma^\mu \vec{D}_\mu \psi$, $\bar{\psi} \overleftarrow{D}_\mu \gamma^\mu$. Only one scalar, two vector and one anti-symmetric (rank 2) tensor gauge fields couple with $\psi, \bar{\psi}$ in the hermitean bilinear Lorentz and local U(2,2) invariant $\frac{1}{2} i \bar{\psi} (\overleftarrow{D}_\mu \gamma^\mu - \gamma^\mu \vec{D}_\mu) \psi$

Physical interpretations, a local U(2,2) "breaking" mechanism and further generalisations are suggested.

1. INTRODUCTION

The most general local phase (gauge) transformation of a single spinless scalar field ϕ is defined by $\phi'(x) = e^{i\theta(x)}\phi(x)$, and it is well-known that the gauge principle of local invariance can introduce electric charge and a minimal electromagnetic coupling for ϕ. Recently it has been shown [1,2,3,4,5] that more general phase (spin-gauge) transformations can introduce more than one kind of charge and electromagnetic coupling for a single spinor field over "*flat*" Minkowski space-time. Similar conclusions have been reached recently by a quite different route [6] based on an assumption that there exist regular solutions of the Dyson equations for standard QED compatible with Lorentz and gauge invariance.

Some authors [2,3] considered local U(4) spin-gauge transformations of a single Dirac bispinor and a Lorentz covariant Dirac algebra. Dirac [7] first introduced the SO(4)-covariant Clifford algebra of 4 generators $\{\alpha_\mu;\ \mu=0,1,2,3\}$ where

$$\tfrac{1}{2}\{\alpha_\mu, \alpha_\nu\} = \delta_{\mu\nu} I, \quad \alpha_\mu^\dagger = \alpha_\mu; \quad \mu,\nu = 0,1,2,3 \tag{1}$$

and $\{\alpha_\mu;\ \mu=0,1,2,3\}$ transforms like a 4-dimensional Euclidean vector under 4-dimensional rotations [8] according to

$$U(R)\alpha_\mu U^{-1}(R) = (R^{-1})_{\mu\nu}\alpha_\nu ; \quad \mu,\nu = 0,1,2,3 \tag{2}$$

J. S. R. Chisholm and A. K. Common (eds.), Clifford Algebras and Their Applications in Mathematical Physics, 371–376.

where $R \in SO(4)$ and $U \in SU(4)$.

In order to verify the Lorentz covariance of his bispinor equation with real Minkowski space-time coordinates, Dirac introduced the Clifford algebra of 4 generators $\{\gamma_o \equiv \alpha_o,\ \underset{\sim}{\gamma} \equiv \alpha_o \underset{\sim}{\alpha}\}$ where

$$\tfrac{1}{2}\{\gamma_\mu,\gamma_\nu\} = g_{\mu\nu}I,\ \bar{\gamma}_\mu \equiv \gamma_o\gamma_\mu^\dagger\gamma_o = \gamma_\mu;\quad \mu,\nu = 0,1,2,3. \tag{3}$$

This Clifford algebra provides the Minkowski space-time metric g with signature +1,-1,-1,-1. The $\{\gamma_\mu;\ \mu=0,1,2,3\}$ transforms like a 4-dimensional Minkowski vector [9] according to

$$S(A)\gamma_\mu\ S^{-1}(A) = (\Lambda^{-1}(A))_\mu{}^\nu\gamma_\nu\ ;\quad \mu,\nu = 0,1,2,3 \tag{4}$$

where $A \in SL(2,C)$, $\Lambda \in SO(3,1)$ and $\bar{S} \equiv \gamma_o S^\dagger \gamma_o = S^{-1}$, $\det S = 1$. This Clifford algebra (3) provides a spin space metric γ_o with signature +1,+1,-1,-1; also

$$S = \exp\left(\frac{i}{8}\,\theta^{\mu\nu}\sigma_{\mu\nu}\right) \tag{5}$$

where $\sigma_{\mu\nu} = \frac{i}{2}[\gamma_\mu,\gamma_\nu]$ and $\Lambda_{\mu\nu} = \delta_{\mu\nu} + \theta_{\mu\nu}$ to first order in θ.

2. U(2,2) SPIN-GAUGE TRANSFORMATIONS

We use the same Clifford algebra (3) and the same Minkowski space-time metric g as in [2,3], but unlike [2,3], we use the more natural pseudo-Euclidean spin space metric γ_o.

We only consider "flat" Minkowski space-time with metric g independent of x, and "flat" spin space with metric γ_o independent of x. However, γ_o is independent of x *only* when γ_o represents the spin space metric otherwise γ_o will become x dependent under a local U(2,2) spin-gauge transformation $\Omega(x)$ according to

$$\gamma_\mu \to \gamma_\mu(x) = \Omega(x)\gamma_\mu\Omega^{-1}(x),\ \Omega \in U(2,2);\quad \mu=0,1,2,3\ . \tag{6}$$

The Clifford algebra structure (3) is preserved by (6) while the metrics g and γ_o remain independent of x since

$$\tfrac{1}{2}\{\gamma_\mu(x),\gamma_\nu(x)\} = g_{\mu\nu}I\ ,\quad \bar{\gamma}_\mu(x) \equiv \gamma_o\gamma_\mu^\dagger(x)\gamma_o = \gamma_\mu(x)\ , \tag{7}$$

and

$$\bar{\Omega}(x) \equiv \gamma_o\Omega^\dagger(x)\gamma_o = \Omega^{-1}(x)\ .$$

Also

$$\Omega(x) = \exp(i\theta^A(x)\gamma_A) \tag{8}$$

where $\{\gamma_A;\ A=1,2,\ldots,16\}$ are basis elements of the Clifford algebra (3) chosen such that $\bar{\gamma}_A = \gamma_A$, and $\{\theta_A;\ A=1,2,\ldots,16\}$ are independent parameters of U(2,2). Hence the Clifford algebra (3) also provides the generators of the local U(2,2) spin-gauge transformation by (8).

3. FIRST SIMPLIFICATION

The first simplification is due to the definition (16) of only *one* derivative for each of ψ and $\bar{\psi}$, and was used by Chisholm and Farwell [1] who needed to introduce only two vector gauge fields to achieve local SL(2,C) spin-gauge invariance with a single complex charge, as if local SL(2,C) $\otimes$ SL(2,C) spin-gauge invariance is effectively local U(1) $\otimes$ U(1) spin-gauge invariance. In standard gauge theories it is conventional to introduce the same number of vector gauge fields as the number of independent parameters to specify an arbitrary gauge transformation, e.g. 6 vector gauge fields for local SL(2,C) gauge invariance.

In local U(2,2) spin-gauge theory, parallel transport for Clifford algebra (3) is described by

$$\gamma_\mu(y) = T(x,y)\gamma_\mu(x)T^{-1}(x,y),\ T(x,y) = \exp\left(-\int_x^y iV_\mu(x')dx'^\mu\right) \qquad (9)$$

and introduces 16 real vector gauge fields $\{V_\mu^A;\ A=1,2,\ldots,16\}$ in

$$V_\mu(x) = V_\mu^A(x)\gamma_A;\quad \mu=0,1,2,3\ . \qquad (10)$$

Under transformation Ω it follows that

$$T(x,y) \to T'(x,y) = \Omega(y)T(x,y)\Omega^{-1}(x)\ . \qquad (11)$$

$$\therefore\ V_\mu(x) \to V_\mu'(x) = \Omega(x)\ (V_\mu(x) + \Omega^{-1}(x)i\partial_\mu\Omega(x))\Omega^{-1}(x);\ \mu=0,1,2,3. \qquad (12)$$

Also under transformation Ω it follows that

$$\psi(x) \to \psi'(x) = \Omega(x)\psi(x)\ . \qquad (13)$$

$$\therefore\ \bar{\psi}(x) \to \bar{\psi}'(x) = \psi'^\dagger(x)\gamma_o = \psi^\dagger(x)\Omega^\dagger(x)\gamma_o = \psi^\dagger(x)\gamma_o\Omega^{-1}(x) = \bar{\psi}(x)\Omega^{-1}(x) \qquad (14)$$

also

$$\bar{\psi}'(x)\gamma_A(x)\psi'(x) = \bar{\psi}(x)\gamma_A\psi(x);\quad A=1,2,\ldots,16 \qquad (15)$$

are hermitean, bilinear, Lorentz covariants and local U(2,2) invariants. Local U(2,2) covariant space-time derivatives are defined on bispinors by

$$\vec{D}_\mu\psi(x)dx^\mu = T^{-1}(x,x+dx)\psi(x+dx)-\psi(x) = (\vec{\partial}_\mu+iV_\mu(x))\psi(x)dx^\mu\ , \qquad (16)$$

$$\bar{\psi}(x)\overleftarrow{D}_\mu dx^\mu = \bar{\psi}(x+dx)T(x,x+dx)-\bar{\psi}(x) = \bar{\psi}(x)(\overleftarrow{\partial}_\mu-iV_\mu(x))dx^\mu$$

to first order in dx^μ. However, Dirac's bispinor equations require only the two derivatives

$$\vec{\not{D}} \equiv \gamma^\mu\vec{D}_\mu = \gamma^\mu(\vec{\partial}_\mu+iV_\mu) \equiv \vec{\not{\partial}}+i\not{V}^A\gamma_A$$
$$\overleftarrow{\not{D}} \equiv \overleftarrow{D}_\mu\gamma^\mu = (\overleftarrow{\partial}_\mu-iV_\mu)\gamma^\mu \equiv \overleftarrow{\not{\partial}}-i\gamma_A\not{V}^A \qquad (17)$$

where only 12 vector gauge fields survive due to algebraic identities

$$\left.\begin{aligned}\gamma^\mu\sigma_{\alpha\beta} &= i(g^\mu{}_\alpha\gamma_\beta - g^\mu{}_\beta\gamma_\alpha) - \epsilon^{\mu\nu}{}_{\alpha\beta}\gamma_5\gamma_\nu \\ \sigma_{\alpha\beta}\gamma^\mu &= -i(g^\mu{}_\alpha\gamma_\beta - g^\mu{}_\beta\gamma_\alpha) - \epsilon^{\mu\nu}{}_{\alpha\beta}\gamma_5\gamma_\nu\end{aligned}\right\} \mu,\alpha,\beta=0,1,2,3 \tag{18}$$

and $\{I, \gamma_\mu, \gamma_5 \equiv \gamma_1\gamma_2\gamma_3\gamma_o, i\gamma_5\gamma_\mu, \sigma_{\alpha\beta}\}$ are hermitean (in the sense $\bar{\gamma}_A(x) \equiv \gamma_o\gamma_A{}^\dagger(x)\gamma_o = \gamma_A(x)$), Lorentz covariant basis elements of the Clifford algebra (3); also $\epsilon^{\mu\nu\alpha\beta}$ is the totally antisymmetric (rank 4) tensor with $\epsilon^{0123} = 1$.

4. SECOND SIMPLIFICATION

The second simplification is due to the Lorentz and gauge invariant, hermitean (in the sense $L^\dagger = L$) bilinear form $L(\psi,\bar{\psi}) \equiv \bar{\psi}[\frac{1}{2}i(\vec{D}-\overleftarrow{D})-m]\psi$ which is the only possible bilinear Lagrangian term dependent on bispinors $\psi,\bar{\psi}$. Here

$$\tfrac{1}{2}i(\vec{\not{D}}-\overleftarrow{\not{D}}) = \tfrac{1}{2}i(\vec{\not{\partial}}-\overleftarrow{\not{\partial}}) - \tfrac{1}{2}\{\gamma_A, \not{V}^A\} \tag{19}$$

and

$$\tfrac{1}{2}\{\gamma_A, \not{V}^A\} = e\not{A} + aA_\mu^{(\mu)} + a\sigma_{\mu\nu}F^{\mu\nu} + g\gamma_5\not{B}^{(5)} \;. \tag{20}$$

Only $3\frac{3}{4}$ gauge vector fields survive in (20)! This is quite different from, but similar to, results found by Zhu Dongpei [3].

5. PHYSICAL INTERPRETATIONS

Physical interpretations for chiral U(2,2) contractions of the form

$$U(1) \otimes \{T_4^\pm \;x)\; [SL(2,C) \otimes T_1{}^D]\} \;,$$

where x) indicates the semi-direct product which occurs in the Poincaré group structure, are as follows:

(i) Local U(1) gauge invariance introduces the well-known electromagnetic vector potential field A_μ and the electronic charge (-e).

(ii) Invariance under local 4-dimensional translations $T_4^\pm$ generated by $\frac{1}{2}(1\pm i\gamma_5)\gamma_\mu$ introduces the well-known electromagnetic (rank 2) antisymmetric tensor $F_{\mu\nu}$ and anomalous magnetic moment a. $F_{\mu\nu} \equiv \partial_\mu A_\nu - \partial_\nu A_\mu$ (provided group parameters $\theta_\mu \equiv \partial_\mu\theta$) are identified consistently and discussed in [4,5] in relation to the well-known Pauli coupling. Local $T_4^\pm$ gauge invariance also introduces a scalar gauge field $A_\mu^{(\mu)}$ which has been consistently identified by $A_\mu^{(\mu)} \equiv \partial_\mu A^\mu$ (provided again $\theta_\mu \equiv \partial_\mu\theta$) in [4,5] in the interests of existence and uniqueness of A_μ and the avoidance of an infra-red catastrophe.

(iii) Local SL(2,C) gauge invariance introduces the (neutral) axial vector gauge field $B_\mu^{(5)}$ and a third coupling constant g which may be used to model parity-violating neutral currents.

6. LOCAL SU(2,2) SYMMETRY "BREAKDOWN"

These 4 types of couplings have been suggested by a quite different argument [6] along with speculations about a "broken" more general (than standard U(1)) gauge theory and the generation of mass spectra. Dongpei [3] points out that there is a natural "breakdown" mechanism

$$\partial_\mu \gamma_\nu(x) = 0 \ ; \quad \mu,\nu = 0,1,2,3 \tag{21}$$

since local U(2,2) covariant space-time derivatives on the Clifford algebra (3) are defined by

$$\begin{aligned} D_\mu \gamma_\nu(x) dx^\mu &= T^{-1}(x,x+dx)\gamma_\nu(x+dx)T(x,x+dx) - \gamma_\nu(x) \\ &= \partial_\mu \gamma_\nu(x) dx^\mu + i[V_\mu(x),\gamma_\nu(x)]dx^\mu \end{aligned} \tag{22}$$

consistent with (9) to first order in dx^μ.

The only derivative which corresponds to those defined by (17) is

$$D_\mu \gamma^\mu(x) = \partial_\mu \gamma^\mu(x) + i[V_\mu(x),\gamma^\mu(x)] \tag{23}$$

which is hermitean since $\overline{\gamma^\mu}(x) = \gamma^\mu(x)$ and $\bar{V}_\mu(x) = V_\mu(x)$.

A Lagrangian field theory for the space-time dependent Clifford algebra (3) will require an additional Lorentz and gauge invariant hermitean bilinear form $tr[D_\mu \gamma^\mu(x)]^2$. A *third simplification* occurs here if the local U(2,2) invariance is "broken" by the constraint

$$\partial_\mu \gamma^\mu(x) = 0 \ , \tag{24}$$

hence

$$\begin{aligned} \frac{1}{16} tr[D_\mu \gamma^\mu(x)]^2 &= -\frac{1}{16} tr[V_\mu(x),\gamma^\mu(x)]^2 \\ &= -(g_5)^2 A_\mu^{(5)} A^{(5)\mu} + g^2 B_\mu B^\mu - a^2 (A_\mu^{(\mu)})^2 - 2a^2 F_{\mu\nu} F^{\mu\nu} \end{aligned} \tag{25}$$

where

$$\left.\begin{aligned} \gamma^\mu V_\mu &= e\not{A} + g_5 \not{A}^5 \gamma_5 + a\not{A}^{(\alpha)}(1 \pm i\gamma_5)\gamma_\alpha - ig(\not{B} + \gamma_5 \not{B}^{(5)}) \\ V_\mu \gamma^\mu &= e\not{A} - g_5 \not{A}^5 \gamma_5 + a(1 \pm i\gamma_5)\gamma_\alpha \not{A}^{(\alpha)} + ig(\not{B} - \gamma_5 \not{B}^{(5)}) \end{aligned}\right\} \tag{26}$$

and

$$-\tfrac{1}{2}\,\epsilon^{\mu\ \nu}_{\ \alpha\ \beta} A_\mu^{(\alpha)} = F^\nu_{\ \beta} \tag{27}$$

for consistency with (20).

Rather than reject any of the terms in (25) as in [3], they may all be acceptable and may even help justify the inclusion of terms which are invented in the interests of existence and uniqueness of A_μ and the avoidance of an infra-red catastrophe.

Further investigation of this kind of "broken" symmetry could be important also in the generation of realistic mass spectra.

Further extension to a Clifford algebra of 6 generators may be needed to include all elementary particle states according to Basri and Barut [10], then an interesting comparison is possible with the standard extension of QED to include "weak" couplings.

REFERENCES

[1] J.S.R. Chisholm, R.S. Farwell: Lecture Notes in Physics **116** (Springer, Berlin 1980) 305-307; Proc. Roy. Soc. London **A377** (1981) 1.

[2] W.J. Wilson: Phys. Letts. **75A** (1980) 156.

[3] Zhu Dongpei: Phys. Rev. **D22** (1980) 2027.

[4] A.O. Barut, J. McEwan: Phys. Letts. **135B** (1984) 172 and Erratum, Phys. Letts. **139B** (1984) 464.

[5] A.O. Barut, J. McEwan; Letts. Math. Phys. **11** (1986) 67.

[6] F.A. Kaempffer: Phys. Rev. **D23** (1981) 918; **D25** (1982) 439, 447.

[7] P.A.M. Dirac: The Principles of Quantum Mechanics 4th Edition, Oxford Univ. Press (1958) Chap. XI.

[8] R.H. Good, Jr.: Rev. Mod. Phys. **27** (1955) 187.

[9] A. Messiah: Quantum Mechanics Vol. II (Wiley, 1965) Chap. XX.

[10] S.A. Basri, A.O. Barut: Int. Journ. Theoret. Phys. **22** (1983) 691.

SPIN(8) GAUGE FIELD THEORY

Frank D. (Tony) Smith, Jr.
P. O. Box 1032
Cartersville, Georgia 30120 U.S.A.

ABSTRACT. The 16-dimensional spinor representation of Spin(8) reduces to two irreducible 8-dimensional half-spinor representations that can correspond to the 8 fundamental fermion lepton and quark first-generation particles and to their 8 antiparticles in a gauge field theory whose gauge group is Spin(8) and whose base manifold is S^4. Numerical values for force strength constants and ratios of particle masses to the electron mass are given. No predictions of transitions or decay rates have yet been made.

1. INTRODUCTION.

Spin(8) gauge field theory, as defined herein, begins with the Yang-Mills principal fibre bundle having gauge group Spin(8) and base manifold S^8 described by Grossman, Kephart, and Stasheff (1984). The base manifold S^8 is reduced to S^4 by a geometric Higgs symmetry breaking mechanism described in Smith (1986). The reduced base manifold S^4 corresponds to space-time. The quaternionic structure of S^4 naturally corresponds to a (3,3,4,3) lattice structure of space-time that has a natural quaternionic structure.

The gauge group Spin(8) decomposes at the Weyl group level into Spin(5), SU(3), Spin(4), and the maximal torus $U(1)^4$.

The Spin(5) component corresponds to a gauge theory of de Sitter gravitation with a cosmological term as described by MacDowell and Mansouri (1977).

The SU(3) component corresponds to a gauge theory of the color force.

The Spin(4) = SU(2)xSU(2) component corresponds to a gauge theory of the weak force with a geometric form of spontaneous symmetry breaking from Spin(4) to SU(2). Details are given in Smith (1986).

The $U(1)^4$ component corresponds to the four components of the photon in the path integral formulation of quantum electrodynamics.

It may be useful to consider the decomposition of Spin(8) in terms of the parameterization of Spin(8) given by Chisholm and Farwell (1984).

J. S. R. Chisholm and A. K. Common (eds.), Clifford Algebras and Their Applications in Mathematical Physics, 377–383.

For spinor matter fields, there is an associated bundle to the principal bundle that is related to the spinor representation of Spin(8). One of the two mirror image 8-dimensional half-spinor spaces, denoted by Q^8+, corresponds to the first-generation fermion particles. The other, denoted by Q^8-, corresponds to the first-generation anti-particles.

Q^8+ can be parameterized by triples of U and D spinors. Equivalently, it can also be parameterized by the octonions, for which a basis can be taken to be (1, O1, O2, O3, O4, O5, O6, O7), the real unit 1 and the 7 octonion imaginary units. The octonion parameterization is particularly useful in considering Q^8+ as the compact manifold RP^1xS^7, where RP^1 corresponds to the real axis and S^7 to the imaginary octonions. Compact manifolds are useful in Spin(8) gauge field theory calculations because they have finite volume and ratios of volumes can be well-defined.

In Spin(8) gauge field theory, the first-generation fermion particles correspond to triples of spinors or octonions as follows:

Fermion:	Triple of Spinors:	Octonion:
electron	UxUxU	O7
green up quark	DxUxU	O3
blue up quark	UxDxU	O2
red up quark	UxUxD	O1
green down quark	UxDxD	O6
blue down quark	DxUxD	O5
red down quark	DxDxU	O4
neutrino	DxDxD	1

The mirror image 8-dimensional half-spinor space corresponds similarly to the first-generation fermion antiparticles. Second and third generation fermions correspond to second-order and third-order direct products of the irreducible half-spinor representations.

2. HISTORICAL BACKGROUND.

Armand Wyler (1971) wrote a paper in which he purported to calculate the fine structure constant to be a = 1/137.03608 and the proton-electron mass ratio to be mp/me = $6pi^5$ = 1836.118 from the volumes of homogeneous symmetric spaces. Although the numerical values were close to experimental data, the physical reasons he gave for using the particular volumes he chose were not clear. Freeman Dyson invited Wyler to the Institute for Advanced Study in Princeton for a year to see if Wyler could develop good physical reasons. However, Wyler was primarily a mathematician and did not produce a convincing physical basis for his numerical calculations. With no clear physical basis, Wyler's results were dismissed by many physicists, such as those writing letters in the November 1971 issue of Physics Today, as mere unproductive numerology.

As far as I know, no further work was done on Wyler's results. It seemed to me that even though Wyler didn't come up with a good physical basis, there might be one. To look for one, I started by trying to study generalizations of complex manifolds to quaternionic manifolds that have the structure of space-time.

Joseph Wolf (1965) wrote a paper in which he classified the 4-dimensional Riemannian symmetric spaces with quaternionic structure. There are just 4 equivalence classes, with the following representatives:

$$T^4 = U(1)^4$$
$$S^2 x S^2 = SU(2)/U(1) \times SU(2)/U(1)$$
$$CP^2 = SU(3)/S(U(2) x U(1))$$
$$S^4 = Spin(5)/Spin(4)$$

Although Wolf's paper was pure mathematics with no attempt at physical application, it seemed to me that the occurrence of the gauge group of electromagnetism U(1), the gauge group of the weak force SU(2), the gauge group of the color force SU(3), and the gauge group of de Sitter gravitation Spin(5) might be physically significant.

Another indication of the possible physical significance of quaternionic structure was the paper of David Finkelstein, J. M. Jauch, S. Schiminovich, and D. Speiser (1963) in which they used quaternionic structure to construct a geometric spontaneous symmetry breaking mechanism producing two charged massive vector bosons and one massless neutral photon.

I then started trying to construct a gauge field theory with a 4-dimensional base manifold having quaternionic structure and a gauge group that would include $U(1)^4$, SU(2)xSU(2) = Spin(4), SU(3), and Spin(5). Such a gauge group should have dimension at least 4+6+8+10= 28. If $U(1)^4$ is taken to be part of a maximal torus, the rank of the gauge group should be at least 4.

As Spin(8) has rank 4 and dimension 28, it is a natural candidate. However, it does not even include U(1)xSU(2)xSU(3) as a subgroup. To have $U(1)^4$, Spin(4), SU(3), and Spin(5) included in it, Spin(8) must be decomposed at the Weyl group level rather than decomposed into subgroups.

Then, as described in Smith (1986), the 28-dimensional adjoint representation of Spin(8), after reduction to 24 dimensions by the geometric Higgs mechanism, corresponds to the gauge bosons; the 8-dimensional vector representation of Spin(8), after reduction to 4 dimensions by the geometric Higgs mechanism, corresponds to space-time; and the 16-dimensional reducible spinor representation of Spin(8) corresponds to the 8 first-generation fermion particles and their 8 antiparticles, each corresponding to a mirror image irreducible 8-dimensional half-spinor representation of Spin(8).

3. CALCULATION OF FORCE STRENGTH CONSTANTS.

The relative strengths of the four forces of gravitation, the color force, the weak force, and electromagnetism should be determined in part by their geometric relationships to the base manifold S^4 and each half-spinor manifold $Q^8_{\pm}$ of Spin(8) gauge field theory.

The other part of their relative strengths should be determined by considering them to be proportional to the ratio of the square of the electron mass to the square of the characteristic mass, if any, associated with the forces. The electron mass me is the only mass term that is not calculable in Spin(8) guage field theory, in which it is a fundamental quantity like the speed of light and Planck's constant. Only gravitation, involving the Planck mass, and the weak force, involving the sum of the squares of the weak vector boson masses, have mass terms.

To calculate the geometric part of the relative force strengths, proceed in three steps.

3.1. Base Manifold Component Of Geometric Part.

Each of the four forces has a natural global action on a part of the base manifold S^4.

Gravitation has gauge group Spin(5), which has a natural global action on Spin(5)/Spin(4) = S^4. Therefore the base manifold component of the geometric part of the strength of gravitation is S^4, and the volume $V(S^4) = 8pi^2/3$.

The color force has gauge group SU(3), which has a natural global action on SU(3)/S(U(2)xU(1)) = CP^2. CP^2 is the same as S^4 except for structure at infinity. Therefore the base manifold component of the geometric part of the strength of the color force is S^4, and the volume $V(S^4) = 8pi^2/3$.

The weak force has gauge group Spin(4) = SU(2)xSU(2), but that is reducible by spontaneous symmetry breaking to SU(2), which has a natural global action on SU(2)/U(1) = S^2. The base manifold component of the geometric part of the strength of one half of the weak force is S^2, which is contained in S^4. $V(S^2) = 4pi$.

Electromagnetism has gauge group $U(1)^4$, the maximal torus of Spin(8), but that is reducible to U(1) by considering each of the four U(1)'s in $U(1)^4$ to be one space-time component of the photon in the Feynman path-integral formulation of quantum electrodynamics. U(1) has a natural global action on U(1) = S^1. S^4 contains S^1. The base manifold component of the geometric part of the strength of one fourth of electromagnetism is S^1, and $V(S^1) = 2pi$.

3.2. Half-Spinor Component of Geometric Part.

Each of the four forces has a natural local action on a part of the half-spinor manifold $Q^8\underline{+} = RP^1xS^7$.

Gravitation has gauge group Spin(5), so that it has a natural local action on an irreducible symmetric bounded domain of type IV5, $D^5\underline{+}$ = Spin(7)/Spin(5)xU(1). $D^5\underline{+}$ has Silov boundary $Q^5\underline{+} = S^4xRP^1$. $Q^8\underline{+}$ contains $Q^5\underline{+}$. The half-spinor component of the geometric part of the strength of gravitation is $Q^5\underline{+} = S^4xRP^1$, and $V(Q^5\underline{+}) = 8pi^3/3$.

The color force has gauge group SU(3), so that it has a natural local action on an irreducible symmetric bounded domain of type I1,3, $D^{1,3}\underline{+}$ = SU(4)/S(U(3)xU(1)) = B^6. $D^{1,3}\underline{+}$ has Silov boundary $Q^{1,3}\underline{+} = S^5$. $Q^8\underline{+}$ contains $Q^{1,3}\underline{+}$. The half-spinor component of the geometric part of the strength of the color force is $Q^{1,3}\underline{+} = S^5$, and $V(Q^{1,3})\underline{+} = 4pi^3$.

Each of the SU(2) gauge groups in the Spin(4) of the weak force has a natural local action on an irreducible symmetric bounded domain of type IV3, $D^3\pm$ = Spin(5)/SU(2)xU(1). $D^3\pm$ has Silov boundary $Q^3\pm = S^2xRP^1$, which is contained in $Q^8\pm$. The half-spinor component of the geometric part of the strength of one half of the weak force is $Q^3\pm = S^2xRP^1$, and $V(Q^3\pm) = 4pi^2$.

Each of the U(1) gauge groups in the $U(1)^4$ of electromagnetism has a natural local action on $U(1) = S^1$. $Q^8\pm$ contains S^1. The half-spinor component of the geometric part of the strength of one fourth of electromagnetism is S^1, and $V(S^1) = 2pi$.

3.3. Dimensional Adjustment Factor of Geometric Part.

Note that in some cases the dimension of the base manifold component differs from the dimension of the half-spinor component. In those cases, Spin(8) gauge field theory requires use of a dimensional adjustment factor. The dimensional adjustment factor is intuitively the hardest part of Spin(8) gauge field theory to understand. A similar factor was used by Wyler (1971) in calculating his value of the electromagnetic fine structure constant, and the difficulty in finding a physical interpretation for it was a major factor in criticisms of Wyler's work (Gilmore (1972)).

Gravitation has a 5-dimensional half-spinor component $Q^5\pm = S^4xRP^1$, which is the Silov boundary of $D^5\pm$. The base manifold component S^4 is 4-dimensional. The gravitational dimensional adjustment factor is the fourth root of the volume of $D^5\pm$, $V(D^5\pm)^{\frac{1}{4}} = (pi^5/2^7 15)^{\frac{1}{4}}$.

The color force has a 5-dimensional half-spinor component $Q^{1,3}\pm = S^5$, which is the Silov boundary of $D^{1,3}\pm$. The base manifold component S^4 is 4-dimensional. The color force dimensional adjustment factor is the fourth root of the volume of $D^{1,3}\pm$, $V(D^{1,3}\pm)^{\frac{1}{4}} = (pi^3/6)^{\frac{1}{4}}$.

One half of the weak force has a 3-dimensional half-spinor component $Q^3\pm = S^2xRP^1$, which is the Silov boundary of $D^3\pm$. The base manifold component S^2 is 2-dimensional. The dimensional adjustment factor for one half of the weak force is the square root of the volume of $D^3\pm$, $V(D^3\pm)^{\frac{1}{2}} = (pi^3/24)^{\frac{1}{2}}$.

The electromagnetism half-spinor component and base manifold component are both S^1 for one fourth of the electromagnetic force, so no dimensional adjustment factor is needed.

3.4. Final Force Strength Calculation.

The geometric part of the force strengths is then: for gravitation, $VG = V(S^4)V(Q^5\pm)/V(D^5\pm)^{\frac{1}{4}} = 3444.0924$; for the color force, $VC = V(S^4)V(Q^{1,3}\pm)/V(D^{1,3}\pm)^{\frac{1}{4}} = 2164.978$; for one half of the weak force, $VW = V(S^2)V(Q^3\pm)/V(D^3\pm)^{\frac{1}{2}} = 436.46599$; and for one fourth of electromagnetism, $VE = V(S^1) = 6.2831853$.

Therefore, mass factors for the weak force and gravitation aside, the relative strengths of the forces are: gravitation, VG/VG = 1; the color force, VC/VG = 0.6286062; the weak force, 2VW/VG = 0.2534577; and electromagnetism, 4VE/VG = 1/137.03608.

When the mass factors are taken into account, Spin(8) gauge field

theory gives the following values for force constants:
fine structure constant for electromagnetism = 1/137.608;
weak Fermi constant times proton mass squared = 1.03 x 10^{-5};
color force constant (at about 10^{-13} cm.) = 0.6286; and
gravitational constant times proton mass squared = 3.4-8.8 x 10^{-39}.
The corresponding experimental values are, respectively:
1/137.03604; 1.02 x 10^{-5}; about 1; and 5.9 x 10^{-39}.

4. PARTICLE MASSES AND KOBAYASHI-MASKAWA PARAMETERS.

Values for particle masses and Kobayashi-Maskawa parameters can also be calculated in Spin(8) gauge field theory. Particle masses are calculated in terms of the electron mass, which is a fundamental constant in Spin(8) gauge field theory. The masses for quarks are constituent masses. The mass ratio of the down quark to the electron is related to the volume of the compact half-spinor space $Q^8_{\pm} = RP^1 x S^7$. The masses of the other leptons and quarks of all three generations come from straightforward consideration of symmetries of their representations as octonions or triples of spinors (for the first generation), as direct products of two octonions or two triples of spinors (for the second generation), and as direct products of three octonions or three triples of spinors (for the third generation).

Spin(8) gauge field theory indicates that there should be three generations of weak bosons, with the second generation having masses around 300 to 400 Gev and the third generation having masses around 18 to 22 Tev. It is from the three generations of weak bosons that Spin(8) gauge field theory gives the Kobayashi-Maskawa parameters.

Details of the calculations can be found in Smith (1985, 1986). The results are as follows:
electron-neutrino mass = 0 (experimentally 0);
down quark constituent mass = 312.8 Mev (experimentally about 350 Mev);
up quark constituent mass = 312.8 Mev (experimentally about 350 Mev);
muon mass = 104.8 Mev (experimentally 105.7 Mev);
muon-neutrino mass = 0 (experimentally 0);
strange quark constituent mass = 523 Mev (experimentally about 550 Mev);
charm quark constituent mass = 1.99 Gev (experimentally about 1.7 Gev);
tauon mass = 1.88 Gev (experimentally 1.78 Gev);
tauon-neutrino mass = 0 (experimentally 0);
charged W mass (first-generation) = 81 Gev (experimentally 81 Gev);
neutral W mass (first-generation) = 99 Gev (experimentally 93 Gev);
charged W mass (second-generation) = 329 Gev;
neutral W mass (second-generation) = 403 Gev;
charged W mass (third-generation) = 17.5 Tev;
neutral W mass (third-generation) = 21.5 Tev;
Planck mass = about 1-1.6 x 10^{19} Gev (experimentally 1.22 x 10^{19} Gev);
Kobayashi-Maskawa-Chau-Keung sin(x)=0.239 (experimentally 0.23);
Kobayashi-Maskawa-Chau-Keung sin(y)=0.0188;
Kobayashi-Maskawa-Chau-Keung sin(z)=0.0046 (experimentally 0.005);
beauty quark constituent mass = 5.63 Gev (experimentally about 5.2 Gev);
truth quark constituent mass = 130 Gev;

No experimental values for the second and third-generation W masses are given, because no experiments have been done at the energies needed to observe them directly. The Kobayashi-Maskawa-Chau-Keung parameter sin(y) is related to the truth quark constituent mass. As discussed in Smith (1986), current experimental results are consistent with a value of sin(y) = 0.019 if the truth quark mass is 130 Gev, but if the truth quark mass is 45 Gev, then sin(y) = 0.05.

CERN has announced that the truth quark mass is about 45 Gev (Rubbia (1984)), but I think that the phenomena observed by CERN at 45 Gev are weak force phenomena that are poorly explained by the standard SU(2)xU(1) model. I further think that current CP-violation experimental results (Wojcicki (1985)) are consistent with a truth quark mass of about 130 Gev and the Kobayashi-Maskawa-Chau-Keung parameters calculated herein from Spin(8) gauge field theory, and that the CERN value of 45 Gev for the truth quark mass is not consistent with those experimental results. As of the summer of 1985, CERN has been unable to confirm its identification of the truth quark in the 45 Gev events, as the UA1 experimenters have found a lot of events clustering about the charged first-generation W mass and the UA2 experimenters have not found anything convincing. (Miller (1985)) I think that the clustering of UA1 events near the charged first-generation W mass indicates that the events observed are nonstandard weak force phenomena.

At this time, I have not yet calculated any transition amplitudes or decay rates.

5. ACKNOWLEDGEMENTS.

I would like to thank David Hestenes, David Finkelstein and members of his Georgia Tech seminar, many participants at the 1985 Clifford Algebra workshop at the University of Kent at Canterbury, and the referee for their help and encouragement.

6. REFERENCES.

Chisholm, J. S. R., and Farwell, R. S. (1984), Il Nuovo Cimento 82A, 145.
Finkelstein, D., Jauch, J., Schiminovich, S., and Speiser, D. (1963), J. Math. Phys. 4, 788.
Gilmore, R. (1972), Phys. Rev. Lett. 28, 462.
Grossman, B., Kephart, T. W., and Stasheff, J. D. (1984), Commun. Math. Phys. 96, 431.
Miller, D. (1985), Nature 317, 110.
Rubbia, C. (1984), talk at A.P.S. D.P.F. annual meeting at Santa Fe.
Smith, F. (1985), Int. J. Theor. Phys. 24, 155.
Smith, F. (1986), to be published in Int. J. Theor. Phys.
Wojcicki, S. (1985), 'Particle Physics' in The Santa Fe Meeting, ed. by T. Goldman and M Nieto (World Scientific, Singapore).
Wolf, J. (1965), J. Math. Mech. 14, 1033.
Wyler, A. (1971), C. R. Acad. Sci. Paris A272, 186.

CLIFFORD ALGEBRAS, PROJECTIVE REPRESENTATIONS AND CLASSIFICATION OF FUNDAMENTAL PARTICLES

A. O. Barut
Department of Physics
University of Colorado
Campus Box 390
Boulder, CO 80309

ABSTRACT. We discuss the use of Clifford Algebras in the classification of elementary particles as an alternative to the use of unitary Lie groups as internal symmetry groups, their physical interpretation and advantages. The relation of Clifford algebras to the projective representations of finite groups is given. We further introduce the concepts of Clifford algebras over the Heisenberg ring and Symplectic Clifford algebras that describe the internal geometry of relativisitic quantum systems.

1. CLIFFORD ALGEBRAS AND PROJECTIVE REPRESENTATIONS OF FINITE GROUPS

In quantum theory we use not the vector representations of symmetry groups of a physical system but rather the representations up to a phase factor, the projective representations (Wigner's theorem[1]), because the overall phase factor of the wave function is not observable. The connection between the projective representations of some very simple finite groups and Clifford algebras is both physically and mathematically interesting and perhaps not generally known.

Consider first a very simple abelian group of symmetries, namely a set of n commuting parity-like operators

$$\Gamma_i^2 = 1, \quad \Gamma_i \Gamma_j = \Gamma_j \Gamma_i, \qquad i,j = 1,2,\ldots n \;. \tag{1}$$

The physical system is assumed to be invariant under n separate exchange operations (e.g. particle -antiparticle exchange, electron-muon exchange, space reflections, ...).

The vector representations of the abelian group (1) are all one dimensional. However, the projective representations of (1) satisfy the group law

$$U(\Gamma_i^2) = 1, \qquad U(\Gamma_i)U(\Gamma_j) = \omega_{ij} U(\Gamma_i \Gamma_j) \;, \tag{2a}$$

hence

$$U(\Gamma_i)U(\Gamma_j) = C_{ij} U(\Gamma_j)U(\Gamma_i) \;; \qquad C_{ij} = \frac{\omega_{ij}}{\omega_{ji}} \;. \tag{2b}$$

J. S. R. Chisholm and A. K. Common (eds.), Clifford Algebras and Their Applications in Mathematical Physics, 385–391.

Equation (2) generate a generalized Clifford algebra of dimension 2^n belonging to the factor system $\{C_{ij}\}$. One can now study all possible phase systems subject to associativity and equivalence[2]. Among these possible phase systems there is one case where all $C_{ij}=-1$, in which case the representations $U(\Gamma_i)$ generate a bona fide Clifford algebra of dimension $2^{n/2}$ or $2^{(n-1)/2}$. Thus even a set of discrete commuting reflection symmetries implies the existence of nontrivial particle multiplets of dimension $2^{n/2}$.

This result has further implications.

Because the representation of a Clifford algrebra coincides with the fundamental spin representation of the orthogonal group, we could, and we generally do, also describe the particle multiplets by the spin representations of the orthogonal Lie groups or by their Lie algebras. Also the complete set of commuting operators can be chosen to be the same in both cases. However, the Lie group has infinitely many representations which must all occur in nature if the symmetry is fundamental, but the Clifford algebra has only one representation. Moreover, there is no internal space on which the continuous Lie group acts. We shall see that these and other considerations favor the Clifford algebras as internal symmetry operations and not the Lie groups.

2. INTERNAL SYMMETRY GROUPS OF FUNDAMENTAL PARTICLES AND CLIFFORD ALGEBRAS

By internal symmetries in particle physics we mean essentially how many different kinds of fundamental particles are postulated as a starting point of a theory to describe the further interactions and bound states formed from these. The space-time symmetry based on the Poincaré group contains the spin and momentum degrees of freedom of the particles together with space parity and time reversal. The Dirac electron theory based on the Clifford algebra C_4 already entails an internal symmetry. Part of the C_4 is identified with the spin degrees of freedom of the Poincaré group, the other part describes the particle-antiparticle exchange symmetry. That is because the Dirac theory provides a reducible representation of the Poincaré group and contains two kinds of particles, the electron and the positron.

In order to describe all the remaining particles, we must separate, if we can, the fundamental particles from the multitude of composite entities. There are approximate symmetries which describe the multiplets of composite systems. For example, a single unitary infinite dimensional representation of the conformal group SO(4,2) contains all the quantum numbers of states of the Hydrogen atom. Such symmetries have been called dynamical groups, because they are the result of the dynamics of the interaction of two basic constituents, the electron and the proton; mathematically, the result of the tensor product of the representations of the fundamental constituents. Symmetries found in atoms, nuclei, or in periodic system of elements,... are all of this type. In particle physics, many of the fundamental particles of the 1950's and 1960's, nucleons, pions, kaons,... are found now to behave like composite systems again. At present, the basic objects generally assumed are the six leptons (e, ν_e, μ, ν_μ, τ, ν_τ) and their antiparticles, and the six quarks (u d s c b t) and their antiparticles (each with three colour degrees of freedom) interacting via a large number of gauge vector mesons and some additional scalar "Higgs" particles to generate the meson and baryon resonances. Most of these basic particles are unobserved or unobservable.

From a mathematical and physical point of view this is not the most economical system. We shall indicate now how Clifford algebras and their tensor products with their unique representations can provide a simple and concise system of particle classification which has also a direct physical interpretation.

It turns out actually that two basic particles states are sufficient to construct all other particle multiplets by successive tensor products: the electron and the neutrino, that is one charged and one neutral spin $\frac{1}{2}$ particle.[3] Physically these are probably the only absolutely stable particles; all other states are unstable and decay eventually into these plus the radiation (i.e. photons). Although the proton is composite, the proton decay is not yet seen. The proton can be viewed as a $(e^+e^+e^-)$- composite which is extremely stable so that it can be used itself as a basic stable constituent to build other composite sytems.

Now the electron-neutrino complex can be described by the tensor product of Dirac Clifford algebra C_4 with a 2-dimensional Clifford algebra C_1 describing the two charge states Q=0 and 1:

$$C_4 \times C_1 \sim C_5 \sim C_4 \oplus C_4 \qquad (3)$$

Stable baryons, like proton, can be made out of three electro neutrino particles. It is convenient to start from a basic baryonic doublet, one with the quantum numbers of proton, p, the other neutral and with the quantum numbers of a neutron, n. This doublet is a second C_5. The Clifford algebra which contains these two C_5 doublets, $(e\nu_e)$ and (pn) is not C_6 but C_7.

From the tensor product of three C_7's we can build successively the next spin-$\frac{1}{2}$ recurrences of leptons and baryons. For leptons the pairs (μ, ν_μ), (τ, ν_τ),... and for baryons the pairs (Λ_s, Λ_c),... The physical reason for this is the observed decay modes, e.g. $\mu^- \to e^-\nu_\mu\bar{\nu}_e$, and the dynamical model for μ as a resonance in the $(e\nu\bar{\nu})$-system.

In order to establish contact with the standard particle classification using Lie algebras, SU(3), SU(4) or SU(6),... we use the fact that Clifford algebras are also closed under commutators, hence form Lie algebras, and the representations of Clifford algebras coincide with the representations of the lowest dimension of the Lie groups. Taking for example the four leptons $\ell = (e\nu_e\mu\nu_\mu)$ and the four baryons $b = (p\ n\ \Lambda_s\ \Lambda_c)$ it is easy to show that all meson states M and all the baryon states B are constructed as

$$M = \ell \otimes \bar{\ell}\ , \qquad B = b \otimes \ell \otimes \bar{\ell}\ . \qquad (4)$$

The resultant multiplet structure obtained by these tensor products of

Clifford algebras is exactly the same as that of the standard quark model using the representations of SU(4).[4] Similar constructions hold at the SU(6) level using six leptons.

Next we consider the symmetries observed in the interactions of hadrons, e.g. in the scattering of mesons and baryons. Here again one has observed approximate invariance of the scattering amplitudes under the unitary Lie groups, like SU(3), SU(6). It can be shown that this invariance can also be described by the representations of finite groups and charge conservation. The finite groups that occur here are interpreted as the exchange of leptonic constituents of the same type when two hadrons collide.[5] Thus all such symmetries are of Clifford type as we discussed in Section 1.

3. CLIFFORD ALGEBRAS OVER HEISENBERG FIELDS

For the complete description of a particle we must include besides the Clifford algebra C_4 the space-time properties, energy-momentum, angular momentum,... of the particle, properties that arise from the Poincaré group. For this purpose we propose to use the concept of a "Clifford algebra $\mathcal{A}$ over the Heisenberg algebra" (or, more generally, over the operator algebra B(H) in a Hilbert space H). Thus important physical quantities, like

$$p = p^\mu \gamma_\mu , \quad x = x^\mu \gamma_\mu , \ldots$$

are in $\mathcal{A}$. Note that the components of p^μ and x^μ in a basis $\{\gamma_\mu\}$ do not commute: $[p^\mu, x^\nu] = ig^{\mu\nu}$.

In $\mathcal{A}$ the states of the particles are given by the product of a set of idempotents (or projection operators). This includes both the projectors in the Clifford algebra and in the Heisenberg algebra. For example in $\mathcal{A}_7$ the general state is given by

$$\Psi = P_Q^\pm \, P_L^\pm \, P_h^\pm \, (p \pm m) dP(\vec{p}) \ . \tag{5}$$

Here $P_Q^\pm$ = projects charged or neutral particles

$P_L^\pm$ = projects baryon or lepton

$P_h^\pm$ = projects into positive or negative helicity states

$(p \pm m)$ = projects particle or antiparticle

$dP(\vec{p})$ = projects momentum $\vec{p}$ in the range between $\vec{p}$ and $\vec{p}+d\vec{p}$.

Note that $p = p^\mu \gamma_\mu$ is the proper time mass operator.

If A is an element of the Clifford algebra representing an observable such that $A^2=1$, then $P_a = \frac{1}{2} (1 \pm A)$ are projection operators, as well as "eigenstates" of the observable A:

$$A \, P_a^\pm = \pm \, P_a^\pm \ . \tag{6}$$

Thus both the observables and states are in the Clifford algebra.

4. A SYMPLECTIC CLIFFORD ALGEBRA

In the theory of the Dirac electron the physical meaning of the dynamical variables associated with the Clifford algebra C_4 (spin) is not as intuitive as the other dynamical variables (x^μ, p^μ). The phase space of the Schrödinger equation has a classical counterpart. This is not so for the spin variables. Many people have tried to model the Dirac particle as a spinning top, but the correct configuration space could not be modelled. It has been possible recently to give to the spin variables a status equal to the (x,p) variables, in fact transform them into a new set of cannonical symplectic variables (Q_μ, P_μ), both in classical theory and in quantum theory. This is best explained on the basis of the new classical model of the electron based on the Lagrangian

$$L = -\frac{\lambda}{2i}(\dot{\bar{z}}z - \bar{z}\dot{z}) + p_\mu(\dot{x}^\mu - \bar{z}\gamma^\mu z) + eA_\mu \bar{z}\gamma^\mu z \quad . \tag{7}$$

Here $z(\tau)$ is a classical internal spinor variable and τ is an invariant time-parameter. We see by the way here the occurence of the Clifford algebra R_4 in a purely classical theory. The variables x_μ, p_μ, A_μ have their usual meanings. This is a symplectic system with the phase space $\Gamma = (\bar{z}, z, x, p)$, $\bar{z}$ and z as well as x and p being two conjugate pairs. The Hamiltonian with respect to τ is $\mathcal{H} = p^\mu\gamma_\mu$ and is a constant of the motion. In order to interpret the equations of the motion and the internal variables we write the solution as $x_\mu = X_\mu + Q_\mu$, where $X_\mu = (p_\mu/m)\tau$ is the coordinate of the center of mass of the electron moving like a relativistic particle (m is the value of the integral of motion $\mathcal{H}$) and Q_μ is an oscillatory relative coordinate. Thus the charge oscillates around a center of mass, a motion whose quantum analog has been called "zitterbewegung" by Schrödinger, and extensively studied recently.[7] Spin is the orbital angular momentum of the oscillatory zitterbewegung around the center of mass. Let P_μ be the relative momentum and the spin tensor be $S_{\mu\nu} = \frac{1}{2}\bar{z}[\gamma_\mu,\gamma_\nu]z$. The internal variables $Q_\mu, P_\mu, S_{\mu\nu}, \mathcal{H}$ form a Poisson bracket algebra which is in one-to-one correspondence with the commutator Lie algebra of the corresponding variables in quantum theory[8]. This symplectic algebra has a very interesting structure:

$$[Q_\mu, Q_\nu] = \frac{1}{m^2}\tilde{S}^{\mu\nu}; \quad [P_\mu, P_\nu] = 4m^2\tilde{S}^{\mu\nu}; \quad [Q_\mu, P_\nu] = -\tilde{g}_{\mu\nu}\frac{\mathcal{H}}{m},$$

$$[P_\mu, \mathcal{H}] = -4m^2 Q_\mu, \quad [\tilde{S}_{\mu\nu}, \mathcal{H}] = 0, \qquad [Q_\mu, \mathcal{H}] = \frac{1}{m}P_\mu$$

$$[Q_\mu, \tilde{S}_{\mu\lambda}] = \tilde{g}_{\mu\nu}Q_\lambda - \tilde{g}_{\mu\nu}Q_\nu); \quad [P_\mu, \tilde{S}_{\mu\lambda}] = \tilde{g}_{\mu\nu}P_\lambda - \tilde{g}_{\mu\lambda}P_\nu$$

$$[\tilde{S}_{\alpha\beta}, \tilde{S}_{\mu\nu}] = (\tilde{g}_{\alpha\mu}\tilde{S}_{\beta\nu} + \tilde{g}_{\beta\nu}\tilde{S}_{\alpha\mu} - \tilde{g}_{\alpha\nu}\tilde{S}_{\beta\mu} - \tilde{g}_{\beta\mu}\tilde{S}_{\alpha\nu}) \tag{8}$$

$$\tilde{g}_{\mu\nu} = g_{\mu\nu} - \frac{1}{m^2}p_\mu p_\nu$$

$$\tilde{S}_{\mu\nu} = S_{\mu\nu} - \frac{1}{m^2}p_\mu p_\alpha S^{\alpha\nu} - \frac{1}{m^2}p_\nu p_\alpha S^{\mu\alpha}$$

In the quantum case, P_μ, Q_μ,... satisfy in addition anticommutation relations.

The classical spinor model of the electron contains the notion of antiparticles which is thus not solely a quantum concept. It further allows one to solve an old outstanding problem of electrodynamics, namely the path integral formulation of the interaction of the electrons starting from classical trajectories. Path integrals over continuous classical spin variables can result in the quantized propagators with discrete spin values.[9]

This symplectic Clifford algebra can be generalized to n degrees of freedoms and seems to be a universal algebra for the internal dynamics of relativistic systems[10] (again the representations of the Clifford algebra coinciding with the spinor representations of SO(n+2)), even further to infinitely many degrees of freedom.

A relativistic wave equation is the boosting of the symplectic Clifford algebra.[8] Representations of (8) of different dimensionality give different wave equations, from Dirac equation all the way to infinite component wave equations.[11]

REFERENCES

1. E. P. Wigner, Group Theory and Its Applications to Quantum Mechanics of Atomic Spectra, Academic Press, N.Y. 1959. See also A. O. Barut and R. Rączka, The Theory of Group Representations and Applications, Polish Scientific Publ., Second Ed. Warsaw 1980, Ch. 13.
2. A. O. Barut and S. Komy, J. Math. Phys. 7, 1903 (1966); A. O. Barut, J. Math. Phys. 7, 1908 (1966).
3. A. O. Barut, in Quantum Theory and Space Time, Vol. 5 (L. Castell at all, editors), C. Hanser Verlag, München, 1983.
4. A. O. Barut and S. Basri, Lett. Nuovo Cim. 35, 200 (1982); S. Basri and A. O. Barut, Intern. J. Theor. Phys. 22, 691 (1983); J. Math. Phys. 26, 1355 (1985).
5. A. O. Barut, Physica 114 A, 221 (1982).
6. A. O. Barut and N. Zanghi, Phys. Rev. Lett. 52, 2009 (1984).
7. A. O. Barut and A. J. Bracken, Phys. Rev. D23, 2454 (1981); Austr. J. Phys. 35, 353 (1982); Phys. Rev. D24, 3333 (1981).
8. A. O. Barut and W. D. Thacker, Phys. Rev. D31, 1386; 2076 (1985).
9. A. O. Barut and I. H. Duru, Phys. Rev. Lett. 53, 2355 (1984); Phys. Rev. (in press).
10. A. O. Barut and A. J. Bracken, J. Math. Phys. 26, 2515 (1985).
11. For the axiomatics of relativistic wave equations (finite or infinite) from the point of view of dynamical groups see A. O. Barut, in Groups, Systems and Many-Body Physics (edit. by P. Kramer et al), Vieweg Verlag (1980); Ch. V1, p.285.

FERMIONIC CLIFFORD ALGEBRAS AND SUPERSYMMETRY

Geoffrey Dixon
315 Boston Post Rd.
Weston MA 02193
USA

ABSTRACT. Fermionic Clifford algebras were originally developed to serve as part of the foundation of a novel approach to unification, we here briefly outline the historical link between fermionic Clifford algebras, supersymmetry, and unification, and we follow this with a new characterization of nilpotent Clifford algebras of which our fermionic variety form a subclass.

1. NILPOTENT CLIFFORD ALGEBRAS

1.1. Historical Context

A subtitle for this paper might be "Eight Years of Clifford Algebra Mutations", but the starting point of these mutations lies in the beauty of ordinary Clifford algebras, and in particular in the fact that the simplest expression of the field equations of the spin-1 electromagnetic field (Maxwell's equations) is found on the Clifford algebra of spin-1 Minkowski space (spin-1 because under the Lorentz group the vectors of Minkowski space, like the electromagnetic potential, transform like vectors; in general I shall refer to integral spin particles and pseudo-orthogonal geometries as bosonic, and half-integral spin particles and symplectic geometries as fermionic).

The question occurred: can one do much the same thing for a fermion field? That is, instead of having fermion fields reside in the spinor or ideal space of the Clifford algebra of a bosonic geometry, as is customary, create a new kind of fermionic Clifford algebra on the generating set of which a fermion field would find a natural home (as the EM field does on the generating set of $R_{1,3}$ and $R_{3,1}$), and from which properties of the fermion field would be derivable. More ambitious still, find such an algebra on which a strong-weak-electromagnetic (-gravity?) unified theory of elementary particles can be developed, taking account of: leptons and quarks, at least the standard symmetry (SU(3)xSU(2)xU(1)), gauge and Higgs fields, charges, symmetry breaking, parity nonconservation, the structure of a lepton-quark family, the multiplicity of families, and so forth. I have in fact been quite successful in this endeavor [1,2,3,4], and "Algebraic Unification", by

J. S. R. Chisholm and A. K. Common (eds.), Clifford Algebras and Their Applications in Mathematical Physics, 393–398.

which I refer to the result, continues to provide a rich field for work in both mathematics and physics, and its predictions are presently in complete accord with experiment. This success required a reformulation and generalization of Clifford algebra theory, and it is this I will present in the next section (elsewhere [1,5] I have presented related reformulations, but here I present a new one which I have found more useful and elegant).

I began by creating a hybrid of a Clifford algebra and a supersymmetry [6] which I termed a super Clifford algebra [7], which is a ZxZ_4-graded algebra. The thought behind this was that as supersymmetry connects bosons and fermions, a super Clifford algebra might connect bosonic and fermionic Clifford algebras, and from the connection the latter might be discovered. This proved to be the case, and my work on super Clifford algebras led to a unified approach to Clifford algebras for both orthogonal and alternating (symplectic) bilinear forms.

The equations characterizing the ordinary Clifford algebra of an n-dimensional pseudo-orthogonal space take the form

$$f_a f_b + f_b f_a = 2g_{ab}e, \tag{1}$$

where the f_a, a=1,...,n, generate the algebra, e is the identity, and $g_{ab}=g_{ba}$ is an array of real coefficients characterizing the space's bilinear form. In such a characterization the symmetry in the indices of g_{ab} must be reflected in the indices on the left-hand-side, and as this is fixed equation (1) cannot form the basis of a unified treatment. I have found that by replacing equation (1), which relates second-order (in the generators) and zeroth order expressions, by an equation which relates third order and first order expressions a more unified treatment may be developed, and one which perfectly suited my requirements as a foundation for a physical theory.

1.2. Nilpotent Clifford Algebras

Let V be an n-dimensional real vector space, and let x and y be vectors in V with components x^a, y^a, a=1,...,n, in some representation. Given an nxn array of real coefficients g_{ab} we can define a bilinear form on V by

$$[x,y] = x^a g_{ab} y^b, \tag{2}$$

where summation is assumed over indices repeated once up and once down. For simplicity we assume $\det(g_{ab})=\pm 1$, and for all a,b, either $g_{ab}=g_{ba}$ or $g_{ab}=-g_{ba}$ (in this case n is even). We define g^{ab} so that

$$g_{ab}g^{bc} = \delta_a^{\ c}, \quad g^{ab}g_{bc} = \delta^a_{\ c}. \tag{3}$$

Let A be a real, associative algebra with 2n linearly independent elements e_a and e^a, a=1,...,n, which generate A as an algebra. Let

x^a, y^a, u^a, v^a be the components of arbitrary vectors in V, and define $x=x^a e_a$, $\hat{x}=e^b g_{ba} x^a \equiv \hat{e}_a x^a$, etc., and let $[x,y]=x^a g_{ab} y^b$, a scalar. If

$$(x+\hat{y})(u+\hat{v})(x+\hat{y}) = (2[x,v]x-[x,x]v)+(2[u,y]\hat{y}-[y,y]\hat{u}), \quad (4)$$

then we refer to A as the nilpotent Clifford algebra (NCA) of V with the bilinear form (2). Without further specifying the g_{ab} one can show

$$e_a e_b = e^a e^b = 0, \quad (5)$$

hence the term nilpotent.

As is the case with ordinary Clifford algebras, bivector terms of an NCA generate elements of the symmetry group of (2). In particular, if the real coefficients $w^a{}_b$ satisfy

$$g_{ab} w^b{}_c = -g_{bc} w^b{}_a, \quad (6)$$

then

$$x \rightarrow W^a{}_b x^b = x^a + w^a{}_b x^b + \tfrac{1}{2} w^a{}_b w^b{}_c x^c + \ldots = \exp(w)^a{}_b x^b \quad (7)$$

is a transformation on the vectors of V leaving (2) invariant. One can then prove using (4) that

$$\exp(w^a{}_b e_a e^b/2) x \exp(w^a{}_b e^b e_a/2) = e_a (W^2)^a{}_b x^b \quad (g_{ab}=g_{ba}) \quad (8a)$$

$$= e_a W^a{}_b x^b \quad (g_{ab}=-g_{ba}). \quad (8b)$$

In the first case the resultant group of transformations is isomorphic to some SO(p,q), in the second case (where n is even) to Sp(n).

Now assume $g_{ab}=g_{ba}$, and in particular that g_{ab} is diagonal, $g_{aa}=1$, $a=1,\ldots,p$, $g_{aa}=-1$, $a=p+1,\ldots,p+q=n$. The ordinary Clifford algebra for this case is denoted $R_{p,q}$, and we denote the NCA by $N_{p,q}$. Define $f_a=e_a+\hat{e}_a$. It is not difficult to prove that

$$f_a f_b + f_b f_a = 2g_{ab} e, \quad (9)$$

where the identity e therefore satisfies

$$e = e_a e^a + e^a e_a \quad \text{(no sum)}, \ a=1,\ldots,n. \quad (10)$$

Similarly, let $t_a=e_a-\hat{e}_a$, then

$$t_a t_b + t_b t_a = -2g_{ab} e. \quad (11)$$

Consequently the subalgebra of $N_{p,q}$ generated by $f_a(t_a)$ is isomorphic to

$R_{p,q}(R_{q,p})$.

For any fixed a define

$$h = e_a e^a - e^a e_a \quad \text{(no sum)}. \tag{12}$$

This element satisfies

$$hf_a = t_a = -f_a h, \tag{13a}$$

$$ht_a = f_a = -t_a h, \tag{13b}$$

$$h^2 = e. \tag{13c}$$

Therefore its inclusion into the subalgebra $R_{p,q}(R_{q,p})$ expands it to $R_{p+1,q}(R_{q+1,p})$, and in fact

$$N_{p,q} = R_{p+1,q} = R_{q+1,p} = N_{q,p}. \tag{14}$$

In particular, the NCA of Minkowski space is

$$N_{1,3} = N_{3,1} = C(4), \tag{15}$$

the complex Dirac algebra (see [1-3]).

I am only going to treat the simplest symplectic case as I believe this is the one relevant to physics. I shall change the notation in this case and replace g_{ab} by ϵ_{ab}, a,b=1,2, $\epsilon_{11}=\epsilon_{22}=0$, $\epsilon_{12}=-\epsilon_{21}=1$. I also replace e_a by S_a, e^a by S^a, and denote the algebra S.

Using (4) one can rearrange and reduce products and demonstrate that a complete basis for S can be given in terms of elements up to the fourth order (ie., elements of the form $S_a \hat{S}_b S_c \hat{S}_d$ and $\hat{S}_a S_b \hat{S}_c S_d$). One can then show that

$$S = R(3) \times R(3). \tag{16}$$

I shall denote elements of S by ordered pairs like (x,y) and (u,v), x,y,u,v in R(3), and define the product (x,y)(u,v)=(xu,yv).

The simplest symplectic NCA is now discussed as I believe this is the one relevant to physics. I shall change the notation in

$$s_1 = 2^{\frac{1}{2}} \begin{bmatrix} & 1 & \\ 0 & 0 & 0 \\ & 0 & \end{bmatrix}, \quad s_2 = 2^{\frac{1}{2}} \begin{bmatrix} & 0 & \\ 0 & 0 & 0 \\ & -1 & \end{bmatrix}, \tag{17}$$

and define the following antiautomorphism on R(3):

$$\overline{\begin{bmatrix} x_{11} & x_{12} & x_{13} \\ x_{21} & x_{22} & x_{23} \\ x_{31} & x_{32} & x_{33} \end{bmatrix}} = \begin{bmatrix} x_{33} & -x_{23} & -x_{13} \\ -x_{32} & x_{22} & x_{12} \\ -x_{31} & x_{21} & x_{11} \end{bmatrix}. \tag{18}$$

We now set

$$S_a = (s_a, s_a^-). \tag{19}$$

For (x,y) in R(3)xR(3) define

$$(x,y)^\wedge = (x^-, y^-) \tag{20}$$

(this is consistent with the previous use of this symbol) so that

$$S_1^\wedge = S^2\epsilon_{21} = -S^2 = (s_1^-, -s_1),\ S_2^\wedge = S^1\epsilon_{12} = S^1 = (s_2^-, -s_2). \tag{21}$$

Two elements of R(3) which play important roles in the study of S are

$$e^{ex} = \begin{bmatrix} 1 & & \\ & 0 & \\ & & 1 \end{bmatrix},\ e^{in} = \begin{bmatrix} 0 & & \\ & 1 & \\ & & 0 \end{bmatrix} \tag{22}$$

(see [1] for the origin of this notation). With these we can define the invariants

$$Q = \tfrac{1}{2}S_a S^a = (e^{ex}, 2e^{in}) \tag{23a}$$

and

$$Q^* = \tfrac{1}{2}S^a S_a = (2e^{in}, e^{ex}). \tag{23b}$$

Equations (9) and (11) have their counterparts in S which involve Q and Q*. Define $F_a = S_a + S_a^\wedge$, $T_a = S_a - S_a^\wedge$, then

$$F_a F_b - F_b F_a = 2\epsilon_{ab}(Q - Q^*), \tag{24a}$$

$$T_a T_b - T_b T_a = 2\epsilon_{ab}(Q + Q^*). \tag{24b}$$

Further, and again like the orthogonal case but with different signs, the three independent elements $F_a F_b + F_b F_a = -(T_a T_b + T_b T_a)$ are, as a Lie algebra, isomorphic to sp(2) from which is generated the symmetry group of the symbol ϵ_{ab}.

As a complete study of the relationship of S to the symplectic geometry has not been made I shall finish by observing that in neither the orthogonal or symplectic case can one use the coefficients g_{ab} or ϵ_{ab} as lowering operators as the e_a and e^a and S_a and S^a are linearly independent. It is possible on the subalgebras generated by the f_a (or t_a) or F_a (or T_a), but in general one can not be too free in performing this operation, and in fact this is important in applications.

REFERENCES

1. G. M. Dixon, Phys. Rev. D28 (1983) 833.

2. G. M. Dixon, Phys. Rev. D29 (1984) 1276.

3. G. M. Dixon, Phys. Lett. 152B (1985) 343.

4. G. M. Dixon, to appear in J. of Physics.

5. G. M. Dixon, Lett. Math. Phys. 5 (1981) 411.

6. V. G. Kac, Comm. Math. Phys. 53 (1977) 31.

7. G. M. Dixon, J. Math. Phys., 19 (1977) 2103.

ON GEOMETRY AND PHYSICS OF STAGGERED LATTICE FERMIONS

Hans Joos
Deutsches Elektronen-Synchrotron DESY
Hamburg

ABSTRACT. This report contains: 1.Introduction. 2.The Problem of Lattice Fermions. 3.Relation between Dirac Fields and Differential Forms. 4.Dirac Kaehler Equation and Clifford Product on a Cubic Lattice. 5.The Symmetry of the Dirac Kaehler Equation on the Lattice. 6.A Physics Problem. 7.Summary and Outlook.

1. INTRODUCTION

In a series of lectures E.Kaehler [1] discussed the possibility of interpreting the equation ('Dirac Kaehler Equation': DKE)

$$(d - \delta + m)\Phi = 0 \tag{1}$$

as a generalization of the Dirac equation. Why have people showed a renewed interest in the DKE at the present time? It was shown [2] [3] [4] that a systematic lattice approximation of Equ. (1) is equivalent to the Kogut Susskind description [5] of lattice fermions! This approximation was done in the spirit of the DeRham isomorphy which maps the differential form Φ, the exterior differentiation d , and the codifferential operator δ on cochains,boundary operator, and coboundary operator on the lattice. How is this fact related to our topic of *CLIFFORD ALGEBRA*? The connection between the DKE and the Dirac equation relies heavily on the introduction of a Clifford product for differential forms.Thus the many important physics questions which one wants to solve with the help of the lattice approximation of the Quantum Chromodynamics(QCD) lead to interesting new problems related to Clifford algebras. It is the aim of my lecture to illustrate this assertion by examples taken from actual research in lattice QCD.

J. S. R. Chisholm and A. K. Common (eds.), Clifford Algebras and Their Applications in Mathematical Physics, 399–423.

2. THE PROBLEM OF LATTICE FERMIONS

The fundamental formula for the calculation of the expectation value of a physical observable Ω is [6] [7]

$$\langle\Omega\rangle = \frac{1}{Z}\int \mathcal{D}[\mathbf{A}]\mathcal{D}[\psi,\overline{\psi}]\Omega[\mathbf{A},\overline{\psi},\psi]e^{-S(\mathbf{A},\overline{\psi},\psi)}. \tag{2}$$

$\mathbf{A}(x)$ denotes the colour gauge potential describing the gluons, $\psi(x),\overline{\psi}(x)$ the quark fields. The action $S = S_g + S_q$ consists of a gluon part

$$S_g = \frac{1}{2g^2}\int dx\ trace(\mathbf{F}^{\mu\nu}\mathbf{F}_{\mu\nu}), \tag{3}$$

$$\mathbf{F}_{\mu\nu} = \partial_\mu\mathbf{A}_\nu - \partial_\nu\mathbf{A}_\mu + i[\mathbf{A}_\mu,\mathbf{A}_\nu], \tag{4}$$

and a part describing quark fields interacting with gluon fields

$$S_q = \int dx\overline{\psi}(x)(D_\mu\gamma^\mu + m)\psi(x), \tag{5}$$

$$D_\mu\psi(x) = (\partial_\mu + i\mathbf{A}_\mu)\psi(x). \tag{6}$$

The expression (2) is supposed to describe averaging $\Omega[...]$ with a normalized measure $\frac{1}{Z}\mathcal{D}[\mathbf{A}]\mathcal{D}[\psi,\overline{\psi}]$ over all field configurations. This procedure is not well defined mathematically. The significance of formula (2) for physics is based mainly on its formal perturbative evaluation, which leads to an understanding of many experimental facts in QCD, and it leads in QED to experimentally confirmed results of highest precision [6]. In order to make Eq. (2) more rigorous mathematically, one considers it for a field theory in Euclidean space time, using imaginary time coordinates. Real time field theory follows by analytic continuation [8]. This first step we have already performed by the definition of an 'Euclidean' action in (3), (5). The real time action would have been imaginary. The next step is to approximate Euclidean space time by a finite lattice. Then the 'path integral' in (2) becomes a finite dimensional integral. It is hoped that the 'thermodynamic limit' to an infinite volume of space time, and a 'renormalized continuum limit' leads to an interpretation of (2) which is acceptable for physics.

A guiding principle for the formulation of the lattice approximation of gauge theories is given by the geometrical meaning of the basic quantities. This leads to the 'Wilson action' for gauge fields on the lattice. Let us denote in a cubic lattice $\overline{\Gamma}$, by $\overline{x} = \overline{b}(\overline{n}^1,...,\overline{n}^d)$ the lattice points, by $[\overline{x},\mu_1,...,\mu_h]$ the lattice h-cells, by $\overline{e}_\mu$ the unit lattice vector (see Fig.1.); we put the lattice constant $\overline{b} = 1$ most of the time. Then the geometric meaning of the gauge potential $\mathbf{A}_\mu$ as connection in the principle bundle of the colour group $\mathcal{P}[R^d, SU(3)]$ leads to the associated lattice description of $\mathbf{A}_\mu$ by finite parallel transports $U(\overline{x})$ of a colour vector along links (1-cells) $[\overline{x},\mu]$:

$$U(\overline{x},\mu) \sim \exp(i\int_{[\overline{x},\mu]}\mathbf{A}_\mu(x)dx^\mu), \qquad U(\overline{x}+\overline{e}_\mu,-\mu) = U^{-1}(\overline{x},\mu). \tag{7}$$

The lattice analog of the field strength is the parallel transport around a plaquette $\mathbf{P} = [\bar{x}, \mu\nu]$

$$U(\mathbf{P}) = U(\bar{x}, \mu\nu) = U^{-1}(\bar{x}, \nu)U^{-1}(\bar{x} + \bar{e}_\nu, \mu)U(\bar{x} + \bar{e}_\mu, \nu)U(\bar{x}, \mu) \sim \exp(i\mathbf{F}_{\mu\nu}(\bar{x})). \quad (8)$$

With these quantities we can define the Wilson action for the SU(n)-gauge theory [9]

$$S_g = -\frac{\beta}{2n} \sum_{\mathbf{P} \in \bar{\Gamma}} (U(\mathbf{P}) + U^{-1}(\mathbf{P}) - 2), \quad (9)$$

which in a formal continuum limit $\bar{b} \to 0$ approaches the gauge action S_g of Eq. (3). Of course, these classical limits mean little for the existence of the renormalized quantum mechanical limit.

Next we have to find the lattice approximation of the quark fields. In a first naive attempt one considers fermion fields as defined on lattice points, and one substitutes differentiation by a difference approximation which respects the Hermite symmetry of the Dirac operator. This 'naive' free lattice fermion action reads then to

$$S_q = \sum_{\bar{x}} (\sum_\mu \frac{1}{2\bar{b}} [\bar{\psi}(\bar{x})\gamma_\mu \psi(\bar{x} + \bar{e}_\mu) - \bar{\psi}(\bar{x} + \bar{e}_\mu)\gamma_\mu \psi(\bar{x})] + \bar{m}\bar{\psi}(\bar{x})\psi(\bar{x})). \quad (10)$$

The propagator related to S_q is in momentum representation

$$G(\bar{p}) = (\sum_\mu \gamma_\mu \frac{1}{\bar{b}} \sin \bar{p}_\mu \bar{b} + \bar{m})^{-1}, \quad (11)$$

with $\bar{p}$ varying in the first Brillouin zone: $-\frac{\pi}{\bar{b}} < \bar{p} \leq \frac{\pi}{\bar{b}}$. However,this propagator poses some serious problems for physics! For lattice spacing $\bar{b} \to 0, G(\bar{p})$ is non-vanishing in 16 different regions, namely for $\bar{p}_\mu \approx 0, \pi,\ \mu = 1, 2, 3, 4$, where it approaches the continuum expression $G_{cont}(p) = (p_\mu \gamma^\mu + m)^{-1}$. Thus the action (10) describes not one but 16 Dirac particles [10].

This spectrum degeneracy is related to the symmetry [11] of the action S_q under the transformations

$$(\hat{M}^H \psi)(\bar{x}) = e^{i\pi(\bar{e}_H, \bar{x})} M^H \psi(\bar{x}), \qquad (\hat{M}^H \bar{\psi})(\bar{x}) = e^{i\pi(\bar{e}_H, \bar{x})} \bar{\psi}(\bar{x}) M^{H\dagger},$$

$$M^H = \prod_{\mu \in H} M^\mu, \qquad M^\mu = i\gamma^5\gamma^\mu, \qquad \bar{e}_H = \sum_{\mu \in H} \bar{e}_\mu. \quad (12)$$

Here we used a multi-index notation $H = (\mu_1, ...\mu_h),\ \ \mu_1 < \mu_2 < ... < \mu_h$. The $\hat{M}$ satisfy the defining relations of the γ- matrices: $\hat{M}^\mu \hat{M}^\nu + \hat{M}^\nu \hat{M}^\mu = 2\delta^{\mu\nu}$. They transform the γ-matrices like

$$(\hat{M}^H)^{-1} \gamma^K \hat{M}^H = e^{i\pi(\bar{e}_H, \bar{e}_K)} \gamma^K, \quad (13)$$

from which the invariance of the action (10) follows immediately. The phase factor $\exp i\pi(\bar{e}_H, \bar{x})$ acts on the Fourier transform $\tilde{\psi}(\bar{p}) = (2\pi)^{d/2} \sum_{\bar{x}} e^{i\bar{p}\bar{x}} \psi(\bar{x})$ by translating the argument

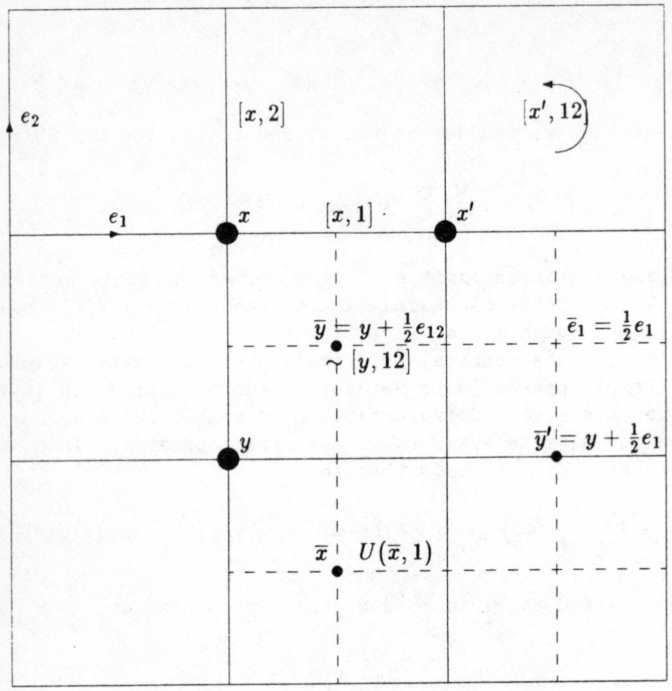

Figure 1: Illustration of the Lattice Notions

$(\hat{M}^H\tilde{\psi})(\overline{p}) = M^H\tilde{\psi}(\overline{p} + \pi\overline{e}_H)$ The group $\{\hat{M}^H\}$ acts as a permutation group of the 16 regions of degeneracy. Therefore it is called sometimes 'spectrum doubling group': SDG.

In order to reduce the unwanted inflation of fields, it was suggested [5] that one use the symmetry of the SDG to separate the action (10) into parts with lower degeneracy. For this we follow a procedure which prepares our later discussion. We imbed the lattice $\overline{\Gamma}$ in a lattice Γ with the double lattice spacing (see Fig.1.):

$$x \in \Gamma : x = b(n^1, ..., n^d) = 2\overline{b}(\overline{n}^1, ..., \overline{n}^d), \qquad e_\mu = 2\overline{e}_\mu,$$

$$\overline{x} \in \overline{\Gamma} : \overline{x} = x + \frac{1}{2}e_{H(x)}, \qquad \phi(x, H) \equiv \chi(\overline{x}) \tag{14}$$

With this notation we transform the naive Dirac fields

$$\psi_a(\overline{x}) = \gamma_{ai}^{H(\overline{x})}\chi_i(\overline{x}), \qquad \chi_i(\overline{x}) = \sum_a (\gamma^{H(\overline{x})})^{-1}_{ia}\psi_a(\overline{x}). \tag{15}$$

It follows from (13) that $\chi_i(\overline{x})$ transforms without an $\overline{x}$-dependent phase $(\hat{M}^H\chi)_i(\overline{x}) = M^H_{ik}\chi_k(\overline{x})$. Since the Dirac-operator $\gamma^\mu\overline{\triangle}^\mu$ commutes with the transformations

$\hat{M} \in SDG$, the action separates in the irreducible components $\chi_i(\overline{x})$. Indeed we get

$$S_q = \sum_{i=1}^{4} \sum_{\overline{x}} \{ \sum_{\mu} \frac{\check{\rho}_{\mu H}}{2\overline{b}} [\overline{\chi}_i(\overline{x})\chi_i(\overline{x}+\overline{e}_\mu) - \overline{\chi}_i(\overline{x}+\overline{e}_\mu)\chi_i(\overline{x})] + \overline{m}\overline{\chi}_i(\overline{x})\chi_i(\overline{x})\}, \tag{16}$$

with a sign function $\check{\rho}_{\mu H}$ explicitely defined in Eq. (19). The lattice fields $\chi_i(\overline{x})$ are called the 'staggered fermion components' of the naive Dirac fields. The restriction to one component $\chi(\overline{x})$ reduces the degeneracy by a factor 4 in 4-dimensional Euclidean space time. In the following sections we are mainly concerned with the geometry and physics of staggered fermion fields.

Our presentation of the problem of lattice fermions might give the impression that spectrum doubling is a more or less superficial problem. This is not correct. Mathematically it was shown by H.B.Nielson and M.Ninomiya [12] and others [13] that it originates deeply in the topological structure of the 'kinetic energy' terms defined on the periodic Brillouin zone. It can be explained by the theorems on 'spectral flow' [14]. The methods applied in physics to deal with the problem are rather arbitrary. One procedure [15] is to add an arbitrary term which breaks the degeneracy in such a way that only one Dirac field survives in the continuum limit. However, this destroys important symmetries of the lattice theory. The point of view we present in this paper is that staggered fermions represent in an appropriate form the spectrum problem [16]. The underlying geometry may give some hints on interesting structure in elementary particle physics.

3. RELATION BETWEEN DIRAC FIELDS AND DIFFERENTIAL FORMS

Before we give a geometric interpretation of staggered lattice fermions by a lattice approximation of the DKE, we want to clarify the relation between the DKE and the Dirac equation in the Euclidean space time continuum. The essential point in this consideration is the introduction of a Clifford product for differential forms[1]. Let us first introduce our notation. (x^μ) are coordinates with respect to an orthogonal basis of Euclidean R^d. We use for the expansion of a complex differential form Φ in the basis generated by the wedge products of dx^μ ('Cartesian basis') the multi-index notation of Eq. (12)

$$\phi = \sum_{h=1,\mu_i}^{d} \frac{1}{h!} \phi_{\mu_1 \ldots \mu_h}(x) dx^{\mu_1} \wedge \ldots \wedge dx^{\mu_h} \equiv \sum_H \phi(x,H) dx^H. \tag{17}$$

The bilinear, associative Clifford algebra $C\mathcal{L}\{\Phi\}$ is generated by the products of dx^μ

$$dx^\mu \vee dx^\nu = dx^\mu \wedge dx^\nu + \delta^{\mu\nu}. \tag{18}$$

It follows for the products of the basis elements dx^H

$$dx^H \wedge dx^K = \hat{\rho}_{H,K} dx^{H \cup K}, \qquad dx^H \vee dx^K = \check{\rho}_{H,K} dx^{H \triangle K},$$

$$\hat{\rho}_{H,K} = 0 \quad for H \cap K \neq \emptyset, \quad \hat{\rho}_{H,K} = (-1)^{\nu} \quad for \quad H \cap K = \emptyset, \quad \check{\rho}_{H,K} = (-1)^{\nu},$$
$$\nu \quad number \quad of \quad pairs \quad (\mu_i^1, \mu_j^2), \quad \mu_i^1 \in H, \quad \mu_j^2 \in K, \quad \mu_i^1 > \mu_j^2. \tag{19}$$

The main automorphism $\mathcal{A} : \mathcal{A}(\Phi \wedge \Psi) = \mathcal{A}\phi \wedge \mathcal{A}\Psi, \quad \mathcal{A}(\Phi \vee \Psi) = \mathcal{A}\phi \vee \mathcal{A}\Psi$ and the main antimorphism $\mathcal{B} : \mathcal{B}(\Phi \wedge \Psi) = \mathcal{B}\Psi \wedge \mathcal{B}\Phi, \quad \mathcal{B}(\Phi \vee \Psi) = \mathcal{B}\Psi \vee \mathcal{B}\Phi$ transform the multi-indexed basis elements as

$$\mathcal{A}dx^H = (-1)^h dx^H, \qquad \mathcal{B}dx^H = (-1)^{h(h-1)/2} dx^H. \tag{20}$$

The contraction operator $e_\mu \neg$, defined linearly as antiderivation

$$e_\mu \neg (\Phi \wedge \Psi) = (e_\mu \neg \Phi) \wedge \Psi + \mathcal{A}\Phi \wedge e_\mu \neg \Psi, \qquad e^\mu \neg dx^\nu = \delta^{\mu\nu}, \quad e^\mu \neg 1 = 0.$$

relates Clifford product and wedge product

$$dx^\mu \vee \Phi = dx^\mu \wedge \Phi + e^\mu \neg \Phi. \tag{21}$$

With this notation the differential operators d and δ can be expressed conveniently expressed by

$$d\Phi = dx^\mu \wedge \partial_\mu \Phi, \qquad \delta\Phi = -e^\mu \neg \partial_\mu \Phi,$$

and because of Eq. (21)

$$(d - \delta)\Phi = dx^\mu \vee \partial_\mu \Phi. \tag{22}$$

∂_μ denoting differentiation of the components $\phi(x, H)$.
The relation between the DKE (1) and the Dirac equation in flat space is a consequence of the isomorphism $dx^\mu \leftrightarrow \gamma^\mu$ between the Clifford algebra $\mathcal{CL}\{\Phi\}$ and the algebra of the Dirac matrices γ^μ which follows from the equivalence of Eq. (18) with the defining relations of the $\gamma^\mu : \gamma^\mu\gamma^\nu + \gamma^\nu\gamma^\mu = 2\delta^{\mu\nu}$. In order to describe this relation, we use an explicit construction of the decomposition of $\mathcal{CL}\{\Phi\}$ into primitive left ideals [2][17] [18]. For this we use in even dimension d the $2^{d/2}$-dimensional, irreducible representation of the $\gamma^H = \gamma^{\mu_1} \cdot\cdot \gamma^{\mu_h}$, to define a new basis $Z = (Z_a^b)$:

$$\Phi = \sum_H \phi(x, H) dx^H = \sum_{a,b} \psi_a^b(x) Z_a^b,$$

$$Z = 2^{-d/2} \sum_H (\gamma^H)^T \mathcal{B} d^H. \tag{23}$$

This basis has the property

$$dx^\mu \vee Z = \gamma^{\mu T} Z, \qquad Z \vee dx^\mu = Z\gamma^{\mu T}, \qquad Z_a^b \vee Z_c^d = Z_a^d \delta_c^b. \tag{24}$$

The transformation between the 'Cartesian'-components $\phi(x, H)$ and the 'Dirac' components ψ_a^b of Φ:

$$\phi(x, H) = 2^{-d/2} trace(\gamma^{H\dagger} \psi(x, H)), \qquad \psi(x) = \sum_H \phi(x, H)\gamma^H, \tag{25}$$

results from the completeness and orthogonality relations of the γ-matrices. Now it follows from Eqs. (22), (24), (13) that the Dirac components of a solution of the DKE satisfy the Dirac equation

$$(\gamma^\mu \partial_\mu + m)\psi^b(x) = 0. \tag{26}$$

In this sense, the DKE in flat space is equivalent to a set of $2^{d/2}$ Dirac equations.
This degeneracy of the DKE is descibed by the following symmetry. Let

$$c(u) = \sum_H u(H)dx^H = \sum_{a,b} u_a^b Z_a^b \tag{27}$$

be a constant differential form: $\partial_\mu c = 0$. Due to the associativity of the $\vee$-multiplication , $\Phi \to \Phi \vee c$, transforms a solution of the the DKE into a solution,

$$0 = ((d - \delta + m)\Phi) \vee c = (dx^\mu \vee \partial_\mu \Phi + m\Phi) \vee c$$
$$= (dx^\mu \partial_\mu (\Phi \vee c) + m(\Phi \vee c). \tag{28}$$

It follows immediately from Eq. (24) that $c(u)$ transforms the Dirac components by the matrix u_c^b:

$$(\Phi \vee c(u))_a^b = \sum_d \psi_a^d u_d^b. \tag{29}$$

In case $(u_a^b) \in SU(4)$, these transformations are called flavour transformations. Eqs (27), (29) describe the flavour transformations by Clifford right multiplication, i.e.by an element of the Clifford group [17].
c(u) may also represent a projection operator. Thus
$\mathbf{P}_a = \frac{1}{4}(1 \pm i dx^{12})(1 \pm dx^{1234})$ projects on the 4 different flavour Dirac components in a representation of the γ-matrices where γ^5, γ^{12} are diagonal.
Finally we want to express the scalar product of differential forms by Dirac components:

$$(\Phi, \Psi)(x) = \sum_H \phi^*(x, H)\psi(x, H) = 2^{-d/2} \sum_{a,b} \phi_a^{*b}(x)\psi_a^b(x),$$

with

$$\psi_a^b(x) = (Z_a^b, \Psi), \qquad \phi_a^{*b}(x) = (\Phi, Z_a^b). \tag{30}$$

The Clifford product allows an algebraic definition of

$$(\Phi, \Psi)_0 = (\Phi, \Psi)dx^{1234} : \qquad (\Phi, \Psi)_0 = (\mathcal{B}\Phi^* \vee \Psi) \wedge dx^{1234}$$

In terms of these expressions, the free action might be written as

$$S_q = \frac{1}{4}\int (\overline{\Phi}, (d - \delta + m)\Phi)_0 = \sum_b \int dx^{1234}\overline{\psi}^b(x)(\gamma^\mu \partial_\mu + m)\psi^b(x), \tag{31}$$

which should be compared with Eq. (5).

4. DKE AND CLIFFORD PRODUCT ON A CUBIC LATTICE

Now we consider the lattice approximation of the DKE with the aim to show that this leads to a geometric interpretation of staggered lattice fermions. Imitating the mapping between DeRham complexes and simplicial complexes [19], we get as lattice correspondence of the differential form Φ

$$\Phi(\mathcal{C}) = \int_{\mathcal{C}} \Phi, \tag{32}$$

where $\mathcal{C}$ is a sum of lattice cells. $\Phi(\mathcal{C})$, a linear functional on linear combinations of cells, can be expressed in a basis $\{d^{x,H}\}$:

$$\Phi = \sum_{x,H} \Phi(x,H) dx^{x,H}, \qquad dx^{x,H}([x',H']) = \delta^{x}_{x'} \delta^{H}_{H'}. \tag{33}$$

This is the lattice analog of (17). Because of Stokes' theorem, the mapping Eq. (32) implies

$$d \to \check{\triangle}, \qquad \check{\triangle}\Phi(\mathcal{C}) = \Phi(\triangle \mathcal{C}), \qquad \delta \to \check{\nabla}, \qquad \check{\nabla}\Phi(\mathcal{C}) = \Phi(\nabla \mathcal{C}).$$

Here $\triangle$ and ∇ are the boundary and the coboundary operator applied to lattice cells. Thus the DKE on the lattice becomes

$$(\check{\triangle} - \check{\nabla} + m)\Phi = 0. \tag{34}$$

In the continuum theory, it was important for the further treatment of the DKE to introduce a Clifford product for differential forms. Now we want to discuss the possibilities of a Clifford product on the lattice. We follow the procedure of algebraic topology and introduce the cup product and the cap product corresponding to the $\wedge$- product and $e^{\mu}\neg$ -operator

$$\begin{aligned} d^{x,H} \wedge d^{y,K} &= \hat{\rho}_{H,K} \delta^{x+e_H, y} d^{x, H\cup K}, \\ e^{\mu} \neg d^{y,K} &= \tilde{\rho}_{\mu,K} d^{x,K-\mu}, \end{aligned} \tag{35}$$

($\tilde{\rho}_{\mu,K}$as in the continuum), and combine both like in Eq. (21) for a Clifford product

$$d^{x,H} \vee d^{y,K} = \check{\rho}_{H,K} \delta^{x+e_H, y} d^{x+e_\Lambda, H\triangle K}, \qquad \Lambda = H \cap K, \qquad H\triangle K = H\cup K - \Lambda. \tag{36}$$

Some examples are illustrated in Fig.(2). The wedge product is associative and satisfies

$$\check{\triangle}(\Phi \wedge \Psi) = \check{\triangle}\Phi \wedge \Psi + \mathcal{A}\Phi \wedge \check{\triangle}\Psi.$$

The important rules for the lattice contraction operator $e_\mu \neg$ are

$$e_\mu \neg (d^{x,H} \wedge d^{y,K}) = (e_\mu \neg d^{x,H}) \wedge T_{e_\mu} d^{x,K} + \mathcal{A} d^{x,H} \wedge (e_\mu \neg d^{x,K}),$$

$$T_{e_\mu} d^{x,H} = d^{x-e_\mu, H}, \qquad e^{\mu}\neg e^{\nu}\neg + e^{\nu}\neg e^{\mu}\neg = 0.$$

However,the Clifford product is not associative, but satisfies

$$(d^{x-e_{H\cap L}, H} \vee d^{y,K}) \vee d^{z-e_{H\cap K}, L} = d^{x,H} \vee (d^{y,K} \vee d^{z,L}). \tag{37}$$

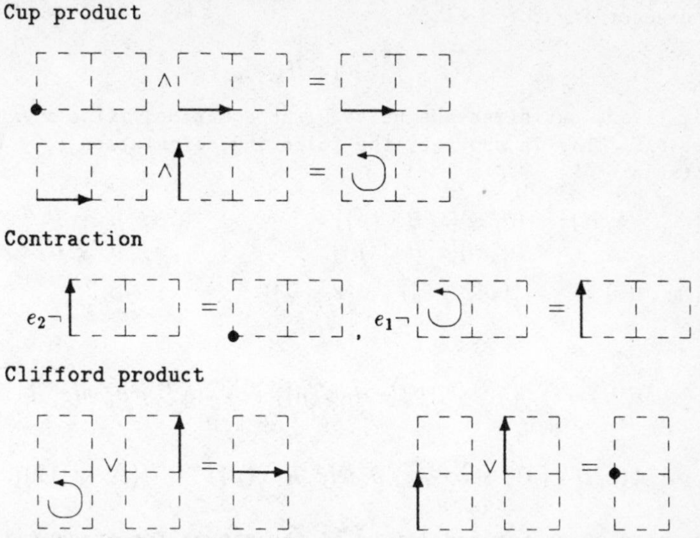

Figure 2: Examples of of operations with elementary cochains of a 2-dimensional lattice: contraction, $\wedge$ product and $\vee$ product. These illustrate the 'matching' conditions.

This is a consequence of the 'matching conditions' which must be satisfied by the $\vee$ -product of elementary cochains $d^{x,H}$ for giving a non-vanishing result.
Is it worthwhile to consider such a Clifford product on the lattice? Formally it helps us to procede with our analogy to the continuum.
Defining, with help of the translation operator T_{e_μ}, difference operators $\partial_\mu^+ = T_{e_\mu} - 1, \partial_\mu^- = 1 - T_{-e_\mu}$ allows a representation of $\check{\Delta}, \check{\nabla}$ very similar to (22):

$$\check{\Delta} = d^\mu \wedge \partial_\mu^-, \qquad \check{\nabla} = -e^\mu \neg \partial_\mu^-, \qquad \check{\Delta} - \check{\nabla} = d^\mu \vee \partial_\mu^-. \tag{38}$$

with $d^H = \sum_x d^{x,H}$. Summing over x,y in Eq. (37) leads to an 'associative' law:
$(d^H \vee d^{x,K}) \vee d^L = d^H \vee (d^{x,K}) \vee d^L)$. From this follows, as in Eq. (28), that $\Phi \to \Phi \vee d^H$ transforms a solution of the lattice DKE into a solution. Thus the Clifford product on the lattice describes an important symmetry of staggered fermions. We shall discuss this in more detail in the following sections.
Now we show that the lattice Dirac Kaehler field $\phi(x,H)$ becomes a staggered fermion field $\chi(\overline{x})$ if we make the identification

$$\phi(x,H) = \phi(\overline{x}, H(\overline{x})) = \chi(\overline{x}), \tag{39}$$

with notations explained in Eq. (14). By this mapping the r-cochains of the 'coarse' lattice $\Gamma : \phi(x,H)$ get identified with lattice fields defined at the points of the 'fine' lattice $\overline{x} \in \overline{\Gamma}$ which are central points of the cells $[x,H]$ (see Fig.1). The proof of the equivalence of Dirac Kaehler fermions with staggered fermions consists by the demonstration that a Dirac field defined according to Eq. (15):

$$\psi_a(\overline{x}) = \gamma_{ai}^{H(\overline{x})}\phi(\overline{x}, H(\overline{x}))$$

i arbitrarily fixed, satisfies the naive Dirac equation, iff $\phi(x,H)$ satisfies the DKE. This is shown by the following calculation. We have the following formula

$$\begin{aligned}(\overline{\triangle}_\mu\psi)(x,H) &= (\partial_\mu^+\psi)(x, H\setminus\{\mu\}) && if\ \mu\ \in H,\\ &= (\partial_\mu^-\psi)(x, H\cup\{\mu\}) && if\ \mu\ \notin H.\\ (\overline{\triangle}_\mu\psi)(\overline{x}) &= \tfrac{1}{2}(\psi(\overline{x}+\overline{e}_\mu) - \psi(\overline{x}-\overline{e}_\mu)).\end{aligned}$$

and hence

$$\gamma^{H\dagger}(\gamma^\mu\overline{\triangle}_\mu\psi)(x,H) = \sum_{\mu\in H}(\gamma^{H\dagger}\gamma^\mu\partial_\mu^+\psi(x,H\setminus\{\mu\}) + \sum_{\mu\notin H}(\gamma^{H\dagger}\gamma^\mu\partial_\mu^-\psi(x,H\cup\{\mu\})$$

$$= \sum_{\mu\in H}\hat{\rho}_{\mu,H}\partial_\mu^+\Phi(x,H\setminus\{\mu\}) + \sum_{\mu\notin H}\tilde{\rho}_{\mu,H}\partial_\mu^-\Phi(x,H\cup\{\mu\}) \;=\; ((\check{\triangle} - \check{\bigtriangledown})\Phi)(x,H).$$

By a similar calculation the equalitity of the action for staggered fermions, Eq. (16), with the Dirac Kaehler action

$$S_q = \frac{1}{4}\sum_{x,H}\{\overline{\Phi}(x,H)((\check{\triangle} - \check{\bigtriangledown} + m)\Phi)(x,H)\} \tag{40}$$

follows.

The DKE is defined geometrically on the coarse lattice. We want to check the energy momentum spectrum. For this we multiply the DKE by the adjoint operator

$$\begin{aligned}&(-(\check{\triangle} - \check{\bigtriangledown} + m)(\check{\triangle} - \check{\bigtriangledown} + m)\Phi\\ &= (\check{\triangle}\check{\bigtriangledown} + \check{\bigtriangledown}\check{\triangle} + m^2)\Phi = (-\partial_\mu^+\partial^{\mu-} + m^2)\Phi = 0.\end{aligned} \tag{41}$$

The iterated DKE is indeed the correct Klein Gordon equation on the lattice. We consider plane wave solutions

$$\phi(x,H) = u(p,H)e^{-ipx}.$$

With this ansatz Eq. (41) becomes

$$(\sum_\mu(\frac{2}{b}\sin\frac{p_\mu b}{2})^2 + m^2)u(p,H) = 0.$$

Because $(\frac{2}{b}\sin\frac{p_\mu b}{2})$ is monotonous in the cut-off momentum range $-\frac{\pi}{b} < p_\mu \leq \frac{\pi}{b}$, the energy momentum spectrum has the same multiplicity as that of the DKE in the continuum. This is in contrast to the spectrum problem which arises from the naive lattice approximation of the Diac equation. J.M.Rabin [20] has given a topological argument why the mapping Eq. (32) does not introduce new zero solutions of the kinetic energy: The zero solutions present harmonic forms, which are in 1-1 correspondence to the homology classes in the continuum and on the lattice according to DeRham's Theorem.

5. THE SYMMETRY OF THE DKE ON THE LATTICE

The transformations of the spinorial Euclidean group $\mathcal{SE}$ and of the flavour group $\mathcal{F}$ generate a symmetry group $\mathcal{G}$ of the DKE in the continuum: $\mathcal{G} \simeq \mathcal{F} \times \mathcal{SE}$ What is left of this symmetry by the lattice approximation? Before we answer this question, we want to describe the continuum symmetry in an appropriate way [3]. The infinitesimal translations $\delta_\mu \phi_a^b(x) = \partial_\mu \phi_a^b(x)$ and rotations $\delta_{\mu\nu}\phi_a^b(x) = (x_\mu \partial_\nu - x_\nu \partial_\mu)\phi_a^b(x) + \frac{1}{4}([\gamma_\mu, \gamma_\nu]\phi)_a^b(x)$ of the spinor components can be expressed directly as operations on the forms

$$\delta_\mu \Phi = \partial_\mu \Phi,$$

$$\delta_{\mu\nu}\Phi = (x_\mu \partial_\nu - x_\nu \partial_\mu)\Phi + \frac{1}{2} S_{\mu\nu} \vee \Phi, \qquad S_{\mu\nu} = dx_\mu \wedge dx_\nu. \tag{42}$$

These transformations generate the unit component of the group $\mathcal{SE}$ which should be supplemented by space reflections. On the other hand, if the Cartesian components transform as O(4) tensors, we get the transformation law

$$\delta_{\mu\nu}^G \Phi = (x_\mu \partial_\nu - x_\nu \partial_\mu)\Phi + \frac{1}{2}(S_{\mu\nu} \vee \Phi - \Phi \vee S_{\mu\nu}). \tag{43}$$

These 'geometric' rotations differ from spinor rotations by a flavour transformation of the form Eq. (29). Thus we may express an element g of the symmetry group of the DKE either as a direct product $g = (f) \otimes (a, s) \equiv (f, a, s)$, with $(f) \in \mathcal{F}, (a, s) \in \mathcal{SE}$, or we compose it by a geometric transformation $(s, a, s) \in G\mathcal{E}$ and another flavour transformation $(\overline{f})$

$$g = (\overline{f}) \circ (s, a, s) = (\overline{f}s, a, s) = [\overline{f}, a, R(s)]. \tag{44}$$

The group multiplication in these different equivalent forms is:

$$(f, a, s) \circ (f', a', s') = (ff', R(s)a' + a, ss'),$$

$$[f, a, R(s)] \circ [f', a', R(s')] = [fsf's^{-1}, R(s)a' + a, R(s)R(s')]. \tag{45}$$

Restricting space time to a lattice implies restrictions of the symmetry group of the DKE. These are most obvious for the geometrical transformations generating the geometrical Euclidean group $G\mathcal{E}$. The contineous translation group gets restricted to the group of lattice translations: $\mathcal{T}_L = \{[a]\}, a = b(n^1 \ldots n^d), n^i \in Z$. The cubic lattice allows only rotations belonging to the symmetry group W_d of the d-dimensional cube. In d = 4 dimensions, W_4 is a group with 384 elements which is generated by rotations $R_{\mu\nu}$ in the (μ, ν) -plane by $\frac{\pi}{2}$, and by a reflection $\Pi : (x^1, x^2, x^3, x^4) \to (-x^1, -x^2, -x^3, x^4)$. These rotation-reflections map r-cells, $r = 1 \ldots 4$, onto r-cells, and commute with the boundary and coboundary operations. Therefore the DKE is invariant under the group $\mathcal{T}_L \ltimes W_4 \simeq G\mathcal{E}_L \subset G\mathcal{E}$ of these transformations; ($\ltimes$ denotes a semi-direct product). The transformation law of staggered fermion fields is

$$([a]\Phi)(\overline{x}, H(\overline{x})) = \Phi(\overline{x} - a^\mu e_\mu, H(\overline{x})), \qquad H(\overline{x} - a^\mu e_\mu) = H(\overline{x}),$$

$$([R]\Phi)(\overline{x}, H(\overline{x})) = \rho(R, H(\overline{x}))\Phi(R^{-1}\overline{x}, H(R^{-1}\overline{x})), \qquad R \in W_4, \tag{46}$$

where the sign $\rho(R,H)$ is the same as in the transformation of the basis differential of the continuum : $Rdx^H = \rho(R,H)dx^{R^{-1}\circ H}$.
The lattice restriction of the flavour transformations is determined by our definition of the Clifford product on the lattice. The symmetry transformation of the DKE

$$d^K\Phi : \Phi \to \epsilon\Phi \vee (d^K)^{-1}, \quad \epsilon = \pm 1,$$

implies for staggered fermion fields

$$(d^K\Phi)(\overline{x}, H(\overline{x})) = \epsilon\check{\rho}_{H(\overline{x}),K}\Phi(\overline{x} + \frac{1}{2}e_K, H(\overline{x} + \frac{1}{2}e_K)). \tag{47}$$

It follows that $(d^K)^2 = [-e_K]$, i.e. the flavour transformations generate translations. These transformations on the lattice appear as the following restrictions of the continuum group $\mathcal{G} : [F, a, R] = [\epsilon\gamma^K, -\frac{1}{2}e_K, 1]$. The group $\mathcal{K}_4 = \{[\epsilon\gamma^K, -\frac{1}{2}e_K + a, 1]\}/\mathcal{T}_L$ is isomorphic to the finite group of 32 elements generated by the γ-matrices. The group $\mathcal{K}_4$ plays an important role in the discussion of Clifford algebras [21]. Now we may state in a crystallographic language [24]: Flavour transformations form a generalized point group isomorphic to the group $\mathcal{K}_4$ which is a non-symmorphic extension of the translation group.
Putting these considerations together, we find the following sub-group $\mathcal{G}_L \subset \mathcal{G}$ which is a symmetry group of the lattice DKE :
$\mathcal{G}_L = \{[\epsilon d_k, -\frac{1}{2}e_K + a, R] \mid a \in \mathcal{T}_L, R \in W_4\}$ with the composition law following from (46)

$$[\epsilon d_K, -\frac{1}{2}e_K + a, R] \circ [\epsilon' d_L, -\frac{1}{2}e_L + a', R']$$
$$= [\epsilon\epsilon'\rho(R, R\circ L)\check{\rho}_{K,R\circ L}d^{K\triangle R\circ L}, -\frac{1}{2}(e_K + Re_L) + Ra' + a, RR']. \tag{48}$$

There are further symmetries of the DKE. The complex DKE is invariant with respect to phase transformations. In the massless case,there are independent phase transformations of the even and odd forms. In the following we restrict ourselves to the discussion of $\mathcal{G}_L$.
Now we try to give a short summary of the representation theory of $\mathcal{G}_L$ [22], [23]. Because $\mathcal{T}_L$ is a normal sub-group of $\mathcal{G}_L$ with $\mathcal{G}_L/\mathcal{T}_L \simeq \mathcal{K}_4 \otimes W_4$, the induction procedure by E.P.Wigner - G.W. Mackey [25] allows the construction of the irreducible, unitary representations ('irreps') of $\mathcal{G}_L$ from the representations of $\mathcal{T}_L$, of the finite group $\mathcal{K}_4$, and of the sub-groups of W_4. We shall give a short glossary of this Wigner- Mackey procedure. Because we apply it to a relatively simple discrete group, we can omit all mathematical sophistication.

Take a group G with a normal sub-group N and the factor group F. An irrep $g \to U(g)$ of G restricted to $N = \{n\}$ decomposes in irreps L^p. Here $p \in \hat{N}$ is a label of the irreps of N. It is the first step of the procedure to analyze this restriction $U(G)|_N$. For this we construct an orthonormal basis of the representation space in which $U(G)|_N$ decomposes explicitly:

$$\{\Phi_{p,a;\eta}\} : \qquad U(n)\Phi_{p,a;\eta} = \sum_{\overline{a}} \Phi_{p,\overline{a};\eta}L^p_{\overline{a},a}(n). \tag{49}$$

η being a degeneracy label. (A direct integral decomposition we treat formally in the same way).
Since N is a normal sub-group of G, $L^p(g^{-1}ng)$ is equivalent to some $L^{p'}(n)$,i.e. G acts as a transformation group of $\hat{N} : p' = gp$. Acting with $U(g^{-1}ng)$ on the basis (49) , we see that p varies over G -orbits $\Theta_j = \{gp_j | g \in G\} \subset \hat{N}$, where p_j denotes an arbitrary but fixed reference point. Indeed, an irrep U(g) is characterized by a single orbit Θ_j. The group $S_j^{(1)}$ leaving p_j invariant: $S_j^{(1)} = \{s | sp_j = p_j, s \in \mathcal{G}\}$ is called the little group of the first kind. N is a normal sub-group of $S_j^{(1)}$. The little group of second kind is defined as the factor group $S_j^{(2)} = S_j^{(1)}/N$. For the further procedure it is important that one can extend always $(L^{p_j}_{a',a}(n))$, up to equivalence, uniquely to a projective representation $\overline{L}^{p_j}(s), \quad s \in S_j^{(1)}$:

$$\overline{L}^{p_j}(s)\overline{L}^{p_j}(t) = \sigma_j(s,t)\overline{L}^{p_j}(st), \qquad s,t \in S_j^{(1)}, \tag{50}$$

such that the multiplier $\sigma_j(s,t)$ only depends on the N-cosets, i.e. on the elements of $S_j^{(2)}$. For the proof we refer to [25]. Since $S_j^{(1)}$ is the stability group of p_j, the sub-space spanned by $\{\Phi_{p_j}\}$ is invariant under the transformations of $S_j^{(1)}$:

$$U(s)\Phi^{j;\chi}_{p_j,a,r} = \sum_{a',r'} = \Phi^{j;\chi}_{p_j,a',r'}\overline{L}^{p_j}_{a',a}(s)D^{\chi}_{r',r}(s). \tag{51}$$

Here $D^{\chi}(s)$ is a projective representation of $S_j^{(2)} = S_j^{(1)}/N$, with the multiplier $\sigma_j^{-1}(s,t)$ .It may be considered as an unfaithful projective representation of $S_j^{(1)}$.
We can now formulate the MAIN THEOREM. [26] All irreducible unitary representations of a group G with a normal sub-group N are characterized by the G-orbits $\Theta^j \subset \hat{N}$, and the irreducible projective representations of the related little groups of second kind $S_j^{(2)} \to D^{\chi}(s)$ with multiplier of the equivalence class $\sigma_j^{-1}(s,t)$.
For the explicit construction we define for the points p of the orbit Θ^j a 'boost' transformation: $p = g(p)p_j, g(p) \in G$ and standardize by equivalence transformations the representation $L^p(n)$ and the basis $\{\Phi^{j;\chi}_{p,a,r}\}$:

$$L^p_{m,m'}(n) = L^{p_j}_{m,m'}(g^{-1}(p)ng(p)),$$

$$U(g(p))\Phi^{j;\chi}_{p_j,a,r} = \Phi^{j;\chi}_{p,a,r}. \tag{52}$$

Because we can write an arbitrary $g \in G$ as a product of boost transformations and an element of $S_j^{(1)}$:

$$g = g(gp)s(g,p)g^{-1}(p), \qquad i.e. \qquad s(g,p) = g^{-1}(gp)gg(p) \qquad \in S_j^{(1)}, \tag{53}$$

it follows from Eq. (51) and (52) that

$$U(g)\Phi^{j;\chi}_{p,a,r} = \sum_{a',r'} \Phi^{j;\chi}_{gp,a',r'} \overline{L}^{p_j}_{a',a}(s(g,p)) D^{\chi}_{r',r}(s(g,p)). \tag{54}$$

One may verify by direct calculation that Eq. (54) defines a representation of G . This representation is irreducible if p is restricted to a single orbit Θ_j, and if $D^{\chi}_{r',r}(s)$ is irreducible. In the construction above the representation of an orbit Θ_j by a reference point p_j and boost transformations g(p) induces some arbitrariness. One convinces oneself easily that the choice of different reference points and boost transformations lead to equivalent representations of G. Similarly, equivalence transformations of $D^{\chi}(s)$ and L^{p_j} lead to equivalent representations.

The iterated application of the Wigner-Mackey procedure leads to a complete classification of the irreps of $\mathcal{G}_L$ by a ' MOMENTUM STAR St_j', a ' FLAVOUR ORBIT $\Theta_{j,k}$ ', and a ' REDUCED SPIN σ'.
In order to show this, we consider in a first application of the Wigner-Mackey procedure the translation group $\mathcal{T}_L$ as normal sub-group of $\mathcal{G}_L$. The 1-dimensional irreps of $\mathcal{T}_L : [a] \to e^{ipa}$ are labelled by 'momenta' $p = (p_1, \ldots p_4)$ varying in the Brillouin zone: $-\frac{\pi}{b} < p_\mu \leq \frac{\pi}{b}$. We denote by the star St_j the orbit of the rotations $R \in W_4$ applied to the momenta $p \to Rp$. Depending on the orientation of p there are 13 qualitatively different stars. For each St_j one may choose a reference point $p_j \in St_j$, boost operators $\Lambda(p)p_j = p$, $\Lambda(p) \in W_4$, and determine the stability group $\overline{S}_j = \{R | Rp_j = p_j\}$.

The little group of the first kind $S^{(1)}_j$ in this application of the Wigner-Mackey procedure is generated by the translations, flavour transformations, and the rotations of $\overline{S}_j$ .The little group of the second kind $S^{(2)}_j \simeq S^{(1)}_j/\mathcal{T}_L$ is generated by $\overline{S}_j$ and the elements of $\mathcal{K}_4$. The representations of $\mathcal{G}_L$, corresponding to Eqs. (52), (53), (54) have the form

$$U[\epsilon d^K, -\frac{1}{2}e_K + a, R]\Phi^{j,\chi}_{p,n} = e^{i(Rp,a-\frac{1}{2}e_K)} \sum_{n'} \Phi^{j,\chi}_{Rp,n'} D^{\chi}_{n'n}(s(g,p)), \tag{55}$$

with

$$p \in St_j, \qquad s(g,p) = [1,0,\Lambda^{-1}(Rp)] \circ [\epsilon d^K, -\frac{1}{2}e_K + a, R] \circ [1,0,\Lambda(p)] \in S^1_j.$$

The extension of the representation of $\mathcal{T}_L$: $[a] \to e^{ip_j a}$ to a representation $\overline{L}(s)$ of $S^{(1)}_j$ is given trivially by $[\epsilon d^K, -\frac{1}{2}e_K + a, R] \to e^{-i(p,a-1/2e_K)}$, $R \in \overline{S}_j$. $D^{\chi}(s)$ is an irrep of $S^{(2)}_j$ which we have to construct now.

This little group of the second kind $S^{(2)}_j$ contains $\mathcal{K}_4$ as a normal sub-group. We apply for the construction of the irreps of $S^{(2)}_j$ the Wigner-Mackey procedure for a second time. For this we have to consider first the irreps of $\mathcal{K}_4$. These are the well known 4-dimensional representation $\epsilon d^K \rightarrow \epsilon\gamma^K$ which we give the label L = 0, and the 16 one dimensional representations of the factor group

$$\mathcal{K}_4: \qquad \epsilon d^K \rightarrow e^{i\pi(e_L,e_K)} \equiv \Gamma^L(\epsilon d^K), \qquad e_L = \sum_{\mu\in L} e_\mu.$$

L is a multi-index like in Eq. (12). For the construction of the representations of $S_j^{(2)}$, we have to consider further the transformations of the irreps of $\mathcal{K}_4$ under the rotations of $\overline{S}_j$:
$\Gamma^L(R^{-1}(\epsilon d^K)R) \simeq \Gamma^{R\circ L}$. In the case L=0, this is an equivalence transformation for all R, therefore
$R\circ(L=0) = (L=0)$. The set of 1-dimensional representations decompose under the rotations of $\overline{S}_j$ in 'flavour orbits' $\Theta_{j,k}, k = 1,\ldots N_j$. For flavour orbits we again may fix a refence point $\overline{L}_k$, choose boost operators $f(L)\overline{L}_k = L, f(l) \in \overline{S}_j$, and determine the stability group
$S_{j,k} = \{R|R\overline{L}_k = \overline{L}_k; R\in S_j\} \subset S_j$ The little group of the first kind $S_{j,k}^{(1)}$ is a semi-direct product of $S_{j,k}$ with $\mathcal{K}_4$, $\mathcal{K}_4$ normal sub-group. The little group of the second kind is $S_{j,k}^{(1)}/\mathcal{K}_4 \simeq S_{j,k}$
With these concepts we can construct the irreps of $S_j^{(2)}$ according to Eqs. (52), (53), (54),

$$\mathcal{U}(\epsilon d^K, R)\Phi^{k,\sigma}_{L,a,n} = \sum_{a',n'} \Phi^{k,\sigma}_{L,a',n'}\tilde{\Gamma}^{\overline{L}_k}_{a',a}(\xi(s,L))\mathcal{D}^{\sigma}_{n',n}(\xi(s,L)), \tag{56}$$

with

$$L\in\Theta_{j,k}, \qquad s = [(\epsilon d^K, R)] \in S_j^{(2)}, \qquad \xi(s,L) = \in S_{j,k}^{(1)}.$$

$\tilde{\Gamma}^0_{a',a}(\xi), \xi\in S_{j,k}^{(1)}$ denotes the extension of the 4-dimensional representation of $\mathcal{K}_4$ to a projective representation of $S_{j,k}^{(1)}$. Here the representations of the rotations $R\in S_{j,k}$ are given by the projective representation of W_4 generated by $R_{\mu\nu} \rightarrow \frac{1}{2}(1+\gamma_\mu\gamma_\nu)$, $\Pi \rightarrow \gamma^4$, restricted to $S_{j,k}$. In the extensions of the 1-dimensional representations: $\tilde{\Gamma}^{\overline{L}_k}_{\overline{a},a}(\xi)$,these rotations,reflections are represented trivially: $R_{\mu\nu} \rightarrow 1$, $\Pi \rightarrow 1$.
The groups $S_{j,k}$ are direct products of the symmetric groups Sym(n) ,n = 2,3,4, the dihedral group D_4, and the hyper-cubic group W_4. Their projective irreps are wellknown, or can be constructed easily. The irreps of $S_{j,k} \ni \xi \rightarrow \mathcal{D}^{\sigma}_{r',r}(\xi)$, which determine the reduced spin σ , may then be labelled by combinations of the primitive characters of these elementary groups.
The discussion of the irreps of $S_j^{(1)}$ together with Eq. (55) completes the construction of the irreps of the symmetry group $\mathcal{G}_L$ of the lattice DKE. The substitution of $D^{\chi}_{n',n}$ in Eq. (55) by $\mathcal{U}(s)$ in Eq. (56) is straightforward. However, it leads to clumsy formulas which we do not want to reproduce here. Instead we want to illustrate the different concepts introduced here by an important example [10]. For a more complete discussion of the representation theory of $\mathcal{G}_L$ we refer to the original publication.

We consider the momentum star of 8 points

$$St_4: \qquad (p_\mu) = (0,0,0,\pm E), (0,0,\pm E,0), ..$$

The stability group of the reference point $p_j = (0,0,0,E)$, $E > 0$, of this star is $\overline{S}_6 \simeq W_3$. In this case there are three types of flavour orbits

(a) The 1-point orbit Γ^0

(b) 1-point orbits $\Theta_{6,k} = \{(e_L)\}, k = 1 \ldots 4$ with $S_{4,k} \simeq W_4$:
$\{(0,0,0,0)\}, \{(0,0,0,1)\}, \{(1,1,1,0)\}, \{(1,1,1,1)\}$.

(c) 3-point orbits $\Theta_{6,k}, k = 5 \ldots 8$, with $S_{4,j} \simeq D_4 \times Z_2$:
$\{(1,0,0,0),(0,1,0,0),(0,0,1,0)\}$, $\{(0,1,1,0), \ldots\}$, $\{(1,0,0,1), \ldots\}$, $\{(0,1,1,1), \ldots\}$.

In this example we have $S_{4,k} \simeq Sym(4) \times Z_2 \simeq W_3$ for L = k = 0,..4, and $S_{4,k} \simeq D_4 \times Z_2$ for k = 5,..8.
For k=0 we have to consider projective representations of W_3 with the multipliers of the spinor representations. There are two 2-dimensional and one 4-dimensional of such representations for Sym(4). This means altogether six reduced spin-parity combinations for this class of representations characterized by St_4 and $\Theta_{4,0}$. We denote them by $(\tilde{2}^\pm), (\tilde{2}'^\pm), (\tilde{4}^\pm)$.
The symmetric group Sym(4), i.e. the proper cubic group, has two 1-dimensional, one 2-dimensional, and two 3-dimensional representations. Together with the parities $\Pi = \pm 1$, this gives 10 spin parity combinations related to the reduced spin group of the orbits $\Theta_{4,k}$ k = 1,..4. We denote them by $(1^\pm)_{W_3}, (1'^\pm)_{W_3}, (2^\pm)_{W_3}, (3^\pm)_{W_3}, (3'^\pm)_{W_3}$.
The dihedral group D_4 has four 1-dimensional representations and one 2-dimensional representaion. This gives again 10 spin parity combinations, however of a different type which belong to the flavour orbits $\Theta_{j,k}$, k = 5,..8. These are denoted by $(1^\pm)_{D_4}, (1'^\pm)_{D_4}, (1''^\pm)_{D_4}, (1'''^\pm)_{D_4}, (2^\pm)_{D_4}$.
Thus we have a complete classification of the 86 irreps of $\mathcal{G}_L$ with momentum star St_4.
The group theoretical analysis of the DKE is one of the main tools for the investigation of its physical significance. We shall illustrate this in the next section.

6. A PHYSICS PROBLEM

How do these different geometric considerations relate to actual calculations of physical quantities? In this section I shall make an attempt to describe in a cursory way the calculation of the meson spectrum. Because of the very crude approximations involved, the results shall not have any real significance. But they should illustrate more directly the physics related to the solution of the lattice fermion problem offered by the Dirac-Kaehler approach.
The aim is to calculate the 'meson propagator' $\langle\overline{\psi}\psi(x)\overline{\psi}\psi(x')\rangle$ with help of the lattice approximation of formula (2). We use the Wilson action, (9), for the gluon field, and of course we use an action of staggered fermion

fields for the description of the quarks:

$$S_q = \sum_{\overline{x}} \{ \sum_{\mu} \check{\rho}_{\mu H(\overline{x})} \frac{1}{2} [\overline{\chi}(\overline{x}) U(\overline{x},\mu) \chi(\overline{x}+\overline{e}_\mu) - \overline{\chi}(\overline{x}+\overline{e}_\mu) U^{-1}(\overline{x},\mu) \chi(\overline{x})] + \overline{m}\overline{\chi}(\overline{x})\chi(\overline{x}) \}$$

$$\equiv \overline{m} \sum_{\overline{x},\overline{x}'} \overline{\chi}(\overline{x}) (\delta_{\overline{x},\overline{x}'} + (\overline{x}|Q[U]|\overline{x}') \chi(\overline{x}). \quad (57)$$

This is the gauge invariant version of the free action Eq. (16), or (40), $\overline{b} = 1$.
The total action $S = S_g + S_q$ is symmetric under the group $\mathcal{G}_L$ if $\chi(\overline{x}) = \Phi(\overline{x}, H(\overline{x}))$ transforms according to Eqs. (46), (47), and $U(\overline{x})$ transforms geometrically under [a,R] and according to

$$U(\overline{x},\mu) \xrightarrow{d^K} U(\overline{x}+\overline{e}_K)$$

under lattice flavour transformations. We shall use this symmetry below for the classification of the different meson states.
There is some arbitrariness in the introduction of a gauge invariant interaction into the Dirac Kaehler action (40)[3]. Different from the 'Susskind' -coupling defined by the action (57), one could consider also a 'coarse' coupling [2] [27]. In this case , gauge fields are defined on the links of the coarse lattice only. The local gauge transformations $\phi(x,H) \rightarrow g(x)\phi(x,H)$ act at the points of the coarse lattice. However, an action based on such a gauge invariant coupling is not invariant under the transformations of $\mathcal{G}_L$. This has serious consequences for the renormalized continuum limit [28].
In the evaluation of the path integral

$$\langle \overline{\chi}\chi(\overline{x})\overline{\chi}\chi(\overline{x}')\rangle = \frac{1}{Z} \int \int \mathcal{D}[U]\mathcal{D}[\chi,\overline{\chi}]\overline{\chi}(\overline{x})\chi(\overline{x})\overline{\chi}(\overline{x}')\chi(\overline{x}') \exp - (S_g + \overline{m}\overline{\chi}(1+Q[U])\chi), \quad (58)$$

we perform first the 'Gaussian integration'[29] with respect to the 'Grassmann' variables $\overline{\chi}(\overline{x}), \chi(\overline{x})$.
This results in

$$\langle \overline{\chi}\chi(\overline{x})\overline{\chi}\chi(\overline{x}')\rangle = \frac{1}{Z} \int \mathcal{D}[U] \frac{1}{\overline{m}^2} (\overline{x}|(1+Q)^{-1}|\overline{x}')(\overline{x}'|(1+Q)^{-1}|\overline{x}) \det \overline{m}(1+Q) e^{-S_g}. \quad (59)$$

For the calculation of the quark determinant $\det \overline{m}(1+Q)$ and the quark propagator
$S(\overline{x},\overline{y};U) = \frac{1}{\overline{m}}(\overline{x}|(1+Q)^{-1}|\overline{y})$, we use as an intermediate step the so-called hopping parameter expansion:

$$S(\overline{x},\overline{y};U) = \frac{1}{\overline{m}} \sum_{L=0}^{\infty} (-1)^L (\overline{x}|(Q)^L|\overline{y}) = \frac{1}{\overline{m}} \sum_{L=0}^{\infty} (\frac{-1}{2\overline{m}})^L \sum_{C_L(\overline{x},\overline{y})} \prod_{l_\mu \in C_L} \rho(l_\mu) U(l_\mu),$$

$$\det(1+Q) = \exp\{ \sum_{L=0}^{\infty} \sum_{x} \frac{1}{L} (\frac{-1}{2\overline{m}})^L \sum_{C_L(\overline{x},\overline{x})} \prod_{l_\mu \in C_L} \rho(l_\mu) U(l_\mu) \}. \quad (60)$$

Here $\mathcal{C}_L(x,y)$ are paths of length L from $\bar{y}$ to $\bar{x}$ composed of the links $l_\mu = [\bar{z},\mu]$,
$\rho(l_\mu) = \breve{p}_{\mu,H(\bar{z})}, \rho(-l_\mu) = -\rho(l_\mu)$. The product of the $U(l_\mu)$ form the parallel transport of the quark colour vector along $\mathcal{C}_L(x,y)$. The expression of $(\bar{x}|Q^L|\bar{y})$ by sums of parallel transports along paths depends crucially on the fact that the matrix $(\bar{x}|Q|\bar{y})$ is only different from zero for neighbouring points $\bar{x},\bar{y}$.
After inserting this expression (60) in Eq. (59) we may perform the integration with respect to the gauge field variables $U(\bar{x},\mu)$ by expanding the Boltzmann factor $e^{-\beta S'_g}$ in β ('strong coupling approximation'). The orthogonality of matrix elements of irreducible group representations with respect to averaging with the Haar measure

$$\int_{SU(3)} d\mu(g) U_{\alpha,\beta}(g) U^\star_{\alpha',\beta'} = \frac{1}{\dim U}\delta_{\alpha,\alpha'}\delta_{\beta,\beta'} \tag{61}$$

allows a systematic evaluation of gauge field averages. Already in zero order we get a non-trivial result. The meson propagator is given by the sum over all strictly parallel quark lines, see Fig.(3).

$$\overline{\chi}\chi(\bar{x})\overline{\chi}\chi(0) = \sum_L (\frac{-1}{4\overline{m}^2})^L B_{\bar{x}}(L). \tag{62}$$

$B_{\bar{x}}(L)$ is the number of double paths from 0 to $\bar{x}$.In this approximation the quark and anti-quark are strongly bound; they can not separate by even one lattice link. The next order in β contains paths as in Fig.(3). This order describes 'internal relative motion' over the distance of one lattice link.
Now we approximate even further the zero order expression (62) by restricting the summation to tree graphs, and omitting paths with closed loops [30]. In order to perform the summation over all tree graphs in (62), we use a combinatorial method introduced by O.Martin [31] for this type of calculation. He remarked that a general tree graph might be composed by a trunk and branches. Branches are $q\overline{q}$ lines which branch off a single quark or anti-quark line. Trunks are tree graphs without branches. In (62) one has to sum only over all trunks , if one renormalizes the mass $\overline{m} \to \frac{1}{2}(\overline{m} + \sqrt{\overline{m}^2+7})$. Setting $\alpha = \overline{m} + \sqrt{\overline{m}^2+7}$, we get instead of Eq. (62)

$$\overline{\chi}\chi(\bar{x})\overline{\chi}\chi(0) = \sum_L (\frac{-1}{\alpha^2})^L T_{\bar{x}}(L). \tag{63}$$

$T_{\bar{x}}(L)$ is the number of trunks from 0 to $\bar{x}$. Trunks are $\overline{q}q-$lines which are not back tracking, otherwise they would have branches. Therefore it is useful to classify trunks according to their last step. $T_{\mu,\bar{x}}(L)$ is the number of trunks from 0 to $\bar{x}$ with last step in μ-direction. We have the recursion relation

$$T_{\mu,\bar{x}}(L+1) = \sum_{\nu\neq-\mu} T_{\nu,\bar{x}-\bar{e}_\nu}(L).$$

Fourier transformation with respect to $\bar{x}$ gives

$$\tilde{T}_\mu(L+1) = \sum_\nu e^{i\overline{p}_\mu}(1-\delta_{\mu,-\nu}\tilde{T}_\nu(L)) = \sum_\nu \mathcal{M}_{\mu,\nu}(\overline{p})\tilde{T}_\nu(L). \tag{64}$$

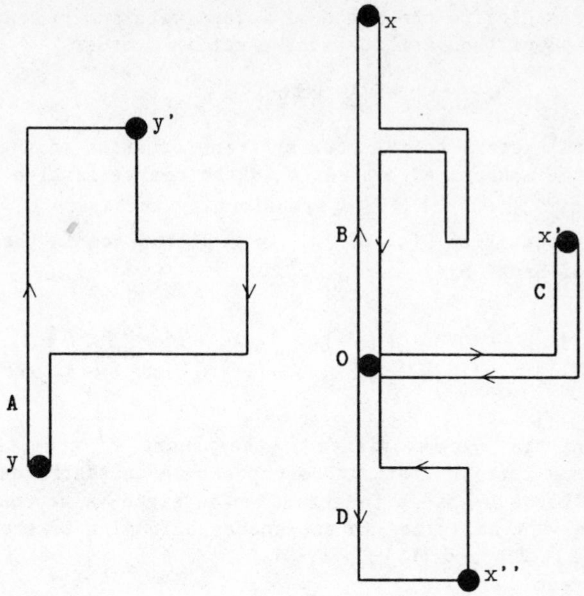

Figure 3: $q\bar{q}$-paths contributing to the meson propagator: A.'General' path from y to y'. B.Trunk with branch from 0 to x. C.Trunk from 0 to x'. D.1-order contribution.

This formula allows the summation over the length L in Eq. (63) for the Fourier transform of the meson propagator $G(\overline{p})$:

$$G(\overline{p}) = \sum_{\overline{x}} e^{i\overline{p}\overline{x}} \langle \overline{\chi}\chi(\overline{x})\overline{\chi}\chi(\overline{0})\rangle = \xi^{\dagger}(1 + \frac{1}{\alpha^2}\mathcal{M}(\overline{p}))^{-1}\xi. \quad (65)$$

$\xi^{\dagger}$, ξ denote the vectors of the final and initial orientations of the paths, which are given by the details of the meson field $\overline{\chi}\chi(x)$. The poles of the propagator , the eigenvalues $\lambda(\overline{p})$ of $\mathcal{M}(\overline{p})$ with $\lambda(\overline{p}) = -\alpha^2$, determine the masses of the mesons. In a way, this eigenvalue problem plays a role similar to the energy eigen-value problem of the Schroedinger equation in non-relativistic quantum mechanics, or the Bethe Salpeter equation in perturbative relativistic field theory. It allows us to illustrate the use of the group theoretical methods developed in the last section.
The symmetry of the lattice DKE can be used to classify the meson states. It follows from Eq. (46) that mesons described by $\overline{\chi}\chi(\overline{x})$ transform by 1-dimensional flavour transformations. In particular the fields $\mu^L(\overline{x}) = \overline{\chi}(\overline{x})e^{i\pi e_L \overline{x}}\chi(\overline{x})$ transform like

$$([\epsilon d^K]\mu^L)(\overline{x}) = e^{i\pi(e_L, e_k)}\mu^L(\overline{x} + \overline{e}_K). \quad (66)$$

The μ^L-propagator matrix $\mathcal{M}^L(p)$ may be determined from Eq. (64) for meson configurations with 'zero space-like momenta', i.e for meson states with quantum numbers belonging to the class of irreps with momentum star St_j, as described at the and of the last section. We get in 0-order

$$\mathcal{M}^L(p) = e^{ip_\mu/2} e^{i\pi(e_L, e_\mu)}(1 - \delta_{\mu,-\nu}). \tag{67}$$

(Reminder: $\overline{p}$ in (64) refers to the Fourier transformation in the fine lattice, whereas the momenta of St_j refer to the coarse lattice.)
$\mathcal{M}^L$ is a 8×8 matrix, $\mu, \nu = \pm 1, \ldots \pm 4$, transforming the space $\mathcal{H} = \{\xi\}$. The 8-dimensional representation D(s) of $S_j^{(1)}$ in $\mathcal{H}$ decomposes in the following irreps (notation of Sect. 5):

$$D(s) \sim 3(1^+)_{W_3} + (2^+)_{W_3} + (3'^+)_{W_3} \quad \text{for} \quad L \sim k = 1,..4,$$
$$D(s) \sim 4(1^+)_{D_4} + (1^-)_{D_4} + (1''^+)_{D4} + (2^-)_{D_4} \quad \text{for} \quad L \sim k = 5,..8.$$

As a consequence of the $\mathcal{G}_L$ -invariance of the theory $\mathcal{M}^L$ commutes with D(s). Schur's lemma implies that $\mathcal{M}^L$ decomposes in sub-matrices which leave the irreducible sub-spaces invariant. The eigen-value condition $\lambda(iE) = -\alpha^2$ can only be satisfied in sub-spaces belonging to the quantum numbers: $(1^+)_{W_3}$, k = 2,4 and $(1^+)_{D_4}$, k = 7,8.
They lead to the result

$$\cosh E = \frac{\alpha^4 - 6\alpha^2 + 7}{2\alpha^2} + 2j, \qquad j = 0, 1, 2, 3 \quad \text{for} \quad k = 4, 8, 7, 2.$$

The next problem would be to associate these states with physical particles. There is the following necessary condition which can be solved by the group theoretical method. The irreps of the continuum group $\mathcal{G}$ determine spin parity, and SU(4)-flavour of physical particles. A state of a lattice meson can be associated to such a particle, if the irrep of $\mathcal{G}$ restricted to $\mathcal{G}_L$ contains the lattice representation of the meson. With help of Mackey's sub-group theorem [26] it is a straight forward calculation to get such 'branching' rules. For details we refer to [23].
A realistic physical discussion requires higher order calculations supplemented by Monte Carlo calculations [34]. In 1-order in β, the matrices $\mathcal{M}^L$ become 80 dimensional. They can be treated successfully with help of the group theoretical methods sketched here [33]. However, a critical physical discussion is beyond the scope of this report. It was only my intention to give a realistic impression of the long way one has to go if one wants to analyze the physics of the DKE and its symmetry in the context of Quantum Chromodynamics.

7. SUMMARY AND OUTLOOK

The consistent quantum mechanical treatment of the standard gauge theoretical models of elementary particles by the method of lattice approximation suggests a description of the fermions by the Dirac Kaehler equation. The reason for this suggestion is the spectrum doubling problem of lattice fermions: The maximal reduction of the number of Dirac particles described by the naive lattice approximation of the Dirac

equation leads to staggered fermion fields, and these fields are equivalent to a systematic lattice approximation of the Dirac Kaehler forms.
However, this natural looking solution of the lattice fermion problem leads to some open physical problems. Since complex Dirac Kaehler forms are equivalent to four Dirac fields, there is an additional degree of freedom, called (Susskind) flavour hidden in the DKE. The physical meaning of the Susskind flavour is completely unclear. One may have the following opinions:

-In spite of the possible solution of the lattice fermion problem by the DKE, the question of the 'Susskind flavour' is a lattice artifact. It should be eliminated (artificially ?!) in the contiuum limit.

-Spectrum doubling is a quantum effect of (chiral) fermion fields, related to the problem of anomalies [13] The Susskind flavour of complex (or real) Dirac Kaehler forms has some physical meaning. It might be related to the quark flavour in QCD [35], to weak isospin [36], to the family structure of elementary particles, etc. However there is not yet a physically relevant theory of such a realistic interpretation of Susskind flavour.

In order to judge on the importance of these aspects of the Dirac Kaehler formalism for elementary particle physics, one has to treat specific problems. Symmetry considerations form always an important link between a dynamical equation and physical states. For this reason we gave a description of the symmetry group of Dirac Kaehler fermions together with a sketch of its representation theory. An application to the strong coupling scheme for the calculation of the meson spectrum in the frame work of QCD was used to illustrate these type of considerations. In the studies of the DKE on the lattice the structure of the Clifford algebra and its extension appeared in a new context: as a non-asssociative algebra of cochains on a cubic lattice, the group $\mathcal{K}_4$[21] as a generalized point group of a non-symmorphic space group, $\mathcal{K}_4$ as a lattice flavour group etc.
There is a general interest in the study of lattice aspects of different physical ideas relevant for elementary particle physics. If fermions are involved, one has to rely on the Dirac Kaehler formalism. Therefore it is a first step in a lattice approximation to formulate such a problem in the continuum in terms of differential forms instead of Dirac spinors. We mention shortly a few examples for such a procedure. Some of these problems are treated more extensively in another context in other contributions to this proceedings. [37]

(i) CLASSICAL SOLUTIONS OF THE DKE IN AN EXTERNAL INSTANTON FIELD. It is generally believed that certain configurations of gauge fields, which are close to solutions of the classical field equations with a topological charge -'instantons'- , play an important role in the dynamics of QCD. These gauge field configurations allow zero mass solutions of the Dirac equation of a given chirality, as it is stated by the Atiyah Singer index theorem.[38] There are attempts to define topological charges also for lattice approximations of gauge fields [39] and to investigate these by numerical calculations. Recent calculations include the effect of these gauge field configurations on the spectrum of fermion systems. From this we derive an interest in the structure of solutions of the DKE

$$(d-\delta)\Phi - i\mathbf{A}\vee\Phi = 0 \tag{68}$$

in a classical SU(2) - instanton field

$$\mathbf{A}(x) = \frac{x^2}{x^2+\lambda^2}\omega \tag{69}$$

with

$$\omega = \frac{1}{x^2}(x^i dx^4 - x^4 dx^i + \epsilon^i_{kl} x^k dx^l)\tau_i \ . \tag{70}$$

The $\tau_i, i = 1,2,3$ are the Pauli matrices. ω is closely related to the Lie algebra valued Maurer Cartan forms of the group SU(2). With help of the Maurer Cartan relations, one gets after a short calculation the solution

$$\Phi = (r^2+\lambda^2)^{-\frac{3}{2}}(1 - r dr \wedge \omega) \vee (1 + dx^{1234}), \qquad r = \sqrt{x^2}. \tag{71}$$

Of course this solution corresponds to the solution of the Dirac equation with positive chirality which is postulated by the index theorem. It would be intersting to see if something of the correlation between the gauge field and the fermion form expressed by Eq. (71) survives the lattice approximation.

(ii)REMARK ON GRAVITATION There are also attempts to investigate the theory of gravitation in lattice approximation [40]. The natural way to include fermions in these considerations leads to the DKE. Since the DKE is defined for a general ('pseudo'-) Riemannian manifold, it describes in a straightforward manner the interaction the interaction with a classical gravitational field. However, it was pointed out by several authors [41], that in curved space there is no longer an equivalence between the Dirac equation and the DKE. The general relativistic DKE postulates an interaction of the flavour degrees of freedom with the gravitational field. In a weak gravitational field described by a metric tensor:
$g_{00} = -(1+V(x)) = -g(x), g_{\mu\nu} = \delta_{\mu\nu} \quad for \quad \mu \neq 0, \nu \neq 0,$
$V(x)$ a static potential, the DKE might be written [42]:

$$-\frac{1}{g}\partial_0\Phi + \sum_{i=1}^{3} dx^{0i} \vee \partial_i\Phi + m dx^0 \vee \Phi$$

$$-\frac{1}{g^2}\partial_i V dx^i \vee e_0 \neg \Phi - \frac{1}{2g}\partial_i V \cdot \Phi \vee dx^{0i} = 0. \tag{72}$$

The first line corresponds to the Dirac equation in a gravitational field.The second line expresses the deviations introduced by the Dirac Kaehler approach. These are proportional to the gradient of V, and hence are small in slowly varying fields. However, in fundamental theories which include gravitation, e.g. like super-symmetric graviatation [43], the gravitational interaction of flavour might open interesting aspects for its physical understanding.

(iii) SUPER-SYMMETRY In the recent attempts of the formulation of fundamental theories, the consideration of super-symmetry plays an important role The Dirac Kaehler formalism represents Dirac fields by a coherent superposition of inhomogeneous differential forms. On the other hand, forms describe in a natural way tensor fields which are bosonic. This similarity strongly suggests the investigation of super-symmetry between bosons and fermions in terms of differential forms. In order to be more specific, one may consider a bosonic form Φ and a fermionic form Ψ which satisfy the DKE. Then the expressions

$$j_\mu^f = (\Psi \vee f, dx_\mu \wedge \overline{\Phi}), \qquad \overline{\Phi} = \mathcal{A}\Phi \tag{73}$$

describe conserved currents depending on the flavour transformations f. It is possible to show [45], that in an explicit 2-dimensional model the charges of this mixed boson fermion currents: $Q = \int j_0^f dx^{123}$ are closely related to the generators of super-symmetry. Other examples are discussed in the contribution by R.W.Tucker. Here we want only to emphasize that the formulation of super-symmetry in differential forms opens the interesting possibility of treating super-symmetric models in lattice approximation with help of the methods developped in Sections 3 -5. [46]
The investigation of geometry and physics of lattice fermions is still a large ,unsettled program.

ACKNOWLEDGEMENT.I would like to thank Dr.A.K.Common for his comments with respect to the final form of the manuscript.

References

[1] E. Kaehler: *Rend.Mat.Ser.V*, **21**(1962)425.

[2] P.Becher, H.Joos: *Z.Phys.C-Particles and Fields* **15**(1982)343.

[3] M.Goeckeler, H.Joos: *Progress in Gauge Field Theory* , (Cargese 83) p.247. G. t'Hooft et.al. (ed.) 1984 Plenum Press New York.

[4] P.Becher: *Proceedings of the Johns Hopkins Workshop on Current Problems in Particle Theory* 7,p.13. G.Domokos, S.Kovesi-Domokos (ed.), World Scientifique Publishing Co. 1983 Singapore.

[5] L.Susskind: *Phys.Rev.***D16**(1976)3031.

[6] P.Becher, M.Boehm, H.Joos: *'Gauge Theories of the Strong and Electroweak Interactions'*. John Wiley and Sons 1984.(B.G.Teubner 1981).

[7] C.Itzikson, J.-B. Zuber: *Quantum Field Theory* , McGraw-Hill, New York, 1980.

[8] K.Osterwalder, R.Schrader: *Comm.Math.Phys.* **42**(1975)281.

[9] K.G.Wilson: *Phys.Rev.* **D10**(1974)2445.

[10] L.H.Karsten, J.Smit: *Nucl.Phys.* **183**(1981)103.

[11] A.Chodos, J.B.Healy: *Nucl.Phys.* **127**(1977)426.
H.S.Charatchandra, H.J.Thun, P.Weisz: *Nucl.Phys.* **192**(1981)205.

[12] H.B.Nielsen,N.Ninomya: *Nucl.Phys.* **185**(1980)20, **193**(1981)173.

[13] D.Friedan: *Comm.Math.Phys.* **85**(1982)482.

[14] M.F.Atiyah, W.K.Patodi, I.M.Singer: *Math.Cambr.Philosoph.Soc.* **79**(1976)71.

[15] K.G.Wilson: *New Phenomena in Sub-Nuclear Physics.* A.Zichichi (ed.), Plenum Press New York, 1977.

[16] A.K.Common: *'The reduction of the Dirac Kaehler Equation to the Dirac equation'*, University of Kent (preprint 1983)

[17] P.Lounesto: Contributionto this Proceedings.

[18] W.Graf: *Ann.Inst.H.Poincare.* Sect.**A29**(1978)85.

[19] I.M.Singer, J.A.Thorpe: *Lecture Notes on Elementary Topology and Geometry* , Springer, New York 1967.

[20] J.M.Rabin: *Nucl.Phys.* **B201**(1982)315.

[21] N.Salingaros: *J.Math.Phys.* **22**(1981)226, **23**(1982)1.
H.W.Braden: *J.Math.Phys.* **26**(1985)613.
We suggest to call this group K_4 which does not have a generally accepted name (H.W.Braden) in honour of this workshop: 'Kent Group'.

[22] M.Schaefer: *'Die Gitterfermiongruppe und ihre Darstellungen'*, Diplomarbeit Hamburg 1985.

[23] H.Joos, M.Schaefer: *'The Representation Theory of the Symmetry Group of Lattice Fermions as a Basis for Kinematics in Lattice QCD'.* (in preparation)

[24] H.Brown, R.Buelow, J.Neubueser, H.Wondrascheck, H.Zassenhaus: *Cristallographic Groups of Four-Dimensional Space* . J.Wiley and Sons,(New York) 1978.

[25] E.P.Wigner: *Ann.Math.* **40**(1939)149.

[26] G.W.Mackey: *Acta Math.* **99**(1958)365. *'The Theory of Group Represenations'*, (Chicago mimeographed Lecture Notes).

[27] P.Becher, H.Joos: *Lett.Nuovo Cim.* **38**(1983)293.

[28] O.Napoli: *Phys.Lett.* **B183**145.
P.Mitra, P.Weisz: *Phys.Lett* **126B**(1983)355.

[29] F.A.Berezin: *The Method of Second Quantization* , Academic Press, New York, 1966.

[30] H.Kluber-Stern, A.Morel, B.Peterson: *Nucl.Phys.* **B215**(1983)527.

[31] O.Martin: *'Large N Gauge Theory at Strong Coupling with Chiral Fermions'*, PhD Thesis Caltech. 68-1048.
Phys.Lett. **114**(1982)152.

[32] M.F.L.Golterman, J.Smit: *Nucl.Phys.* **B245**(1984)61.

[33] M.Mehamid, private communication.

[34] J.P.Gilchrist, G.Schierholz, H.Schneider, M.Teper:
Phys.Lett. **136B**(1984), *Nucl.Phys.* **B248**(1984)29. and literature quoted there.

[35] T.Banks et.al.: *Phys.Rev.* **D15**(1977)1111.

[36] D.Hestenes: *Found.Phys.* **12**(1982)153. See also: Contribution to this Proceedings.

[37] See especially the contributions by D.Hestenes, and R.W.Tucker.

[38] M.F.Atiyah, I.M.Singer: *Ann.Math.* **87**(1968)485, 546; **93**(1971)1, 119, 139.
E.F. Corrigan, D.B.Fairlie, S.Templeton, P.Goddard: *Nucl.Phys.* **B140**(1978)31.
R.Jackiw, C.Rebbi: *Phys.Rev.* **D16**(1977)1052.

[39] M.Luescher: *Comm.Math.Phys.* **85**(1982)29.
I.A.Fox, M.L.Laursen, G.Schierholz: *Phys.Rev.Lett.* **54**(1985)749.
I.A.Fox et.al., M.Goeckeler: *Phys.Lett.* **158B**(1985)332.

[40] J.Cheeger,W.Mueller,R.Schrader: *'Lattice Gravity or Riemann Structure on Piecewise Linear Spaces'*, in *Unified Theories of Elementary Particle*, P.Breitenlohner, H.P.Duerr (ed.). Springer Verlag (Berlin, Heidelberg, New York),1982.

[41] T.Banks, Y.Dothan, D.Horn: *Phys.Lett.* **117B**(1982)413.

[42] J.Krueger: *Klassische Loesungen der Dirac Kaehler Gleichung*, (Diplomarbeit Hamburg 1985)

[43] J.Wess, J.Bagger: *Supersymmetry and Supergravity*, Princeton University Press 1982.

[44] C.G.Callan, R.Dashen, D.J.Gross: *Phys.Rev.* **D17**(1978)2717.
C.Vafa, E.Witten: *Comm.Math.Phys.* **95**(1984)257.

[45] H.Joos: *Revista Brasileira de Fisica, M.Schoenberg Vol.* 1984, p.169.

[46] S.Elitzur, E.Rabinovici, A.Schwimmer: *Phys.Lett.* **119B**(1982)165.
H.Aratyn, A.H.Zimerman: *Phys.Lett.* **137B**(1984)392, *Z.Phys.C. - Particles and Fields*,
C27(1985)536.

A SYSTEM OF VECTORS AND SPINORS IN COMPLEX SPACETIME AND THEIR APPLICATION TO ELEMENTARY PARTICLE PHYSICS.

Jaime Keller
División de Estudios de Posgrado, F.Q.
Universidad Nacional Autónoma de México,
Apartado 70-528, 04510 México, D.F.

ABSTRACT.

Using a systematic notation we give a unified presentation of the spinors space and the vector space C^4, of complex spacetime, as a single mathematical system, as well as the real spacetime subalgebras and the three dimensional cut-offs. This allows a compact Clifford analysis of the relationship between the vector space and space-time spinors and their symmetries. The system is applied to construct a theory of elementary particles and their interaction fields, which is unified and contains the standard SU(3)xSU(2)xU(1) gauge symmetry.

I. INTRODUCTION.

Most applications to physics of vector Clifford algebras involve spinors and multivectors in spaces with five dimensions or less. Historically, a Clifford algebra of two dimensional space was introduced at the beginning of last century by Wessel and thoroughly discussed by Hamilton in his famous formula $i^2 = j^2 = k^2 = ijk = -1$ to study (surprisingly, because the duality between surfaces and vectors in R^3 was not known) three dimensional rotations and, independently, by Pauli in the study of the spin algebra $\mathcal{P}$. The multivectors for three dimensional space $\mathcal{P}_c$ were introduced, at the beginning of the century, either as complex quaternions by Hamilton or as the standard three dimensional space vector algebra, through the definition of axial and polar vectors, by Gibbs. The case of four $\mathcal{D}$ and five (complex four $\mathcal{D}_c$) dimensions were developed by Dirac who also introduced the use of spinors. Table I presents these algebras in a systematic form, where each one corresponds to the even part of the next higher dimensional multivector algebra. ($\mathcal{D}_c \sim C(4)$, $\mathcal{D} \sim H(2)$, $\mathcal{P}_c \sim C(2)$).

In the following three sections we present the multivector formalism for these cases, the spinors and the spinor construction of multivectors and, finally, their application to the dual geometry theory of elementary particles and their interaction fields. Some uses of Clifford multivector algebras can be found in Proca, 1930a, 1930b; Sauter 1930a, 1930b; Juvet, 1930, 1932; Eddington, 1936; Mercier,

J. S. R. Chisholm and A. K. Common (eds.), Clifford Algebras and Their Applications in Mathematical Physics, 425–434.

1934, 1935; Sommerfeld, 1939; Ravsevskii, 1957; Riez, 1946, 1953, 1958; Teitler, 1965a, 1965b, 1965c, 1966a, 1966b; Hestenes, 1966, 1975; Salingaros and Dresden, 1975; Greider, 1980; Casanova, 1970, 1976; Boudet, 1971, 1974, 1985, Quilichini, 1971; Keller, 1981, 1982a, 1982b, 1984, 1985; and papers in the present volume.

TABLE I.

DIMENSIONS		SPACETIME MULTIVECTOR					
		S	V	BiV	Tri V	Tetra V	Penta V
0	R	$\mathbb{1}$					
1	C	$\mathbb{1}$	$\mathbf{i}_1 = \gamma_{12}$				
2	P	$\mathbb{1}$	$e_h = \gamma_{h3}$	$\gamma_{12} = e_{12}$			
3	P_c	$\mathbb{1}$	$\gamma_i^{P_c} = \gamma_{01}$	γ_{ij}	$\gamma_5 = \gamma_{ijk}^{P_c}$		
4	$\mathcal{D}$	$\mathbb{1}$	$\gamma_\mu = \gamma_\mu$	$\gamma_{\mu\nu}$	$\gamma_{\mu\nu\lambda}$	$\gamma_5 = \gamma_{0123}$	
5	$\mathcal{D}_c$	$\mathbb{1}$	$\gamma_A = \mathbf{i}_2\gamma_\mu,$	γ_{AB}	γ_{ABC}	γ_{ABCD}	$\mathbf{i}_2 = i$

The $\gamma_i^{P_c}$ are isomorphic to twice the (complex) Pauli algebra or $\mathcal{D}_{even} = (P_c)^{(2)}$. S = scalar, V = vector, BiV = bivector, etc.

II. VECTROS AND SPINORS IN COMPLEX SPACE-TIME AND THEIR SUBSPACES.

1. VECTORS.

The multivectors are generated by the antisymmetric, Grassmann, outer product Λ of a basis set $\{\gamma_\mu\}$ in N dimensions

$$\gamma_{AB} = \gamma_A \wedge \gamma_B = \frac{1}{2}(\gamma_A\gamma_B - \bar{\gamma}_B\bar{\gamma}_A), \tag{1}$$

where

$$\gamma_A = \gamma_{\mu\nu\ldots\lambda} \text{ and } \bar{\gamma}_A = (-\gamma_\mu)(-\gamma_\nu)\ldots(-\gamma_\lambda); \text{ Also } \tilde{\gamma}_A = \gamma_{\lambda\cdots\nu\mu}. \tag{2}$$

The corresponding Clifford algebra is constructed using the Grassman algebra and an inner (dot) product;

$$\gamma_A \cdot \gamma_B = \frac{1}{2}(\gamma_A\gamma_B + \bar{\gamma}_B\bar{\gamma}_A), \tag{3}$$

to define the total, or geometric, product:

$$\gamma_A\gamma_B = \gamma_A \cdot \gamma_B + \gamma_A \wedge \gamma_B .$$

The metric of space-time $R^{1,3}$, ($\mu = 0,1,2,3$) is defined through the product

$$g_{\mu\nu} = \gamma_\mu \cdot \gamma_\nu = \text{diag}\ (1,-1,-1,-1). \tag{4}$$

If the multivector algebra C^N is considered as the complexification of $R^{m,n}$ $(N = m+n)$, we require the concept of absolute value squared $|\gamma_A|^2 = \gamma_A \cdot \gamma_A^*$, where γ_A^* is a multivector with all coefficients being the conjugate of those of γ_A. We can write formally for $\mathcal{D}_c \sim C^4$

$$\mathcal{D}_c = \mathcal{D} + i\mathcal{D} = \mathcal{D}_R + \mathcal{D}_I\ ; \quad \mathcal{D}_R^* = \mathcal{D}_R \text{ and } \mathcal{D}_I^* = -\mathcal{D}_I \quad \text{and}$$

$$\mathcal{D} = \mathcal{D}_+ + \mathcal{D}_-\ ;\ \mathcal{D}_+ = \overline{\mathcal{D}}_+ \text{ and } \mathcal{D}_- = -\overline{\mathcal{D}}_-\ , \text{ also } \mathcal{D}_- = \gamma_o \mathcal{D}_+\ .$$

Also, if $\gamma_5 = \gamma_{0123}$, $\mathcal{D} = \mathcal{D}_1 + \gamma_5\mathcal{D}_1$ is the main U(1) operation of the algebra. See table II, where $\mathcal{D}_1$ corresponds to the upper half of it.

We can construct projection operators P_A, with any of the multivectors $\gamma_A : \gamma_A^2 = 1$ (except $\gamma_A = 1$)

$$P_A^+ = (1 + \gamma_A)/2 \quad \text{and} \quad P_A^- = (1 - \gamma_A)/2;$$

$$P_A^+P_A^- = P_A^-P_A^+ = 0; \quad P_A^+P_A^+ = P_A^+\ , \ P_A^-P_A^- = P_A^- \quad \text{and} \quad P_A^+ + P_A^- = 1.$$

For an algebra of dimension N the number p of independent P_A is p = integer (N/2).

TABLE II. Spacetime Geometry

Geometric Type	Multivector				Operator	
Scalar		1			$\mathbf{1}$ →	$\mathbf{1}$
Vectors	γ_0	γ_1	γ_2	γ_3	γ_α	**PT**
Surfaces		$\gamma_0\gamma_1$	$\gamma_0\gamma_2$	$\gamma_0\gamma_3$	$\gamma_{\alpha\beta}$ →	$\mathcal{L} = \begin{cases} R \to \gamma_{ij} \\ L \to \gamma_{0i} \end{cases}$
		$\gamma_2\gamma_3$	$\gamma_3\gamma_1$	$\gamma_1\gamma_2$		
Volumes	$\gamma_1\gamma_2\gamma_3$	$\gamma_2\gamma_3\gamma_0$	$\gamma_3\gamma_0\gamma_1$	$\gamma_0\gamma_1\gamma_2$	$\gamma_{\alpha\beta\gamma}$	$\mathbf{P^D T^D}$
Hypervolumes		$\gamma_5 = \gamma_0\gamma_1\gamma_2\gamma_3$			γ_5 →	**D**

Space-time multivectors, their symmetries and their properties as operators.

2. SPINORS.

We will define the spinors of any set of multivectors $\mathcal{D}_c$ as a vectorial space L such that:

Definition: if $\chi \in L$ and $M \in \mathcal{D}_c$ then $M\chi \in L$, (8)
with dual spinor space L^+ :
Definition: if $\chi^+ \in L^+$ and $M \in \mathcal{D}_c$ then $\chi^+ M \in L^+$ (9)

in such a way that we obtain $\mathcal{D}_c$ from the "outer" spinor product

$\chi\chi^+ \in \mathcal{D}_c$ and the "inner" spinor product $\chi^+_\beta \chi_\lambda = C_{\beta\lambda}$. (10)

Proof: if $M = M^{\alpha\beta}\chi_\alpha\chi^+_\beta$, then $M\chi_\lambda = M^{\alpha\beta}\chi_\alpha\chi^+_\beta\chi_\lambda = C_{\beta\lambda}M^{\alpha\beta}\chi_\alpha = \chi_\sigma \in L$

and correspondingly for the $\chi^+_\alpha \in L^+$. The $M^{\alpha\beta}$ are the matrix representation of the $M \in \mathcal{D}_c$. ($\mathcal{D}_c$ = any multivector Clifford algebra).

3. a) CLASSIFICATION OF $L_{\mathcal{D}_c}$ SPINORS.

Define the main projectors Q_R and Q_L and name the two spinor subspaces generated by the Q on the $L_{\mathcal{D}_c}$ left handed L and right handed R, such that

$$L_{\mathcal{D}} = L_R + L_L . \tag{11}$$

For $\mathcal{D}_c$ = complex spacetime we need two projection operators, which will be either

($i\gamma_5$ and $i\gamma_{12}$) or ($i\gamma_{12}$ and γ_o) and

$$Q_{R/L} = (1 \pm i\gamma_5)/2 \text{ (chiralty projectors).} \tag{12}$$

The other commonly employed projectors, are the

mass projector: $m_{+/-} = (1 \pm \gamma_o)/2$, and the (12)

spin projectors: $S_{\uparrow/\downarrow} = (1 \pm i\gamma_1\gamma_2)/2$. (14)

The relations used to construct the starting multivectors will be:

$$\gamma_o i\gamma_5\chi_L = -(\gamma_o\chi_L) = -i\gamma_5(\gamma_o\gamma_L) \tag{15}$$

or $\gamma_o i\gamma_5\chi_R = (\gamma_o\chi_R) = -i\gamma_5(\gamma_o\chi_R)$, (16)

$$(i\gamma_5)(i\gamma_{12}\chi_L) = i\gamma_{12}i\gamma_5\chi_L = -(i\gamma_{12}\chi_L) \tag{17}$$

or $(i\gamma_5)(i\gamma_{12}\chi_R) = i\gamma_{12}i\gamma_5\chi_R = +(i\gamma_{12}\chi_R)$. (18)

In the following: ch = {L,R}, S = {↑,↓} and m = {+,-}.

3. b) THE ELEMENTS OF $\mathcal{D}_c$ AS A LINEAR COMBINATION OF SPINOR PRODUCTS $\chi\chi^+ \in \mathcal{D}_c$.

In the spin chiral representation we can construct $i\gamma_5$ as:

$$i\gamma_5 \equiv \sum_S (\chi_{SR}\chi^+_{SR} - \chi_{SL}\chi^+_{SL})\, e^{i\theta_5} \quad \text{and} \tag{19}$$

$$i\gamma_{12} \equiv \sum_{ch} (\chi_{\uparrow ch}\chi^+_{\uparrow ch} - \chi_{\downarrow ch}\chi^+_{\downarrow ch})e^{i\theta_{12}}$$

$$\gamma_o \equiv \sum_S (\chi_{SL}\chi^+_{SR} + \chi_{SR}\chi^+_{SL})e^{i\theta_o} \;; \tag{20}$$

in this representation $\theta_{12} = \theta_o = \theta_5 = 0$, (usually $\theta_o = \pi$!).

From the commutation and anticommutation relations with γ_5 and γ_{12} the vector basis set is found to be

$$\begin{aligned}
\gamma_o &= \chi_{\uparrow L}\chi^+_{\uparrow R} + \chi_{\downarrow L}\chi^+_{\downarrow R} + \chi_{\uparrow R}\chi^+_{\uparrow L} + \chi_{\downarrow R}\chi^+_{\downarrow L} \;,\\
\gamma_1 &= -\chi_{\uparrow L}\chi^+_{\downarrow R} - \chi_{\downarrow L}\chi^+_{\uparrow R} + \chi_{\uparrow R}\chi^+_{\downarrow L} + \chi_{\downarrow R}\chi^+_{\uparrow L} \;,\\
\gamma_2 &= i(-\chi_{\uparrow L}\chi^+_{\downarrow R} + \chi_{\downarrow L}\chi^+_{\uparrow R} + \chi_{\uparrow R}\chi^+_{\downarrow L} - \chi_{\downarrow R}\chi^+_{\uparrow L}) \;,\\
\gamma_3 &= -\chi_{\uparrow L}\chi^+_{\uparrow R} + \chi_{\downarrow L}\chi^+_{\downarrow R} + \chi_{\uparrow R}\chi^+_{\uparrow L} - \chi_{\downarrow R}\chi^+_{\downarrow L} \;.
\end{aligned} \tag{21}$$

3\. c) MATRIX REPRESENTATIONS.

If we define the column matrix of the basis $\chi_\alpha \in L$ and the row matrix of the $\chi^+_\alpha \in L^+$, the multivectors will be represented by the coefficients obtained from (21) as $\gamma_A \to M^{\alpha\beta}_A$, the usual matrices used in quantum mechanics.

The standard matrix representation of quantum mechanics corresponds to the choice of γ_o and $i\gamma_{12}$ as projection operators and the corresponding spinors

$$\gamma_o\chi_{S+} = \chi_{S+} \quad \text{and} \quad \gamma_o\chi_{S-} = -\chi_{S-} \tag{22}$$

3\. d) THE MOST COMMONLY USED SUBSPACES OF C^4 AND OF $R^{1,3}$.

First rewrite the five vector basis of $\mathcal{D}_c$ in the form

$$\gamma_v = \{i\gamma_{123},\ i\gamma_0, i\gamma_{01}, i\gamma_{02}, i\gamma_{03}\} \;; \tag{23}$$

and the D_c multivector algebra $\mathcal{D}_c = \{1, \gamma_u, \gamma_{uv}, \gamma_{muv}, \gamma_{uvmx}, i\}$

in order that the even part of it corresponds to $\mathcal{D} = R^{1,3}$;

$$\mathcal{D} = \{1, \gamma_{mv} = \{\gamma_{\mu\nu}\} + \{\gamma_{o\mu}\} + \{\gamma_{\mu\nu\lambda}\},\ \gamma_{uvwx} = \{\gamma_5, \gamma_\mu\}\ \}$$

The curly brackets give the equivalence in terms of the standard spacetime basis γ_μ. The standard reduction chain of even subalgebras:

$$\mathcal{D}_c \to \mathcal{D} \to P_c \to P \to C^1 \to R^1$$

is straightforward, (Table 1), $\mathcal{D}_c$ and $\mathcal{D}$ share the same spinor basis, P_c

and P have two common representations, in our system they use either the spinor basis $(\chi_{\uparrow L},\chi_{\downarrow L})$ or $(\chi_{\uparrow R},\chi_{\downarrow R})$. The spinor basis of C^1 and R^1 is trivial.

It is important to emphasize here that all rotations in our representations of $\mathcal{D}_c$ are quantized in the standard formulation of quantum mechanics (Keller 1985) and that changes in the spinor basis $\chi_\alpha \to \chi_\alpha + \varepsilon_\alpha^\beta \chi_\beta$ correspond to changes, using (10), in the multivector space. In general $\varepsilon_\alpha^\beta = \varepsilon_\alpha^\beta(\underline{x})$.

4. COVARIANT VECTOR AND SPINOR DERIVATIVES.

Following Hestenes (1966) define a differentail operator $\Box_i$ by:

$$\Box_i \phi = \partial_i \phi \tag{24}$$

where ϕ is a scalar and $\Box_i$ maps scalars into scalars. And for a vector field γ_j : $\Box_i \gamma_j = - L_{ij}^k \gamma_k$ (25)

and for multivectors A and B $\Box_i(AB) = (\Box_i A)B + A\Box_j B$ (26)

$$\Box_i(A+B) = \Box_i A + \Box_i B \ . \tag{27}$$

In general if $\underline{a} = a_j \gamma^j$, then $\Box_i \underline{a} = (\partial_i a_j + a_k L_{ij}^k)\gamma^j$. (28)

Hestenes (1966) uses this operator to discuss problems in general relativity (see also Hestenes and Sobczyk (1984) in this context). For our spinor spaces L and L^+ we can define:

$$\Box_i \chi^\alpha = K_{i\beta}^\alpha \chi^\beta \quad \text{and} \quad \Box_i \chi^{\alpha+} = - K_{i\beta}^\alpha \chi^{\beta+} \tag{29}$$

where the $K_{i\delta}^\rho$ are related with the L_{ij}^k using the (representation dependent) expansion (10) of the $\gamma^k = M_{\alpha\beta}^k \chi^\gamma \chi^{\beta+}$.

III. THE MULTIVECTOR BASIS OF THE DIRAC FORM OF THE THEORY OF ELEMENTARY PARTICLES.

An observer in reference system S (S') associates an energy momentum vector p (p') to an electron (in fact to any "elementary" particle with mass $m_o > 0$)

$$p^\beta \gamma_\beta = p'^\alpha \gamma'_\alpha \ , \tag{30}$$

where the basis vectors of S' and S are related through a Lorentz Boost

$$\gamma'_\alpha = L\gamma_\alpha L^{-1} \ ; \quad LL^{-1} = L^{-1} L = 1 \ . \tag{31}$$

Observer S' is taken to be that where $p' = m_o c\gamma'_o$; then postmultiplying (30) by L we obtain an equation in multivector form relating the particle's system to the observer's system S:

$$p^\beta \gamma_\beta L = m_o c L \gamma_o \ . \tag{32}$$

Introduce, (see Hestenes 1966), the Schrödinger operator $\hat{p}^\beta$

$$\tilde{p}^\beta L \equiv \hbar \partial^\beta L\mathbf{I} = p^\beta L \quad \text{with } \mathbf{I}^2 = -1 \tag{33}$$

to obtain the multivector Dirac equation

$$-\gamma_\beta \partial^\beta L = m_o c L \gamma_o \mathbf{I} : \tag{34}$$

here $\hbar = 1$ and $\mathbf{I}$ is some rotation plane. Hestenes (see article in this book) proposes $\mathbf{I} = \gamma_{12}$.

The solution $L = LQ_o$ to the multivector equation (34), where $Q_o = A \exp(-\mathbf{I}\underline{p}\cdot\underline{x}/\hbar)$, can be "gauged"

$$L \to \Psi = A \exp(-\mathbf{I}\{\underline{p}\cdot\underline{x} + \phi(\underline{x})\}/\hbar) , \tag{35}$$

where the more general gauge "angle" is

$$\phi(\underline{x}) = \phi_{\text{scalar}}(\underline{x}) + \gamma_5\phi_{PS}(\underline{x}) + \gamma_\mu\gamma_\nu\partial^\nu\Omega^\nu(\underline{x}) . \tag{36}$$

The scalar part is usually interpreted as corresponding to the electromagnetic field, the pseudoscalar part we interpret as corresponding to the weak and color fields, and the bivector part to the gravitational field. See Keller 1984, 1985.

That is, the interaction fields are given as boundary condition to represent both the rest of the physical world, and the physical effect of the particle on itself. Then the electromagnetic interaction appears as a (complex) phase factor ϕ which will produce an "extra" energy-momentum

$$\partial^\mu e^{-\mathbf{I}\phi}\,\mathbf{I} = \frac{eA_\mu}{c}\, e^{-\mathbf{I}\phi} ;$$

A_μ is the usual electromagnetic field vector. The weak and color fields produce an extra vector -"axial" vector energy-momentum and the gravitational field changes the local, fiducial, frame $\gamma_\mu \to \gamma''_\mu(\underline{x})$. The gravitational interaction arises because, in order to compensate such a gauge transformation, a vierbein is needed:

$$f_j = (f^o e^{-\partial\Omega})_j \tag{37}$$

where the f^o are locally Lorentzian tetrads

$$g_{\mu\nu} = g^o_{\alpha\beta} f^\alpha_\mu f^\beta_\nu = \left[g e^{-2\,\Box\Omega} \right]_{\mu\nu} \tag{38}$$

defining a (gauge invariant) gravitational "field"

$$\phi^\nu_\mu = \partial_\mu\Omega^\nu + \partial^\nu\Omega_\mu - \delta^\nu_\mu\partial^\alpha\Omega_\alpha ; \tag{39}$$

this "field" will obey, for self-consistency, the "field" equation

$$\Box^2\phi = 4G\pi(\mathbf{T} - \tfrac{1}{2}\, gT) \tag{40}$$

where $\mathbf{T}$ is the energy-momentum stress tensor of the total sources. The origin of the electroweak interaction will be discussed below.

The standard form of the Dirac equation is obtained (see Casanova 1976) using a spinor u such that $\gamma_o u = u$ and $\mathbf{I}\, u = iu$, to define $\Psi u = \psi$ and $\Psi\gamma_o \mathbf{I} = -i\psi$; here ψ is now a particular spinor projected out of Ψ and

$$i\gamma_\mu\partial^\mu\psi = m_o c\psi \quad \text{or} \quad D\psi = -im_o c\psi ; D^2 = \partial_\mu\partial^\mu . \tag{41}$$

For a massless particle $D\psi_o = 0$. The ψ_o (more generally ψ) obeys the Klein-Gordon equation

$$\partial_\mu \partial^\mu \Phi = 0 \quad \text{or} \quad (-\partial_\mu \partial^\mu \Phi = (m_o c)^2 \Phi) \tag{42}$$

from which the Dirac solution is obtained using the Dirac operator D:

$$\psi_o = (D + m_o ci)\Phi \ . \tag{43}$$

Now we use multivectors to generalize (41) and to develop the theory of Symmetry Constrained Dirac Particles (Keller 1984, 1985). For this purpose we generalize the Dirac construction to a differential operator D valued in the (complex) multivector algebra $\mathcal{D}_c$:

$$DD^* = D^*D = \partial^\mu \partial_\mu \tag{44}$$

The Klein-Gordon equation operator ($c = \hbar = 1$)

$$(\partial^\mu \partial_\mu + m^2) = (D^* + mi)(D - mi) \tag{45}$$

requires $-D^*m + mD = 0$; that is either

{"A" : $D^* = D$ and $m \neq 0$ (Dirac's)} or {"B" : any D obeying (44) if $m = 0$ } (46)

Let us restrict ourselves to case "B" (massless particles) and a D where we change (one or) several of the vectors γ_μ into a more general element γ_A. A hint comes from the special role of $i\gamma_5$ in elementary particle physics and from the general solution Ψ above. Then, if we define a set d of coefficients $\{t^d_\mu\}$, for the construction of a diracon operator D_d: $a_d(\mu)\gamma_\mu \partial^\mu$,

$$D = \gamma_\mu \partial^\mu \rightarrow D_d = (\cos[(n + t^d_\mu)\tfrac{\pi}{2}] + i\gamma_5 \sin[(n + t^d_\mu)\tfrac{\pi}{2}])\gamma_\mu \partial^\mu \text{ and } \psi_d = D^*_d \Phi. \tag{47}$$

With the choice of n and t^d_μ integers we obtain a set of diracon massless fields with definite chiralty $i\gamma_5 \psi_d = \pm\, \psi_d$. In that case $a_d(\mu) = \pm\, 1$ or $\pm\, i\gamma_5$, provided we also restrict $t^d_\mu = 0$ or 1 in order not to mix different chiralties. Each n represents a FAMILY of fields (Keller 1984):

Field	t(0)	t(1)	t(2)	t(3)	Future Identification
ν	0	1	1	1	neutrino
	0	0	1	1	
d	0	1	0	1	quark down 3 "colors"
	0	1	1	0	
	0	0	0	1	
$\bar{u}$	0	0	1	0	(anti) quark up 3 "colors"
	0	1	0	0	
e	0	0	0	0	electron

IV. SPACETIME DUAL GEOMETRY THEORY OF GAUGE FIELDS AND OF COMPOSITE ELEMENTARY PARTICLES.

The experimentally observed elementary particles are not those corresponding to the "bare" diracon fields described in the preceeding section (with possible exception of the neutrinos) but massive leptons (composed of left and right handed ℓ-fields) and baryons (3 quarks in a colorless combination) or mesons (2 quarks in a colorless combination), in interaction among themselves and with the, rest of the world, radiation fields.

There are two main possibilities in our spinor system to represent the collection of spinor fields $\chi_\alpha^{(d)}$ of composite particles 1) as a row Ψ of $\chi_\alpha^{(d)}$ or 2) as a column ψ of $\chi_\alpha^{(d)}$. Each one of these possibilities has its own advantages. The practical use of Ψ is that it can be operated on the right by the elements of γ_A if Ψ always contains 4 columns (in $\mathcal{D}_c$). The use of ψ is standard in most of the elementary particle literature.

The composite wave function Ψ can be symbolically written $\Psi = a_{d_1}^{\alpha}\chi^{d_1}\chi_\alpha^{+}$ where the linear combination of diracon fields $a_{d_1}^{\alpha}\chi^{d_1}$ is placed by χ_α^{+} ($\alpha = 1.4$), in the α-th column of Ψ. This is the procedure to construct the massive electron wave equation in (32) and used by Casanova (1976) to study the baryon and meson multiplets.

The standard procedure of constructing a supercolumn spinor ψ is used by Keller 1985 to give an explicit formulation of SU(2) x U(1) electroweak interactions, $SU(3)_{color}$ chromodynamics and a unified presentation of SU(3) x SU(2) x U(1) in terms of the gauging (36) of the diracon fields. In the latter theory $\mathbb{I} = \gamma_5$ in equation (32) corresponding to a plane of $\mathcal{D}_c$ in (23).

V. CONCLUSIONS.

The systematic use of multivectors and spinors Clifford algebras shows itself to be basic solving several of the major questions of today's physics: the unification of gravitation and electroweak and color interaction, the explanation of leptons and quarks, their symmetries and their properties, like confinement and the existent families of fermions; providing a new start for a general quantum mechanical formulation of the theory of the elementary particles and their interaction fields.

In this paper we have tried to use standard notations, see for example the source article (Greenberg, 1982) and also to conform our results to the now accepted standard model (Fritzch and Minkowski 1984, Georgi and Glashow, 1974, Georgi 1975, Salam 1968 and Weinberg 1967).

VI. REFERENCES.

Boudet, R., C.R. Acad. Sci. Paris Ser. A **272**, 767, (1971).
Boudet, R., C.R. Acad. Sci. Paris Ser. A **278**, 1063, (1974).
Boudet, R., C.R. Acad. Sci. Paris Ser. II, **300**, 157 (1985).
Casanova, G., C.R. Acad. Sci. Paris Ser. A **270**, 1202 (1970).

Casanova, G., L'algebre Vectorielle, Que Sais-je, No. 1657, (Presses Universitaires de France, Paris, 1976).
Eddington, A.S., Relativity Theory of Protons and Electrons (McMillan, New York, 1936). (And see references therein).
Fritzch, H. and Minkowski, P., Ann. Phys. N.Y. 93, 193 (1974).
Georgi, H. and Glashow, S.L., Phys. Rev. Lett. 32, 438 (1974).
Georgi, H., Particles and Fields, (C.E. Carlson, Ed., AIP, New York, 1975).
Greenberg, O.W., Am.J.Phys. 50, 1074-89 (1982).
Greider, T.K., Phys. Rev. Lett. 44, 1718 (1980).
Hestenes, D., Spacetime Algebra (Gordon & Breach, New York 1966).
Hestenes, D., J. Math. Phys. 16, 556 (1975), and references therein.
Hestenes, D. and Sobczyk, G., Clifford Algebra to Geometric Calculus, (D. Reidel Publishing Co., Dordrecht 1984).
Juvet, G., Commun. Math. Helv. 2, 225 (1930).
Juvet, G., Bull. Soc. Neuchateloise S. Nat. 57, 127 (1932).
Keller, J., Rev. Soc. Quim. Mex. 25, 28 (1981).
Keller, J., Int. J. Theor. Phys. 21, 829 (1982).
Keller, J., in Proceedings of the Mathematics of the Physical Spacetime, Mexico 1981, J. Keller Ed., p.117 (University of Mexico, 1982).
Keller, J., Int. J. Theor. Phys., 23, 818. (1984).
Keller, J., Paper presented at the 3rd. Loyola Conf. on Quantum Theory & Gravitation", New Orleans 1985., Int. J. Ther. Phys. 25,..., (1985).
Mercier, A., Actes de la Soc. Helv. des Sci. Nat. Zurich, 278 (1934).
Mercier, A., These U. de Geneve 1935, and Archives des Sciences Physiques et Naturelles 17, 278 (Switzerland 1935).
Proca, A., C.R. Acad. Sci. Paris 190, 1377 (1930).
Proca, A., C.R. Acad. Sci. Paris 191, 26 (1930).
Proca, A., J. Phys. VII, 1, 236 (1930).
Quilichini, P., C.R. Serie B 273, 829 (1971).
Ravsevskii, P.K., Trans. Am. Math. Soc. 6, 1 (1957).
Riesz, M., Comptes Rendus du Dixieme Congres des Mathematiques des Pays Scandinaves, vol. 123 (Copenhagen 1946).
Riesz, M., Comptes Rendus du Douzieme Congres des Mathematiques des Pays Scandinaves, Vol. 241 (Lund, 1953).
Riesz,M., Lecture series No. 38, University of Maryland (1958).
Salam, A., Proc. 8th Nobel Symp. (Stockholm), N. Svatholm, ed., p. 367 (Almquist and Wiksell, Stockholm 1968).
Salingaros, N. and Dresden, M., Phys. Rev. Lett. 43 (1979).
Sauter, F., Z. Phys. 63, 803; Z. Phys. 64, 295 (1930).
Sommerfeld, A., Atombau und Spektrallinien, Vol. II, p. 217, (Braunschweig 1939).
Teitler, S., Nuovo Cimento Suppl. 3, 1 (1965a).
Teitler, S., Nuovo Cimento Suppl 3, 15 (1965b).
Teitler, S., J. Math. Phys. 6, 1976 (1965c).
Teitler, S., J. Math. Phys. 7 1730 (1966).
Teitler, S., J. Math. Phys. 7, 1739 (1966).
Weinberg, S., Phys. Rev. Lett. 19, 1264 (1967).

SPINORS AS COMPONENTS OF THE METRICAL TENSOR IN 8-DIMENSIONAL RELATIVITY

Dieter W. Ebner
Physics Department
University of Konstanz
P.O. Box 5560
D-775 Konstanz
W-Germany

ABSTRACT. It is assumed that each point of space-time is a microscopic $S^3 \times S^1$ (S^3 = 3-sphere). This space is metrically disturbed, so that an isometry is obtained only when a spatial rotation is linked with half a rotation of the S^3. This disturbed space is identified with the physical vacuum having an SO_3-spin-structure.

1. KALUZA-KLEIN THEORIES[1][2]

In the 5-dimensional Kaluza (1921) theory, the world is a <u>5-dimensional Riemannian space</u> (coordinates t,x,y,z,χ; signature +----; topology $R^4 \times S^1$; $o \leqq \chi \leqq 2\pi$) and is described by a metrical tensor γ_{ij}.

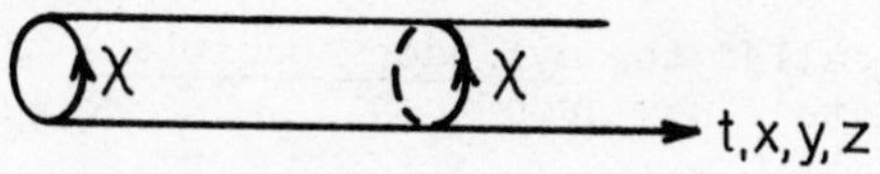

Figure 1. 5-dimensinal Kaluza-theory with cylinder condition.

What we see macroscopically as an event (point in space at time t) is in fact a small circle. For the metrical tensor the <u>5-dimensional Einstein vacuum equations</u>:

$$(1.1) \quad R_{ij} - \frac{1}{2}\gamma_{ij}R = o$$

are postulated. (R_{ij}=Ricci-tensor, R=Ricci-scalar.) It is the great achievement of Kaluza to have shown that (1.1) contains the usual 4-dimensional Einstein equations ($i,j=1,\ldots 5$; $\mu,\nu=1,\ldots 4$)

J. S. R. Chisholm and A. K. Common (eds.), Clifford Algebras and Their Applications in Mathematical Physics, 435–443.

(1.2a) $R_{\mu\nu} - \frac{1}{2} g_{\mu\nu} R = T_{\mu\nu}$,

(1.2b) $T_{\mu\nu} = F_{\mu\alpha} F^{\alpha}{}_{\nu} + \frac{1}{2} F_{\alpha\beta} F^{\alpha\beta} g_{\mu\nu}$

and Maxwell's equations

(1.2c) $F_{\mu\nu;\alpha} = o$,

(1.2d) $F_{\mu\nu} = A_{\nu,\mu} - A_{\mu,\nu}$.

Here $g_{\mu\nu}$ is the metrical tensor of space-time coordinates t,x,y,z; (,) is the partial derivative, (;) is the covariant derivative with respect to $g_{\mu\nu}$; A_μ is the electromagnetic 4-vector potential; $F_{\mu\nu}$ is the electromagnetic field-strength tensor containing the electric and magnetic field; $T_{\mu\nu}$ is the energy-momentum-tensor describing energy and momentum density of the electromagnetic field A_μ according to Maxwell's theory; $R_{\mu\nu}$ and R are the Ricci-tensor and Ricci-scalar of $g_{\mu\nu}$. The tensors $g_{\mu\nu}$ and A_μ are constructed from γ_{ij} in a suitable way called projection, explained below.

Thus, Kaluza's theory is a successful unification of gravity with electromagnetism. Eq. (1.2) describes empty space containing a gravitational field $g_{\mu\nu}$ (gravitons) and an electromagnetic field A_μ (photons).

A general Riemannian space would look like Fig. 2 rather than Fig. 1.

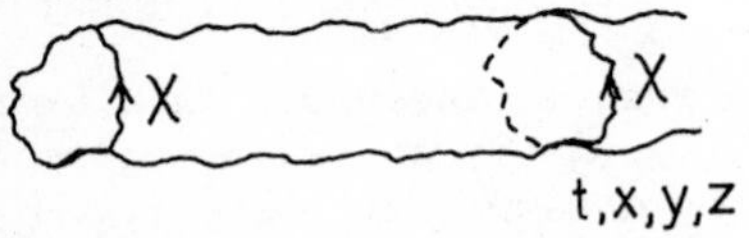

Figure 2. Kaluza-Klein model without cylinder condition.

It was Kaluza's assumption (called the cylinder condition) that there is a system of closed χ-coordinate lines such that

(1.3a) γ_{ij} is independent of χ,

(1.3b) $\gamma_{55} = \text{const} = 1$

as suggested in Fig. 1. (Equations (1.2) are obtained only if (1.3) is assumed.) The electromagnetic 4-vector potential A_μ is directly identified with components of the metrical tensor:

(1.4a) $A_\mu = \gamma_{\mu 5} = \gamma_{5\mu}$,

dimensional factors being put equal to unity. Eq. (1.4a) can also be formulated in a coordinate independent way, using the Killing vector fields tangential to the χ-lines. (1.3a) is essential for the identification of A_μ as a space-time field.

The coordinate χ with property (1.3) are not unique, permitting the transformations

(1.5a) $$x'^{\mu} = f^{\mu}(x^{\sigma})$$

(1.5b) $$\chi' = \chi + \lambda(\chi^{\sigma}).$$

Since (1.4a) (1.4b) entail

(1.6) $$A_\mu' = A_\mu + \lambda,_\mu ,$$

(1.5b) can be identified with the gauge-transformations of electrodynamics. This is the second part of the great success of Kaluza's theory: An internal symmetry of physics (gauge-transformations, leading to the conservation of electric charge according to Noether's theorem) and external (i.e. space-time) symmetries are formulated simultaneously as geometric symmetries in the 5-dimensional world. Eq.(1.5a) is a symmetry of Einstein's equations. In special cases it is Poincaré-invariance, leading to the conservation of energy and momentum.

The metrical 4-tensor $g_{\mu\nu}$ cannot be identified directly with the components of the metrical 5-tensor: $g_{\mu\nu} \neq \gamma_{\mu\nu}$. The correct projection, leading to a $g_{\mu\nu}$ independent of the gauge-freedom (1.5b), and having the usual transformation properties under (1.5a), independent of λ in (1.5b), is obtained according to Fig. 3: The 5-line element

$$d\sigma^2 = \gamma_{ij}\, dx^i\, dx^j$$

is directly identified with the ordinary 4-line element

$$ds^2 = g_{\mu\nu}\, dx^\mu\, dx^\nu$$

($ds=d\sigma$) for a displacement dx^i which is orthogonal to the χ-lines, leading to

(1.4b) $$g_{\mu\nu} = \gamma_{\mu\nu} + A_\mu A_\nu .$$

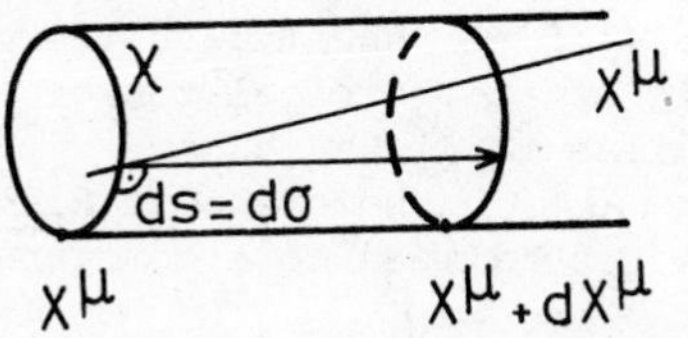

Figure 3. Geometric motivation for the projection of the metrical 5-tensor.

In Kaluza's theory there is no explanation for the cylinder condition. When this condition is postulated only approximately, a cosmological explanation could be attempted, similar to the explanation of the approximate flatness of space by thermodynamic equilibrium in the early stages of the universe. Only when small derivations from the cylinder condition are observed, the theory would be appreciated as verified. Thus far it is considered only as a formal unification of Maxwell's and Einstein's theory, which could be the result of a mathematical coincidence.

Theories discarding the cylinder condition (1.3) are called Kaluza-Klein-theories. Here, the γ_{ij} are Fourier-decomposed

$$(1.5) \qquad \gamma_{ij} = \sum_{n=-\infty}^{\infty} \gamma_{ijn}(x^{\sigma})\, e^{in\chi}$$

leading thus to an infinity of space-time fields (particles) γ_{ijn}.

We mention that the known quantization procedures applied to Kaluza-Klein theories, and indeed to all Einstein theories, lead to fundamental difficulties such as unremovable divergencies (non-renormalization), and non-uniqueness of the light cone. Furthermore, a semi-classical treatment of quantization seems to indicate that Kaluza-Klein models are unstable shrinking down to a point within Planck time (E.Witten's instability[3]).

2. GLASHOW-SALAM-WEINBERG (GSW)-Theory[4][5]

In the last few years, there has been progress in the unification of forces in physics. The GSW-theory is a unified gauge theory of electromagnetic and weak forces, which is receiving currently considerable verification at the CERN. The theory is based on the group

$$(2.1) \qquad U_1 \times SU_2.$$

A gauge-field is a connection in a principal fibre bundle[6][2] with space-time as the base-manifold and with a group, for example (2.1), as the typical fibre. Also particle fields (sections in a tensor bundle) are considered. A local gauge-theory is obtained by imposing certain field-equations which are form-invariant under the local action of the group. Assuming the ordinary Yang-Mills type of field equations, the gauge-fields can be formulated as components of the metrical tensor in a higher dimensional Kaluza-theory, similar to (1.4a). The S^1 is thereby replaced by the group manifold (2.1), being a homogeneous Riemannian space. The cylinder condition is thus generalized to a homogeneity condition.

In these generalized Kaluza-theories there is no geometrical interpretation for the particle fields. Such an interpretation seems in particular difficult for spinor fields. Spinors have the puzzling property that they assume a factor -1 when space has made a complete rotation about 2π. This property does not seem to be explainable in a pure geometrical model, and usually spinors must be assumed ad hoc.

3. AN 8-DIMENSIONAL KALUZA-KLEIN THEORY HAVING A SPIN STRUCTURE

We consider a Riemannian space with topology

(3.1) $\mathbb{R}^4 \times S^1 \times S^3.$

A special example is the "flat" case

(3.2) $M^4 \times S^1 \times S^3.$

We call it the pre-vacuum. In (3.2), M^4 is flat Minkowski-space, S^3 is the metrical sphere and x implies metrical orthogonality. The symmetry (isometry) group of the pre-vacuum is

(3.3) $ISO(3,1) \times U_1 \times SO_4.$

(ISO(3,1) is the Poincaré-group.) By considering deviations from (3.2), the pre-vacuum gets filled with 8-gravitons. In analogy to (1.4a)(1.4b), the components of the 8-graviton can be interpreted as ordinary 4-gravitons, as $U_1 \times SU_2$ gauge-fields and as spinor-fields. As coordinates we choose x^μ (space-time), χ (for the S^1) and θ_1, θ_2, ϕ (spherical coordinates for the S^3). The metrical tensor γ_{ij} (i,j=1,...8) is Fourier-decomposed

(3.4) $\gamma_{ij} = \Sigma\, \gamma_{ijnll'm}(x^\sigma)\, e^{in\chi}\, Y_{ll'm}(\theta_1, \theta_2, \phi),$

where Y are the spherical harmonics of the S^3, obtaining an infinity of space-time particle fields $\gamma_{ijnll'm}$. It is usual to assume that only a few of them have been recognized experimentally because they have a sufficient low mass.

By filling the pre-vacuum, its symmetry gets broken, in the same way as translation invariance of the Euclidean plane E^2 gets broken by inserting a particular point. We construct such a disturbance, a system of gravitational waves, so that

(3.5) $SO_4 \simeq SU_2 \times SU_2$

gets linked with the subgroup of spatial rotations

(3.6) $SO_3 \subset ISO(3,1).$

The pre-vacuum together with this disturbance will be identified with the physical vacuum. We will show that it has spin-structure. The linkage is constructed in such a way that a full rotation of space-time($2\pi \in SO_3$) corresponds to a half rotation of the S^3 ($-1 \in SU_2$). When we make a rotation by 2π, only in our blurred macroscopic 4-dimensional view have we returned to the previous position. In the more refined 8-dimensional view, all microscopic 3-spheres (S^3) attached to each point have made only half a rotation. Only by making a further turn, everything has settled.

We explain this terms of a simpler model.

Figure 4. Pre-vacuum. Independent translations and rotations.

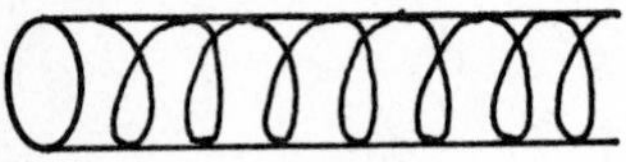

Figure 5. Vacuum. Isometry only if rotations and translations are linked.

Fig. 4 shows the pre-vacuum in the form of an exact cylinder. It has translation and rotation invariance ($\mathbb{R} \times U_1$). Fig. 5 shows a disturbance like grooves on a screw. An isometry is obtained only if the translation is linked in a suitable way with the rotations, so that we have the symmetry breaking

$$\mathbb{R} \times U_1 \longrightarrow \mathbb{R}\,. \tag{3.7}$$

The symmetry of physics is the symmetry of the vacuum. As physics progresses, the vacuum gets emptier and its symmetry group increases. In pre-Newtonian physics, the gravitational field of the earth was counted as part of the vacuum. The symmetry at that time was the group of Euclidean motions in the horizontal plane. As an example a pendulum clock does not function in the same way when in the vertical position or when tilted with respect to this direction. Later on it was recognized that the gravitational field in our laboratory is produced by the accidental presence of the earth and should not be counted as part of the vacuum. The vacuum has got emptier and its symmetry has increased. When we rotate the clock, we have to rotate every other system which might interact with the clock. In this example we must not forget to rotate the earth (about the centre of the chosen rotation), in order to get the required symmetry. To rotate an electron without rotating all the microscopic 3-spheres, which might interact with the electron, is like rotating the clock without rotating the earth.

4. EMBEDDING SPACE FORMULATION TO OBTAIN LINEARITY

Instead of the spherical coordinates θ_1, θ_2, ϕ of the S^3, we use an embedding space $\xi^6, \xi^7, \xi^8, \xi^9$ with Euclidean topology. The embedding space,

Fig. 6, is also a Riemannian space.

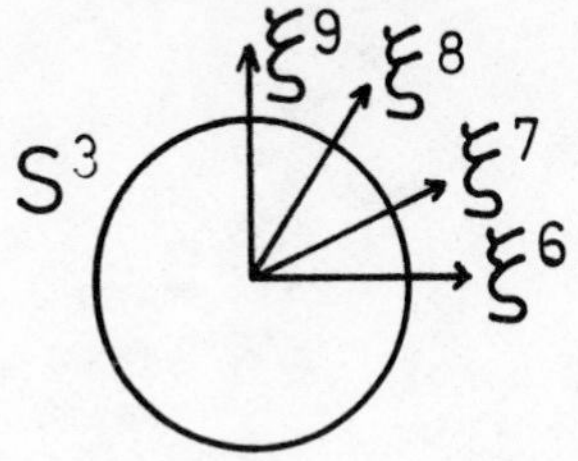

Figure 6. Embedding space for the microscopic S^3.

It is used only to introduce the topology of the S^3. The Riemannian metric has relevance only in so far as it induces a metric on the physical points

$$(4.1) \qquad (\xi^6)^2 + (\xi^7)^2 + (\xi^8)^2 + (\xi^9)^2 = 1.$$

Points not satisfying (4.1) do not correspond to an "event" of the world, which according to our theory is an 8-dimensional manifold. All physical point belonging to a particular microscopic $S^3 \times S^1$, given by (4.1), correspond to an event in the usual macroscopic 4-dimensional space-time

To make contact with the usual spinor-formulation, we consider the following complex column

$$(4.2) \qquad \phi = \begin{pmatrix} \phi_1 \\ \phi_2 \end{pmatrix} = \begin{pmatrix} \xi^6 + i\xi^7 \\ \xi^8 + i\xi^9 \end{pmatrix}$$

to denote a point of the embedding space. Physical points satisfy

$$(4.1') \qquad \phi^\dagger \phi = 1.$$

It is immaterial whether we use the real $\xi^6, \ldots \xi^9$ or the complex column ϕ to denote a physical point of the embedding space.

Rotations of the S^3 are obtained by

$$(4.3) \qquad \phi' = B\,\phi, \quad B \in SU_2 .$$

In this way we have found a "spinorial" formulation of rotations of the S^3. We do not obtain all rotations of the S^3, but only those corresponding to one of the SU_2-factors contained in SO_4 according to (3.5). For the emergence of a spin-structure it is essential that SO_4 contains an $SU_2 \in SO_4$, not just an SO_3, which is only locally isomorphic to SU_2. The $\simeq$ sign in (3.5) is necessary because -1 is a common element of both SU_2-factors, as shown in Fig. 7, which would not be the case in a direct product.

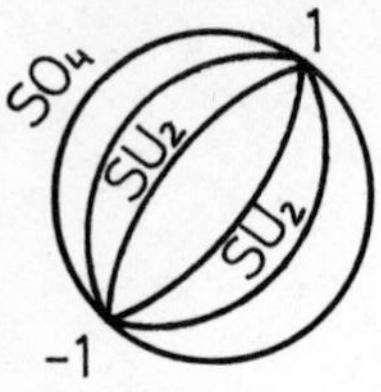

Figure 7. The group SO_4 with two invariant subgroups SU_2.

The general element $B \in SU_2$ can be written as

$$B = \cos(\theta/2) + in_a\sigma_a \sin(\theta/2), \tag{4.4}$$

with $n_a n_a=1$, σ_a = Pauli-matrices.

We are now in a position to give the linkage of spatial rotations with half rotations of the S^3, which is the most fundamental relation in this theory, and which was discussed in §3 in a qualitative way.

A spatial rotation can be given by θ, the angle of rotation, and by n_a, the axis of the rotation. The corresponding half rotation of the S^3 is then simply given by (4.4). Indeed by (4.3) (4.2), the transformation B of (4.4) can be interpreted as a rotation of the s^3. A complete rotation $\theta=2\pi$, gives half a rotation (B=-1) of the S^3.

We can express this fundamental relation also in terms of the real coordinates $\xi^6,\dots\xi^9$ of the embedding space of the S^3. It reads

$$\xi' = D\xi, \tag{4.5}$$

where D is given by the following matrix:

$$\begin{pmatrix} \cos(\theta/2) & -n_z\sin(\theta/2) & n_y\sin(\theta/2) & -n_x\sin(\theta/2) \\ n_z\sin(\theta/2) & \cos(\theta/2) & n_x\sin(\theta/2) & n_y\sin(\theta/2) \\ -n_y\sin(\theta/2) & -n_x\sin(\theta/2) & \cos(\theta/2) & n_z\sin(\theta/2) \\ n_x\sin(\theta/2) & -n_y\sin(\theta/2) & -n_z\sin(\theta/2) & \cos(\theta/2) \end{pmatrix}$$

Now we are able to identify the spinors in our theory. We denote the metrical tensor of the embedding space again by γ_{mn}, (m,n = 1, ...9). It can be shown, that the following column of complex linear combinations of components of the metrical tensor, which is constructed in analogy to (4.2), namely

$$\Gamma = \begin{pmatrix} \Gamma_1 \\ \Gamma_2 \end{pmatrix} = \begin{pmatrix} \gamma_{56}+i\gamma_{57} \\ \gamma_{58}+i\gamma_{59} \end{pmatrix} \tag{4.6}$$

has the transformation properties of a spinor under spatial rotations. This can be seen by the following arguments. Spatial rotations leave the components $\gamma_{56},\gamma_{57},\gamma_{58},\gamma_{59}$ of the metrical tensor unaffected. However, since a spatial rotation is linked with a rotation of the S^3, given by (4.5), the components 6,7,8,9 transform, the components 5 still remaining unchanged. More exactly, γ_{5m}, m=6,7,8,9 transforms as a vector under (4.5). The linear combinations Γ, given by (4.6), being analogous to (4.2), transform similarly to (4.3) (4.4), i.e. as a spinor under spatial rotations.

Whereas the "spinors" (4.2) are coordinate spinors only, denoting a point in the 8-dimensional world, the spinors Γ, given by (4.6), describe a physical field, namely the "gravitational" field, in the 8-dimensional world. However, it must be admitted that the spinor Γ also contains redundant (unphysical) information. Two metrical tensors γ_{mn} of the embedding space might induce the same metrics on the submanifold of physical points, given by (4.1), and thus are physically equivalent.

The theory was presented here in an intuitive and qualitative form. A detailed mathematical exposition will be given in reference [7].

References:

(1) H.C.Lee,(Ed.), *An Introduction to Kaluza-Klein Theories, World* Scientific (1984).

(2) C.van Westenholz, *Differential Forms in Mathematical Physics*, Amsterdam, North-Holland (1978).

(3) E.Witten, Nucl.Phys.B.195 (1982),481.

(4) K.Huang, *Quarks, Leptons & Gauge-Fields*,World Scientific (1982).

(5) D.Ebner, *Glahow-Salam-Weinberg Theory in Curved Space-Time*, Lecture Notes, University of Konstanz (1981).

(6) W.Drechsler, M.E. Mayer, *Fibre Bundle Techniques in Gauge Theories*, Lecture Notes in Physics 67 (Berlin, 1977).

(7) D.Ebner, Natural Origin of SO_3-spinors in an Eight-Dimensional Kaluza-Theory Exhibiting an Additional $U_1 \times SU_2$ Internal Symmetry', submitted to *Annals of Physics*.

MULTIVECTOR SOLUTION TO HARMONIC SYSTEMS

William M. Pezzaglia Jr.
Physics Dept., California State University
Sacramento, CA 95819
U.S.A.

ABSTRACT. The real Clifford Algebra $R_{3,1}$ or R(4) in the (-+++) metric allows the formulation of a new pedagogical solution to the classical harmonic oscillator. The formulation provides examples of physical quantities which can be usefully described by multivector aggregates, i.e. sums of unlike geometries. The equation of motion of the harmonic oscillator is "geometrized" by being factored into a first order multivector differential equation. This is analogous to factoring the (relativistic) Klein-Gordon equation into the linear Dirac equation by the use of Dirac matrices. The solution is a special "geometric phasor", which is analogous to the complex phasor of standard theory, however the timelike vector e_0, replaces the usual imaginary "i". The geometric phasor is reinterpreted as a vector in a particular two-dimensional subalgebra which has the properties of a geometric phase space. Clifford Algebraic velocity (timelike bivector) and position are found to be naturally perpendicular in the phase space.

A single multivector conservation equation containing all the conserved quantities of the system is derived directly from the multivector equation of motion. The formulation is a classical discrete mechanical analog of the relativistic quantum results put forward by Greider[3] and Ross[2]. A continuum generalization allows for a R(4) formulation Dirac theory. The R(4) algebra [of spacetime] used here is one dimension less than the C(4) [five dimensional] algebra used in standard Dirac theory.

1. INTRODUCTION/NOTATION

The philosophical theme that this work follows is that all physical quantities can be represented by sums of geometric elements. One of the most important aspects is the representation of spinors by multivector aggregates. These multivector hybrids are the sum of unlike geometries (e.g. vector plus scalar). In this paper, a natural multivector aggregate in the classical domain is derived. The Geometric Phasor is the solution to the geometrized harmonic oscillator, and provides the discrete analog to the continuous wave theories of

J. S. R. Chisholm and A. K. Common (eds.), Clifford Algebras and Their Applications in Mathematical Physics, 445–454.

Greider[3], Ross[2] & Teitler[1].

The four dimensional (16 element) Clifford Algebra has four generators e_μ (μ=0,1,2,3) which are interpreted to be the unit basis vectors in Minkowski space-time. The assumed metric is $g_{\mu\mu}$=(-+++), which has the representation of R(4), not requiring any external commutative "i" in the representation. The generators are thus isomorphic to real Dirac Γ matrices (i.e. in the Majorana representation, 4 by 4 matrices with real number entries only), with the same multiplication rule, $\{e_\mu, e_\nu\}=2g_{\mu\nu}$.

The scalar (1), pseudoscalar ($\varepsilon=e_0e_1e_2e_3$) along with the 4 vectors, 4 trivectors and 6 bivectors make up the 16 element (four-dimensional) Clifford Algebra [isomorphic to R(4), 4 by 4 real matrices]. A general linear combination of these 16 elements is called a multivector, which will usually be represented in this paper by a Greek or script (italic) Latin letter.

2. GEOMETRIZATION OF HARMONIC EQUATION OF MOTION

Most introductory books on field theory derive the concept of a field from a system of coupled harmonic oscillators. Hence the harmonic problem is of great pedagogical importance. Here the harmonic problem is reformulated[4] and extended by the use of Clifford Algebra. The reader may be surprised that there is more to be discovered about such a well known problem, yet the inclusion of geometry will offer new perspectives, solutions, insights and different interpretations. Further, the new ideas put forward in this discrete mechanical realm can be extended into continuum mechanics, although a complete discussion of the latter is beyond the scope of this paper.

Consider the second order differential equation of motion,

$$(\partial_t^{\,2} + \omega^2)X = -(e_0\partial_t + \omega)(e_0\partial_t - \omega)X = m^{-1}F \tag{1}$$

where $X=(X^1e_1 + X^2e_2 + X^3e_3)$ is the position vector and F is the external driving force vector. The differential operator was factored utilizing the time unit basis vector e_0 (where $e_0{}^2 = -1$) and the unique ability of Clifford Algebra to consistently add a scalar (ω) to a vector (e_0).

We define the geometric phasor, $\mathscr{X}$, which is seen to be a solution to a first order multivector differential equation of motion,

$$\mathscr{X} = (e_0\partial_t - \omega)X = e_0V - \omega X \ , \tag{2}$$
$$(e_0\partial_t + \omega)\mathscr{X} = - m^{-1}F \ . \tag{3}$$

The geometric phasor is a multivector aggregate, containing the physical observables of the system, the position and velocity of the "particle". The two quantities are kept separate by their distinctly different geometric interpretation, i.e. the position is associated with the vector portion, while the velocity is associated with the (timelike) bivector portion,

$$e_0 V = \vec{V} = V^1 e_0 e_1 + V^2 e_0 e_2 + V^3 e_0 e_3 \ . \tag{4}$$

The association of the velocity with a timelike bivector is geometrically desirable since velocity is the change in position with respect to time and should therefore have the geometric properties of both position and time. Also, it can be argued that both spacelike vectors and timelike bivectors of four-dimensional space will be observed as three-vectors with a three-dimensional perception[5].

In contrast, consider the analogous formulation using standard Gibbs vectors. The factorization now requires the external imaginary "i"; consequently the phasor $\mathcal{X}$ would be a complex Gibbs (three) vector,

$$-(\partial_t^{\ 2} + \omega^2) = (i\partial_t + \omega)(i\partial_t - \omega) \ , \tag{5}$$

$$\mathcal{X} = (i\partial_t - \omega)\vec{X} = i\vec{V} - \omega\vec{X} \ . \tag{6}$$

The real part of $\mathcal{X}$ is the position (multiplied by ω) while the imaginary part is the velocity. Although this standard phasor resembles the multivector one, they are distinctly different. The algebra of i (which commutes with all elements) is different from that of e_0 (which anticommutes with other vectors). For one dimensional motion, the position and velocity vectors are always collinear. Consequently, the homogeneous phasor (i.e. no driving force) for this system can be conveniently expressed in the polar form,

$$\mathcal{X} = (2E/m)^{\frac{1}{2}}\, e_1 \exp(\omega\, t\, e_0) \ . \tag{7}$$

General two and three dimensional solutions can be constructed by the superposition of one-dimensional solutions of orthogonal polarization (e.g. replace e_1 by e_2 and add a phase shift).

3. MULTIVECTOR CONSERVATION EQUATION

A single multivector conservation equation which contains all the conserved quantities of the harmonic system is derived directly from the multivector equation of motion. This is another illustration of useful general multivector aggregates in classical physics. It is also a discrete mechanical analog of the relativistic quantum results put forward by Greider[3] and Ross[2]. The method of interpretation follows theirs; the type of geometry defines the quantity. The advantage of this classical result (over their quantum result) is that here we have source term (a driving force). An interaction provides an unambiguous interpretation of conserved quantities.

The transposed multivector equation of motion is obtained by an automorphism which inverts all basis vector elements ($e_\mu \to -e_\mu$), and reverses the order of all elements. The equation is

$$-\mathcal{X}(e_0\partial_t - \omega) = -\, m^{-1}F \ , \tag{8}$$

where the derivative operates to the left. The conservation equation is obtained by multiplying equation (3) on the left by $-\mathcal{x}$, and adding to equation (8) multiplied on the right by $\mathcal{x}$,

$$\partial_t \mathcal{g} = -(-\mathcal{x}F + F\mathcal{x}) = [\mathcal{x},F] \ , \tag{9}$$

$$\mathcal{g} = -\mathcal{x} e_0 \mathcal{x} = e_0 \ \mathcal{x}^2 \ . \tag{10}$$

The scalar, trivector and pseudoscalar parts of multivector quantity $\mathcal{g}$ are identically zero. The vector part is proportional to the energy "E" of the system, while the bivector part is proportional to the angular momentum "M" of the system.

$$\mathcal{g} = 2m^{-1}\{E\ e_0 + \tfrac{1}{2}\omega[M^1 e_2 e_3 + M^2 e_3 e_1 + M^3 e_1 e_2]\} \tag{11}$$

$$E = (V^2 + \omega^2 X^2)m/2 \tag{12}$$

$$M = mX \wedge V \ . \tag{13}$$

In short, the multivector quantity $\mathcal{g}$ contains all the quantities of the harmonic system that would be conserved in the absence of any driving force. In the case of non-zero driving force[9], the right side of equation (9) gives,

$$-(-\mathcal{x}F + F\mathcal{x}) = [\mathcal{x},F] = 2(e_0\ V \cdot F - \omega\ X \wedge F) \ . \tag{14}$$

The bivector part is the torque on the system, demonstrating that the bivector part of quantity $\mathcal{g}$ is the angular momentum of the system (i.e. the torque is the time derivative of the angular momentum). Further, this agrees with Greider's result[7]. The timelike vector part is the rate at which work is being done on the system, demonstrating that the timelike vector part of quantity $\mathcal{g}$ is the energy (also consistent with Greider's result[6] for relativistic quantum theory, where in addition he associated the spacelike vectors with linear momentum). The absence of spacelike vectors in the conserved quantity $\mathcal{g}$ might be interpreted as either non-conservation of linear momentum for the harmonic oscillator, or perhaps interpreted as meaning the total momentum of the system is zero. At this time, no meaning can be ascribed to the absence of both scalar and pseudoscalar portions.

4. GEOMETRIC INTERPRETATIONS

Within the R(4) Clifford Algebra (of spacetime geometry) there are a number of geometric subspaces. In standard physics, subspaces are usually formed from subsets of basis vectors. The unique property of Clifford Algebra is that a geometric subspace can be made from, for example, a vector and a plane (if the elements anticommute). Here we explore one such subspace, which is a Clifford Algebra analog of "phase space". Connected with this idea is the general theme of the proper association of physical quantities with mathematical (geometric) elements.

Consider that the two elements e_1 and e_0e_1 anticommute, thus form a two dimensional geometric subalgebra, with the metric (+,+). Recalling that the "x" position of a particle is associated with the element e_1, and the "x" velocity (V_x) with the element e_0e_1, this two dimensional space can be interpreted as a plot of velocity versus position. Thus this subspace is the geometric analog of phase space. The geometric phasor [for one dimensional motion, i.e. equation (7)] can be reinterpreted as a vector in this two dimensional "phase space". For one-dimensional motion, the point in phase space described by the phasor traverses a circle as time progresses. The angular position of the point on the circle (i.e. the classical phase) has the geometry of the plane of the two-dimensional phase space, corresponding to the geometric element which is the product of the two phase space generators: $e_0e_1e_1 = e_0$.

Standard theory would have in general a six-dimensional phase space, formed from the mutually perpendicular axes of the three positions and their associated momenta (equivalently velocities). However, it is obvious one cannot get a six dimensional space out of the intrinsically four dimensional R(4) Clifford Algebra. Explicitly, the set of elements (e_1, e_0e_1, e_2, e_0e_2, e_3, e_0e_3) do not mutually anticommute, hence cannot be the basis for a six dimensional Clifford Algebra (e.g. e_1 and e_0e_2 commute), thus they cannot serve as a set of generators. This failure is related to the fact that to have a geometric space, physical significance must be attributed to rotations between the axes. For the two-dimensional phase space, a rotation between the axes corresponds to a change of phase. However, for a six-dimensional phase space, no physical significance can be attributed to a rotation between, for example, the X position and the Y velocity. This geometric problem is reflected by the algebraic property that e_1 and e_0e_2 commute.

Several times there has been more than one quantity associated with a certain geometric element. For example, the timelike basis vector e_0 was associated with the coordinate time ($x^0 = t$), as well as the conserved quantity energy ($p^0 = E$). Similarly, spacelike bivectors have been associated with angular position coordinates, as well as with the conserved quantity angular momentum. In each case there are two physical quantities associated with a given geometric element. One quantity corresponds to the "coordinate variable" of classical mechanics, while the other the "canonical conjugate variable". The latter quantity is conserved when the coordinate variable is cyclic. Table I provides a summary of the correspondence between physical quantities and geometric elements. It also classifies the quantities as either coordinate or conjugate. The table is incomplete, containing only those quantities that are unambiguous. Note that force has been associated with timelike bivectors. This may appear inconsistent with equation (1) which contains only vectors. This ambiguity is resolved if the geometry of e_0 is associated with inertial mass (i.e. relativistic mass, which is really energy). Thus while F is a timelike bivector, (F/m) is a vector (in four-space). One must be cautious in the non-relativistic domain to interpret whether a mass term is truly a rest mass (scalar) or an inertial mass (hence a timelike vector). Note also, the three-dimensional perception of physical quantities will be different. For example, force, velocity, position and momentum will all be experienced in a three-dimensional perception as (three) vectors, while angular position, angular momentum, torque and angular velocity will be perceived as pseudovectors. The relationship between four-dimensional geometries and the three-dimensional perception of these quantities is described in a previous work[5].

5. CONTINUUM MECHANICS

The classical field can be regarded as an ensemble of harmonic oscillators. Will the use of only geometric elements change the way in which classical (and quantum) fields are represented? Consider the classical spin 0^+ charged field, described by the Klein-Gordon equation,

$$(\Box^2 - m^2)\ \Phi(x) = 0\ , \tag{15}$$

where the function $\Phi(x) = \exp(ip\cdot x)$, is a (complex) scalar. Normally the solutions for a scalar charged field utilize the external imaginary "i" as the generator of quantum phase, considered to be a necessary feature for charged states. Since previously we found the unit time vector e_0 to be the generator of classical phase, we consider geometric solutions of the form,

$$\Phi(x) = \exp(-e_0\ p\cdot x)\ \Gamma\ . \tag{16}$$

The independent geometric factor Γ determines the "spin" of the field. Ignoring for the moment factors of e_0 (which can be absorbed as a phase factor) there are 8 independent spin states. They group into 4 multiplets: $\Gamma = 1$ is a spin 0^+ scalar, $\Gamma = e_i$ is a vector triplet hence spin 1^-, and the geometric duals ε, εe_i represent pseudoscalar 0^- and pseudovector 1^+ fields respectively. These states are found to be charged without the use of the external "i". Of course a complete discussion of these ideas is a subject in itself, hence will appear in later papers[10]. However, the basic ideas are summarized below.

We factor the Klein-Gordon equation into a linear form, just like we did the harmonic oscillator. We define the multivector field $\Psi = (\Box + m)\Phi$, which satisfies a first order multivector differential equation,

$$(\Box - m)\ \Psi(x) = (e_\mu \partial^\mu - m)\ \Psi(x) = 0\ , \tag{17}$$

$$\Psi(x) = (\Box + m)\ \Phi(x)\ . \tag{18}$$

Although equation (17) looks like the Dirac Equation, it doesn't necessarily describe a half integral spin particle. The spin depends upon the geometric structure of the potential $\Phi(x)$, i.e. factor Γ in equation (16) above. Assuming the spin 0^+ case (a three-dimensional scalar, $\Gamma = 1$) the field is found to be the sum of a scalar and time-like vector [both of which are perceived as scalars in a three-dimensional perception[5]]. The multivector eigenfunctions can be expressed (in normalized form),

$$\Phi_p(x) = (2\pi)^{-3/2} \, [2E(E+m)]^{-1/2} \, \exp(-e_0 \, p\cdot x) \, , \tag{19a}$$

$$\Psi_p(x) = (2\pi)^{-3/2} \, [2E(E+m)]^{-1/2} \, [\vec{P} + (E+m)] \, \exp(-e_0 \, p\cdot x) \, , \tag{19b}$$

$$\vec{P} = P^i \, e_0 e_i \, ; \qquad (i \neq 0) \tag{19c}$$

which have the important property that they form a Hilbert Space with the inner product being defined,

$$\langle \Psi_p \mid \Psi_k \rangle = \int d^3X \, \Psi_p^{\dagger}(x) \, \Psi_k(x) \, , \tag{20a}$$

$$e_i^{\dagger} = e_i \quad (i = 1,2,3) \, ; \qquad e_0^{\dagger} = -e_0 \quad (i=0) \, . \tag{20b}$$

In this we mean that the inner product of orthogonal states is zero, the states form a complete set, and the norm of an eigenfunction (for spin 0) is a scalar. The latter property is an important difference from the formulations of Greider[3] and Ross[2] whose eigenfunctions (using a commutative "i") have a multivector norm. Explicitly, for their spin zero charged particle, their wavefunction and norm are, respectively,

$$\Psi p(x) = (2\pi)^{-3/2} \, [m/E]^{1/2} \, [\tfrac{1}{2}(1 + ip/m)] \, \exp(i \, p\cdot x) \, , \tag{21a}$$

$$\int d^3X \, e_0 \, \Psi_p^{\dagger}(x) \, \Psi_k(x) = (-i + p/m) \, . \tag{21b}$$

This has the disadvantage of divorcing the theory from standard Bra-Ket notation, and hence the great body of standard formulation. The theoretical consequences of this departure are at this time unclear.

Another special property of the multivector wavefunction is that its time evolution satisfies a right-sided path integral formula,

$$\Psi(X',t') = \int d^3X \, \Psi(X,t) \, F(X',t';X,t) \, , \tag{22a}$$

$$F(X',t';X,t) = \sum_{\text{paths}} \exp(+e_0 \, S) \, , \tag{22b}$$

where S is the Classical action (of a free particle). Note the appearance again of e_0 as the generator (replacing the commutative "i" of standard theory). Modifying the action by including the (well-known) electromagnetic interaction generates the form of the quantum interaction. From this route, the interacting field equation is found to be

$$(\Box - m) \, \Psi(x) = - A \, \Psi(x) \, e_0 \, , \tag{23a}$$

$$\Psi(x) = (\Box + m) \, \Phi(x) - A \, \Phi(x) \, e_0 \, . \tag{23b}$$

Antiparticles are described by the dual geometry. For example, if $\Psi(x)$ describes a particle, then $\Psi\varepsilon$ describes an antiparticle with the opposite charge [easily verified by multiplying equation (23a) on the right by ε, moving ε through e_0 will give a sign change for the interaction].

Substituting equation (23b) into (23a), the equation for the potential function $\Phi(x)$ is

$$(\Box^2 - m^2 - A^2)\ \Phi(x) = -\ \{\Box, A\}\ \Phi(x)\ e_0\ , \tag{24a}$$

$$= 2A\cdot\Box\Phi(x)\ e_0 + (\Box\cdot A + \Box\wedge A)\ \Phi(x)\ e_0\ . \tag{24b}$$

For a non-zero magnetic field, it is clear that the potential $\Phi(x)$ must be a multivector, e.g. scalar (spin 0^+) plus spacelike bivector (spin 1^+). The mixture of spin 0 and 1 (of the same parity) is interpreted to be a spin half particle. Note the geometric structure is similar (nearly identical) to the multivector form of the spinor used by Greider[3] and Ross[4], although showing the equivalence is beyond the scope of this paper.

6. SUMMARY

One of the basic themes in the development of the above is in the spirit of Occam's razor, to describe the maximum amount of phenomena with the least amount of elements or abstraction. Standard theory (using the external "i") is formulated in the 5-dimensional Clifford Algebra C(4). The theory presented here is formulated in one lower dimension, R(4), a 4-dimensional Clifford Algebra. For the most part the imaginary "i" has been replaced (not arbitrarily) by the timelike vector e_0. The replacement is not trivial, for e_0 has a different algebra (it doesn't commute with other vectors while "i" does). In the case of the classical harmonic oscillator, the use of only R(4) provides more concrete interpretation for the familiar concepts of phase space and phasors. The geometric phasor also provides us with a simple example of a classical object that is best described by a multivector aggregate.

When these ideas are extended into continuum mechanics, it is found that electromagnetic charge and Dirac particles can be described by the use of only R(4) with the metric (-+++). Indeed, if the field is interacting, it necessarily **must** have mixed 0 and 1 spin (i.e. spin half). It is interesting to note that most authorities are of the opinion that an external (non-geometric) "i" is needed for (charged) quantum theory. In particular a recent work by Salingaros[8] argues that there is an "intrinsic connection between the spin and the necessity for having the complex form of the Clifford Algebra in Minkowski spacetime". His proof does not apply in this case because of the assumption that momentum eigenfunctions necessarily have idempotent structure. This is not the case, although a more complete exposition of the continuum theory will be needed to address this and related issues.

TABLE I PHYSICAL QUANTITIES CLASSIFIED AS COORDINATE OR CONJUGATE AND THEIR CORRESPONDENCE WITH GEOMETRIC ELEMENTS OF R(4)

GEOMETRIC ELEMENT			SYMBOL	VARIABLES	
NAME	#	TYPE	Γ	COORDINATE	CONJUGATE
VECTOR	1	TIMELIKE	e_0	Time[a]	Energy
	3	SPACELIKE	e_i	Position	Linear Momentum
BIVECTOR	3	TIMELIKE	e_0e_i	Force[c]	Velocity
	3	SPACELIKE	εe_0e_i	Angular Position	Angular Momentum
TRIVECTOR	3	TIMELIKE	εe_i	Torque[b]	Angular Velocity
	1	SPACELIKE	$e_1e_2e_3$	Volume	--------[d]

[a]Equivalently, the "phase"
[b]Torque, being the time derivative of angular momentum, would thus have the same number of degrees of freedom and transformation properties as a timelike trivector.
[c]Force, being the time derivative of momentum, would thus have the same number of degrees of freedom and transformation properties as a timelike bivector.
[d]Mass density could be represented by a spacelike trivector.

7. REFERENCES

1. S. Teitler, Nuovo Cim. Suppl. 3,1 (1965) and 3,15 (1965).

2. M. Ross, 'Geometric Algebra in Classical and Quantum Physics', Ph.D. Thesis, Univ. of Calif., Davis, 1980. Available from University Microfilms International, 300 N. Zeeb Road, Ann Arbor, Michigan 48106 (U.S.A.).

3. K. Greider, Found. of Phys. 14,467 (1984).

4. W. Pezzaglia, 'A Clifford Algebra Multivector Reformulation of Field Theory'. Ph.D. Thesis, Univ. of Calif., Davis, 1983. Available from University Microfilms International, 300 N. Zeeb Road, Ann Arbor, Michigan 48106 (U.S.A.).

5. W. Pezzaglia, op. cit., Chapter 11, Section G 'The Connection Between Four and Three Space'.

6. K. Greider, op. cit. equation (2b).

7. K. Greider, op. cit. equation (2c).

8. N. Salingaros, 'The Clifford Algebra of Differential Forms' in Acta Applicandae Mathematicae 4, 271 (1985)

9. M. Ross (private communication, December 1980).

10. W. Pezzaglia, 'MULTIVECTOR RELATIVISTIC QUANTUM MECHANICS: Part I Representation of Spin 0, 1 and ½' and 'Part II Charged Dirac Theory Without the Imaginary i, Spinors, or Other Abstract Quantities', to be submitted for publication later this year.

THE IMPORTANCE OF MEANINGFUL CONSERVATION EQUATIONS IN RELATIVISTIC QUANTUM MECHANICS FOR THE SOURCES OF CLASSICAL FIELDS

F.G. Morris† and K.R. Greider
University of California
Davis, CA 95616
U.S.A.

ABSTRACT. Relativistic quantum theory for a free particle, such as the Dirac theory, yields continuity equations for tensor bilinear densities that lead to conservation laws for the physical properties of that particle. It is shown here that the same continuity equations serve as the source equations for classical fields produced by the particle and described by the Clifford algebra C_4. The continuity equations must yield physically meaningful conservation laws in order to produce classical fields which also are physically meaningful and correct. Specifically, the charge-current conservation equation leads directly to the correct classical Maxwell electromagnetic equations if the sign of the charge is indefinite. Similarly, the energy-momentum conservation equation leads to a classical theory of gravity if the sign of the energy tensor is positive-definite. The equations obtained, although very different from those of general relativity, agree with all the current gravitation experiments.

1. INTRODUCTION

1.1. Classical Fields from Quantum Laws

The proposal that classical laws can be mathematically derived from quantum laws, i.e., that the quantum laws are primary and that the classical domain should be derivable from them, is not an unusual or particularly controversial point of view. The contribution made here, however, is that the classical fields for electromagnetism and gravity arise naturally from the quantum conservation laws for elementary Dirac particles. Specifically, we propose that classical electromagnetism and gravity fields have their sources in the continuity equations for the charge-current density and the energy-momentum tensor density, respectively, of each elementary charged particle. Further, the total classical field is obtained via a superposition of the fields from an assembly of a finite number of elementary particles.

†deceased

J. S. R. Chisholm and A. K. Common (eds.), Clifford Algebras and Their Applications in Mathematical Physics, 455–463.

For this procedure to be successful, consistent and meaningful the conservation laws for a single elementary particle must be physically meaningful and consistent; otherwise one obtains classical laws (by superposition) which would be unphysical. Specifically, we require for a single charged elementary particle that the charge-current density be indefinite (the charge may have either sign) and that the energy-momentum tensor describes a particle (or field) of positive-definite energy density.

The connections between quantum and classical physics presented here is a brief summary of a Ph.D. thesis by one of the authors[1] and earlier unpublished work[2]. A C_4 Clifford algebra description of the quantum conservation laws with a physically meaningful charge-current density and energy-momentum tensor density has been published elsewhere[3].

In what follows we denote the four generators of the abstract (non-matrix) algebra C_4 as $\underset{\sim}{e}_\mu$ (μ=0,1,2,3) with

$$\underset{\sim}{e}_\mu \underset{\sim}{e}_\sigma + \underset{\sim}{e}_\sigma \underset{\sim}{e}_\mu = 0 \quad (\mu \neq \sigma) \quad . \tag{1}$$

The $\underset{\sim}{e}_\mu$ are interpreted in the standard geometric way, as the unit basis vectors in space-time, with the metric (-+++). That is,

$$\underset{\sim}{e}_1^2 = \underset{\sim}{e}_2^2 = \underset{\sim}{e}_3^2 = -\underset{\sim}{e}_0^2 = +1 \tag{2}$$

where 1 is a real number and not a matrix.

1.2. Poincaré's Lemma and the Construction of Classical Fields

If one has a conservation equation for the density of a particular physical quantity (i.e., charge-current, energy-momentum, and so on)

$$\partial_\mu v^\mu(x) = 0 \quad , \tag{3}$$

one can write it as the inner product (divergence)[4] of a C_4 vector $\underset{\sim}{v} = v^\mu \underset{\sim}{e}_\mu$:

$$\underset{\sim}{\Box} \cdot \underset{\sim}{v} = \partial_\mu v^\mu = 0 \ . \tag{4}$$

The identity[4]

$$\underset{\sim}{\Box} \cdot (\underset{\sim}{\Box} \cdot \underset{\sim}{v}) \equiv 0 \tag{5}$$

implies that there exists a bivector function $\underset{\sim}{B}(x)$ such that

$$\underset{\sim}{\Box} \cdot \underset{\sim}{B}(x) = C\underset{\sim}{v}(x) \tag{6}$$

where C is a scalar constant. The bivector $\underset{\sim}{B}(x)$ may be calculated directly from a theorem, often called Poincaré's Lemma[1,5] :

$$\underset{\sim}{B}(x) = C \int_0^1 \underset{\sim}{x} \wedge \underset{\sim}{v}(sx) s^2 ds + \underset{\sim}{\Box} \cdot [\underset{\sim}{e}^5 \underset{\sim}{u}(x)] . \tag{7}$$

The proof of this theorem is given in Refs. 1 and 5. In Eq. (7), $\underset{\sim}{u}(x)$ is an arbitrary differentiable vector function to be specified later, and direct calculation shows that Eq. (6) follows when one takes the divergence of Eq. (7).

Eq. (7) gives a formal method of obtaining the classical field equations for $\underset{\sim}{B}(x)$ in terms of $\underset{\sim}{v}(x)$ as a consequence of the conservation equation for the physical density $\underset{\sim}{v}(x)$, instead of the normal procedure which reverses the causal order (i.e., obtains the conservation equation from the field equation).

2. CLASSICAL ELECTROMAGNETISM - THE MAXWELL THEORY

2.1. The Conserved Current

This section gives a particularly useful example since all the results are already well-known. The C_4 treatment here reproduces that of Hestenes[4] and obtains nothing new, except for the point of view (obtaining the electromagnetic field equations from the charge-current conservation equation). The conservation equation is

$$\partial_\mu j^\mu = 0 \tag{8}$$

where the physical requirements on j^μ are that it must be the components of a 4-vector, is indefinite (i.e., have either sign) and is time-like. Equation (8) can be written in C_4 as a divergence,

$$\underset{\sim}{\Box}\cdot\underset{\sim}{j}(x) = 0 \ . \tag{9}$$

The equation analogous to Eq. (6) is

$$\underset{\sim}{\Box}\cdot\underset{\sim}{F} = -\frac{4\pi}{c}\underset{\sim}{j} \tag{10}$$

where c is the velocity of light and $\underset{\sim}{F}$ is the electromagnetic field bivector

$$\underset{\sim}{F} = \underset{\sim}{\vec{E}} + \underset{\sim}{e}_5\underset{\sim}{\vec{B}} \ . \tag{11}$$

$\underset{\sim}{\vec{E}}$ and $\underset{\sim}{\vec{B}}$ are the electric and magnetic fields, respectively, and $\underset{\sim}{e}_5$ is the unit pseudoscalar. The notation here follows that of Hestenes[4]: $\underset{\sim}{\vec{E}}$ and $\underset{\sim}{\vec{B}}$ are time-like bivectors

$$\underset{\sim}{\vec{E}} = E_i\underset{\sim}{e}_i\underset{\sim}{e}_o \tag{12a}$$

$$\underset{\sim}{\vec{B}} = B_i\underset{\sim}{e}_i\underset{\sim}{e}_o \tag{12b}$$

where i=1,2,3.

2.2. Construction of the Field Bivector from Poincaré's Lemma

The general procedures outlined in the Introduction can be applied directly to the electromagnetic case. One replaces the general four-current density $\underset{\sim}{v}(x)$ of Eqs. (4) and (6) by the physical charge-current density $\underset{\sim}{j}(x)$ to obtain Eqs. (9) and (10). From Poincaré's Lemma, the bivector electromagnetic field is written in terms of the charge current density $\underset{\sim}{j}$:

$$\underset{\sim}{F}(x) = \frac{4\pi}{c} \int_0^1 \underset{\sim}{x} \wedge \underset{\sim}{j}(sx) s^2 ds + \underset{\sim}{\Box} \cdot [\underset{\sim}{e}^5 \underset{\sim}{u}(x)] \ . \tag{13}$$

The arbitrary vector function $\underset{\sim}{u}(x)$ can now be chosen[1,5] so that the homogenous Maxwell equations are satisfied

$$\underset{\sim}{\Box} \wedge \underset{\sim}{F} = 0 \ . \tag{14}$$

The choice $\underset{\sim}{\Box} \wedge \underset{\sim}{F}=0$ can be argued from physical considerations owing to the experimental absence of magnetic charges. (In the presence of both magnetic and electric charges for an elementary particle, one may still obtain $\underset{\sim}{\Box} \wedge \underset{\sim}{F} = 0$ by making use of a duality transformation on the field $\underset{\sim}{F}$ and on the charge-current sources[6]). Equations (10) and (14) yield

$$\underset{\sim}{\Box F} = - \frac{4\pi}{c} \underset{\sim}{j} \tag{15}$$

which is the familiar form of Maxwell's equations in the C_4 formalism[3,4,7]. (The more familiar four separate Maxwell equations are obtained by multiplying Eq. (15) on the left by $\underset{\sim}{e}_o$, as shown explicitly by Hestenes[4]). In the usual manner, one multiplies the reversion of Eq. (15) on the right by $\underset{\sim}{F}$, Eq. (15) on the left by $\underset{\sim}{F}$ and adds the two results to obtain

$$- \frac{1}{8\pi} \underset{\sim}{F \Box F} = - \frac{1}{8\pi} \partial_\mu (\underset{\sim}{F} \underset{\sim}{e}_\mu^{-1} \underset{\sim}{F}) = \frac{1}{c} \underset{\sim}{F} \cdot \underset{\sim}{j} \tag{16}$$

where the Lorentz force $\underset{\sim}{f}$ is identified with $\frac{1}{c} \underset{\sim}{F} \cdot \underset{\sim}{j}$:

$$\underset{\sim}{f} = \frac{1}{c} \underset{\sim}{F} \cdot \underset{\sim}{j} = \frac{d\underset{\sim}{p}}{d\tau} \ . \tag{17}$$

The more familiar time and 3-space form of Eq. (17) is obtained by multiplying on the left by $\underset{\sim}{e}_o$[4]. One finds the time component of the Lorentz force

$$f_o = -\frac{1}{c} \underset{\sim}{\vec{E}} \cdot \underset{\sim}{\vec{j}} \tag{18}$$

and the 3-vector space components

$$\underset{\sim}{\vec{f}} = \frac{1}{c} [j^o \underset{\sim}{\vec{E}} + \underset{\sim}{e}_5 \underset{\sim}{\vec{B}} \wedge \underset{\sim}{\vec{j}}] = \frac{1}{c} [j^o \underset{\sim}{\vec{E}} + \underset{\sim}{\vec{j}} \times \underset{\sim}{\vec{B}}] . \tag{19}$$

To complete the theory, Eq. (14) implies that $\mathbf{F}$ may be derived from a 4-vector potential $\mathbf{a}$. That is

$$\mathbf{F} = \boldsymbol{\Box}\wedge\mathbf{a} \tag{20}$$

because of the identity $\boldsymbol{\Box}\wedge(\boldsymbol{\Box}\wedge\mathbf{a}) \equiv 0$, similar to Eq. (5).

This completes our brief review of classical electromagnetism for which more complete details have been given earlier by Hestenes[4]. The main purpose here was to obtain the main results of Eq. (15) and also Eqs. (16) and (17) from the conservation equation, Eq. (9).

3. CLASSICAL GRAVITATION

3.1. The Conserved Current

In this section we will derive a Clifford algebra theory of gravitation that parallels as closely as possible the development of electromagnetism of the previous section. Instead of starting with the charge-current conservation equation, Eq. (9), we now begin with the conservation equation for the energy-momentum tensor. In tensor coefficient language one has

$$\partial_\mu T^{\mu\sigma} = 0 \tag{21}$$

where the physical conditions on $T^{\mu\sigma}$ are that it is symmetric and T^{oo} is positive-definite[3]. (For a free particle $T^{\mu\sigma}$ must be proportional to the product of the 4-momentum p^μ and the same 4-momentum p^σ, i.e., $T^{\mu\sigma} \propto p^\mu p^\sigma$). In the C_4 formalism, it is evident that Eq. (21) becomes

$$\partial_\mu T^{\mu\sigma}\mathbf{e}_\sigma = 0 \; . \tag{22}$$

To obtain a divergence equation in C_4 using the gradient operator $\boldsymbol{\Box}$, like Eq. (9) for the charge-current density, it is clear that the gradient vector $\boldsymbol{\Box}$ cannot be in the same C_4 algebra as $\mathbf{e}_\sigma$ since $T^{\mu\sigma}$ is symmetric. Thus we are forced to introduce a second C_4 algebra with generators $\{\underline{\underline{e}}_\mu\}$ that commute with the generators $\{\mathbf{e}_\sigma\}$. Equation (22) may be written in the product $C_4{\times}C_4$ double algebra

$$\underline{\underline{\Box}}\cdot\boldsymbol{\mathbb{T}} = \boldsymbol{\Box}\cdot\boldsymbol{\mathbb{T}} = 0 \tag{23}$$

where $\boldsymbol{\mathbb{T}}$ is a vector-vector entity in the double algebra (denoted by the double vertical lines);

$$\boldsymbol{\mathbb{T}} = T^{\mu\sigma}\,\underline{\underline{e}}_\mu\mathbf{e}_\sigma = T^{\mu\sigma}\,\mathbf{e}_\sigma\underline{\underline{e}}_\mu \; . \tag{24}$$

The $C_4{\times}C_4$ double algebra is not new to physics. It is the algebra used to describe two interacting particles in QED and was introduced earlier by Eddington[8] (his E-F numbers). The main difference here is that the product algebra $C_4{\times}C_4$ is required (due to the symmetric $T^{\mu\sigma}$) for the energy-momentum tensor of a single particle instead of two distinct particles. Considering Eq. (23) as the basic conservation equation [analogous to Eq. (9) for electromagnetism], one may construct a bivector-vector function $\mathbb{E}$ in the double algebra. It follows that

$$\square \cdot \mathbb{E} = -\kappa \mathbb{T} \tag{25}$$

3.2. Construction of the Field Bivector-Vector from Poincaré's Lemma

The procedures given in the Introduction and in the electromagnetic case can be applied also to the double algebra for gravitation. One replaces the four-current density $v(x)$ of Eqs. (4) and (5), or $j(x)$ of Eqs. (9) and (10) by the vector-vector gravitational source $\mathbb{T}(x)$, Eq. (24). From Poincaré's Lemma one obtains the bivector-vector gravitation field $\mathbb{E}$ in terms of $\mathbb{T}$

$$\mathbb{E}(x) = -\kappa \int_0^1 x \wedge \mathbb{T}(sx) s^2 ds + \square \cdot (e_5 \, \mathbb{u}) \tag{26}$$

analogous to Eq. (13) for electromagnetism. Again, following electromagnetic theory, we assume no gravitational "magnetic" sources (or eliminate them via a duality transformation) and postulate that the arbitrary function $\mathbb{u}(x)$ can be chosen such that

$$\square \wedge \mathbb{E} = 0 \ . \tag{27}$$

Equations (25) and (27) yield the gravitational equivalent of Maxwell's equations

$$\square \mathbb{E} = -\kappa \mathbb{T} . \tag{28}$$

Following what has become standard procedure we use the reversion of Eq. (28) to obtain

$$-\frac{1}{2\kappa}(\mathbb{E} \, \square \mathbb{E}) = -\frac{1}{2\kappa}\partial_\sigma(\mathbb{E} \; e^\sigma \; \mathbb{E}) = \frac{1}{2}(\mathbb{E}\mathbb{T} - \mathbb{T}\mathbb{E}) \tag{29}$$

which is the analogous to Eq. (16) for electromagnetism. The dimensions of this equation are (force/volume), so by analogy with electromagnetism (the Lorentz force), it is natural to propose that the right hand side of Eq. (29) is the force density of a field generated by the energy-momentum tensor $\mathbb{T}$.

Again, in analogy with electromagnetism, one obtains a vector-vector potential function $\mathbb{h}$ (the gravitational potential) from Eq. (27).

$$\mathbb{E} = \square \wedge \mathbb{h} . \tag{30}$$

Further details on this potential function and its relation to $\mathbb{T}$ may be found in Ref. 1.

The gravitation analogy to the electromagnetic Lorentz force density of Eq. (29) must be a 4-vector in one C_4 algebra and not in the double algebra. Therefore, it is necessary to contract the double $C_4 \times C_4$ algebra into a single C_4 algebra. This procedure may be carried out in a number of ways[1] but there is one physically consistent method that preserves Lorentz covariance of the equation in both vector algebras, and is independent of which of the two vector algebras one chooses to express the gravitational "Lorentz" force.

The contraction operation that is used was introduced by Temple[9] in 1930 and used subsequently by Dirac[10]. It utilizes the "interchange operator" $\mathbb{I}$ to contract the double $C_4 \times C_4$ algebra into one C_4 algebra in a symmetric, Lorentz covariant way. The interchange operator is the sum of the products of Clifford "double algebra" multivectors of the same rank and of the same space-time direction

$$\mathbb{I} = \frac{1}{4}[1 + \underset{\sim}{e}_0 \underset{\approx}{e}_0 + \underset{\sim}{e}_1 \underset{\approx}{e}_1 + \cdots + \underset{\sim}{e}_5 \underset{\sim}{e}_3 \underset{\approx}{e}_5 \underset{\approx}{e}_3 + \underset{\sim}{e}_5 \underset{\approx}{e}_5] \ . \tag{31}$$

When $\mathbb{I}$ brackets any multivector $\underset{\approx}{M}$ in one of the C_4 algebras, it changes it into $\underset{\sim}{M}$ in the other. Applying the symmetric contraction operation to the gravitational "Lorentz force" of Equation (29) one finds the gravitational force, equivalent to Equations (18) and (19) for electromagnetism.

The resulting gravitation force is considerably more complicated than its electromagnetic counterpart, since there are four times the number of terms in the gravitation "Lorentz force" as compared to the electromagnetic case. The reason for the more complicated gravitation force is not surprising, since the gravitation source $\underset{\sim}{\mathbb{T}}$ is more complicated than the electromagnetic source $\underset{\sim}{j}$.

We consider the special case of the force obtained for a large, massive, spherically symmetric source of mass M at rest, in order to compare our results with the well-known gravitation experiments. For this central force problem, the form of the source energy-momentum tensor $\underset{\sim}{\mathbb{T}}_s$ is

$$\underset{\sim}{\mathbb{T}}_s = Mc^2 \, \delta^3(\vec{\underset{\sim}{r}}) \, \underset{\sim}{e}_o \underset{\approx}{e}_o \tag{32}$$

where $\vec{\underset{\sim}{r}}$ is the radius vector from the center of the source. The force on a test particle of 4-momentum $\underset{\sim}{u} = u^o \underset{\sim}{e}_o + u^i \underset{\sim}{e}_i$ and rest mass m is[1,2]

$$f^o = - \frac{\alpha m}{r} (\vec{\underset{\sim}{u}} \cdot \vec{\underset{\sim}{r}}) \, u^o \tag{33}$$

for the time component. The 3-space force is

$$\vec{\underset{\sim}{f}} = - \frac{\alpha m}{r^3} [u_o^2 \vec{\underset{\sim}{r}} + \vec{\underset{\sim}{u}}^2 \vec{\underset{\sim}{r}} - (\vec{\underset{\sim}{u}} \cdot \vec{\underset{\sim}{r}}) \vec{\underset{\sim}{u}}] \tag{34}$$

where α is the usual gravitation parameter, $\alpha = GM/c^2$. In the low velocity limit $u_o \to ce_o$ and $\vec{u} \to 0$, so that one obtains the Newtonian limit

$$f^o = 0 \tag{35}$$

and

$$\vec{f} = -\frac{GMm}{r^3}\vec{r} \quad . \tag{36}$$

The comparison with the classic tests of gravitation can also be made. One can see immediately from Equation (33) the red-shift result for light, $\vec{u} = \vec{c}$, and the bending of light by a massive object from Equation (34) for which $|u_o| = |\vec{u}| = c$, and $\vec{u}\cdot\vec{r} = 0$. What is most surprising is that it turns out also that the perhelion shift gives the same result as Einstein's general relativity. The fact that these equations agree with all the gravitation experiments is already known and acknowledged in the literature, since Equations (33) and (34) are identical to a gravitation theory of G.D. Birkhoff[11] in 1943 and the results are discussed elsewhere[1,6]. The Birkhoff theory never gained much favor and has been dismissed due to several untenable physical assumptions he made[6]. It is a very different theory from that presented here. The general gravitation force equations and the beginning physical assumptions are not the same as ours, but by coincidence, Birkoff's theory reduces to our results of Equations (33) and (34) in the static limit.

4. DISCUSSION

The only novel feature of the electromagnetic theory of Section 2 is its starting point, namely we regard the conserved charge-current density as fundamental, and that the equations for the electromagnetic fields, the vector potential and the Lorentz force follow from the original continuity equation. The classical Maxwell theory can be obtained directly by superposition of the currents and fields of an assembly of such particles.

In the gravitation theory of Section 3, although agreement with all experimental tests of gravitational phenomena is the same as in the Einstein theory, it is obvious that unlike the Einstein theory, the present theory is incomplete. Ours is a linear theory in which no account is taken of the interaction of the energy-momentum of the field with the energy-momenta of the source and the "test" particles. Work is progressing at present to develop a complete theory that incorporates the non-linear effects in a natural way.

It is important also to point out that the linear theory described here is not the same as the linear approximations to the Einstein theory or the so-called "post-Newtonian approximations"[6] for which one does not obtain the correct value of the perhelion shift observed by experiment and predicted by the Einstein theory and the theory here.

Finally, the present theory is not a metric theory as in general relativity. We consider the gravitational force as a real force in flat Minkowski space-time, completely analogous to the Lorentz force for electromagnetism. However, it may very well be the case, and there is strong evidence for it, that a complete theory which includes the non-linear interactions with the gravitational field back on the source particles will have to abandon the concept of a global flat Minkowski space-time. At least this has been the experience of others who begin with a linear theory in Minkowski space-time[12].

However, the main advantage of our approach with gravitation as well as with electromagnetism is that it is a unifying approach to the description of both classical fields. First, it is unifying on a mathematical level due to the fact that only one formalism - the Clifford algebra - is required for the description of the fields, the currents that produce the fields and the forces on test particles in the fields. This approach is both mathematically powerful and physically intuitive. Second, it is unifying in that the fields and the forces for both gravitation and electromagnetism arise from the conservation laws for the energy-momentum and charge-current, respectively, of elementary particles, expressed also in the same Clifford algebra formalism. In this way one obtains an important and novel connection between the relativistic quantum theory of elementary particles and the classical fields they create.

5. REFERENCES

1. F. Morris, Classical Fields Derived from Quantum Sources, Ph.D. thesis, University of California, Davis (1983).
2. K. Greider, Lecture Notes on Clifford Algebra, University of California, Davis (1980), unpublished.
3. K. Greider, Found. Phys. **14**, 467 (1984) and Phys. Rev. Lett. **44**, 178 (1980).
4. D. Hestenes, Space-Time Algebra (Gordon and Breach, New York, 1966).
5. H. Flanders, Differential Forms with Applications in Physics (Academic Press, New York, 1963).
6. C. Misner, K. Thorne and J. Wheeler, Gravitation (W.H. Freeman, San Francisco, 1972).
7. M. Riesz, Clifford Numbers and Spinors, Lecture Series No. 38 (University of Maryland, College Park, 1958), Chaps. I-IV.
8. A.S. Eddington, Relativity Theory of Protons and Electrons (Macmillan, New York, 1936).
9. G. Temple, Proc. Roy. Soc. **A127**,342 (1930).
10. P.A.M. Dirac, Principles of Quantum Mechanics, 3rd edition (Oxford University Press, 1947) p. 221.
11. G.D. Birkhoff, Proc. Nat. Acad. Sci. **29**, 231 (1943).
12. R.P. Feynman, Lectures on Gravitation, (1963) unpublished.

ELECTROMAGNETIC THEORY AND NETWORK THEORY USING CLIFFORD ALGEBRA

E. Folke Bolinder
Chalmers University of Technology
Division of Network Theory
S-412 96 Gothenburg
Sweden

ABSTRACT. Clifford algebra was introduced in electromagnetic theory around 1930 and in network theory in 1959. While it is necessary to use Clifford algebra in quantum mechanics in splitting up the Schrödinger equation to get the Dirac equation, it is not necessary to use Clifford algebra in electromagnetic theory or network theory. However, as Clifford algebra is the natural tool to use in connection with, for example, the Minkowski model of Lorentz space, it is quite useful in studying problems dealing with partially polarized waves (the Stokes vector), active, lossy, and noisy networks, etc. Some applications of Clifford algebra in electromagnetic theory and network theory will be discussed.

1. INTRODUCTION

In 1901 Ricci and Levi-Civita published the first systematic description of tensor algebra and some of its applications [1]. Tensors have been used extensively in electromagnetic theory [2,3] and the pioneer work of G. Kron in network theory is well-known [4]. In the special case of skew-symmetric tensors Clifford algebra is a sufficient and effective tool.

2. CLIFFORD ALGEBRA

Clifford algebra was introduced by W.K. Clifford in 1876 and first published by him in 1878 [5]. Such an algebra was also suggested by R. Lipschitz in 1880 [6]. See also Reference 7.

Clifford algebra is an associative and non-commutative algebra over the real or complex field. It can be built on the simple postulate that the square of a vector is equal to its quadratic form:

$$x^2 = x \cdot x \quad , \qquad x = \sum_{i=1}^{n} x_i e_i \quad , \tag{1}$$

where $x \cdot y$ is the scalar product. With

J. S. R. Chisholm and A. K. Common (eds.), Clifford Algebras and Their Applications in Mathematical Physics, 465–483.

$$x \cdot y = \sum_{i,k}^{n} g_{ik} x_i y_k \quad , \quad g_{ik} = g_{ik} = e_i \cdot e_k \tag{2}$$

the Clifford algebra can also be defined by the postulate

$$e_i e_k + e_k e_i = 2g_{ik} \quad . \tag{3}$$

Thus, for Euclidean 3-space:

$$e_i e_k + e_k e_i = 2\delta_{ik} = \begin{cases} 1 & i=k \\ 0 & i \neq k \end{cases} \quad ; \quad x \cdot x = x_1^2 + x_2^2 + x_3^2 \tag{4}$$

and for Lorentz 4-space: $e_o^2 = 1$, $e_i^2 = -1$, $i = 1,2,3$

$$e_i e_k = -e_k e_i \quad , \quad i \neq k \quad ; \quad x \cdot x = x_0^2 - x_1^2 - x_2^2 - x_3^2 \quad . \tag{5}$$

We can write $(x+y) \cdot (x+y) = (x+y)^2$ so that

$$xy + yx = 2x \cdot y \quad . \tag{6}$$

Also,

$$xy = \frac{1}{2}(xy + yx) + \frac{1}{2}(xy - yx) = x \cdot y + x \wedge y \quad , \tag{7}$$

where $x \wedge y$ is the outer product [8,9]. Two vectors are orthogonal if $x \cdot y = 0$. Then $xy = -yx$, so that x and y anticommute. A simple bivector, spanned by the vectors x and y, simplifies with an orthogonal basis to:

$$F = x \wedge y = \sum_{i<k} (x_i y_k - x_k y_i) e_i e_k \quad . \tag{8}$$

We obtain an algebra C generated by the basis elements e_1, e_2, ..., e_n of the vector space E. The algebra contains all products of the basis elements and also all linear combinations of such products. It is called a Clifford algebra C with the metric g_{ik}, Eq(3). The elements of C are called Clifford numbers. See M. Riesz [8] and D. Hestenes [9].

3. SPECIAL FORMS OF CLIFFORD ALGEBRA

For simplicity the following notations will be used: CE_n^e means "the even subalgebra of an n-dimensional Clifford algebra with Euclidean metric", CO_n means "an n-dimensional Clifford algebra without metric" (all $g_{ik}=0$), and "L" in CL_n means "Lorentz metric".

3.1. The system of complex numbers, CE_2^e

A geometric interpretation of the complex numbers seems to have been made for the first time by C. Wessel in 1797. This is an even subalgebra of C for n = 2, and it is spanned by the elements 1 and e_1e_2.

3.2. Grassmann's exterior algebra, CO_n

The word "exterior" means that the algebra is not restricted to three dimensions, but is valid for any number of dimensions. Grassmann presented his "Ausdehnungslehre" in two books in 1844 and 1862. In 1853 he wrote a paper on different multiplications [10], and in 1877, just before he passed away, he commented on a paper on quaternions [11] and wanted to show how quaternions fitted into his exterior algebra [12]. According to the remarks by E. Study [13] the attempt was not too successful.

3.3. Hamilton's quaternion algebra over the real field, CE_3^e

Hamilton was struck by the idea on quaternions in 1843 and published two books on quaternion algebra and its applications in 1853 and 1866 (the latter one posthumously). It is an even subalgebra of C for n = 3, and it is spanned by the elements 1, $i = e_3e_2$, $j = e_1e_3$, and $k = e_2e_1$.

3.4. Hamilton's quaternion algebra over the complex field (biquaternions), CE_3.

Hamilton's biquaternions are quaternions with complex coefficients. They have to be distinguished from Clifford's biquaternions which are composed of the sum of a quaternion and another quaternion premultiplied by the pseudoscalar. Hamilton's algebra is spanned by the elements 1, $i = e_3e_2$, $j = e_1e_3$, $k = e_2e_1$, $l = e_1e_2e_3$, $li = e_1$, $lj = e_2$, and $lk = e_3$.

3.5. Vector algebra

In 1881 - 1884 J.W. Gibbs and O. Heaviside independently created a simple vector algebra. Gibbs was influenced by Grassmann's exterior algebra, and Heaviside by Hamilton's quaternion algebra. The cross product of the vector algebra is related to the outer product by means of the formula

$$a \times b = -i(a \wedge b) = -a \cdot (ib) = (ib) \cdot a \qquad (9)$$

where $i = e_1e_2e_3$ is the pseudoscalar. See Hestenes [9].

3.6. Spinor algebra

This algebra seems to have originated with E. Cartan in 1913. The word "spinor" was later introduced when it was found that the algebra was useful in quantum mechanics. Cartan published two volumes on spinors in 1938 [14] and R. Brauer and H. Weyl wrote a fundamental paper on n-dimensional spinors in 1935 [15].

3.7. Space-time algebra, CL_{3+1}

With the arrival of relativity theory many papers appeared in which relativity theory was used in the electromagnetic theory. The basic relativity papers from 1895 to 1920 by Lorentz, Einstein, Minkowski, (Sommerfeld), and Weyl have been published by Dover Publications, Inc. [16]. The space-time Clifford algebra found an important application in splitting up the Schrödinger equation to get the Dirac equation for the electron [17, 18].

3.8. Space algebra, CE_3

D. Hestenes [9] considers this algebra to be the even subalgebra of the space-time algebra mentioned above. The six bivectors in the space-time algebra are converted into three vectors and three bivectors in the space algebra by selecting a specific time-like vector. The space algebra can be used in representing the Pauli matrix algebra [17].

3.9. The algebra of differential forms

The Grassmann exterior algebra, which has no inner product, was used by E. Cartan in building his calculus with exterior differential forms and "moving frames". He wanted coordinate-free formulas [19]. So did D. van Danzig [20]. A basic paper was published by E. Kähler [21], who used the algebra of differential forms in studying Maxwell's equations. Two interesting publications on exterior differential forms and electromagnetic theory were presented by G. Deschamps [22, 23]. D. Hestenes [9] and D. Hestenes and G. Sobczyk [24] claim that the use of Clifford algebra, which integrates the inner and outer products into a single operation, is superior to the use of exterior differential forms. The calculation of an example by using the two methods would be very interesting.

Note: Matrix algebra was introduced by A. Cayley in 1858.

4. APPLICATIONS OF CLIFFORD ALGEBRA IN ELECTROMAGNETIC THEORY

4.1. Differential operators

We introduce the differential operator

$$\square = \sum_{i=1}^{n} e_i \frac{\partial}{\partial x_i} = \sum_{i=1}^{n} e_i \partial_i \quad . \tag{10}$$

Thus, the gradient operator is

$$\square u = \sum_{i=1}^{n} e_i \partial_i u \quad . \tag{11}$$

4.2. Maxwell's equations

We need eight equations to express Maxwell's equations in component form, four in vector form, two in tensor or exterior differential forms and one in complex quaternion form [25, 26], complex 4D form [27], or Clifford algebra form. With Clifford algebra we get the following simple expression for the Maxwell's equations:

$$\square F = s \tag{12}$$

where $\square$ is the differential operator in Eq(10), F is the general field bivector, and s is the four-current vector. Inspired by a paper by A. Proca on the use of Clifford algebra in splitting the Schrödinger equation [28], G. Juvet, A. Schidlof, and A. Mercier made a basic study on the use of Clifford algebra in electromagnetic theory [29-31]. Later on M. Riesz [8], D. Hestenes [9], and P. Lounesto [32] studied Maxwell's theory by means of Clifford algebra.

4.3. The Maxwell-Minkowski stress-energy tensor [33]

By using the Einstein summation convention the Maxwell-Minkowski stress-energy tensor can be written:

$$x'_i = S_i^{\ k} x_k = \left[F_{ir} F^{kr} - \frac{1}{4} F_{rs} F^{rs} \delta_i^{\ k} \right] x_k \quad . \tag{13}$$

See, for example, Reference 34.

In Clifford algebra Eq(13) can be written [30]:

$$x' = Sx = -\frac{1}{2} FxF \quad . \tag{14}$$

It can be shown that Eq(14) is a Lorentz transformation multiplied by a scalar factor:

$$x'^2 = \frac{1}{4} FxF^2xF = \frac{1}{4} Fx(\alpha+\beta Q)xF = \frac{1}{4} F(\alpha-\beta Q)x^2F =$$

$$= \frac{1}{4} F^2(\alpha-\beta Q)x^2 = \frac{1}{4}(\alpha^2+\beta^2)x^2 = \lambda^2 x^2 \quad , \tag{15}$$

where

$$\lambda = \frac{1}{2}\sqrt{\alpha^2+\beta^2} \tag{16}$$

and $F^2 = \alpha + \beta Q$ is a quasiscalar with α and β real numbers, and the unit pseudoscalar $Q = e_0e_1e_2e_3$, so that $Q^2 = -1$. Note: $Qx = -xQ$, $QF = FQ$. We split F into two totally orthogonal parts, a time bivector F_T (area A) and a space bivector F_S (area B), $x = x_T + x_S$.

Then $$F^2 = (F_T + F_S)^2 = \alpha + \beta Q$$

so that $$\alpha = A^2 - B^2 \quad , \quad \beta = \pm\, 2AB.$$

Thus $$\lambda = \frac{1}{2}(A^2 + B^2) \quad , \tag{17}$$

and Eq(14) is a Lorentz transformation multiplied by a scalar factor.

What is $x' = Sx$ geometrically?

$$x' = -\frac{1}{2} FxF = -\frac{1}{2}(F_T + F_S)(x_T + x_S)(F_T + F_S)$$

$$= \frac{1}{2}(F_T^2 - F_S^2)(x_T - x_S), \text{ because } x_SF_S = -F_Sx_S \ , \ x_TF_T = -F_Tx_T \ ,$$

$$x_TF_S = F_Sx_T \ , \ \text{etc.}$$

Thus, $$x' = \frac{1}{2}(F_T^2 - F_S^2)(x_T - x_S) = \frac{1}{2}(A^2 + B^2)(x_T - x_S) = \lambda(x_T - x_S) \ .$$

We find that $x' = Sx$ geometrically means a reflection in F_T followed by a multiplication by λ. λ is given by Eq(17) where A is the area of F_T and B the area of F_S.

A ray (pencil) of bivectors is defined by the expression

$$\omega F = F\omega \ ,$$

where ω is a quasiscalar. By putting $\omega = \alpha + \beta Q$ and $\bar{\omega} = \alpha - \beta Q$ we get

$$\omega FxF\omega = \omega\bar{\omega}FxF = (\alpha^2 + \beta^2)FxF = \text{const}(FxF) \ .$$

Riesz also treated the lightlike case. He probably intended to publish this material on the Maxwell-Minkowski stress tensor in Chapter 5 of his Maryland Notes [8]. On February 8, 1959, Riesz told the author that Chapter 5 would also contain a study of the four-dimensional Schilling figure and a proof of the Pascal theorem in space-time. A first outline was dictated to the author [33].

Before leaving the electromagnetic theory two additional papers should be mentioned [35, 36]. The first, by Lewis in 1910, describes three- and four-dimensional ($x_4 = ict$) vector analyses in Grassmann's spirit with applications to electrodynamic theory, and the second paper, by Wilson and Lewis in 1912, is dealing with the non-Euclidean geometry in two-, three-, and four-dimensional Lorentz space, with applications to mechanics and electromagnetics.

5. APPLICATIONS OF CLIFFORD ALGEBRA IN NETWORK THEORY

5.1. Some unsolved network problems in the 1940's.

In the 1940's some of the prominent unsolved problems in network theory were the following:

1. Optimum design of tapered lines
2. Synthesis of networks yielding specified transients
3. Explicit expressions for elliptic filters
4. Synthesis of networks without hidden transformers
5. Synthesis of hyperelliptic filters.

The first problem was solved approximately by the author in 1950 [37] by using Fourier transforms. By using the Minkowski model of Lorentz space and Clifford algebra a more accurate solution may be obtained. The second problem was solved by Manuel V. Cerrillo at the Research Laboratory of Electronics, MIT, in 1951-52 [38] by approximating the real part of the transfer function. This approximation made it possible to obtain reliable approximations of both the time and the frequency functions in the Laplace transform pair. The extended problem of studying transient propagation in three dimensions could probably lead to the use of Clifford algebra (4D radar).

The third problem has recently been solved by the author by using the Minkowski model of 3D Lorentz space [39]. See Section 5.3.

The fourth and fifth problems have not been solved yet.

5.2. Network synthesis

In network analysis we apply a known signal to a known network and look at the output signal. If it is not what we want, we change the network by cut-and-try until we get what we want. In network synthesis we specify the input and output signals and want to find first the transfer function and then, after approximating it by a rational function, we want to find the network elements. The synthesis method used today is the insertion loss method, originally created in 1937-1942 by Norton and Darlington in the United States, Cauer, Piloty, and Bader in Germany, and Cocci in Italy. The insertion loss power ratio P_L is defined as the ratio of the power absorbed in the load resistance R_L when it is directly connected to the generator resistance R_1, to the power absorbed in R_L after a lossless filter has been inserted between R_L and R_1:

$$P_L = \frac{1}{1-|\Gamma|^2} \quad ; \quad |\Gamma|^2 = \frac{P_L - 1}{P_L} \quad , \tag{18a,b}$$

where Γ is the reflection coefficient at the input of the filter.

In the insertion loss theory from 1937-42, Γ was found from P_L in Eq(18b) by a splitting up procedure. Then the element values of the lossless network were found by realizing the elements successively by inverse element transformations in the complex reflection coefficient plane, the Smith Chart, $\Gamma = \Gamma_r + j\Gamma_i$, or the complex impedance plane, $Z = R + jX$. See, for example, Weinberg [40].

5.3. Outline of a new insertion loss method based on the use of the Minkowski model of Lorentz space and Clifford algebra

In studying, for example, lossy and noisy two-port networks it is better to use power quantities. If we "couple" the insertion power ratio P_L to the timelike coordinate in a 3D Lorentz space by means of the following expression

$$P_3 = 2P_L - 1 \tag{19}$$

we can use the invariant quadratic form

$$P_3^2 - P_2^2 - P_1^2 = 1 \tag{20}$$

to derive the formula

$$P_1^2 + P_2^2 = 4P_L(P_L - 1) \ . \tag{21}$$

By using analytic continuation and zeros of $P_L = 0$ and $P_L - 1 = 0$ which guarantee a realizable network we can find $P_1(\omega) + jP_2(\omega)$. From this expression and Eq(19) we find the coordinates of a point $P = (P_1, P_2, P_3)$ on the Minkowski model of 3D Lorentz space (ML - 3). Figure 1 shows how this point can be projected down in the Smith Chart by a stereographic projection with the projection center at the top of the lower hyperboloid. The connections between the ML-3 and the Poincaré (P-2) and Cayley-Klein (CK-2) models of two-dimensional hyperbolic space, first shown by M. Riesz [41], are also shown in the figure. Thus the Smith Chart can be interpreted to be a P-2 model. The following simple expressions are obtained for the transformations between ML-3, "Spaceland" and the Smith Chart, "Flatland", and vice versa:

Flatland ⇒ Spaceland

$$\begin{cases} P_1 = 2\Gamma_r P_L = \dfrac{2\Gamma_r}{1-|\Gamma|^2} & \text{(22a)} \\[2ex] P_2 = 2\Gamma_i P_L = \dfrac{2\Gamma_i}{1-|\Gamma|^2} & \text{(22b)} \\[2ex] P_3 = 2P_L - 1 = \dfrac{1+|\Gamma|^2}{1-|\Gamma|^2} & \text{(22c)} \end{cases}$$

Spaceland ⇒ Flatland

$$\begin{cases} \Gamma_r = \dfrac{P_1}{P_3+1} & \text{(23a)} \\[2ex] \Gamma_i = \dfrac{P_2}{P_3+1} & \text{(23b)} \end{cases}$$

In the 3D Lorentz space, transformations through the different network elements correspond to rotations. For example, an ideal transformer is

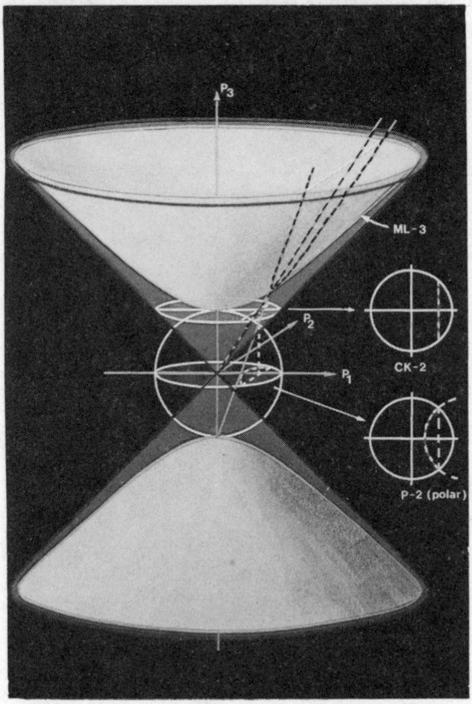

Fig.1. The Minkowski model of three-dimensional Lorentz space with its submodels.

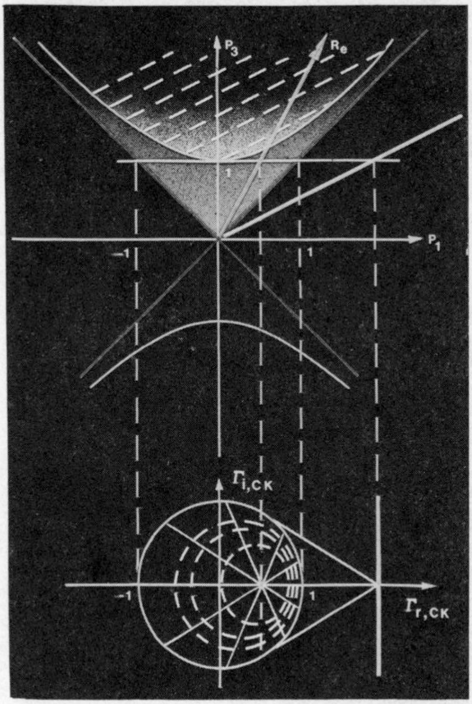

Fig.2. The hyperbolic rotation.

a hyperbolic rotation, a piece of a transmission line is an elliptic rotation, and series inductors and shunt capacitors are parabolic rotations. Thus, from the point P on ML-3 we can split off a network element by performing an inverse rotation. The rotation is performed until the coefficients of the highest ω-powers are zero. Then a new point with simpler P_1, P_2, and P_3 expressions is obtained. The arc-length passed will give the value of the network element. After all network elements have been taken out, the end point will be the "matching point" (0,0,1) = (0,0) in the Smith Chart, corresponding to the unit load resistance. Matrix algebra can be used in performing the rotations, but Clifford algebra is a more convenient tool, because we can then write all operations on a single line and no clumsy matrices are needed. Also, programs can be written in Clifford algebra. See Mikkola and Lounesto [42].

5.4. Rotations in the three-dimensional Lorentz space

5.4.1. General theory. In his Notes [8,p160] M. Riesz states the following basic theorem: "Let E be a real vector space of Euclidean or

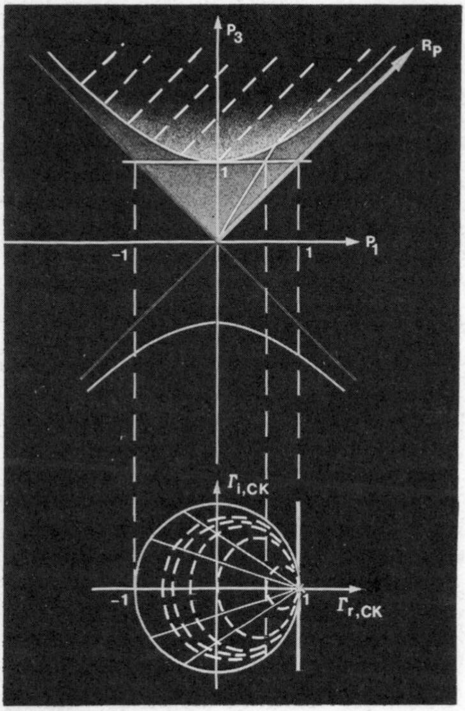

Fig.3. The elliptic rotation.

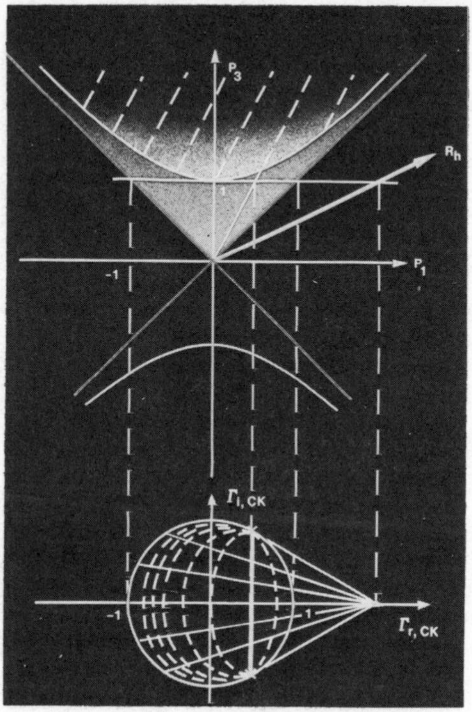

Fig.4. The parabolic rotation.

Lorentz signature and C the corresponding Clifford algebra. To every isometry L of E, connected with the identity, there exists a bivector F in C such that

$$x' = Lx = e^{-\frac{F}{2}} x e^{\frac{F}{2}} \tag{24}$$

for every vector x. This property does not belong to any other signature." The right hand side of Eq(24) must be interpreted in the Clifford sense. The signature indicates the signs in the invariant quadratic form.

5.4.2. <u>The hyperbolic rotation in ML-3</u>. The hyperbolic rotation is performed around a spacelike vector R_h (outside the lightcone) in timelike parallel planes, all Lorentz orthogonal to R_h. The planes cut the hyperboloid in hyperbolas; this is the reason for the name of the rotation. See Fig.2. From Eq(24) we get:

$$P' = e^{-\Psi I_h} P e^{\Psi I_h} , \tag{25}$$

where

$$I_h^2 = 1 \quad , \quad I_h = e_3 e_1 \quad , \quad e^{\Psi I_h} = \cosh\Psi + \sinh\Psi \cdot I_h \quad . \tag{26}$$

I_h commutes with e_2 and anticommutes with e_3 and e_1. By expanding Eq(25) we obtain

$$P' = (P_1 \cosh 2\Psi + P_3 \sinh 2\Psi)e_1 + P_2 e_2 + \tag{27}$$

$$+ (P_1 \sinh 2\Psi + P_3 \cosh 2\Psi)e_3 \quad .$$

We now introduce the two lightlike vectors ℓ_1 and ℓ_2:

$$\ell_1 = e_3 + e_1 \quad ; \quad \ell_2 = e_3 - e_1 \tag{28}$$

and obtain

$$P' = (P_3 + P_1)e^{2\Psi} \frac{\ell_1}{2} + P_2 e_2 + (P_3 - P_1)e^{-2\Psi} \frac{\ell_2}{2} \quad . \tag{29}$$

Thus, the hyperbolic rotation with the hyperbolic angle 2Ψ corresponds to a multiplication of ℓ_1 by $e^{2\Psi}$ and ℓ_2 by $e^{-2\Psi}$.

5.4.3. The elliptic rotation. The elliptic rotation is performed around a timelike vector R_e (inside the lightcone) in spacelike parallel planes all Lorentz orthogonal to R_e. The planes cut the hyperboloid in ellipses; this is the reason for the name of the rotation. See Fig.3. From Eq(24) we get

$$P' = e^{-\theta I_e} P e^{\theta I_e} \tag{30}$$

where $$I_e^2 = -1 \quad , \quad I_e = e_1 e_2 \quad , \quad e^{\theta I_e} = \cos\theta + \sin\theta I_e \quad . \tag{31}$$

I_e commutes with e_3 and anticommutes with e_1 and e_2. By expanding Eq(30), using Eq(31), we get

$$P' = (P_1 \cos 2\theta + P_2 \sin 2\theta)e_1 + (-P_1 \sin 2\theta + P_2 \cos 2\theta)e_2 + P_3 e_3 \quad . \tag{32}$$

5.4.4. The parabolic rotation. The parabolic rotation is performed around a lightlike vector R_p (on the lightcone) in lightlike parallel planes all Lorentz orthogonal to R_p. The planes cut the hyperboloid in parabolas; this is the reason for the name of the rotation. See Fig.4. From Eq(24) we get:

$$P' = e^{-\frac{1}{2} w I_P} P e^{\frac{1}{2} w I_P} \quad , \tag{33}$$

where

$$I_p^2 = 0 \quad , \quad I_p = (e_3+e_1)e_2 \quad , \quad e^{\frac{1}{2}wI_p} = 1 + \frac{1}{2}wI_p \quad . \tag{34}$$

By expanding Eq(33), using Eq(34), we get:

$$P' = (1-\frac{1}{2}wI_p)\,P\,(1+\frac{1}{2}wI_p) = P + \frac{w}{2}(PI_p - I_pP) - \frac{w^2}{4}I_pPI_p \quad . \tag{35}$$

Here, $\frac{1}{2}(PI_p - I_pP) = TP$ is a skew-symmetric transformation, defined by $P\cdot(TP) = 0$.
Also,

$$-\frac{1}{2}I_pPI_p = \frac{1}{4}[(PI_p-I_pP)I_p - I_p(PI_p-I_pP)] = T^2P \quad . \tag{36}$$

Thus,

$$P' = (1+wT+\frac{1}{2}w^2T^2)P \quad . \tag{37}$$

In the same way as in the case of the hyperbolic rotation we utilize lightlike planes by putting $\ell_1 = e_3 + e_1$ and $\ell_2 = e_3 - e_1$. The P-vector

$$P = P_1e_1 + P_2e_2 + P_3e_3 \tag{38}$$

will get the following appearance:

$$P = (P_3+P_1)\frac{\ell_1}{2} + P_2e_2 + (P_3-P_1)\frac{\ell_2}{2} \quad . \tag{39}$$

In order to perform the parabolic rotation given by Eq(35) we must determine how the operators T and T^2 operate on $\frac{\ell_1}{2}$, e_2, and $\frac{\ell_2}{2}$.
We obtain:

$$Te_2 = \frac{1}{2}(e_2I_p - I_pe_2) = e_2I_p = e_2(e_3+e_1)e_2 = e_2e_3e_2 + e_2e_1e_2$$

$$= e_3 + e_1 = \ell_1$$

because $e_3^2 = 1$, $e_2^2 = -1$, $e_1^2 = -1$, $e_1e_2 = -e_2e_1$, etc.

and $-I_pe_2 = e_2I_p$

$$T^2e_2 = \frac{1}{2}\left[(Te_2)I_p - I_p(Te_2)\right] = \frac{1}{2}(\ell_1I_p - I_p\ell_1) = 0$$

because $I_p\ell_1 = \ell_1I_p$.

If we continue in the same way with the rest of the operations we get the following tables, valid for the two lightlike planes $I_p = (e_3+e_1)e_2$ and $I_p = (e_3-e_1)e_2$:

Table 1a

P	TP	T^2P
$\frac{\ell_1}{2}$	0	0
e_2	ℓ_1	0
$\frac{\ell_2}{2}$	e_2	ℓ_1
$I_p = (e_3+e_1)e_2$		

Table 1b

P	TP	T^2P
$\frac{\ell_1}{2}$	e_2	ℓ_2
e_2	ℓ_2	0
$\frac{\ell_2}{2}$	0	0
$I_p = (e_3-e_1)e_2$		

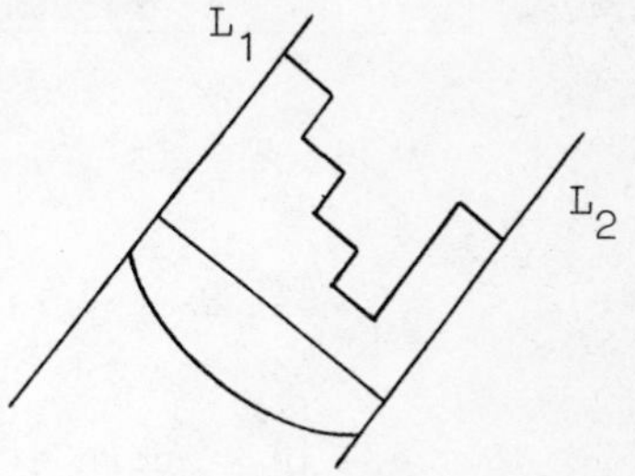

Fig 5. Curves with equal lengths in a light-like plane.

Table 1a is used in connection with a series inductor, for example, and Table 1b used with a shunt conductor.

In the three-dimensional Lorentz space two vectors are orthogonal if they are symmetric to the lightcone. This means that a lightlike vector is orthogonal to itself! This peculiar fact has many interesting repercussions. Two planes through a lightlike vector (and not tangent to the lightcone) form the angle 0 (or π). This is the dual theorem to the theorem that the distance between two points on a lightlike line is zero. Also, in a lightlike plane all curves between two parallel light-like lines have the same lengths if they don't cut any other parallel line more than once. This is the key to the understanding of the rôle of the parabolic transformation in network synthesis. Thus, in Fig.5 all curves between L_1 and L_2 have the same lengths, also the straight line, so that a parabolic arc can be measured by the length of the straight line Euclidean perpendicular to L_1 and L_2. As a matter of fact we can even measure in the Euclidean sense. Example: The simplest low-pass filter is the so-called Butterworth-1 filter which consists of a simple element, a series inductor, for example. Here $P_L = 1+\omega^2$ and the point on ML-3 has the coordinator $P_1 = 2\omega^2$, $P_2 = 2\omega$, and $P_3 = 1+2\omega^2$. The parabolic arclength from the point to the matching point (0,0,1) is equal to the P_2-coordinate, so we find directly from $P_2 = 2\omega$ that $L_1 = 2$.

A rotation in a lightlike tangent plane is a sort of shear. The constant angles Ψ and θ in the hyperbolic and elliptic cases correspond in the parabolic case to an arc with constant length. With this in mind we can treat all three kinds of rotation on equal basis. It is indeed quite fascinating that mankind stumbled into this very peculiar and singular parabolic transformation already at the start of network theory around 1916 when Karl Willy Wagner in Germany and Georg Campbell in the United States invented the first filters.

In general, when the lightlike plane does not pass through the matching point (0,0,1), the following formula can be used for the length of the arc between two points P and P' on the hyperboloid:

$$\Delta P_2 = \left|\frac{P_2'-P_2}{P_3'\pm P_1'}\right| = \left|\frac{P_2-P_2'}{P_3\pm P_1}\right| \qquad (+\equiv C\ ;\ -\equiv L) \tag{40}$$

5.4.5. Example. Butterworth-1 with impedance transformation.

$$P_L = \cosh^2 a(1+\omega^2) \quad ; \quad a = \frac{1}{2}\ln\frac{R_L}{R_1}$$

$$P_1^2 + P_2^2 = 4\cosh^4 a(1+s)(1-s)(\tanh a+s)(\tanh a-s)$$

$$P_1 + jP_2 = 2\cosh^2 a(1-j\omega)(\tanh a+j\omega)$$

in accordance with the splitting procedure. Thus,

$$P = (2\omega^2\cosh^2 a + \sinh 2a)e_1 + 2\omega e^{-a}\cosh a + (2\omega^2\cosh^2 a + \cosh 2a)e_3 .$$

Eq(39) yields

$$P = (4\omega^2\cosh^2 a + e^{2a})\frac{\ell_1}{2} + 2\omega e^{-a}\cosh a + e^{-2a}\frac{\ell_2}{2} .$$

We now take out an ideal transformer by multiplying $\ell_1/2$ by $e^{-2\Psi_1}$ and $\ell_2/2$ by $e^{2\Psi_1}$ [see Eq(29)] and an inductor L_1 by using Eq(37) with $w = -\omega L_1$ (-sign because we perform an inverse transformation) and by using Table 1a:

$$P_a = e^{-2\Psi_1}(4\omega^2\cosh^2 a + e^{2a})\frac{\ell_1}{2} + 2\omega e^{-a}\cosh a e_2 + e^{2\Psi_1}e^{-2a}\frac{\ell_2}{2} -$$

$$-\omega L_1 4\omega e^{-a}\cosh a\,\frac{\ell_1}{2} - \omega L_1 e^{2\Psi_1}e^{-2a}e_2 + \omega^2 L_1^2 e^{2\Psi_1}e^{-2a}\frac{\ell_1}{2} .$$

From the e_2-terms we get

$$L_1 = 2 \quad ; \quad e^{2\Psi_1} = e^a\cosh a$$

so that $P_a = e^{2a-2\Psi_1}\frac{\ell_1}{2} + e^{2\Psi_1-2a}\frac{\ell_2}{2}$.

We now take out a second ideal transformer:

$$P_b = e^{-2\Psi_2+2a-2\Psi_1}\frac{\ell_1}{2} + e^{2\Psi_2+2\Psi_1-2a}\frac{\ell_2}{2}$$

with $e^{2\Psi_2} = e^{2a-2\Psi_1} = \frac{e^a}{\cosh a}$;

P_b simplifies to $P_b = \frac{\ell_1}{2} + \frac{\ell_2}{2} = e_3$ (the matching point) .

The lossless two-port network in the Butterworth-1 case will thus consist of an ideal transformer, a series inductor (L_1=2), and another ideal transformer. The load resistance equals unity.

5.5. Some additional applications of the new insertion loss power ratio method

Clifford algebra was introduced in network theory in 1959 [43]. Since then the author has applied this tool to many different network problems, some mentioned above. See also Reference 44. Among the problems the following can be mentioned:

1. Broadband matching
2. Invariants, passivity ↔ stability, two-state switching
3. Microwave theory, stub filters, step lines, gyrators, mixed filters, non-commensurate filters
4. Lossy and noisy networks. Losses: Bolinder, 1958 [45](CK-3). Noise: Bolinder, 1958 [46]
5. Digital filters
6. Active filters .

Transformations through lossy two-port networks can be performed by means of the lightcone in the Minkowski model of four-dimensional space. The network theory described above is obtained from a cross section of this lightcone, when we normalize the power quantities with respect to the real power. If, instead, we normalize with respect to the reactive power the theory is converted into a method for active network synthesis. The close parallel between the Darlington passive network synthesis procedure [47] and the Szentirmai active network synthesis procedure [48] has been clearly pointed out by Tuttle [49]. Finally, if we normalize with respect to the total power, we obtain a sphere which, for example, can be considered to be the Poincaré sphere in polarization theory.

Transformations in connection with noisy two-port networks also require a four-dimensional Lorentz space. The four-dimensional Stokes vector, introduced by G.G. Stokes in 1852 [50] to represent partially polarized light, can here find an application in network theory. See Reference 46.

Finally, a pioneer geometric paper on the representation of waveguides and waveguide junctions was published by Deschamps in 1952 [51]. Geometric representations, the complex plane C, the sphere S, the spaces C^2, E^3, and ML-4, were introduced for the "state" of the waves at a point in a waveguide, for the transformation of a state through a linear junction, and for any hermitian form of the complex vector which represents the state.

6. CONCLUSIONS

A survey has been made of some applications of Clifford algebra in electromagnetic theory and network theory. Although Clifford algebra is by no means a necessary tool to use in these fields it is a very handy tool especially in connection with geometric pictures (models). The use of Clifford algebra is bound to increase now, when computer software programs on the Clifford algebra operations start to appear.

7. REFERENCES

[1] Ricci M.M.G., Levi-Civita T. (1901), 'Méthodes de calcul différential absolu et leurs applications', Math Ann, 54, 125-201.

[2] Weyl, H. (1921), Raum, Zeit, Materie, 4th ed, Springer, Berlin; in English: Dover, New York.

[3] Post, E.J. (1962), Formal Structure of Electromagnetics, North-Holland Publ Co.

[4] Kron, G. (1959), Tensors for Circuits, Dover Publ, Inc, New York, N.Y.

[5] Clifford, W.K. (1878), 'Applications of Grassmann's extensive algebra', American J of Math Pure and Applied, 1, 350-358. 'On the classification of geometric algebras' Math Papers by William Kingdon Clifford, MacMillan and Co 1882, 397-401.

[6] Lipschitz, R. (1880), 'Principles d'un calcul algébrique qui contient comme espèces particulières le calcul des quantités imaginaires et des quaternions', CR Acad Sc Paris, 91, 619-621, 660-664. Also: (1887) Bull des Sciences Math, 2 series, 9, 115-120.

[7] Study, E., Cartan, E. (1908), 'Nombres complexes', L'Encyclopédie des Sciences Math, 1, 329-468 (Clifford algebra 463-466).

[8] Riesz, M. (1958), Clifford Numbers and Spinors, Lecture Series, No 38, The Institute for Fluid Dynamics and Appl. Math, Univ. of Maryland, Maryland.

[9] Hestenes, D. (1966), Space-Time Algebra, Gordon and Breach, New York, N.Y.

[10] Grassmann, H. (1854), 'Sur les différents genres de multiplication', J für die reine und angewandte Math, 49, 123-141.

[11] Dillner, G. (1876), 'Versuch einer neuen Entwicklung der Hamilton'schen Methode, gennant "Calculus of quaternions",' Math Ann, 11, 168-193.

[12] Grassmann, H. (1877), 'Der Ort der Hamilton'schen quaternionen in der Ausdehnungslehre', Math Ann, 12, 375-386.

[13] Study, E. (1904), 'Der Ort der Hamilton'schen quaternionen in der Ausdehnungslehre (von H. Grassmann)', Hermann Grassmanns Gesammelte Math u Phys Werke, 2, Part 1, (herausgegeben von E. Study, G. Scheffers und F. Engel, Teubner, Leipzig).

[14] Cartan, E. (1938), Leçons sur la Theorie des Spineurs, I et II, Actualités Scientifiques et Industrielles Nos 643 et 701, Hermann

et Cie, Paris.

[15] Brauer, R., Weyl, H. (1935), 'Spinors in n dimensions', Am J of Math, 57, 425-449.

[16] Einstein, A., Lorentz, H.A., Minkowski, H., and Weyl, H. (1923), The Principle of Relativity, Methuen and Co, translated by Dover Publs, Inc.

[17] Rojansky, V. (1938), Introductory Quantum Mechanics, Prentice-Hall, Inc, New York. (Dirac: Chapter XIV, Pauli: Chapter XIII).

[18] Riesz, M. (1946), 'Sur certaines notions fondamentales en théorie quantique relativiste', Dixième Congres des Mathematiciens Scandinaves, 123-148.

[19] Cartan, E. (1971), Les Systèmes Différentiels Exterieurs et Leurs Applications Géométriques, Hermann, Paris (Lectures 1936-37). Also, Cartan, E. (1924), Ann Ecole Normale, 41, 1.

[20] Van Danzig, D. (1934), 'The fundamental equations of electromagnetism, independent of metrical geometry', Proc Amsterdam Acad, 37, 421-427.

[21] Kähler, E. (1937), 'Bemerkungen über die Maxwellschen Gleichungen', Abh der Hamburgischen Universität, 12, 1 Heft, 1-28.

[22] Deschamps, G.A. (1970), 'Exterior differential forms', Mathematics Applied to Physics, Chapter III, (ed E. Roubine), Springer Verlag, Berlin-Heidelberg-New York.

[23] Deschamps, G.A. (1981), 'Electromagnetics and differential forms', Proc IEEE, 69, 676-696.

[24] Hestenes, D., Sobczyk, G. (1984), Clifford Algebra to Geometric Calculus, D. Reidel Publ Co, Dordrecht, Holland.

[25] Silberstein, L. (1912), 'Quaternionic form of relativity', Phil Mag, 23, 790-809.

[26] Silberstein, L. (1914), The Theory of Relativity, McMillan, London.

[27] Bolinder, E.F. (1957), 'The classical electromagnetic equations expressed as complex four-dimensional quantities', J. Franklin Inst, 263, 213-223.

[28] Proca, A. (1930), 'Sur l'équation de Dirac', J de Physique, Ser 7, T1, 235-248. Also: CR Paris, T190, séance 16.6.1930, 1377-1379, and T191, séance 7.7.1930, 26-29.

[29] Juvet, G. (1930), 'Opérateurs de Dirac et équations de Maxwell',

Comment Math Helv, 2, 225-235.
Juvet, G. (1932), 'Les nombres de Clifford et leurs applications à la physique mathématique', Congrès int des mathematiciens de Zurich, 306-307.

[30] Juvet, G., Schidlof, A. (1932), 'Sur les nombres hypercomplexes de Clifford et leurs applications à l'analyse vectorielle ordinaire, à l'électromagnétisme de Minkowski et à la théorie de Dirac', Bull Soc Neuchateloise Sc Nat, 57, 127-147.

[31] Mercier, M. (1935), 'Expression des équations de l'électromagnétisme au moyen des nombres de Clifford', Thesis No 953, University of Genève.

[32] Lounesto, P. (1979), Spinor Valued Regular Functions in Hyper-Complex Analysis, Helsinki Univ of Techn, Inst of Math, Report HTKK-MAT-A154, 1-79.

[33] Riesz, M. (1959), Private communication, April 11, 1959, Indiana University, Dept of Math, Bloomington, Indiana.

[34] Pauli, W. (1958), Theory of Relativity, Pergamon Press (Eq 222, p 85).

[35] Lewis, G.N. (1910), 'On four-dimensional vector analysis, and its application in electrical theory', Contrib from Res Lab of Phys Chem of the Mass Inst of Techn, No 58, 165-181.

[36] Wilson, E.B., Lewis, G.N. (1912), 'The space-time manifold of relativity. The non-Euclidean geometry of mechanics and electromagnetics', Proc Am Acad of Arts and Sciences, 48, No 11, 389-507.

[37] Bolinder, E.F. (1951), Fourier Transforms in the Theory of Inhomogeneous Transmission Lines, Trans Royal Inst of Techn, No 48, 1-84. Acta Polytechnica Elec Eng Series 3. Also: Proc IRE, 38, 1354, 1950.

[38] Cerrillo, M.V. et al (1952), 'Basic existence theorems in network synthesis', I) M.V. Cerrillo, Quart Prog Rep, RLE, MIT, Jan 15, 1952, 71-95, - II) M.V. Cerrillo, QPR, RLE, MIT, April 15, 1952, 80-89. - III) M.V. Cerrillo, E.A. Guillemin, Techn Rep 233, RLE, MIT, June 4, 1952, - IV) M.V. Cerrillo, E.F. Bolinder, Techn Rep 246, RLE, MIT, Aug 15, 1952.

[39] Bolinder, E.F. (1985), 'Basic study of elliptic low-pass filters by means of the Minkowski model of Lorentz space', Proc of the Symp on Math Theory in Network Systems, Stockholm, Sweden, June 1985 (to be published).

[40] Weinberg, L. (1962), Network Analysis and Synthesis, McGraw Hill Book Co.

[41] Riesz, M. (1943), 'An instructive picture of non-Euclidean geometry. Geometric excursions into relativity theory (in Swedish), Acta Physiogr Soc, Lund, NF 53, No 9, 1-76.

[42] Mikkola, R., Lounesto, P. (1983), 'Computer-aided vector algebra', Int J Math Educ Sci Technol, 14, No 5, 573-578.

[43] Bolinder, E.F. (1959), 'Impedance, power, and noise transformations by means of the Minkowski model of Lorentz space', Ericsson Techniques, No 2, 1-35.

[44] Bolinder, E.F. (1982), 'Unified microwave network theory based on Clifford algebra in Lorentz space', Proc 12th European Microwave Conf, Helsinki, 24-35.

[45] Bolinder, E.F. (1957), 'Impedance transformations by extension of the isometric circle method to the three-dimensional hyperbolic space', J Math and Physics, 36, 49-61.

[46] Bolinder, E.F. (1959), 'Theory of noisy two-port networks', J. Franklin Inst, 267, 1-23, Tech Report 344, Res Lab of Electronics, MIT, June 1958.

[47] Darlington, S. (1939), 'Synthesis of reactance 4-poles which produce prescribed insertion loss characteristics', J Math Phys, 30, 257-353.

[48] Szentirmai, G. (1973), 'Synthesis of multiple-feedback active filters, Bell Syst Techn J, 52, 527-555.

[49] Tuttle, D.F. (1974), 'A common basis for the Darlington and Szentirmai network syntheses', Proc IEEE, 62, 389-390.

[50] Stokes, G.G. (1852), 'On the composition and resolution of streams of polarized light from different sources', Trans Cambr Phil Soc, 9, 399-416. - Math and Phys Papers by G.G. Stokes, 3, 233-258, Cambr Univ Press, 1922.

[51] Deschamps, G. (1952), 'Geometric viewpoints in the representation of waveguides and waveguide junctions', Proc Symp on Modern Network Synthesis, Polytechn Inst of Brooklyn, 277-295.

REMARKS ON CLIFFORD ALGEBRA IN CLASSICAL ELECTROMAGNETISM

B. Jancewicz
Institute of Theoretical Physics
University of Wrocław
Cybulskiego 36
50-205 Wrocław
Poland

ABSTRACT. Using bivectors for description of the magnetic field, uniting electric vector and magnetic bivector into a single quantity, and employing the formalism of Clifford algebra reveals the intrinsic structure of electromagnetic phenomena, simplifies the methods of solving equations, allows one to visualize solutions and sometimes gives new ones.

1. DESCRIPTION OF THE MAGNETIC FIELD

A deeper understanding of phenomena stems from consequent use of bivectors for magnetic quantities, namely the magnetic field and the magnetic induction. One may even fit the definition of magnetic induction to its bivector nature starting from the Lorentz force $\vec{F}_m = q\vec{v}\times\vec{B}$ (in SI units). When written with the use of bivector $\overset{\frown}{B} = I\vec{B}$ (here I is the unit right-handed trivector) it takes the form $\vec{F}_m = -q\vec{v}\cdot\overset{\frown}{B}$ with the inner product between $\vec{v}$ and $\overset{\frown}{B}$. If $\vec{v}\,||\,\overset{\frown}{B}$ (physically this manifests by the fact the motion of a test charge q is planar) then $\vec{v}\cdot\overset{\frown}{B} = \vec{v}\overset{\frown}{B}$ with the Clifford product between $\vec{v}$ and $\overset{\frown}{B}$ and $\vec{F}_m = -q\vec{v}\overset{\frown}{B}$. From this we obtain

$$\overset{\frown}{B} = -q^{-1}\vec{v}^{-1}\vec{F}_m = -q^{-1}\vec{v}^{-1}\wedge F_m = q^{-1}\vec{F}_m\wedge\vec{v}^{-1}.$$

This formula can serve for determining the bivector of magnetic induction when the test charge q performs a planar motion (but not along a straight line) in the magnetic field. It is analogous to the formula $\vec{E} = q^{-1}\vec{F}_e$ for the electric field. Of course it is valid only for small q and $\vec{v}$ when the field can be treated as homogeneous.

There exists a possibility of introducing a concept analogous to the force lines known for the electric field: *force surfaces* for a bivector field are oriented surfaces to which the bivector field is tangent at each point and has the same orientation. The existence problem for such surfaces is governed by the Frobenius theorem [1] and for the magnetic field it gives

J. S. R. Chisholm and A. K. Common (eds.), Clifford Algebras and Their Applications in Mathematical Physics, 485–493.

$$\frac{\partial \vec{D}}{\partial t} + \vec{j} \perp \vec{H} \quad \text{or equivalently} \quad \frac{\partial \vec{D}}{\partial t} + \vec{j} \parallel \widetilde{H}$$

as the necessary and sufficient condition [2] . We show force surfaces on Figures 1-4 for simple magnetic fields.

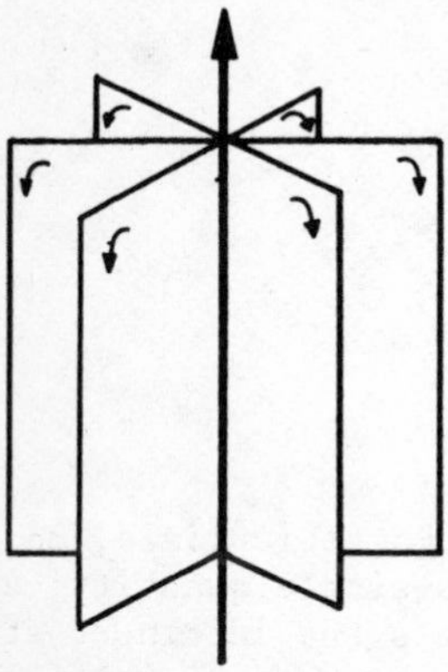

Fig. 1. Force surfaces around a linear current.

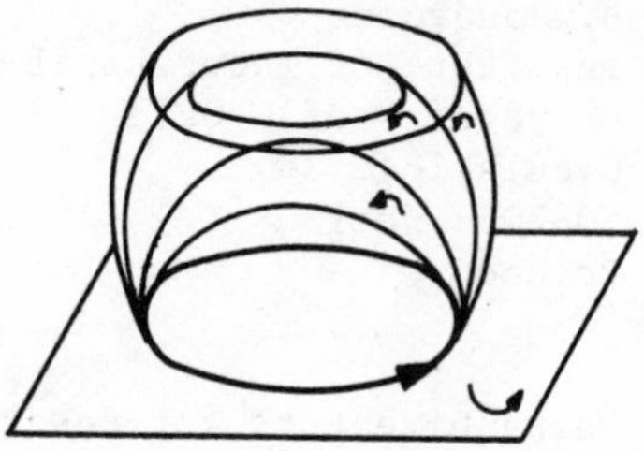

Fig. 2. Force surfaces around a circular current.

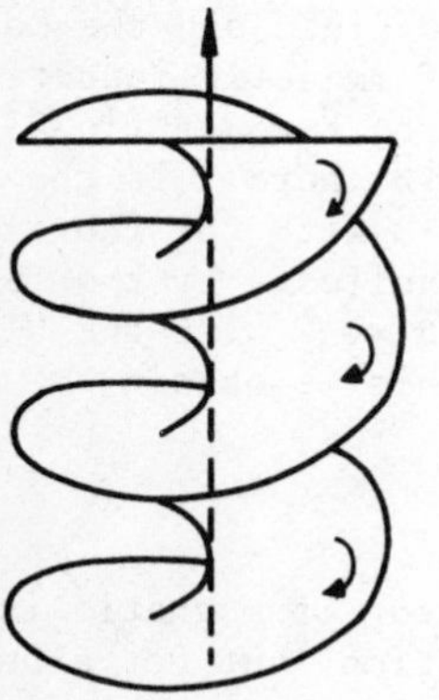

Fig. 3. One force surface for superposition of field shown on Fig.1 with homogeneous one.

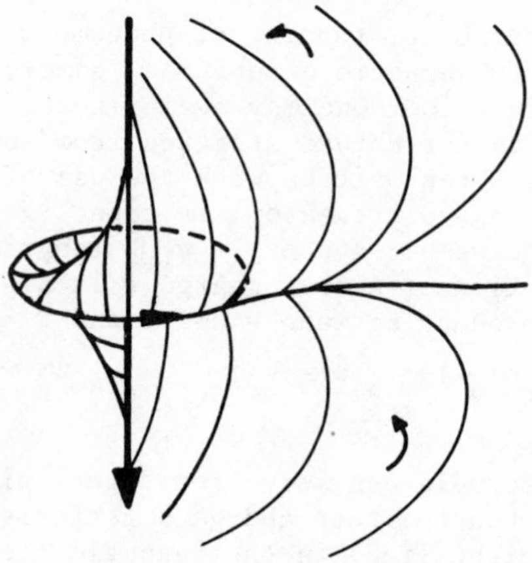

Fig. 4. One force surface for superposition of fields of Fig. 1 and 2.

One may draw the following conclusions from the examples. The force surfaces of the magnetic field have their boundaries on the currents. Orientation of the force surfaces is consistent with that of the currents at the boundaries. In this way the bivector language reveals close physical connection between the magnetic field and electric currents as its sources.

It also helps one to judge which analogy between the electromagnetic quantities is more proper: the Lorentz-Abraham one which links E with B and D with H, or the Hertz-Heaviside one that is E↔H and D↔B. It is well known that the Lorentz force formula and the Maxwell equations support the Lorentz-Abraham analogy and that only boundary conditions between two media support the Hertz-Heaviside one. But when written with the use of bivectors:

$$\vec{E}_{1t} = \vec{E}_{2t}, \qquad \hat{B}_{1t} = \hat{B}_{2t},$$

$$\vec{D}_{1n} - \vec{D}_{2n} = \sigma\vec{n}, \qquad \hat{H}_{1n} = \hat{H}_{2n} = \vec{i} \wedge \vec{n}$$

with the usual meaning of symbols, they also take side of Lorentz and Abraham.

2. MAXWELL EQUATION AND ITS FUNDAMENTAL SOLUTION

As was shown in [3] , p.29, when discussing Maxwell equations within the Pauli algebra it is useful to form the Clifford number $\vec{E}+\hat{B}$. But this is practical only in empty space. In the presence of material medium one has to take into account its electric permittivity ε and magnetic permeability μ. The SI units help in forming the proper products of $\vec{E}$ and $\hat{B}$ with some powers of ε and μ to obtain the same physical dimension. The best possibility is $f = \sqrt{\varepsilon}\,\vec{E}+(1/\sqrt{\mu})\hat{B}$ which has the dimension $\sqrt{J/m^3}$ that is square root of the energy density. In this respect it resembles the wave function from quantum mechanics which has the dimension of square root of the probability density.

With the usual notation f^+ for the reversion of Clifford numbers we obtain

$$ff^+ = w + \frac{1}{u}\vec{S}$$

where $w = \frac{1}{2}(\varepsilon E^2 + \frac{1}{\mu}B^2)$ is the energy density for the electromagnetic field, $\vec{S}$ is the Poynting vector and $u=1/\sqrt{\varepsilon\mu}$ is the phase velocity for the electromagnetic waves in the medium.

Having denoted $\vec{e}=\sqrt{\varepsilon}\,\vec{E}$ and $\hat{b}=(1/\sqrt{\mu})\hat{B}$ one may write the Maxwell equations in the following single formula :

$$\mathcal{D}f + \vec{e}\mathcal{D}\ln\sqrt{\varepsilon} + b\mathcal{D}\ln\sqrt{\mu} = \tilde{j} \tag{1}$$

where $\mathcal{D} = \vec{\nabla} + \sqrt{\varepsilon\mu}\,\frac{\partial}{\partial t} = \vec{\nabla} + \frac{1}{u}\frac{\partial}{\partial t}$ is the Dirac operator and $\tilde{j}=(1/\sqrt{\varepsilon})\rho+ - \sqrt{\mu}\vec{j}$. The logarithmic derivative terms stem from differentiation of the products $(1/\sqrt{\varepsilon})\vec{e}$ and $\sqrt{\mu}\hat{b}$ in which both factors can, in general, be functions of time and position. We call (1) simply the *Maxwell equation*.

If the material medium is homogeneous in space and constant in time, the equation becomes simply

$$\mathcal{D}f = \tilde{j} \ . \tag{2}$$

It is interesting to apply the method of fundamental solutions for solving it when the charge and current density is given. Recall that the *fundamental solution* of the equation (2) is a distribution g (in our case a Clifford number valued distribution) such that $\mathcal{D}g(\vec{r},t)=\delta^4(\vec{r},t)$ with the Dirac delta on the right hand side. Then

$$f(r,t) = \int_{R^3}\int_{R} g(\vec{r}-\vec{r}',t-t')\,\tilde{j}(\vec{r}',t')\,d^3r'dt' + f_o(\vec{r},t) \tag{3}$$

is a solution of (2) where f_o is a free solution, that is solution of (2) with $\tilde{j}=0$. It is easy to check that when D is a fundamental solution for the d'Alembert operator $\square = \Delta - \frac{1}{u^2}\frac{\partial^2}{\partial t^2}$, then g can be written as $g = (\vec{\nabla} - \frac{1}{u}\frac{\partial}{\partial t})D$. In this way for the retarded (-) and advanced (+) fundamental solutions $D^{(\pm)}$ for the d'Alembert operator,

$$D^{(\pm)}(\vec{r},t) = -\frac{1}{4\pi r}\,\delta(t \pm \frac{r}{u}),$$

one obtains the Clifford number valued distributions:

$$g^{(\pm)}(\vec{r},t) = \frac{1}{4\pi}\left[\frac{\vec{r}}{r^3}\,\delta(t \pm \frac{r}{u}) + \frac{1}{ur}(1 \mp \frac{\vec{r}}{r})\,\delta'(t \pm \frac{r}{u})\right]$$

for the Dirac operator, that is for the Maxwell equation (2).

The physically interesting and relatively simple applications are: electric dipole with the moment $\vec{p}$ changing in time, for which

$$\tilde{j}_e = -\frac{1}{\sqrt{\varepsilon}}\,(\vec{p}(t)\cdot\vec{\nabla})\,\delta^3(\vec{r}\,) - \sqrt{\mu}\,\frac{d\vec{p}(t)}{dt}$$

and the magnetic dipole with the moment $\overset{\frown}{M}$ for which

$$\tilde{j}_m = -\sqrt{\mu}(\overset{\frown}{M}(t)\cdot\vec{\nabla})\,\delta^3(\vec{r}\,).$$

In these cases one obtains respectively

$$f_e(\vec{r},t) = \frac{1}{4\pi\sqrt{\varepsilon}}\left[\frac{3\vec{n}(\vec{n}\cdot\vec{p})-\vec{p}}{r^3} + \frac{3\vec{n}(\vec{n}\cdot\dot{\vec{p}})-\dot{\vec{p}}}{ur^2} - \frac{\vec{n}(\vec{n}\wedge\ddot{\vec{p}})}{u^2r} - \vec{n}\wedge\left(\frac{\dot{\vec{p}}}{ur^2} + \frac{\ddot{\vec{p}}}{u^2r}\right)\right] ,$$

$$f_m(\vec{r},t) = \frac{\sqrt{\mu}}{4\pi}\left[\frac{2\overset{\frown}{M}-3\vec{n}(\vec{n}\cdot\overset{\frown}{M})}{r^3} + \frac{2\dot{\overset{\frown}{M}}-3\vec{n}(\vec{n}\cdot\dot{\overset{\frown}{M}})}{ur^2} + \frac{\vec{n}(\vec{n}\cdot\ddot{\overset{\frown}{M}})}{u^2r} - \vec{n}\cdot\left(\frac{\dot{\overset{\frown}{M}}}{ur^2} + \frac{\ddot{\overset{\frown}{M}}}{u^2r}\right)\right] ,$$

where $\vec{n} = \vec{r}/r$ and the dipole moments $\vec{p}$ and $\overset{\frown}{M}$ and their derivatives depend on time through the difference t-r/u.

The far fields, i.e. the terms prevailing for $r \to \infty$, have the form

$$f_{e,f}(\vec{r},t) = \frac{1}{4\pi\sqrt{\varepsilon}u^2 r}\,(1+\vec{n})(\ddot{\vec{p}} \wedge \vec{n}) \ ,$$

$$f_{m,f}(\vec{r},t) = \frac{\sqrt{\mu}}{4\pi u^2 r}\,(1+\vec{n})(\ddot{\overset{\frown}{M}}\cdot\vec{n}) \ .$$

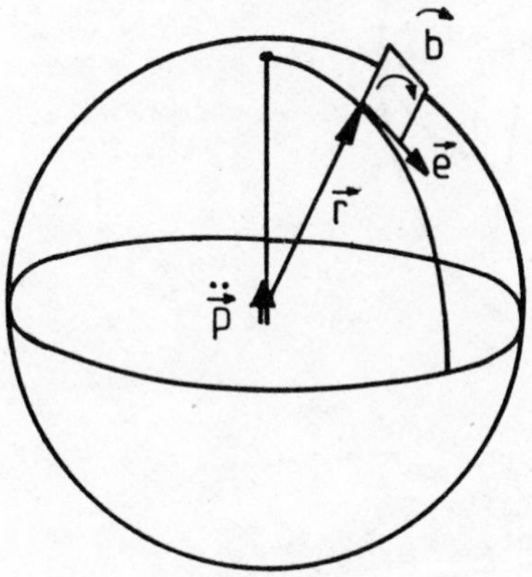

Fig. 5. Far field of electric dipole.

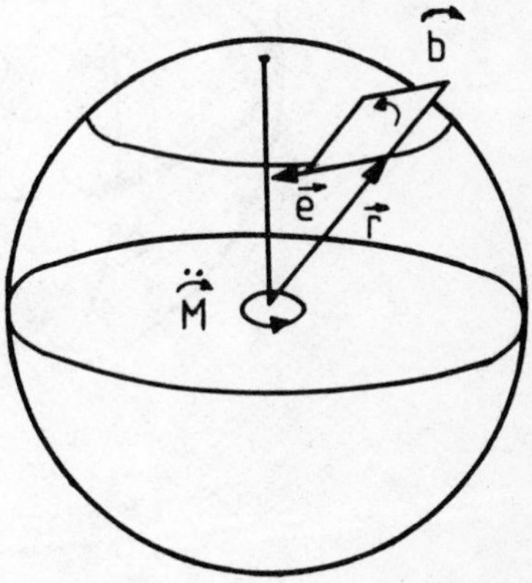

Fig. 6. Far field of magnetic dipole.

We illustrate them on figures 5 and 6 using rectangles for magnetic bivectors. Notice that the whole electromagnetic fields of the two figures are perpendicular to each other. This is caused by the perpendicularity of the moments $\vec{p}$ and $\overset{\frown}{M}$ generating the fields. Both fields have axial symmetry.

For the special time dependence

$$\vec{p}(t) = e^{-\frac{1}{2}\overset{\frown}{\omega}t}\,\vec{p}_o\,e^{\frac{1}{2}\overset{\frown}{\omega}t} \ ,$$

that is for the uniform rotation with angular velocity $\overset{\frown}{\omega}$ and initial moment $\vec{p}_o \| \overset{\frown}{\omega}$ we obtain

$$f_{e,f}(\vec{r},t) = \frac{k^2}{4\pi\sqrt{\varepsilon}r}\,(1+\vec{n})\left\{\vec{n} \wedge \left[\vec{p}_o\, e^{\overset{\frown}{\omega}(t-\frac{r}{u})}\right]\right\}$$

where $k=\omega/u$. This gives a spherical wave with the elliptic polarization. In particular the polarization is circular in direction perpendicular to $\overset{\frown}{\omega}$ and linear in direction parallel to $\overset{\frown}{\omega}$, see fig. 7. One may notice that the ellipse of electric oscillations is similar to image of the circle $t \to \vec{p}_o e^{\overset{\rightarrow}{\omega}t}$ seen from the point in which the field is measured.

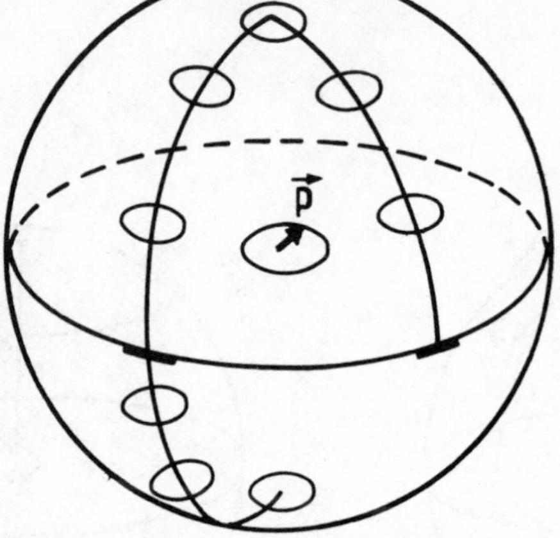

Fig. 7. Changes of the electric vector for rotating electric moment.

3. DESCRIPTION OF POLARIZATIONS

The Clifford algebra is particularly useful for exploring the plane wave solutions of the homogeneous version of eq.(2). One may compactly write down superpositions of two circularly polarized waves running in opposite directions. When the two waves have the same orientation (left or right) one obtains the expression

$$f_s(\vec{r},t) = e^{\pm I[\omega t - \vec{n}(\vec{k}\cdot\vec{r})]}\ f_o\ , \tag{4}$$

where $\vec{n}=\vec{k}/k$, $\omega=uk$ and I is the unit trivector. The initial Clifford number f_o and the field f anticommute with $\vec{n}$ which is equivalent to the transversality condition $\vec{e}\perp\vec{n}\perp\vec{b}$. When the superposed waves have opposite orientation one obtains the wave

$$f_r(\vec{r},t) = e^{\pm I(\vec{n}\omega t\ -\ \vec{k}\cdot\vec{r})}\ f_o\ . \tag{5}$$

The two kinds of solutions (4) and (5) posses interchanged type of time and space dependences. The time dependence in (4) is given by the factor $e^{I\omega t}$ which,when acting as an operator on cliffords,implements the Larmor-Reinich transformation [4] :

$$\vec{e} \to \vec{e}\cos\omega t - \vec{b}\sin\omega t, \qquad \vec{b} \to \vec{b}\cos\omega t + \vec{e}\sin\omega t,$$

called also duality rotation [5] . The space dependence in (4) is governed by the factor $e^{I\vec{n}(\vec{k}\cdot\vec{r})}$ with the bivector exponent. Due to the anticommutativity of f_o with $\vec{n}$ this factor implements rotation in the plane perpendicular to $\vec{n}$. This implies that the space diagram for the electric vector is a helix - therefore we say that the wave f_s has a *spiral polarization.*

The space dependence in solution (5) is the Larmor-Reinich transformation and the time dependence is rotation, thus the tip point of the electric vector in a fixed place traces a circle - this is the reason we have chosen the expression: the wave f_r has a *round polarization.*

New quantities may be introduced for describing these polarizations, namely the bivector of *roundness:*

$$\tilde{\Omega} = \frac{1}{2w}\left(\vec{e}\wedge\frac{\partial\vec{e}}{\partial t} + \vec{b}\wedge\frac{\partial\vec{b}}{\partial t}\right),$$

and trivector of *spirality:*

$$s = \frac{1}{2w}\left[\tilde{b}\wedge(\vec{\nabla}\cdot\tilde{b}) - \vec{e}\wedge(\vec{\nabla}\wedge\vec{e})\right] .$$

One obtains

$$\tilde{\Omega} = \pm I\vec{n}\omega\ , \qquad s = \pm\frac{1}{uw}I(\vec{k}\cdot\vec{S}) \qquad \text{for } f_r,$$

$$\tilde{\Omega} = \pm\frac{1}{uw}\omega I\vec{S}\ , \qquad s = \pm kI \qquad \text{for } f_s.$$

So the roundness for the round wave is equal to the bivector of angular velocity of the electric field, and the spirality for the spiral wave is equal to the unit trivector of the helix orientation multiplied by its pitch.

The ordinary running wave with the elliptic polarization can be also written in the single formula:

$$f(\vec{r},t) = (1+\vec{n})\vec{n}\wedge\left[\vec{h}e^{\tilde{i}(\vec{k}\cdot\vec{r}-\omega t)}\right]$$

where $\tilde{i}$ is a unit bivector arbitrarily directed with respect to $\vec{n}=\vec{k}/k$, $\vec{h}$ is a vector parallel to $\tilde{i}$ and perpendicular to $\vec{n}$. The expression in square bracket is rotation of $\vec{h}$ in the plane of $\tilde{i}$ by the angle $\phi=\vec{k}\cdot\vec{r}-\omega t$. Applying the knowledge about orthogonal projections (see [6] , Sec.1.2) we see that the diagram of electric oscillations is projection of the circle $\phi\to\vec{h}e^{\tilde{i}\phi}$ onto the plane perpendicular to $\vec{n}$, i.e. the ellipse.

This suggests that different kinds of elliptic polarizations for running waves can be described by the relative position of $\tilde{i}$ with respect to $\vec{n}$ or of the unit vector $\vec{i} = I\tilde{i}$ with respect to $\vec{n}$. All possible positions obviously form a sphere (see fig.8) which is analogous to the Poincaré sphere.(The similarity between figures 7 and 8 is not so great as it seems: the ellipses of fig.7 are tangent to the sphere whereas the ellipses of fig.8 are parallel to the plane of equator.)

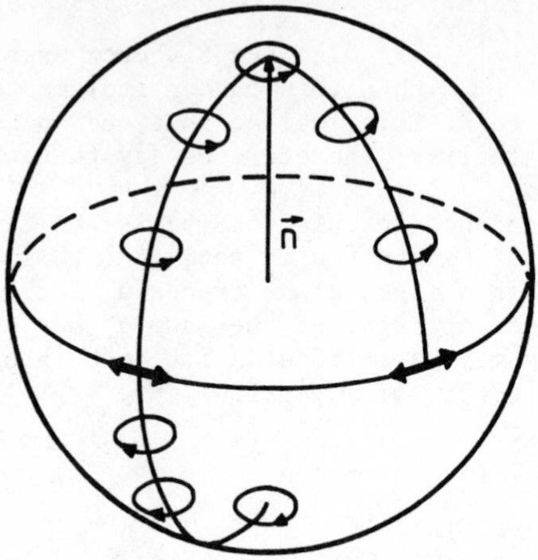

Fig.8. Sphere describing possible elliptic polarizations

4. PLANE WAVE IN PLANE-STRATIFIED MEDIUM

It would be interesting to find solutions periodic in time of eq. (1) when $\tilde{j}=0$ but ε and μ are not constant. It is very difficult to do this in the general case. It was able to solve it when u is constant, ε depends on position in one direction only (that is for plane-stratified inhomogeneous medium) and the electromagnetic field changes along the same direction - let it be z axis. Then eq.(4) takes the form

$$\vec{e}_3 \frac{\partial f}{\partial z} + \sqrt{\mu\varepsilon(z)} \frac{\partial f}{\partial t} + \vec{e}\,\vec{e}_3 \frac{1}{\sqrt{\varepsilon}} \frac{\partial\sqrt{\varepsilon}}{\partial z} = 0.$$

This equation has no elementary function solutions for arbitrary dependence $\varepsilon(z)$. The most convenient function is $\mu\varepsilon(z)=\alpha^2/(\omega z+\alpha c)^2$ where α is dimensionless constant. Then the solution with round polarization is

$$f = \left(\frac{\alpha c}{\omega z + \alpha c}\right)^{\frac{1}{2}}[(\vec{e}_3+1)\vec{A}(\cos\phi + \frac{\alpha}{\beta}\,\vec{\imath}\,\sin\phi) + (\vec{e}_3-1)\,\frac{\vec{A}}{2\beta}\sin\phi]e^{-\vec{\imath}\omega t} \qquad (6)$$

where $\vec{\imath}=\vec{e}_1\vec{e}_2$, $\vec{A}$ is a constant vector orthogonal to $\vec{e}_3$, $\beta=\sqrt{\alpha^2-1/4}$, $\phi=\beta\ln\frac{\omega z+\alpha c}{\alpha c}$, for $\alpha>\frac{1}{2}$, or

$$f = \left(\frac{\alpha c}{\omega z+\alpha c}\right)^{\frac{1}{2}}\{\vec{A}[\alpha x^{\beta} - (1 + 2\beta)x^{-\beta}(\alpha + \beta\vec{\imath})] +$$

$$+ \vec{e}_3\vec{A}[2\alpha x^{-\beta}(\alpha\tilde{i} - \beta) - (\frac{1}{2} + \beta)x^{\beta}\,\tilde{i}]\}e^{-\tilde{i}\omega t} \quad (7)$$

where $x = \dfrac{\alpha c}{\omega z+\alpha c}$, $\beta = \sqrt{\frac{1}{4}-\alpha^2}$ for $\alpha < \dfrac{1}{2}$.

This means that when the changes of ε are slow enough (that is $\alpha > \frac{1}{2}$) the solution has some elements of periodicity (i.e. the presence of trigonometric functions) in the space dependence. For too rapid changes of ε the solution is aperiodic.

When α tends to infinity, which implies that ε tends to constant function, the solution (6) tends to ordinary plane wave with circular polarization. When α tends to zero which approximates discontinuous change on the interface between two media, the solution (7) reproduces the well known Fresnel formulas.

References

[1] R. Sulanke and P. Wintgen: *Differentialgeometrie und Faserbündel*, Deutscher Verlag der Wiss., Berlin 1972.
[2] B. Jancewicz: *Eur.J.Phys.* 1(1980)179.
[3] D. Hestenes: *Space-Time Algebra*, Gordon and Breach, New York 1966
[4] I. Larmor: *Collected Papers*, London 1928.
G.I. Rainich: *Trans.Am.Math.Soc.* 27(1925)106.
[5] N. Salingaros: *Found.Phys.* 14(1984)777.
[6] D. Hestenes and G. Sobczyk: *Clifford Algebra to Geometric Calculus*, D. Reidel, Dordrecht 1984.

QUATERNIONIC FORMULATION OF CLASSICAL ELECTROMAGNETIC FIELDS AND THEORY OF FUNCTIONS OF A BIQUATERNION VARIABLE

K. Imaeda
Faculty of Science
Okayama University of Science
Okayama 700
Japan

ABSTRACT. A quaternionic formulation of electromagnetic fields by using complex-quaternions (biquat) is presented. Also, the solutions of Maxwell's equations have been given using regular functions of a biquat variable.

1. BIQUATERNIONS, BIQUATS, MAXWELL'S EQUATIONS AND REGULAR CONDITIONS FOR A FUNCTION OF A BIQUAT VARIABLE.

1.1. Biquaternions

Complex quaternions are powerful mathematical tools to describe classical electrodynamics and to obtain solutions of Maxwell's equations in vacuum by means of regular functions of a biquat variable.[1,2,3]

A complex quaternion (we hereafter use "biquaternion") Z is defined as follows:

$$(1.1) \qquad Z = \sum_{\mu=0}^{3} e_\mu (x_\mu + iy_\mu) = X + iY = x_0 + \underline{x} + i(y_0 + \underline{y})$$

where e_μ satisfy the following equations:

$$(1.2) \qquad e_0 = 1,\ e_k^2 = 1,\ e_1 e_2 = -e_2 e_1,\ \text{etc.},\ i = \sqrt{-1},$$

$$\underline{x} = e_1 x_1 + e_2 x_2 + e_3 x_3\ ,\ \underline{y} = e_1 y_1 + e_2 y_2 + e_3 y_3$$

where k = 1, 2, 3 and $\underline{x}$ and $\underline{y}$ are the "vector part" of X and Y respectively, x_0, and y_0 are the "real part" and the "imaginary part" respectively of the scalar part of Z.[3]

1.2. Biquats.

When the imaginary part of a biquaternion is zero we call it

J. S. R. Chisholm and A. K. Common (eds.), Clifford Algebras and Their Applications in Mathematical Physics, 495–500.

specifically "biquat", distinguishing it from a biquaternion Z but, denoting it by $X = \sum_{\mu=0}^{3} e_\mu x_\mu$.

We define (i) hyperconjugate, (ii) complex-conjugate, (iii) double-conjugate to

$$Z = x_0 + \underline{x} + i(y_0 + \underline{y})$$

as

(i) $Z^\dagger = x_0 - \underline{x} + i(y_0 - \underline{y})$,

(ii) $Z^* = x_0 + \underline{x} - i(y_0 + \underline{y})$,

(iii) $\overline{Z} = (Z^*)^\dagger = (Z^\dagger)^*$.

1.3. Maxwell's equations

Maxwell's equations in vacuum can be cast into a biquaternion form: [2,3]

(1.3) $$DF^*(X) = 4\pi I, \quad F(X) = E + iH = \sum_{\mu=0}^{3} e_\mu(E_\mu + iH_\mu)$$

$$D = \partial_0 - \sum_{k=1}^{3} e_k\partial_k, \quad \partial_\mu = \partial/\partial x_\mu ,$$

$$E_0 = H_0 = 0, \; I = e_\mu i_\mu = i_0 + \underline{i}.$$

Equation (1.3) is equivalent to the following equations:

$$\mathrm{div}\, \underline{E} = -4\pi i_0, \; \mathrm{div}\, \underline{H} = 0,$$

$$\partial_0\underline{E} - \mathrm{curl}\, \underline{H} = -4\pi \underline{i}, \quad \partial_0\underline{H} + \mathrm{curl}\, \underline{E} = 0.$$

1.4. Regularity conditions

Extending the regularity conditions for functions of a quaternion variable [1] to that of a biquat variable, we find, using a functional derivative definition of the regularity conditions [2], the following conditions:

(1.4) $$D\Phi(X) = 0,$$

where D is defined by (1.3) and the function $\Phi(X)$ is a function of a biquat variable $X = \Sigma e_\mu x_\mu$ ($\mu = 0,1,2,3$), $x_\mu \in R$, and

(1.5) $$\Phi(X) = u_0 + \underline{u} + i(v_0 + \underline{v}),$$

where u_μ, v_μ are real scalar function of x_μ, and twice differentiable by x_μ. The $\Phi(X)$ will be called left regular iff $\Phi(X)$ satisfies (1.4). Writing in components, we have:

$$(1.6) \qquad \partial_0 u_0 - \text{div}\, \underline{u} = 0, \quad \partial_0 \underline{u} - \text{grad}\, u_0 + \text{curl}\, \underline{v} = 0,$$

$$\partial_0 v_0 - \text{div}\, \underline{v} = 0, \quad \partial_0 \underline{v} - \text{grad}\, v_0 - \text{curl}\, \underline{u} = 0.$$

Comparing (1.3) and (1.4) provided I = 0 and identifying $\Phi(X) = F^*(X)$ together with the conditions ($E_0 = H_0 = 0$), $\Phi(X)$ becomes identical with the electromagnetic field $F(X)$ defined by (1.3) with I = 0.

Thus, an electromagnetic field quantity and regular functions of a biquat variable are intimately connected.

2. SOLUTIONS OF MAXWELL'S EQUATIONS.

2.1. Initial condition method.

A straight forward method to connect $F^*(X)$ to a left D-regular function by using (1.3) and (1.4) is to impose the "vector condition" to a left D-regular function $\Phi(X)$ as an initial condition. Then, if $\Phi(X)$ satisfies $u_0 = v_0 = 0$, then $\Phi(X)$ satisfies the condition for all the time.

2.2. Potential function method.

We now introduce another practical method which derive a potential function of an electromagnetic field.

Let $F^*(X)$ be defined by:

$$(2.1) \qquad F^*(X) = (1/2)[\Phi(X)D]$$

where $\Phi(X)$ is a left D-regular function satisfying (1.4). Then, $F^*(X)$ satisfies Maxwell's equations as well as the vector condition, as will be shown below.

$$DF^*(X) = D(1/2)[\Phi(X)D] = (1/2)[D\Phi(X)]D = 0.$$

From eq. (1.2), the scalar part of (2.1) is: $E_0 = \partial_0 u_0 - \text{div}\, \underline{u} = 0$, $H_0 = \partial_0 v_0 - \text{div}\, \underline{v} = 0$, by virtue of the first eq. of (1.6). Thus, $F^*(X)$ given by (2.1) satisfies Maxwell's equations with I = 0.

Taking complex conjugate of (1.4) and using $D^* = D$, we find:

$$(2.2) \qquad \Phi^*(X)D = 0,$$

where we used the relation $(A\, B)^* = B^*A^*$. Adding $\Phi^*(X)D = 0$ to the rhs of eq. (2.1), we have

$$F^*(X) = (1/2)[\Phi(X) + \Phi^*(X)]D = U(X)D, \tag{2.3}$$

where U(X) is the real part of

$$\Phi(X) = U(X) + iV(X) = u_0 + \underline{u} + i(v_0 + \underline{v})$$

Writing (2.3) in component functions of F(X), we get:

$$\begin{aligned} E_0 &= \partial_0 u_0 - \text{div}\,\underline{u}, \quad H_0 = 0, \\ \underline{E} &= -\partial_0\underline{u} + \text{grad}\,u_0\ , \ \underline{H} = \text{curl}\,\underline{u}. \end{aligned} \tag{2.4}$$

Using the first equation of (1.6), we find that u_0, $\underline{u}$ satisfy the Lorentz condition:

$$E_0 = \partial_0 u_0 - \text{div}\,\underline{u} = 0$$

and play the role of the scalar and vector potentials respectively of the electromagnetic field.

3. GENERATION OF REGULAR FUNCTIONS

Let $G(Z) = u(z_0, z) + ev(z_0, z)$, $e^2 = 1$, $e \neq 1$, be a regular function of a "hyperbolic variable" $Z = z_0 + ez$ [4] satisfying the following conditions:

$$\begin{aligned} &\partial_0 u - \partial v = 0, \\ &\partial_0 v - \partial u = 0, \end{aligned} \qquad \partial = \partial/\partial z,\ \partial_0 = \partial/\partial z_0 \tag{3.1}$$

Replace z_0, Z and e by the following expressions:

$$\begin{aligned} e &= [e_1x_1 + e_2x_2 + e_3x_3]/x,\ x = [x_1^2 + x_2^2 + x_3^2]^{1/2}, \\ z_0 &= x_0,\ Z = X = x_0 + e_1x_1 + e_2x_2 + e_3x_3. \end{aligned} \tag{3.2}$$

Then,

$$\Phi(X) = \Box G(X),\quad \Box = DD^\dagger = D^\dagger D \tag{3.3}$$

is a left as well as right D regular function:

$$D\Phi(X) = \Phi(X)D = 0. \tag{3.4}$$

4. TRANSFORMATION OF FIELD QUANTITIES

Let G(Z) and H(Z) be analytic functions of a hyperbolic

variable $Z = z_0 + ez$, then $G(H(Z))$ is also an analytic function of Z.

Applying this process described by (3.1) and (3.2), to G(Z) and H(Z) we obtain functions G(X) and H(X) of a biquat variable X which are a bothside D-regular function:

$$\Phi(X) = \Box G(X), \quad \Psi(X) = \Box G(H(X)):$$

(4.1)

$$D\Phi(X) = \Phi(X)D = 0, \; D\Psi(X) = \Psi(X)D = 0.$$

$\Phi(X)$ and $\Psi(X)$ are related by the mapping function H(X) through the change of variable $\Psi(X) = \Phi(H(X))$.

We may apply the mapping to a field function obtained from an analytic function of a hyperbolic variable through the above mentioned process using equations (2.1) and (2.3). However, since both $\Phi(X)$ and $\Psi(X)$ are both-side D regular we cannot obtain field functions. We can obtain field functions simply multiplying an arbitrary constant biquaternion K from the right hand side by $\Phi(X)$ and $\Psi(X)$. Then, $\Phi(X)K$ and $\Psi(X)K$ are left D-regular functions but not right D-regular functions. Thus,

$$F(X) = [\Box\Phi(X)K]$$

$$F'(X) = [\Box\Psi(X)K]D = [\Box\Phi(H(X))K]D$$

are field functions satisfying Maxwell's equations.

The mapping is a kind of exten ion of the conformal mapping of a function of a complex (or a hyperbolic) variable to that of a biquat variable.

4. REFERENCES

[1] R. Fueter: Comm. Math. Helv., 7, 307 (1934-35), 8, 371 (1936-37)

[2] 'The relation of Maxwell's equations to the theory of functions of a biquaternion variable':
K. Imaeda: Prog. Theor. Phys., 5, 133(1950). The Memoir of the Faculty of Liberal Arts and Education, Yamanashi Univ., 2, 111(1951). The latter is included in 'Quaternionic Formulation of Classical Electrodynamics and Theory of Functions of a Biquaternions Variable', Report-FPL-83, Feb. 1983.
Other references are:
D. Hestenes: Space-Time Algebra, Gordon and Breach, New York (1966), (And earlier references on the quaternionic form of Maxwell's equations).

R. Hermann: Spinors, Clifford and Cayley Algebras, Math Sci Press, Brooklyn, Ma (1974).

D. Hestenes and G. Sobczyk: Clifford Algebra to Geometric Calculus, D. Reidel Pub. Co., Dordrecht (1984).

[3] The e_i (i = 1,2,3) satisfy the same relations as Pauli matrices. As to the geometric interpretation: see D. Hestenes Space-Time Algebra loc. cit.

[4] D. Riabouchinsky: Compt. Rend., **19**, 1139 (1924).

M. Goto: Journal Inst. Electrical Engineers of Japan (in Japanese) 742 (1929) and 275 (1929).

COMPARISON OF CLIFFORD AND GRASSMANN ALGEBRAS IN APPLICATIONS TO ELECTROMAGNETICS

Georges A. Deschamps
Department of Electrical and Computer Engineering
University of Illinois
1406 W. Green St.
Urbana, IL 61801

and

Richard W. Ziolkowski
Electronics Engineering Department
Lawrence Livermore National Laboratory
P.O. Box 5504, L-156
Livermore, CA 94550

ABSTRACT

This paper presents some points of comparison between Grassmann and Clifford algebras in their applications to electromagnetics. The Grassmann algebra applies directly to (exterior) differential forms leading to Cartan's calculus. Forms of various degrees correspond most naturally to electromagnetic quantities and their relations are expressed by means of the exterior differential operator d. These relations are conveniently represented either in space or space-time by flow diagrams. They reveal the existence of potentials (Poincaré's Lemma), the independence from coordinate systems, and the role of the metric. All of these properties are briefly reviewed. A calculus for functions having their values in a Clifford algebra can also be developed based on the Dirac operator $\mathbf{D}$, whose square is a generalized Laplacian. Besides its role in Dirac's theory of the electron, this Clifford calculus can also be adapted to electromagnetics, although less directly than with Cartan's calculus. The point of departure is the role of a metric in a Clifford algebra. This can be remedied by introducing the Hodge star operator in the Grassmann algebra, an operation that is needed to express the property of matter usually represented by means of ϵ and μ.

1. INTRODUCTION

When Maxwell discovered the law of electromagnetics, the problem of expressing them in a convenient form arose. As for any law in Physics, the ultimate representation of the quantities of interest is by ordinary numbers, the components of these quantities. These numbers, however, may be grouped and acquire a mathematical structure, together with rules to manipulate them. This makes their handling easier and sometimes reveals unexpected properties.

J. S. R. Chisholm and A. K. Common (eds.), Clifford Algebras and Their Applications in Mathematical Physics, 501–515.

At first in Maxwell's presentation of electromagnetics two such systems, besides the use of components, were in competition; one based on vectors, the other on Hamilton's quaternions. Influenced by his friendship with P. E. Tait, Maxwell at first used quaternions. After a heated debate between proponents of the two systems, the formalism based on vectors was adopted, mostly due to the efforts of W. Gibbs and O. Heaviside. Together with the appropriate calculus this formalism dominates the literature of electromagnetics as well as that of most of physics. Later, following the development of Riemannian geometry, tensor calculus was introduced and applied, among other topics, to relativity and electromagnetics.

Recently, during the past few decades, a calculus due to E. Cartan proved a powerful tool in differential geometry and was shown to apply to Electromagnetics, to Classical Mechanics, and to several other fields in physics. At the same time Clifford algebras, which are generalizations of quaternions were used, with the appropriate calculus, for the same purpose. In particular the Clifford algebra associated with Minkowski space-time augmented with the differential structure defined by the Dirac operator turned out to play a major role in Dirac's relativistic theory of the electron and led to the discovery of the positron.

The purpose of this paper is to discuss the relative merits of the two approaches when applied to Electromagnetics. Both approaches can be introduced through their relations to the classical vector calculus and (if this is done correctly!) they are obviously equivalent. The preference for one notation is mostly a matter of convenience, ease in performing computation, simplicity and conciseness in expressing the results. To some extent it is a matter of taste and esthetic, hence partly subjective.

A thorough comparison of the two formalisms would require complete presentations of electromagnetics by the two methods. In this short article we can only sketch such presentations and refer to available publications for further details.

Considering that several papers in this workshop review in detail the Clifford-Dirac approach and remark on its application to electromagnetics, we shall emphasize in our discussion the application of Cartan's calculus to that same topic. A reference perhaps best known to physicists is the book *Gravitation* by Misner, Thorne and Wheeler [6]. A very readable introduction to differential forms is [3]. Most modern texts on differential geometry contain at least some allusions to Electromagnetism. We shall here use the notations of [2] with the warning that a few of them are not the conventional ones.

2. CARTAN CALCULUS AND ELECTROMAGNETICS

The following is a list of the properties that make Cartan calculus applicable to electromagnetics.

2.1 *Electromagnetic quantities are represented by differential forms of various degrees.* For instance, the electric field on R^3, classically a vector $\vec{E}$ function of position $\vec{r} = (x, y, z)$, defines the work done on a unit charge when it describes a curve C, by the integral

$$\int_C \vec{E} \cdot d\vec{r}. \tag{2.1}$$

The integrand $\vec{E} \cdot d\vec{r} = X dx + Y dy + Z dz$ is a one-form which we shall denote by E (dropping the vector sign). The integral over C is denoted by $E|C$ following the notations in [2]. Note that (x, y, z) do not have to be cartesian coordinates, in which case (X, Y, Z) are not components of $\vec{E}$. These components however are not usually needed. Other quantities: current $\vec{J}$, charge ρ, vector $\vec{D}$ and $\vec{B}$, potential $\vec{A}$, etc. are associated with differential forms of various degrees matching the dimensions of their domains of integration. The letters conventionally used are adopted for the corresponding forms [2]. Thus to the vector $\vec{J}$ is associated the 2-form $J = n \cdot \vec{J} ds$, which integrated over a surface S gives the current $J|S$ through S. Other forms in space and in space-time are discussed in [2].

2.2 *Differential forms are multiplied according to the rules of a Grassmann algebra.* A differential p-form may be written

$$\alpha = \sum a_J dx^J \tag{2.2}$$

where J is an ordered p-index $J = (j_1, j_2, \ldots, j_p)$, $j_1 < j_2 < \ldots < j_p$. The Grassmann (or exterior) product of two forms α and β is denoted by $\alpha \wedge \beta$. [The wedge can be omitted when no confusion results.] It can be computed by distributivity from

$$dx^J \wedge dx^K = dx^{JK}. \tag{2.3}$$

When J and K have at least one common index $dx^{JK} = 0$. Otherwise the juxtaposition JK is a $(p+q)$ index which can be ordered by successive exchanges of neighbors, using the rule

$$dx^i \wedge dx^j = -dx^j \wedge dx^i. \tag{2.4}$$

For instance, the one forms E and H corresponding to $\vec{E}$ and $\vec{H}$ have a product $E \wedge H$, the Poynting 2-form, which integrated over a surface S gives the power flow through S.

2.3 *Change of variable. Pullback.* If the coordinates (x, y, z) are functions of (u, v, w)

$$f: \quad (u, v, w) \mapsto (x, y, z) \tag{2.5}$$

a differential form $\alpha(x, y, z)$ is transformed into $\alpha'(u, v, w)$ by expressing x, y, z, dx, dy, dz in terms of u, v, w. To evaluate α' one makes use of the differential of scalar functions such as

$$dx = x_u du + x_v dv + x_w dw \tag{2.6}$$

and performs products according to the rules of exterior algebra. The resulting form α' is called the pullback of α by f and is denoted by $f^*\alpha$. If α is integrated over a domain $\mathcal{D}$ whose image under f is $f_*\mathcal{D}$ we have

$$\alpha|\mathcal{D} = (f^*\alpha)|(f_*\mathcal{D}). \tag{2.7}$$

2.4 *The exterior differential d.* Given a p-form

$$\alpha = a dx^J \tag{2.8}$$

where p is an p-index, its exterior differential is defined as

$$d\alpha = da \wedge dx^J \tag{2.9}$$

where $da = \sum_i \partial_i a \, dx^i$. This differential has the following properties.

a) It can take the place of the operations curl, grad, div if one translates scalars and vectors into differential forms as shown in [2].

b) It is applicable in spaces of any dimension.

c) The differential of a product of a p-form α by any form β satisfies a modified Leibnitz rule

$$d(\alpha \wedge \beta) = (d\alpha) \wedge \beta + (w\alpha) \wedge d\beta = (d\alpha) \wedge \beta + (-)^p \alpha \wedge d\beta \tag{2.10}$$

where w is the main automorphism: $w\gamma = (-)^p \gamma$ for a p-form γ.

d) Applying d twice in succession to any form gives 0:

$$d \circ d = 0. \tag{2.11}$$

2.5 *Stokes' Theorem.* This fundamental theorem, simply expressed by

$$\omega|\partial\mathcal{D} = d\omega|\mathcal{D}, \tag{2.12}$$

relates differential and integral formulations of many equations of electromagnetics. The boundary of $\mathcal{D}$ is denoted by $\partial\mathcal{D}$. To every pair of a form and its differential corresponds such an integral relation.

2.6 *Poincaré Lemma.* Property 4d can be expressed by saying that the differential $d\alpha$ of any form α is *closed* (i.e., $d(d\alpha) = 0$). If $d\alpha = 0$ holds in a contractible domain the given form α is *exact*, i.e., there exists a form β such that $d\beta = \alpha$. This form β is not unique since it can be modified by addition of any closed form (Gauge transformation). This is the basis for the introduction of potentials.

2.7 *Hodge's Star Operation.* When the manifold carrying the differentials forms is Riemannian a metric is defined over its cotangent space T^*M (space, or space-time for electromagnetics). The metric may be specified by any standard method: a scalar product $g(x, y)$ or an isomorphism $E \to E^*$: $x \mapsto \bar{x}$ with the dual E^* or E such that $x \cdot y = \bar{x}(y)$. It is convenient to use Hodge's *star operator* which maps the space of p-covectors Γ_p onto the space of (n-p)-covectors Γ_{n-p}, expressing the metric (and space orientation) by the conditions

$$\begin{aligned} x \cdot y &= *^{-1}(x * y) \\ *1 &= \omega \end{aligned} \tag{2.13}$$

where ω is the unit volume n-form. An inner product (α, β) of two p-forms is defined as the integral of the n-form $\alpha * \beta$, short for $\alpha \wedge (*\beta)$.

2.8 *Adjointness with respect to the inner product.* The linear operator L^* is the adjoint of L, L being linear, if

$$(L\alpha, \beta) = (\alpha, L^*\beta) \tag{2.14}$$

for any α and β. This defines the *co-differential*

$$d^* = *^{-1}\, d * w \tag{2.15}$$

equal to the adjoint of d and the *co-multiplier*

$$\alpha^* = - *^{-1}\, \hat{\alpha} * w \tag{2.16}$$

equal to the adjoint of the operator $\hat{\alpha}$, exterior product from the left by the one-form α ($\hat{\alpha} : \xi \mapsto \alpha \wedge \xi$). Note that δ is often used for the codifferential instead of d^*. The operator α^* has the remarkable property of being an antiderivative:

$$\alpha^*(\beta \wedge \gamma) = (\alpha^*\beta) \wedge \gamma + (-)^p \beta \wedge (\alpha^*\gamma), \tag{2.17}$$

p being the degree of β. The systematic use of α^* is convenient: for instance the Lie derivative for the flow defined by a vector field $a = \bar{\alpha}$, or the one form α is

$$L_a = \alpha^* \circ d + d \circ \alpha^*. \tag{2.18}$$

The effect of α^* on β is to give the product $\alpha \cdot \beta = \beta|a$. Note that no confusion arises from using star superscripts for both the pullback and the adjoint operations if one pays attention to the objects to which they are applied.

2.9 *Flow Diagrams for Electromagnetics*† The equations that relate the various quantities in Electromagnetics: fields, sources, potentials, etc., are displayed as *flow diagrams* in Tables I, II, and III. Maxwell's equations separate into two sets represented respectively in Tables I and II. In both of these tables forms of the same degree are placed on horizontal lines and the horizontal arrows represent the operation ∂_t ($-i\omega$ for time harmonic fields). The oblique arrows, inclined at 45°, represent the exterior differential operator $\mathbf{d}$ with respect to space. These arrows point downward in Table I and upward in Table II. This convention results from a reversed ordering in Tables I and II of the degrees associated with each horizontal line. It has been chosen for no other reason than to make Table III, the superposition of Tables I and II, more readable. In each table all of the equations have the form $\beta = \mathbf{d}\alpha$, where α is a p-form, β a (p+1)-form. Since a pullback commutes with $\mathbf{d}$, all of these equations are covariant; i.e., they have the same expression in *any system of coordinates.* To join Tables I and II into Table III the relation between the two sets of quantities must be expressed. This is accomplished by transcribing the usual constitutive relations $\vec{D} = \epsilon\vec{E}$ and $\vec{B} = \mu\vec{H}$ into differential form notation:

$$\begin{aligned} D &= \boldsymbol{\epsilon} E = \epsilon \star E \\ B &= \boldsymbol{\mu} H = \mu \star H, \end{aligned} \tag{2.19}$$

the scalar factors ϵ, μ having become the weighted star operators $\boldsymbol{\epsilon} = \epsilon\star$, $\boldsymbol{\mu} = \mu\star$, $\star$ being the Hodge star operator in R^3. Note that in vacuum $\boldsymbol{\epsilon} = \boldsymbol{\mu} = \star$. The complete set of equations superimposing Tables I and II is given in Table III.

† The authors would like to thank the IEEE Publishing Services for their permission to incorporate Tables I-IV of Ref. 2 into this paper.

TABLE I MAXWELL-FARADAY EQUATIONS

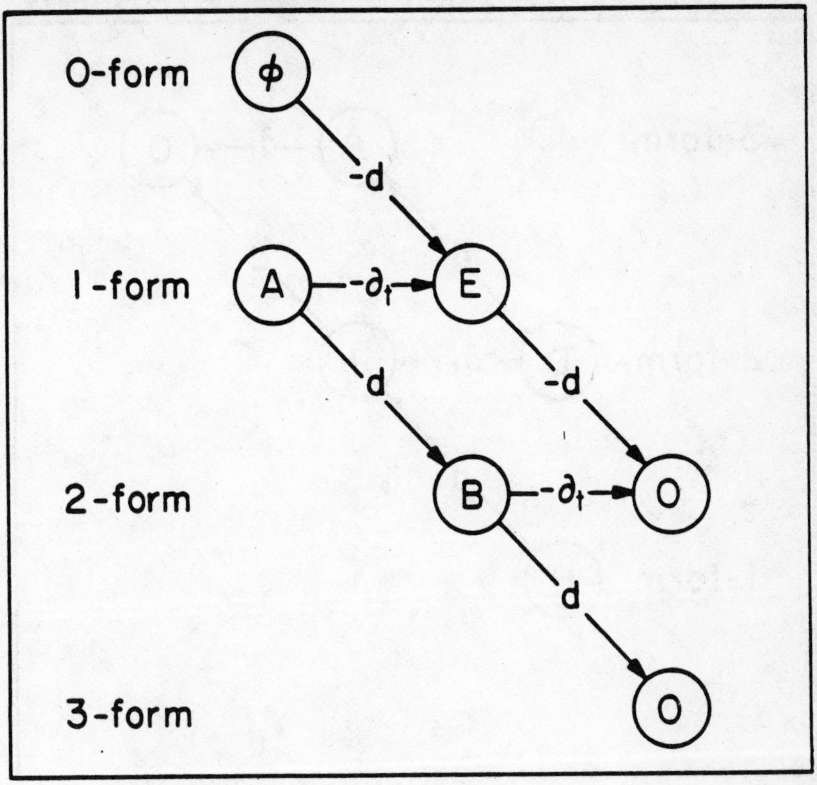

SPACE-TIME FORMULATION

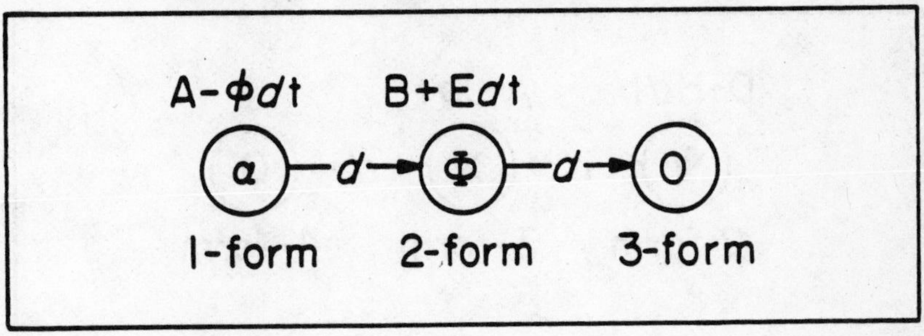

TABLE II MAXWELL-AMPERE EQUATIONS

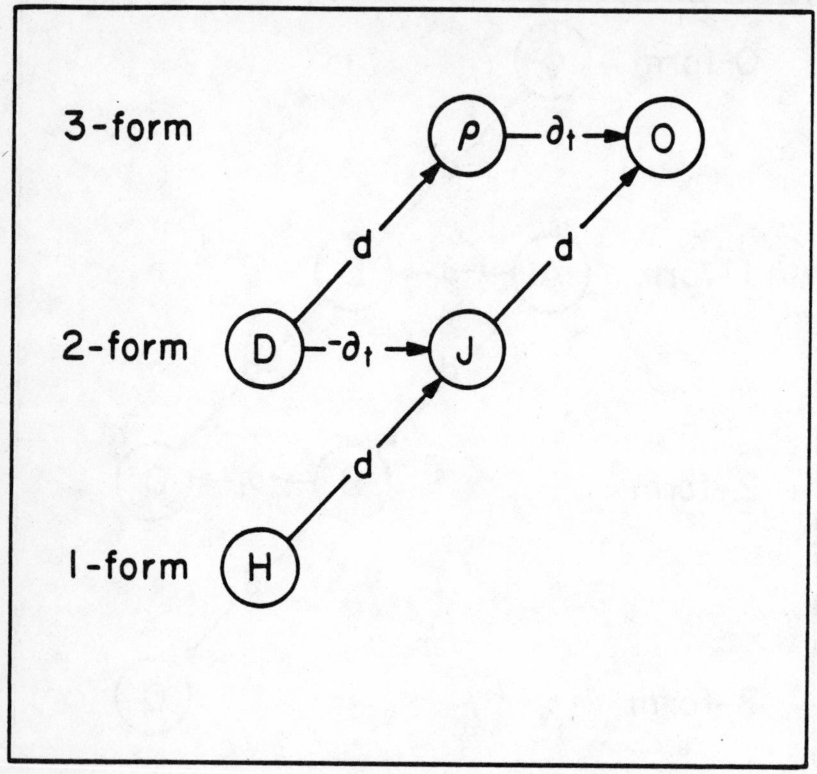

SPACE-TIME FORMULATION

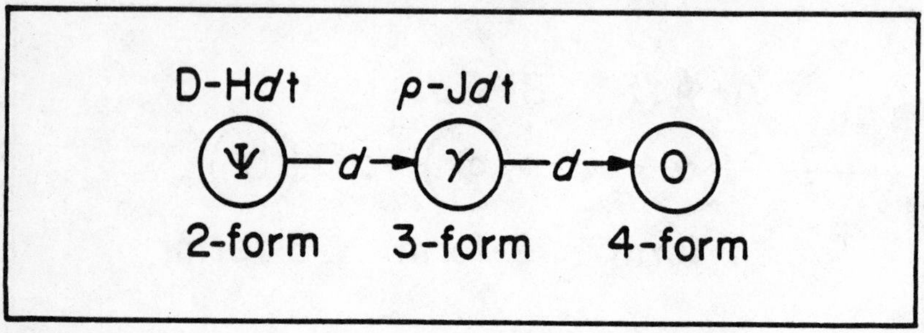

The constitutive relations are represented by the vertical arrows. A bar across an arrow means a negative sign. The quantity in any circle is the sum of those contributed by the arrows leading to it. The curved arrows represent the inverse operators L and G. For time harmonic fields $L = -(\Delta + k^2)$ and G is convolution by $e^{ikR}/4\pi R$.

TABLE III. Electromagnetics Flow Diagram

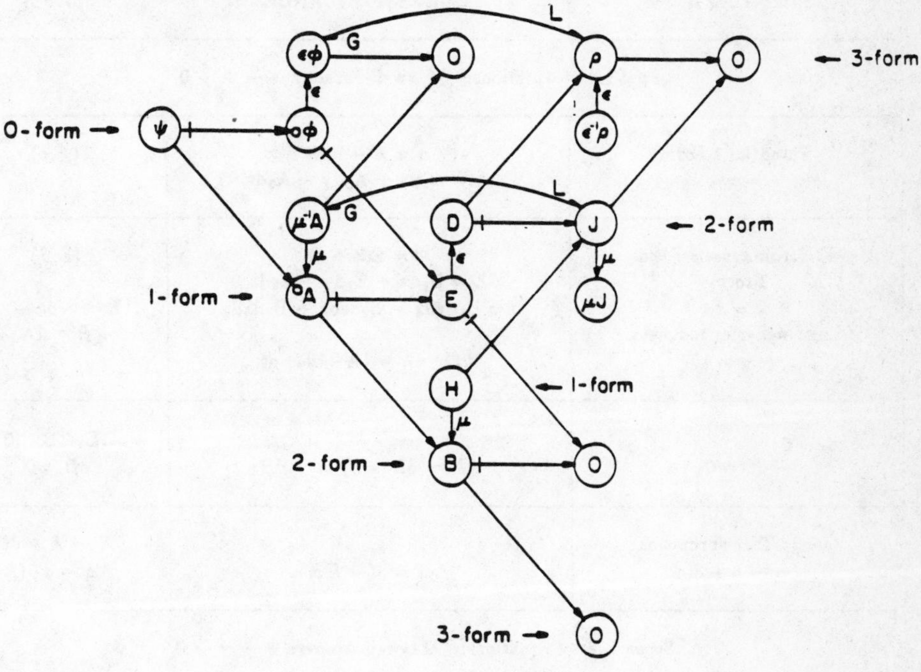

A radical simplification of the equations occurs when they are expressed in space-time as shown in the bottom of Tables I and II where pairs of quantities in columns of Tables I and II have been combined into single differential forms:

$$\begin{aligned} \alpha &= A - \phi\, dt \\ \Phi &= B + E\, dt \\ \gamma &= \rho - J\, dt. \end{aligned} \tag{2.20}$$

The operator d now indicates a differential with respect to space and time: $d = \mathbf{d} + dt\,\partial_t$. This conversion from the formulation in space (R^3) to the one in space-time (R^{3+1}) is described in detail in Table IV. Introducing the star operator for the Minkowski metric, these space-time diagrams of Tables I and II are simply connected by the relation

TABLE IV

RELATIONS BETWEEN THE THREE AND FOUR DIMENSIONAL FORMULATIONS OF THE EQUATIONS OF ELECTROMAGNETICS

Formulation in Space-Time R^{3+1}	CORRESPONDANCE	Formulation in Space R^3
	First set of equations: Maxwell-Faraday $\alpha \xrightarrow{d} \Phi \xrightarrow{d} 0$	
Potential 1-Form α	$\alpha = \mathbf{A} - \phi\, dt$ $\mathbf{A} = A_1 dx + A_2 dy + A_3 dz$	$(\mathbf{A}, \phi)$
Electromagnetic Field 2-form $\Phi = d\alpha$ (unit: weber or magnetic charge g = 137 e)	$\Phi = \mathbf{E}dt + \mathbf{B}$ $\mathbf{E} = E_1 dx + E_2 dy + E_3 dz$ $\mathbf{B} = B_1 dydz + B_2 dzdx + B_3 dxdy$ $d\alpha = \mathrm{d}\mathbf{A} - (\dot{\mathbf{A}} + \mathrm{d}\phi)\, dt$	$(\mathbf{E}, \mathbf{B})$ $\mathbf{E} = -\mathrm{d}\phi - \dot{\mathbf{A}}$ $\mathbf{B} = \mathrm{d}\mathbf{A}$
$d\Phi = 0$	$d\Phi = (\mathrm{d}\mathbf{E} + \dot{\mathbf{B}})\, dt + \mathrm{d}\mathbf{B}$	$\mathrm{d}\mathbf{E} + \dot{\mathbf{B}} = 0$ $\mathrm{d}\mathbf{B} = 0$
Gauge Transformation $\alpha \to \alpha + df$	$df = \mathrm{d}f + \dot{f}\, dt$	$\mathbf{A} \to \mathbf{A} + \mathrm{d}f$ $\phi \to \phi - \dot{f}$
	Second set of equations: Maxwell-Ampere $\Psi \xrightarrow{d} \gamma \xrightarrow{d} 0$	
Field 2-form Ψ, Charge-Current 3-form γ: $d\Psi = \gamma$ (unit: Coulomb or electron charge: e)	$\Psi = \mathbf{D} - \mathbf{H}\, dt$ $\mathbf{D} = D_1 dydz + D_2 dzdx + D_3 dxdy$ $\mathbf{H} = H_1 dx + H_2 dy + H_3 dz$ $\gamma = \rho - \mathbf{J}\, dt$ $\rho = q\, dxdydz$ $\mathbf{J} = J_1 dydz + J_2 dzdx + J_3 dxdy$	$(\mathbf{D}, \mathbf{H})$ $\mathrm{d}\mathbf{D} = \rho$ $\mathrm{d}\mathbf{H} - \dot{\mathbf{D}} = \mathbf{J}$ $(\rho, \mathbf{J})$
Continuity eq. $d\gamma = 0$	$d\gamma = -(\dot{\rho} + \mathrm{d}\mathbf{J})\, dt$	$\mathrm{d}\mathbf{J} + \dot{\rho} = 0$

The differential d applies only to space coordinates, while *d* applies also to time. The dot means a time derivative.

$$\Psi = *\Phi. \tag{2.21}$$

Alternatively one may use the codifferential d^*, eliminate Ψ, and introduce the source 3-form $\Gamma = *\gamma = j - q\,dt$ to obtain:

$$\begin{aligned} d\alpha &= \Phi \qquad d\Phi = 0 \\ d^*\Phi &= \Gamma \qquad d^*\Gamma = 0. \end{aligned} \tag{2.22}$$

These results are represented by the flow diagram in Table V. The inverse operators L and G now represent respectively the D'Alembertian: $\Box = \Delta - \partial^2_{ct}$, and convolution by $\delta(t - R/c)/4\pi R$. The Lorentz gauge condition becomes $d^*\alpha = 0$. It is satisfied if $\alpha = d^*\Pi$ where Π is the Hertzian potential two-form. The whole theory of Hertz potentials and gauge transformations[10] is radically simplified by using differential forms. For instance, Nisbet's gauge transformation of the third kind may be expressed simply as $\alpha \longrightarrow \alpha + d\Lambda$.

TABLE V SPACE–TIME ELECTROMAGNETICS FLOW DIAGRAM

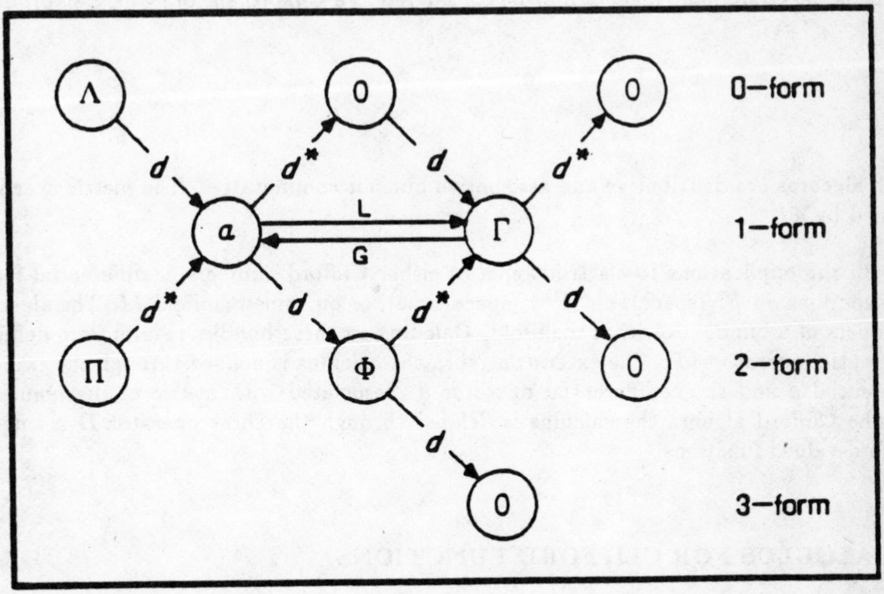

3. COMPARISON OF THE GRASSMANN AND CLIFFORD ALGEBRAS

Both the Grassmann and the Clifford algebras have for elements the multivectors Γ constructed on a vector space V of dimension n. They differ by the product designated respectively by $\wedge$ for the former (exterior, or wedge product) and $\vee$ for the latter (Clifford, or vee product). One can assume that V is referred to an orthonormal basis $e^1, \ldots, e^n$ which is extended to a basis $\{e^J\}$ of Γ, where the multiindex J runs over ordered subsets of $N = (1, 2, \ldots, n)$.

In the Clifford algebra the rule for forming products is such that the relation

$$\left(\sum_{i=1}^{n} x_i e^i\right)^2 = e^0 |x|^2 \tag{3.1}$$

holds and the resulting algebra is associative. The Clifford algebra multiplication rule

$$\begin{aligned} e^i \vee e^j &= e^i \wedge e^j + e^i \cdot e^j \\ e^i \cdot e^j &= g^{ij} \end{aligned} \tag{3.2}$$

for the 1-vectors in $\Gamma_1 = V$ can be extended to elements of Γ. [Ref. 1, p. 61]. An exterior algebra can be defined on the same space Γ of multivectors by means of the exterior product $\wedge$. Thus the Grassmann algebra may be considered as a Clifford algebra where all products $e^i \vee e^i = g^{ii} = 0$:

$$e^{ij} = -e^{ji}, \qquad e^{ii} = 0. \tag{3.3}$$

Both algebras are distributive and associative but not commutative. The metric over V is defined by g^{ij}.

In the applications to electromagnetics either Clifford numbers or differential forms are functions on R^3 (space), on R^{3+1} (space-time), or on some manifold M. The algebras are fibers of a bundle over that manifold. Calculus on these bundles results from defining differential operators. For the exterior algebra the calculus is defined through the exterior differential d and the codifferential operator d^* associated with metric of the manifold. For the Clifford algebra the calculus is defined through the Dirac operator $\mathbf{D}$ acting on Clifford-valued functions.

4. CALCULUS FOR CLIFFORD FUNCTIONS

Clifford numbers constructed on a vector space $V = R^n$, endowed with a metric, form a linear space Γ of dimension 2^n. In the Clifford algebra where functions $u : \Gamma \to \Gamma : x \mapsto y$ are considered, partial derivatives of the components of y with respect to the components of x can be taken as the coefficients of a Clifford function. Let us consider the case of Clifford numbers that represent electromagnetic quantities over space and time. A point in space-time is represented by a Clifford number $x = \sum_{i=1}^{4} x_i e^i$. A field is a function $u(x)$ whose value at x is a multivector

$$y(x) = \sum_J y_J \, e^J. \tag{4.1}$$

The Dirac differential operator

$$\mathbf{D} = \sum_{i=1}^{4} e^i \, \partial_i \tag{4.2}$$

where the ∂_i, partial derivative operators with respect to x_i, are treated as numbers making $\mathbf{D}$ into a symbolic Clifford number which can be represented by a 4×4 matrix operator as done by Dirac. One may form the product $\mathbf{D} \vee y$ according to the Clifford algebra:

$$\mathbf{D} \vee y = \sum (\partial_i y_J) e^i \vee e^J. \tag{4.3}$$

Usually $\mathbf{D} \vee y$ is simply written $\mathbf{D}y$. The result decomposes into two parts: one which comes from the terms where $i \notin J$ (i.e., i not an element of J) and one with terms $i \in J$. Because of the correspondences

$$e^i \longleftrightarrow dx^i \qquad \text{and} \qquad e^J \longleftrightarrow dx^J, \tag{4.4}$$

the result may be expressed as follows. For the first term

$$\sum (\partial_i y_J) e^i \wedge e^J = \sum (\partial_i y_J) \, dx^{iJ} = dy. \tag{4.5}$$

The second term can be shown to be the codifferential of y, $d^* y$. We may write this result in operator form

$$\mathbf{D} = d + d^* \tag{4.6}$$

and verify immediately that

$$\mathbf{D}^2 = d\,d^* + d^*\,d = -\Delta \tag{4.7}$$

where Δ is the Laplacian. The associativity of $\vee$ is used to show that if u satisfies $\mathbf{D}u = 0$ then $\mathbf{D}(\mathbf{D}u) = 0$. This short discussion suggests that Clifford-valued functions (such as spinors, the four component ψ of Dirac) could be replaced by differential forms.[8,9]

Consider now the electromagnetic field with $\Phi = B + E\,dt$ in 4 dimensions. As shown by the flow diagram in Table V, it satisfies

$$d\Phi = 0 \qquad \text{and} \qquad d^*\Phi = \Gamma. \tag{4.8}$$

These relations can be combined to give the Clifford algebra form of Maxwell's equations:

$$\mathbf{D}\Phi = J \tag{4.9}$$

which reduces to $\mathbf{D}\Phi = 0$ in free-space.

5. CONCLUSION

The authors are obviously biased the favor of the Cartan calculus approach and they are not alone in this preference! It must be acknowledged, however, that convenience in computation depends largely on the background of the person doing the computation. It cannot be evaluated fully objectively and the judgment about it may evolve as science develops. In some fields the Clifford approach is superior. Perhaps a combination of the two formalisms into a single algebra (Kähler-Atiyah) with two products could be superior to either. In the case of electromagnetics this remains to be demonstrated by concrete examples. At this point, however, and for classical electromagnetics, the Cartan calculus seems particularly well suited.

REFERENCES

1. C. Chevalley (1946), *Theory of Lie Groups*, Princeton University Press.

2. G. Deschamps (1981), 'Electromagnetics and Differential Forms,' *Proc. IEEE*, Vol. 69, pp. 676-696.©1981 IEEE

3. H. Flanders (1963), *Differential Forms with Applications to the Physical Sciences*, New York, Academic Press.

4. E. Kähler (1938), 'Bemerkungen uber die Maxwellschen Gleichungen,' *Hamburg Abhandlungen*, Vol. 12, 1-28.

5. A. Mercier (1935), 'Expression des equations de l'Electromagnetisme au moyen des nombres de Clifford,' *Thesis*, Geneva.

6. C. W. Misner, K. S. Thorne, and J. A. Wheeler (1973), *Gravitation*, W. H. Freeman and Co.

7. M. Riesz (1958), 'Clifford numbers and Spinors,' *Lecture Series No. 38*, The Institute for Fluid Dynamics and Applied Mathematics, University of Maryland.

8. W. Graf (1978), 'Differential Forms as Spinors', *Ann. Inst. H. Poincaré*, Vol. 29(1), pp. 85-109.

9. E. Kähler (1960), 'Innerer und äusserer Differentialkalkül', *Abh. Deutsch. Akad. Wiss. Berlin Kl. Math. Phys. Tech.*, no. 4, 32 pp.

10. A. Nisbet (1955), 'Hertzian Electromagnetic Potentials and Associated Gauge Transformations', *Proc. Roy. Soc. London*, Vol. A231, pp. 250-263.

SYMPLECTIC CLIFFORD ALGEBRAS

A. CRUMEYROLLE
Université P.Sabatier
Mathématiques
118, route de Narbonne
31062 TOULOUSE Cedex
FRANCE

ABSTRACT. We give a construction of symplectic Clifford algebras according to an algebraic process analogous to the one used in orthogonal Clifford algebras. We define Clifford and spinors groups and symplectic spinors. We develop two applications : first a geometric approach to the Fourier transform, second a deformation theory for the algebras associated with a symplectic manifold.

INTRODUCTION. The nice results obtained in the orthogonal Clifford algebras and spinors, naturally suggest some parallel developments by means of mathematical objects called symplectic Clifford algebras and symplectic spinors.

The ancestral origin of these topics is probably the "Weyl algebra".

Considerer the algebra generated by the variables $p_1, p_2, \ldots p_r$, $q_1, q_2, \ldots q_r$, with the relations : $p_i p_j = p_j p_i$, $q_i q_j = q_j q_i$, $p_i q_j - q_j p_i = \delta_{ij}$; we get a non commutative algebra, linearly isomorphic to a polynomial algebra.

The classical correspondences :

$$p_j \longrightarrow \partial/\partial x^j\ , \quad q_j \longrightarrow i\lambda x^j, \quad 1 \longrightarrow i\lambda, \quad \lambda \in \mathbb{C}^*,$$

allows one to describe the Weyl algebra like differential operators with constant coefficients.

Several works by A. Weil, Shale and al---, pointed out the role of the "metaplectic group" and covering groups for the symplectic group.

Historically these subjects were approached first through the functional analysis and nice algebraic structures were often hidden behind the scholarship !

My aim was devoted to develop an algebraic approach, such that the metaplectic group becomes like a spinor group. Ten years ago I started to construct enlarged Weyl algebras and I gave the crucial formula $\exp(ta^2)x\exp(-ta^2) = x + 2\,tF(a,x)a$ (cf. below), connec-

J. S. R. Chisholm and A. K. Common (eds.), Clifford Algebras and Their Applications in Mathematical Physics, 517–529.

ting the exponentials and the symplectic transvections. This formula is the "open sesame" of the theory, the one which permits the construction of symplectic spinor groups.

Using our approach, in despite of technical difficulties, we can for example explain the geometric meaning of the Maslov index, build a new deformation theory for the Poisson algebra, give a new approach to the Fourier transform, etc...

If one compares the methods and results in this way, a very deep parallelism appears with the orthogonal case. There are some technical problems, because the symplectic Clifford algebra is infinite dimensional, so that convergence questions appears on the way. Also I find advantageous to my purpose the introduction of a parameter h which represents the "Planck constant", but h is considered as either an infinitely small parameter or an indeterminate. I'll try to show both algebraic and geometric ideas.

1. BASIC TECHNICS AND GENERALITIES

1.1. Weyl algebra

If E is a symplectic vector space ($\dim E = n = 2r$) over the field $\mathbb{K}$ ($\mathbb{K} = \mathbb{R}$ or $\mathbb{C}$), we denote by hF the symplectic form (h is either an indeterminate or a $\mathbb{K}$-valued parameter which in mechanical applications corresponds to the Planck's constant, about 10^{-27} C.G.S.)

We define the symplectic Clifford algebra, in the strict sense (Weyl algebra), $C_S(hF)$, as the quotient of the tensor algebra $\otimes E$ by the two-sided ideal I generated by the elements :

$$x \otimes y - y \otimes x - hF(x,y), \quad x, y \in E. \tag{1}$$

We can interpret $C_S(hF)$ either as a $\mathbb{K}$-algebra (h is a $\mathbb{K}$-valued parameter) or an algebra over the formal power series ring $\mathbb{K}[[h]]$ (h is an indeterminate).

It is easy to see that algebra has the analogous elementary properties to the orthogonal Clifford algebra :

a) Universal property : If $(A,*)$ is a $\mathbb{K}$-algebra $\rho_F : \otimes E \longrightarrow \otimes E/I = C_S(hF)$, u a linear operator $E \longrightarrow (A,*)$, such that :

$u(x) * u(y) - u(y) * u(x) = hF(x.y)$,

there exists $\hat{u} : C_S(hF) \longrightarrow A$, with : $u = \hat{u} \circ \rho_F$.

b) Trivial center.

c) Linear isomorphism with the symmetric algebra $\vee E$.

We write the basis table :

$$(e_i)^k (e_j)^\ell = (e_j)^\ell (e_i)^k + h\ell k\, F(e_i,e_j)(e_j)^{\ell-1}(e_i)^{k-1} + \dots$$
$$\dots + (hF(e_i,e_j))^p\, C^p_\ell C^p_k\, p!\,(e_j)^{\ell-p}(e_i)^{k-p} + \dots \tag{2}$$

$p \leqslant \ell$, $p \leqslant k$, $(e_1, e_2, \ldots e_n)$ is a frame of E.

1.2. Enlarged formal symplectic Clifford algebra $\check{C}_S(hF)$.

Here h is an indeterminate.

We write a formal symplectic power series $\hat{u}$:

$$\hat{u} = \Sigma \lambda_{h_1, h_2 \ldots h_r, k_1, k_2, \ldots k_r} (e_1)^{h_1} (e_2)^{h_2} \ldots (e_r)^{h_r} (e_{1^*})^{k_1} (e_{2^*})^{k_2} \ldots (e_{r^*})^{k_r}$$

$$= \sum_{H\,K^*} \lambda_{HK^*}\, e^H e^{K^*}, \quad (e_\emptyset = 1)$$

with an infinity of terms, $\{e_\alpha, e_{\beta *}\}$ is a symplectic frame. We obtain an associative algebra.

1.3. Enlarged (non formal) symplectic Clifford algebra $C_S(hF)^\alpha$.

$h \in \mathbb{R}$ or $\mathbb{C}$.

We choose from some rank :

$$|\lambda_{HK^*}| < \frac{\sigma(\hat{u})\ \rho(\hat{u})^{|H|+|K^*|}}{(H!\ K^*!)^{1/2+\alpha}} \tag{3}$$

$H! = h_1!\ h_2! \ldots h_r!$, $K^*! = k_1!\ k_2! \ldots k_r!$, $|H| = \Sigma h_i$, $|K^*| = \Sigma k_i$,

$\alpha > 0$ is a fixed real number.

These conditions are independent of the selected frame.

These conditions determine an algebra over $\mathbb{K}$.

<u>Remark</u> :

$\exp(t(e_\alpha)^2)$, $\exp(t(e_{\alpha^*})^2)$ belong to $\check{C}_S(hF)$ if $t \in \mathbb{K}$, and satisfy (3) with $\alpha = 0$. The convergence of $\hat{u}\hat{v}$ requires an upperbound for $\rho(\hat{u})$, $\rho(\hat{v})$: we can take $(2h)^{-1/2}$; this upperbound is very large, the order of 10^{13} C.G.S.

2. SYMPLECTIC CLIFFORD GROUPS AND SYMPLECTIC SPINOR GROUPS.

1.1. Definition 1.

We call symplectic formal Clifford group G_S^H, the subgroup of the elements $\gamma \in \check{C}_S(hF)^*$ such that :

$$\gamma x \gamma^{-1} \in E \otimes \mathbb{K}[[h]] = E_H, \quad \forall x \in E_H.$$

One can see that G_S^H is a group, that $p : \gamma \to p(\gamma)$ with $p(\gamma)(x) = \gamma x \gamma^{-1}$ is a homomorphism from G_S^H into $Sp(n, \mathbb{K}[[h]])$, group of

$\mathbb{K}[[h]]$ - linear isomorphism σ of E_H such that $hF(\sigma x, \sigma y) = hF(x,y)$, and $\ker p = \mathbb{K}[[h]]^*$. The elements $\exp(ta^2)$, $t \in \mathbb{K}[[h]]$, $a \in E_H$, give :

$$\exp(ta^2)x \exp(-ta^2) = x + 2hF(a,x)ta \quad (4)$$

and belong to G_S^H, for any t.

When $\mathbb{K} = \mathbb{R}$ or $\mathbb{C}$, we can consider h as an arbitrary fixed element of $\mathbb{K}$ in the second member of (4), and $p \circ \exp(ta^2)$, $a \in E$, furnishes according to (4) a symplectic transvection of E, and the set of the images of these exponential elements $Sp(n,\mathbb{K})$, a classical result.

We conclude with the following definition:

1.2. Definition 2.

We call symplectic Clifford group G_S the quotient of the subgroup of elements of G_S^H with form :

$$\pi_i(\lambda_i h) \exp(t_i(a_i)^2),\ \lambda_i(h) \in \mathbb{K}[[h]]^*,\ t_i \in \mathbb{K},\ a_i \in E,$$

by the subgroup of the elements $(1+u)$, $u \in \mathbb{K}[[h]]$.

One can consider $\ker(p)$ in $\mathbb{K}^*$ and write the exact sequence :

$$1 \longrightarrow \mathbb{K}^* \longrightarrow G_S \xrightarrow[p]{} Sp(n, \mathbb{K}) \longrightarrow 1. \quad (5)$$

As $Sp(n, \mathbb{K})$ is a connected group, any element of $Sp(n, \mathbb{K})$ is a finite product of symplectic transvections belonging to an arbitrary small neighbourhood of the identity.

Thus we can consider G_S generated by exponentials satisfying the condition $|\rho(\hat{u})| < (2h)^{-1/2}$, the set of these exponentials constitutes in G_S a local Lie group Γ ; we can construct that local Lie group Γ by means of a neighbourhood $\mathcal{U}$ of 0 in the Lie algebra L_h obtained from the local Lie algebre L, fixing h. By some standard results in Lie group theory [1], G_S possesses a Lie group structure.

1.3. Definition .

Let E' be the complexified space of E, with symplectic form hF ; the subgroup of the elements γ of G_S' such that $\gamma x \gamma^{-1} \in E$, $\forall x \in E$ with $|N(\gamma)| = 1$ is called the metaplectic group $M_p(r)$.

We have defined $N(\gamma) = \beta(\gamma)\gamma$; β is the principal anti-involution of the Clifford algebra such that $\beta|_{E'} = i\mathrm{Id}$, $N(\gamma) \in \mathbb{C}^*$; if $\gamma \in G_S'$ is the Clifford group for E'. The group $M_p(r)$ gives rise to the exact sequence :

$$1 \longrightarrow S^1 \longrightarrow M_p(r) \longrightarrow S_p(n, \mathbb{R}) \longrightarrow 1 \quad . \quad (6)$$

More particularly, if we choose $N(\gamma) = 1$, we obtain the group $Sp_2(r)$, and the exact sequence :

$$1 \longrightarrow \mathbb{Z}_2 \longrightarrow Sp_2(r) \longrightarrow Sp(n, \mathbb{R}) \longrightarrow 1 \ . \quad (7)$$

$Sp_2(r)$ is called spinor symplectic group of order 2. There exist also $Sp_q(r)$, $Sp_\infty(r)$, spinor groups of order q, ∞ [2,a].

3. SYMPLECTIC SPINORS.

3.1. Non formal case :

This case corresponds to $C_S(hF)^{\alpha}$.
Consider $\mathcal{J}m = \{\hat{u} \in C_S(hF)^{\alpha},\ \hat{u} = \Sigma\, \lambda_{HK^*}\, e^H e^{K^*},\ K^* \neq \emptyset\}$
$\mathcal{J}_m$ is a left maximal ideal and the quotient space $C_S(h_F)^{\alpha}/\mathcal{J}_m$ is a natural representation space for $C_S(hF)^{\alpha}$, this representation is irreducible because $\mathcal{J}m$ is maximal. This quotient is a spinor symplectic space, it is linearly isomorphic to a subspace of the symmetric algebra in $e_1, e_2, \ldots, e_r$.

3.2. Formal case :

This case corresponds to $\check{C}_S(hF)$, but it is more convenient replacing $\mathbb{K}((h))$ by $\mathbb{K}((h))$, the set of rational fractions constructed over $\mathbb{K}((h))$.

We define $\mathcal{J}m$ as we did in 3.1 and we obtain an irreducible representation in the quotient

$$C_S(hF)/\mathcal{J}_m, \text{ written } C_S(hF)\Phi^* .$$

It is possible to prove that $C_S(hF)$ is a "simple" algebra (direct sum of isotropic simple algebras).

If $u\Phi^* = \lambda_H e^H \Phi^*$ is any symplectic spinor, $v \longrightarrow vu\Phi^* = vu (\mathrm{mod}\ \mathcal{J}m)$ is the spinor representation of the formal symplectic Clifford algebra.

There are some analogies with the orthogonal case.

<u>Remark</u> :

If we consider the quotient $\overline{H}$ of the algebra $\mathbb{K}[h]$ of the polynomials in h by the ideal of multiples of h^p, p fixed integer, $p \geqslant 2$, $\overline{H}$ is a p-dimensional algebra over $\mathbb{K}$. If we replace $\mathbb{K}[[h]]$ by $\overline{H}$ we obtain truncated symplectic Clifford algebras which look extremely interesting in physical applications, because h could be considered as an "infinitesimal" quantity.

4. SYMPLECTIC SPINORS AND FOURIER TRANSFORM.

4.1. A Review

Let E be a real vector space with dimension $n = 2r$, possessing a symplectic form F. If J is an endomorphism of E such that $J^2 \equiv -\mathrm{Id}$, J is called a complex structure.

<u>Definition 1</u> :

J is adapted to F, with signature (p,q), p+q=r, if there exists a nondegenerate hermitian form $\eta : E \times E \to \mathbb{C}$, with signature (p,q), such that $F = -\mathrm{Im}(\eta)$.

The real part $\mathcal{R}(\eta(x,y)) = (x,y)$ defines a quadratic form with (2p, 2q) signature.

It is easy to establish that if
$(Jx,y) = F(x,y)$, $(Jx,Jy) = (x,y)$, $F(Jx,Jy) = F(x,y)$,
then J is orthogonal and symplectic.

We call lagrangian a maximal totally isotropic space of E, according F (with real dimension r): $F(x,y) = 0$.

Proposition 1 :

Let L be a lagrangian, non isotropic for the pseudo-metric $(\cdot,\cdot)$; if I is adapted to F, J(L) is a supplementary lagrangian of L, orthogonal to L.

Proposition 2 :

If $E = L \oplus L'$, L and L' being both lagrangian subspaces and g a scalar product with (p,q) signature over L, there exists over E a unique complex structure J, adapted to F with (p,q) signature, such that J(L) = L', and the hermitian form of imaginary part (-F), restrained to L, is g.

We can introduce a symplectic frame (e_α, e_{β^*}) adapted to the decomposition $E = L \oplus L'$ with
$g(e_\alpha,e_\alpha) = 1, \quad \alpha = 1,\ldots,p, \; g(e_\alpha,e_\alpha) = -1, \quad \alpha = p+1,\ldots,p+q$
$g(e_\alpha,e_{\beta^*}) = 0$
$J(e_\alpha) = \chi_\alpha e_{\alpha^*}, \; J(e_{\alpha^*}) = -\chi_\alpha e_\alpha, \qquad \chi_\alpha = \pm 1$.

$\chi_\alpha = 1$ if $\alpha = 1,\ldots p$, $\chi_\alpha = -1$ if $\alpha = p+1,\ldots p+q$.

$$\varepsilon_\alpha = \frac{e_\alpha - i\chi_\alpha e_{\alpha^*}}{\sqrt{2}}, \quad \varepsilon_{\alpha^*} = \frac{e_\alpha + i\chi_\alpha e_{\alpha^*}}{\sqrt{2}}.$$

$J(\varepsilon_\alpha) = i\varepsilon_\alpha, \; J(\varepsilon_{\alpha^*}) = -i\varepsilon_{\alpha^*}$.

4.2. We choose first h = 1, but only as a question of convenience. It is easy to prove the interesting result :

$$p \circ \exp(-\tfrac{\pi}{2} \Sigma \chi_\alpha \varepsilon_\alpha \varepsilon_{\alpha^*}) = J, \quad [2,a].$$

If we put

$$U = \frac{1}{2} \sum_\alpha \chi_\alpha (e_\alpha)^2, \; V = \frac{1}{2} \sum_\alpha \chi_\alpha (e_{\alpha^*})^2, \quad W = \frac{r}{2} - \sum_\alpha e_\alpha e_{\alpha^*}, \tag{8}$$

then

$$[U,V] = -W, \; [U,W] = -2U, \; [V,W] = 2V, \tag{9}$$

where the bracket $[U,V]$, for example, is UV - VU (products in the Clifford algebra. The reader can recognize the Lie algebra of $SL(2,\mathbb{R})$ (or SO(2,1)).

If we consider the Lie algebra of O(p,q), it is the set of

linear combinations with real coefficients of the $\chi_\beta e_\alpha e_{\beta^*} - \chi_\alpha e_{\beta^*} e_\alpha$, and $O(p,q)$, $SL(2,\mathbb{R})$ are in $Sp(2r,\mathbb{R})$ centralizer subgroups of each other

If h is now any number, we identify the vector e_α and the variable q^α, thus a symplectic spinor becomes a formal sequence of the q^α, $(\alpha = 1, 2,\dots,r)$.

Using the famous "correspondence principle" of the quantum mechanics, the formula $e_{\alpha^*}(e_\alpha)^k = (e_\alpha)^k e_{\alpha^*} - hk(e_\alpha)^{k-1}$ and the definition above of the symplectic spinors, show that if e_α acts like product by q^α, e_{α^*} acts like the derivation $-h\, \partial/\partial q^\alpha$; using systematically this correspondence principle,

U becomes $\frac{1}{2} g_{\alpha\beta} q^\alpha q^\beta$ (beltramian) or $\frac{1}{2}\rho^2$,

V becomes $\frac{1}{2} g^{\alpha\beta} \dfrac{\partial^2}{\partial q^\alpha \partial q^\beta}$ (laplacian) or $\frac{1}{2}\Delta$,

W becomes $\frac{r}{2} - q^\alpha\, \partial/\partial q^\alpha$ (the last term is a Liouville field).

We have called $g_{\alpha\beta}$, $g^{\alpha\beta}$ the components of the pseudo-metric tensor.

<u>Proposition 3</u> :

$\tilde{J} = \exp(-\frac{\pi}{2}(U+V))$ is the lift of J in $Sp_2(r)$.

In terms of infinitesimal operators J corresponds to

$$-\frac{\pi}{4}\left(g_{\alpha\beta} q^\alpha q^\beta + g^{\alpha\beta} \frac{\partial^2}{\partial q^\alpha \partial q^\beta}\right)$$

(modulo a constant factor, it is the sum of the laplacian and the beltramian).

4.3. The Fourier transform and the complex structure J.

We choose now $h = \frac{1}{2\pi i}$ for easy comparison of our results with some classical analytic results.

Infinitesimally J comes from $-i\pi^2(U+V)$ or

$$-\frac{\pi i}{4}\left(2\pi\rho^2 - \frac{\Delta}{2\pi}\right).$$

With these new notations $\tilde{J}$ is writen

$$\tilde{J}_1 = \exp\left(-\frac{\pi i}{4}\left(2\pi\rho^2 - \frac{\Delta}{2\pi}\right)\right). \qquad (10)$$

Here we are using some results pointed out in [3]. A symplectic spinor corresponds to a sequential function f of the q^α, $\mathcal{F}^{p,q}$ is the Fourier transform :

$$\mathcal{F}^{p,q}(f)(y) = \int_{\mathbb{R}^{p+q}} f(x) e^{-2\pi i(x.y)} dx \quad . \qquad (11)$$

We can give the very simple result :

Proposition 4 :

Modulo the scalar constant factor $\exp(-\frac{\pi i}{4}(p-q))$, the Fourier transform $\mathfrak{F}^{p,q}$ can be identified with the lifting $\tilde{J}_1$ of J in $Sp_2(p+q)$, acting in the symplectic spinor space by left product.

Consider $f(x) = \exp(-\pi(x\circ x))$ (Hermite function); this function corresponds to $\exp(-\pi\,\Sigma(e_\alpha)^2)\Phi^* = \exp(-2\pi U)\Phi^*$ in the spinor space; immediately we obtain its Fourier transform

$$\exp\frac{\pi i(p-q)}{4}\ \exp(-\frac{\pi i}{4}(2\pi\rho^2 - \frac{\Delta}{2\pi}))\ \exp(-2\pi U)\Phi^* = \exp(-2\pi U\Phi^*),$$

without any integral calculus.

The classical formula

$$\chi_\alpha\, \mathfrak{F}(\frac{\partial f}{\partial q^\alpha}) = 2\pi i q^\alpha\, \mathfrak{F}(f)$$

is nothing but

$$\chi_\alpha \tilde{J}_1 e_{\alpha^*} \tilde{J}_1^{-1} = e_\alpha \quad ,$$

equivalent to :

$$J(e_{\alpha^*}) = -\chi_\alpha e_\alpha \quad .$$

The commutation property given in 4.2. between the elements of $O(p,q)$ and $SL(2,\mathbb{R})$ involves the classical property of the Fourier transform which commutes with the action of the isometries $O(p,q)$.

The previous calculations are formal or not formal. If we are interested in convergence problems the algebras $C_S(hF)^\alpha$ are useful, so also are the truncated algebras. The formal point of view is extremely easy to work. (Compare this with the Fourier transform of distributions).

4.4. Sesquilinearity and Parseval-Plancherel theorem.

The constant h is changed into $\frac{1}{2\pi i}$ again.

We take : $E_\alpha = \exp(-\frac{i\pi}{4})e_\alpha$, $E_{\alpha^*} = \exp(-\frac{i\pi}{4})\, e_{\alpha^*}$,

$E_\alpha E_{\beta^*} - E_{\beta^*}E_\alpha = \frac{1}{2\pi}\,\delta_{\alpha\beta}$, and consider E_α, E_{α^*}, as "real" elements.

From the evident isomorphism between the Clifford algebras constructed over the (e_k) and the (E_k) we consider now a Fourier transform determined by the product according to

$$\exp\frac{\pi i(p-q)}{4}\,\tilde{J}_1 \ ,$$

where $\tilde{J}_1$ corresponds to J with $J(E_\alpha) = \chi_\alpha E_{\alpha^*}$, $J(E_{\alpha^*}) = -\chi_\alpha E_\alpha$.

$$F(E_\alpha, E_{\gamma^*}) = \frac{\delta\alpha\gamma}{4\pi} \ ,$$

β is the principal anti-involution with

$$\beta(E_\alpha) = -iE_{\alpha^*},\quad \beta(E_{\alpha^*}) = iE_\alpha,$$

and the symplectic spinors are constructed with the E_α.

Naturally we write :

$$\beta(u\Phi^*) = \Phi\beta(u)$$

$$\beta(\overline{u\Phi^*}) = \Phi\beta(\overline{u})$$

where (-) denotes complex conjugation.

We are writing symbolically, in double classes :

$$\Phi\ \beta(\overline{u})v\ \Phi^* = \Phi\mathcal{H}(u\Phi^*,\ v\Phi^*)\ \Phi^* .$$

$\mathcal{H}(u\Phi^*, v\Phi^*)$ is a scalar, with

$$\mathcal{H}(e^H\Phi^*,\ e^K\Phi^*) = 0, \text{ if } K \neq H$$

$$\mathcal{H}(e^H\Phi^*,\ e^H\Phi^*) = |h|^H\ H!$$

$\mathcal{H}$ is a sesquilinear definite positive hermitian form. $\mathcal{H}$ is invariant under the $Sp_2(r)$ action and particularly under $\tilde{J}_1$ (symplectic Parseval-Plancherel theorem).

Remark :

The preceding spinor representations correspond to the Schrödinger representation. Using $(\varepsilon_\alpha,\varepsilon_{\beta^*})$ instead of (e_α,e_{β^*}) we can construct spinors corresponding to the Bargmann-Fock representation.

Observe that the definition $\mathcal{H}$ possesses the symplectic invariance.

5. THE LAPLACIAN AND THE SYMPLECTIC GEOMETRY.

We now put h = 1.

$\sum_\alpha x_\alpha(\chi_{\alpha^*})^2$ corresponds to Δ, symplectic spinors being polynomials of q^α, $\alpha = 1,\dots,r$.

Taking the exponential $\gamma = \exp(\Sigma\chi_\alpha(e_{\alpha^*})^2 t)$, we obtain a symplectic transvection.

$\Delta u\Phi^* = 0$ is equivalent to $\gamma u\Phi^* = u\Phi^*$.

Proposition 5 :

The harmonic symplectic spinors are the spinors invariant under the natural action of the one parameter group :

$$t \longrightarrow \exp(\Sigma\chi_\alpha(e_{\alpha^*})^2 t).$$

Consider the quotient of $\mathcal{C}_S(hF)$ by the bilateral ideal (S) generated by $\Sigma(e_\alpha)^2-1$.(elliptic signature).

Any homogeneous polynomial of the (e_α) is written:

$$Q(e_\alpha) = P_o(e_\alpha) + (\Sigma(e_\alpha)^2)P(e_\alpha)$$

$$= P_o(e_\alpha)+(\Sigma(e_\alpha)^2)P_1(e_\alpha)+(\Sigma(e_\alpha)^2)^2P_2(e_\alpha)+(\Sigma(e_\alpha)^2)^3P_3(e_\alpha)+\dots$$

where $\Sigma(e_\alpha)^2$ doesn't divide $P_0, P_1, P_2, P_3 \ldots$

Applying $\gamma(1) = \exp(\Sigma(e_\alpha)^2)$ to $P_i(e_\alpha)$, according to

$$\gamma(1)\; P_i(e_{\alpha^*})(\gamma_1)^{-1} ,$$

taking then the quotient by (S), one see that $P_i(e_{\alpha^*})$ is invariant, because $c\ell(\gamma_1)$, mod(S) = 1.

We thus obtain very quickly a classical result about the polynomials which are written with the help of harmonic factors $P_i(q^\alpha)$ and $\Sigma(q^\alpha)^2$. In particular if q^α is real, the restriction to the unit sphere S_{n-1} of any polynomial in n variables is a sum of restrained harmonic polynomials.

The symplectic spinors give after quotient by (S), the Fourier series of several variables (q^α replace e_α).

6. DEFORMATION THEORY OF THE ALGEBRAS ASSOCIATED WITH THE PHASE-SPACE.

6.1. We recall briefly the basic definitions about the deformations of associative algebras and Lie algebras. A is an associative algebra over $\mathbb{K}$, h an indeterminate; $\mathbb{K}((h))$ is defined above. We put

$$A_h = A \otimes \mathbb{K}((h)) .$$

If there exists a bilinear application $f_h : A_h \times A_h \longrightarrow A_h$ such that :

(1) $f_h(a,b) = ab + hF_1(a,b) + h^2F_2(a,b) + \ldots$, a, b $\in$ A with :

(11) $f_h(f_h(a,b),c) = f_h(a,f_h(b,c))$ (associativity), we have a one-parameter family of deformations of A.

With a Lie algebra L, we must postulate :

(1) and (11)' $f_h(a,b) = -f_h(b,a)$

$$(111)\ \sum_{p.c} f_h(f_h(a,b),c) = 0 \quad \text{(Jacobi identity).}$$

Here are some examples (associative algebras) :

a) the orthogonal Clifford algebra obtained with the metric hg, where g is a particular metric which is a deformation of the exterior algebra.

b) the symplectic Clifford algebra (Weyl algebra) obtained with the form hF (cf. above), which is a deformation of the symmetric algebra.

c) The Moyal product over a symplectic space $\mathbb{R}^{2r}$; if $f_1, f_2 \in C^\infty(R^{2r})$.

$$f_1 \underset{M}{*} f_2 = \sum_0^\infty \frac{h^i}{i!} F^{i_1 j_1} \ldots F^{i_r j_r} (\partial_{i_1, \ldots i_r} f_1)(\partial_{j_1 \ldots j_r} f_2),$$

which is a deformation of the usual product $f_1 f_2$.

6.2. (V,F) is a symplectic manifold, dim V = n = 2r, we consider the associative algebra $C^\infty(V,\mathbb{C})$ and the Poisson algebra with the bracket

$$\{f,g\} = \frac{1}{2} F^{ij}(\partial_i f)(\partial_j g). \qquad (12)$$

$C_S(V,F)$ is the bundle of symplectic Clifford algebras (formal or not formal) with the structural group $Sp(n,\mathbb{R})$, reducible to $U(r)$, $C_S'(V,F)$ is the complexified bundle.

∇ is a particular symplectic torsion free connection (with $\nabla F = 0$).

The basic idea is the existence of linear isomorphism between symmetric algebra and symplectic Clifford algebra, working in the sheaves of the cross-sections of the two bundles ("Kähler-Atiyah" correspondence).

First we replace F by hF. If $x_o \in V$ and $(x^1,x^2,\dots x^n)$ are local coordinates with $(x^1,x^2,\dots,x^n)_o = 0$, let $\tilde{f}$ be the germ of $f \in C^\infty(x_o)$. We denote by $\Sigma = \overset{**}{V} T^*_{x_o}(V)$ the algebra of formal series constructed over the cotangent space at x_o.

We define $\hat{f} \in \Sigma$, by :

$$\hat{f} = f(x_o) + (\nabla_i f)_{x_o} dx^i + \dots S(\nabla_{\ell_1\ell_2\dots\ell_k} f)_{x_o} dx^{\ell_1} \vee dx^{\ell_2} \vee \dots \vee dx^{\ell_k} + \dots ;$$

$\vee$ denotes the symmetric product and S the symmetrisation operator.

We can make the hypotheses (H)

$(\nabla_{(\ell_1\ell_2\dots\ell_k)}\tilde{f})_{x_o} \leqslant Aa^k$, $k \in \mathbb{N}$, A, a are constants depending on x_o and f.

$\varphi : \tilde{f} \longrightarrow \hat{f}$ is injective,

These hypotheses are possible if we suppose for V some convenient analyticity property ; they will permit the construction of an algorithm that we shall preserve, even though these hypotheses are loose.

First consider an hermitian frame $\mathcal{R}$ coming from an orthonormed symplectic frame (e_α, e_{β^*})

$$\varepsilon_\alpha = \frac{e_\alpha - ie_{\alpha^*}}{\sqrt{2}}, \quad \varepsilon_{\alpha^*} = \frac{e_\alpha + ie_{\alpha^*}}{\sqrt{2}}, .$$

The choice of $\mathcal{R}$ determines a linear isomorphism between $C_S'(hF)$ and the symmetric algebra ; we define $i_{\mathcal{R}}$ to be this isomorphism.

Let $\hat{\hat{f}}$ be $i_{\mathcal{R}}(\hat{f}) = i_{\mathcal{R}}(\varphi(\tilde{f})) = \varphi_1(\tilde{f})$.

O denotes the multiplicative rule in the symplectic Clifford algebra. we define :

$$f * g = \varphi_1^{-1}(\hat{\hat{f}} \;\mathrm{O}\; \hat{\hat{g}}) \qquad (13)$$

and obtain a deformation $*$ of the associative algebra of the germs satisfying the hypotheses (H) above. The associativity results from a structural transfer.

The explicit evaluation of $f * g$ leads to :

$$\tilde{f} * \tilde{g} = \tilde{f}\tilde{g} + hF^{\alpha^*\beta}(\nabla_{\alpha^*}\tilde{f})(\nabla_{\beta}\tilde{g}) + \frac{h^2}{2!}F^{\alpha_1^*\beta_1}F^{\alpha_2^*\beta_2}(\nabla_{(\alpha_1^*\alpha_2^*)}\tilde{f})(\nabla_{(\beta_1\beta_2)}\tilde{g}) +$$

$$+ \frac{h^3}{3!}F^{\alpha_1^*\beta_1}F^{\alpha_2^*\beta_2}F^{\alpha_3^*\beta_3}(\nabla_{(\alpha_1^*\alpha_2^*\alpha_3^*)}\tilde{f})(\nabla_{(\beta_1\beta_2\beta_3)}\tilde{g}) + \dots$$

$$\dots + h^2\,\alpha_2(\tilde{f},\tilde{g}) + h^3\,\alpha_3(\tilde{f},\tilde{g}) + \dots ; \qquad (14)$$

$\alpha_2, \alpha_3, \dots$ contain curvature terms (null at $x = x_o$ and null in the flat case);

$$h[f,g] = \tilde{f} * \tilde{g} - \tilde{g} * \tilde{f}$$

gives a non formal deformation of the Poisson algebra N of the preceding germs at x_o.

6.3. The formula (14) gives

$$x_1 \longrightarrow (\tilde{f}_{x_o} * \tilde{g}_{x_o})(x_1),\ x_1 \text{ being close to } x_o.$$

If $x_1 = x_o$, varying x_o we obtain the same algorithm (14), with $\alpha_2, \alpha_3 \dots = 0$.

Forgetting (H) we can consider formal deformations of $C^\infty(V,\mathbb{C})$ and $N(V,\mathbb{C})$, then we have :

Proposition 6.

There exist formal deformations over every symplectic manifold, for the associative algebra $\mathbb{C}^\infty(V,C)$ and the Poisson algebra $N(V,\mathbb{C})$.

6.4. Consider the real case $f \in C^\infty(V,\mathbb{R})$. The preceding method works for the Poisson algebra, because in $f * g - g * f$ the odd terms are real and the even terms (in h^{2k}, $k > 1$) are purely imaginary then ;

$$f * g - g * f,\ \text{mod}\ h^{2k},\ k \geqslant 1,$$

gives a real bracket if $f, g \in C^\infty(V,\mathbb{R})$.

Proposition 7.

The sufficient Vey's hypothesis, $H^3(V,\mathbb{R}) = 0$ for obtaining deformation of the Poisson algebra is invalid.

We can develop an analogous calculation in the real context if (V,F) is a symplectic manifold containing real global lagrangian fields: the structural group $Sp(n,\mathbb{R})$ reduces to $O(r) \times O(r)$, thus providing a deformation for $C^\infty(V,\mathbb{R})$; this method works over the phase space $T^*(M)$.

However, we can observe that our hypothesis justifies a global correspondence between the bundle of the symplectic Clifford algebras and the bundle of the symmetric algebras. But if we consider a cross-section in the sheaf of the symmetric algebras, we can canonically put in it a symplectic Clifford algebra structure, also the proposition 6 is also valid for $C^{\omega}(V,\mathbb{R})$. De Wilde and Lecomte, using the Lichnerowicz approach and cohomological methods also obtained this result [4].

6.5. What is the relation with the Moyal product ?

Using the universal property of symplectic Clifford algebras, it is easy to prove that our deformation is isomorphic to the Moyal deformation in the flat case. Notice that this isomorphism does not respect the grading. Over a manifold, it is more difficult to show that deformation is sent onto the Moyal deformation by the help of a homomorphism.

Our deformation has an universal property.

The reader can find in [2,c] more copious development of these topics, in particular reference to quantum mechanics.

REFERENCES

1. Bourbaki : Groupes et algèbres de Lie. Chapitres 2 et 3. Hermann - Paris, 1972.

2. A.Crumeyrolle :
a)'Algèbre de Clifford symplectique'..etc... J.Math.pures et appl, 56, 1977, p.205-230.
b)'Classes de Maslov, fibrations spinorielles symplectiques et transformation de Fourier'. Ibid, 58, 1979, p.111-120.
c)'Deformations d'algèbres associées à une variété symplectique'... Ann. Inst. Henri Poincaré, vol 35, n°3, 1981, p.175-194.
d)'Planck's constant and symplectic geometry.' Colloquium on differential geometry, Debrecen 1984.
e)'Structures symplectiques, structures complexes, spineurs symplectiques et transformation de Fourier.' Preprint. Toulouse 1985.

3. R.Howe : 'On the role of the Heisenberg group in harmonic analysis' Bull. of the Am. Math. Soc. vol. 3, n°2, Sept.1980.

4. M.DeWilde et P.Lecomte :
'Existence of star-products and of formal deformations of the Poisson Lie algebra of arbitrary symplectic manifolds'. Letters in Math.Physics. 7, 1983, 487-496.

WALSH FUNCTIONS, CLIFFORD ALGEBRAS AND CAYLEY-DICKSON PROCESS

Per-Erik Hagmark
State Institute for Technical Education
Helsinki, Finland

Pertti Lounesto
Helsinki University of Technology
Espoo, Finland

ABSTRACT. The present paper scrutinizes how the sign of the product of two elements in the basis for the Clifford algebra of dimension 2^n can be computed by the Walsh functions of degree less than 2^n. In the multiplication formula the basis elements are labelled by the binary n-tuples, which form an abelian group Ω, which in turn gives rise to the maximal grading of the Clifford algebra. The group of the binary n-tuples is also employed to the Cayley-Dickson process.

1. WALSH FUNCTIONS

Consider n-tuples $\underline{a} = a_1 a_2 \ldots a_n$ of binary digits $a_i = 0,1$. For two such n-tuples $\underline{a}$ and $\underline{b}$ the sum $\underline{a} \oplus \underline{b} = \underline{c}$ is defined by termwise addition modulo 2, that is,

$$c_i = a_i + b_i \mod 2.$$

These n-tuples form a group so that the group characters are <u>Walsh functions</u> [4]

$$w_{\underline{a}}(\underline{b}) = (-1)^{\sum_{i=1}^{n} a_i b_i}.$$

The Walsh functions have only two values ± 1, and they satisfy $w_{\underline{k}}(\underline{a} \oplus \underline{b}) = w_{\underline{k}}(\underline{a}) w_{\underline{k}}(\underline{b})$, as group characters, and $w_{\underline{a}}(\underline{b}) = w_{\underline{b}}(\underline{a})$. The Walsh functions $w_{\underline{k}}$, labelled by binary n-tuples $\underline{k} = k_1 k_2 \ldots k_n$ can be ordered by integers $k = \sum_{i=1}^{n} k_i 2^{n-i}$.

1.1. Sequency order

In applications one often uses the <u>sequency order</u> of the Walsh functions

J. S. R. Chisholm and A. K. Common (eds.), Clifford Algebras and Their Applications in Mathematical Physics, 531–540.

$$\tilde{w}_{\underline{k}}(\underline{x}) = (-1)^{k_1 x_1 + \sum_{i=2}^{n} (k_{i-1}+k_i) x_i},$$

for instance, in spectral analysis of time series, signal processing, communications and filtering [6], [8]. In the sequency order the index $\underline{k}$ is often replaced by an integer $k = \sum_{i=1}^{n} k_i 2^{n-i}$ and the argument $\underline{x}$ by a real number on the unit interval $x = 2^{-n} \sum_{i=1}^{n} x_i 2^{i-1}$ (Fig. 1 and Fig. 2).

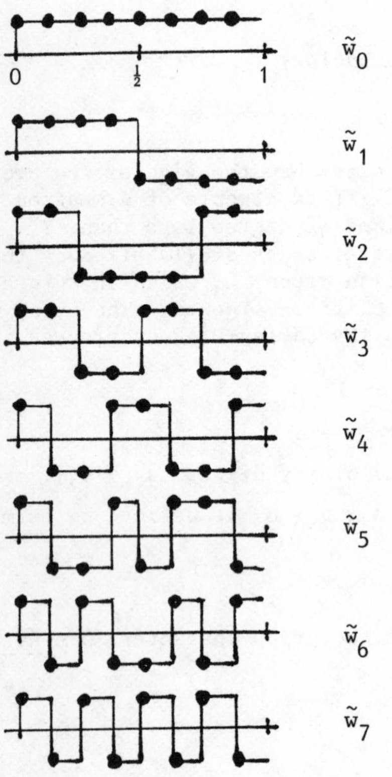

Figure 1. The first 8 Walsh functions

$$\tilde{w}_k(x) = (-1)^{k_1x_1+(k_1+k_2)x_2+(k_2+k_3)x_3}$$

with $k = 4k_1 + 2k_2 + k_3$ and $x = \frac{1}{8}(x_1 + 2x_2 + 4x_3)$. Observe that the number of zero crossings per unit interval equals k.

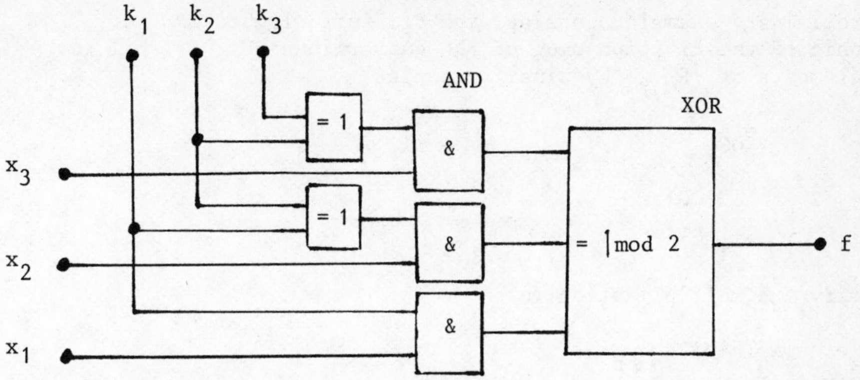

Figure 2. The first 8 Walsh functions in hardware, $\tilde{w}_k(x) = (-1)^f$.

1.2. Gray code

The passage to the sequency order is related to the Gray code g defined by

$$g(\underline{k})_1 = k_1,\ g(\underline{k})_i = k_{i-1} + k_i \mod 2, \quad i = 2,\ldots,n.$$

The formula $\tilde{w}_{\underline{a}}(\underline{x}) = w_{g(\underline{a})}(\underline{x})$ reorders the Walsh functions. The Gray code is a single digit change code, that is, the codes of two consecutive integers differ only in one bit. (Table I).

TABLE I. The Gray code for $k < 8$.

k	$\underline{k}$	$g(\underline{k})$
0	0 0 0	0 0 0
1	0 0 1	0 0 1
2	0 1 0	0 1 1
3	0 1 1	0 1 0
4	1 0 0	1 1 0
5	1 0 1	1 1 1
6	1 1 0	1 0 1
7	1 1 1	1 0 0

The Gray code is a group isomorphism among the binary n-tuples, that is, $g(\underline{a} \oplus \underline{b}) = g(\underline{a}) \oplus g(\underline{b})$. The inverse h of the Gray code is obtained by

$$h(\underline{a})_i = \sum_{j=1}^{i} a_j \mod 2.$$

2. BINARY REPRESENTATIONS OF CLIFFORD ALGEBRAS

As a preliminary example, consider the Clifford algebra $R_{0,2}$, isomorphic to the division ring of the quaternions H. Relabel the basis elements of $R_{0,2}$ by binary 2-tuples

$$\begin{array}{ll} 1 & e_{00} \\ e_1, e_2 & e_{10}, e_{01} \\ e_{12} & e_{11} \end{array}$$

and verify the multiplication rule

$$e_{\underline{a}} e_{\underline{b}} = w_{\underline{a}}(h(\underline{b})) e_{\underline{a} \oplus \underline{b}} .$$

For an alternative representation reorder the basis elements by the formula

$$\tilde{e}_{\underline{a}} = e_{g(\underline{a})} \quad \text{or} \quad e_{\underline{a}} = \tilde{e}_{h(\underline{a})}$$

to get the correspondences

$$\begin{array}{ll} 1 & \tilde{e}_{00} \\ e_1, e_2 & \tilde{e}_{11}, \tilde{e}_{01} \\ e_{12} & \tilde{e}_{10} . \end{array}$$

This yields the multiplication rule

$$\tilde{e}_{\underline{a}} \tilde{e}_{\underline{b}} = \tilde{w}_{\underline{a}}(\underline{b}) \tilde{e}_{\underline{a} \oplus \underline{b}} .$$

2.1. Clifford multiplication

In general, consider the Clifford algebra $R_{0,n}$ with n generators $e_1, e_2, \ldots, e_n$ such that

$$e_i^2 = -1 \quad \text{for} \quad i = 1,2,\ldots,n$$
$$e_i e_j = -e_j e_i \quad \text{for} \quad i \neq j.$$

Theorem 1. If a real 2^n-dimensional algebra A has the multiplication rule

$$e_{\underline{a}} e_{\underline{b}} = w_{\underline{a}}(h(\underline{b})) e_{\underline{a} \oplus \underline{b}}$$

between the basis elements labelled by the binary n-tuples, then A is isomorphic to the Clifford algebra $R_{0,n}$.

Proof. It is sufficient to show that A is associative, has unit element and is generated by n anticommuting elements with square −1.

The element $e_{\underline{0}} = e_{000..00}$ is the unit, since $e_{\underline{a}}e_{\underline{0}} = w_{\underline{a}}(h(\underline{0}))e_{\underline{a}\oplus\underline{0}} = w_{\underline{a}}(\underline{0})e_{\underline{a}} = +e_{\underline{a}}$ and similarly $e_{\underline{0}}e_{\underline{a}} = +e_{\underline{a}}$. The following n basis elements

$$e_{100...00}, e_{010...00}, \ldots, e_{000...01}$$

generate by definition all of A. Each generator has square $-e_{\underline{0}}$; in particular for the i:th generator $e_{\underline{a}}$

$$\underset{}{\underline{a}} = \underset{1\quad\ \ i\quad\ \ n}{00\ldots010\ldots00}, \quad h(\underline{a}) = \underset{1\quad\ \ i\quad\ \ n}{00\ldots011\ldots11}$$

and so $w_{\underline{a}}(h(\underline{a})) = -1$, from which one concludes that $e_{\underline{a}}e_{\underline{a}} = w_{\underline{a}}(h(\underline{a}))e_{\underline{a}\oplus\underline{a}} = -e_{\underline{0}}$. In a similar manner one finds that the generators anticommute with each other.

Finally, A is associative, since for three arbitrary basis elements $e_{\underline{a}}, e_{\underline{b}}, e_{\underline{c}}$ the condition $(e_{\underline{a}}e_{\underline{b}})e_{\underline{c}} = e_{\underline{a}}(e_{\underline{b}}e_{\underline{c}})$ is equivalent to

$$w_{\underline{a}}(h(\underline{b}))w_{\underline{a}\oplus\underline{b}}(h(\underline{c})) = w_{\underline{a}}(h(\underline{b}\oplus\underline{c}))w_{\underline{b}}(h(\underline{c})),$$

which is a consequence of $w_{\underline{a}\oplus\underline{b}}(\underline{x}) = w_{\underline{a}}(\underline{x})w_{\underline{b}}(\underline{x})$ and $w_{\underline{a}}(\underline{x}\oplus\underline{y}) = w_{\underline{a}}(\underline{x})w_{\underline{a}}(\underline{y})$ and h being a group isomorphism. □

It is convenient to assume the correspondences

$$e_i = e_{\underset{1\quad\ \ i\quad\ \ n}{00\ldots010\ldots00}} \qquad \text{for } i = 1,2,\ldots,n$$

between the ordinary and binary representations of the generators of the Clifford algebra $R_{0,n}$. Then the basis elements of $R_{0,n}$ are labelled by the binary n-tuples $\underline{a} = a_1a_2\ldots a_n$ as follows

$$e_{\underline{a}} = e_1^{a_1}e_2^{a_2}\ldots e_n^{a_n}, \quad a_i = 0,1.$$

Since the Gray code is a group isomorphism among the binary n-tuples, we can reorder the basis of the Clifford algebra $R_{0,n}$ by

$$\tilde{e}_{\underline{a}} = e_{g(\underline{a})}.$$

This reordering results in a simple multiplication formula:

Corollary. The product of the basis elements of the Clifford algebra $R_{0,n}$ is given by

$$\tilde{e}_{\underline{a}}\tilde{e}_{\underline{b}} = \tilde{w}_{\underline{a}}(\underline{b})\tilde{e}_{\underline{a}\oplus\underline{b}}.$$

Proof.

$$\tilde{e}_{\underline{a}}\tilde{e}_{\underline{b}} = e_{g(\underline{a})}e_{g(\underline{b})} = w_{g(\underline{a})}(h(g(\underline{b}))e_{g(\underline{a})\oplus g(\underline{b})}$$

$$= w_{g(\underline{a})}(\underline{b})e_{g(\underline{a}\oplus\underline{b})} = \tilde{w}_{\underline{a}}(\underline{b})\tilde{e}_{\underline{a}\oplus\underline{b}} . \quad \square$$

If you choose the signs in $e_{\underline{a}}e_{\underline{b}} = \pm e_{\underline{a}\oplus\underline{b}}$ in some other way, you get other algebras than $R_{0,n}$. For instance, the Clifford algebra $R_{p,q}$ over the quadratic form $x_1^2+\ldots+x_p^2 - x_{p+1}^2 - \ldots - x_{p+q}^2$ has the multiplication formula

$$e_{\underline{a}}e_{\underline{b}} = (-1)^{\sum_{i=1}^{p} a_i b_i} w_{\underline{a}}(h(\underline{b}))e_{\underline{a}\oplus\underline{b}} .$$

Of course, this might be also written without Walsh functions

$$e_{\underline{a}}e_{\underline{b}} = (-1)^{\sum_{i=p+1}^{n} a_i b_i} (-1)^{\sum_{i>j} a_i b_j} e_{\underline{a}\oplus\underline{b}} ,$$

a formula essentially obtained by Brauer and Weyl [2]. See also Refs. [1] and [3] for a related definition of the product on the Clifford algebras (based on sums of multi-indices).

2.2. An iterative process to form Clifford algebras

Clifford algebras can be obtained by a method analogous to the Cayley-Dickson process. Consider pairs (u,v) of elements u and v in the Clifford algebra $R_{p,q}$. Define a product for two such pairs

$$(u_1,v_1)(u_2,v_2) = (u_1u_2 \pm v_1v_2', u_1v_2 + v_1u_2')$$

where $u \to u'$ is the main involution of $R_{p,q}$. This results in an algebra isomorphic to the Clifford algebra

$R_{p+1,q}$

or

$R_{p,q+1}$

according to the $\pm$ sign. This iterative process could be repeated by noting that $(u,v)' = (u',-v')$.

For more details on the Clifford algebras see Refs. [9] and [10].

3. SOME CLIFFORD-LIKE ALGEBRAS

All the above algebras are special cases of the following. Let A be a real linear space of dimension 2^n. Label a basis for A by binary n-tuples $\underline{a}$ to get the basis elements $e_{\underline{a}}$. Then define a multiplication between the basis elements $e_{\underline{a}}$ and extend it to all of A by linearity. The definition is of the form

$$e_{\underline{a}} e_{\underline{b}} = \pm e_{\underline{a} \oplus \underline{b}}$$

for a certain choice of signs. Then the algebra A is a direct sum of the one-dimensional subspaces $U_{\underline{a}}$, spanned by $e_{\underline{a}}$, satisfying

$$U_{\underline{a}} U_{\underline{b}} \subset U_{\underline{a} \oplus \underline{b}} .$$

In other words A is an algebra graded by the abelian group of binary n-tuples Ω. This grading is maximal [7], and these algebras will be called Clifford-like algebras. Next we shall study some Clifford-like algebras.

3.1. Cayley-Dickson process

Consider a generalized quaternion ring Q with $i^2 = \gamma_1$, $j^2 = \gamma_2$ and $k^2 = \gamma_1\gamma_2$, where $\gamma_1, \gamma_2 = \pm 1$. The conjugation-involution $u \to u^L$ of Q is given by

$$i^L = -i, \quad j^L = -j, \quad k^L = -k.$$

Introduce a multiplication in the 8-dimensional real linear space $Q \times Q$ by the formula

$$(u_1, v_1)(u_2, v_2) = (u_1 u_2 + \gamma_3 v_2^L v_1, v_2 u_1 + v_1 u_2^L)$$

where $\gamma_3 = \pm 1$. Inducing an anti-involution $(u,v)^L = (u^L, -v)$ of $Q \times Q = CD(\gamma_1, \gamma_2, \gamma_3)$ makes it possible to repeat this Cayley-Dickson process to get an algebra $CD(\gamma_1, \gamma_2, \ldots, \gamma_n)$, where $\gamma_i = \pm 1$. In fact, the Cayley-Dickson process could be started with R to give $CD(\gamma_1)$ and $Q = CD(\gamma_1, \gamma_2)$.

Example. $CD(-1) \simeq C$, $CD(-1,-1) \simeq H$, and $CD(-1,-1,-1) \simeq 0$, the real 8-dimensional alternative division algebra of octonions [10]. □

The algebras $CD(\gamma_1, \gamma_2, \ldots, \gamma_n)$ obtained by the Cayley-Dickson process are simple flexible algebras of dimension 2^n [11]. Every element of such an algebra satisfies a quadratic equation with real coefficients.

3.2. Binary representation of Cayley-Dickson process

The algebras formed by the Cayley-Dickson process are Clifford-like algebras. For instance, choose a basis of $CD(\gamma_1) = R \times R$

$$e_0 = (1,0), \quad e_1 = (0,1)$$

and introduce the multiplication rule

$$e_{\underline{a}} e_{\underline{b}} = \gamma_1^{a_1 b_1} e_{\underline{a} \oplus \underline{b}} \quad (\underline{a} = a_1, \ \underline{b} = b_1).$$

The involution is given by

$$e_0^L = (1,0) = e_0, \quad e_1^L = (0,-1) = -e_1$$

or in a condensed form $e_{\underline{a}}^L = (-1)^{a_1} e_{\underline{a}}$.

Theorem 2. A Clifford-like algebra A, dim A = 2^n, with multiplication rule

$$e_{\underline{a}} e_{\underline{b}} = f(\underline{a},\underline{b}) e_{\underline{a} \oplus \underline{b}}$$

$$f(\underline{a},\underline{b}) = (-1)^{\sum_{i=1}^{n-1}((S_i(\underline{a})+S_i(\underline{b})+S_i(\underline{a}\oplus\underline{b}))b_{i+1}+S_i(\underline{b})a_{i+1})} \times \prod_{i=1}^{n} \gamma_i^{a_i b_i},$$

where $S_i(\underline{a})$ is the maximum of a_j for $1 \leq j \leq i$, is isomorphic to $CD(\gamma_1, \gamma_2, \ldots, \gamma_n)$. The anti-involution is

$$e_{\underline{a}}^L = (-1)^{S_n(\underline{a})} e_{\underline{a}} .$$

Proof. The first case of the mathematical induction is proved in the example above.

Assume that the statement holds up to the n:th step, and apply the Cayley-Dickson process. If the new basis elements are denoted

$$e_{a_1 a_2 \ldots a_n a_{n+1}} = \begin{cases} (e_{\underline{a}}, 0), & a_{n+1} = 0 \\ (0, e_{\underline{a}}), & a_{n+1} = 1 \end{cases}$$

or $e_{\underline{a} a_{n+1}} = e_{a_1 a_2 \ldots a_n a_{n+1}}$ in short, then

$$e_{\underline{a}0} e_{\underline{b}0} = (e_{\underline{a}},0)(e_{\underline{b}},0) = (e_{\underline{a}} e_{\underline{b}},0) = f(\underline{a},\underline{b})(e_{\underline{a}\oplus\underline{b}},0) = f(\underline{a},\underline{b}) e_{\underline{a}\oplus\underline{b}0}$$

$$e_{\underline{a}1} e_{\underline{b}0} = (0,e_{\underline{a}})(e_{\underline{b}},0) = (0,e_{\underline{a}} e_{\underline{b}}^L) = (-1)^{S_n(\underline{b})} f(\underline{a},\underline{b}) e_{\underline{a}\oplus\underline{b}1}$$

$$e_{\underline{a}0} e_{\underline{b}1} = (e_{\underline{a}},0)(0,e_{\underline{b}}) = (0,e_{\underline{b}} e_{\underline{a}}) = f(\underline{b},\underline{a}) e_{\underline{a}\oplus\underline{b}1}$$

$$e_{\underline{a}1} e_{\underline{b}1} = (0,e_{\underline{a}})(0,e_{\underline{b}}) = (\gamma_{n+1} e_{\underline{b}}^L e_{\underline{a}},0) = \gamma_{n+1} (-1)^{S_n(\underline{b})} f(\underline{b},\underline{a}) e_{\underline{a}\oplus\underline{b}0} .$$

These four equations can be condensed into one equation

$$e_{\underline{a}a_{n+1}} e_{\underline{b}b_{n+1}} = f(\underline{a},\underline{b})^{1-b_{n+1}} f(\underline{b},\underline{a})^{b_{n+1}}$$
$$\times\ \gamma_{n+1}^{a_{n+1}b_{n+1}} (-1)^{a_{n+1} S_n(\underline{b})} e_{\underline{a}\oplus\underline{b}(a_{n+1}\oplus b_{n+1})} ,$$

where

$$f(\underline{b},\underline{a}) = f(\underline{a},\underline{b})(-1)^{S_n(\underline{a})+S_n(\underline{b})+S_n(\underline{a}\oplus\underline{b})} ,$$

which is a consequence of $(e_{\underline{a}}e_{\underline{b}})^L = e_{\underline{b}}^L e_{\underline{a}}^L$. Thus we have proved the desired multiplication rule in the case $n+1$. The induced anti-involution is also of the assumed type:

$$e_{\underline{a}0}^L = (e_{\underline{a}}^L, 0) = (-1)^{S_n(\underline{a})} e_{\underline{a}0}$$

$$e_{\underline{a}1}^L = (0, -e_{\underline{a}}) = -e_{\underline{a}1}$$

or in a condensed form

$$e_{\underline{a}a_{n+1}}^L = (-1)^{\max(S_n(\underline{a}),a_{n+1})} e_{\underline{a}a_{n+1}} . \quad \square$$

The algebra $CD(\gamma_1,\gamma_2,\ldots,\gamma_n)$ is generated by an n-dimensional vector space, whose elements

$$x_1 e_{100\ldots00} + x_2 e_{010\ldots00} + \ldots + x_n e_{000\ldots01}$$

have squares $(\gamma_1 x_1^2 + \gamma_2 x_2^2 + \ldots + \gamma_n x_n^2)e_{\underline{0}}$. As opposed to the Clifford algebras different orderings of the parameters γ_i in $CD(\gamma_1,\gamma_2,\ldots,\gamma_n)$ may result in non-isomorphic algebras in case $n > 3$.

Another construction relating Clifford algebras and Cayley-Dickson algebras is found in Ref. [12].

For more details of the algebraic extensions of the group of binary n-tuples Ω see Ref. [5].

REFERENCES

[1] E. Artin: Geometric Algebra. Interscience, New York, 1957.

[2] R. Brauer, H. Weyl: 'Spinors in n dimensions.' Amer. J. Math. 57 (1935), 425-449.

[3] R. Delanghe, F. Brackx: 'Hypercomplex function theory and Hilbert modules with reproducing kernel'. Proc. London Math. Soc. (3) 37

(1978), 545-576.

[4] N.J. Fine: 'On the Walsh functions'. Trans. Amer. Math. Soc. 65 (1949), 372-414.

[5] P.-E. Hagmark: 'Construction of some 2^n-dimensional algebras'. Helsinki University of Technology, Mathematical Report A177, 1980.

[6] H.F. Harmuth: Sequency Theory, Foundations and Applications. Academic Press, New York, 1977.

[7] A.K. Kwasniewski: 'Clifford- and Grassman-like algebras - old and new'. J. Math. Phys. 26 (1985), 2234-2238.

[8] M. Maqusi: Walsh Analysis and Applications. Heyden, London, 1981.

[9] A. Micali, Ph. Revoy: Modules Quadratiques. Cahiers Mathématiques 10, Montpellier, 1977.

[10] I.R. Porteous: Topological Geometry. Van Nostrand-Reinhold, London, 1969. Cambridge University Press, Cambridge, 1981.

[11] R.D. Schafer: 'On the algebras formed by the Cayley-Dickson process'. Amer. J. Math. 76 (1954), 435-446.

[12] G.P. Wene: 'A construction relating Clifford algebras and Cayley-Dickson algebras'. J. Math. Phys. 25 (1984), 2351-2353.

Z(N)-Spin Systems and Generalised Clifford Algebras

T.T. Truong
Institut für Theoretische Physik, Freie Universität Berlin
Arnimallee 14, D-1000 Berlin 33, W.-Germany

ABSTRACT. We show that Generalised Clifford Algebras (G.C.A.) arise naturally in the formulation of the statistical mechanics of spin systems which generalise the well-known Ising model. A special one-parameter automorphism exhibits remarkable properties and leads to the introduction of lattice fermion fields with "exotic" statistics, nowadays thought to be relevant in some physical situations (such as charge fractionalisation).

The use of Clifford Algebras (C.C.A.) in physics has been widespread since the discovery by P.A.M. Dirac of the relativistic wave equation.[1] Its occurence in statistical mechanics as key method to formulate and solve the two-dimensional Ising model[2] is perhaps no accident since in recent years the role of the Clifford Group (C.G.) emerges as a crucial procedure to calculate the correlation functions of the Ising model[3]. It is natural thus to raise the question whether there exists generalizations of the Ising model for which a generalized version of the Clifford algebra would play an important role. In this paper we discuss a concrete example of spin system namely the so-called Potts model[4] whose algebraic structure is conveniently described by a generalization of the Clifford Algebra introduced by K. Yamazaki[5] hereafter denoted by (G.C.A.).

We first introduce the q-state Potts model on a square lattice. On each lattice site n is a classical spin variable S_n taking the values: $1, \xi, \ldots \xi^{(q-1)}$ where $\xi = \exp 2i\pi/_q$. Between neighboring sites n and m we have a pair interaction defined by the energy

$$\beta E_{nm} = \{K_o + K \sum_{j=1}^{q-1} S_n^j S_m^{j+1}\} \qquad (1)$$

where K_o and K are coupling constants and $\beta = 1/kT$ with k being the Boltzmann constant and T the temperature. The ob-

J. S. R. Chisholm and A. K. Common (eds.), Clifford Algebras and Their Applications in Mathematical Physics, 541–548.

ject in statistical mechanics is the computation of the partition function of the model, namely the sum over all spin configurations of

$$Z(q) = \sum_{\{\text{Conf}.S_n, S_m\}} \exp \sum_{n,m} \beta E_{nm} . \tag{2}$$

Now for q = 2 we recover the known Ising model. In the following we shall discuss for the sake of simplicity the case q = 3. Many of the features encountered here are valid for q > 3 but a detailed account for q > 3 is beyond the limitations of the paper and shall be reported elsewhere.

Generalized Clifford Algebras have been the subject of numerous studies both from the mathematical and physical points of view. We have tried to give as many references as possible[8] but this list is certainly not complete. For our limited purpose here we shall only make use of the following definition. A G.C.A. is the polynomial algebra over the complex numbers generated by the p elements: $\Gamma_1, \Gamma_2, \ldots, \Gamma_p$ subjected to the relations:

$$\Gamma_i^q = I \quad , \quad \Gamma_i \Gamma_j = \xi \Gamma_j \Gamma_i \quad i<j \tag{3}$$

for i,j = 1,2,...,p. The relevant G.C.A. for the three-state Potts model is generated by two generators called Ω and M, following the work of L. Mittag & M. Stephen[7] and represented by the 3 x 3 matrices:

$$\Omega = \begin{vmatrix} 1 & 0 & 0 \\ 0 & j & 0 \\ 0 & 0 & j^2 \end{vmatrix} \qquad M = \begin{vmatrix} 0 & 1 & 0 \\ 0 & 0 & 1 \\ 1 & 0 & 0 \end{vmatrix} \tag{4}$$

with $j = \exp 2i\pi/3$ and the relations:

$$\Omega^3 = M^3 = I, \; M\Omega = j\Omega M, \; M\Omega = j^2 \Omega M . \tag{5}$$

If we associate to each site r a pair $\Omega_r, M_r)$ then we can express the partition function Z(q=3) as the trace of a transfer matrix defined as follows; for a square lattice of N' rows and N sites per row:

$$Z(q=3) = \text{Trace}(V_1 V_2)^{N'} \tag{6}$$

in appropriate normalisation, where we have:

$$V_1 = \prod_{r=1}^{N} (I + \omega(M_r + M_r^+)) ; \quad \omega = e^{-2K} , \tag{7a}$$

$$V_2 = \exp K \sum_{r=1}^{N}{}' \{ \Omega_r \Omega_{r+1}^+ + \Omega_r^+ \Omega_{r+1} \} \tag{7b}$$

A similar structure exists also in the Ising model[2] where the roles of M and Ω are played by the Pauli matrices σ^x and σ^z. We assume as usual periodic boundary conditions. Although analytically many symmetries and automorphic properties are known for Z(q=3), and by indirect mapping the model is known to be soluble at criticality where $\omega = \omega_c$ [6]) a diagonalization of the transfer matrix remains to be found. With this goal in mind, we recall that the diagonalization of the Ising model is based on the observation that both V_1 and V_2 are in fact linear automorphisms of the "Majorana spinor" associated with the basic Clifford Algebra. This fact helps to reduce the original diagonalization down to a diagonalization of a 2N x 2N matrices, which is basically a product of 2 dimensional rotations. The interested reader can find the details in ref. (2) and (3).

It is thus natural to ask whether such an approach applied to the G.C.A. for q=3 would shed new light on the structure of the transfer matrix which might eventually lead to its diagonalization.

Following L. Mittag and M. Stephen[7]) let us construct a "Z(3)-Majorana" spinor with 3-components:

$$\pi_r = M_1^+ \times \ldots \times M_{r-1}^+ \times \Omega_r \times I_{r+1} \times \ldots \times I_N \tag{8a}$$

$$Q_r^+ = M_1^+ \times \ldots \times M_{r-1}^+ \times M_r^+ \Omega_r \times I_{r+1} \times \ldots \times I_N \tag{8b}$$

$$\pi_r^+ Q_r = M_1^+ \times \ldots \times M_{r-1}^+ \times \Omega_r M_r \times I_{r+1} \times \ldots \times I_N \,. \tag{8c}$$

The hermitian adjoint is the triplet $(\pi_r^+, Q_r, Q_r^+ \pi_r)$. It is customary to denote $M_1 x \ldots x M_{r-1} \times I_r x \ldots x I_N \equiv \mu_r$ as a "disorder" spin on the dual lattice. Some commutators are:

$$\begin{aligned}
Q_r \pi_s &= j\, \pi_s Q_r \quad r \geq s, & Q_r \pi_s &= j^2 \pi_s Q_r \quad r < s\\
\pi_r \pi_s &= j^2 \pi_s \pi_r \quad r > s, & Q_r Q_s &= j^2 Q_s Q_r \quad r < s\\
\pi_r^+ Q_s &= j^2 Q_s \pi_r^+ \quad r > s, & \pi_r^+ Q_s &= j\, Q_s \pi_r^+ \quad r \leq s;
\end{aligned} \tag{9}$$

others are obtained by taking their hermitian adjoint and we have the basic relations

$$Q_r^3 = \pi_r^3 = I \tag{10}$$

which generalize the concept of "Majorana" spinor to 3 component objects.

We now study the first automorphism induced by the matrix $V_1 = V_1(\omega)$. Since it is simply the tensor product of local operators we can immediately obtain its inverse

$$V_1^{-1}(\omega) = V_1(\underline{\omega}) \frac{1-\omega}{1+\omega-2\omega^2} \,, \tag{11a}$$

$$\underline{\omega} = - \frac{\omega}{1+\omega} \,. \tag{11b}$$

Therefore we can deduce the transformation:

$$V_1^{-1}\begin{vmatrix}\pi_r\\ Q_r^+\\ \pi_r^+Q_r\end{vmatrix} V_1 = V_1(\omega)\begin{vmatrix}\pi_r\\ Q_r^+\\ \pi_r^+Q_r\end{vmatrix} = \begin{vmatrix}\lambda_1 & j\lambda_2 & \lambda_2^+\\ j^2\lambda_2^+ & \lambda_1 & \lambda_2\\ \lambda_2 & j\lambda_2^+ & \lambda_1\end{vmatrix}\begin{vmatrix}\pi_r\\ Q_r^+\\ \pi_r^+Q_r\end{vmatrix},$$

where w is a real number: (12)

$$\lambda_1 = \frac{1+\omega+\omega^2}{1+\omega-2\omega^2}, \quad \lambda_2 = (1-j^2)\omega\frac{1-j\omega}{1+\omega-2\omega^2}, \quad \lambda_2^+ = \text{c.c.}\lambda_2 . \tag{13}$$

The matrices can be simply expressed as:

$$V(\omega) = \{\lambda_1 I + \lambda_2 M\,\Omega + \lambda_2^+\Omega^+M^+\} \tag{14}$$

and form a one-parametric subgroup with the rule:

$$V_1(\omega)V_1(\omega') = V_1(\omega'') . \tag{15}$$

The matrix elements of $V_1(\omega'')$ are defined by the functions λ_1'', λ_2'', $\lambda_2^{+''}$ which are expressed in terms of the functions λ_1, λ_2, λ_2^+ and λ_1', λ_2', $\lambda_2'^+$; defining $V_1(\omega)$ and $V_1(\omega')$ by

$$\lambda_1'' = \lambda_1\lambda_1' + \lambda_2\lambda_2'^+ + \lambda_2^+\lambda_2', \quad \lambda_2'' = \lambda_1\lambda_2' + \lambda_2\lambda_1' + \lambda_2^+\lambda_2'^+ \tag{16}$$

$$\lambda_2''^+ = \text{c.c.}\lambda_2''$$

the corresponding relation for the ω, ω' and ω'' parameters is:

$$\omega'' = \frac{\omega+\omega'+\omega\omega'}{1+2\omega\omega'} . \tag{17}$$

Let us remark that for the q > 3 Potts model we find a generalized form of eq. (17), namely

$$\omega'' = \frac{\omega+\omega'+(q-2)\omega\omega'}{1+(q-1)\omega\omega'} . \tag{18}$$

But we know that all the ω must have the same functional form, say with respect to a new parameter cK^*, where c is simply a scaling factor. Equation (18) implies that we search for a mapping of the group law of eq. (15) onto the additive group in the variables K^*:

$$K''^* = K^* + K'^* ;$$

then we must solve the functional equation:

$$\omega(K^*+K'^*) = \omega(K''^*) = \frac{\omega(K^*)+\omega(K'^*)+(q-2)\omega(K^*)\,\omega(K'^*)}{1+(q-1)\omega(K^*)\omega(K'^*)} . \tag{18a}$$

The solution is readily obtained from ref. (9):

$$\omega(K^*) = \frac{1 - \exp(-2cK^*)}{1 - (q-1)\exp(-2cK^*)} . \tag{19}$$

For q = 3, let us set c = 1 and write down the functions λ_1 and λ_2:

$$\lambda_1(\omega) = \frac{1}{3}(1 + 2\cosh 2K^*), \tag{20a}$$

$$\lambda_2(\omega) = \frac{1}{3}\{(1-j^2)\sinh 2K^* + \frac{2}{3}(1-j)^2\sinh^2 K^*\}, \tag{20b}$$

$\lambda_2^+(\omega)$ being obtained by complex conjugation of $\lambda_2(\omega)$. Note that for $K^* = 0$, $\lambda_1 = 1$ and $\lambda_2 = 0$ and $V_1(\omega(K^*))$ reduces to the identity. So we have obtained a kind of canonical parametrisation for our one-parameter subgroup of matrices $V_1(\omega(K^*))$. We recall that for q = 2 (Ising Model) $V_1(\omega)$ being a rotation matrix, its "canonical" parameter K^* has the meaning of an imaginary angle[2),]

$$V_1(\omega) = \begin{vmatrix} \lambda_1 & \lambda_2^+ \\ \lambda_2 & \lambda_1 \end{vmatrix} \tag{21}$$

where

$$\lambda_1 = \frac{1+\omega^2}{1-\omega^2}, \quad \lambda_2 = \frac{2i\omega}{1-\omega^2}; \tag{22a}$$

or using eq. (19):

$$\lambda_1 = \cosh 2K^*, \quad \lambda_2 = i \sinh 2K^* . \tag{22b}$$

The determinant is obviously

$$\mathrm{Det}\, V_1(\omega) = \lambda_1^2 - \lambda_2 \lambda_2^+ = 1 . \tag{23}$$

For q = 3, $V_1(\omega)$ is also an element of SL(3) since

$$\mathrm{Det}\, V_1(\omega) = \lambda_1^3 + \lambda_2^3 + \lambda_3^3 - 3\lambda_1\lambda_2\lambda_3^+ = 1. \tag{24}$$

It is remarkable that it has the canonical hessian form of a cubic form in three variables. (In fact if $\lambda_2^{\pm}$ were real function then the parametrization would have been realized by the so-called generalized hyberbolic functions of order 3[10),13)] instead of eq. (19).

We now come to the central point of our discussion: the introduction of fermion variables. In previous works[11] the fermion variables are defined heuristically first for generalized integrable Heisenberg chains and later for the Potts model in the hamiltonian limit. In both cases the continuum limit was taken leading to quantum field theories in one space dimension. As shown by T.D. Schultz, D. Mattis and E.H. Lieb[12] the fermion variables simplify considerably the discussion of the solution in the spinor language of L. Onsager and B. Kaufman[2]. It is thus useful to recast the three-state Potts model transfer matrix in the fermion description.

Since the fermion operators are obtained in the Ising model as eigenvectors of the $V_1(\omega)$ matrix, it is natural to extend this construction to the Potts model. The characteristic equation of the matrix $V_1(\omega)$ is

$$-\lambda^3 + 3\lambda_1(\lambda-1)\lambda + 1 = 0 \tag{25}$$

where λ is the eigenvalue; it has three roots:

$$\lambda, \lambda', \lambda'' = \frac{1-\omega}{1+2\omega}, \; 1 \; , \; \frac{1+2\omega}{1-\omega} \; . \tag{26}$$

The corresponding eigenvectors are denoted by $\psi_1^+(r)$, $\psi_2^+(r)$, $\psi_3^+(r)$,

$$V_1^{-1} \begin{vmatrix} \psi_1^+(r) \\ \psi_2^+(r) \\ \psi_3^+(r) \end{vmatrix} V_1 = \begin{vmatrix} \lambda \; \psi_1^+(r) \\ \lambda' \; \psi_2^+(r) \\ \lambda'' \; \psi_3^+(r) \end{vmatrix} \tag{27}$$

where

$$\begin{aligned} \psi_1^+(r) &= 3^{-1}\{ \pi_r + Q_r^+ + j\,\pi_r^+ Q_r \} \\ \psi_2^+(r) &= 3^{-1}\{ \pi_r + j^2 Q_r^+ + j^2 \pi_r^+ Q_r \} \\ \psi_3^+(r) &= 3^{-1}\{ \pi_r + jQ_r^+ + \pi_r^+ Q_r \} , \end{aligned} \tag{28}$$

obeying the following commutation rules:

$$\begin{aligned} \psi_\alpha^+(r)\;\psi_\beta^+(r') &= (1-\delta_{rr'})\, j^{2\,\mathrm{sign}(r'-r)}\, \psi_\beta^+(r')\,\psi_\alpha^+(r) + \\ &+ \delta_{rr'}\,\varepsilon_{\alpha\beta\gamma}\psi_\gamma(r) . \end{aligned} \tag{29}$$

$$\psi_\alpha^+(r)\;\psi_\beta(r') = (1-\delta_{rr'})\, j^{\mathrm{sign}(r'-r)}\, \psi_\beta(r')\psi_\alpha^+(r) + \delta_{rr'}\delta_{\alpha\beta}\psi_\alpha^+(r)\psi_\alpha(r')$$

reflecting the features of "fractional" statistics. These operators are, however, fermion-like because they obey Pauli's principle:

$$\{\psi_\alpha^+(r)\}^2 = \{\psi_\alpha(r)\}^2 = 0 \tag{30}$$

as can be directly checked from the spinor commutation relations eq. (9).

Before discussing the automorphism generated by V_2, let us follow L. Mittag and M. Stephen[7] and observe that

$$V_2 = D^{-1} V_1 D \tag{31}$$

where D is the so-called duality transformation, i.e.

$$D^{-1} \begin{vmatrix} \pi_r \\ Q_r^+ \end{vmatrix} D = \begin{vmatrix} Q_r \\ \pi_{r+1}^+ \end{vmatrix}, \tag{32}$$

valid for all sites of the lattice except the last one r = N. Observe that D^2 is simply a translation by one lattice spacing for π_r and Q_r^+. Basically V_2 operates as follows:

$$V_2^{-1} \begin{vmatrix} \pi_r \\ Q_{r-1}^+ \\ \pi_r^+ Q_{r-1} \end{vmatrix} V_2 = V_2(\omega) \begin{vmatrix} \pi_r \\ Q_{r-1}^+ \\ \pi_r^+ Q_{r-1} \end{vmatrix} = \begin{vmatrix} \lambda_1 & \lambda_2^+ & j^2\lambda_2 \\ \lambda_2 & \lambda_1 & j^2\lambda_2^+ \\ j\lambda_2^+ & j\lambda_2 & \lambda_1 \end{vmatrix} \begin{vmatrix} \pi_r \\ Q_{r-1}^+ \\ \pi_r^+ Q_{r-1} \end{vmatrix}. \tag{33}$$

$V_2(\omega)$ forms also a one-parameter subgroup, its determinant is one and it has the same eigenvalues as $V_1(\omega)$, and the eigenvectors are the linear combinations (up to arbitrary phase factor):

$$\begin{aligned} \Phi_1^+(r) &= 3^{-1}\{\pi_r + j\, Q_{r-1}^+ + j\, \pi_r^+ Q_{r-1}\} \\ \Phi_2^+(r) &= 3^{-1}\{\pi_r + j^2 Q_{r-1}^+ + \pi_r^+ Q_{r-1}\} \\ \Phi_3^+(r) &= 3^{-1}\{\pi_r + Q_{r-1}^+ + j^2 \pi_r^+ Q_{r-1}\} \end{aligned} \tag{34}$$

associated with the eigenvalues $\lambda, \lambda', \lambda''$. We may view them as "dual" fermion variables with fractional statistics, since they fulfil again the Pauli principle:[11]

$$\{\Phi_\alpha(r)\}^2 = \{\Phi_\alpha^+(r)\}^2 = 0 . \tag{35}$$

Note that they are bilocal in the original "spinor" variables between neighboring sites, thus might be considered as fermion variables defined on the respective bond of the lattice.

The passage from fermions to dual fermions is clearly not linear as in the case of the Ising model. This is due to the fact that the

Z(3)-symmetry implies already a built-in interaction as can be seen from eq. (29). Thus the diagonalization of the Potts model necessitates a new approach which should replace the Fourier transformation used in the diagonalization of the Ising model.

In conclusion we have brought about a fermionic structure with "exotic" (or fractional) statistics to the transfer matrix of the Potts model. We have also indicated how this structure generalizes the usual fermionic structure of the Ising model, the underlying role being played by a Generalized Clifford Algebra[8] instead of a simple Clifford Algebra. We expect that this fermionic formulation will be more flexible to handle and may eventually lead to a complete diagonalization of the transfer matrix. Work along this direction is in progress.

1. P.A.M. Dirac, Proc. Roy. Soc. A117, 610 (1928)
2. L. Onsager, Phys. Rev. 65, 117 (1944)
   B. Kaufman, Phys. Rev. 76, 1232 (1949)
3. M. Sato, T. Miwa and M. Jimbo, Publ. Research Institute for Math. Sciences (Kyoto Univ.) 16, 531 (1980)
4. R.B. Potts, Proc. Camb. Phil. Soc. 48, 106 (1952)
5. K. Yamazaki, J. Fac. Sci. Univ. Tokyo Sect. I,10, 147 (1964)
6. H.N.V. Temperley and E.H. Lieb, Proc. Roy. Soc. A322, 251 (1971),
   R.J. Baxter, J. Phys. C: Solid State Phys. 16, L445 (1973)
7. L. Mittag and M. Stephen, J. Math. Phys. 12, 441 (1971)
8. A.O. Morris, Quart. J. Math. Oxford) 18, 7 (1967), 19, 1289 (1968)
   I. Popovici and C. Gheorghe, C.R. Acad. Sci. (Paris) 262A,682(1966)
   There are extensive results obtained by A. Ramakrishnan and his collaborators on applications of G.C.A. in physics and mathematics, essentially reported in numerous publications as well as in the book "L-matrix theory or grammar of Dirac matrices", Tata Mc.Graw Hill, New Dehli, India, 1972.
   R. Jagannathan and N.R. Ranganathan, Rep. Math. Phys. 5,229 (1974)
9. J. Aczel, in "Lectures on Functional Equations and Applications", p. 80, Acad. Press, New York (1966)
10. W. Magnus, F. Oberhettinger and R.P. Soni, in "Bateman Project Vol. III", p. 212 (1958)
11. T.T. Truong and H.J. de Vega, Phys. Lett. 151B, 136 (1985)
    P. Bhattacharyya and S. Wadia, J. Phys. A: Math. Gen. 18, L303 (1985)
12. T.D. Schultz, D. Mattis and E.H. Lieb, Rev. Mod. Phys. 36, 856 (1964)
13. K.A. Kwasniewski, contribution to this workshop.

GENERALIZED CLIFFORD ALGEBRAS AND SPIN LATTICE SYSTEMS

A. K. Kwaśniewski
Institute of Theoretical Physics
University of Wrocław
50-205 Wrocław, Cybulskiego 36
Poland

ABSTRACT. Generalized Clifford algebras are shown to be a very efficient and natural tool for formulating and solving problems in statistical mechanics of lattice systems with Z_n symmetry.

While the importance of usual $C_m^{(2)}$ Clifford algebras for physics is nowadays not in question - the generalized $C_m^{(n)}$ Clifford algebras are not that familiar to physicists though they are potentially a powerfull tool in formulating and solving various problems of physics.
In this talk we want to show how advantageous is the use of these generalized algebras in Potts models.
Generalizations of Clifford algebras were introduced by the authors of [1-4] and the first application to physics known to us dates from 1969 [5].
Another application one may find in [6] which is several years later, and then, in 1978 these algebras were fruitfully applied by Yoneya [7] to the study of Z_n lattice gauge models.
On the way, in 1971 Mittag and Stephen [8] have, in a sense, reinvented these $C_{2p}^{(n)}$ algebras as an inevitable outcome of the algebraic analysis of the transfer matrices for Potts models.

It seems that the authors of [8,7] then followed by [9-11] did not realized that as a matter of fact they had been using the $C_{2p}^{(n)}$ algebras of [4].

Some recent applications are due to the author [12].

We start our discussion of Potts models with a general statement. Namely, the transfer matrix technique for a statistical system with the most general translational invariant and globally symmetric Hamiltonian on the lattice - does generate appropriate algebras of operators which form an algebra of the type of algebra extension [2] of that naturally assigned to the lattice - grading group.
For the Z_n cyclic group as a symmetry group this very algebra becomes, in the known cases, the generalized Clifford algebra.

Our talk is devoted to both standard and planar [13] Potts models

J. S. R. Chisholm and A. K. Common (eds.), Clifford Algebras and Their Applications in Mathematical Physics, 549–554.

and the main result of the considerations to follow is the statement that the complete partition functions of Potts models (for finite torus lattices) are polynomials in generalized "cosh" functions.
Before we proceed with the derivation of this result we need some preliminary knowledge and notation. The generalized Clifford algebra $C_{2p}^{(n)}$ [3,4] is generated by the matrices satisfying:

$$\omega\gamma_i\gamma_j = \gamma_j\gamma_i \;;\quad i<j, \qquad \gamma_i^n = 1 \qquad i,j=1,\dots,2p \tag{1}$$

where $\omega = \exp\left\{\frac{i2\pi}{n}\right\}$.

Similarily to the usual Clifford algebra case one finds - up to equivalence - the only one faithful and irreducible representation of these generators in terms of generalized Pauli matrices [4] according to:

$$\begin{aligned}
\gamma_1 &= \sigma_3 \otimes I \otimes I \otimes \dots \otimes I \otimes I \\
\gamma_2 &= \sigma_1 \otimes \sigma_3 \otimes I \otimes \dots \otimes I \otimes I \\
&\dots\dots\dots\dots \\
\gamma_p &= \sigma_1 \otimes \sigma_1 \otimes \sigma_1 \otimes \dots \otimes \sigma_1 \otimes \sigma_3 \\
\gamma_{p+1} &= \bar{\gamma}_1 = \sigma_2 \otimes I \otimes I \otimes \dots \otimes I \otimes I \\
\gamma_{p+2} &= \bar{\gamma}_2 = \sigma_1 \otimes \sigma_2 \otimes I \otimes \dots \otimes I \otimes I \\
&\dots\dots\dots\dots \\
\gamma_{2p} &= \bar{\gamma}_p = \sigma_1 \otimes \sigma_1 \otimes \sigma_1 \otimes \dots \otimes \sigma_1 \otimes \sigma_2
\end{aligned} \tag{2}$$

where, for n odd

$$\sigma_1 = \left(\delta_{i+1,j}\right), \qquad \sigma_2 = \left(\omega^j \delta_{i+1,j}\right), \qquad \sigma_3 = \left(\omega^i \delta_{ij}\right), \tag{3}$$

while for n even

$$\sigma_1 = \left(\delta_{i+1,j}\right), \qquad \sigma_2 = \left(\xi^{2i+1} \delta_{i+1,j}\right), \quad \sigma_3 = \left(\xi \sigma_1^{n-1} \sigma_2\right), \tag{4}$$

with ξ - the primitive 2n-th root of unity. Matrix elements are labelled by indices $i,j \in Z_n' = \{0,1,\dots,n-1\}$ from the additive cyclic group.

In the following we show that transfer matrices of all Potts models are just polynomials in these generalized γ's.
We shall also need the generalized "cosh" functions which we define to be from now on

$$f_i(x) = \sum_{k=0}^{\infty} \frac{x^{nk+i}}{(nk+i)!}, \qquad i \in Z_n' \tag{5}$$

where x might be an element of any finite dimensional, associative algebra with unity.
Most of identities satisfied by cosh and sinh functions generalize to the case of f_i's and here we list some of them:

$$f_i(x) = \frac{1}{n}\sum_{k=0}^{n-1} \omega^{-ki}\exp\{\omega^k x\} \qquad i\in Z'_n , \tag{6}$$

$$\sum_{i=0}^{n-1} f_i(x)f_{k-i}(x) = f_k(2x) , \qquad k\in Z'_n , \tag{7}$$

$$\frac{d}{dx} f_i(x) = f_{i-1}(x) , \qquad x\in\mathbb{C} , \qquad i\in Z'_n , \tag{8}$$

$$\sum_{k=0}^{n-1} f_k(x)f_{-k}(x) = f_o(2x) , \qquad x\in\mathbb{R} , \tag{9}$$

$$\sum_{k=0}^{n-1} f_{k+r}(x)\, f_{-k}(x) = f_r(2x) \qquad x\in\mathbb{R} , \quad r\in Z'_n \tag{10}$$

and so on, where for complex values of x generalized "cosh" functions satisfy $\overline{f_s(x)} = f_s(\bar{x})$. One may develop further this type of generalization introducing generalized "cth" functions:

$$t_k(x) = f_k(x)\, f^{-1}_{k-1}(x) , \qquad k\in Z'_n, \qquad x\in\mathbb{C} . \tag{11}$$

The use of f_i's for a system with Z_n symmetry is inevitable as can be seen already from the following two examples.

Example I.

Consider the one dimensional open chain with Z_n symmetry in the zero external field for which the partition function is defined to be

$$\mathcal{Z}_N(a) = \sum_{\mu_1,\dots,\mu_N\in Z_n} \exp\left\{a \sum_{i=1}^{N-1} \mathrm{Re}(\mu_i\mu_{i+1})\right\} , \tag{12}$$

where $Z_n = \{\omega^i,\ i\in Z'_n\}$.

Calculations similar to those [14] for n=2 yield

$$\mathcal{Z}_N = [n\chi_o(a)]^{N-1} \tag{13}$$

$$\text{where } \chi_o = n \sum_{i\in Z'_n} f_i^2(a/2). \tag{14}$$

Example II.

The problem is to calculate the partition function of the one dimensional periodic chain $/\mu_{N+1}=\mu_1/$ in an external magnetic field, namely:

$$\mathcal{Z}_N(a,b) = \sum_{\mu_1,\ldots,\mu_N \in Z_n} \exp\left\{\frac{a}{2}\sum_{i=1}^{N} \mathrm{Re}(\mu_i\mu_{i+1}) + \frac{b}{2}\sum_{i=1}^{N} \mathrm{Re}\mu_i\right\} . \tag{15}$$

One finds rather easily the transfer matrix for the problem and, after the dual "a*" parameter has been introduced [15]

$$\mathcal{Z}_N = [\det \hat{a}(a)]^{\frac{N}{2}} \,\mathrm{Tr}[\exp\{a^*(\sigma_1 + \sigma_1^+)\} \exp\{b(\sigma_3 + \sigma_3^+)\}]^N , \tag{18}$$

where $\hat{a}(a) = \sum_{k=0}^{n-1} \lambda_k \sigma_1^k$ with $\lambda_k = \exp\{a\mathrm{Re}\omega^k\}$ is at the same time the interaction matrix for the planar Potts model.
Exponentials in (18), due to $\sigma_i^n = 1$ property, give rise to $f_i(a^*)$'s and $f_i(b)$'s, resulting in the conclusion that $\mathcal{Z}_N$ is a polynomial in these f_k's.
The same statement is valid for Potts models which we now discuss in detail.
The transfer matrix for both standard and planar Potts models [13] is represented as a product M = BA, where, in the case of standard Potts model [8,16]

$$B = \prod_{k=1}^{p} \exp\left\{ \frac{1}{n} \sum_{i=0}^{n-1} (Z_k^+ Z_{k+1})^i \right\} \tag{19}$$

and $$A = \prod_{k=1}^{p} \left(e^a + \sum_{i=1}^{n-1} X_k^i \right) , \tag{20}$$

while for the planar model

$$B = \prod_{k=1}^{p} \exp\left\{ Z_k^+ Z_{k+1} + Z_{k+1}^+ Z_k \right\} , \tag{21}$$

$$A = \prod_{k=1}^{p} \left(\sum_{i=0}^{n-1} \lambda_i X_k^i \right) \quad \text{with} \quad \lambda_i = \exp\{a\mathrm{Re}\, \omega^i\} , \tag{22}$$

where

$$X_k = I \otimes \ldots \otimes I \otimes \sigma_1 \otimes I \otimes \ldots \otimes I \qquad (\text{ p - terms}) \tag{23}$$

$$Z_k = I \otimes \ldots \otimes I \otimes \sigma_3 \otimes I \otimes \ldots \otimes I \qquad (\text{ p - terms}) \tag{24}$$

σ's being placed at the k-th site.

Again, introducing the dual parameter "a*" one gets instead of (22), of which (20) is a special case

$$A = [\det \hat{a}(a)]^{\frac{p}{n}} \exp\left\{ a^* \sum_{k=1}^{p} (X_k + X_k^+) \right\}, \tag{25}$$

and again due to the property: $X_k^n = Z_k^n = 1$, we conclude that

$\mathcal{Z} = \operatorname{Tr}(BA)^q$ is a polynomial in $f_i(a^*)$'s and $f_j(b)$'s.

The use of generalized Clifford algebras allows one to learn still more about the transfer matrix for Potts models.
Here as an illustration we shall quote the corresponding results for the planar, n - odd, Potts model [16].
Namely, one may prove that

$$\mathcal{Z} = \operatorname{Tr} M^q = \sum_{k=0}^{n-1} \operatorname{Tr}\left\{ [B_k B_k^+ A]^q \, V_k \right\} \tag{26}$$

where the projection matrices V_k are defined to be:

$$V_k = \frac{1}{n} \sum_{i=0}^{n-1} \omega^{-ki} \, U^i, \qquad k \in Z_n', \tag{27}$$

where

$$U = \prod_{k=1}^{p} \gamma_k^{-1} \bar{\gamma}_k \qquad \text{and} \qquad V_k V_1 = \delta_{k1} V_k,$$

while

$$B_k = \exp\left\{ b \sum_{\alpha=1}^{p-1} \bar{\gamma}_\alpha^{-1} \gamma_{\alpha+1} \right\} \exp\left\{ b\omega^k \, \bar{\gamma}_p^{-1} \gamma_1 \right\}. \tag{28}$$

If one also takes into account that $X_k = \omega^{-1} \gamma_k^{-1} \bar{\gamma}_k$, one readily finds that M^q is a polynomial in generalized γ's.

As the formula for $\operatorname{Tr}(\gamma_{i_1} \dots \gamma_{i_s})$ is known [16] there is some hope that this might be the way to get an exact expression for the complete partition function of any of the discussed models.
Also in the case of the one-dimensional periodic chain (Example 2) one has a convenient representation of the transfer matrix:

$$M^N = [\det \hat{a}(a)]^{\frac{N}{n}} \exp\left\{ \sum_{\alpha \in \Gamma} N c_\alpha e_\alpha \right\} \tag{29}$$

where $\Gamma = Z_n' \oplus Z_n'$ is the maximally grading group of $C_2^{(n)}$ generalized Pauli algebra with $\{e_\alpha\}_{\alpha \in \Gamma}$ its canonical basis [3] while c_α ($c_o = 0$) are some functions of a^* and b parameters. One arrives at (29) using the Baker-Hausdorf formula for two exponential factors of M.
Naturally e_α, $\alpha \in \Gamma - \{0\}$ form the Lie algebra; (for details see [17]).
The expression (29) might be a starting point for other approaches to

the problem of calculating the complete partition function.

ACKNOWLEDGMENTS:

The author expresses his thanks to T.T. Truong for inspiring discussions during the Workshop, and to the Organizers for their hospitality.

REFERENCES

[1] Morinaga K. et all. *J.Sci.Hirishima Univ.* (A) 16 (1952) 13-41
[2] Yamazaki K. *J.Fac.Sc.Univ.Tokyo*, Sect.I, vol 10 (1964), 147
[3] Popovici J. et all. *C.R.Acad.Sc.Paris* 262, 1966, 682-685
[4] Morris A.O. *Quart.J.Math.Oxford* (2) 18 (1967), 7-12
[5] Ramakrishnan A. et all. *J.Math.Phys.Sci.Madras* 3 (1969) 307
Ramakrishnan A. *Proceedings on the Conference on Clifford Algebras, Its Generalizations and Applications Matscience*, Madras 1971
[6] Santhanam T.S. *Foundations of Physics* 7, (1977), 121
[7] Yoneya T. *Nuclear Physics* B 144 (1978) 195-218
[8] Mittag L. Stephen M.J. *J.Math.Phys.* 12, 441 (1971)
[9] Alcaraz F.C. Koberle R. *J.Phys.A: Gen.* 13 (1980) L153-L160
ibid. 14 (1981) 1169-1192
[10] Bashilov Y.A. Pokrovsky S.V. *Commun.Math.Phys.* 76 (1980) 129-141
[11] Gehlen G.V. Rittenberg V. UGVA - DPT 1984/10 - 447
[12] Kwaśniewski A.K. *J.Math.Phys.* 26 (1985) 2234
[13] Domb C. *J.Phys.*, 1974, A 7 p. 1335
[14] Thompson C.J. *Mathematical Statistical Mechanics*, MacMillan, New York, 1971
[15] Kwaśniewski A.K. *Communications of JINR* E17-85-86 Dubna 1985
[16] Kwaśniewski A.K. *J.Phys.A* (in press)
[17] Popovici I. et all. *Rev.Roum.Math.Pures et Appl.* 1966, XI p.989

CLIFFORD ALGEBRA, ITS GENERALISATIONS AND PHYSICAL APPLICATIONS

A. Ramakrishnan
Alladi Centenary Foundation
62, Luz Church Road
Madras-600 004
India

ABSTRACT An operation is described which generates Dirac matrices from Pauli matrices. Repeating this operation leads to matrix representations of higher dimensional Clifford algebras. It is also used to construct matrices satisfying ω-commutation relations where ω is a root of unity. Further generalisations and the use of these results in forming Lie algebras of physical interest is discussed.

This is a brief summary of the results obtained with my collaborators on Clifford Algebra and its generalisations during the years 1967-79 at MATSCIENCE, Madras, and embodied in a monograph [1] and the proceedings of conferences [2,3].

My interest started with the observation of the intriguing placement of the Pauli matrices in the Dirac matrices, leading to the question: "Is there a logical manner in which one goes from 2×2 anti-commuting matrices to 4×4 or higher dimensions anticommuting matrices?" or more brashly: "Why did not Pauli discover the Dirac matrices?"

The answer came in 1967 with a sudden realisation that a simple algorithm, which I called the σ-operation, leads to this extension of anticommutation to higher dimensional matrices from the basic Pauli set [4]. Starting with the 2×2 matrix

$$L_3 = \begin{bmatrix} \lambda_3 & \lambda_1 - i\lambda_2 \\ \lambda_1 + i\lambda_2 & -\lambda_3 \end{bmatrix}$$

with three real parameters λ_i embedded in it, we have

$$L_3^2 = (\lambda_1^2 + \lambda_2^2 + \lambda_3^2)I = \Lambda_1^2 I ,$$

$$L_3 = \sum \lambda_i \alpha_i ,$$

$$\alpha_i \alpha_j = -\alpha_j \alpha_i \quad (i \neq j) ,$$

$$\alpha_1^2 = \alpha_2^2 = \alpha_3^2 = 1 ,$$

J. S. R. Chisholm and A. K. Common (eds.), Clifford Algebras and Their Applications in Mathematical Physics, 555–558.

where

$$\alpha_1 = \begin{bmatrix} 0 & 1 \\ 1 & 0 \end{bmatrix}, \quad \alpha_2 = \begin{bmatrix} 0 & -i \\ i & 0 \end{bmatrix}, \quad \alpha_3 = \begin{bmatrix} 1 & 0 \\ 0 & -1 \end{bmatrix} .$$

We define the σ-operation as follows: replace any one of the λ's, *but only one*, by the matrix L_3 and relabel the other two as λ_4 and λ_5 attaching unit matrices. We then obtain for example

$$L_5 = \begin{bmatrix} \lambda_5 I & L_3 - i\lambda_4 I \\ L_3 + i\lambda_4 I & -\lambda_5 I \end{bmatrix} .$$

Writing this as

$$L_5 = \sum \lambda_1 \alpha_i$$

we obtain five mutually anticommuting Dirac matrices,

$$\alpha_i \alpha_j = -\alpha_j \alpha_i \ (i \neq j), \ \alpha_i^2 = 1$$

$$L_5^2 = (\Lambda_1^2 + \lambda_4^2 + \lambda_5^2) I = \Lambda_2^2 I .$$

By successive applications of the σ-operation we obtain the matrix L_{2n+1} of dimension $2^n \times 2^n$ with 2n+1 parameters, such that

$$L_{2n+1}^2 = \Lambda_n^2 I = (\Lambda_{n-1}^2 + \lambda_{2n}^2 + \lambda_{2n+1}^2) I$$

$$L_{2n+1} = \sum \alpha_i \lambda_i, \ \alpha_i \alpha_j = -\alpha_i \alpha_j \quad (i \neq j).$$

The matrix L_{2n+1} has only two eigenvalues : $\pm\Lambda_n$ and therefore its eigenvectors are degenerate. To resolve the degeneracy we consider the n commuting matrices

$$L_{2n+1}, \begin{bmatrix} L_{2n-1} & \\ & L_{2n-1} \end{bmatrix}, \ldots, \begin{bmatrix} L_3 & & \\ & \ddots & \\ & & L_3 \end{bmatrix}$$

with eigenvalues $\pm\Lambda_n, \pm\Lambda_{n-1}, \ldots, \pm\Lambda_1$ and require an eigenvector to be a simultaneous eigenvector of the n matrices.

It is noted that the number of parameters increases by two as the dimension gets doubled; there is a 'shell' structure for L_{2n+1}, with two parameters in each shell and three in the first. We call a shell 'saturated' if both the parameters therein are non zero and unsaturated if one of them is zero. The symmetry operation on the eigenvalues by various L-matrices have been studied in detail in a series of papers. It is noted that the Dirac Hamilton is unsaturated in the second shell with the identification

$$\lambda_1 = p_1, \quad \lambda_2 = p_2, \quad \lambda_3 = p_3, \quad \lambda_4 = 0, \quad \lambda_5 = m$$

where p_1, p_2, p_3 are the components of momentum and m the mass of the particle; $\pm\Lambda_1/\Lambda_1$ and $\pm\Lambda_2$ are recognised to be Helicity and Energy respectively.

The most remarkable feature of the σ-operation is that it is applicable even to the generalisation of the anticommutation relation to the ω-commutation relation where ω is the mth root of unity. In the lowest representation we have as in the Pauli case only *three* mutually ω-commuting matrices with the representation of $\alpha_1, \alpha_2, \alpha_3$ as

$$\begin{bmatrix} 0 & 1 & & \\ & 0. & 1 & \\ & & \cdot & \\ & & \cdot 0 & 1 \\ 1 & \cdots & & 0 \end{bmatrix}, \quad \begin{bmatrix} 0 & \omega & & \\ & 0. & \omega^2 & \\ & & \cdot & \\ & & \cdot 0 & \omega^{m-1} \\ 1 & \cdots & & 0 \end{bmatrix}, \quad \begin{bmatrix} 1 & & & \\ & \omega & & \\ & & \omega^2 & \\ & & & \ddots \\ & & & & \omega^{m-1} \end{bmatrix},$$

and L_3 is m×m matrix

$$\sum \lambda_i \alpha_i \ ,$$

where

$$\alpha_1\alpha_2 = \omega\alpha_2\alpha_1 \ ,$$
$$\alpha_2\alpha_3 = \omega\alpha_3\alpha_2 \ ,$$
$$\alpha_1\alpha_3 = \omega\alpha_3\alpha_1 \ .$$

The commutation relations are 'ordered' since $\omega \neq \omega^{-1}$ except when $\omega = \pm 1$.

Performing the σ-operation we can generate 2n+1 ω-commuting matrices

$$L_{2n+1} = \sum \lambda_i \alpha_i$$

with

$$L_{2n+1}^m = \Lambda_n^m I = (\lambda_1^m + \lambda_2^m + \ \dots \ + \lambda_{2n+1}^m)I \ .$$

The ramifications of the eigenvectors, eigenvalues and the symmetry operations have been studied in great detail.

It was soon realised that, with L-matrices as building blocks, suitable linear combinations can be chosen such that we obtain Lie Algebraic matrices, the eigenvalues of which have physical interpretation as in Gell-Mann's unitary symmetry. This has led to the generalisation of the well-known Gellmann Nishijima relation

$$Q = I_z + \frac{1}{2} Y$$

where Q is the charge, Y the hypercharge and I_z the z component of the isotopic spin of the elementary particle. It also gives the relation

$$I_{k\ell} = \frac{1}{2}(S_k - S_\ell)$$

where S_k, S_ℓ are eigenvalues of commuting Lie Algebraic operators which are linear combinations of generalised Clifford Algebraic operators.

A very elegant mathematical extension of ω-commutation results if we relax the commutation rule to include higher powers of ω, that is, if we require

$$\alpha_i\alpha_j = \omega^{t_{ij}}\alpha_j\alpha_i \ ,$$

where $\omega^m = 1$ and the 2n+1 $\{\alpha_i\}$ are N×N matrices with $N = m^n$. It is to be noted that the product of the 2n+1 matrices is a constant and therefore we have only 2n independent matrices. There is also the subtle but important feature that the factor $\omega^{\frac{1}{2}}$ occurs in L_2 when m is even. In "Pauli matrices", the fourth root of unity occurs though we are dealing with m=2.

We define a product transformation connecting two sets α_i' and α_i:

$$\alpha_i' = \alpha_1^{u_{i,1}} \alpha_2^{u_{i,2}} \ldots \alpha_{2n}^{u_{i,2n}} .$$

If the α_i are characterised by a matrix T with elements t_{ij} and L' by T', it was proved that

$$T' = U T \tilde{U} ,$$

where U is matrix with elements U_{ij}, the exponents in the product transformations. A 'canonical' form of the T matrix is

$$t_{ij} = +1 , \quad t_{ji} = -1 , \quad t_{ii} = 0 ,$$

when we have ω-commutation among the 2n matrices.

REFERENCES

1. *L-Matrix Theory or the Grammar of Dirac Matrices* (Tata-McGraw Hill, New Delhi, India, 1972).
(This contains the list of papers published in the Journal of Mathematical Analysis and Applications, 1967-72.)

2. *Topics in Numerical Analysis*, Ed. J.H. Miller, Academic Press, 132 (1976).
(Proceedings of the conference on numerical analysis held in Dublin, Ireland, 1974.)

3. *Symmetries in Science*, Ed. Bruno Gruber and R.S. Millman, Plenum Press, New York and London, 323 (1979).
(Proceedings of the Einstein Centennial Celebration Science Symposium at Southern Illinois University, Carbondale, Illinois, U.S.A., 1979.)

4. A. Ramakrishnan, 'The Dirac Hamiltonian as a member of a hierarchy of matrices', *J. Math. Anal. and Applications*, **20**, 9-16, 1967.

APPLICATION OF CLIFFORD ALGEBRAS TO $*$-PRODUCTS

Jacques HELMSTETTER
Institut Fourier
Université de Grenoble
B.P. 74
38402 Saint Martin d'Hères
France

ABSTRACT. Some theorems involving symmetric bilinear forms and Clifford algebras may be "translated" to theorems involving symplectic forms and Moyal-products of distributions. The translation of some properties of Clifford groups is carried out.

I will explain a method which enables one to "translate" certain assertions concerning Clifford algebras into assertions concerning $*$-products. For the proofs of these assertions, I give the reference [H] . For the properties and meaning of the $*$-products, I give the reference [BFFLS] ; briefly speaking, the $*$-multiplication was first a deformation of the ordinary multiplication of functions, in order to get a new (non commutative) multiplication isomorphic to the multiplication of operators associated by quantum mechanics to these functions. You will find here a non-classical presentation of Clifford groups, which gives an essential importance not to the group itself, but to its closure (for any sensible topology) ; this presentation is explained in [H] and [SMJ] .

Let E be a real vector space of dimension n , provided either with a non degenerate symmetric bilinear form Φ , or with a symplectic structure ; in this latter case n must be even $(n = 2m)$, and Φ will denote the symplectic bilinear form on the <u>dual</u> space E^* . In the first case I shall use the exterior algebra $\wedge E$, and deform its exterior multiplication in order to get a Clifford multiplication (which I will also note like a $*$-multiplication) such that, for all f and g in E ,

(1a) $$f * g + g * f = \Phi(f, g) ;$$

in the second case I shall use the space $\mathcal{S}'E$ of complex-valued tempered distributions on E , and deform its multiplication in order to get a $*$-multiplication such that, for all f and g in E^*,

(1b) $$f * g - g * f = i\Phi(f, g) \quad (i = \sqrt{-1}) ;$$

J. S. R. Chisholm and A. K. Common (eds.), Clifford Algebras and Their Applications in Mathematical Physics, 559–564.

of course both ordinary and $*$-multiplications in $\mathcal{S}'E$ are only defined for privileged couples of factors ; this already accounts for the fact that not all assertions about Clifford algebras may be "translated". For the sake of simplicity I shall allow notations like $\int f(x)h(x)dx$, for any $f \in \mathcal{S}'E$ and any $h \in \mathcal{S}E$ (that is, h is a "rapidly decreasing" function on E).

It is already known how to construct canonically a Clifford multiplication on ΛE , that satisfies (1a) (see for instance [K]) ; I shall give here another description of this construction. But first let us remember how we may define the exterior product $f \wedge g$ of two elements f and g of ΛE , when we consider f , g and $f \wedge g$ as linear forms on the dual space ΛE^* . Let the mapping M $(E \oplus E \to E)$ be defined by $M(x, y) = x+y$, and let ΛM^* be the corresponding homomorphism $\Lambda E^* \to \Lambda(E \oplus E)^*$; some people call ΛM^* the coproduct of ΛE^* ; since $\Lambda(E \oplus E)$ is canonically isomorphic to the (twisted) tensor product $\Lambda E \otimes \Lambda E$, we may consider $f \otimes g$ as an element of $\Lambda(E \oplus E)$ and we may write the following formula, which defines $f \wedge g$ as a linear form on ΛE^* :

$$\langle f \wedge g, \varphi \rangle = \langle f \otimes g, \Lambda M^*(\varphi) \rangle ;$$

here φ is any element of ΛE^* ; if f , g and φ are respectively in $\Lambda^p E$, $\Lambda^q E$ and $\Lambda^{p+q} E^*$, this formula is equivalent to the classical definition of the exterior product of a p-linear skew-symmetric form on E^* and a q-linear one. Now remember that $\Lambda^2(E \oplus E)^*$ contains a subspace canonically isomorphic to $(E \otimes E)^*$; I denote by Φ' the image of Φ in $\Lambda^2(E \oplus E)^*$; Φ' is a <u>symplectic</u> form on $E \oplus E$; in the algebra $\Lambda(E \oplus E)^*$ I consider the exponential of $\Phi'/2$; $\exp(\Phi'/2)$ is a finite sum, because the exterior power Φ'^p vanishes whenever $p > n$ (dimension of E) . With the same notations as above, the Clifford multiplication on ΛE may be defined as follows :

(2a) $\quad \langle f * g, \varphi \rangle = \langle f \otimes g, \Lambda M^*(\varphi) \wedge \exp(\Phi'/2) \rangle$.

Indeed if f and g are in E , then (2a) implies

(3a) $\quad f * g = f \wedge g + \Phi(f, g)/2$;

whence (1a) .

At first it is not evident how to "translate" (2a) to the symplectic case, for it is difficult to "translate" $\langle f, \varphi \rangle$ when f and φ are functions or even distributions on E and E^* respectively, unless f or φ is a polynomial ; to overcome this difficulty, I use the Fourier transformation $\mathfrak{F}$; I suppose E and E^* are provided with dual Lebesgue measures, and for all integrable functions f on E ,

$$\mathfrak{F}f(y) = (2\pi)^{-m} \int f(x) \cdot \exp(-ix \cdot y) \cdot dx ;$$

the following definition

(4b) $\langle f, \varphi\rangle = (2\pi)^{-m} \int \mathfrak{F} f(y) \cdot \varphi(y)\, dy$

may be used for any $f \in \mathcal{S}'E$ and any $\varphi \in \mathcal{S}E^*$; if f is a polynomial, it does not give the ordinary value of $\langle f, \varphi\rangle$, but that of $\langle f(ix), \varphi\rangle$; however this has little importance here.

Now I can write the "translation" of (2a) , which defines the ∗-product of 2 elements f and g of $\mathcal{S}'E$:

(2b) $\int \mathfrak{F}(f * g)(y) \cdot \varphi(y) dy = (2\pi)^{-m} \iint \mathfrak{F} f(y) \cdot \mathfrak{F} g(z) \cdot \exp \frac{-i}{2} \Phi(y, z) \cdot \varphi(y+z) dy dz$;

here φ is any element in $\mathcal{S}E^*$; notice that Φ interferes in (2b) as a quadratic form on $(E \oplus E)^*$, whence a symmetric bilinear form Φ' on $(E \oplus E)^*$. As already mentioned, (2b) fails to define a ∗-product for all f and g ; so that it is defined for sufficiently many couples (f, g) , we must "force" the 2nd member of (2b) to be defined, for instance by introducing a factor $\exp Q_n(y, z)$, where Q_n is a sequence of non-degenerate negative quadratic forms, that converges to **0** ; but anyhow the existence of $f * g$ is always to be proved. If f and g are in E^*, then (2b) implies

(3b) $f * g = fg + i\Phi(f, g)/2$;

whence (1b) .

For what follows, it is not necessary to understand that (2b) is just another definition of the Moyal product (see [BFFLS]) ; but if you want to realize this, just notice that the Poisson bracket P(f, g) may be defined in this way :

$$\int \mathfrak{F} P(f, g)(y) \cdot \varphi(y) dy = -(2\pi)^{-m} \iint \mathfrak{F} f(y) \cdot \mathfrak{F} g(z) \cdot \Phi(y, z) \varphi(y+z)\, dy\, dz \ ;$$

and remember that the Moyal multiplication is the "symbolic exponential" of iP/2 . I recall that the Moyal multiplication admits the constant function 1 as unit element, that it is associative (with hypotheses of existence), and that the complex conjugation γ is an anti-automorphism :

(5) $\gamma(f * g) = \gamma(g) * \gamma(f)$.

Since (2a) defines a multiplication isomorphic to that of the Clifford algebra of Φ , we may consider the Clifford group $\Gamma(\Phi)$ as a subset of ΛE ; its closure in ΛE is well known ([SMJ] or [H]) ; it is the set XE of all elements $x \wedge \exp u$, where exp u is the exterior exponential of an element u of $\Lambda^2 E$, and x is a decomposable element of ΛE (a scalar, or a vector, or an exterior product of vectors). If $g = x \wedge \exp u$ as above, S(g) will denote the support of x (the subspace of $e \in E$ such that $e \wedge x = 0$) ; if $S(g) \neq 0$, the knowledge of g only determines the skew-symmetric bilinear form defined by u on the subspace of E^* that annihilates S(g) . I shall denote by γ the principal anti-automorphism of the Clifford multiplication ; thus (5) is also valid in ΛE and $\gamma(f) = f$ if $f \in E$. I recall that to every u is associated an element U in the Lie algebra of the orthogonal group $O(\Phi)$:

(6a) $U(f) = u * f - f * u$ if $f \in E$.

Since the group $O(\Phi)$ operates naturally on the whole space ΛE , I may write $G(f)$ for any $G \in O(\Phi)$ and any $f \in \Lambda E$. I recall the following results.

THEOREM 1. -

(a) If $g \in XE$ (the closure of the Clifford group), then $\gamma(g) \in XE$, and $g * \gamma(g)$ and $\gamma(g) * g$ are equal scalars, that will be noted $N(g)$.

(b) The Clifford group $\Gamma(\Phi)$ is the set of all g in XE such that $N(g) \neq 0$; the restriction of N to $\Gamma(\Phi)$ is a homomorphism.

(c) If $g \in \Gamma(\Phi)$, there exists an orthogonal transformation $G \in O(\Phi)$ such that

$$g * f * g^{-1*} = G(f) \text{ if } g \text{ or } f \text{ is even,}$$
$$= -G(f) \text{ if } g \text{ and } f \text{ are odd ;}$$

here f is any element of ΛE , and g^{-1*} is the $*$-inverse of g , that is, $\gamma(g)/N(g)$. The kernel of $G+I$ is the subspace $S(g)$ defined above. The mapping $g \mapsto G$ is a homomorphism, the image of which is $O(\Phi)$, the kernel of which is the group of invertible scalars.

(d) Suppose that $S(g) = 0$; for instance $g = \exp u$; and let U be defined by (6a) ; then

(7) $$G = (I+U/2)(I-U/2)^{-1} ,$$

(8) $$U = 2(G-I)(G+I)^{-1} ,$$

(9a) $$N(\exp u) = \text{dét}(I-U/2) = \text{dét}(I+U/2) .$$

(e) If an element g of $\Gamma(\Phi)$ approaches an element of XE which is not in $\Gamma(\Phi)$, then $N(g)$ tends to 0 .

Let us attempt to "translate" this first theorem. Instead of the elements $\lambda \exp u$, where λ is a scalar and $u \in \Lambda^2 E$, we must use the functions $\lambda \exp iu$, where u is a (real) quadratic form on E ; for reasons which will appear later, λ must be a real or a purely imaginary number, other than 0 . But the theorem of the stationary phase shows that in the closure of those functions $\lambda \exp iu$, all the following distributions g are to be found ; for any h in $\mathcal{S}E$,

$$\int g(x)h(x)\,dx = \lambda \exp \frac{is\pi}{4} \int_S \exp iu(x) \cdot h(x)\,dx \ ;$$

such a distribution g depends on λ and u (which are chosen as above), and on a subspace S of E , of arbitrary dimension s $(0 \leq s \leq n)$; notice that the knowledge of g only determines the restriction of u to S ; this subspace S will be noted $S(g)$. The factor $\exp is\pi/4$ is here a new feature ; its presence accounts for the fact that λ may pass from real values to purely imaginary values ; it may be related to problems of Maslov indices.

Let M_*E be the set of all such distribution g ; M_*E is the Maslov bundle of $E \oplus E^*$, provided with its natural symplectic structure of cotangent bundle over E ; indeed it is proved in [W] and in [H] that M_*E is a principal bundle over the set of lagrangian (maximal completely singular) subspaces of $E \oplus E^*$, for the structural group of invertible real or purely imaginary numbers. This recalls the Chevalley bundle (see [Ch]), which is the set X_*E of non null elements of XE ; consider now the natural quadratic form Q on $E \oplus E^*$ (that is, $Q(x, y) = \langle x, y \rangle$), and the set of all maximal completely Q-singular subspaces of $E \oplus E^*$; X_*E is a principal bundle over this set, for the structural group of invertible (real) scalars.

There are other facts which show that M_*E is the good "translation" of X_*E . For instance the equality

(10b) $\quad \mathfrak{F}(M_*E) = M_*E^*$

is just the "translation" of

(10a) $\quad \mathfrak{F}(X_*E) = X_*E^*$,

provided that I explain what $\mathfrak{F}$ means in (10a) ; choose dual elements ω and ω^* in $\Lambda^n E$ and $\Lambda^n E^*$ respectively ; for any f in ΛE , $\mathfrak{F}f$ is the inner product of ω^* and f ; thus $\mathfrak{F}$ is the well known transformation which maps every $\Lambda^p E$ onto $\Lambda^{n-p} E^*$. Many facts show that the Fourier transformation is the "translation" of this $\mathfrak{F}$; for instance (4b) is the "translation" of

(4a) $\quad \langle f, \varphi \rangle = \int (\varphi \wedge \mathfrak{F} f) \qquad (f \in \Lambda E\ ,\ \varphi \in \Lambda E^*)$;

I explain at once that, for any φ in ΛE^*, $\int \varphi$ is the scalar λ such that $\lambda \omega^*$ is the component of φ in $\Lambda^n E^*$.

Now I will "translate" the first theorem.

THEOREM 2. -

(a) If $g \in M_*E$, its complex conjugate $\gamma(g)$ is also in M_*E , and $g * \gamma(g)$ and $\gamma(g) * g$ are equal constant functions, the value of which will be noted N(g) ; however for exceptional elements g of M_*E , these $*$-products do not exist, and then we let $N(g) = \infty$.

(b) Let $\Gamma(\Phi)$ be the set of all g in M_*E such that $N(g) \neq \infty$; $\Gamma(\Phi)$ is a $*$-group ; the mapping N determines a homomorphism of $\Gamma(\Phi)$ onto the group of invertible positive numbers ; $f * g$ and $g * f$ exist whenever $g \in \Gamma(\Phi)$ and $f \in \mathcal{S}'E$.

(c) If $g \in \Gamma(\Phi)$, there exists a symplectic transformation $G \in Sp(\Phi)$ such that

$$g * f * g^{-1*} = f \circ G^{-1}$$

for all $f \in \mathcal{S}'E$. The image of $G+I$ is $S(g)$. The mapping $g \mapsto G$ is a homomorphism, the image of which is the symplectic group $Sp(\Phi)$, the kernel of which is the group of constant real or purely imaginary functions.

(d) Suppose that $S(g) = E$; for instance $g = \exp iu$; let U be the endomorphism of E such that

(6b) $\qquad -f \circ U = iu * f - f * iu$ for all $f \in E^*$;

U is in the Lie algebra of $Sp(\Phi)$; then the formulas (7) and (8) are still valid; but instead of (9a) we have

(9b) $\qquad N(\exp iu) = |\text{dét}(I-U/2)|^{-1} = |\text{dét}(I+U/2)|^{-1}$.

(e) If an element g of $\Gamma(\Phi)$ approaches an element of M_*E which is not in $\Gamma(\Phi)$, then $N(g)$ tends to ∞. This fact, which agrees with formula (9b), shows that the "translation" of the null element of XE is an element ∞, which symbolizes the exceptional apparition of a strictly divergent integral, and that we should let $ME = M_*E \cup \{\infty\}$.

Actually the group $\Gamma(\Phi)$ has 2 connected components; the metaplectic group $Mp(\Phi)$ (that is, the 2-covering group of $Sp(\Phi)$) may be defined as the intersection of its neutral connected component with the kernel of N; the metaplectic group becomes the "translation" of the spinor group; the Segal-Shale-Weil representation (a unitary irreducible representation of $Mp(\Phi)$) is the "translation" of the spinorial representation of the spinor group. For more details I refer to [H].

REFERENCES

[BFFLS] BAYEN, FLATO, FRONSDAL, LICHNEROWICZ, STERNHEIMER, 'Quantum mechanics as a deformation of classical mechanics', Letters in Math. Physics, vol. 1, 1977.

[Ch] Claude CHEVALLEY, The algebraic theory of spinors, Columbia University Press, 1954.

[H] Jacques HELMSTETTER, 'Algèbres de Clifford et algèbres de Weyl', 'Algèbres de Weyl et *-produits', Cahiers Mathématiques de Montpellier, n°25 (1982) et n°34 (1985).

[K] Erich KAEHLER, 'Der innere Differentialkalkül', Rendiconti di matematiche e delle sue applicazioni, vol. 21, 1962.

[SMJ] M. SATO, T. MIWA, M. JIMBO, 'Holonomic quantum fields', Publ. R.I.M.S., vol. 14, 1978.

[W] Alan WEINSTEIN, Lectures on symplectic manifolds, Regional conference series in Math. n°29, Am. Math. Soc., 1977.

- o -

ON REGULAR FUNCTIONS OF A POWER-ASSOCIATIVE HYPERCOMPLEX VARIABLE

Mari Imaeda
I.F.U.P.,31700,Cheswick Place,Solon,Ohio,44139,U.S.A..

ABSTRACT. The paper first discusses several regularity conditions of functions of a 2^N-dimensional hypercomplex variable, each of which is defined on different subspace of the 2^N-dimensional space. The power-associative law is assumed but not the associative or the alternative laws for the algebra. Some such regular functions are then constructed: a set of regular functions are constructed from functions of a complex variable; polynomial and exponential functions are constructed using the generating function. The Fourier representation of the regular function is briefly discussed. Some analysis obtained for the associative case does carry over to the nonassociative and nonalternative cases. The results of this paper apply to functions of a Clifford variable as well as that of a Cayley-Dickson variable.

1. INTRODUCTION

The present work (1) stems from previous studies on quaternions and octonions (2,3) where it was realized that the theory of functions of these variables may be extended, at a reasonable ease, to that of a nonassociative and nonaltenative higher dimensional hypercomplex variable. The theory developed here holds for a Cayley-Dickson

J. S. R. Chisholm and A. K. Common (eds.), Clifford Algebras and Their Applications in Mathematical Physics, 565–572.

variable (4) as well as for a Clifford variable. We do not assume associative or alternative laws throughout this paper, however we do assume a form of anticommutativity for the basis elements, i_k, of the vector space:

$$i_j\, i_k + i_k\, i_j = -2\,\delta_{jk}, \quad i,j = 1,2,\ldots,2^N-1, \quad (1)$$

where N is a positive integer.

Our main development is the regular (or monogenic) functions of these variables. Since Fueter (5) defined the regularity of a function of a quaternion variable, there have been some modifications of the "definition" to that of the functions of a biquaternion, an octonion and a Clifford variable (2,3,6). The significance of the regularity of some such functions were explicitly found in the formulation of Maxwell's equations, as well as of some gauge theories. We review, along these lines, the regularity of a function of a 2^N-dimensional hypercomplex variable (section 2) and construct some regular functions of such variable (sections 3 and 4). We also derive the Fourier representation of regular functions (section 5).

2. CONDITION OF REGULARITY

We may define more than one regularity condition for a function, F(X), of 2^N-dimensional hypercomplex variable, X. Suppose the space, V^n, where $n=2^N$, is spanned by the basis $(1,i_1,i_2,\ldots,i_{n-1})$. F(X) is left D_m- regular in a subspace $V^m \subset V^n$ $(m \leqslant n = 2^N)$ if

$$D_m F(X) = 0 \;, \quad (2)$$

and is D_m-right regular in $V^m \subset V^n$, if

$$F(X) D_m = 0 \;,$$

where $D_m := \partial_o + \sum_{k=1}^{m-1} i_k \partial_k$, $\partial_o := \frac{\partial}{\partial x_o}$, $\partial_k := \frac{\partial}{\partial x_k}$,

and $X := x_o + \sum_{k=1}^{m-1} i_k x_k$.

Similarly, F(X) is $\bar{D}_m$-left regular if $\bar{D}_m F(X)=0$, where

$$\bar{D} := \partial_o - \sum_{k=1}^{m-1} i_k \partial_k \ .$$

By integrating equation (2) over the m-dimensional subspace, V^m, and using Gauss's theorem one can express the above conditions in an integral form. The above definition modifies the usual "full" regularity condition for the function of a 2^N-dimensional hypercomplex variable.

3. SPECIAL REGULAR FUNCTIONS

We now construct some specific regular functions; these are obtained by direct generalization of the method used for functions of octonions (2). We describe a method of constructing functions, regular in an m-dimensional hypersurface, from functions of a complex variable which satisfy the Cauchy-Riemann conditions. The method applies for constructing functions of a Clifford variable as well as that of a Cayley-Dickson variable. Let $G(z) := u(z) + iv(z)$ be a function of a complex variable, where $z := x_o + ix$, and satisfies the Cauchy-Riemann conditions, $u_o = v_x$, and $u_x = - v_o$, where subscripts o and x refer to the differentiation with respect to the variables x_o and x, respectively. Now define a function, G(X), with

$$X := x_o + \sum_{k=1}^{m-1} i_k x_k \ , \text{ by replacing}$$

$$x := \Big(\sum_{k=1}^{m-1} x_k^2 \Big)^{1/2} \ , \text{ and } i := x^{-1} \Big(\sum_{k=1}^{m-1} i_k x_k \Big).$$

Here, i is regarded as a function of x_k ; however, $i^2 = -1$ is always assured by the anticommutation relation of equation (1). Note also that $[i, i_k] \neq 0$. Now define the following operators :-

$$D := \partial_o + \sum_{k=1}^{m-1} i_k \partial_k \ , \quad \text{and} \quad \Box := D\bar{D} = \sum_{\mu=0}^{m-1} \partial_\mu^2 \ .$$

Then the following relations may be proved by means of the first principle of induction.

Proposition : Define

$$u^{(r)} := \Box^r u \ , \qquad iv^{(r)} := \Box^r (iv) \ ,$$

and

$$w^{(r)} := \Box^{r-1}(DF(X)) = \Box^{r-1} D(u+iv) \ ,$$

where r is a positive integer. Then,

$$u^{(r)} = (m - 2r) (x^{-1} u^{(r-1)}_x) \ ,$$

$$v^{(r)} = (m - 2r) (x^{-1} v^{(r-1)}_x) \ ,$$

and

$$w^{(r)} = (m - 2r) (x^{-1} w^{(r-1)}_x) \ ,$$

where the subscript x refers to the differentiation with respect to x. In proving these relations (for details, see ref.(1)), one may note the following relations.

$$u^{(r-1)}_{oo} + u^{(r-1)}_{xx} = - 2 (r - 1) x^{-1} u^{(r-1)}_x \ ,$$

$$v^{(r-1)}_{oo} + v^{(r-1)}_{xx} = - 2 (r - 1) x^{-1} v^{(r-1)}_x \ ,$$

$$w^{(r-1)}_{oo} + w^{(r-1)}_{xx} = - 2 (r - 1) x^{-1} w^{(r-1)}_x \ .$$

The following theorem now follows directly from the proposition.

Theorem : Let m=2r, an even integer, and G(X) the function of an m-dimensional hypercomplex variable constructed as above. Then

$$\square^{r} G(X) = 0 ,$$

and the function, F(X), defined by

$$F(X) := \square^{r-1} G(X)$$

is left D-regular.

It is brought to the author's attention by the referee that a similar result has been obtained for the associative case (7,8).

4. POLYNOMIAL AND EXPONENTIAL FUNCTIONS

Consider the polynomial functions associated with the generating function,

$$K^{n}(X,\underline{t}) := (\, x_o\underline{t} + (\underline{x}.\underline{t})\,)^{n} = (\, \sum_{k=1}^{m-1} t_k\,(x_k + i_k x_o)\,)^{n} ,$$

where $\underline{t} := \sum_{k=1}^{m-1} i_k t_k$, $\underline{x} := \sum_{k=1}^{m-1} i_k x_k$,

and $\underline{x}.\underline{t} = \sum_{k=1}^{m-1} x_k t_k$.

The polynomials, $Pn_1n_2....n_{m-1}$ (X), are defined by

$$K^{n}(X,\underline{t}) := \sum_{(n_1,..,n_{m-1})}^{n} n!\, Pn_1n_2....n_{m-1}(X)\, t_1^{\,n}\, t_2^{\,n} \,....\, t_{m-1}^{\,n},$$

$$n := \sum_{k=1}^{m-1} n_k ,$$

where the summation is over all possible combinations of $n_1,.,.n_{m-1}$.

By a straightforward calculation, one can show that K^n is left and right $\bar{D}$-regular and also that the polynomials, $Pn_1n_2.....n_{m-1}(x)$, are $\bar{D}$-regular. These polynomials are the extension of the polynomials for quaternions which were first studied by Fueter. One may write the polynomials more explicitly as follows:-

$$Pn_1....n_{m-1}(x_o,x_1,...,x_{m-1}) = (n!)^{-1} \sum_{(Kr)} (x_k + i_k x_o)..(x_k + i_k x_o)$$

where, n_1 = number of times 1 appears in the series $(k_1,...k_n)$,
and n_2 = " 2 " " , and
the summation is over all possible permutations of the series $(k_r) = k_1,...k_n$.

One can show the following relations of the polynomials by direct calculations:

$$\frac{\partial^n}{\partial x_1^{r_1} \partial x_2^{r_2} .. \partial x_{m-1}^{r_{m-1}}} Pn_1n_2...n_{m-1}(X) = \delta n_1 r_1 \, \delta n_2 r_2 \, .. \, \delta n_{m-1} r_{m-1},$$

and $\partial_k (Pn_1n_2...n_{m-1}(X)) = Pn_1...n_k-1..n_{m-1}(X)$,

where $n := \sum_{k=1}^{m-1} n_k = \sum_{k=1}^{m-1} r_k$.

Various other properties of these polynomials follow trivially from the definition.

We can now define exponential functions which are regular as a consequence of the above polynomial functions. Define

$$\phi(X) := e^{i(x_o\underline{t} + (\underline{t}.\underline{x}))} = \sum_{n=0}^{\infty} (n!)^{-1} (i)^n (x_o\underline{t} + (\underline{x}.\underline{t}))^n$$

$$= \sum_{n=0}^{\infty} (n!)^{-1} (i)^n K^n .$$

One can easily show that these are left and right $\bar{D}$-regular.
Note that the anticommutation relation was not used in the construction of the polynomial and exponential functions.

5. FOURIER REPRESENTATION

Consider the following function;

$$\phi(X) = \int_{-\infty}^{+\infty} \cdots \int e^{i(x_o \underline{t} + (\underline{x}.\underline{t}))} A(\underline{t})\, dt_1 dt_2 \ldots dt_{m-1} ,$$

where $A(\underline{t})$ is a function of $t_1, t_2, \ldots, t_{m-1}$, and $\phi(X)$ is left $\bar{D}$ regular. Putting $x_o = 0$ in $\phi(X)$, we have

$$\phi(X) \Big|_{x_o=0} = \phi(\underline{x}) = \int \cdots \int e^{i(\underline{x}.\underline{t})} A(\underline{t})\, dt_1 dt_2 \ldots dt_{m-1} ,$$

which is a Fourier integral. $A(\underline{t})$ is then expressed as

$$A(\underline{t}) = ((2\pi)^{m-1})^{-1} \int_{-\infty}^{+\infty} \cdots \int e^{-i(t.\underline{\tau})} \phi(\underline{\tau})\, d\tau_1\, d\tau_2 \ldots d\tau_{m-1}.$$

Inserting $A(t)$ in $\phi(X)$, we have

$$\phi(X) = ((2\pi)^{m-1})^{-1} \int_{-\infty}^{+\infty} \cdots \int e^{i(x_o \underline{t} + \underline{t}(\underline{x} - \underline{\tau}))} \phi(\underline{\tau})$$

$$\times \; d\tau_1 d\tau_2 \ldots d\tau_{m-1} \;\; dt_1 dt_2 \ldots dt_{m-1}.$$

6. CONCLUSIONS

The nonassociativity of certain algebras has been shown to play important roles in the fundamental understanding of nature. However, it is not at all understood whether the nonalternativity of any algebra will take part in the formulation of a physical theory; and thus our theory developed here remains a mathematical enquiry, for the present. But due to the "more-general" nature of this theory it applies to functions of a quaternion, an octonion, and even of a Clifford variable.

ACKNOWLEDGEMENT This work was in collaboration with Professor K.Imaeda, Okayama Univ.of Science, Japan.

REFERENCES

(1) Imaeda,K, Imaeda,M, Bull.Okayama Univ.Sci.,20A,133 (1985).

(2) Imaeda,K, Imaeda,M, Bull.Okayama Univ.Sci.,19A, 93 (1984).

(3) Imaeda,K, Nuovo Cim.,32,139 (1976);
also the contributed talk of this workshop.

(4) For Cayley-Dickson algebras, see for example, Schafer,R.D, An Introduction to Nonassociative Algebras, Academic Press (1966).

(5) Fueter,R, Comm.Math.Helvetici,7,307 (1935).

(6) Delanghe,R, Math.Ann.,185,91 (1970);
also the review talk of this workshop.

(7) Sce,M, Lincei.Rend.Sc.Fis.Mat.e Nat.,23,220 (1957).

(6) Ryan,J, Complex Variables:Theory & applications,1,119 (1982).

ON A GEOMETRIC TOROGONAL QUANTIZATION SCHEME

J. TIMBEAU
Université Paul Sabatier
Mathématiques
118, route de Narbonne
31062 TOULOUSE Cedex
FRANCE

ABSTRACT. The aim of this article is to present a method for geometric quantization over a pseudo-riemannian manifold making use of torogonal and spinor structures, with orthogonal polarizations.

0. INTRODUCTION.

The Kostant-Souriau [7] theory cannot be applied to classical phase spaces with non vanishing second Stiefel-Whitney class, since in this case, metaplectic structures and half-forms do not exist ; moreover the quantizing Hilbert space is built from sections in a complex line bundle as the tensor product of the prequantum bundle and the bundle of the half-forms, with a partial connection.

To cure these defects, in the Hess quantization method [4], given a symplectic manifold, considering the metaplectic and toroplectic groups, which are respectively the two coverings and the central extension of the symplectic group, it is possible to construct a complex line bundle arising from a toroplectic structure, equipped with an ordinary connection.

In our torogonal quantization approach, given a pseudo-riemannian manifold, making use of the spinor group in the Popovici sense, we may construct a torogonal group and, from the corresponding structures, namely spinor and torogonal structures, we introduce the torogonal prequantization notion. Then, we construct also a line bundle provided with an ordinary connection satisfying a generalized quantum dogma for the first real Chern spinor class.

In the case of a symplectic manifold, considering a subordinate definite positive metric we find again the Hess results.

1. TOROGONAL AND SPINOR STRUCTURES.

1.1. Torogonal and spinor groups

We denote by (E,Q) a 2n-dimensional real vector space provided with

J. S. R. Chisholm and A. K. Common (eds.), Clifford Algebras and Their Applications in Mathematical Physics, 573–582.

a metric Q non-degenerate, and the subscript "prime" represents the complexification (E',Q') and utilises complex elements ; $O(Q)$ and $O(Q')$ are the orthogonal groups ; $C(Q)$ and $C(Q') \simeq C(Q)'$ the Clifford Algebras ; if $C^*(Q)$ is the unities group of $C(Q)$, then G is the Clifford group relative to the representation φ from $C^*(Q)$ to $C(Q)$ defined by $\varphi_g(t) = gtg^{-1}$ for $g \in C^*(Q)$ and $t \in C(Q)$; we consider $\varphi' : G' \longrightarrow O(Q')$.

Therefore $G^c = \varphi'^{-1}(O(Q))$ represents the "real" elements of G'; with the spinor norm N such that $N(g) = \beta(g)g$, where β is the principal anti-automorphism of $C(Q)$, we can introduce the central extension of $O(Q)$ by the 1-unitary group $S^1 = U(1)$: $\Delta(Q) = N^{-1}(S^1) \cap G^c$ named the torogonal real group and the spinor group : pin $Q = \mathrm{Ker}(N|\Delta(Q))$.

We have the following commutative exact diagram of Lie groups :

$$(\mathfrak{D}) : \quad \begin{array}{ccccccccc}
 & & 1 & & 1 & & 1 & & \\
 & & \downarrow & & \downarrow & & \downarrow & & \\
1 & \longrightarrow & S^o=\{\pm 1\} & \longrightarrow & \text{pin } Q & \xrightarrow{\sigma} & O(Q) & \longrightarrow & 1 \\
 & & \downarrow & & \downarrow & & \| & & \\
1 & \longrightarrow & S^1 & \longrightarrow & \Delta(Q) & \xrightarrow{\delta} & O(Q) & \longrightarrow & 1 \\
 & & \downarrow {\scriptstyle \cdot^2} & & \downarrow {\scriptstyle N} & & \downarrow & & \\
1 & \longrightarrow & S^1 & = & S^1 & \longrightarrow & 1 & & \\
 & & \downarrow & & \downarrow & & & & \\
 & & 1 & & 1 & & & &
\end{array}$$

where δ and σ denote the restrictions of φ' and $\cdot^2$ the squaring map.

1.2. Clifford bundles.

If (ξ,Q) represents a 2n-real vector bundle ξ with a metric Q, over a paracompact manifold M, we denote by (E,Q) the typical-fibre and by O_ξ the principal $O(Q)$-bundle.

By the universal property of Clifford algebras we define : the extension morphism : ext : $O(Q) \longrightarrow \mathrm{Aut}_C(C(Q'))$ from $O(Q)$ to the automorphisms group relative to $C(Q')$; the corresponding extension of O_ξ which is an $\mathrm{Aut}_C(C(Q'))$ - principal bundle ; and the associated algebraic bundle named Clifford bundle $C(\xi',Q') = O_\xi \times_{O(Q)} C(Q')$.

1.3. Generalized S.C.structure.

Given a complex algebraic bundle η over M, with typical-fibre a complex central simple algebra A, if ρ is a simple typical representation from A to a complex space vector S, we have the Skolem-Noether diagram :

$$1 \longrightarrow C^* \longrightarrow GL_C(S) \xrightarrow{f} \mathrm{Aut}_C(A) \longrightarrow 1$$

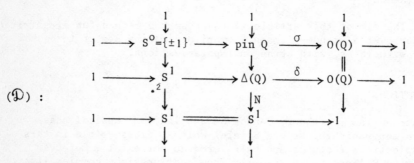

where $f(u) = \rho^{-1} \circ \mathrm{Int}(u) \circ \rho$ for $u \in GL_C(S)$; Int denote the inner automorphism.

Using Čech Cohomology with coefficients in sheaves of germs of C^∞ functions on M with values in not necessarily abelian groups, from the cohomological sequence induced by this diagram and by the following :

$$0 \longrightarrow Z \longrightarrow C \xrightarrow{m} C^* \longrightarrow 1$$

where $m(z) = \exp(2i\pi z)$ for $z \in C$ we obtain the Bocksteins diagram :

$$H^1(M, GL_C(S)) \xrightarrow{f_*} H^1(M, Aut_C(A)) \xrightarrow{B_f^1} H^2(M, C^*) \xrightarrow{B_m^2} H^3(M, Z)$$

where f_* denote the cohomological extension functor and B_f^1 and B_m^2 the Bockstein operators.

Therefore we can define the obstruction class $W_3(\eta) = B_m^2 o B_f^1[\eta]$ in the cohomology group $H^3(M,Z)$, associated with the equivalence class of η.

<u>Proposition</u> :

The following three assertions are equivalent :

(i) : there exists a complex vector bundle $\mathcal{T}$ with typical-fibre S and an algebraic bundle isomorphism from η to the bundle of vector bundle endomorphisms of $\mathcal{T}$, inducing on each point of M a simple representation ; such a structure is named : generalized S.C.structure.

(ii) : $f_*[\mathcal{T}] = [\eta]$

(iii) : $W_3(\eta) = 0$.

In particular case of S.C.structures over (ξ,Q) (introduced by G. Karrer [6]) : $\eta = C(\xi',Q')$ and the corresponding cohomology class $W_3(\xi)$ is independent of the metric.

1.4. Torogonal structure.

A torogonal structure over (ξ,Q) is a δ-lifting of 0_ξ i.e. a principal $\Delta(Q)$-bundle Δ_ξ together with a principal bundle morphism also denoted by : $\delta : \Delta_\xi \longrightarrow 0_\xi$.

The Clifford structures are define in the same way as φ or φ'-liftings of 0_ξ.

<u>Proposition</u> :

The notions of S.C.structure, torogonal structure and "real" Clifford structure are equivalent.

1.5. Spinor-structure.

A spinor structure over (ξ,Q) is a σ-lifting of 0_ξ i.e. a principal pin Q-bundle pin_ξ together with a principal bundle morphism

$\sigma : pin_\xi \to O_\xi$.

The cohomology class $W_2(\xi,\sigma) = B^1_{2\sigma}[O_\xi]$ represents the spinor obstruction in the cohomology group $H^2(M,S^o)$. Considering the exact sequence :

$$0 \longrightarrow Z \longrightarrow Z \xrightarrow{\text{mod}_2} Z_2 \longrightarrow 0$$

and the isomorphism $j : S^o \xrightarrow{\simeq} Z_2$ we can obtain :

Proposition :

For all metrics: $W_3(\xi) = B^2_{\text{mod}_2} \circ j_*(W_2(\xi,\sigma))$.

Therefore a spinor structure induces a torogonal structure.

Remark :

Since $O(p) \times O(q)$ is a retract of $O(p+q)$ we can consider the Sylvester decomposition : $\xi = \xi^+ \oplus \xi^-$ and the corresponding Stiefel-Whitney classes ; expliciting $W_2(\xi,\sigma)$ we find the counterpart of the Karoubi results obtained by method of algebraic topology [5].

2. TOROGONAL PREQUANTIZATION.

Henceforth we consider tangential torogonal or spinor structures over a 2n-dimensional paracompact pseudo-riemannian manifold (M,Q) ; in this case ξ is the tangent bundle TM and O_ξ is the principal O(Q)-bundle O_M.

2.1. Notion of pseudo-connection.

From the torogonal diagram ($\mathfrak{D}$) we obtain : $\Delta(Q) \simeq S^1 \times_{S^o} pin\, Q$.

Whence we have the Lie algebras decomposition :
$L\Delta(Q) = iR \oplus LO(Q)$ because pin Q is a two-covering of O(Q). The projection $\mu : L\Delta(Q) \longrightarrow iR$ is equivariant with respect to the adjoint action of $\Delta(Q)$.

Given a torogonal structure $\delta : \Delta_M \longrightarrow O_M$ a μ-pseudo-connection (in the Hess sense [4]) is a one form $\tilde{\gamma}$ on Δ_M with values into iR, equivariant, and such that $\tilde{\gamma}(A^*) = \mu(A)$ where A^* denotes the vertical vector field corresponding to $A \in L\Delta(Q)$.

The local representations of μ-pseudo-connections are similar to those of ordinary connections ; in particular, from the Cartan structure equation, a μ-pseudo-connection $\tilde{\gamma}$ possesses a curvature two-form $d\tilde{\gamma}$.

2.2. Definition.

A torogonal prequantization over (M,Q) provided with a real closed two-form ω is a torogonal structure $\delta : \Delta_M \longrightarrow O_M$ together with a μ-pseudo-connection $\tilde{\gamma}$ satisfying the curvature condition $d\tilde{\gamma} = \pi^*(i\omega)$ where $\pi : \Delta_M \longrightarrow M$ is the natural projection.

Assuming the torogonal prequantization existence, given a $\Delta(Q)$-cocycle $\Delta_{k\ell} : U_k \cap U_\ell \longrightarrow \Delta(Q)$ lifting the $O(Q)$-cocycle $g_{k\ell} : U_k \cap U_\ell \longrightarrow O(Q)$ i.e $\delta \circ \Delta_{k\ell} = g_{k\ell}$, where $\mathcal{U} = \{U_i\}_{i\in I}$ is a suitable choice of principal coordinate representations of Δ_M and O_M, it is possible to choose cochains, in general not cocycles : $z_{k\ell} = \exp(i\theta_{k\ell})$ and $\tilde{g}_{k\ell}$ taking values into S^1 and pin Q respectively, with $\Delta_{k\ell} = z_{k\ell}.\tilde{g}_{k\ell}$.

Therefore $\tilde{g}_{jk}\,\tilde{g}^{-1}_{ik}\,\tilde{g}_{ij}$ represents the two obstruction spinor cocycle and the local representation of $\tilde{\gamma}$ implies : $\tilde{\gamma}_\ell - \tilde{\gamma}_k = id\theta_{k\ell}$.

The curvature condition means that : $\theta_{ij} + \theta_{jk} - \theta_{ik}$ just represents the Čech cohomology class corresponding to the De Rham cohomology class $[\omega]$; denoting by :

$I_R : H^2_R(M,C) \xrightarrow{\sim} H^2(M,C)$ this De Rham isomorphism we have :

<u>Theorem</u> :

There exists a torogonal prequantization if and only if the cohomology class : $\exp_*(I_R[i\omega]).\,W_2(M,\sigma)$ or $m_*(I_R[\frac{\omega}{2\pi}]).\,W_2(M,\sigma)$ vanishes into $H^2(M,S^1)$.

<u>Corollary</u> :

If a spinor structure exists, this condition becomes : the real cohomology class $[\frac{\omega}{2\pi}]$ is integral.

2.3. Generalization of the Kostant-Souriau procedure.

If $[\frac{\omega}{2\pi}]$ is integral we can construct a classical prequantization [7] i.e. a principal S^1-bundle L^* provided with an ordinary hermitian connection γ such that $\pi_L^*(i\omega) = d\gamma$ for the natural projection $\pi_L : L^* \longrightarrow M$.

Assuming the spinor structure existence $\sigma : pin_M \longrightarrow O_M$ we consider the ψ-extension Δ_M of the fibre-product : $L^* \times_M pin_M$ which is a principal $S^1 \times$ pin Q-bundle, associated to the group morphism $\psi : S^1 \times \text{pin } Q \longrightarrow \Delta(Q)$; there exists a μ-pseudo-connection $\tilde{\gamma}$ defined by : $\psi^*(\tilde{\gamma}) = p_1^*(\gamma)$ denoting by $\psi : L^* \times_M pin_M \longrightarrow \Delta_M$ the principal extension morphism and by $p_1 : L^* \times_M pin_M \longrightarrow L^*$ the first projection.

Therefore $(\Delta_M, \tilde{\gamma})$ is a torogonal prequantization, which is said to be the amalgamation of the Kostant-Souriau prequantization (L^*, γ)

and the spinor structure pin_M.

2.4. Application to the 2n-symplectic manifold (M,ω).

If $(M, \mathfrak{J}, Q)$ represents a subordinate almost hermitian structure where Q is positive definite, therefore pin Q is a classical spinor group. From the primary decomposition relative to $\mathfrak{J}$ we define the canonical cohomology class $C_1(M,\omega)$ into the cohomology group $H^2(M,Z)$ and the second Stiefel-Whitney class : $W_2(M) = mod_{2*}(C_1(M,\omega)$ into $H^2(M,Z_2)$. Then recall [9] :

Proposition :

There exists a spinor structure if and only if $W_2(M) = 0$ i.e. $\frac{1}{2}C_1(M,\omega)$ is integral ; this condition is equivalent to : there exists also a metaplectic structure.

Therefore we have the following :

Integrality criterion :

A torogonal prequantization over a symplectic manifold (M,ω) exists if and only if the real cohomology class : $\frac{1}{2}C_1(M,\omega) + \left[\frac{\omega}{2\pi}\right]$ is integral ; this is also the existence condition of a toroplectic prequantization [4].

Having done the prequantization, the quantization depends on a new auxiliary structure, namely :

3. ORTHOGONAL POLARIZATIONS AND ASSOCIATED CONNECTIONS.

3.1. Definition :

A complex vector subbundle P of the complexified tangent bundle $(TM)^c$ is said to be an orthogonal polarization if : it is totally maximal isotropic with rank n ; $P \cap \overline{P}$ is a "real" vector bundle with constant rank equal to the index of Q and we have the Sylvester decomposition : $P = (P \cap \overline{P}) \oplus P^{\varepsilon(Q)})$ for the pseudo-hermitian extension $Q'(z_1,\overline{z_2})$ of Q where $\varepsilon(Q)$ is an invariant for Q.

Remark :

This polarization can be involutive in the Frobenius sense i.e. $[P,P] \subset P$; and regular, relatively to the associated foliations.

By complexification, the orthogonal group O(Q) operates on the vector totally isotropic maximal spaces, so that this set is an homogeneous space, whence :

3.2. Existence theorem :

The orthogonal polarizations P bijectively correspond to the reductions O_P of the principal O(Q)-bundle O_M, from structure group O(Q) to isotropy group S_P for a typical vector totally isotropic maximal space denoted also by P.

Remark :

S_P is said to be the spinoriality group in the Crumeyrolle sense [1,2].

This group S_P is a semi-direct product of a diagonal group : $GL(r,R) \times U(n-r)$ (linear and unitary groups) where r = index (Q) by an almost nilpotent subgroup.

3.3. For the restriction group morphism : $ind_P : S_P \longrightarrow GL_C(P)$ defined by $ind_P(u) = u|P$ the ind_P-extension of O_P is naturally isomorphic to the whole frame bundle R_P of P ; this also implies the existence of a natural line bundle isomorphism : $\overset{n}{\Lambda} P \simeq O_P \times_{detoind_P} C$ where $\overset{n}{\Lambda} P$ is the outer product.

Then we can define the first Chern spinor class :

$$C_1(P) = B_m^1[\overset{n}{\Lambda} P] \text{ into } H^2(M,Z) \ [3].$$

Remark :

If we consider the metalinear group $ML_C(P)$ as the two coverings of the linear group $GL_C(P)$, a P-metalinear-structure is a lifting of R_P and the half-P-forms are the sections of the associated line bundle.

3.4. Definition :

A P-connection (named spin-euclidean connection in [2]) is a principal connection over the reduction O_P.

We will denote by ω_P and Ω_P the connection and curvature forms taking values into the Lie algebras LS_P which is a semi-direct product of a diagonal Lie algebras $g\ell(r,R) \times LU(n-r)$ (linear and anti-hermitian algebras) by a determined ideal.

Now we can bring together again these ingredients to construct a :

4. TOROGONAL QUANTIZATION.

Denoting $pin(P) = \sigma^{-1}(S_P)$ and $\Delta(P) = \delta^{-1}(S_P)$ we get an analogue $(\mathfrak{D}_P)$ of the torogonal diagram $(\mathfrak{D})$; the morphisms being denoted with an additional subscript P. We may identify : $\Delta(P) \simeq S^1 \times_{S^o} pin(P)$ whence :

$L\Delta(P) = iR \oplus LS_P$ with projection $\mu_P : L\Delta(P) \longrightarrow iR$.

Given a torogonal structure $\delta : \Delta_M \longrightarrow O_M$ we define a polarized torogonal structure $\delta^{-1}(O_P) = \Delta_P$ getting the following diagram of principal bundles :

$$\begin{array}{ccc} \Delta_P & \hookrightarrow & \Delta_M \\ \delta_P \downarrow & & \downarrow \delta \\ O_P & \hookrightarrow & O_M \end{array}$$

Moreover the μ-pseudo connection $\tilde{\gamma}$ on Δ_M restricts to a μ_P-pseudo connection $\tilde{\gamma}_P$ onto Δ_P.

Proposition :

For some P-connection ω_P :
$\tilde{\beta}_P = \tilde{\gamma}_P + \delta_P^*(\omega_P)$ is a principal connection on Δ_P.

In a second step, there exists a group morphism :

$\chi^P : \mathrm{pin}(P) \longrightarrow C^*$ satisfying : $(\chi^P)^2 = \det \circ \mathrm{ind}_P \circ \sigma_P$ and $|\chi^P|^2 = |\det \circ \mathrm{ind}_{P\cap\bar{P}} \circ \sigma_P|$ where $\mathrm{ind}_{P\cap\bar{P}} : S_P \longrightarrow GL_C(P\cap\bar{P})$.

Then we can define a one dimensional representation : $\nu^P : \Delta(P) \longrightarrow C^*$ by $\nu^P(z.\hat{g}) = z\chi^P(\tilde{g})$ which satisfies :

$$(\nu^P)^2 = N_P.\ \det \circ \mathrm{ind}_P \circ \delta_P\ , \qquad (1)$$

$$|\nu^P|^2 = |\det \circ \mathrm{ind}_{P\cap\bar{P}} \circ \delta_P|\ . \qquad (2)$$

Thus we obtain the ν^P-extension of Δ_P and the associated complex line bundle : $L^P = \Delta_P \times_{\nu^P} C$ together with the linear connection ∇^P over L^P associated with $\tilde{\beta}_P$, the curvature of which is : $i\omega + \frac{1}{2}\mathrm{Tr}(\text{curvature } \nabla|P)$, where ∇ denote the linear connection associated to ω_P ; ω the curvature of the pseudo-connection $\tilde{\gamma}$ and Tr is the trace operator.

The first relation (1) implies that : $C_1(L^P) = I_R\left[\frac{\omega}{2\pi}\right] + \frac{1}{2}C_1(P)$. Whence, the line bundle L^P yielded by this prodedure satisfies this generalized quantum dogma.

Now, if we consider the sections s of L^P i.e. the tensors $\tilde{s}$ over Δ_P with values into C, the type of which is $\nu^P(\Delta(P))$, we can construct a sesquilinear bundle morphism :

$<,>_P : L^P \times_M L^P \longrightarrow R_P \times_{|\det|} C$ into the line bundle of densities of weight 1 over $P \cap \bar{P}$. (due to the second relation (2)).

This local scalar product is compatible with the connection ∇^P and when Q is definite, it is an ordinary hermitian structure onto L^P. Finally, $(L^P, \nabla^P, <,>_P)$ represents the quantum bundle arising from the polarized torogonal prequantization and complies with the quantum dogma.

The representation space is obtained from the sheaf of germs of sections in L^P, which are covariant constant along P and determines the quantizing Hilbert space.

Remark :

We can find again the results on the quantization of phase space [8] making use of transverse involutive polarizations. In the particular case of classical prequantization (L^*,γ), if $[\frac{\omega}{2\pi}]$ is integral, and from a spinor structure $\sigma : pin_M \longrightarrow O_M$ we can define a polarized spinor structure $\sigma^{-1}(O_P) = pin_P$ and the corresponding half-P-forms bundle : $\sqrt{\Lambda^n P} = pin_P \times_{\chi} {}_P C$; therefore $L^P = L \otimes \sqrt{\Lambda^n P}$ and the connection ∇^P and the scalar product are also obtained by tensoring the corresponding objects on L and $\sqrt{\Lambda^n P}$.

Moreover, in the second particular case of a symplectic manifold (M,ω), we construct a subordinate torogonal structure associated with the primary decomposition relative to $(M, \mathfrak{J}, Q)_P$ and from an almost hermitian connection, we get a quantum bundle L^P such that :

$C_1(L^P) = I_R\left[\frac{\omega}{2\pi}\right] + \frac{1}{2}\, C_1(M,\omega)$.

Obviously, the right hand side of this relation has to be an integral class, but it is not necessary that $I_R\left[\frac{\omega}{2\pi}\right]$ and $\frac{1}{2}\, C_1(M,\omega)$ are separately integral.

If both of them are only half-integral, as in the example of the Kählerian manifold $P_{n-1}(C)$ with odd n, arising in energy levels quantization of the n-dimensional harmonic oscillator, then a torogonal quantum bundle exists too, while the Kostant-Souriau theory does not apply.

REFERENCES

1. A. Crumeyrolle : 'Spin fibrations over manifolds and generalized twistors'. Proceedings of Symposia A.M.S. Stanford (1973). Part I, p.69.
2. A. Crumeyrolle : Periodica Math Hungarica 6, (2), 1975, p.143-171
3. A. Crumeyrolle : 'Classes caractéristiques réelles spinorielles' C.R. Acad. Sc. Paris, Série A, t.283, p.359.
4. H. Hess : 'On a geometric quantization scheme generalizing those of Kostant-Souriau and Czyz'. Proceedings Clausthal, Germany, 1978.
5. M. Karoubi : Ann. Scient. Ecole normale Sup. tome 1, fasc.2, 1968, p.166-177.
6. G. Karrer : Ann. Acad. Sci. Fennicae, série A, 1, 1973, n°521.
7. B. Kostant : 'On the definition of quantization' - Colloq. Int. du C.N.R.S., Aix en Provence, 1974.

8. J. Timbeau : 'Quantification des systèmes classiques sur les variétés pseudo-riemanniennes'. Thèse 3ème cycle, 1976, Toulouse

9. J. Timbeau : 'Structure spinorielle sur une variété presque symplectique'. C.R. Acad. Sciences, 279, série A, 1974, p.273.

INDEX

Ahlfors 213, 319
Algebra
  bispinor 353-9
  Cayley-Dickson 565
  central K- 106
  Dirac 236, 363, 371
  Fierz 353, 356
  Grassmann 467, 501
  Hamilton 467
  K- 106
  Kac-Moody 310
  Kemmer-Duffin 299
  Neveu-Schwarz 310
  non-associative 295
  Pauli 323, 487
  Poisson 518, 528
  Poisson bracket 389
  separable 67
  simple 276
  spinor 49, 56, 467
  vector 467
  Virasoro 308
  Weyl 517
Analytic
  functions 118
  functionals 119
Angular momentum 448
Anomalous magnetic moment 374
Anti-automorphism 574
Antiderivation 248
Antiparticle 281, 298, 451
Antisymmetric tensor 49
Artinian ring 62
Asymptotic field 489
Automorphism 447
  interior 70
  orthogonal 71

Barrier (N=3) 316
Baryon states 388
Baltramian 523
Bessel functions 155
Bianchi identity 230
Bilinear form
  degenerate 93
Bilinear Lagrangian 374
Biquaternion 64, 495
Biregular 160
  plane wave 150, 157
  polynomial 149, 155
Birkhoff 462
Bispinor 363
  density 353, 358
  field 371-3
Bivector 7, 150, 365-7
  magnetic induction 485
  normal 241
  timelike 447
  -vector 460
Big bang 269
Bochnar-Martinelli representation 162
Boundary
  behaviour 153
  value 149
Bra-ket notation 451
Broad band matching 479
Bundle
  of geodesics 230
  orthonormal frame 182
Bureš 211
Butterworth filter 477

Cartan
  calculus 501
  classification 88
  product 208
Cartan-Dieudonné theorem 72
Cartesian components 404
Casimir operator 116, 150
  superspin 316
Cauchy
  integral formula 14, 46, 210
  theorem 11, 45
  transform 127
Cauchy-Riemann
  equations 13, 133, 201, 338

generalised operators 160, 201
Cayley-Dickson
algebra 565
proccess 537
Cell division 269
Central
charge 313
extension 96
force problem 461
quadratic 106
Characteristic matrix 275
Chern class 573, 5779
Charge conjugation 297
Chevalley Bundle 563
Chiral projection 342
Circle multiplication 55

Classical
action 451
field 450, 455
mechanics 449
Clifford algebra 7, 79, 160, 351, 365, 371, 387, 445, 452, 465, 501
basis 370, 372
classification 52, 80
complex 9, 120, 150, 220, 326, 445, 459
contraction 94
degenerate 61
direct prodouct 279, 459
even subalgebra 79
fermionic 393, 394
generalised 541, 549
geometric representation 50
hermitean 245
matrix 169, 172, 173
metric on 361
nilpotent 394, 396
non-degenerate 61
non-universal 50
over Heisenberg field 388
real 9, 79, 326
R(4) 445, 452
semi simple 90
simple 279
space-time dependent 375
super 394
symplectic 389, 396, 519
tensor product 387-8
universal 50
vector 425
Clifford analysis 150, 159
Clifford bundle 177, 574
Clifford group 64, 72, 73, 94, 168, 520
closure 562
even 64
homomorphism 80
Clifford-like algebras 537
Clifford multiplication 251
Clifford number 512
Coarse coupling 415
Coderivative 230
second 230
Co-differential 505
Cohomology
Čech 575
Color 301
Commutation relations 300
Commutator
product 229
Co-multiplier 505
Complex
geometry 39, 40, 45
numbers 10, 335
structure 521, 523
Conformal invariance 213, 276
Killing vector 234
Conformal mapping 499
Conjugate variable 449
Conjugation matrix 363, 365
Conservation 267, 453, 461
equations 455
quantum law 456
Conserved current 188
convection 281
quantity 451
Continuity equation 461, 468
Correspondence principle 523
Contraction 461
Coudazzi-Mainardi equation 229
Covariant derivative 364
spinor 177
U(2,2) 373
Crummeyrolle 36
Curl 13, 16
Curvature
condition 577
extrinsic 230, 243
form 579
intrinsic 243
scalar 194
tensor 238
Cyclic
group 549
variable 449

Cylinder condition 142

Darboux system 153
Decomposition
of spinors 367
Deformation theory 526
star 526
Delanghe 207
Density matrix 274
Derivative
exterior 16
of extensor 228
vector 12, 228
de Sitter model 238-9
Differential
complexform 16, 178, 204, 345, 403, 468, 501
multiform 247, 255
operator 468
operators, infinite order 155
second 230
Dipole
electric 488
magnetic 488
Dirac 455, 461
adjoint equation 267
algebra 236, 275, 293, 297, 325, 387, 396
components 404
current 333, 341
equation 46, 133, 193, 267, 274, 280, 330, 337, 347, 353-60, 445, 465
matrices 16, 347, 445
operator 116, 149, 206, 253, 501, 512
relation 250
see also "Kähler-Dirac"
Dirac-Feynmann Gell-Mann 273, 280
Direct sum
decomposition 62
Discriminant 69
Distribution 153
Divergence 13, 16
properties 313
Division
agebra 191
ring 29
Dodson 219
Domain
of biregularity 162
of holomorphy 159
Double covering 95
Dual 324
geometric 450
symmetry 243
Duality
rotation 491
transformation 458
Dynkin diagram 25
Dyson equations 371

Edington 276, 459
Eigenfunction 451
momentum 452
Einstein
equations 435
model 236
non-symmetric theory 253
theory 462
Einstein-Yang-Mills equations 177
Eisenhart 231
Elasticity operator 115
Electrodynamics 296
Electromagnetic 501
field 19, 364, 497
gauge transformation 334
interaction 451
tensor 374
theory 465
two-form 16
vector potential 374
Elementary particle states 376
Embedding
class 236
space 440
Electron 363
charge 374
classical model 389
Electrostrong interaction 363, 369
Elementary particles 363, 425, 430, 456
Energy 448, 469
Energy-momentum
spectrum 408
tensor 19
Entropy function 273
Equation of Motion
charged particle 21
Euclidean operator 116
Euler Poisson-Darboux equations 149
Expectation value 400
Extensor 228
Exterior
bundle 178
derivative 346, 504
product 276, 345

Fegan 202, 214
Fermion 364, 546
  families 189
  operators 245, 282
Filter
  active 479
Fixed point 171
  elliptic 172
  hyperbolic 172
  loxodromic 172
  parabolic 172
Flag
  plus pole model 277
Flatland 472
Flavour
  group 409
  orbit 412
  transformation 404
Flow diagram 506
Force strength constant 377
Fourier
  representation 571
  transform 518, 521
Fractional statistics 546
Fritzjohn's formula 155
Fueter 133, 208, 566
  equation 39, 40, 45
  operator 45

Gårding 209
Gauge field 360, 400, 438
  axial vector 375
  scalar 374
  tensor 371
  transformation 365, 437, 505, 511
  vector 371
Gauge group 360
  electromagnetic 334
  electroweak 341
  U(1) 374
  (*see also* "Lattice", "Spin gauge")
Gauge theory 363
Gegenbauer
  equations 150
  polynomials 155
Geodesic 230
Geometric
  algebra 227
  calculus 5, 228
  function theory 10, 14, 22
  representation 22, 50, 56
  subalgebra 449
  supspace 448
Geometrical transformation 409
Geometry
  extrinsic 228
  intrinsic 228
Geon 243
Gibbs' vector 447
Glashow-Salam-Weinberg 363, 366, 438
Grade 8
Gradient 241
Grading
  group 549
  involution 50, 54, 59
Graded tensor product 62
Grassmann 7, 303, 501
  mutually dual algebras 249
  products 247
  (*see also* "Algebra")
Gravitation 182, 296, 426, 365, 455, 460
  Clifford algebra theory 459
Gravitino 197
Graviton 439
Gray code 533
Green's formula 267
Group
  cohomology 575
  conformal 82, 235
  cyclic 549
  de Sitter 99
  dynamical 386
  extended Galilei 100
  even 69, 73
  Galilei 94
  grading 549
  Heisenberg 96
  internal symmetry 386
  Lorentz 49
  metaplectic 517, 520, 563
  of idempotents 69
  of isometries 231
  orthogonal 71
  Picard 68
  pin 774, 95
  Poincaré 99, 269
  pseudo-unitary 79
  quadratic extension 70
  rotation 116
  SO(4) 204
  spin 64, 93, 95, 116, 119, 149
  spinor 74, 80, 89, 573
  spinor symplectic 520
  spinorial Euclidean 409

spinoriality 579
SU(1,1) 370
symmetry 409
sympletic 84, 517
symplectic Clifford 520
toroplectic 573
U(2,2) 371
unimodular orthogonal 71
unitary symplectic 88
(*see also* "Clifford group", "Gauge group")
Group representations 149
infinitesmal 121
spin 121, 149

Hamilton-Jacobi equation 338
Harmonic function 220
oscillator 449
symplectic spinor 525
Hartog's theorem 160
Heat operator 115
Helicity 366
Helmholtz operator 115
Hertz potential 511
Hestenes 28, 36, 236, 457
Heterotic string 318
Hierarchy of algebras 274
Hilbert space 273, 451
Hitchin 213-4
Hodge
dual 276
map 178
operator 208, 501
Hodge de Rham operator 180, 194
Holomorphic functions 159
several complex variables 159
Homogeniety condition 438
Homogeneous
differential operator 153
polynomial 153-4
Hopping parameter expansion 415
Hurwitz (*see also* Radon)
condition 41
pair 39, 41, 44
problem 39
Hyperbolic
functions 545
rotations 370
space 472
variable 498
Hypercomplex
function theory 14
variables 162

Hyperfunctions 119
Hypersurface 236

Ideal 80, 325, 330
descending chain 64
indecomposable 64
left 29, 64, 274, 364, 404
minimal 29, 49, 64, 253, 274, 330, 364
primitive 404
two-sided 80
Idempotent 29, 49, 62, 83, 183, 330, 388, 450
decompositions 62
group 69
lifting 63, 64
mutually annihilating 29, 62
primitive 29, 62, 274, 276
Imaginary, unit 447
Index
wrt cycle 46
Infrared catastrophe 374
Inner differential 266
calculus 265
Inner product 451, 505
Insertion loss 472
power ratio 471
Instanton field 419
Integral
formula 210, 219, 220
geometry 149
operators 161
representations 159
Interaction field equation 451
Interaction field 431
Interior product 59
Internal symmetry 386
Invariants 479
Involution 30, 167, 179, 191
Irreducible representation
unitary 410
Ising model 542
Isometry 59, 231
Isotropy 196

Jacobian 237
Jacobson radical 61
Jordan-Wigner
algebra 278
operator 280

Kähler 3, 331
(-Dirac) equation 185, 214, 257, 261, 280, 290, 399, 406, 409

Kähler-Atiyah algebra 250
Kaluza-Klein 40, 44, 296, 435
Killing
  current 188
  equations 231
  vector 194, 230, 234
Klein-Gordon
  equation 445, 450
  operator 116
Korteweg-de Vries 311
Kustaanhelmo 27

Lagrangian 309, 522
  field theory 363, 383
Laplace-Beltrami operator 119, 150, 180, 194
Laplace equation 115
Laplacian 13, 523
  m-dimensional 149
Larmor-Reinich 490
Lattice
  approximation 400
  system 309
  fermion 400
  free lattice fermion 401
  gauge theory 400
  Kähler-Dirac equation 406
Legendre polynomials 117
Leibnitz product rule 231
Leray residue 210, 219
Lie
  bracket, generalised 231, 233
  derivative 194, 232
Light cone 475
Light like plane 476
Liouville 273, 281
  field 523
Local gauge transformation 363, 368
Local interaction 295
Lorentz (*see* "group Lorentz")
  covariance 373
  force 339, 458, 462
  3-dimensional space 471
  transformation 21, 276, 333
Lounesto 206
Lyapounov 273
Magnetic
  charge 458
  field 486
Majorana representation 446
Manifold 178, 227
Mapping, projective 11
Mass spectra 375
Massless field 208
  spin half 219
Maslov bundle 563
Maxwell 246, 457, 462, 469
  equations 15, 393, 436, 458, 468-9, 487, 495
Meson
  mass 417
  propagator 414
  state 387
Metaplectic structure 573
Metric 445
  form 239
  tensor 232
Microwave theory 479
Minkowski space 236, 370, 372, 446, 452, 465, 479
Möbius transformation 167, 170
Module
  indecomposable 62
  quadratic K- 67, 103
  principal indecomposable 61, 63
  quadratic 67
Moisil 208
Momentum 448
  star 412
Monomorphism 51
Moyal product 528, 561
Multi-index notation 401
Multiplet 450
Multispinor 365
Multivector 9, 247, 445
  aggregate 446
  conservation 445
  differential equation 450
  eigenfunction 450
  norm 451

Networks 465
  active 465
  analysis 471
  lossy 465, 479
  noisy 465, 479
  two-port 479
  synthesis 471
Neutral current 375
Neutrino 363
Newtonian limit 462
Nilpotent
  ideals 62
  radical 61
Non-alternative 565
Non-associative 565

Non-linear
Dirac equation 360
sigma model 360
Non-symmorphic extension 410
Norm 74, 366, 451
spinorial 74, 80
n-tuple 531
Null space 62

Occam's razor 452
Octonian 23, 296, 537
w-commutation relations 555, 557
Orthonomal
base 50
frame 239
subset 50
Outer differential 266

Parallel transport 364, 373
Parastatistics 301
Parity violation 375
Parseval-Plancheral 525
Partially polarised wave 465
Particle
classification 387
field 438
mass 377
Partial differential equation 149
Partial polarisation 479
Partition function 551
Pascal theorem 470
Path integral 400, 451
Pauli
coupling 380
matrices 550, 555
principle 552
Pauli-Lubanski vector 316
Penrose R. 211, 214, 219
distribution 289
foliation 290
Perihelion shift 462
Periodic chain 553
Periodicity-8 theorem 52
Phase space 273, 445, 449, 526
algebraic spinor 276
distribution function 280
quantisation 581
relativistic 278
superoperator 280
Phase transformation 363
Phasor 445
geometric 446
Physical observables 353
Physical concepts 295
Plane elliptic system 150
Planck constant 524
Plane-stratified 472
Poincaré
lemma 456, 460, 505
sphere 491
Point nature 295
Poison bracket 389
Polarisation
elliptic 491
orthogonal 578
round 491
spiral 491
Polynomial, left regular 136
Porteous 36
Position 446
Post-Newtonian approximation 462
Potential 450
4-vector 436
scalar 498
vector 498
Potts model 541, 549
Power-associative 565
Pre-quantisation 577
Pre-vacuum 439
Primitive
Dirac number 274
elements 274, 276
matrix 276
(*see also* "Idempotent")
Probability current 280
Product
geometric 7
inner 7, 275
outer 7, 275, 323
Projection 229, 436
momentum 277
operator 227, 342, 353, 358, 369, 388
spin 277
Projective
representation 385
transformation 17
Propagator
poles 417
Pseudo-connection 576
Pseudo-euclidean 39
Hurwitz pair 41
space 79, 82, 236
Pseudo-quaternion structure 87
Pseudo-Riemannian 82, 178, 576

Pseudoscalar 9, 18, 328, 450
  field 227
  unit 323
Pseudovector field 369, 450
Pullback 503

Quadratic form 67
Quantisation, phase space 581
Quantum electrodynamics 371
Quantum mechanics 353
  interpretation 321
  relativistic 347
Quark 301, 365
Quasi-conformal mapping 40
Quasi-distribution function 273
Quasi-particle operator 250
Quaternion 79, 81, 106, 467
  scalar product 84
  valued function 167

Radon-Hurwitz
  number 29, 62
  sequence 50, 52
Rediscoveries of Clifford Algebra 2
Red-shift 462
Reduced spin 412
Reflection Coefficient 471
Reisner-Nordstrom 187
Regular 498
  exponential 569
  function 167 , 496, 565
  left- 135, 566
  mapping 220
  polynomial 136
  right- 566
  spinor fields 206
Relativistic
  quantum mechanics 455
  wave equation 390
Relativity 468
  8-dimensional 435
Renormalisation 295
Renormalised continuum limit 400, 415
Representation, matrix 429
Reversion 19, 31, 277
Ricci identity 230
Riemann 245
  curvature 229
  surface 163
Riemannian connection 228
Riesz M. 36, 211, 219, 274, 322
  formula 221

Rigid motion 64
Robertson-Walker 360
Rotation
  elliptic 473, 475
  hyperbolic 473-4
  parabolic 473, 475
Roundness 491

Scalar product 266
  pseudo-euclidean 39
  pseudo-quaternion 85
Schilling figure 470
Schönberg spinor 279
Schroedinger operator 115
Schwarz distributions 118
Schwarzchild 187
  manifold 243
  model embedded 240
Semi-direct product 98
Semi-Grassmannian 89
Sequency order 531
Sesquilinear
  bundle 580
  form 84
Several complex variables 159
Shape operator 229
  σ-operation 555, 557
Skolen-Noether theorem 70
Soliton 40
Souček 219
Soup-plate trick 28
Source
  classical fields 455
  equation 455
Spaceland 472
Space-time
  algebra 15, 332, 468
  complex 425
  creation and annihitation 278
  geometry 448
  split 17, 250
Special functions 149
Spectrum doubling group 402
Spherical harmonics 115, 150, 154
  of S_3 445
Spherical mean 149, 152
  oriented 151
Spherical monogenics 125, 149, 150
  inner 126
  left 125
  outer 126
Spin

connection 364, 579
current 188
-half 273, 280, 458
metric 277, 372
structure 439
-zero 452
Euclidean operator 116, 129
Spin gauge theory 363, 365, 371
local U(4) 371
spin (8) 377
U(2,2) 371-2
Spin group, *see* "Group"
Spinor 299, 427, 435, 442, 445, 543
actions 195
algebra 49, 56
algebraic 183, 274, 276, 280, 282, 285
classification 428
contravariant 279
covariant derivative 177, 191
Dirac 276, 327, 335, 364
expansion inversion theorem 358
field 189
forms 238
group, *see* "Group"
parallel 257
parallel transport 364
Pauli 330
phase space 279
pure 196, 257
regularisation 27
representations 49, 327, 335
rotations 441
spaces 56, 253
structure 285, 291, 575, 581
symplectic 523
Spirality 491
Stein 202, 208
Stiefel-Whitney class 573, 578
Stokes
4-dimensional vector 479
theorem 12, 510
Stress tensor 183, 469
Strong coupling approximation 416
Structure factors 355-6
Subalgebra, even 18
Submodule, irreducible 38
Superbracket 249
Supercharge 250
Supergravity 313
Superoperator 250, 273, 282
Supersymmetry 177, 197, 250, 279, 300, 303, 313, 394, 420
Superstring 293, 306, 313
closed 318
Supervector 273
Susskind coupling 415
Symmetry
breaking 140, 268, 371, 375, 439-40
external 437
internal 437
principle 365
transformation 296
vacuum 440
Symplectic 245
Clifford algebra 389
decomposition 29, 40, 45
manifold 573
power series 519
spinor 517, 521
transvection 520
truncated Clifford algebra 521

Takabayashi 277
Tangent algebra 227
Temple 461
Tensor 279
density 455
field 228
non-local 272
outer product 273
representation 271
Theordorescu 208
Thermodynamics 273
Topology 68, 268
change 319
Grothendieck 76
Torogonal
pre-quantisation 577
polarised structure 579
quantisation 573
structure 575
Torque 448
Transposition 447
Translation 233

Unification, algebraic 393
Unified theory
elementary particles 393
electromagnetism and gravity 463
spin gauge 369

Vacuum-state 278
Vector
algebra 467

basis 367
light-like 475
manifold 227
normal 237, 241
pure 167
-vector entity 459
Vector field invariant 121
Vector-valued forms 204
Velocity 446
Vey's hypothesis 528

Walsh fraction 531
Wave guide 479
junction 479
Wedge product 404
Weinberg-Salam 340, 363
(*see als* "Glashow-Salam-Weinberg")
Weiss 202, 208
Weyl 518
representation 277
solution 243
Wigner-Moyal 274, 281
Wilson action 401
Witt decomposition 277
Wyler 378

Yang-Mills 181, 309, 438
maximal super 313

Zitterbewegung 340, 389